Conducting Polymers for Advanced Energy Applications

Conducting Polymers for Advanced Energy Applications

Edited by

Ram K. Gupta

CRC Press
Taylor & Francis Group
Boca Raton London New York

CRC Press is an imprint of the
Taylor & Francis Group, an **informa** business

First edition published 2022
by CRC Press
6000 Broken Sound Parkway NW, Suite 300, Boca Raton, FL 33487-2742

and by CRC Press

2 Park Square, Milton Park, Abingdon, Oxon, OX14 4RN

Library of Congress Cataloging-in-Publication Data

Names: Gupta, Ram K., editor.
Title: Conducting polymers for advanced energy applications / edited by Ram K. Gupta.
Description: First edition. | Boca Raton, FL : CRC Press, 2022. | Includes
   bibliographical references and index. | Summary: "This book details the
   use of conducting polymers and their composites in supercapacitors,
   batteries, photovoltaics, and fuel cells, covering nearly the entire
   spectrum of energy area under one title. It covers a range of advanced
   materials based on conducting polymers, the fundamentals, and chemistry
   behind these materials for energy applications. This work provides new
   direction to scientists, researchers, and students in materials science
   and polymer chemistry seeking to better understand the chemistry behind
   conducting polymers and how to improve their performance for use in
   advanced energy applications"—Provided by publisher.
Identifiers: LCCN 2021036380 (print) | LCCN 2021036381 (ebook) | ISBN
   9780367713355 (hbk) | ISBN 9780367713409 (pbk) | ISBN 9781003150374 (ebk)
Subjects: LCSH: Conducting polymers. | Electric power supplies to
   apparatus--Materials. | Solar cells—Materials. | Fuel cells—Materials.
   | Supercapacitors—Materials.
Classification: LCC TK2945.P65 C72 2022 (print) | LCC TK2945.P65 (ebook)
   | DDC 620.1/9204297—dc23/eng/20211006
LC record available at https://lccn.loc.gov/2021036380
LC ebook record available at https://lccn.loc.gov/2021036381

ISBN: 978-0-367-71335-5 (hbk)
ISBN: 978-0-367-71340-9 (pbk)
ISBN: 978-1-003-15037-4 (ebk)

DOI: 10.1201/9781003150374

Typeset in Times
by KnowledgeWorks Global Ltd.

Dedication

*I would like to dedicate this book to my wife Rajani Gupta, my daughter
Anjali Gupta, and my niece Payal Gupta for their
love, motivation, and encouragement.*

Contents

Preface

Traditional polymers such as polyethylene, polyvinyl chloride, and polypropylene are insulating in nature and are mostly used to insulate electrical cables and electronic devices. Conducting polymers are a unique class of polymers that can transport charge through conjugated double bonds. Conducting polymers such as polyaniline, polythiophene, and polypyrrole have attracted considerable research interest since their discovery due to their unique electrical, electrochemical, optical, and mechanical properties. Their facile synthetic routes and processability provide additional advantages for many applications. The morphology and chemical structures and thus the properties of conducting polymers can be easily tuned using various methods. In particular, their tunable electrochemical properties make them very suitable for energy applications.

Energy is one of the most important parts of modern society for its proper functioning and advancement. In the last few decades, there has been substantial scientific development that heavily requires lightweight and high-performance energy devices. Energy devices such as supercapacitors, batteries, solar cells, and fuel cells are used in many areas such as electronic devices, appliances, defense, automobiles, and biomedicals, and conducting polymers find their wide applications in these devices. Some of the issues such as low energy density and electrochemical stability of conducting polymers can be improved by making nanocomposites with other materials such as 2D materials, metal oxide, and metal sulfides. Scientists and researchers have developed many strategies to synthesize nanocomposites of conducting polymers with unique morphology and improved electrochemical properties. For example, nanocomposite of polyaniline with graphene facilitates charge transport and provides mechanical strength, high energy density, and stable electrochemical properties for their applications in energy devices.

The main purpose of this book is to provide current, state-of-the-art knowledge of the conducting polymers and their composites for supercapacitors, batteries, photovoltaics, and fuel cells, covering almost the entire spectrum in the energy area under one title. This book covers a range of advanced materials based on conducting polymers, the fundamentals, and the chemistry behind these materials for energy applications.

Ram K. Gupta

Editor

Ram K. Gupta, PhD, is an Associate Professor at Pittsburg State University. Dr. Gupta's research focuses on conducting polymers and composites, green energy production and storage using nano-materials, optoelectronics and photovoltaics devices, organic-inorganic hetero-junctions for sensors, bio-based polymers, flame-retardant polymers, bio-compatible nanofibers for tissue regeneration, scaffold and antibacterial applications, corrosion inhibiting coatings, and bio-degradable metallic implants. Dr. Gupta has published over 230 peer-reviewed articles, made over 280 national, inter-national, and regional presentations, chaired many national/international meetings sessions, edited many books, and written several book chapters. He has received over two and a half million dollars for research and educational activities from many funding agencies. He is serving as Editor-in-Chief, Associate Editor, and editorial board member of numerous journals.

Contributors

Jiji Abraham
Department of Chemistry
Vimala College
Kerala, India

Charles Oluwaseun Adetunji
Applied Microbiology, Biotechnology and
    Nanotechnology Laboratory
Department of Microbiology
Edo University Iyamho
Auchi, Nigeria

Frank Ofori Agyemang
Department of Materials Engineering
College of Engineering
Kwame Nkrumah, and University of Science
    and Technology
Kumasi, Ghana

A.R. Ajitha
Department of Chemistry
Newman College
and
International and Interuniversity Centre for
    Nanoscience and Nanotechnology
Mahatma Gandhi University
Kerala, India

H. Akhina
Post-Graduate Department of Chemistry
MSM College
and
International and Interuniversity Centre for
    Nanoscience and Nanotechnology
Mahatma Gandhi University
Kerala, India

Daniel Nframah Ampong
Department of Materials Engineering
College of Engineering
Kwame Nkrumah, and University of Science
    and Technology
Kumasi, Ghana

György Bánhegyi
Polygon Consulting
Budapest, Hungary

Suddhasatwa Basu
Department of Chemical Engineering
Indian Institute of Technology
Delhi, India
and
CSIR-Institute of Minerals and Materials
    Technology
Bhubaneswar, India

Shikha Chander
Department of Chemistry
University of Technology
Jaipur, India
and
St. Francis Degree and P.G. College for Women
Hyderabad, India

Xuecheng Chen
Faculty of Chemical Technology and
    Engineering
West Pomeranian University of Technology
Szczecin, Poland
and
School of Environment and Chemical
    Engineering
Shenyang University of Technology
Shenyang, China

Dipak Kumar Das
Department of Chemistry
GLA University
Mathura, India

Deepak P. Dubal
Centre for Materials Science
School of Chemistry and Physics
Queensland University of Technology
Brisbane, Australia

Modou Fall
Laboratory of Organic Physical Chemistry and
    Environmental Analyses (LCPOAE)
Department of Chemistry
Faculty of Sciences and Techniques
Cheikh Anta Diop University
Dakar-Fann, Senegal

Diariatou Gningue-Sall
Laboratory of Organic Physical Chemistry and
    Environmental Analyses (LCPOAE)
Department of Chemistry
Faculty of Sciences and Techniques
Cheikh Anta Diop University
Dakar-Fann, Senegal

Katarzyna Grochowska
Centre for Plasma and Laser Engineering
The Szewalski Institute of Fluid-Flow
    Machinery
Polish Academy of Sciences
Gdańsk, Poland

Amadou Bélal Guèye
Laboratory of Organic Physical Chemistry and
    Environmental Analyses (LCPOAE)
Department of Chemistry
Faculty of Sciences and Techniques
Cheikh Anta Diop University
Dakar-Fann, Senegal
and
School of Energy Materials
International and Inter University Center
    for Nanoscience and Nanotechnology
    (IIUCNN)
School of Chemical Sciences
Mahatma Gandhi University
Kerala, India

Ram K. Gupta
Department of Chemistry
Kansas Polymer Research Center
Pittsburg State University
Pittsburg, Kansas, United States

P. Hema
Department of Chemistry
Amrita Viswavidyapeedom
Amrithapuri Campus
Kerala, India

Rudolf Holze
State Key Laboratory of Materials-Oriented
    Chemical Engineering
School of Energy Science and Engineering
Nanjing Tech University
Nanjing, China
and
Chemnitz University of Technology
Institut für Chemie
AG Elektrochemie
Chemnitz, Germany
and
Institute of Chemistry
St. Petersburg State University
St. Petersburg, Russia

Abel Inobeme
Department of Chemistry
Edo University Iyamho
Auchi, Nigeria

Jayapriya Jayaprakash
Department of Applied Science and
    Technology
Alagappa College of Technology
Anna University
Chennai, India

Leta Tesfaye Jule
Department of Physics
College of Natural and Computation Science
and
Centre for Excellence in Indigenous Knowledge
Innovative Technology Transfer and
    Entrepreneurship
Dambi Dollo University
Dembi Dolo, Ethiopia

Anuj Kumar
Department of Chemistry
GLA University
Mathura, India

Parmod Kumar
J.C. Bose University of Science and Technology
Faridabad, India

Vinod Kumar
Department of Physics
College of Natural and Computation Science
and
Centre for Excellence – Indigenous Knowledge
Innovative Technology Transfer and
    Entrepreneurship
Dambi Dollo University
Dembi Dolo, Ethiopia

Amir Ershad Langroudi
Color and Surface Coatings Group
Polymer Processing Department
Iran Polymer and Petrochemical Institute
    (IPPI)
Tehran, Iran

Ajay Lathe
Department of Chemistry
Mahatma Phule ASC College
Panvel, India

Momath Lo
Laboratory of Organic Physical Chemistry and
    Environmental Analyses (LCPOAE)
Department of Chemistry
Faculty of Sciences and Techniques
Cheikh Anta Diop University
Dakar-Fann, Senegal

Kalpana Madgula
SAS Nanotechnologies LLC
Wilmington, Delaware, USA

S. Malavika
Department of Chemistry
Amrita Viswavidyapeedom
Amrithapuri Campus
Kerala, India

Meenu Mangal
Department of Chemistry
Poddar International College
Jaipur, India

Hanna J. Maria
Laboratory of Organic Physical Chemistry and
    Environmental Analyses (LCPOAE)
Department of Chemistry
Faculty of Sciences and Techniques
Cheikh Anta Diop University
Dakar-Fann, Senegal

John Tsado Mathew
Department of Chemistry
Ibrahim Badamasi University
Lapai, Nigeria

Kwadwo Mensah-Darkwa
Department of Materials Engineering
College of Engineering
Kwame Nkrumah, University of Science and
    Technology
Kumasi, Ghana

Gopika G. Nair
Department of Physics
CMS College (Autonomous)
Kerala, India

Saranya Narayanasamy
Department of Applied Science and
    Technology
Alagappa College of Technology
Anna University
Chennai, India

Vanya Nayak
Department of Biotechnology
Indira Gandhi National Tribal University
Amarkantak, India

Olugbemi T. Olaniyan
Laboratory for Reproductive Biology and
    Developmental Programming
Department of Physiology
Edo University Iyamho
Auchi, Nigeria

Adrian Olejnik
Centre for Plasma and Laser Engineering
The Szewalski Institute of Fluid-Flow
    Machinery
Polish Academy of Sciences
and
Gdańsk University of Technology
Gdańsk, Poland

Avinash R. Pai
International and Inter University Center for
    Nanoscience and Nanotechnology
Mahatma Gandhi University
Kerala, India

Anil M. Palve
Department of Chemistry
Mahatma Phule ASC College
Panvel, India

Hamidreza Parsimehr
Color and Surface Coatings Group
Polymer Processing Department
Iran Polymer and Petrochemical Institute
    (IPPI)
Tehran, Iran

Shruthy D. Pattathil
Gulbarga University
Karnataka, India
and
St. Francis Degree and P.G. College
    for Women
Hyderabad, India

Venkata Sreenivas Puli
Smart Nanomaterials Solutions LLC
Orlando, Florida, USA and
Materials and Manufacturing Directorate
Wright Patterson Air Force Base
Ohio, USA

Krishnaraj Ramaswamy
Centre for Excellence – Indigenous
    Knowledge
Innovative Technology Transfer and
    Entrepreneurship
and
Department of Mechanical Engineering
Dambi Dollo University
Dembi Dolo, Ethiopia

Deepti Rawat
Miranda House College
University of Delhi
Delhi, India

Arunima Reghunadhan
School of Energy Materials
Mahatma Gandhi University
Kerala, India

S. Rijith
Post Graduate and Research Department of
    Chemistry
DST-FIST Supported Department
Sree Narayana College
Kollam
Affiliated to University of Kerala
Kerala, India

Mohamed Lamine Sall
Laboratory of Organic Physical Chemistry and
    Environmental Analyses (LCPOAE)
Department of Chemistry
Faculty of Sciences and Techniques
Cheikh Anta Diop University
Dakar-Fann, Senegal

Sreedha Sambhudevan
Department of Chemistry
Amrita Viswavidyapeedom
Amrithapuri Campus
Kerala, India

P.K. Sandhya
School of Chemical Sciences
Mahatma Gandhi University
and
Post Graduate Research Department of
    Chemistry
Sree Sankara College
Kerala, India

S. Sarika
Post Graduate and Research Department of
    Chemistry
DST-FIST Supported Department
Sree Narayana College
Kollam
Affiliated to University of Kerala
Kerala, India

U.N. Mohamed Shahid
Department of Chemical Engineering
School of Engineering
University of Petroleum and Energy Studies
Dehradun, India

Beer Pal Singh
Department of Physics
Chaudhary Charan Singh University
Meerut, India

Jay Singh
Department of Chemistry
Institute of Science
Banaras Hindu University
Varanasi, India

Kshitij R.B. Singh
Department of Chemistry
Government V.Y.T. Post Graduate Autonomous
    College
Durg, India

Manohar Singh
Department of Physics
Chaudhary Charan Singh University
Meerut, India

Ravindra Pratap Singh
Department of Biotechnology
Indira Gandhi National Tribal University
Amarkantak, India

Rahul Singhal
Department of Physics and Engineering
    Physics
Central Connecticut State University
New Britain, Connecticut, USA
and
Shivaji College
University of Delhi
Delhi, India

Katarzyna Siuzdak
Centre for Plasma and Laser Engineering
The Szewalski Institute of Fluid-Flow
    Machinery
Polish Academy of Sciences
Gdańsk, Poland

R.K. Soni
Department of Chemistry
Chaudhary Charan Singh University
Meerut, India

Felipe M. de Souza
Department of Chemistry
Kansas Polymer Research Center
Pittsburg State University
Pittsburg, Kansas, USA

Muhammad Rizwan Sulaiman
Department of Chemistry
Kansas Polymer Research Center
and
Plastic Engineering
Kansas Polymer Research Center
Pittsburg State University
Pittsburg, Kansas, USA

V.S. Sumi
Department of Chemistry
Government College
Kerala, India

Preema C. Thomas
Department of Physics
CMS College (Autonomous)
Kerala, India

Sabu Thomas
School of Energy Materials
International and Inter University Center
    for Nanoscience and Nanotechnology
    (IIUCNN)
and
School of Chemical Sciences
Mahatma Gandhi University
Kerala, India

Kelsey Thompson
Department of Chemistry
Kansas Polymer Research Center
Pittsburg State University
Pittsburg, Kansas, USA

Nobel Tomar
J.C. Bose University of Science and Technology
Faridabad, India

Shrestha Tyagi
Department of Physics
Chaudhary Charan Singh University
Meerut, India

Vinod Kumar Vashistha
Department of Chemistry
GLA University
Mathura, India

Abbaraju Venkataraman
Department of Chemistry
Gulbarga University
Karnataka, India

Yuping Wu
State Key Laboratory of Materials-Oriented
    Chemical Engineering
School of Energy Science and Engineering
Nanjing Tech University
Nanjing, China

Ghulam Yasin
Institute for Advanced Study
Shenzhen University
Shenzhen, China

1 Introduction

Conductive Polymers from the Nobel Prize to Industrial Applications

György Bánhegyi
Polygon Consulting, Budapest, Hungary

CONTENTS

1.1 THE BEGINNINGS

The discovery and the industrial realization of silicon-based electronics (beginning with the transistor) led to extremely rapid development in almost all areas of life, and all this advancement was based on a deep understanding and manipulation of the silicon single crystal. This was one of the leading themes of the "newborn" materials science, which required close cooperation between theoretical and experimental physicists, chemists, and all kinds of engineers. This led Moore, one of the founders of Intel in 1965, to predict that the spatial density of transistors in a unit space will double every two years. This conjecture became a "law", and the actual doubling rate even surpassed the predicted value. Nevertheless, there are physical limits of miniaturization as the molecular dimensions are approached. Nowadays, the quasi-monopolistic position of silicon-based electronics is challenged not only by the compound semiconductors (such as GaAs or InP) but also by molecular electronics, graphene, or carbon nanotube-based systems. Simultaneously with microelectronics there emerged another industry that profoundly rearranged our everyday life: the production and processing of synthetic polymers. These new synthetic materials have been and mostly still are insulators and are utilized as such. The obvious benefits of easy processability and free-shape formation made the goal of synthesizing conductive polymers attractive. It has been well known from the band theory of solid-state physics that the prerequisite of metallic conduction is the existence of overlapping valence and conduction bands without an energy gap in-between. Under these conditions, the outermost electrons of the atoms become *delocalized*, mobile.

As polymers are mostly organic compounds, thinking started from delocalization in *conjugated* systems, where single and double (or triple) bonds follow each other in an *alternating* manner. In such systems, the so-called σ-bonds located between the adjacent atoms form a planar framework and the π-bonds formed by the overlap of *p*-orbitals, oriented perpendicularly to this plane, create *delocalized* molecular orbitals above and below the σ-frame, but exhibit zero density in the plane itself, which is a nodal plane. With the increasing number of atoms, the number of these delocalized orbitals increases, and the highest occupied molecular orbital (HOMO) gets closer to the lowest unoccupied molecular orbital (LUMO). This is reflected by the fact that the electronic transition between them occurs at lower energy: butadiene absorbs in the UV range, while carotene with its extended conjugated system in the visible range. One would expect that in an infinite conjugated chain, the *band gap* diminishes, and the material becomes a conductor. There is, however, a phenomenon called the *Peierls distortion*, which predicts a spontaneous (fluctuation induced) "clustering" of the uniform C–C distances into atom-pairs of alternating bond length. This spontaneous breakdown of symmetry in 1D metals with exactly half-filled band structure is somewhat like the Jahn–Teller effect observed in polyatomic systems with degenerate electronic structures. (For the historical aspects of the Peierls distortion in 1D metals see Pouget (2016).) This kind of limitation does not exist for 2D organic "metals", like graphite, so no wonder that some of the earliest attempts to develop organic semiconductors and conductors were associated with the development of carbon fibers, namely the monitoring of conductivity changes during the carbonization of polyacrylonitrile. Just to mention one example: a close correlation was observed between the differential thermal analysis (DTA) signals, the electron paramagnetic resonance (EPR) bandwidth, the conductivity, and the activation energy of conductivity during the pyrolysis of oriented PAN–carbamide complexes (Hedvig, Kulcsár et al. (1968)). Other early developments include acene-quinone radical polymers (Pohl, Opp (1962)), poly(phthalocyanines) (Felmayer, Wolf (1958)), and other similar systems, which involved partial carbonization and/or the inclusion of metal atoms in the main chain.

The Nobel Prize presentation of Heeger (Heeger (2001)) nicely summarizes the history of polyacetylene itself. Acetylene polymerization by Ziegler–Natta catalysts in 1974 (both with cis and trans structures, depending on the polymerization temperature) yielded a material with a metallic luster, which, however, did not exhibit metallic conduction – only after "*doping*" by oxidative or reductive agents, which was observed in the late 1970s. This "doping" is not an exact analogy of the process used in silicon-based semiconductors, but the effects are similar, so the name was retained. Doped polyacetylene was a semiconductor even in the sense that its conductivity increased rather than decreased with temperature, unlike in metals.

A relatively early conference proceedings volume (Kuzmany, Mehring et al. (1985)) gives a vivid picture of the turbulent research going on in this field in the 1980s. Here the author quotes just some items of general interest, without mentioning the author's details; these can be found in the book itself. One essential step forward was the development of polyacetylene synthesis using precursor polymers (the so-called "Durham route"). The synthesis involved monomers obtained by Diels–Alder chemistry, containing cyclobutene moieties, resulting in non-crystalline linear polymers. In later steps, the fluorine-containing auxiliary parts of the molecule were eliminated by heat, resulting in cis-polyacetylene, which could be isomerized to all-trans polyacetylene (see Figure 1.1). If the pyrolytic treatment is

cis-polyacetylene

trans-polyacetylene

FIGURE 1.1   The schematic structure of all-cis polyacetylene and all-trans polyacetylene.

performed simultaneously with stretching, an ordered structure can be obtained. Much effort was devoted to understanding the conduction mechanism in doped polyacetylene. It turned out that the "*soliton model*", a single unpaired electron, an excess electron, or a hole that travels along the polyacetylene chain, plays a surprisingly small role in the actual conduction process. At low doping levels, phonon-assisted charge carrier *hopping* (including variable range hopping, VRH) between localized states is dominant, while in strongly doped states, the inter-fibrillar resistances determine the conduction level. In the intermediate range, tunneling-controlled charge transport occurs between regions of higher and lower conductivities (see also Bakhshi (1995)). The extensive use of Raman spectroscopy to elucidate the fine structure (bond length, cis-trans distribution, defects, etc.) of these low-polarity compounds, which, however, have large polarizability started at this time and became even more widespread with the development of carbon nanostructures. Considerable attention was devoted to aromatic polymers such as polyaniline, polythiophene, polypyrrole, and poly(phenylene-vinylene), which will be the subject of the next subchapter.

1.2   THE DEVELOPMENT OF CONDUCTIVE POLYMERS WITH AROMATIC MAIN CHAIN

In addition to the processability problems of polyacetylene, there was another big obstacle before the practical use of this class of conductive polymers (hereafter CPs): instability in air. This led to an intense study of other, aromatic polymers, which – though still not melt processable – proved to be more stable under ambient conditions. These include *polyphenylene vinylene*, *polyaniline* (PANI), *polypyrrole* (PPy), *polythiophene* (PTh), and related structures (see Figure 1.2).

The reason for the higher stability of these polymers in the air is due to the robustness of the aromatic units as compared to the polyconjugated linear aliphatic chain. In contrast to polyacetylene polymerized originally by standard polymerization methods (using Ziegler–Natta catalysts) the CPs mentioned above are usually synthesized from the monomer either by *chemical oxidation* or by *electrochemical* methods (Inzelt (2008)). Both methods have advantages and disadvantages, which are largely complementary. The wet chemical process can yield large amounts of the product, which can be collected easily. The reaction is relatively complicated, and it is not easy to control the doping level (doping is frequently done in a separate step). Roughly the converse is true for the electrochemical process: it can be easily performed, the doping level can be controlled, but the amount of the product is small and cannot be easily removed from the electrode. The use of one or another method is decided by the targeted application. If the powder product is to be compounded with e.g., plastics, the chemical route is suggested. If, however, we want to apply a coating onto a metal product or want to combine the CP with carbonaceous materials (as in, say, supercapacitors), the electrochemical synthesis is much better. In aqueous solution-based processes the most frequently used oxidative agents are Fe^{3+} salts, Sb(V) compounds, hydrogen peroxide, persulfates, perchlorates, etc. The product is usually a powder that is not soluble either in water or in organic solvents, and the powder cannot be melt-processed. Various strategies have been used to render the insoluble CP chains soluble either in the precursor or in the final form, such as using long or bulky aliphatic side-groups (mainly for polypyrrole and polythiophene), copolymerization, ionic substituents (e.g., in polyaniline), solubilizing agents, and carriers (see e.g., Bhattacharya, De (2007)). The main obstacle in rendering the CPs soluble is the high rigidity of their main chains and the strong intermolecular interactions between them. If bulky side-groups or copolymerization are used, as a rule, the orderliness of the structure is also reduced, resulting in lower in-chain conductivity and loss of coplanarity. The same modifications tend to increase the number of polymerization defects too. In some cases, water-soluble polymers (e.g., chitosan) are grafted with conductive branches (e.g., polyaniline). Although these approaches increase the solubility, the CP sequences tend to micro-phase separate (into layered structures or otherwise) because of the strong interaction of the highly polarizable aromatic groups (stacking). In several technologies, instead of trying to make the CP soluble, various kinds of dispersions are produced (see e.g., Hussin, Gan et al. (2017)). In fact,

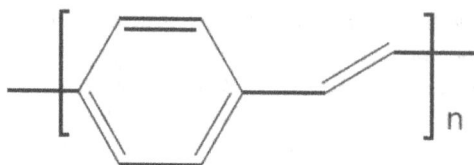

Polyphenylene vinylene

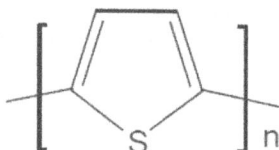

Polythiophene

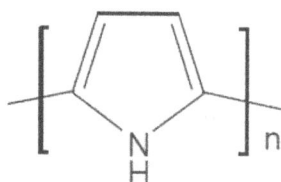

Polypyrrole

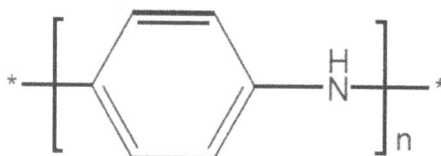

Polyaniline

FIGURE 1.2 The chemical structure of a few conductive polymers with aromatic main chain.

one of the leading developers of polyaniline considers proper dispersion the key to success in this technology (Wessling (1998)). We return to his reasons for doing so in the next subchapter dealing with conduction phenomena.

Polyacetylene did not develop to the commercial scale, but the aromatic main chain polymers did. It is interesting to observe that some early developers of polyaniline and other CPs and their composites (as e.g., Ormecon in Germany or Panipol in Finland) disappeared from the market, which continues to grow steadily, consolidates, and slips over to the hands of major polymer manufacturers and/or compounders, coating manufacturers (SABIC, Lubrizol, Solvay, PolyOne, 3M, McDermid, etc.). A special subgroup of polythiophenes, namely *poly(3,4-ethylenedioxythiophene)* (PEDOT), in combination with *polystyrene sulfonate* (PSS), synthesized from aqueous polystyrene sulfonate solution and 3,4-ethylenedioxythiophene in the presence of oxidizing agents (developed by Bayer AG and marketed under the trade name of Baytron P, where P stands for polymer, or Baytron M, where M stands for monomer – the latter for *in situ* polymerization) began a life on its own as a transparent, thermoelectric, conductive polymer salt, that can be solution processed into transparent, conductive coatings (see e.g., Karri, Srinivasan (2019)).

1.3 CONDUCTION, DOPING, AND PROCESSING

The understanding of the conduction mechanism of CPs always started from the 1D linear molecular structure, mostly using quantum chemical considerations. Figure 1.3 (adapted from Heeger (2001)) shows schematically the hierarchy of various *quasiparticles* playing a role in CPs.

In trans polyacetylene, unpaired electrons could be detected, which had not been detectable before the cis-trans isomerization, probably due to isolated π electrons, called neutral *solitons*. They do not have charge but have ½ spin. Upon contact with acceptors or donors, they form positively or negatively *charged solitons*, respectively. In aromatic main chain polymers, reactions with acceptors or donors change shorter or longer sequences of aromatic units into quinoidal structures, and this combination of charge and unpaired electron forms the *polarons* with positive or negative charges. This reaction causes changes in various physical properties of the material (e.g., color), other than conductivity. The acceptor may extract the unpaired electron from the polaron, resulting in a *bipolaron* with a double *positive* charge, or the donor may add a further electron to the negatively charged polaron resulting in a *bipolaron* with a double *negative* charge. The relative occurrence of polarons and bipolarons in a polyaromatic system with a non-degenerate ground state depends on the level of doping. At lower doping levels polarons are dominating, while at higher doping levels the concentration of bipolarons increases. For a more detailed exposition of the conduction mechanism of single chains of CPs, see e.g., Le, Kim et al. (2019). Polyaniline requires even more attention, as here not only the state of oxidation but also the protonated/deprotonated forms (salt and base forms) must be considered (see e.g., Molapo, Ngandili et al. (2012)). Increasing states of oxidation lead from *leucoemeraldine* to *emeraldine* and then to *pernigraniline*, which are in the (protonated) salt form at low pH and in the base form at high pH. Of these forms only the emeraldine salt is conductive. Interestingly the conductivity of PANI hardly depends on the molecular weight (Yilmaz, Kücükyavuz (2009)), which is explained by the fact that the rate of inter-chain charge transfer is higher than the lifetime of the charge carrier within a single chain. In contrast to polyacetylene, this is a rare case when not the inter-chain charge transfer is the rate-determining step in conduction.

In addition to the main chain structure the type and level of doping (i.e., the addition of oxidizing or reducing agents, electron acceptors, or donors) also strongly influence the conductivity level of conjugated main chain polymers. The most frequently used types of doping include vapor phase, solution, and electrochemical doping by various chemicals, such as halogens, halogenated group-V elements (like P, As) and their anions, alkali metals, sulfonic acid derivatives, sulfites, inorganic Fe^{3+}, and other salts (Le, Kim et al. (2019)).

In some cases, even the addition of relatively simple compounds, such as methane sulfonic acid or hydrogen chloride, is enough to achieve the required change (non-redox doping by protonation). Sulfonic acid doping is used for PANI and PPy – both have nitrogen functions that can be protonated. In the case of PPy some of the sulfonic acid doped salts are soluble in cresol, dimethyl sulfoxide, while the emeraldine salt form of PANI is soluble in strong aprotic solvents as N-methyl pyrrolidone (NMP) and some other solvents, allowing solution processing (Cho, Song et al. (2007)). PPy needs a substantial amount of dodecylbenzosulfonic acid (DBSA) to become soluble – more

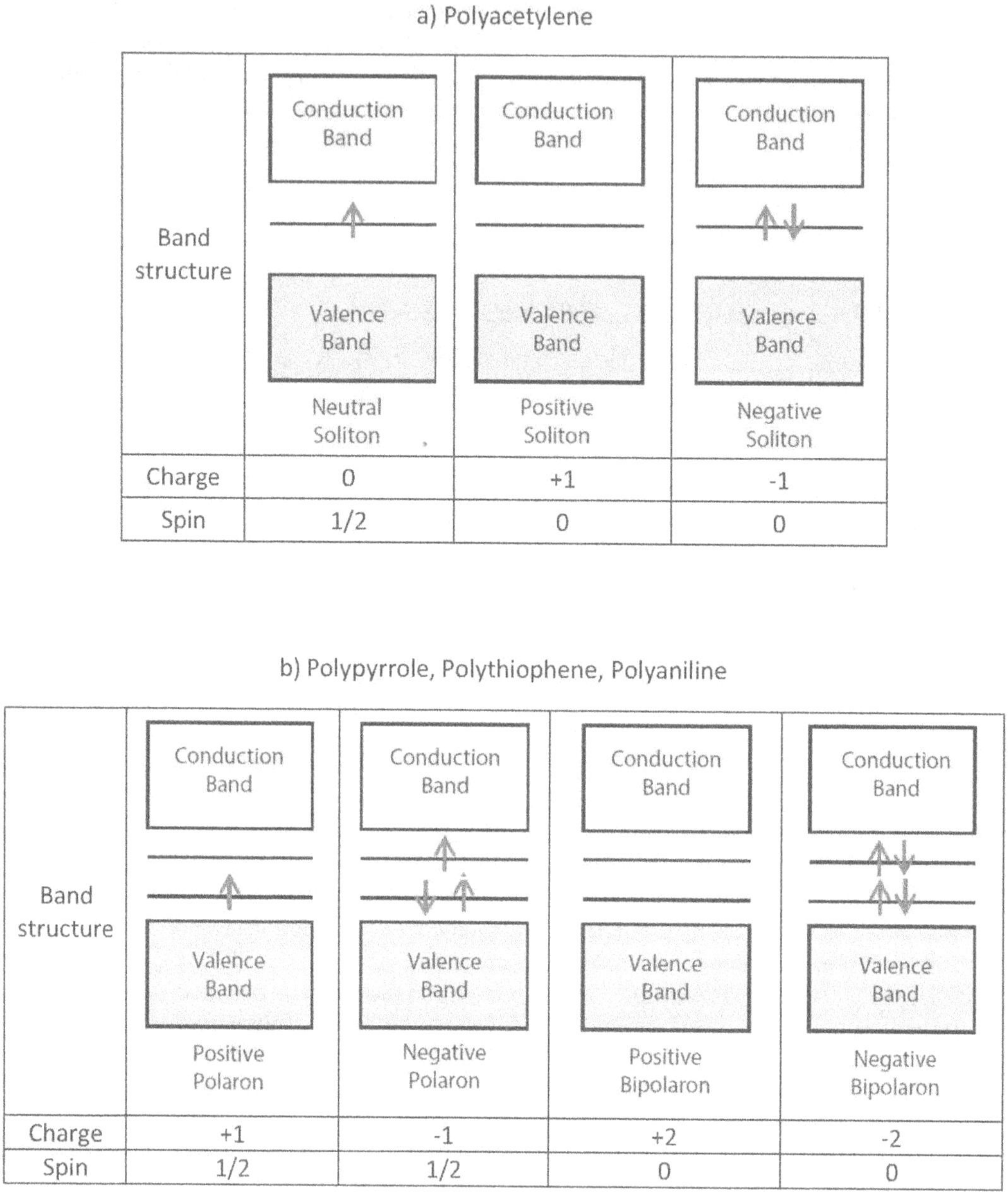

a) Polyacetylene

Band structure	Conduction Band	Conduction Band	Conduction Band
	Valence Band	Valence Band	Valence Band
	Neutral Soliton	Positive Soliton	Negative Soliton
Charge	0	+1	-1
Spin	1/2	0	0

b) Polypyrrole, Polythiophene, Polyaniline

Band structure	Conduction Band	Conduction Band	Conduction Band	Conduction Band
	Valence Band	Valence Band	Valence Band	Valence Band
	Positive Polaron	Negative Polaron	Positive Bipolaron	Negative Bipolaron
Charge	+1	-1	+2	-2
Spin	1/2	1/2	0	0

FIGURE 1.3 Schematic representation of various quasi-particles playing a role in the electric conduction of conjugated polymers, together with their position in the band structure of the polymer. (Adapted after Heeger (2001).)

than that required for doping, so DBSA is a kind of co-solvent, not only a dopant. The extra DBSA molecules surround the salt-like doped PPy molecule and make it soluble in less polar solvents – but this is a micellar solution rather than a true solution. Solution-cast PPy films are smoother than electrodeposited ones. The stability of PPy-DBSA solutions is limited – the polymer tends to precipitate by oxidative gelation. Some of the CPs can be prepared in the fibrillar form either spontaneously, by adjusting the polymerization conditions (PANI, (Huang, Kaner (2007)), or by using *electrospinning* (blends of PANI with more soluble polymers, PPy-DBSA solutions (Cho, Song et al. (2007)). The use of electrospinning technology to obtain fiber mats from CPs has been reviewed among others by Yanilmaz, Sarac (2014). Due to the inherent rigidity, relatively low molecular weight, and poor solubility of typical CPs the realization of electrospinning is not easy – as it has a complex set of requirements with respect to conductivity and solution rheology. Typically, either the CP must be modified to make it more soluble, or it must be combined with other, non-conducting polymers that exhibit higher solubility and flexibility. These tricks reduce the conductivity of the active component, so in some cases further (carbon- or metal-based) additives are used to improve the electrical contacts between the polymer molecules. A review devoted to various techniques for obtaining nanofibers from CPs (Long, Li et al. (2011)) is available, from which electrospinning is only one.

Recently an interesting non-redox solid-state doping technique was suggested for soluble polythiophenes (Kroon, Hofmann et al. (2019)) based on latent acid generators, which can be dissolved together with the polymer to be doped, then, after removing the solvent, can be activated by heat. Chemical doping from the vapor phase can achieve homogeneous doping; if done slowly, the process is reversible (sublimation, desorption). Electrochemical doping can be well controlled and is almost fully reversible and repeatable. Photo-doping is a transient phenomenon, as the formation of charge carriers is caused by the photons, and they disappear by recombination. This may be advantageous, if well controlled, switchable conduction is the goal. In the case of chemical and electrochemical doping, the conductivity level reaches a plateau value (saturation) and may even decrease afterward. When performing the doping process, care must be taken, as the crystalline structure and other (say, mechanical) properties of the polymer may also change.

In addition to coating and fiber (or fiber-mat), formation *patterning* is an important technological process if we want to use the CPs in microelectronics. The application of standard electron- and *photolithographic* techniques (positive and negative *photoresists*) to CP coatings (PANI, PPy) started relatively early (Angelopoulos (2001)), which remained important up till today (Ouyang, Xie et al. (2015)), later followed by *3D printing* technologies, mainly for PEDOT:PSS (Yuk, Lu et al. (2020)). Soft lithography or embossing can also be used for preparing conductive nanofibers (Long, Li et al. (2011)). UV-photolithographic patterning of a negative photoresist can be combined with the *in situ* polymerization of CPs in the structures obtained in the first, photolithographic step (Abargues, Rodriguez-Cantó et al. (2012)). UV light exposure may also change the conductivity of PEDOT–tosylate polymers by changing the polymerization conditions (Edberg, Iandolo et al. (2016)), allowing a $1:10^6$ change in the conductivity according to a pre-determined pattern. Patterned post-polymerization functionalization of CP films (PPy, PTh, and PANI derivatives) obtained by electrochemical methods also allows creating fine structures (Mohammed (2018)) useful as sensor materials. Patterning is not limited to flat (2D) structures. Electrochemical preparation of well-defined spherical microcontainers from PPy in the presence of naphthalene sulfonic acid has been described (Bajpai, He et al. (2004)), and these micro containers can be "designed" to fit into flat patterns created on stainless steel substrate covered with insulating plastic patterns created by microstamping or plasma etching. Other techniques, such as screen printing (Weng, Shepherd, (2010)) and LBL deposition (layer-by-layer deposition, see e.g., Lee, Ryu et al. (2012)), are also available. Various printing technologies such as gravure, transfer, offset, flexography, and nano-imprinting, used to apply CP coatings on textiles are reviewed by Onggar, Kruppke et al. (2020). Directional electrochemical synthesis allows the preparation of CP nanowires (Long, Li et al. (2011)).

It should be noted that the conductivity values quoted in various review papers are always the order of magnitude estimates and rarely refer to homogeneous materials, as in most cases only fibrous or granular agglomerates can be manufactured. In the strict sense of the word, these materials are composites consisting of agglomerates of conducting chains separated by less ordered domains, byproducts, and sparsely distributed voids. The electrical properties of composite materials (both high-frequency limiting permittivity and low-frequency limiting conductivity) can be described at several levels. The first level is based on continuum electrodynamics, and even at this level several models have been suggested (Bánhegyi (1986)). They can be divided into two subgroups: matrix-inclusion type formulae, which are asymmetrical with respect to the phase indices, and statistical mixture type formulae, which are symmetrical, provided that the shape factors assigned to the components are identical. (The shape of the phases is assumed to be characteriz-able by ellipsoids, usually rotational ellipsoids – oblate or prolate ellipsoids.) Prolate ellipsoids reasonably describe fiber or needle-like entities, while oblates approximate well flake-like struc-tures. Equations are available for randomly oriented and parallel subunits and for single-phase or skin-core structures (the latter are approximated by confocal ellipsoids). Skin-core structures can be well utilized to describe structural units that have much higher or much lower surface conductiv-ity than that of the core component. One sub-group of the statistical mixture formulae, based on effective medium theories, predict a phenomenon observable in composite samples composed of phases with widely differing conductivities: *percolation* (Stauffer, Aharony (1994)). The concept was originally developed to describe the flow in porous media, but it can be used for several other phenomena, from insulator-conductor composites to phase transitions. Effective medium theories describe well the observed fact that the critical volume fraction (denoted as *percolation threshold*), where an insulator/conductor composite begins to conduct, decreases in the following order: sphere > oblate > prolate. This approach is, however, not enough for quantitative predictions, as it does not consider the aggregation of the conducting elements in the insulating matrix. The description can be improved distribution of shape factors is introduced, and the aggregates are treated as elongated particles. There is no theory, however, to describe the shape factor distribution vs. the filler volume fraction.

In addition to using strongly non-spherical filler particles, one way of decreasing further the percolation threshold is to utilize the *double percolation* concept. It means that the conductive component (be it CP, metal, or carbon-based conductor) is dispersed in a polymer component which adheres strongly it surface (thus can be easily dispersed in it), and this composite phase is distrib-uted in another polymeric phase that is incompatible with the first polymer and does not wet well the conductive component. With carefully selected components and phase ratios, the first compos-ite phase rendered conductive by the conducting component forms a continuous phase within the second component, so the percolation can happen at a much lower volume fraction of the conduc-tive component. As typical examples, one can mention carbon black composites with two polymer components (Zhang, Chen (2019)), or polyolefin blends containing polyaniline (Yang, Rannou et al. (1998)). This author remembers that when he was involved in the development of a static dissipa-tive composite based on short carbon fibers at the Zoltek Plc (now belonging to the Toray group), improving the conductivity by compounding an elastomer-based masterbatch to the insulating PP matrix and by combining more than one conductive component, e.g., among others, PANI emeral-dine salt, was also tried. This line was, however, not pursued further in that case, as the combination of carbon fiber and conductive carbon black proved to be more effective.

The next level of describing the percolation phenomenon (which plays a decisive role not only in "pure" CP samples but also in the description of blends and composites treated in the next sub-chapter) is the Monte Carlo simulation of percolative networks in two or three dimensions (see e.g., Liu, Regenauer-Lieb (2011)). The critical exponents that describe the change of conductivity near the percolation threshold are closely related to the *fractal structure* and the non-integer *frac-tal dimensions* of the formed clusters formed. First site–percolation and bond-percolation mod-els were developed on 2D (square, trigonal, hexagonal, etc.) lattices, which were later extended to three dimensions, and finally, with improving computational possibilities realistic simulations of e.g., carbon nanotube–filled systems became possible (Tarlton, Sullivan et al. (2017)) where self-avoiding and clustering must be considered. It is, however, argued that even the most sophisticated percolation theories do not take into account the interactions between the conducting units (be they CPs or carbon nanoparticles), leading to the formation of *dissipative structures*, which, at pres-ent, cannot be treated exactly, although a tentative theoretical treatment considering wetting, flow, and spontaneous agglomeration has been advanced (Wessling (1995)). Unfortunately, this approach has been largely neglected, although it could be useful in handling not only the conductivity of composites but also the rheological ones. It has been known for a long time that the conductiv-ity curves of composites filled with conducting particles (be they carbon-, metal-, or conjugated polymer-based) are not well reproducible in repeated heating and cooling cycles (Gul (1996)). It has been usually explained by the temperature-dependent inter-particle distance and by the strongly nonlinear dependence of inter-particle contact resistance between the conducting particles if the charge transfer is determined by *tunneling* (see also Derosa, Michalak (2014)). Therefore, in more recent approaches, where the continuum-based theories expounded above are combined with local phenomena and clustering, special attention is paid to the contact resistance (see e.g., Folorunso, Hamam et al. (2019)). The same phenomenon explains the so-called PTC (positive thermal coefficient) effect in polymer composites filled with conducting fillers (for a recent fundamental study see e.g., Asare, Evans et al. (2016)), and the strong dependence of the electrical properties on the processing conditions. As an example for the latter, here the author just mentions the complex impedance measurements that were performed on LDPE films containing conductive carbon black (Cabelec 969 of Cabot Corporation) that were blow-molded with different speeds to yield films of different thicknesses (30, 80, and 140 µm), perpendicularly to the film direction (Bánhegyi, Hedvig et al. (1991)). The thinner the film, the higher the shear force, orienting the conducting carbon black agglomerates along the blowing direction but breaking up the contacts perpendicularly to the film. With decreasing thickness, the onset of an almost purely resistive behavior shifted to lower frequencies, and the specific low-frequency limiting impedance (the analog of specific volume resistance) increased accordingly by orders of magnitude. This explains why it is so complicated to manufacture semiconducting composites based on conducting and insulating components with repeatable properties on an industrial scale. A phenomenon that complicates further the reproducible measurement of conductivities in such systems is that of micro-breakdown. Thin insulating polymeric layers between CP particles or carbon-based particles tend to break down due to local fields that exceed the breakdown strength of these polymers – nevertheless, these local breakdown events do not lead to an avalanche-like runaway process; only the local interfacial resistance is reduced. Fortunately, these micro-breakdown events occur only during the first electrical loading of the structure and disappear thereafter – unless thermal cycling creates new contacts.

1.4 COPOLYMERS, BLENDS, AND COMPOSITES

As general-purpose thermoplastics, the various properties of CPs can be adjusted by three main methods: by *copolymerization* (i.e., by the combination of more kinds of monomers), by *blending* (i.e., by the combination of various, independent polymer chains), and by adding *fillers* (solid particles, flakes, or needles, or a combination thereof). Linear copolymers can be random or block-like (these are two extremes of a whole spectrum of block-length distribution), while branched copolymers can be graft-copolymers if the branch is grown from the main chain (see Figure 1.4). Post-polymerization modification of polymeric chains may also result in copolymers if e.g., the monomer side chains are modified randomly.

The blending of two polymers is not easy, even in the case of flexible polymers, as the mixing entropy is negligibly small (smaller than in polymer solutions and much smaller than in the mixtures of two low molecular compounds). In fact, polymer pairs which are thermodynamically compatible and mix homogeneously are exceedingly rare even among thermoplastics (only if the mixing enthalpy is negative, i.e., if there is an energetically favorable interaction between them). Even partially miscible systems are, however, called blends and are considered to be "compatible" in the technical sense of the word if they can be mixed, be processed, and yield products with acceptable mechanical properties. In the case of thermoplastic polymers, the main goal was to achieve intermediate properties between the components without copolymerization, which is technically more demanding. In the case of CPs which (as shown in the previous sub-paragraph) are usually non-soluble and cannot be melted, the main goal was to develop blends of CPs and thermoplastics that can be processed by high-throughput technologies like injection molding without losing too much conductivity. This approach proved to be only partially successful, as due to a loss of inter-connectedness, the conductivity of thermoplastics filled with CPs is usually lower than that

Alternating
— A — B — A — B — A — B — A — B — A — B —

Random
— A — A — A — B — A — B — B — A — B — A —

Block
— A — A — A — A — A — B — B — B — B — B —

Graft
— A — A — A — A — A — A — A — A — A — A —
          | |
          B   B
          | |
          B   B
          | |
          B   B
          | |

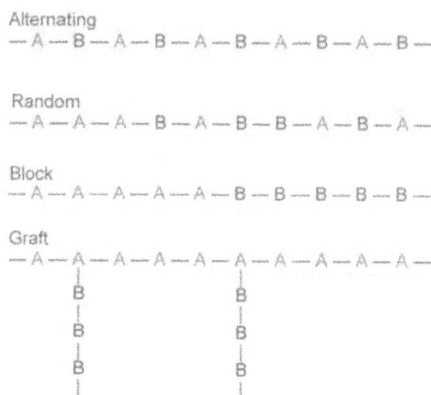

FIGURE 1.4 Schematic view of various kinds of copolymers, where A and B can be any of the monomers shown in Figure 1.2, their substituted analogs, or (especially in the case of block and graft copolymers) any other monomer. In the case of block and graft copolymers the chains or blocks containing non-CP monomers can be polymerized by different mechanisms than the CP grafts or blocks.

of the pure conductive component, so trade-offs had to be found between conductivity and processability. The blending approach is especially popular in electrospinning (Yanilmaz, Sarac (2014)). In fact, the blends of thermoplastics (or thermosets) with conductive plastics can also be considered filled polymers, where the insulating plastic in the matrix, the CP, is the filler. A separate, rapidly growing field is the development of hydrogels with CP additives, which is being widely used in the biomedical and in the sensor field (Sharma, Kumar et al. (2016)).

The last group of multi-component materials to be mentioned in relation to conducting polymers are composites that can be either statistical mixtures of granular materials and CPs or may contain even more components (e.g., thermoplastic, or thermoset matrix, CPs, and other conductive components, namely metals, carbon-based conductors, magnetic materials, etc.). Such combinations can advantageously combine the properties of various conductive components, such as carbon black, carbon micro- and nanofibers, nanotubes, graphene structures, fullerenes, metal nanoparticles, nanowires, and semiconducting nanodots. Moreover, some of these nanoparticles can be generated simultaneously with the CP component, or even together with a non-conducting matrix or hydrogel. This co-generation of hybrid nanostructures allows the realization of intricate fractal structures which could not have been prepared by mixing previously synthesized components. These structures play a key role in electrode materials used e.g., in up-to-date battery electrodes or supercapacitors.

Let us take a closer look at each category. A good recent review is available on the chemical and electrochemical copolymerization of CPs (Jadoun, Riaz (2019)). This reference contains a table with several examples of copolymers together with the achieved conductivity values. In general, the copolymers (if they are real copolymers, not just mixtures of the two homopolymers) exhibit better solubility or higher dispersion stability in solvents than any of the pure components. Here we quote some typical examples.

The combination of pyrrole and aniline (Hammad, Noby et al. (2018)) is relatively straightforward: the oxidative polymerization mechanism of these monomers is similar, so the growing chain can incorporate both kinds of monomers. The morphology, diffraction properties, and porosity of the copolymer depend on the details of synthesis (mixing conditions and sequence, acid added, medium water, or supercritical CO_2). Poly(aniline/pyrrole) copolymers can also be obtained by stepwise (subsequent) polymerization (He, Qin et al. (2018)), where PANI nanofibers are synthesized first and pyrrole is polymerized onto it *in situ*. It changes the smooth fiber surfaces into a rough surface that is used to optimize the electro-rheological properties of the powder. In this case or in other systems, where the pyrrole monomer is polymerized in the presence of pre-formed PANI fibers or dispersions, it is hard to decide whether a surface grafting of the second component occurs to the surface or it is simply a physical mixture of the two polymers, as the interface properties cannot be studied morphologically or by other means. The copolymerization of aniline and pyrrole can be performed in the presence of inorganic nanoparticles (e.g., SiO_2, see Emran, Ali et al. (2017)), where the properties of the final product (in this case, used as an adsorbent) is modified by the presence of the nano-filler.

It is somewhat easier to prepare real copolymers of thiophene and pyrrole in a common solvent (e.g., in acetonitrile, a polar, aprotic solvent, which dissolves not only the monomers but also the anhydrous Fe(III) compound used as an oxidizing agent, forming a homogeneous system (Khademi, Pourabbas et al. (2018)). Electrochemical copolymerization was also performed in the same publication on stainless steel electrodes. The differential scanning calorimetry (DSC) analysis indicated a single glass transition with a Tg value lower than any of the homopolymers – indicating a single-phase and increased mobility in the copolymer.

The preparation of block-copolymers between conductive and other polymers allows the unique combination of both properties, e.g., thermal responsivity and conduction in PNIPAM (poly(N-isopropyl acrylamide)–PANI block copolymers (Abel, Riberi et al. (2019)). In such cases, the synthetic process becomes somewhat more complicated as the PNIPAM block is prepared by free radical polymerization, the PANI block by redox processes. Therefore, a PNIPAM block is synthesized first, and the PANI block is grown onto this end-group. An earlier review (McCullough, Matyjaszewski (2010)) describes a whole series of controlled radical polymerization methods that can be used to incorporate conducting blocks into non-conducting polymers. One reason for the interest in such synthetic techniques is the spontaneous structure formation in block copolymers (lamellar, cylindrical, globular, interpenetrating network) that could have been achieved otherwise by more complicated methods, such as lithographic techniques discussed earlier. By the proper selection of the comonomers, the composition and the casting solvents highly ordered structures can be realized, especially in thin coatings, which are very useful e.g., in sensors. Another review (Maity, Dawn (2020)) describes similar synthetic techniques to obtain graft copolymers of CPs. In the field of grafting, three basic techniques are distinguished: "grafting through" (using macro-monomer – monomer mixtures), "grafting from" (initiation from the main chain + monomer), and "grafting to" (coupling pre-formed sidechains to the main chain). Grafting of conductive chains onto electrospun non-conductive fiber mats is one solution to circumvent the difficulties in the electrospinning on inherently conducting polymers (Blachowicz, Ehrmann (2020)). Graft copolymers are especially looked for in the biomedical field, so the "main chains" to which the graft is attached are frequently cellulose derivatives (e.g., chitosan), but other hydrogels, such as polyacrylamide, are also used. In other cases, the CP is the main chain, and solubilizing side chains (e.g., PEO (poly-ethylene oxide)) are attached to it to improve solubility or dispersibility in an aqueous environment (Such a "shell", however, reduces the probability of charge transfer from one conductive chain to the other.) Aqueous solubility can be achieved in the opposite direction if conductive chains are grown onto PEO blocks. Ionic grafting is also possible if e.g., the counterion in the emeraldine salt is a macro-anion. The PEDOT:PSS pair mentioned above is another example of this approach.

Template polymerization using soft templates, as micelles, emulsions, or hard templates as scaffolds, colloidal particles, channels (porous membranes), and reactive templates is also very popular, especially in the energy-related applications (electrode materials, supercapacitors) as these again allow the synthesis of conductive nanostructures with well-defined morphologies (see e.g., Pan, Qiu et al. (2010)). Template polymerization with hard and soft templates is also a widely used technique for preparing nanofibers from CPs (Long, Li et al. (2011)). Block copolymers, with selective swelling properties with respect to the monomer of the conductive monomer, can also be used as templates for the synthesis of CP nanostructures (Lee, Chang et al. (2006)).

Although, as we have seen, the borderlines are not always clear, let us look at polymer blends based on CPs. It is possible to blend two CPs if a common solvent is found. Tikish, Kumar et al.

(2018)) prepared e.g., PANI/PPy blends were prepared by solution-blending in DMSO solvent from PANI and PPy polymers prepared separately. The x-ray diffraction properties of the blend were determined by the majority phase: semicrystalline for PANI and amorphous for PPy. The DSC investigation proved that the blends were compatible with Tg values situated between those of the components.

Having realized the difficulties of processing CPs, blending with thermoplastics and thermosets processable by conventional technologies was one of the first intensively researched areas to improve the processability without losing too much in conductivity. Wessling, in his cited publications (Wessling (1995, 1998)), discusses experiments on PANI-based systems, in which he was actively involved, dating back to the 1980s, and comes to the conclusion that the key to success is dispersibility. He also emphasizes that the fibrillar structure of PANI (resulting in a low percolation threshold) is only apparent; in many cases, the "fibers" consist of globular subunits that rearrange in shear fields, so good dispersibility should be designed into the CP already during the synthesis. These conclusions are corroborated point by point in an experimental study (Zilberman, Siegman et al. (1998)) where commercial and laboratory synthesized PANI emeraldine salts with two different sulfonates were compounded with thermoplastics of different polarities, and not only the conductivities and morphologies but also the calculated cohesive energy densities regulating the interfacial interaction between the components were compared. Not unexpectedly, as the PANI emeraldine salt has a relatively high cohesive energy density (in the order of 20–24 $[J/cm^3]^{0.5}$), the less polar matrices such as linear low-density polyethylene (LLDPE) or polystyrene (PS) could disperse the PANI particles less effectively than copolyamides or polycaprolactone. As well-known e.g., thermoplastic/carbon black mixtures, if the conductive filler loading is identical, higher conductivities (or lower percolation thresholds) could be obtained for semicrystalline matrices than for amorphous ones, as the conducting particles (carbon black or PANI) are accumulated at the crystalline–amorphous interface, ensuring better contacts. A relatively early review (De Paoli, Gazotti (2002)) summarizes various strategies used until that date to obtain CP blends. If the monomer swells the matrix polymer (e.g., in the case of the pyrrole–PVC pair), the polymerization process may be initiated by electrochemical methods. In a similar way, the polymerization reaction may be initiated by chemical oxidation if the mixture of the insulating polymer, the monomer of the conductive component, and the oxidizing agent are dissolved, and the solvent is evaporated. In other cases, the insulating polymer/conductive monomer pair is dissolved in a solvent, and the cast film is subsequently exposed to another solution containing an oxidizing agent. These are diffusion-controlled processes. A lot of efforts were devoted to the melt-compounding of polymerized CPs with thermoplastics. Mechanical mixing of solutions of thermoplastics and CPs in common solvents yield blend films upon evaporating the solvent. If the base form of the polymer dissolves better, the blend film may be exposed to acid dopant afterward. When using melt blending the highest temperature and residence time must be controlled to avoid conductivity loss. Good dispersion can sometimes be achieved even if the conductive component does not dissolve completely, but a fine aqueous or organic dispersion is available, which can be mixed with the non-conductive component. In such cases sometimes the formation of a percolative network is even more efficient than for a homogeneous mixture.

For the sake of curiosity, we mention a recent paper (Daviddi, Chen et al. (2019)) in which a combination of series of up-to-date nano-imaging technologies was used to understand the macro-, micro-, and nanoscale properties of PTh/PMMA composites at various compositions. Of these techniques, the use of nano-cyclic voltammetry is of special relevance to detect the conductivity differences between the insulating and conducting domains. Nanoscale investigation of the conducting domains helped to clarify that the electrochemical properties of the composite electrode determined under macroscopic conditions are determined not by the intrinsic properties of the PTh domains (which are close to the bulk phase) but by the electrode/composite interface. Such high-resolution test methods may contribute to a better understanding of the agglomeration properties of CPs in an insulating matrix.

The last, and perhaps the largest group, is that of CP composites containing other conductive components as well (first of all, carbon-based conductors, metals, metal oxides), or CPs combined with thermoplastic or thermoset polymers and other conductive components (listed above). One motivation for preparing ternary composites with a single insulating and multiple conductive phases is to improve the connectivity between the conducting particles and to reduce the percolation threshold. Such composites can be prepared by *ex situ* methods, i.e., by mixing the previously prepared components, or by *in situ* methods when the CP is synthesized in the presence of the other component. Nowadays, almost always the second group of methods is used, as the mechanical mixing of the previously prepared CPs is complicated by their rigid structure and the *in situ* preparation allows not only better distribution of the phases but also the application of other chemicals, leading to better interfacial adhesion or other tailored properties (as e.g., the LBL combination of carbon nanostructures with one or more kinds of CPs (Hussain, Abbas et al. (2013)).

Among metal-based CPs composites noble metal composites are favored due to the high stability of the metallic component (Folarin, Sadiku et al. (2011)). The polymerization of monomers in the presence of pre-formed metal nanoparticles is not too effective; *in situ* synthesis in polymer solution is better, but perhaps the best is to oxidize the monomer by the metal ions, so both the polymer and the metal particles are formed simultaneously. The electrochemical method can be two steps or a single step: either first, the metal nanoparticles are deposited onto the electrode followed by the electrochemical polymerization of the monomer, or the polymer formation is performed in a suspension of nanoparticles. The combination of noble metal nanoparticles and CPs improves the catalytic activity of the former and protects the particles providing a kind of scaffold. Due to the plasmon effect metal particles absorb in the visible spectrum depending on the particle size, thus such composites can be used e.g., for preparing optical filters. Further applications will be mentioned in the next subchapter.

CPs combined with graphene, graphene oxide, chemically modified graphenes, graphene nanoplatelets, carbon nanotube, and sometimes also containing metal oxides (mostly transition metal oxides) are widely used in supercapacitors (Bose, Kuila et al. (2012), see also the next subchapter). The lamellar graphene sheets can be utilized either parallel with, or perpendicularly to, the electrical field. The most convenient way to prepare such electrodes is the electrochemical polymerization of conductive monomers on carbon-based electrodes, e.g., graphene paper. Metal oxide nanoparticles can be prepared within the carbon network by sol-gel methods. The application of such combined solutions facilitates the transfer of ions from the liquid phase to the electrodes. CPs are also used for the non-covalent modification of carbon nanotubes (Bilalis, Katsigiannopoulos et al. (2014) and Zhou, Fang et al. (2019)). It is well known that carbon nanotubes cannot be dispersed easily because of the inter-fibrillar interactions resulting in bundle formation. This can be reduced by chemical modification (functionalization), which, however, degrades the conductivity by disrupting conjugation, using surfactants that wrap the nanotubes but separate the conducting particles from each other and by combining the nanotubes with *in situ* formed CPs. These molecules reduce inter-fibrillar interactions but, being conductive themselves, do not destroy electrical contacts. In general, it has been observed that the *in situ* polymerization of conductive monomers with carbon-based fillers reduces the agglomeration tendencies of the latter, resulting in more homogeneous mixtures not only in the case of carbon nanotubes but also conductive carbon black (Xie, Zhang et al. (2019)). The thickness of the wrapping layers around the carbon nanofillers can be tuned by selecting the synthesis conditions (Zhang, Xu et al. (2011)). Graphene–CP composites exhibit a strong chemoresistance – i.e., the resistivity depends on the presence of certain chemicals in the surrounding atmospheres, which is a kind of doping effect that can be utilized in chemosensors (Zamiri, Haseeb (2020)). When using vapor-grown carbon fibers (VGCF) followed by oxidative polymerization of e.g., pyrrole, it is possible to prepare carbon fiber/ PPy nanocomposite layers with nanometer thickness (Sharma, Kim et al. (2009)). It is also possible to fabricate coaxial fibrous structures from two different CPs (e.g., PPy nanotube covered by the polymerization products of aniline or phenylenediamine), which yield high conductivity and, after carbonization, special, nitrogen-doped carbon nanotubes (Stejskal, Sapurina et al. (2015)).

Composites based on metal oxides, sulfides, and other inorganic compounds (clays, zeolites) have also been intensively studied (Maity, Biswas (2006)). Well-dispersed metal or semimetal oxides (SiO_2, SnO_2, TiO_2, MnO_2, ZrO_2, Al_2O_3) provide a template for the polymerization of conductive monomers initiated by oxidizing agents (e.g., $FeCl_3$), and some of these oxides with more oxidation states, or which can be doped with metal ions of different charge (e.g., SnO_2 doped with Sb), can contribute to the electronic conduction process too. Layered oxides (V_2O_5), sulfides (MoS_2), silicates (as montmorillonite), and other compounds (FeOCl, molybdenum bronze, phosphates) can be intercalated by the monomers or protonated monomers of the conducting polymer compounds, and the polymerization process can be initiated, which results in partial or complete exfoliation of the layered mineral. In each case, strong adsorption of the polymer molecules to the particles or platelets is observed. Multi-layer or multi-component metal-oxide and CP dispersions have also been made by solvent and monomer exchange. It is hard, or almost impossible, to predict the changes in conductivity in the presence of inorganic nanoparticles, but e.g., in the case of N-vinyl carbazole, which, by itself is only a semiconducting polymer, significant and consistent improvement in the conductivity was observed. Nevertheless, other properties such as optical, catalytic, or electrorheological properties can be advantageously changed; therefore the exploration of ever newer combinations is actively pursued. The addition of $AgNO_3$ or water-soluble gold compounds to pyrrole or aniline allows not only direct oxidative polymerization but also indirect photopolymerization both in the presence and in the absence of layered silicates (see e.g., Saad, Jlassi et al. (2017), which describes a wide variety of clay-conductive polymer composites), where the noble metal ions play the role of the oxidizing agent. Systematic variation of the substituents in PTh combined with metal and metal-oxide nanoparticles allows the tailored manipulation of the bandgap and broadens the possibility to use such components as photodegradation catalysts in the treatment of wastewater (Ansari, Khan et al. (2015)). This synergistic interaction can be explained by the fact that the energy gap in PTh is smaller than in the metal oxides, and the excited PTh can transfer the excitation to the metal oxide. Nano-fibrous PANI-MoS_2 with advanced electrochemical properties could be synthesized in two steps using hydro-thermal methods: first, MoO_x-nanofibers were prepared in ammonium polymolybdate-aniline solution in the presence of water-soluble persulfates, which were transformed into MoS_2-PANI nanofiber in aqueous thiourea solution in an autoclave (Ganesha, Veeresh et al. (2020)). A somewhat similar effect was observed in mixed metal (Cu, Ni, Zn) oxide–PANI nanocomposites prepared by low-temperature sol-gel technology combined with oxidative polymerization for anticorrosive purposes (Kumar, Boora et al. (2020)): the bandgap of the synthesized composite was smaller than those of the pure oxide or the PANI component. In addition to its excellent anticorrosive properties, the material exhibited extremely high heat resistance (above 600°C) and unique optoelectronic properties.

The electrochemical methods of preparing CP-semiconductor (oxide, sulfide, selenide, and carbon-based) hybrid assemblies have been thoroughly reviewed by Janáky and Rajeshwar (2015) with special attention to the applied current-voltage programs, monomers, solvents, electrolytes, electrode materials, and ways of incorporating inorganic nanoparticles to the polymer component during electro-synthesis. The review deals in detail with the combined application of photochemical and electrochemical methods, where the illumination is used to increase the conductivity of the semiconductor component during synthesis.

A whole chapter has been devoted to the effects of special sample morphologies (hollow nanospheres, layered nanostructures, nano-belts) of CPs (by themselves or combined with carbon or metal-oxide based additives) on the detection possibilities of toxic materials in electrochemical sensors in a recent monograph on Advanced Functional Materials (Ameen, Akhtar et al. (2015)).

1.5 APPLICATIONS

This book focuses on energy-related applications, such as supercapacitors, photovoltaics, electrode materials in batteries, fuel cells, electrocatalysts in water electrolysis, and hydrogen storage applications, but there are other applications closely related to energy conversion or energy saving, such as thermoelectrics (harvesting waste energy), electroactive polymers (direct conversion of electric energy to mechanical energy), antistatic, static dissipative, and EMI (electromagnetic interference) shielding materials used in various fields of electronics and electric industry, and sensors which are used in energy-related fields but are equally important in environmental protection and the biomedical industry. Corrosion protection and tribological (low-friction and wear-reducing) coatings are also directly coupled to the energy industry. As specialized paragraphs are devoted to several fields listed above, the author will discuss them briefly and will spend some more time with the more "exotic" applications. This subchapter shows extensive overlap with the previous one, as the various composite structures mentioned there are mostly used in these applications; in some cases, they were developed exactly for these purposes.

Supercapacitors are charge- and energy-storing devices that bridge the gap between electrolyte capacitors (they can store much more specific energy) and batteries (they deliver the charge much faster). In these systems, the charge separation occurs at a high specific surface carbon/liquid electrolyte interface, usually without chemical reaction (non-Faradaic mechanism – a difference from batteries). These systems are called electric double-layer capacitors (EDLC). In some other cases, fast, reversible electron transfer also occurs on an electroactive component present at the interface (Faradaic mechanism). These latter systems are called pseudocapacitors (PCs). The capacitance increment is higher in pseudocapacitors, but the cyclability is worse and the internal resistance is higher. Of course, both mechanisms can be present in a given system simultaneously, which can be studied by impedance spectroscopy, cyclic voltammetry, and other experimental techniques. One main development direction in this field is the increase of the specific surface of the carbon materials using a wide range of nanostructures (Bose, Kuila et al. (2012)). Carbon nanotubes and graphenes are frequently combined with CPs and/or metal oxides to improve the efficiency, the continuity of electron, and ion transfer. The multiplicity of conduction mechanisms and contacts results in a more balanced behavior (Peng, Zhang et al. (2008)). Various generations of supercapacitor structures are outlined in Table 1.1. Both CPs and transition metal oxides increase the specific capacitance as compared to porous carbon alone. The combination of these additives with the traditional (or new) carbon nanostructures and electrolytes can help to enjoy the advantages of both worlds, so it is up to the applied research projects run at commercial companies to improve the efficiency and the

TABLE 1.1

Various Generations of Supercapacitor Electrode Systems (Modified and Completed after Peng, Zhang, et al. (2008))

Material Group	Subgroup	Charge Accumulation Principle	Developed Since
Carbon	Porous carbon	Electric double-layer	Late 1950s
	Carbon fiber	Non-Faradaic	
	Carbon nanofiber		
Metal Oxide	Transition metal oxides	Pseudocapacitor	1970s
	Semimetal oxides	Faradaic	
Conducting Polymer	PANI		
	PPy		
	PTh		
Composite	Metal oxide + carbon	Combined	1990s
	Conducting polymer + carbon		
	Ternary (metal oxide + conducting polymer + carbon)		

Physicochemical properties	Carbon materials	Metal oxides	Conducting polymers
Non-faradaic capacitance	☺ ☺ ☺ ☺	☺ ☺	☺ ☺
Faradaic capacitance	☻	☺ ☺ ☺ ☺	☺ ☺ ☺ ☺
Conductivity	☺ ☺ ☺ ☺	☺	☺ ☺ ☺ ☺
Energy density	☺	☺ ☺ ☺	☺ ☺
Power density	☺ ☺ ☺	☺	☺ ☺
Cost	☺ ☺	☺ ☺ ☺	☺ ☺
Chemical stability	☺ ☺ ☺ ☺	☺	☺ ☺ ☺
Cycle life	☺ ☺ ☺ ☺	☺ ☺	☺ ☺
Easy fabrication process	☺ ☺	☺	☺ ☺ ☺
Flexibility	☺ ☺	☻	☺ ☺ ☺

Very high, ☺ ☺ ☺ ☺; high, ☺ ☺ ☺; medium, ☺ ☺; low, ☺; very Low, ☻.

FIGURE 1.5 Comparative advantages and disadvantages of various materials used for supercapacitor manufacturing. (Taken from Shown, Ganguly et al. (2015), used under the terms of the Creative Commons Attribution License.)

economic aspects of these devices. The advantage of using CPs vs. metal oxides as "power boosting" components is their relatively low price and synthesis. Some relative strengths and weaknesses of various materials used in supercapacitor fabrication are schematically shown in Figure 1.5.

The advantages of CPs over carbon and metal-oxide-based electrodes are conspicuous if developing flexible supercapacitors (Shown, Ganguly et al. (2015)). In these systems both the electronic and mechanical properties are utilized – both alone and combining CPs with carbon nanostructures or metal oxide nanoparticles. It is worth noting that in addition to combining carbon nanofibers with CPs, it is possible to use non-conducting but flexible fibers, such as cellulose nanofibers, to prepare CP-based nanocomposites for energy storage applications (Liew, Walsh et al. (2017)).

Composites of CPs and metal oxide nanoparticles are widely used as *photovoltaics* and *electrochromic* materials (Ameen, Akhtar et al. (2013), Ates, Karazehir et al. (2012)). Dye-sensitized solar cells (DSSC), containing a dye-sensitized anode and an electrolyte, are relatively cheap versions of solar cells that can be manufactured by roll-printing techniques, utilize transparent photosemiconductor nanostructures made of TiO_2, ZnO, SnO_2, and further oxides combined with the three main CPs (PANI, PPy, PTh), but other photosensitive polymers are also used, such as polyvinyl carbazole, polyazulene, polyindole, and others. As the CPs are themselves colored and photoactive; therefore they can play the role of the dye in oxide–polymer nanocomposites, although oxide-dye-conducting polymer combinations are also used. CPs, as PANI doped with various anions, are also used as counter-electrode instead of Pt.

Lithium batteries came to the fore in energy storage when the development of *electromobility* – first of all, electric cars – became an urgent task due to the carbon emission problems and the foreseeable depletion of fossil fuel sources. Traditional Li-batteries are usually based on mixed Li-oxide anodes and graphite cathodes. CPs appeared mostly as binders (Eliseeva, Kamenskii et al.

(2020), Nguyen, Kuss (2020)) or as electrode component materials (Sengodu, Deshmukh (2015), Tajik, Beitollahi et al. (2020)). As electrode components, one big advantage of CPs is that they can easily and reversibly absorb the "swelling–deswelling" effects caused by Li uptake and release, which tends to destroy more rigid anode or cathode materials like carbon, silicon, or mixed oxides. Lately, water-soluble binders such as CMC (carboxymethyl cellulose) are frequently used instead of the formerly preferred PVDF (polyvinylidene fluoride), which requires non-aqueous electrolytes. Adding PEDOT:PSS to CMC further improves the flexibility and charge transfer properties of the binder material.

Recently much effort was devoted to *lithium-sulfur batteries* due to their high energy density, where the cathode is sulfur and the anode is lithium metal. Various CPs, such as PANI, PTh, and PPy, are used as coating layers, conductive hosts, sulfur-containing compounds, separator modifier/ functional interlayers, binders, and current collectors (Hong, Liu et al. (2020)). The use of CPs improves the cyclability of the batteries due to the robust, continuous conductive structure. Better cyclability can be explained through the interaction between the CPs and the polysulfide ions generated in the batteries, which play a "shuttling" role in the electrochemical processes. PANI can be well used for wrapping and embedding sulfur particles (alone or in combination with carbon nanotubes, graphene, graphene oxide, etc.) or as a conductive medium that allows both ionic and electronic charge transfer. Starting from a chlorinated aniline derivative it is even possible to produce PANI containing covalently bound sulfur. The synthetic techniques mentioned in the previous subchapter allow e.g., the deposition of sulfur and conductive PPy nanolayers, which confine the movement of polysulfide ions but ensure good overall conduction properties. Using template polymerization, PPy-fiber hosts can be synthesized for sulfur particles, while pyrrole-infused graphene can be melt impregnated to yield stable conducting structures. PPy has also been used as a component of the separator film between the cathode and the anode and as current collector films, in combination with sulfur. PTh is mostly used in the PEDOT:PSS form – again combining excellent ionic and electronic conductivity. Utilizing the block copolymer synthetic possibilities discussed earlier it is also possible to develop polythiophene – polysulfide block copolymers that go from insulating to conducting with increasing (substituted) thiophene content.

In *fuel cells* that convert chemical energy to electricity CPs play a special role as contact-improving additives within the proton-conducting membrane used in hydrogen- and methanol-based fuel cells, as well as in bipolar plates, which distribute the gases and liquid movement in the fuel cell and help to maintain the heat balance (Kausar (2017)). In the proton-conducting membrane the combined proton and electronic conduction is the main advantageous property, while in the bipolar plates the key is the low corrosivity as compared e.g., to stainless steel structures. CPs, in combination with the traditional Pt-based catalysts, also play an important role in fuel cells as *electrocatalysts,* while in combination with oxide semiconductors as TiO_2 they improve the efficiency of *photocatalysts* in solar energy conversion and in environmental protection (Ghosh, Maiyalagan et al. (2016)).

In *hydrogen storage*, which is important in fuel cells, in hydrogen-fueled cars but also as transitory energy accumulation technology, CPs play a dual role: if prepared appropriately, they can adsorb a large amount of gases (including hydrogen), but as systems that can be reversibly oxidized and reduced, they can store the hydrogen also chemically (proton-doping/de-doping, see e.g., Mahato, Jang et al. (2020)).

Thermoelectrics that can convert a temperature difference into DC voltage or can be used as *Peltier* modules for creating temperature difference (mostly used for cooling) are intensively researched and developed, as they can be used for *energy harvesting* (utilization of waste heat). It can be shown that the product of the so-called figure of merit (z) of thermoelectric material is proportional to the electrical conductivity and to the square of the Seebeck coefficient and inversely proportional to the heat conductivity of the material (Snyder, Snyder (2017)). It means that we need good electrical conductivity and low heat conductivity – while the two change in the same direction in metals (Wiedemann-Franz law). In semiconductors, however, it can be shown that the zT product (where T is the absolute temperature) exhibits a maximum at a certain charge carrier concentration.

So far, the most successful thermoelectrics (with a zT product around 1) have been compound semiconductors made of heavy elements, such as Bi_2Te_3, SnSe, PbTe, and so on. Best organic thermoelectrics based on CPs still do not reach this level (zT around 0.1–0.5, see Culebras, Gómez et al. (2014)), but the improvement is spectacular: while zT factor of inorganic thermoelectrics improved only by 3–4 in the past few decades, this improvement has been about three orders of magnitude for CP-based thermoelectric, so it is worth continuing. Copolymerization and composite formation are further possibilities to increase the figure of merit (Yao, Che et al. (2015)), and the fabrication and patterning methods (already mentioned above in the previous subchapters) add further variables to optimize the properties of the Seebeck or Peltier modules made of organic thermoelectrics (Kamarudin, Shamir et al. (2013)).

There are two applications where CPs change not (only) the electrical properties of the medium they are embedded in but also their mechanical behavior: electroactive polymers (EAPs) and electrorheological fluids. EAPs, used in artificial muscles and in sensors/activators, exhibit considerable shape and size change in the electrical field (which should not be mixed with piezoelectrics, where the change is relatively small). EAPs are divided into subgroups, one of which are conducting polymers (Kaneto (2016)). PANI, PPy, and PEDOT:PSS have all been used for this purpose. PPy is studied most as it can be deposited in film form onto metal films and these films exhibit acceptable mechanical resilience. PANI, on the other hand, can be processed in film form in aprotic solvents as NMP or DMSO, the cast film exhibits electroactive properties at low pH<3, which is acidic and not amenable to an electronic environment. Moreover, the elongation at the break of the PANI films is much lower than in the case of electro-deposited PPy films. PEDOT:PSS exhibits small deformation and less excellent mechanical properties than the other two CPs, but it has great stability and repeatability under electrochemical conditions. The electroactive behavior of CPs is explained by the fact that in the as-prepared state they are usually oxidized, doped (swollen) with anions, and are relatively stiff because of the ionic crosslinks. If reduced, the network shrinks but can be reversibly oxidized back to the swollen state. Depending on the size of the anion the process can be anion- or cation-controlled. *Electrorheological fluids* are insulating liquids with dispersed, polarizable particles that tend to form filaments under the effect of an electrical field, thus stiffening the structure, resulting in a considerable increase of viscosity and change of the dynamic rheological properties. These polarizable particles can be conducting polymers (Lu, Han et al. (2018)). The advantage of these systems vs. metal oxide powders is that their density differs less from the dispersing medium, and thus settling causes fewer problems. All three CP types are used, as well as blends of CPs with non-conducting polymers and CP composites with insulating filler nanoparticles, such as halloysite or montmorillonite particles. As opposed to conducting applications too high conductivity should be avoided, as high leakage currents cause problems in the operation of the system. For this reason, e.g., PANI is not used in the emeraldine salt form but in the base form, where the polarizability is still considerable, but the conductivity is lower. Various intricate core–shell structures have also been synthesized to cover the conducting core with insulating shells to reduce the leakage current. Main applications include vibration control, brakes, clutches, and dampers.

Various metal- and carbon-based polymer composites have long been used as *antistatic, static dissipative,* and *EMI shielding* compounds and coatings, so it is natural that the inherently conductive polymers have also been tried in these applications. The distinction between these categories is somewhat arbitrary but is usually based on their specific surface and bulk volume resistances. Antistatic compounds usually reduce dust deposition on plastic surfaces by preventing static charge accumulation (about 10^9 ohms), static dissipative compounds remove static charges by conduction if connected to the ground (between 10^9 and 10^5 ohms), and EMI shielding compounds (10^3 to 10^0 ohm) attenuate electromagnetic waves by 30–70 dB. The simplest antistatic surfaces do not need conducting components; it is enough to render the surface hydrophilic (e.g., by adding migrating additives) as the water adsorbed from the ambient atmosphere increases the surface conductivity enough to reduce dust accumulation. Static dissipation requires higher conductivity and several melt-blends of CPs and thermoplastics are in this range, so static dissipative films or injection molded articles can be prepared (see e.g., Gill, Ehsan et al. (2020)). Here the main question is always how can one reduce the amount of the more expensive CP component and how can one get a stable conductivity value on the production line. (These problems have been treated in the subchapter devoted to CP blends and composites. Depending on the doping level and structure of the CP particles used, the percolation threshold for antistatic and ESD applications is usually 5–20%, which can be reduced further by special techniques, see e.g., Wessling (1989)). CP nanofiber mats (Long, Li et al. (2011)) can be used as antistatic or ESD layers by themselves. Aqueous PEDOT:PSS dispersions applied onto insulating fiber mats or plastic surfaces also exhibit the right conductivity range (Onggar, Kruppke et al. (2020), Tseghai, Mengistie et al. (2020)). Electromagnetic interference shielding (especially in GHz) has become an intensively researched subject not only because of the potential health effects of mobile communication but also because of its military importance. As shielding efficiency consists of three components, namely absorption, reflection, and multiple reflections (Lyu, Liu et al. (2018)), it is related to the much-searched stealth technologies used in military fighters. As electromagnetic radiation has both electrical and magnetic components, the composite approach is preferred. High electrical conductivity is not enough, and good magnetic absorption is also needed, so ferrite or other magnetically active components must be added to the CPs, carbon, or metal additives. The cited review investigates the various solutions in groups: thermoplastic with carbon fillers, metal nanofillers, conductive polymer nanofillers, porous carbon nanofillers, and combinations thereof. The acceptable shielding effectiveness in the GHz range starts from 20 dB but in exceptional cases, 60–80 dB can also be achieved. The critical evaluation shows that the combination of various types of fillers brings synergism and that morphology plays a dominant role. Wanasinghe, Aslani et al. (2020) also compared a wide range of polymer composites and nanocomposites used for EMI shielding the GHz range, and the experimental shielding effectiveness/frequency curves show that these systems are highly individual; in some cases the curves are smooth, and in other cases they show one or more maxima (sometimes very sharp ones). Therefore such systems must be developed carefully and individually for each application. At present, it is art rather than science – although there is a significant amount of science in it. Besides EMI shielding and ESD protection, there were other useful applications of CPs in electronics beginning from the relatively early stage of development, including lithography (conducting resins, charge dissipation), metallization, corrosion protection, interconnections and wiring, and as parts of semiconductor devices (diodes, transistors) (Angelopoulos (2001)). Electrospun and printed CP nanowires (both by themselves and as blends) were studied as potential components of field-effect transistors (FET) (Luzio, Canesi et al. (2014)).

Conducting polymers, due to their electrochemical activity and doping sensitivity, are eminently good materials in *sensor* applications. The main areas where both conducting polymers and their composites can be used include *gas sensors*, *electrochemical sensors*, *temperature sensors*, and *biosensors*.

In electrical nose applications used for gas and vapor sensing (Arshak, Moore et al. (2004)) CPs and e.g., carbon-based conducting particles are dispersed in an insulating polymer matrix and, if the gas to be sensed swells the polymer matrix, the resistivity of the near-to-percolation composite increases. (The physics behind this change is similar to the PTC effect mentioned in the subchapter devoted to PC composites). In this case, only the conductivity of the PC component is utilized. If a conducting polymer layer itself is the sensor, the presence of the sensed gas changes the conductivity of the polymer by changing the interchain conductivity, the proton conductivity (by H-bond formation), and the electronic intrachain conductivity itself. A separate review has been devoted to gas sensors based on CP-inorganic nanocomposites (Yan, Yang, et al. (2020)). As CPs by themselves exhibit relatively low conductivity and, besides oxidative or reductive components, because they are sensitive to organic vapors and water, their sensitivity and selectivity are not too good. Therefore the addition of inorganic additives (metals, metal oxides, and carbon-based nano-additives, such as graphenes or carbon nanotubes) help to alleviate these problems. They can accelerate response; improve conductivity (mostly by adding carbon-based nanofillers), sensitivity, and selectivity; and due to their high specific surface allow a wide variety of surface modifications. The tricks used to achieve these ends include Schottky heterojunction formation at the filler/polymer interface (in metal-oxide/CP composites) and tailored morphology (porosity). The constantly improving arsenal of prepared organic/inorganic hybrids and metal-oxide nanotubes allows the fabrication of ever newer flexible nanosensors. Noble metal/CP nanocomposites utilize the high adsorption ability and catalytic activity of these metals, which modulate the electrical conductivity too. Moreover, metal- or metal-oxide "decorated" nanotubes and graphenes can also be synthesized, which conveniently combine the advantages of both subsystems in ternary CP nanocomposites. Ion sensing (Sethumadhavan, Rudd et al. (2018)) can be done using the electrochemical principle (see below), but anions can be directly detected as they influence the doping level and conductivity of CPs and cations indirectly. It should be noted that not only the conductivity but also the UV-Vis spectrum of CPs is influenced by the presence of ions, which can be utilized in ion sensing.

Electrochemical sensors based on CPs or their double or triple composites are mostly prepared by electrochemical methods, either on metal surfaces or using the conductivity of the carbon component (nanotube or graphene layer) deposited on any substrate (El Rhazi, Majid et al. (2018)). These detectors can be used not only for gases but also for detecting metal ions and even bioactive molecules. In addition to metals, metal oxides, and carbon nanoparticles CP composites can incorporate other additives, such as dye molecules, metal-oxide frameworks (MOF), enzymes, and other biologically active molecules or complexes (Naveen, Gurudatt et al. (2017)). It is even possible to prepare MIP (molecularly imprinted polymer) layers on electrochemical sensors. These and entrapped enzymes allow very specific biomolecule detection. Non-covalent functionalization of carbon nanotubes with CPs (Zhou, Fang et al. (2019)) is also utilized in *biosensors*. Functionalization of the CP layer allows building covalent or Van der Waals binding of specific macromolecules to the electrode surface.

Besides biosensors, there are, however, further *biomedical* applications of conducting polymers, which will be just listed here as examples. Electrically conducting, biocompatible hydrogels based on CPs, and conventional hydrogel materials that exhibit I–V characteristics depending on the degree of swelling can be prepared e.g., by freezing–thawing method without using potentially toxic crosslinkers (Gotovtsev, Badranova et al. (2019)). Using hierarchical nanostructures (at the micron, nanometer, and Angstöm size) in polyaniline containing gels based on phythic acid (a six-fold dihydrogenphosphate ester of inositol, a naturally occurring non-polymeric molecule), it is possible to obtain with rugged, scalable conductivity at very low PANI content (Pan, Yu et al. (2012)) – this resembles the techniques used in decreasing the percolation threshold in conductor/insulator composites. In this case ionic crosslinks develop between the phosphate groups and the NH groups of the polyaniline component. CP composites are also used in implants, which are highly sophisticated applications (drug delivery devices, minimally invasive electronics, electrodes for neural interfaces) (see Liu, Yin et al. (2020)). CPs and their various composites have been used to prepare 3D scaffolds that were used to monitor or stimulate cellular activity not only in cell cultures but also *in vivo* (Khan, Cantù et al. (2019)). It is worth noting that PANI fibers (similar to other polycations) exhibit measurable antibacterial activity (Shubba, Kalpana et al. (2016)). Recently, much effort was devoted to developing biocompatible and even biodegradable conductive polymers based on electroactive macromonomers, as PANI by itself has toxic properties (da Silva, de Torresi (2019)). Other biomedical applications of CPs and their composites include scaffold materials, tissue engineering, wound healing, food packaging, actuators, and artificial nerves (Kaur, Adhikari, et al. (2015)).

The last topic to be mentioned is that of *corrosion-resistant* and *low-friction* coatings, which are both indirectly related to energy saving by increasing the useful lifetime of various machine parts and metal objects. CPs and their composites are environmentally more benign than, say, chromate coatings and receive considerable attention (e.g., Ates (2016), Xu, Zhang (2019)). Polyaniline, polypyrrole, as well as polycarbazole coatings have shown protection efficiency in excess of 90% on steel and its alloys, and also on Al, Zn, and their alloys. It should be noted, however, that CP coatings are frequently combined with epoxy or other topcoats as they tend to be porous.

In contrast to corrosion resistance, the advantageous *tribological (friction reducing)* properties were not expected from the structure of the conjugate. Nevertheless, it was observed that, e.g., polyaniline is advantageous as a friction-reducing additive in special greases (Cao, Xia (2017)), but the tribological properties of PANI brush on the gold surface were also found to be beneficial in a high-density data storage system (Yoshida, Fujinami et al. (2013)). Polypyrrole coatings on natural fibers (Wang, Lin et al. (2005)) reduced friction but excellent tribological properties were also observed on bamboo fiber-filled PA6 composites covered by polypyrrole coating (Niu, He et al. (2017)). These and other observations not cited here show that it may be worth studying and understand the reasons for this behavior.

1.6 CONCLUSION

We hope that even this short introductory paragraph shows that the material science of conductive polymers and their composites developed from a curiosity toward a wide, rapidly developing, continuously ramifying subject where, in addition to basic science, there is a lot of technology. Although we are far from the conscious design and tailored manufacturing of CP structures, the empirical material is vast and several rules of thumb could be established. The advantages and disadvantages of various types of CPs and their composites have been largely clarified. Due to the space limitations, we could not do more than pointing to some of the most important and actual tendencies. Even the number of review articles (which we tried to cite) is so large that we can only hope that the interested reader will find the rest of the data in the references and the following specialized paragraphs.

REFERENCES

Abargues R., Rodríguez-Cantó P. J., García-Calzada R., Martínez-Pastor J. (2012) Patterning of Conducting Polymers Using UV Lithography: The in-Situ Polymerization Approach, Journal of Physical Chemistry C, **116**, 17547–17553, https://doi.org/10.1021/jp303425g

Abel S.B., Riberi K., Rivarola C.R., Molina M., Barbero C.A. (2019) Synthesis of a Smart Conductive Block Copolymer Responsive to Heat and Near Infrared Light, Polymers, **11**, 1744, 1–13, https://doi.org/10.3390/polym11111744

Ameen S., Akhtar M.S., Song M., Shin H.S. (2013) Metal Oxide Nanomaterials, Conducting Polymers and Their Nanocomposites for Solar Energy, Chapter 8, in: Solar Cells – Research and Application Perspectives, Morales-Acevedo A. (ed.), InTechOpen, ISBN: 978-953-51-1003-3

Ameen S., Akhtar M.S., Seo H.-K., Shin H.S. (2015) Development of Toxic Chemicals Sensitive Chemiresistors Based on Metal Oxides, Conducting Polymers and Nanocomposites Thin Films, Chapter 1, in: Advanced Functional Materials, Tiwari A., Uzun L. (eds.), Scrivener Publishing, Salem, ISBN 978-1-118-99827-4

Angelopoulos M. (2001) Conducting Polymers in Microelectronics, IBM Journal of Research and Development, **45**, 57–75, https://doi.org/10.1147/rd.451.0057

Ansari M.O., Khan M.M., Ansari S.A., Cho M.H. (2015) Polythiophene Nanocomposites for Photodegradation Applications: Past, Present and Future, Journal of Saudi Chemical Society, **19**, 494–504, https://doi.org/10.1016/j.jscs.2015.06.004

Arshak K., Moore E., Lyons G.M., Harris J., Clifford S. (2004) A Review of Gas Sensors Employed in Electronic Nose Applications, Sensor Review, **24**, 181–198, https://doi.org/10.1108/02602280410525977

Asare E., Evans J., Newton M., Peijs T., Bilotti E. (2016) Effect of Particle Size and Shape on Positive Temperature Coefficient (PTC) of Conductive Polymer Composites (CPC) – A Model Study, Materials and Design, **97**, 459–463, https://doi.org/10.1016/j.matdes.2016.02.077

Ates M. (2016) A Review on Conducting Polymer Coatings for Corrosion Protection, Journal of Adhesion Science and Technology, **30**, 1510–1536, https://doi.org/10.1080/01694243.2016.1150662

Ates M., Karazehir T., Sarac A.S. (2012) Conducting Polymers and Their Applications, Current Physical Chemistry, **2**, 224–240, https://doi.org/10.2174/1877946811202030224

Bajpai V., He P., Dai L. (2004), Conducting-Polymer Microcontainers: Controlled Syntheses and Potential Applications, Advanced Functional Materials, **14**, 145–151, https://doi.org/10.1002/adfm.200304489

Bakhshi A.K. (1995) Electrically Conductive Polymers: From Fundamental to Applied Research, Bulletin of Materials Science, **18**, 469–495, https://doi.org/10.1007/BF02744834

Bánhegyi G. (1986) Comparison of Electrical Mixture Rules for Composites, Colloid and Polymer Science, **264**, 1030–1050, https://doi.org/10.1007/BF01410321

Bánhegyi G. Hedvig P., Petrovic Z.S., Karasz F.E. (1991) Applied Dielectric Spectroscopy of Polymeric Composites, Polymer – Plastics Technology and Engineering, **30**, 183–225, https://doi.org/10.1080/03602559108020136

Bhattacharya A., De A. (2007) Conductive Polymers in Solution – Progress Toward Processibility, Journal of Macromolecular Science Part C Polymer Reviews, **C39**, 17–56, https://doi.org/10.1081/MC-100101416

Bilalis P., Katsigiannopoulos D., Avgeropoulos A., Sakellariou G. (2014) Non-Covalent Functionalization of Carbon Nanotubes with Polymers, RSC Advances, **4**, 2911–2934, https://doi.org/10.1039/C3RA44906H

Blachowicz T. Ehrmann A. (2020) Conductive Electrospun Nanofiber Mats, Materials, **13**, 152, 1–17, https://doi.org/10.3390/ma13010152

Bose S., Kuila T., Mishra A.K., Rajasekar R., Kim N.H., Lee J.H. (2012) Carbon-Based Nanostructured Materials and their Composites as Supercapacitor Electrodes, Journal of Materials Chemistry, **22**, 767–784, https://doi.org/doi:10.1039/C1JM14468E

Cao Z., Xia Y. (2017) Corrosion Resistance and Tribological Characteristics of Polyaniline as Lubricating Additive in Grease, Journal of Tribology, **139**, 061801, 1–7, https://doi.org/10.1115/1.4036271

Cho S.H., Song K.T., Lee J.Y. (2007), Recent Advances in Polypyrrole, Chapter 8, in: Handbook of Conducting Polymers 3rd ed., Theory, Synthesis, Properties and Characterization, Skotheim T.A., Reynolds J.R. (eds.), CRC Press, Boca Raton, ISBN 978-1-4200-4358-7

Culebras M., Gómez C.M., Cantarero A. (2014) Review on Polymers for Thermoelectric Applications, Materials, **7**, 6701–6732; https://doi.org/10.3390/ma7096701

da Silva A.C., de Torresi S.I.C. (2019) Advances in Conducting, Biodegradable and Biocompatible Copolymers for Biomedical Applications, Frontiers in Materials, **6**, 98, 1–9, https://doi.org/10.3389/fmats.2019.00098

Davidddi E., Chen Z., Massani B.B., Lee J., Bentley C.L., Unwin P.R., Ratcliff E.L. (2019) Nanoscale Visualization and Multiscale Electrochemical Analysis of Conductive Polymer Electrodes, ACS Nano, **13**, 13271–13284, https://doi.org/10.1021/acsnano.9b06302

De Paoli M.-A., Gazotti W.A. (2002) Conductive Polymer Blends: Preparation, Properties and Applications, Macromolecular Symposia, **189**, 83–104, https://doi.org/10.1002/masy.200290008

Derosa P.A., Michalak T. (2014) Polymer Mediated Tunneling Transport Between Carbon Nanotubes in Nanocomposites, Journal of Nanoscience and Nanotechnology, **14**, 3696–3702, https://doi.org/10.1166/jnn.2014.7973

Edberg J., Iandolo D., Brooke R., Liu X., Musumeci Ch., Andreasen J.W., Simon D., Evans D., Engquist I., Berggren M. (2016) Patterning and Conductivity Modulation of Conductive Polymers by UV Light Exposure, Advanced Functional Materials, **26**, 6950–6960, https://doi.org/10.1002/adfm.201601794

El Rhazi M., Majid S., Elbasri M., Salih F.E., Oularbi L., Lafdi K. (2018) Recent Progress in Nanocomposites Based on Conducting Polymer: Application as Electrochemical Sensors, International Nano Letters, **8**, 79–99, https://doi.org/10.1007/s40089-018-0238-2

Eliseeva S.N., Kamenskii M.A., Tolstopyatova E.G., Kondratiev V.V. (2020) Effect of Combined Conductive Polymer Binder on the Electrochemical Performance of Electrode Materials for Lithium-Ion Batteries, Energies, **13**, 2163, 1–24, https://doi.org/10.3390/en13092163

Emran K.M., Ali S.M., Al-Oufi A.L.L. (2017) Synthesis and Characterization of Nano-Conducting Copolymer Composites: Efficient Sorbents for Organic Pollutants, Molecules, **22**, 772, 1–14, https://doi.org/10.3390/molecules22050772

Felmayer W., Wolf I. (1958) Conductivity and Energy Gap Measurements of Some Relatives of Phthalocyanine, Journal of the Electrochemical Society, **105**, 141, https://doi.org/10.1149/1.2428778

Folarin O.M., Sadiku E.R., Maity A. (2011) Polymer-Noble Metal Nanocomposites: Review, International Journal of the Physical Sciences, **6**, 4869–4882, https://doi.org/10.5897/IJPS11.570

Folorunso O., Hamam Y., Sadiku R., Ray S.S., Joseph A.G. (2019), Parametric Analysis of Electrical Conductivity of Polymer-Composites, Polymers, **11**, 1250, 1–20, https://doi.org/10.3390/polym11081250

Ganesha H., Veeresh S., Nagaraju Y.S., Vandana M., Ashokkumar S.P., Vijeth H., Devendrappa H. (2020) Growth of 3-Dimensional MoS$_2$-PANI Nanofiber for High Electrochemical Performance, Materials Research Express, **7**, 084001, 1–9, https://doi.org/10.1088/2053-1591/ab9e30

Ghosh S., Maiyalagan T., Basu R.N. (2016) Nanostructured Conducting Polymers for Energy Applications: Towards a Sustainable Platform, Nanoscale, **8**, 6921–6947, https://doi.org/10.1039/c5nr08803h

Gill Y.Q., Ehsan H., Irfan M.S., Saeed F., Shakoor A. (2020) Synergistic Augmentation of Polypropylene Composites by Hybrid Morphology Polyaniline Particles for Antistatic Packaging Applications, Materials Research Express, **7**, 015331, 1–10, https://doi.org/10.1088/2053-1591/ab61b5

Gotovtsev P.M., Badranova G.U., Zubavichus Y.V., Chumakov N.K., Antipova C.G., Kamyshinsky R.A., Presniakov M. Yu., Tokaev K.V., Grigoriev T.E. (2019) Electroconductive PEDOT:PSS-Based Hydrogel Prepared by Freezing-Thawing Method, Heliyon, **5**, e02498, 1–8, https://doi.org/10.1016/j.heliyon.2019.e02498

Gul E.V. (1996) Structure and Properties of Conducting Polymer Composites, CRC Press, Boca Raton, ISBN 9789067642040

Hammad A.S., Noby H., Elkady M.F., El-Shazly A.H. (2018) In-situ Polymerization of Polyaniline/Polypyrrole Copolymer using Different Techniques, IOP Conf. Series: Materials Science and Engineering, **290**, 012001, 1–8, https://doi.org/10.1088/1757-899X/290/1/012001

He K., Qin C., Wen Q., Wang C., Wang B., Yu S., Hao C., Chen K. (2018), Facile Fabrication of Polyaniline/ Polypyrrole Copolymer Nanofibers with a Rough Surface and Their Electrorheological Activities, Applied Polymer Science, **135**, 6289, https://doi.org/10.1002/app.46289

Hedvig P., Kulcsár S. Kiss L. (1968) Electrical and Paramagnetic Properties of Pyrolyzed Polyacrylonitrile Prepared by Radiation Polymerization in Urea Channels, European Polymer Journal, **4**, 601–609, https://doi.org/10.1016/0014-3057(68)90058-X

Heeger A.J. (2001) Nobel Lecture: Semiconductive and Metallic Polymers: The Fourth Generation of Polymeric Materials, Reviews of Modern Physics, **73**, 681–700, https://doi.org/10.1103/RevModPhys.73.681

Hong X., Liu Y., Li Y., Wang X., Fu J., Wang X. (2020) Application Progress of Polyaniline, Polypyrrole and Polythiophene in Lithium-Sulfur Batteries, Polymers, **12**, 331, 1–27, https://doi.org/10.3390/polym12020331

Huang J., Kaner R.B. (2007), Polyaniline Nanofibers: Syntheses, Properties, and Applications, Chapter 7, in: Handbook of Conducting Polymers 3rd ed., Theory, Synthesis, Properties and Characterization, Skotheim T.A., Reynolds J.R. (eds.), CRC Press, Boca Raton, ISBN 978-1-4200-4358-7

Hussain S.T., Abbas F., Kausar A., Khan M.R. (2013) New Polyaniline/Polypyrrole/Polythiophene and Functionalized Multiwalled Carbon Nanotube-Based Nanocomposites: Layer-by-Layer in situ Polymerization, High Performance Polymers, **25**, 70–78, https://doi.org/10.1177/0954008312456048

Hussin H., Gan S.N., Mohamad S., Phang S.W. (2017) Synthesis of Water-soluble Polyaniline by Using Different Types of Cellulose Derivatives, Polymers & Polymer Composites, **25**, 515–519, https://doi.org/10.1177/096739111702500702

Inzelt Gy. (2008) Conductive Polymers: A New Era in Electrochemistry, Springer Verlag, Berlin, https://doi.org/10.1007/978-3-540-75930-0

Jadoun S., Riaz U. (2019) A Review on the Chemical and Electrochemical Copolymerization of Conducting Monomers: Recent Advancements and Future Prospects, Polymer-Plastics Technology and Materials, **59**, 484–504, https://doi.org/10.1080/25740881.2019.1669647

Janáky C., Rajeshwar K. (2015) The Role of (Photo)Electrochemistry in the Rational Design of Hybrid Conducting Polymer/Semiconductor Assemblies: From Fundamental Concepts to Practical Applications, Progress in Polymer Science, **43**, 96–135, https://doi.org/10.1016/j.progpolymsci.2014.10.003

Kamarudin M.A., Shamir S.R., Datta R.S., Long B.D., Sabri M.F.M., Said S.M. (2013) A Review on the Fabrication of Polymer-Based Thermoelectric Materials and Fabrication Methods, The Scientific World Journal, **2013**, 713640, 1–17, https://doi.org/10.1155/2013/713640

Kaneto K. (2016) Research Trends of Soft Actuators Based on Electroactive Polymers and Conducting Polymers, Journal of Physics: Conference Series, **704**, 012004, 1–9, https://doi.org/10.1088/1742-6596/704/1/012004

Karri S.N., Srinivasan P. (2019) Synthesis of PEDOT:PSS using Benzoyl Peroxide as an Alternative Oxidizing Agent for ESD Coating and Electro-Active Material in Supercapacitor, Materials Science for Energy Technologies, **2**, 208–215, https://doi.org/10.1016/j.mset.2019.01.008

Kaur G., Adhikari R., Cass P., Bown M., Gunatillake P. (2015) Electrically Conductive Polymers and Composites for Biomedical Applications, RSC Advances, **5**, 37553–37567, https://doi.org/10.1039/c5ra01851j

Kausar A. (2017) Overview on Conducting Polymer in Energy Storage and Energy Conversion System, Journal of Macromolecular Science, Part A, **54**, 640–653, https://doi.org/10.1080/10601325.2017.1317210

Khademi S., Pourabbas B., Foroutani K. (2018) Synthesis and Characterization of Poly(Thiophene-Copyrrole) Conducting Copolymer Nanoparticles via Chemical Oxidative Polymerization, Polymer Bulletin, **75**, 4291–4309, https://doi.org/10.1007/s00289-017-2264-z

Khan M.A., Cantù E., Tonello S., Serpelloni M., Lopomo N.F., Sardini E. (2019) A Review on Biomaterials for 3D Conductive Scaffolds for Stimulating and Monitoring Cellular Activities, Applied Sciences, **9**, 961, 1–18, https://doi.org/10.3390/app9050961

Kroon R., Hofmann A.I., Yu L., Lund A., Müller Ch. (2019) Thermally Activated in-situ Doping Enables Solid-State Processing of Conducting Polymers, Chemistry of Materials, **31**, 2770–2777, https://doi.org/10.1021/acs.chemmater.8b04895

Kumar H., Boora A., Yadav A., Rahul R. (2020) Polyaniline-Metal Oxide-Nano-Composite as a Nano-Electronics, Optoelectronics, Heat Resistance and Anticorrosive Material, Results in Chemistry, **2**, 100046, 1–7, https://doi.org/10.1016/j.rechem.2020.100046

Kuzmany H., Mehring M., Roth S. (Eds.) (1985) Electronic Properties of Polymers and Related Compounds, Springer Verlag, Berlin, ISBN 3-540-15722-0

Le T.-H., Kim Y., Yong H. (2019) Electrical and Electrochemical Properties of Conducting Polymers, Polymers, **9**, 150, 1–32, https://doi.org/10.3390/polym9040150

Lee D.H., Chang J.A., Kim J.K. (2006) Block Copolymer as a Template for Electrically Conductive Nanocomposites, Journal of Materials Chemistry, **16**, 4575–4580, https://doi.org/10.1039/b611456c

Lee J., Ryu L., Youn H.J. (2012) Conductive Paper Through LbL Multilayering with Conductive Polymer: Dominant Factors to Increase Electrical Conductivity, Cellulose, **19**, 2153–2164, https://doi.org/10.1007/s10570-012-9781-6

Liew S.Y., Walsh D.A., Chen G.Z. (2017) Conducting Polymer Nanocomposite-Based Supercapacitors, in: Conducting Polymer Hybrids, x Kumar V., Kalia S., Swart H.C. (eds.), Springer Nature, Cham, https://doi.org/10.1007/978-3-319-46458-9_9

Liu J., Regenauer-Lieb K. (2011) Application of Percolation Theory to Microtomography of Structured Media: Percolation Threshold, Critical Exponents, and Upscaling, Physical Review E, **83**, 016106, 1–13, https://doi.org/10.1103/PhysRevE.83.016106

Liu Y., Yin P., Chen J., Cui B., Zhang C., Wu F. (2020) Conducting Polymer-Based Composite Materials for Therapeutic Implantations: From Advanced Drug Delivery System to Minimally Invasive Electronics, International Journal of Polymer Science, **2020**, 5659682, 1–16, https://doi.org/10.1155/2020/5659682

Long Y.-Z., Li M-M., Gu C., Wan M., Duvail J.-L., Liu Z., Fan Z. (2011) Recent Advances in Synthesis, Physical Properties and Applications of Conducting Polymer Nanotubes and Nanofibers, Progress in Polymer Science **36**, 1415–1442, https://doi.org/10.1016/j.progpolymsci.2011.04.001

Lu Q., Han W.J., Choi H.J. (2018) Smart and Functional Conducting Polymers: Application to Electrorheological Fluids, Molecules, **23**, 2854, 1–24, https://doi.org/10.3390/molecules23112854

Luzio A., Canesi E.V., Bertarelli C., Carioni M. (2014) Electrospun Polymer Fibers for Electronic Applications, Materials, **7**, 906–947, https://doi.org/10.3390/ma7020906

Lyu L., Jiurong Liu J., Liu H., Liu C., Lu Y., Sun K., Fan R., Wang N., Lu N., Guo Z., Wujcik E.K. (2018) An Overview of Electrically Conductive Polymer Nanocomposites toward Electromagnetic Interference Shielding, Engineered Science, **2**, 26–42, https://doi.org/10.30919/es8d615

Mahato N., Jang H., Dhyani A., Cho S. (2020) Recent Progress in Conducting Polymers for Hydrogen Storage and Fuel Cell Applications, Polymers, **12**, 2480, 1–40, https://doi.org/10.3390/polym12112480

Maity A., Biswas M. (2006) Recent Progress in Conducting Polymer, Mixed Polymer-Inorganic Hybrid Nanocomposites, Journal of Industrial and Engineering Chemistry, **12**, 311–351.

Maity N., Dawn A. (2020) Conducting Polymer Grafting: Recent and Key Developments, Polymers, **12**, 709, 1–23, https://doi.org/10.3390/polym12030709

McCullough L.A., Matyjaszewski K. (2010) Conjugated Conducting Polymers as Components in Block Copolymer Systems, Molecular Crystals and Liquid Crystals, *Proceedings of the Xth International Conference on Frontiers of Polymers and Advanced Materials: Emerging Technologies and Business Opportunities*, 521, 1, 1–55, https://doi.org/10.1080/15421401003719951

Mohammed M.Q. (2018) Patterned functionalisation of conducting polymer films, PhD Thesis, University of Leicester, https://leicester.figshare.com/articles/thesis/Patterned_functionalisation_of_conducting_polymer_films/10243865/files/18491702.pdf, accessed on 30.12.2020.

Molapo K.M., Ndangili P.M., Ajayi R.F., Mbambisa G., Mailu S.M., Njomo N., Masikini M., Baker P., Iwuoha E.I. (2012) Electronics of Conjugated Polymers (I): Polyaniline, International Journal of Electrochemical Science, **7**, 11859–11875.

Naveen M.H., Gurudatt N.G., Shim Y.B. (2017) Applications of Conducting Polymer Composites to Electrochemical Sensors: A Review, Applied Materials Today, **9**, 419–433, https://doi.org/10.1016/j.apmt.2017.09.001

Nguyen V.A., Kuss C. (2020) Review—Conducting Polymer-Based Binders for Lithium-Ion Batteries and Beyond, Journal of the Electrochemical Society, **167**, 065501, 1–15, https://doi.org/10.1149/1945-7111/ab856b

Niu F., He R., Li J. (2017) The Influence of Polypyrrole Coatings on the Mechanical and Friction and Wear Properties of Bamboo Fiber Filled PA6 Composites, Surface and Interface Analysis, **50**, 111–116, https://doi.org/10.1002/sia.6345

Onggar T., Kruppke I., Cherif C. (2020) Techniques and Processes for the Realization of Electrically Conducting Textile Materials from Intrinsically Conducting Polymers and Their Application Potential Polymers, **12**, 2867, 1–46, https://doi.org/10.3390/polym12122867

Ouyang S., Xie Y., Wang D., Zhu D., Xu X., Tan T., Fong H.H. (2015) Surface Patterning of PEDOT:PSS by Photolithography for Organic Electronic Devices, Journal of Nanomaterials, **2015**, 603148, 1–9, https://doi.org/10.1155/2015/603148

Pan L., Qiu H., Dou C., Li Y., Pu L., Xu J., Shi Y. (2010) Conducting Polymer Nanostructures: Template Synthesis and Applications in Energy Storage, International Journal of Molecular Sciences, **11**, 2636–2657, https://doi.org/10.3390/ijms11072636

Pan L., Yu G., Zhai D., Lee H.R., Zhao W., Liu N., Wang H., Tee B. C.-K., Shi Y., Cui Y., Bao Z. (2012) Hierarchical Nanostructured Conducting Polymer Hydrogel with High Electrochemical Activity, Proceedings of the National Academy of Sciences, **109**, 9287–9292, https://doi.org/10.1073/pnas.1202636109

Peng C., Zhang S., Jewell D., Chen G.Z. (2008) Carbon Nanotube and Conducting Polymer Composites for Supercapacitors, Progress in Natural Science, **18**, 777–788, https://doi.org/doi:10.1016/j.pnsc.2008.03.002

Pohl H.A., Opp D.A. (1962), The Nature of Semiconduction in Some Acene Quinone Radical Polymers, Journal of Physical Chemistry, **66**, 2121–2126, https://doi.org/10.1021/j100817a012

Pouget J.-P. (2016), The Peierls Instability and Charge Density Wave in One-Dimensional Electronic Conductors, Comptes Rendus Physique, **17**, 332–356, https://doi.org/10.1016/j.crhy.2015.11.008

Saad A., Jlassi K., Omastová M., Chehimi M.M. (2017) Clay/Conductive Polymer Nanocomposites, Chapter 6 in: Clay-Polymer Nanocomposites, Jlassi K., Chehimi M.M., Thomas S.(eds.), Elsevier, Amsterdam, https://doi.org/10.1016/B978-0-323-46153-5.00006-9

Sengodu P., Deshmukh A.D. (2015) Conducting Polymers and Its Inorganic Composites for Advanced Li-ion Batteries: A Review, RSC Advances, **5**, 42109–42130, https://doi.org/10.1039/C4RA17254J

Sethumadhavan V., Rudd S., Switalska E., Zuber K., Teasdale P., Evans D. (2019) BMC Materials, **1**, 4, 1–14, https://doi.org/10.1186/s42833-019-0001-7

Sharma A.K., Kim J.-H., Lee Y.S. (2009) An Efficient Synthesis of Polypyrrole/Carbon Fiber Composite Nano-Thin Films, International Journal of Electrochemical Science, **4**, 1560–1567.

Sharma K., Kumar V., Kaith B.S., Kalia S., Swart H.C. (2016) Conducting Polymer Hydrogels and Their Applications, in: Conducting Polymer Hybrids, Kumar V., Kalia S., Swart H.C. (eds.), Springer International Publishing, https://doi.org/10.1007/978-3-319-46458-9_7

Shirakawa H., Hiroki K. (N.D.) Fundamentals of Conductive Polymers, Material Matters™ BASICS, Vol. **8.**, Sigma-Aldrich, Japan, https://www.sigmaaldrich.com/content/dam/sigma-aldrich/docs/Aldrich/General_Information/1/material-matters-basics-vol8.pdf, accessed on 12.12.2020.

Shown I., Ganguly A., Chen L.-C., Chen K.-H. (2015) Conducting Polymer-Based Flexible Supercapacitor, Energy Science and Engineering, **3**, 2–26, https://doi.org/10.1002/ese3.50

Shubba L.N., Kalpana M., Rao P.M. (2016) Synthesis, Characterization by AC Conduction and Antibacterial Properties of Polyaniline Fibers, Der Pharmacia Lettre, **8**, 214–219.

Snyder G.J., Snyder A.H. (2017) Figure of Merit ZT of a Thermoelectric Device Defined from Materials Properties, Energy and Environmental Science, **10**, 2280–2283, https://doi.org/10.1039/C7EE02007D

Stauffer D., Aharony A. (1994) Introduction to Percolation Theory, 2nd ed., Taylor and Francis, New York, ISBN 9780748402533

Stejskal J., Sapurina I., Trchová M., Šeděnková I., Kovářová J., Kopecká J., Prokeš J. (2015) Coaxial Conducting Polymer Nanotubes: Polypyrrole Nanotubes Coated with Polyaniline or Poly(p-Phenylenediamine) and Products of their Carbonization, Chemical Papers, **69**, 1341–1349, https://doi.org/10.1515/chempap-2015-0152

Tajik S., Beitollahi H., Nejad F.G., Shoaie I.S., Khalilzadeh M.A., Asl M.S., Le Q.V., Zhang K., Jang H.W., Shokouhimehr M. (2020) Recent Developments in Conducting Polymers: Applications for Electrochemistry, RSC Advances, **10**, 37834–37856, https://doi.org/10.1039/d0ra06160c

Tarlton T., Sullivan E., Brown J., Derosa P.A. (2017) The Role of Agglomeration in the Conductivity of Carbon Nanotube Composites near Percolation, Journal of Applied Physics, **121**, 085103, 1–9, https://doi.org/10.1063/1.4977100

Tikish T.A., Kumar A., Kim J.Y. (2018) Study on the Miscibility of Polypyrrole and Polyaniline Polymer Blends, Advances in Materials Science and Engineering, **2018**, 3890637, 1–5, https://doi.org/10.1155/2018/3890637

Tseghai G.B., Mengistie D.A., Malengier B., Fante K.A., Van Langehove L. (2020) PEDOT:PSS-Based Conductive Textiles and Their Applications, Sensors, **20**, 1881, 1–18, https://doi.org/10.3390/s20071881

Wanasinghe D., Aslani F., Ma G., Habibi D. (2020) Review of Polymer Composites with Diverse Nanofillers for Electromagnetic Interference Shielding, Nanomaterials, **10**, 541, 1–46, https://doi.org/10.3390/nano10030541

Wang L., Lin T., Wang X., Kaynak A. (2005) Frictional and Tensile Properties of Conducting Polymer Coated Wool and Alpaca Fibers, Fibers and Polymers, **6**, 259–262, https://doi.org/10.1007/BF02875651

Weng B., Shepherd R.L., Crowley K., Killard A.J., Wallace G.G. (2010) Printing Conducting Polymers, Analyst, **135**, 2779–2789, https://doi.org/10.1039/C0AN00302F

Wessling B. (1989) Post Polymerization Conductive Polymers Processing – an Integral Part of a New Materials Science of Conductive Polymers, in: Springer Series in Solid-State Sciences 91 Electronic Properties of Conductive Polymers (ICP), 3rd Edition, Kuzmany H., Mehring M., Roth S. (eds.), Springer, Berlin, ISBN 978-3-642-83835-4

Wessling B. (1995) Critical Shear Rate – the Instability Reason for the Creation of Dissipative Structures in Polymers, Zeitschrift für Physikalische Chemie, **191**, 119–135, https://doi.org/10.1524/zpch.1995.191.Part_1.119

Wessling B. (1998) Dispersion as the Key to Processing Conductive Polymers, in: Handbook of Conductive Polymers, 2nd Revised & Expanded Edition, Skotheim T.A., Elsenbaumer R.L., Reynolds J.R. (eds.), Marcel Dekker Inc, New York.

Xie Y., Zhang S.-H., Jiang H.-Y., Zheng H., Wu R.M., Chen H., Gao Y.-F., Huang Y.-Y., Bai H.-L. (2019) Properties of Carbon Black-PEDOT Composite Prepared via in-situ Chemical Oxidative Polymerization, e-Polymers, **19**, 61–69, https://doi.org/10.1515/epoly-2019-0008

Xu H., Zhang Y. (2019) A Review on Conducting Polymers and Nanopolymer Composite Coatings for Steel Corrosion Protection, Coatings, **9**, 807, 1–22, https://doi.org/10.3390/coatings9120807

Yan Y., Yang G., Xu J.-L., Zhang M., Kuo C.-C., Wang S.-D. (2020) Conducting Polymer-Inorganic Nanocomposite Based Gas Sensors: A Review, Science and Technology of Advanced Materials, **21**, 768–786, https://doi.org/10.1080/14686996.2020.1820845

Yang J.P., Rannou R., Planés J., Pron A., Nechtshein M. (1998) Preparation of Low Density Polyethylene-Based Polyaniline Conducting Polymer Composites with Low Percolation Threshold via Extrusion, Synthetic Metals, **93**, 169–173, https://doi.org/10.1016/S0379-6779(97)04093-9

Yanilmaz M., Sarac A.S. (2014) A Review: Effect of Conductive Polymers on the Conductivities of Electrospun Mats, Textile Research Journal, **84**, 1325–1342, https://doi.org/10.1177/0040517513495943

Yao Q., Chen L., Qu S. (2015) Conducting Polymer-Based Nanocomposites for Thermoelectric Applications, Chapter 6, in: Fundamentals of Conjugated Polymer Blends, Copolymers and Composites: Synthesis, Properties and Applications, Saini P. (ed.), Wiley and Sons, New York, https://doi.org/10.1002/9781119137160.ch6

Yilmaz F., Kücükyavuz Z. (2009) The Influence of Polymerization Temperature on Structure and Properties of Polyaniline, e-Polymers, **9**, 5, https://doi.org/10.1515/epoly.2009.9.1.48

Yoshida S., Fujinami S., Esashi M. (2013) Investigation of Mechanical and Tribological Properties of Polyaniline Brush by Atomic Force Microscopy for Scanning Probe-Based Data Storage, e-Journal of Surface Science and Nanotechnology, **11**, 53–59, https://doi.org/10.1380/ejssnt.2013.53

Yuk H., Lu B., Lin S., Qu K., Xu J., Luo J., Zhao X. (2020), Nature Communications, **11**, 1604, 1–8, https://doi.org/10.1038/s41467-020-15316-7

Zamiri G., Haseeb A.S.M.A. (2020) Recent Trends and Developments in Graphene/Conducting Polymer Nanocomposites Chemiresistive Sensors, Materials, **13**, 3311, 1–24, https://doi.org/10.3390/ma13153311

Zhang B., Xu Y., Zheng Y., Dai L., Zhang M., Yang J., Chen Y., Chen X., Zhou J. (2011) A Facile Synthesis of Polypyrrole/Carbon Nanotube Composites with Ultrathin, Uniform and Thickness-Tunable Polypyrrole Shells, Nanoscale Research Letters, **6**, 431, 1–9, https://doi.org/10.1186/1556-276X-6-431

Zhang Q.-H., Chen D.-J. (2019) Percolation Threshold and Morphology of Composites of Conducting Carbon Black/Polypropylene/EVA, Journal of Materials Science, **39**, 1751–1757, https://doi.org/10.1023/B:JMSC.0000016180.42896.0f

Zhou Y., Fang Y., Ramasamy R.P. (2019) Non-Covalent Functionalization of Carbon Nanotubes for Electrochemical Biosensor Development, Sensors, **19**, 392, 1–21, https://doi.org/10.3390/s19020392

Zilberman M., Siegman A., Narkis M. (1998) Melt-Processed Electrically Conductive Polymer/Polyaniline Blends, Journal of Macromolecular Science Part B, **37**, 301–318, https://doi.org/10.1080/00222349808220474

LIST OF ABBREVIATIONS

1D   one dimensional
2D   two dimensional
3D   three dimensional
CMC   carboxymethyl cellulose
CP   conductive polymer

DBSA	dodecylbenzosulfonic acid
DMSO	dimethyl sulfoxide
DSC	differential scanning calorimetry
DSSC	dye-sensitized solar cell
DTA	differential thermal analysis
EAP	electroactive polymer
EDLC	electric double-layer capacitor
EMI	electromagnetic interference
EPR	electron paramagnetic resonance
FET	field-effect transistor
HOMO	highest occupied molecular orbital
LBL	layer by layer
LDPE	low-density polyethylene
LLDPE	linear low-density polyethylene
LUMO	lowest unoccupied molecular orbital
MIP	molecularly imprinted polymer
MOF	metal-oxide framework
NMP	N-methyl pyrrolidone
PAN	polyacrylonitrile
PANI	polyaniline
PANPY	poly(aniline-co-pyrrole) copolymer
PC	pseudocapacitor
PTC	positive thermal coefficient
UV	ultraviolet
VRH	variable range hopping

2 Materials and Chemistry of Conducting Polymers

Kalpana Madgula[1], Shruthy D. Pattathil[2,3],
Venkata Sreenivas Puli[4,5], and Abbaraju Venkataraman[6]

[1]SAS Nanotechnologies LLC,
Wilmington, Delaware, USA

[2]Gulbarga University, Kalaburagi, Karnataka, India

[3]St. Francis Degree and P.G. College for Women,
Begumpet, Hyderabad, Telangana, India

[4]Materials and Manufacturing Directorate, Wright Patterson
Air Force Base, Ohio, USA

[5]Smart Nanomaterials Solutions LLC, Orlando, Florida, USA

[6]Department of Chemistry, Gulbarga University,
Kalaburagi, Karnataka, India

CONTENTS

2.1   INTRODUCTION

A polymer is a large molecule composed of repeating units known as monomers [1]. Polymers play a crucial role from biological structures such as DNA and proteins to plastics like polyethylene owing to their wide range of properties [2]. Polymers were earlier known for their insulating properties, but over the last few years, many organic polymers known as conducting polymers (CPs) with good electrical conductivities were discovered and researched extensively [3]. With the discovery of integrated circuits, there was a need for materials with smaller dimensions and the focus was shifted to CPs [4]. A breakthrough in this field was with the serendipitous discovery of trans-polyacetylene by Hideki Shirakawa and the group. It was observed that the conductivity of polyacetylene increased by almost ten million times on exposure to halogen vapors. This was followed by a collaborative work with Alan Heeger and Alan MacDiarmid, which made them realize the potential of this material. The trio was awarded Nobel Prize for chemistry in the year 2000 for their collaborative work. This pioneering work by the Nobel Laureates paved the way for a whole new area of research [5, 6]. CPs or pi-conjugated polymers possess mechanical properties like traditional polymers along with conductivity like metals, such a class of polymers are known as intrinsically conducting polymers (ICPs) to achieve the desired properties [7]. Polyacetylene, polyaniline (PANI), polypyrrole (PPy), etc. are commonly studied CPs. These polymers have mechanical properties as conventional polymers and conductivity in the metallic or semiconductor range, making them ideal materials to replace conventional inorganic conductors that require continuous etching and lithographic steps during fabrication. Technological applications include biosensors, gas sensors, supercapacitors, and photovoltaic applications [8–10].

CPs are divided majorly into two ways based on the nature of polymer and dopant and their mechanism of conduction [11]. A necessary condition for CPs to be intrinsically conducting is the presence of conjugated structure (or segments) of alternate double and single bonds coupled with atoms (e.g. N, S) providing π-orbitals for continuous overlap. In contrast to the free-electron movement in metals, the polymers should have charge carriers as well as orbital systems (e.g. polymer backbone) that allow free movement of these charge carriers (or delocalization). CP consists of localized and delocalized states and this delocalization is responsible for the development of charge carriers such as polarons (radical cation or radical anion) and bipolarons (dication or anion) which are responsible for conduction in CPs [12]. In an undoped state, the conductivity of CPs lies in between that of insulators and semiconductors with a moderate bandgap of 2–3 eV and as the concentration dopant increases, conductivity also increases [13].

Their general classification and chemical structure are shown in Figures 2.1 and 2.2(a) [12, 14]. Most of the organic polymers lack intrinsic charge carriers and hence charge carriers created by the process, "doping" i.e. introducing electron acceptors or electron donors into the polymer chain creating defects in the form of polaron, bipolaron, or soliton, which then act as charge carriers as shown in the Figure 2.2(b) for the poly(3,4-ethylenedioxythiophene) (PEDOT) [14]. By the process of doping, the conductivity of CP can be enhanced from the range of (semiconductor) insulator to that of close to "metallic conducting regime." The dopants can be introduced into the CP matrix by various synthetic (or chemical) strategies discussed in the following section (Chemistry of CPs) [15]. Doping agents are also being classified based on their nature – (i) *neutral doping agents* (for example, iodine, I_2, Br_2, AsF_2, H_2SO_4, and $FeCl_3$) that can be converted into negative or positive ions with or without chemical modification during the process of doping; (ii) *ionic dopants* (e.g. $LiClO_4$,

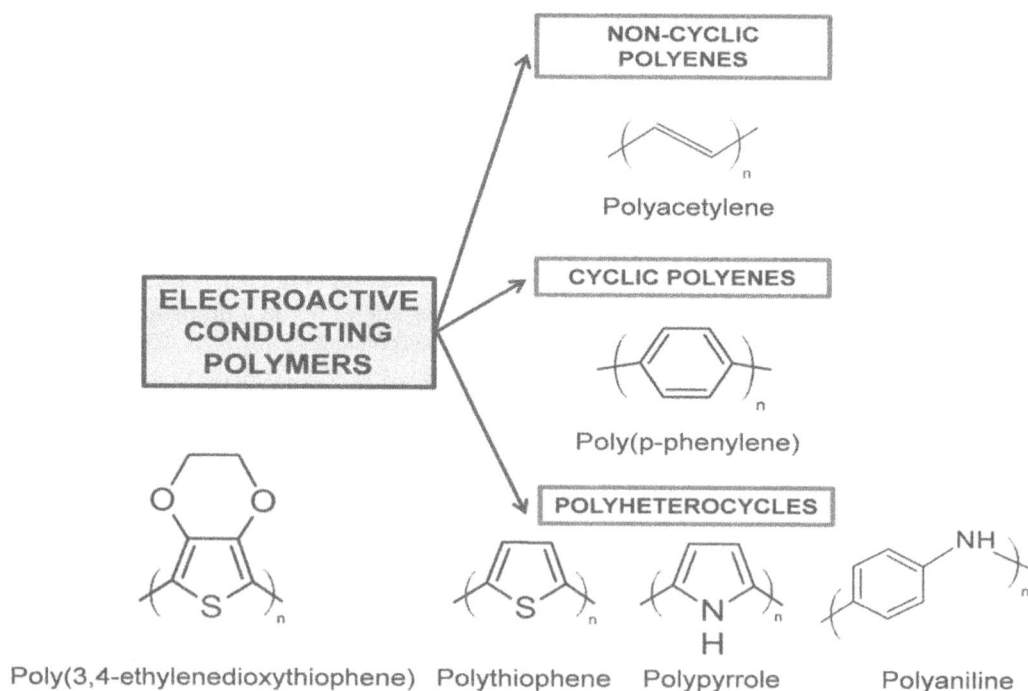

FIGURE 2.1 General classification and chemical structure of electroactive conducting polymers – noncyclic, cyclic, and polyheterocyclic CPs. (Copyright 2015. The Royal Society of Chemistry. Reproduced with permission from Ref. [12].)

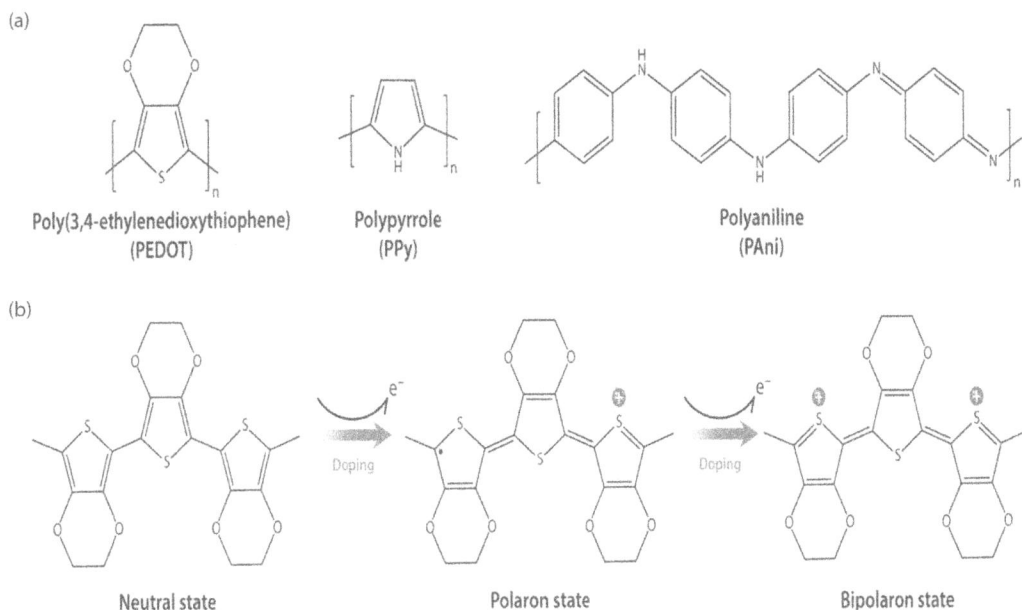

FIGURE 2.2 (a) Chemical structure of few CPs, e.g. poly(3,4-ethylenedioxythiophene)(PEDOT), polypyrrole (Ppy), and polyaniline (PANI). (b) Polaron and bipolaron state in PEDOT. (Copyright 2019. Reproduced with permission from Ref [14].)

$FeClO_4$.) that are either oxidized or reduced by an electron with the polymer and the counter ion remains with polymer making the system overall neutral; (iii) *organic dopants* that are anionic (e.g. CF_3COOH, CF_3SO_3Na) and are incorporated into polymers during anodic deposition of the polymer from aqueous electrolytes; (iv) *polymeric dopants* such as PVS, PPS, PS-co-MA having amphiphilic anions and are functionalized polymeric electrolytes. Finally, (v) *metallic oxide dopants*, belonging to the oxides of cerium (CeO_2), titanium (TiO_2), tin (SnO_2), tungsten (WO_3), cobalt (CoO_3), and similar metal oxides can behave as electron donors due to the ability of their oxygen atoms to donate their lone pair of electrons and help in localization of charge in the polymeric chain resulting into increase in conductivity of the host polymer [13, 16, 17]. Similarly, in the recent reviews and reports published in the literature, CPs have been combined with *carbon-based materials* [18–20] like graphene oxide (GO), reduced graphene oxide (rGO), carbon nanotubes (CNTs), or transition metal chalcogenides (e.g. MoS_2, $MoSe_2$, $NbSe_2$, $MoTe_2$, or WTe_2) or 2D materials for excellent mechanical and electrochemical applications (e.g. in supercapacitors and lithium (Li-) ion batteries) or/and with doped or undoped borocarbonitrides (e.g. BxCyNz) [21, 22] for advanced applications.

2.2 CHEMISTRY OF CONDUCTING POLYMERS

CPs can be synthesized by chemical or electrochemical polymerization. Liquid phase chemical oxidative polymerization involves oxidation of the monomer by oxidizing agents such as ferric chloride, ammonium persulphate to form radical cation, which couples with another monomer or radical cation to form a dimer. The chain growth step involves the repetition of reoxidation and coupling to form CPs [23]. Other than liquid-phase chemical oxidative polymerization, vapor phase polymerization is also reported in the literature. During this process, the oxidant is covered onto a substrate and the monomer vapor reaches the substrate and undergoes polymerization to form a layer of the polymer [24]. The main advantage of liquid-phase chemical oxidative polymerization is that CPs can be prepared in bulk quantities, whereas in vapor phase polymerization, thin CP films can be deposited onto a wide range of substrates. Electrochemical polymerization can be carried out using different methods such as galvanostatic (constant current), galvanodynamic (pulsed current), potentiostatic (constant potential), or potentiodynamic (cyclic voltammetry or pulsed potential) methods [25, 26]. This method is carried out in the solution containing monomer, on applying anodic potential, the monomer gets oxidized onto the substrate electrode, the polymer chain usually carries a positive charge which is counterbalanced by negative ions. The major advantage of electrochemical polymerization is that the yield, morphology, and electrochemical properties of the polymer can be controlled by nucleation and growth process.

Methods for synthesizing CP composites are classified under three categories based on the procedure: ex-situ synthesis, in situ synthesis, and one-pot synthesis. In ex situ synthesis, they are produced separately, and then the composite is obtained by blending two or more individual components. For in situ synthesis, at least one component of the composite is synthesized in the presence of another one. One main advantage of this method is that the interface in-between components can be controlled at the molecular level to achieve a synergistic effect. In the one-pot synthesis method, the monomer and other components can interact with each other forming composite in one single step [9, 27, 28].

CPs containing heterocycles as shown in Figure 2.3 are formed from monomers containing heteroatom, e.g. A = N, S, O, etc., which is coupled with a conjugated system of alternate double and single bonds within the aromatic ring. The formation of polyheterocycle can be divided into three steps: monomer oxidation to form monomer radical cation; two monomer radical cations are coupled to form dimer di(or radical)cation; and finally by chain propagation step, exist in aromatic or quinoid-like forms. Similarly, CPs such as PANI are well known for their mechanical stability, controllable conductivity with acid/base modifications, and environmental stability [29]. The chemical structure of PANI with mechanistic details comprising possible resonance structures, dimer, and polymeric forms are represented in Figure 2.4 [13]. CP composites attracted considerable interest in recent years because of their numerous applications in various electric and electronic devices.

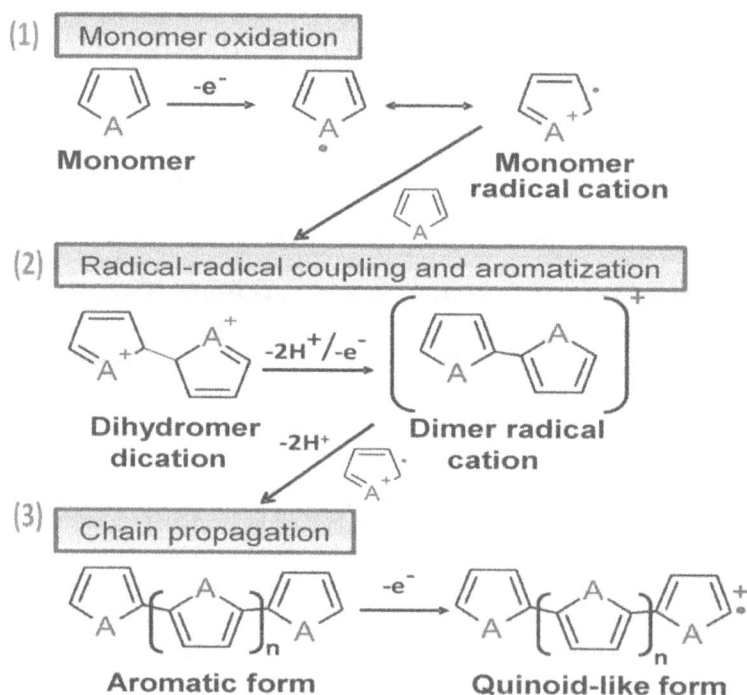

FIGURE 2.3 The mechanism of formation of radical cations from a polyheterocycle in three steps; A is heteroatom coupled with conjugated system. (Copyright 2015. Reproduced with permission from Ref [12].)

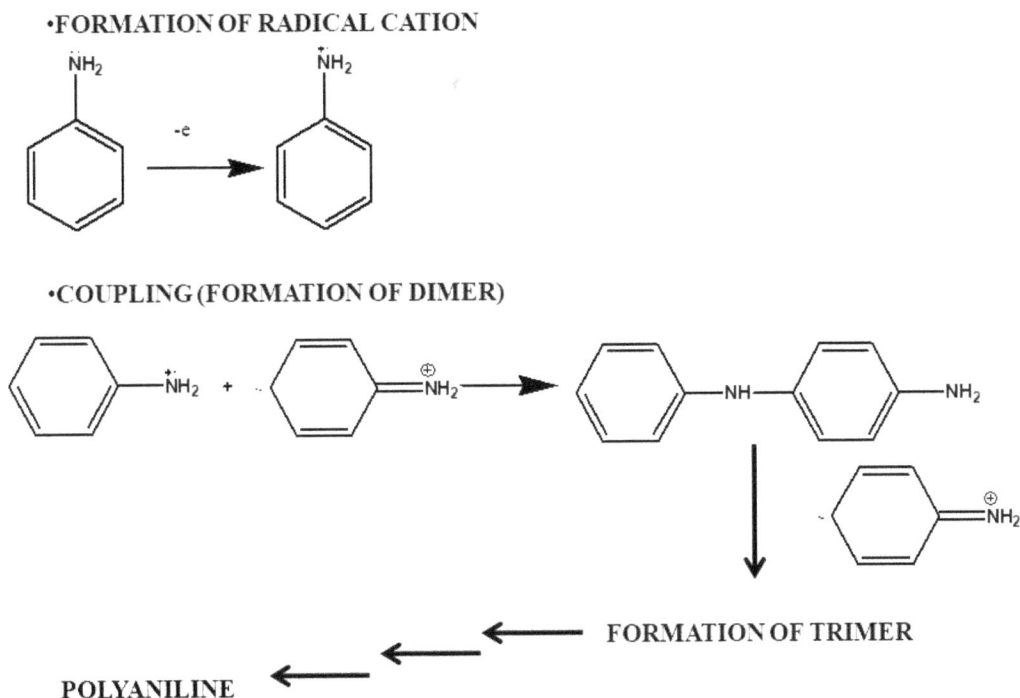

FIGURE 2.4 Mechanism of CP formation. Polyaniline (PANI), aniline monomer forming a radical cation, with possible resonance structures; formation of dimer and polymeric forms.

CP composites with some suitable compositions of one or more insulating materials led to desirable properties [30]. These materials are especially important owing to their bridging role between the world of nanoparticles or other nanostructure varieties, such as fibers, films [31], and that of CPs. This polymer can be used for metallic parts in components, such as sensors, capacitors, displays, and light-emitting diodes (LEDs) [32]. One-dimensional PANI nanostructures, including nanowires, nanorods, and nanotubes, have been studied [33, 34]. Recent studies are focused on the study of composites based on conducting PANI/metal oxide nanoparticles. The properties of these nanocomposites are quite different from PANI alone. By varying the particle size, shape, composition, and extent of dispersion of nanoparticles, the properties can be changed and tuned to the desired applications [35, 36]. As one of the important metal oxides, copper oxide (CuO) is frequently used as anode material for Li-ion batteries owing to its high capacity, safety, and low conductivity [37, 38]. Further, PANI/CuO nanocomposite with their well-known electrical properties [39] and antimicrobial activity are explored for sensor applications.

2.3 CP-BASED MATERIALS AND THEIR APPLICATIONS IN BRIEF

Energy shortage is one of the major issues faced by society for sustainable development. To resolve this issue, there is an urgent need for renewable energy resources or the development of new materials for low-cost efficient energy storage technology. With the development of science and technology, research in this field is progressing rapidly to achieve the desired goal. Energy storage devices, such as batteries, fuel cells, and supercapacitors are being investigated thoroughly. The development of portable devices has also increased the demand for these energy storage devices. Batteries have high energy storage capacity but slow power delivery or uptake. Conventional dielectric capacitors have high storage output but less energy storage capacity. Supercapacitors, ultracapacitors, or

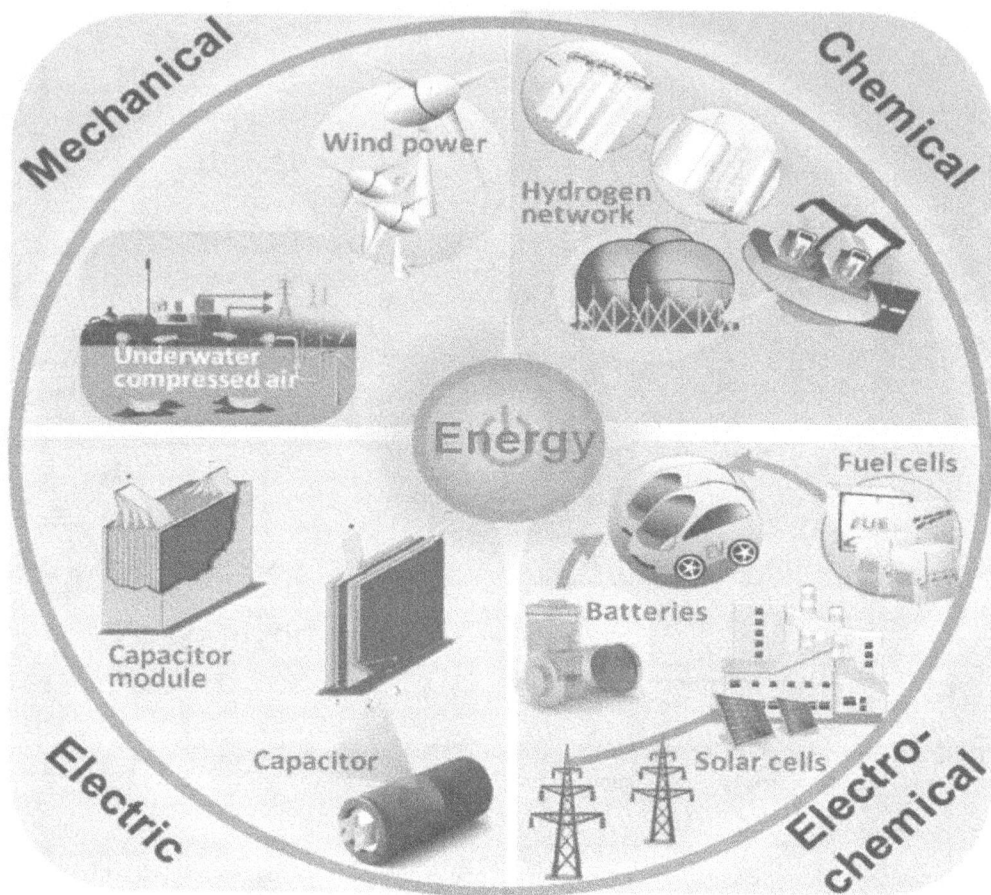

FIGURE 2.5 Different types of energy storage systems based on inherent properties of conducting polymer composites for sustainable and advanced applications. (Copyright 2018. Reproduced with permission from Ref. [43]. Published by The Royal Society of Chemistry.)

electrochemical capacitors can resolve the drawbacks faced by batteries and conventional dielectric capacitors as they possess higher power delivery [9, 40, 41]. Figure 2.5 demonstrates different types of energy storage systems based on inherent properties (mechanical, electrical, chemical, and electrochemical) and capabilities of CP composites for sustainable and advanced applications.

Supercapacitors – with the growing population, use of electric vehicles, portable devices, there is an increasing demand for energy sources or energy storage systems and replacement of conventional energy sources has become the need of the hour. There is a growing demand for supercapacitors in recent times as they store 1000 times more energy than the conventional capacitor. A supercapacitor consists of two electrodes, a separator, and an electrolyte including aqueous electrolyte, organic electrolyte, and ionic liquid which are sealed in strong cases to avoid leakage for safety concerns. Electrodes here play a crucial role in the property of supercapacitors.

Typically, electrical energy is stored and supplied in various forms of energy devices, which includes: batteries, EDLCs, fuel cells, capacitors [electrolytic capacitors, solid-state capacitors, and electromechanical capacitors, electrostatic capacitors (dielectric parallel plate capacitors: ceramic capacitors, ceramic-glass composites, and ceramic-polymer composites, etc.)]. Depending on the type of storage mechanism, supercapacitors can be classified into three types: electrochemical double-layer capacitors (EDLCs), pseudocapacitors, and hybrid capacitors [42]. The performance

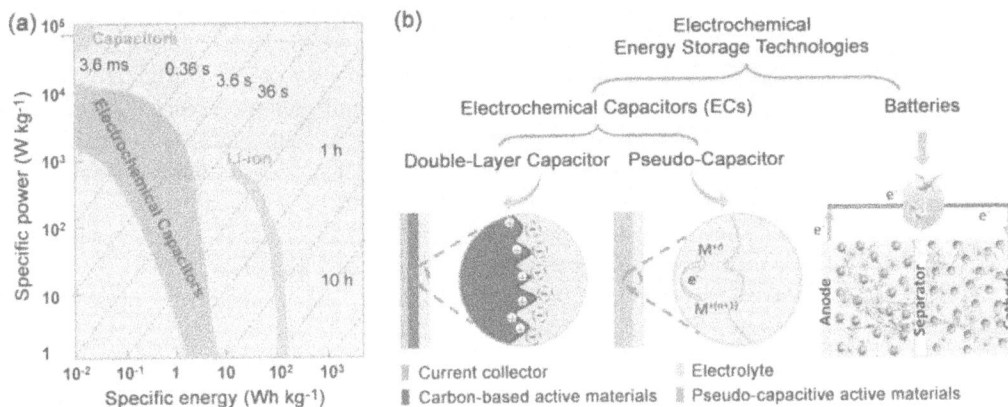

FIGURE 2.6 (a) Ragone plot for energy storage systems. (b) Types of energy storage systems: EDLC's, pseudocapacitor, and batteries. (Copyright 2018. Reproduced with permission from Ref. [43]. Published by John Wiley and Sons.)

metrics of various energy storage materials were enlisted by using Ragone plot (Figure 2.6) [43] that shows the relation between energy density (how far an electric car can go on a single charge) and power density (how fast the car can go). In general, electrical energy storage capacitors are the passive electronic components widely used in electronic circuits, which can store energy in the form of an electrostatic field.

The energy storage mechanism in EDLCs is based on the electrostatic interaction between ions on the surface of the active electrode materials and electrolytes. For pseudocapacitors, the energy storage is based on reversible Faradaic redox reactions on the surface and in the bulk. The properties of hybrid capacitors lie in between EDLCs and pseudocapacitors. Materials derived from carbon such as activated carbon, CNTS, and graphene, which have high surface area are widely used materials for EDLC and also transition metal oxides/hydroxides and together with CPs are commonly used as materials for pseudocapacitors. CPs have garnered more interest from researchers when compared to transition metal oxides due to their high specific capacitance, good conductivity, good environment, stability, ease of synthesis, low cost, and good flexibility; also CP's electrical conductivity is similar to metals and mechanical properties to that of a polymer. The widely used CP's for pseudocapacitors are PPy, PANI, polythiophene (PTh), and its derivatives, such as PEDOT. The source of energy for CPs is from doping and de-doping processes resulting in high specific capacitances, such as 620 F g^{-1} for PPy, 750 F g^{-1} for PANI, and 485 F g^{-1} for PTh [9, 42, 44, 45]. However, CPs-based supercapacitors face a major challenge in practical applications as they face poor cycle stability due to continuous volume change occurring during the doping–dedoping process. To improve the cycle stability and energy storage capacity of CPs-based supercapacitors, researchers have prepared various binary and ternary composites of CP with other active materials like metal hydroxides and carbon-based materials [40].

Many investigations have shown that carbon-based composites of CPs are ideal materials for superconductors. Though they have low capacitance levels, they show good conductivity, cycle stability, possess good specific surface area, and so on, making them a potential candidate for composite material. Recently, CPs integrated with CNT and graphene [46] have been reviewed a lot. CNT and graphene cannot be directly used as high-performance electrodes because of their low accessible surface area due to their compactness. Recently, Sharma et al. proposed a novel method for the fabrication of a ternary composite of cellulose/graphite/PANI. This setup showed stable cycle stability and high capacitance values [47]. Wang et al. developed a pizza-like ternary nanostructure of MoS_2/polypyrrole/PANI, which showed good specific capacitance of 1273 F g^{-1} at 0.5 A g^{-1} [48].

Conductivity, sensors and antimicrobial applications – the emeraldine salt form of PANI nanofibers synthesized using ammonium persulphate as oxidizer were characterized by Ultraviolet-Visible (UV-Vis), Fourier Transform Infrared Spectroscopy (FTIR), X-ray Diffraction (XRD), and Transmission Electron Microscopy (TEM) studies which confirmed the expected structure of the polymer and the formation of PANI nanofibers. From dielectric and conductivity studies, it is observed that there is an increase in alternating current (AC) conductivity for the PANI sample at higher frequencies. It was evident that PANI nanofibers and PANI–Au nanocomposite fibers are reported to be effective against *Klebsiella sps* and *Staphylococcus sps* and are anticipated to be good candidates for the fabrication of biomedical and biosensor devices [49–54].

CPs are extensively used in the field of sensing gases, heavy metal ions, and explosives. The change in pH, color, the flow of current, conductivity, potential, optical properties on the interaction of the CPs with the analyte is used as a measure for detection. Excellent electrical, mechanical properties, optical, and conducting properties resistant to corrosion, low cost, and ease of synthesis have made them an excellent choice in the field of sensing. Electrochemical sensors are gaining more importance these days owing to the excellent electrical properties of CPs. The basic principle of an electrochemical sensor is that the chemical reaction taking place when the analytes interact with the electrode is converted into electrical signals that exhibit changes in current, potential, and conductivity [55]. Poly(3,4-ethylenedioxythiophene)-graphene oxide (PEDOT-GO) nanocomposite was developed by the electrochemical deposition followed by electrochemical reduction of nanocomposite for selective detection of dopamine [56].

Another field of application of CPs that is gaining a lot of interest these days is the detection of explosives by fluorescence technique. The development of sensors for the detection of explosives is a major concern associated with security at airports, borders, public places, mining, etc. In recent years, there has been a lot of increase in terrorist activity making homeland security a major concern. Currently, the methods available for the detection of explosives include trained canines, mass spectrometry, electron capture, metal detectors, and so on. Every method has its advantages; their use is not without a problem. Detection of explosives via a change in optical properties is gaining attention, especially by fluorimetry because of its ease of handling, sensitivity and has wider linear ranges compared to absorbance-based methods. CPs have recently been used compared to small molecule sensors due to their extended pi-conjugation leading to extended exciton migration pathway and good interaction between explosives and polymer backbone [41].

Swager et al. proposed that binding of the analyte to one site resulted in quenching at all emitting sites of an entire conjugated polymer molecule, which is not observed in small molecules. This amplifying effect is termed as molecular wire effect. Deepti et al. used phenylene vinylene crosslinked polymer for trace detection of Royal Demolition Explosive (RDX), trinitrotoluene (TNT), and pentaerythritol tetranitrate (PETN) [57]. Lakshmidevi et al. used PANI doped with camphor sulphonicacid for trace detection of picric acid and p-nitrotoluene [58]. Patil et al. synthesized p-toluene sulphonicacid doped PANI for the detection of high-energy materials such as RDX, PETN, and hexanitrohexaazaisowurtzitane, also called HNIW and CL-20 [59]. Here, the conjugated polymer due to its π-electron-rich nature acts as electron-rich species, and explosives due to the presence of nitro ($-NO_2$) groups acts as electron-deficient species. When explosives interact with CPs, they abstract the electron/energy, thereby altering the fluorescence emission of the polymer.

2.4 CONCLUSIONS AND CURRENT TRENDS

So far, many researchers have advanced the field of CP nanocomposites by introducing sophisticated methods or fabrication strategies, novel properties, and advanced applications by combining CPs with a variety of suitable functional materials to arrive at improved physical and chemical properties. The CPs that are insulators in their pristine form can behave close to the metallic conductors when doped with appropriate fillers or dopants. Even though theoretical modeling of CP composites for experimentation is complex, the fundamental understanding of their complex structure,

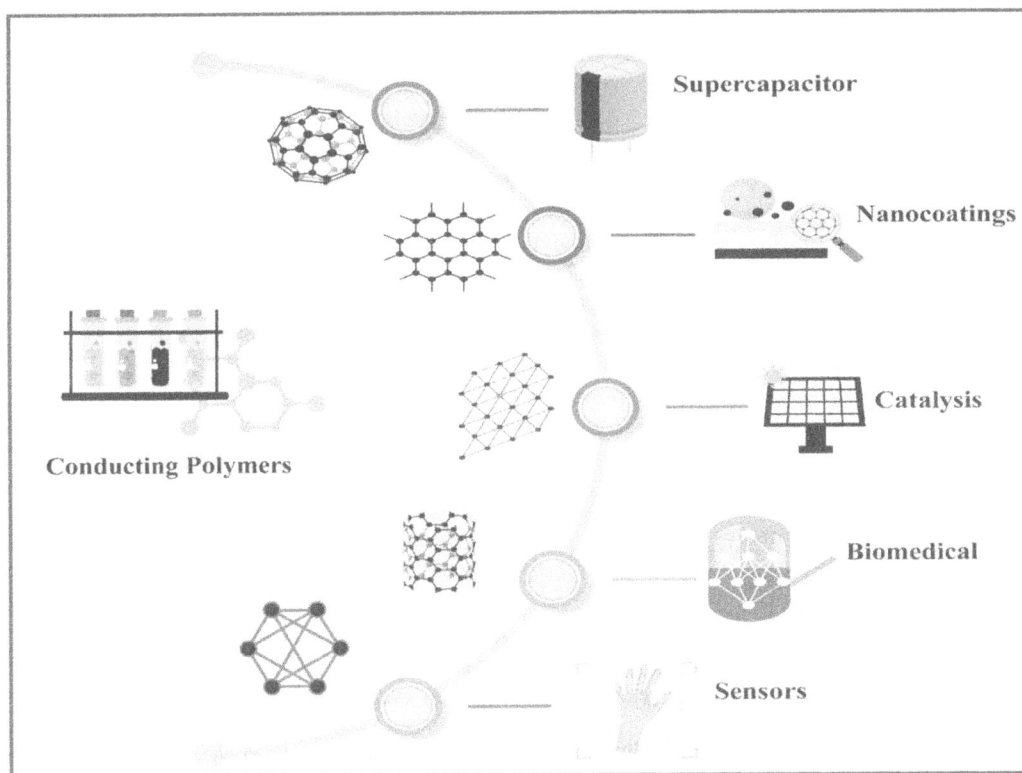

FIGURE 2.7 Application of CPs in various fields. (Copyright 2021. Reproduced with permission from Ref. [13]. Published by the Royal Society of Chemistry.)

mechanism, and improved conductivity from tuning the properties at an appropriate temperature, and processing conditions are further necessary. The wide array of CPs and the CP composite applications are depicted in Figure 2.7.

The economic cost of corrosion is tremendous and estimated at around 4% of the US gross national product. Corrosion causes waste of valuable resources, industrial plant and machinery failure, loss or contamination of products, reduction in efficiency, costly maintenance, etc. In addition to the economic cost, corrosion also jeopardizes human safety and the surrounding environment due to structural failure. Even though corrosion is inevitable, its costs and detrimental impact can be significantly reduced by the use of anticorrosive coatings. Thus, anticorrosive coatings are in great demand, and the market is estimated at around 22 billion. The most effective anticorrosive coatings currently available in the market contain heavy metal pigments, such as chromates, zinc, zinc compounds, and lead compounds. However, because of health, safety, and environmental concerns, the replacement of heavy metal-based anticorrosion pigment is highly desired. The currently available metal-free anticorrosive coatings are not as effective as metal-based coatings. Additionally, if there is a scratch or mechanical damage on the surface of the coating, the exposed metal starts corroding immediately making the anticorrosive coating ineffective. Also, most of the effective anticorrosive coatings in the market are solvent-based, and so there are environmental concerns with regards to their volatile organic compounds (VOCs) as well.

Recent studies on CP nanocomposites are extended to the field of industrial coatings where the toxic chromate coatings are being replaced with less toxic, sustainable, and alternative to the existing functional additives [60–70] in the coatings that protect metals from corrosion. Owing to the strict environmental and regulatory concerns, chromium(VI) compounds are banned by the current REACH (Registration, Evaluation, Authorisation, and Restriction of Chemicals, EU) and

TSCA (Toxic Substance Controlled Act, US) legislation. For example, in the past few years, CPs are potentially considered for controlling metallic corrosion especially in the case of metals, and alloys of iron (or steel) and aluminum. The composites made of multiple layers of CPs (e.g. PANI, PPy) and their composites or the CPs incorporated or encapsulated with corrosion inhibitors provide corrosion resistance by multiple mechanisms such as forming an ionic barrier or reacting with the corrosive ions (reduction) due to the change in electrochemical potential during corrosion, which triggers the release of encapsulants (or corrosion inhibitors) from the polymer matrix or capsules. Morphology, compatibility of CP composites with coating materials, and controlled or triggered release of inhibitors (by reduction) by generating a protective metal oxide layer (primer) below the topcoat are effective methodologies adopted for corrosion inhibition CP-based anticorrosive coatings.

ACKNOWLEDGMENTS

KM acknowledges the support of her supervisor, Sumedh P. Surwade, CEO and Founder, SAS Nanotechnologies LLC, for working toward polymer nanocomposites as self-healing materials and AVR expresses his thanks to UGC, New Delhi for the BSR faculty fellowship [F.4-5(11)/2019(BSR)].

REFERENCES

1. MacDiarmid, A. G. "Synthetic metals": A novel role for organic polymers (nobel lecture). *Angew. Chem., Int. Ed.* 40, 2581 (2001).
2. Swaroopa Rani, H., Basavaraj, S., Basavaraj, C., Huh, D. S., Venkataraman, A. A new approach to soluble polyaniline and its copolymers with toluidines. *J. Appl. Polym. Sci.* 117, 1350 (2010).
3. Nezakati, T., Seifalian, A., Tan, A., Seifalian, A. M. Conductive polymers: Opportunities and challenges in biomedical applications. *Chem. Rev.* 118, 14, 6766–6843 (2018). doi: 10.1021/acs.chemrev.6b00275.
4. Seymour, R. B., *Conductive Polymers*, Plenum Press, New York (1981).
5. Zhang, W. J., Feng, J., MacDiarmid, A. G., Epstein, A. Synthesis of oligomeric anilines. *J. Synth. Met.* 84, 119–120 (1997).
6. Surwade, S. P., Manohar, N., Manohar, S. K. Origin of bulk nanoscale morphology in conducting polymers, *Macromolecules* 42, 1792–1795 (2009).
7. Guo, X., Facchetti, A. The journey of conducting polymers from discovery to application. *Nat. Mater.* 19, 922–928 (2020).
8. Lakard, B. Electrochemical biosensors based on conducting polymers: A review. *Appl. Sci.* 10, 6614 (2020). https://doi.org/10.3390/app10186614.
9. Zhao, C., Jia, X., Shu, K., Yu, C., Wallace, G. G., Wang, C. Conducting polymer composites for unconventional solid-state supercapacitors. *J. Mater. Chem. A* 8, 4677–4699 (2020). https://doi.org/10.1039/C9TA13432H.
10. Ary, R. M., Ahmed, I., Aziz, S. B., Sozan, N. A., Brza, M. A. Conducting polymers for optoelectronic devices and organic solar cells: A review. *Polymers* 12, 11, 2627 (2020). https://doi.org/10.3390/polym12112627.
11. MacDiarmid, A. G. *Prix Nobel*, (2001). https://www.nobelprize.org/prizes/chemistry/2000/macdiarmid/lecture/
12. Pérez-Madrigal, M. M., Armelin, E., Puiggalí, J., Alemán, C. Insulating and semiconducting polymeric free-standing nanomembranes with biomedical applications. *J. Mater. Chem. B* 2015, 3, 5904. doi: 10.1039/C5TB00624D.
13. Namsheer, K., Rout, C. S., Conducting polymers: A comprehensive review on recent advances in synthesis, properties and applications. *RSC Adv.* 11, 5659–5697 (2021). doi: 10.1039/D0RA07800J.
14. Yuk, H., Lu, B., Zhao, X. Hydrogel bioelectronics. *Chem. Soc. Rev (RSC)* 48, 1642 (2019). doi: 10.1039/C8CS00595H.
15. (a) Sai Ram, M., Palaniappan, S. J. Benzoyl peroxide oxidation route to polyaniline salt and its use as catalyst in the esterification reaction. *Mol. Catal.* 201, 289–296 (2003). (b) Sun, Z., Geng, Y., Li, J., Wang, X., Jing, X., Wang, F. Catalytic oxidization polymerization of aniline in an H2O2-Fe2+ system. *J. Appl. Polym.* Sci. 72, 1077–1084 (1999). (c) Liu, W., Kumar, J., Tripathy, S., Senecal, K. J., Samuelson,

L. Enzymatically Synthesized Conducting Polyaniline. *J. Am. Chem. Soc.* 121, 71–78 (1999). (d) Moon, D. K., Osakada, K., Maruyama, T., Yamamoto, T. Preparation of polyaniline by oxidation of aniline using H2O2 in the presence of an iron(II) catalyst. *Makromol. Chem.* 193, 1723–1728 (1992). (e) Dias, H. V. R., Wang, X., Rajapakse, R. M. G., Elsenbaumer, R. L. A mild, copper catalyzed route to conducting polyaniline. *Chem. Commun.* 976–978 (2006). (f) Wang, Y., Jing, X., Kong, Polyaniline nanofibers prepared with hydrogen peroxide as oxidant. *J. Synth. Met.* 157, 269–275 (2007).

16. Graham, M. A., Mammone R. J, Kaner R. B., Lord, P. The concept of "doping" of conducting polymers: The role of reduction potentials. *Philos. Trans. R. Soc. A* 314, 3–15 (1985).
17. Ma, Z., Shi, W., Yan, K., Pan, L., Logo, G. Y. D. Doping engineering of conductive polymer hydrogels and their application in advanced sensor technologies. *Chem. Sci.* 10, 6232–6244 (2019). https://doi.org/10.1039/C9SC02033K.
18. Simotwo, S. K., Delre, C., Kalra, V. Supercapacitor electrodes based on high-purity electrospun polyaniline and polyaniline–carbon nanotube nanofibers. *ACS Appl. Mater. Interfaces*, 8, 21261–21269 (2016).
19. Zhu, X., Hou, K., Chen, C., Zhang, W., Sun, H., Zhang, G., Gao, Z. Structural-controlled synthesis of polyaniline nanoarchitectures using hydrothermal method. *High Perform. Polym.* 27, 207–216 (2015).
20. Wang, R., Han, M., Zhao, Q., Ren, Z., Guo, X., Xu, C., Hu, N., Lu, L. Hydrothermal synthesis of nanostructured graphene/polyaniline composites as high-capacitance electrode materials for supercapacitors. *Sci. Rep.* 7, 1–9 (2017).
21. Chen, I. W. P., Chou, Y. C., Wang, P. Y. Integration of ultrathin MoS2/PANI/CNT composite paper in producing all-solid-state flexible supercapacitors with exceptional volumetric energy density. *J. Phys. Chem. C* 123, 17864–17872 (2019).
22. Gopalakrishnan, K., Sultan, S., Govindaraj, A., Rao, C. N. R. Supercapacitors based on composites of PANI with nanosheets of nitrogen-doped RGO, BC1.5N, MoS2 and WS2. *Nano Energy*, 12, 52–58 (2015).
23. Tan, Y., Ghandi, K. Kinetics and mechanism of pyrrole chemical polymerization *Synth. Met.* 175, 183–191 (2013).
24. Cheng, N., Zhang, L., Joon Kim, J., Andrew, T. L. Vapor phase organic chemistry to deposit conjugated polymer films on arbitrary substrates. *J. Mater. Chem. C* 5, 5787–5796 (2017).
25. Wang, C., Zheng, W., Yue, Z., Too, C. O., Wallace, G. G. Buckled, stretchable polypyrrole electrodes for battery applications. *Adv. Mater.* 23, 3580–3584 (2011).
26. Davies, A., Audette, P., Farrow, B., Hassan, F., Chen, Z., Choi, J. Y., Yu, A. Graphene-based flexible supercapacitors: Pulse-electropolymerization of polypyrrole on free-standing graphene films. *J. Phys. Chem. C* 115, 17612–17620 (2011).
27. Choi, H., Ahn, K.-J., Lee, Y., Noh, S., Yoon, H. Free-standing, multilayered graphene/polyaniline-glue/graphene nanostructures for flexible, solid-state electrochemical capacitor application. *Adv. Mater. Interfaces*, 2, 1500117 (2015).
28. Chen, J., Wang, Y., Cao, J., Liu, Y., Zhou, Y., Ouyang, J.-H., Jia, D. Facile co-electrodeposition method for high-performance supercapacitor based on reduced graphene oxide/polypyrrole composite film. *ACS Appl. Mater. Interfaces*, 9, 19831–19842 (2017).
29. Huang, J., Kaner, R. B. A general chemical route to polyaniline nanofibers. *J. Am. Chem. Soc.* 126, 3, 851–855 (2004).
30. Hardaker, S., Gregory, R. New approaches to the study of polyaniline. *Synth. Met.* 84, 743–746 (1997).
31. (a) Li, W., Wang, H.-L. Oligomer-assisted synthesis of chiral polyaniline nanofibers. *J. Am. Chem. Soc.* 126, 2278–2279 (2004). (b) Zhang, X., Goux, W. J., Manohar, S. K. Synthesis of polyaniline nanofibers by "Nanofiber Seeding". *J. Am. Chem. Soc.* 126, 4502–4503 (2004). (c) Li, D., Huang, J., Kaner, R. B. Polyaniline nanofibers: A unique polymer nanostructure for versatile applications. *Acc. Chem. Res.* 42, 135–145 (2009). (d) Chiou, N.-R., Epstein, A. Polyaniline nanofibers prepared by dilute polymerization. *J. Adv. Mater.* 17, 1679–1683 (2005).
32. Xing, S., Zhao, C., Jing, S., Wang, Z. Morphology and conductivity of polyaniline nanofibers prepared by "seeding" polymerization. *Polymer (Guild)* 47, 7, 2305–2313 (2006).
33. Zhou, Y., Freitag, M., Hone, J., Staii, C., Johnson, A. T., Pinto, N. J., MacDiarmid, A. G. Fabrication and electrical characterization of polyaniline-based nanofibers with diameter below 30 nm. *Appl. Phys. Lett.* 83, 18, 3800 (2003).
34. Surwade, S. P., Madgula, K. Stimuli-responsive micro-reservoirs for release of encapsulants, US Patent App. 16/513,220, 2020. Application number-16513220;US020200016564A120200116 (storage.googleapis.com).
35. Wang, H., Romero, R. J., Mattes, B. R., Zhu, Y., Winokur, M. J. Effect of processing conditions on the properties of high molecular weight conductive polyaniline fiber. *J. Polym. Sci. Part B Polym. Phys.* 38, 194–204 (1999).

36. Sanches, E.A., Soares, J. C., Iost, R. M., Marangoni, V. S., Trovati, G., Batista, T., Mafud, A. C., Zucolotto, V. and Mascarenhas, Y. P. Structural characterization of emeraldine-salt polyaniline/gold nanoparticles complexes, *J. Nanomater.* 2011, 1–7.

37. Liu, A., Bac, L. H., Kim, J. S., Kim, B. K. and Kim, J.-C. Synthesis and characterization of conducting polyaniline-copper composites. *J. Nanosci. Nanotechnol.* 13, 11, 7728–7733 (2013).

38. Sharma, R., Malik, R., Lamba, S., Annapoorni, S. Metal oxide/polyaniline nanocomposites: Cluster size and composition dependent structural and magnetic properties. *Bull. Mater. Sci.* 31, 3, 409–413 (2008).

39. Subha, L. N., Kalpana, M., Madhusudhan Rao, P. *Proceedings of the National Seminar on Frontiers in Chemical Research and Analysis*, St. Francis College for Women, Hyderabad, T.S., 24–25 July, 2015, pp.1–4. https://www.researchgate.net/profile/Kalpana-Madgula/publication/295400974_Study_of_chemi cally_synthesized_polyaniline_copper_oxide_Nanocomposites/links/580e03e908aebfb68a501b7f/ Study-of-chemically-synthesized-polyaniline-copper-oxide-Nanocomposites.pdf.

40. Meng, Q., Cai, K., Chen, Y., Chen, L. Research progress on conducting polymer based supercapacitor electrode materials, *Nano Energy*, 36, 268–285 (2017). https://doi.org/10.1016/j.nanoen.2017.04.040.

41. Madgula K., Shubha L. N. (2020) Conducting Polymer Nanocomposite-Based Gas Sensors. In: Thomas S., Joshi N., Tomer V. (eds) *Functional Nanomaterials. Materials Horizons: From Nature to Nanomaterials.* Springer, Singapore. https://doi.org/10.1007/978-981-15-4810-9_16; 10.1039/ c5cs00496a.

42. Huang, S., Zhu, X., Sarkar, S., Zhao, Y. Challenges and opportunities for supercapacitors. *APL Mater.* 7, 100901 (2019). https://doi.org/10.1063/1.5116146.

43. Wang, J., Xu, C., Jiang, H., Li, C., Zhang, L., Lin, J., Shen, Z. X. *Adv. Sci.* 5, 1700322 (2018). https://doi. org/10.1002/advs.418.

44. Singhal, R., Thorne, D., LeMaire, P. K., Martinez, X., Zhao, C., Gupta, R. K., Uhl, D., Scanley, E., Broadbridge, C. C., Sharma, R. K. Synthesis and characterization of CuS, CuS/graphene oxide nanocomposite for supercapacitor applications. *AIP Advances* 10, 035307 (2020). https://doi. org/10.1063/1.513271355.55.

45. Al Dream, J., Zequine, C., Siam, K., Kahol, P. K., Mishra, S. R., Gupta, R. K. Electrochemical properties of graphene oxide nanoribbons/polypyrrole nanocomposites. *C* 5, 2, 18 (2019). https://doi.org/10.3390/ c5020018.

46. Zhang, M., Wang, X., Yang, T., Zhang, P., Wei, X., Zhang, L., Li, H. Polyaniline/graphene hybrid fibers as electrodes for flexible supercapacitors. *Synth. Met.* 268, 116484 (2020).

47. Sharma, K., Pareek, K., Rohan, R., Kumar, P. Flexible supercapacitor based on three-dimensional cellulose/graphite/polyaniline composite. *Int. J. Energy Res.* 43, 604–611 (2019).

48. Wang, K., Li, L., Liu, Y., Zhang, C., Liu, T. Constructing a "Pizza-Like" MoS2/polypyrrole/polyaniline ternary architecture with high energy density and superior cycling stability for supercapacitors. *Adv. Mater. Interfaces* 3, 1600665 (2016).

49. Shubha, L. N., Kalpana, M., Madhusudana Rao, P. Synthesis, characterization by AC conduction and antibacterial properties of polyaniline fibers. Scholars Research Library. *Der Pharmacia Lettre* 8, 1, 214–219 (2016) (http://scholarsresearchlibrary.com/archive.html).

50. Ramanathan, K., Bangar, M. A., Yun, M., Chen, W., Myung, N. V., Mulchandani, A. Bioaffinity sensing using biologically functionalized conducting-polymer nanowire. *J. Am. Chem. Soc.* 9, 127, 2, 497 (2005).

51. Wena, Y., Xu, J., Li, D., Liu, M., Kong, F., He, H., Novel electrochemical biosensing platform based on poly(3,4-ethylenedioxythiophene):poly(styrenesulfonate) composites. *Synth. Met.* 162, 1308–1314 (2012).

52. Zare, E. N., Makvandi, P., Ashtari, B., Rossi, F., Motahari, A., Perale, G. Progress in conductive polyaniline-based nanocomposites for biomedical applications: A review. *J. Med. Chem.* 63, 1, 1–22 (2020). doi: 10.1021/acs.jmedchem.9b00803.

53. Parthiban, E., Kalaivasan, N., Sudarsan, S. A study of magnetic, antibacterial and antifungal behaviour of a novel gold anchor of polyaniline/itaconic acid/Fe3O4 hybrid nanocomposite: Synthesis and characterization. *Arab. J. Chem.* 13, 4751–4763 (2020).

54. Poyraz, S., Cerkez, I., Huang, T. S., Liu, Z., Kang, L., Luo, J., Zhang, X. One-step synthesis and characterization of polyaniline nanofiber/silver nanoparticle composite networks as antibacterial agents. *ACS Appl. Mater. Interfaces* 6, 20025–20034 (2014).

55. Naveen, M. H., Gurudatt, N. G., Shim, Y.-B. Applications of conducting polymer composites to electrochemical sensors: A review. *Appl. Mater. Today* 9, 419–433 (2017). https://doi.org/10.1016/j. apmt.2017.09.001.

56. Wang, W., Xu, G., Cui, X. T., Sheng, G., Luo, X. Enhanced catalytic and dopamine sensing properties of electrochemically reduced conducting polymer nanocomposite doped with pure graphene oxide. *Biosens. Bioelectron.* 58, 153–156 (2014). https://doi.org/10.1016/j.bios.2014.02.055.
57. Gopalakrishnan, D., Dichtel, W. R. Direct detection of RDX vapor using a conjugated polymer network. *J. Am. Chem. Soc.* 135, 22, 8357–8362 (2013). https://doi.org/10.1021/ja402668e.
58. Venkatappa, L., Ture, S. A, Yelamaggad, C. V, Narayanan Naranammalpuram Sundaram, V., Martínez-Máñez, R., Abbaraju, V. Mechanistic insight into the turn-off sensing of nitroaromatic compounds employing functionalized polyaniline. *ChemistrySelect* 5, 21, 6321–6330 (2020). https://doi.org/10.1002/slct.202001170.
59. Patil, V. B, Ture, S. A, Yelamaggad, C. V, Nadagouda, M. N, Venkataraman, A. Turn-off fluorescent sensing of energetic materials using protonic acid doped polyaniline: A spectrochemical mechanistic approach. *Zeitschrift für Anorg. und Allg. Chemie* 647, 4, 331–340 (2021). https://doi.org/10.1002/zaac.202000321.
60. Sun, M., Ma, Z., Li, A., Zhu, G., Zhang, Y. Anticorrosive performance of polyaniline/waterborne epoxy/poly(methylhydrosiloxane) composite coatings. *Prog. Org. Coat.* 139, 105462 (2020).
61. Wang, H., Qi, Q., Zhang, Y., Chen, S., Dong, B., Zhu, S., Hu, Q., Guo, Z. Anticorrosive epoxy nanocomposite coatings filled with polyaniline-functionalized silicon nitride particles. *Ind. Eng. Chem. Res.* 59, 16649–16659 (2020).
62. Wemmert, S., Ketter, R., Rahnenfuhrer, J., Beerenwinkel, N., Strowitzki, M., Feiden, W., Hartmann, C., Lengauer, T., Stockhammer, F., Zang, K. D., Meese, E., Steudel, W. I., Von Deimling, A., Urbschat, S. Patients with high-grade gliomas harboring deletions of chromosomes 9p and 10q benefit from temozolomide treatment. *Neoplasia* 7, 883–893 (2005).
63. Souto, L. F. C., Soares, B. G. Polyaniline/carbon nanotube hybrids modified with ionic liquids as anti-corrosive additive in epoxy coatings. *Prog. Org. Coat.* 143, 105598 (2020); Chen, Z., Zhang, G., Yang, W., Xu, B., Chen, Y., Yin, X., Liu, Y. Superior conducting polypyrrole anti-corrosion coating containing functionalized carbon powders for 304 stainless steel bipolar plates in proton exchange membrane fuel cells. *Chem. Eng. J.* 393, 124675 (2020).
64. Garcia-Cabezon, C., Garcia-Hernandez, C., RodriguezMendez, M. L., Martin-Pedrosa, F. A new strategy for corrosion protection of porous stainless steel using polypyrrole films. *J. Mater. Sci. Technol.* 37, 85–95 (2020).
65. Muthusamy, P., Konda Kannan, S. K. High efficient corrosion inhibitor of water-soluble polypyrrole–sulfonated melamine formaldehyde nanocomposites for 316L stainless steel. *J. Appl. Polym. Sci.* 138, 49952 (2021).
66. García Rueda, F. C., González, J. T. Electrochemical polymerization of polypyrrole coatings on hard-anodized coatings of the aluminum alloy 2024-T3. *Electrochim. Acta* 347, 136272 (2020).
67. Ren, B., Li, Y., Meng, D., Li, J., Gao, S., Cao, R. Encapsulating polyaniline within porous MIL-101 for high-performance corrosion protection. *J. Colloid Interface Sci.* 579, 842–852 (2020).
68. Yao, Y., Sun, H., Zhang, Y., Yin, Z. Corrosion protection of epoxy coatings containing 2-hydroxyphosphonocarboxylic acid doped polyaniline nanofibers. *Prog. Org. Coat.* 139, 105470 (2020).
69. Badi, N., Khasim, S., Pasha, A., Alatawi, A. S., Lakshmi, M. Silver nanoparticles intercalated polyaniline composites for high electrochemical anti-corrosion performance in 6061 aluminum alloy-based solar energy frameworks. *J. Bio- and Tribo-Corrosion* 6, 1–9 (2020).
70. Li, J., He, Y., Sun, Y., Zhang, X., Shi, W., Ge, D. Synthesis of polypyrrole/V2O5 composite film on the surface of magnesium using a mild vapor phase polymerization (VPP) method for corrosion resistance. *Coatings* 10, 402 (2020).

3 Conducting Polymers for Supercapacitors

Deepak P. Dubal[1], Xuecheng Chen[2,3],
Yuping Wu[4], and Rudolf Holze[4,5,6]
[1]Centre for Materials Science, School of Chemistry and Physics,
Queensland University of Technology,
Brisbane, Queensland, Australia
[2]Faculty of Chemical Technology and Engineering,
West Pomeranian University of Technology, Szczecin,
Szczecin, Poland
[3]School of Environment and Chemical Engineering,
Shenyang University of Technology, Shenyang, China
[4]State Key Laboratory of Materials-Oriented Chemical Engineering,
School of Energy Science and Engineering, Nanjing Tech University,
Nanjing, Jiangsu Province, China
[5]Chemnitz University of Technology, Institut für Chemie,
AG Elektrochemie, Germany
[6]Institute of Chemistry, St. Petersburg State University,
St. Petersburg, Russia

CONTENTS

DOI: 10.1201/9781003150374-3

3.1 INTRODUCTION

3.1.1 Electrochemical Energy Conversion and Storage

When appreciating the numerous attractive advantages of electric energy and its application, a major drawback must be kept in mind: Electric energy must be used at the very moment it becomes available. Terms like "must be consumed or generated" are in obvious contradiction to the first law of thermodynamics: Energy can neither be generated nor annihilated. Nevertheless, this terminology is ubiquitous, no vain attempt will be made here to change this. The unwelcome consequence of ignoring this fundamental drawback becomes quite visible on every scale, even in a large power grid scale when there is a mismatch caused by too many users drawing electric energy from the grid, causing grid instabilities initially in terms of fluctuating frequency and voltage and in even more severe case brownouts and finally blackouts. Certainly, the specific technical details of every such incident are more complicated than a quotation of the first law, but they are somehow based on it and thus closely related to it. Similar problems arise when attempts are made to force-feed more energy into the grid than is currently needed: Instability follows. Numerous attempts and procedures to match supply and demand better than it is done currently have been developed and even made into reality, but a perfect match will most likely never happen. In addition, situations with demand exceeding supply will happen in every system from time to time. Accordingly, energy storage has to be considered at all levels, from the tiny sensor needing electricity at regular intervals for reporting data by telemetry all the way up to international electric grids covering whole continents. Certainly, on the latter level, with highly developed interconnections, this grid by itself helps to match supply and demand, but the need for storage remains. The numerous options of electric energy storage are textbook content [1–3]. The need for such storage on every scale has been stressed frequently [4–6].

Electric energy can be stored without conversion into some other form of energy (and retrieved, too) using inductivities and capacitors (Mode 1 in Figure 3.1). With a conversion step, energy is stored as chemical energy in the electrode and/or the electrolyte solution when electrochemical energy storage (EES) and conversion are considered (Mode 2 in Figure 3.1). These basic facts are sketched below in Figure 3.1.

Electrochemistry provides both options: In supercapacitors running in the electrochemical double layer storage mode option 1 applies; in a redox flow battery option 2.

Available devices for EES and the logically associated conversion processes operate on a wide scale of rates, i.e. speed of conversions. Charging a secondary battery tends to be a slightly slower process (not only from the eyes of the car operator waiting for his battery-driven vehicle to get its battery recharged), whereas charging or discharging a capacitor can be a very fast process. Accordingly, the roles these systems play in EES are different. For grid stabilization, i.e. power quality maintenance (commonly called primary operating reserve or – with reference to steam turbines and electromechanical generators – spinning reserve [1]), fast response of both charge and discharge of a storage device within fractions of a second is needed. A slightly slower response and operation on a few minutes scale is typical of the secondary operating reserve, and even slower operation on a scale of 15 min or more has been found for tertiary operating reserve. Electrochemical systems tend to be expensive, thus their use has to be considered carefully with respect to both technological

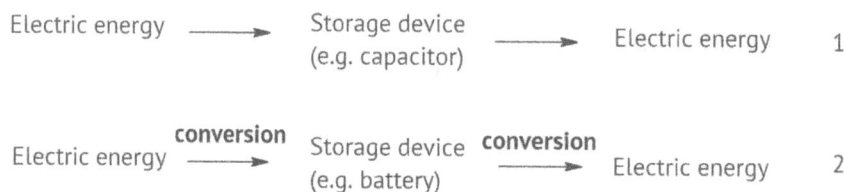

Electric energy ⟶ Storage device (e.g. capacitor) ⟶ Electric energy   1

Electric energy —**conversion**→ Storage device (e.g. battery) —**conversion**→ Electric energy   2

FIGURE 3.1 Basic modes of electrical energy storage.

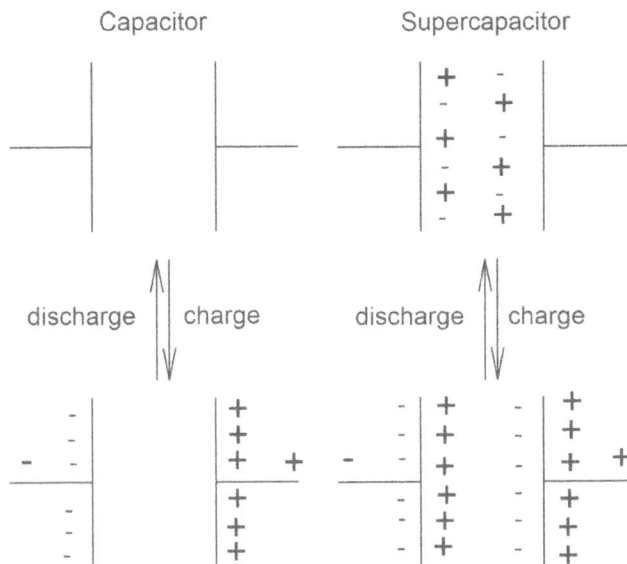

FIGURE 3.2   Charge distribution during operation in a conventional capacitor and in a supercapacitor.

as well as economic criteria. Secondary batteries will be most likely – except for very special circumstances like island or remote operation – economically unattractive for long-term storage. For primary regulation, their limited current capability in the charging mode makes them unlikely candidates for very fast response applications, although the almost legendary overload capability of lead–acid batteries has kept them busy in uninterrupted power supply applications in a wide range of installed capacity as a backup for power stations in insulated grids. Broadly speaking, electrochemical systems are attractive options for secondary and primary operating reserves.

Capacitors have been around for many decades in electrical engineering and electronics, but their storage capacity has been too small to make them competitive for electrical energy storage. This situation has changed dramatically with the advent of the supercapacitor. The very large surface area of activated carbons in contact with an electrode solution showing the capacitive behavior of the electrochemical double layer [7] yields a large double-layer capacitance, and two electrodes assembled as half cells in an electrochemical cell are already a complete supercapacitor, an electrochemical double layer capacitor (EDLC) supercapacitor. Initially, carbon-based electrode materials were used, preferably utilizing the EDLC for charge and thus energy storage [8, 9]. The pathway from the conventional dielectric capacitor to the supercapacitor is sketched in Figure 3.2.

Introductory overviews are available [10, 11], in numerous monographs presumably all aspects from basic functional principles to materials and applications are treated [12–18]. Tremendous power densities have been achieved, whereas energy densities are still disappointingly low. Numerous attempts to improve the energy density without sacrificing power density and stability have been reported [19]. Selective overviews addressing some aspects of possible approaches to increased storage capabilities of electrode materials and subsequently increased energy densities of full devices are available [20]. Starting with the capacitor-like response of some metal oxide electrode materials (RuO_2, MnO_2, etc.), the use of such compounds showing fast and highly reversible redox reactions has been examined [21–25], designation of this behavior as pseudocapacitive has been suggested by Gileadi and Conway [23]. This has extended the previous scheme beyond the double layer and the electrolytic capacitor to the supercapacitor, as shown schematically in Figure 3.3.

ICPs (for more, see below) also show redox reactions making them possible candidates for use as active mass in supercapacitor electrodes, as suggested by Arbizzani et al. [26, 27] and Conway et al. [22].

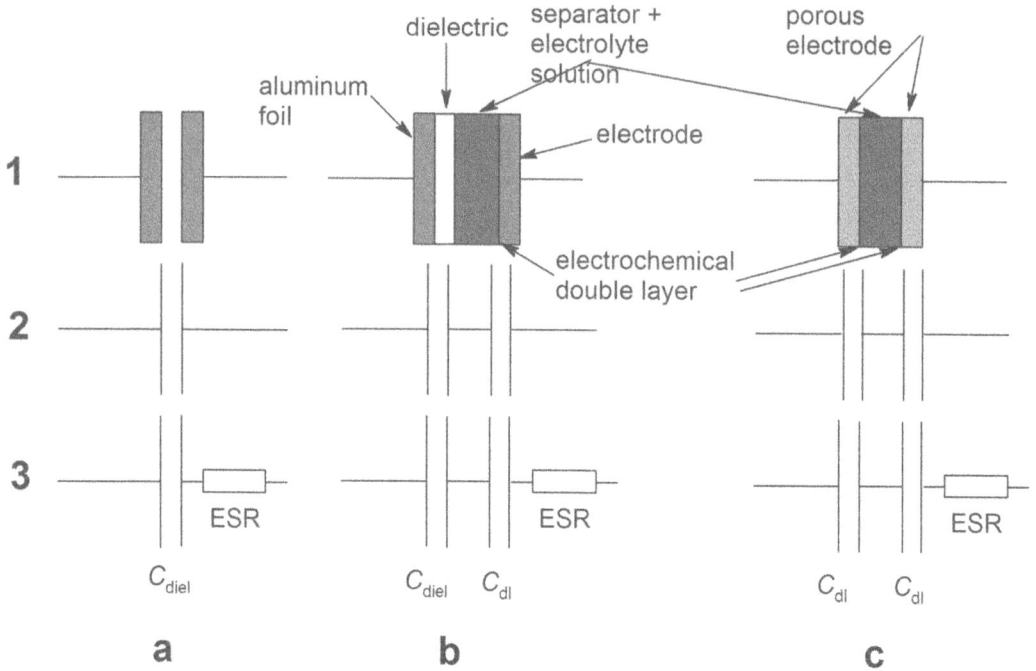

FIGURE 3.3 Schematic cross sections (1), simplified (2) and extended (3) equivalent circuits of (a) dielectric capacitor, (b) electrolytic capacitor and (c) electrochemical double layer capacitor. Inductive contributions/ elements are not shown, they are of minor importance in most supercapacitor applications. C_{diel}: capacitance established between two metallic electrodes separated by a dielectric medium, C_{dl}: capacitance of the electrochemical double layer, ESR: electric series resistance.

3.1.2 POLYMERS IN MATERIALS SCIENCE AND ELECTROCHEMICAL ENERGY TECHNOLOGY

Polymers are commonly known as electric insulators; some polymers are among the most highly insulating materials used in electrical engineering. The absence of mobile charge carriers, the wide band gap between highest or frontier fully occupied molecular orbitals (highest occupied molecular orbital, HOMO) and the lowest empty orbitals (lowest unoccupied molecular orbital, LUMO) and the low charge carrier mobility are the main reasons [28, 29]. Mixing such insulating polymeric materials with a conductive one like metal powders or fibers, carbon or acetylene black yields a composite which shows electronic conductivity once a minimum amount of conductive material sufficient to create continuous and coherent conduction pathways (percolation threshold) in the still insulating polymeric matrix has been passed. These composites are called conducting polymers, more precisely extrinsically conducting polymers. Less precisely, they are called filled polymers in technology; they are beyond the scope of this chapter. For applications requiring flexibility of both the device components as well as of the complete device, flexible materials may gain importance as active masses, beyond the well-established use of polymeric materials as binders and separators; for an overview, see [30, 31].

Structural as well as molecular orbital energy properties of some molecules enable creation of mobile charge carriers by chemical and/or electrochemical reactions (oxidation or reduction) making the initially insulating or very poorly conducting material a conducting one. Because this conductivity is not caused by added conducting substances, the materials are called *intrinsically conducting polymers* (ICPs). They have been around for quite some time [32, 33], but only the discovery that a polymer (polyacetylene) can be changed into a conducting form reported by Shirakawa et al. [34] established ICPs as a class of materials beyond the lab curiosity where they have been staying mostly before. Polyacetylene itself has not turned out to be particularly attractive for electrochemists, it showed only poor performance as an electrode material in secondary batteries [35]. But the wide variety of other ICP materials, mostly containing heteroatoms such as N, S and O, which were developed shortly thereafter, provided many options for electrochemical investigations and developments.

3.2 POSSIBLE APPLICATIONS OF INTRINSICALLY CONDUCTING POLYMERS

Early in the history of ICPs, after the discovery of their redox behavior (see e.g. [36–39]), their use in secondary batteries as active material was suggested in some studies [39–42]. Because of the fundamental similarities between secondary batteries and supercapacitors (for an introduction, see [43]), their use in the latter devices for electrochemical energy conversion and storage has been proposed following the discovery of the suitability of several metal oxides as redox-active storage materials [23, 44, 45]; early overviews are available [40, 46–49]. More recently, reports selectively touching on some aspects and examples of the use of ICPs in supercapacitors have been published [50–52]. Since ICPs can be easily deposited onto an electronically conducting support acting also as current collector, they can be used as active mass without any additive or binder [48]. Because of the frequently observed change of the electronic conductance of the ICP as a function of electrode potential [53], in most reported studies, some form of carbon has been added resulting in a composite material [54–56]. One particularly attractive option appears to be the combination of an ICP with a metal chalcogenide yielding a composite with performance data significantly superior to the respective data of the single components [54, 57]. Putting these examples into a wider perspective, the following applications of ICPs in supercapacitors can be envisaged (Figure 3.4).

A large number of original research reports on these topics and on specific ICPs with different morphologies and modes of preparation, etc. are available. In addition, a number of reviews only dealing with highly selected aspects of specific ICPs are also available. Hence, this chapter is not intended to be another addition (incomplete anyway) to the already existing pool of reviews, rather it will focus on general features, trends and possible future options in research and development. After a general overview on ICPs with particular attention to their behavior in supercapacitor electrodes, the applications indicated above will be inspected more closely.

3.3 ICPs – THE MATERIALS

Figure 3.5 provides an overview of some currently known and investigated ICPs. The molecular structure shown in the figure does not necessarily show the conducting form.

The most frequently studied ICPs are briefly introduced in the following section. Given the many different monomers already examined, a complete presentation of all ICPs is beyond the scope of this chapter. Because many features of ICPs, in particular with respect to their application in supercapacitors, are very similar, such an attempt appears to be not necessary anyway. The section ends with considerations affecting all ICPs.

Taxonomy of ICP applications

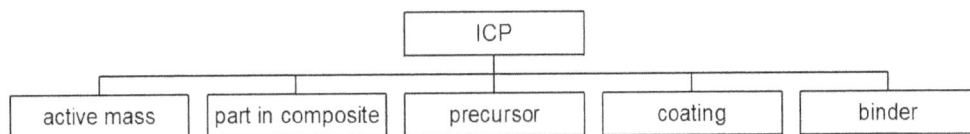

FIGURE 3.4 Taxonomy of ICP applications in supercapacitors.

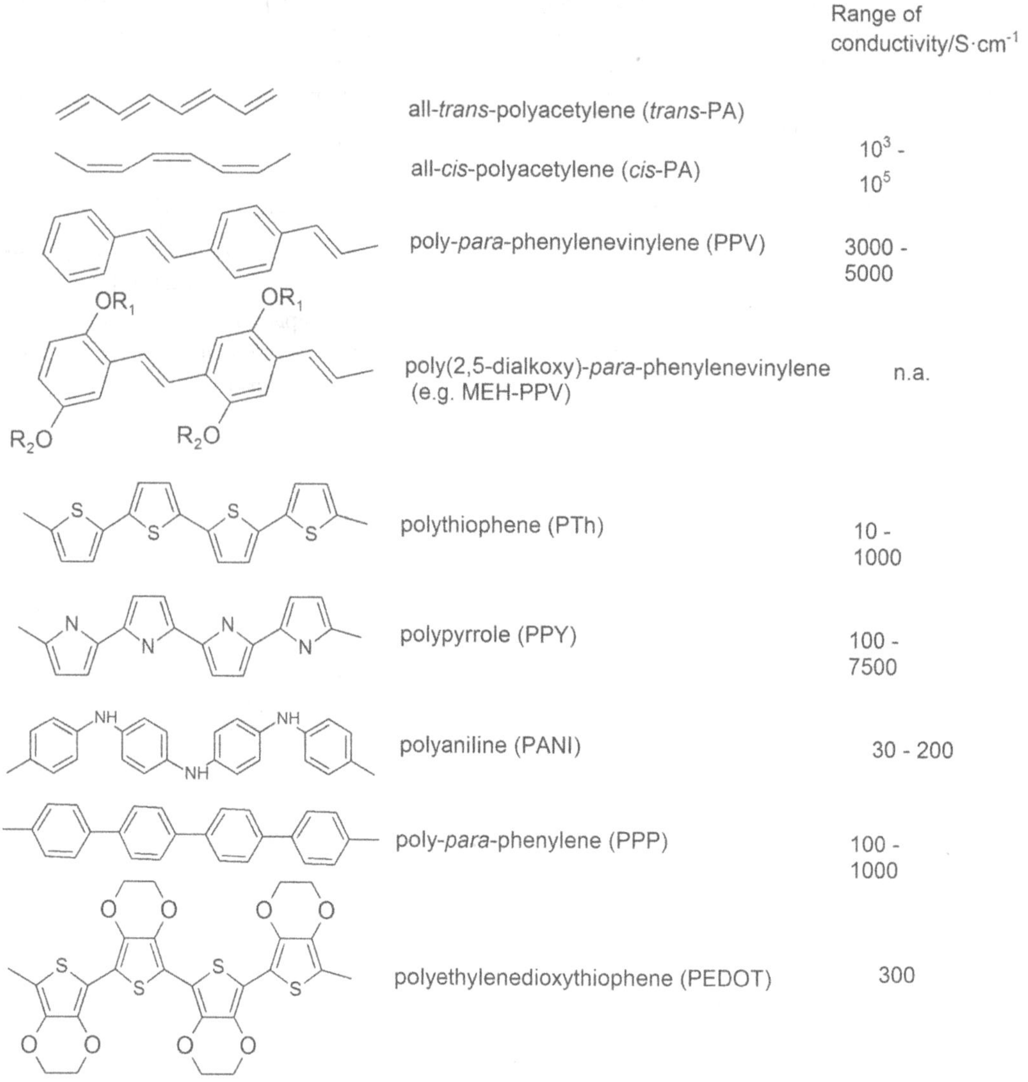

		Range of conductivity/S·cm^{-1}
	all-*trans*-polyacetylene (*trans*-PA)	10^3 - 10^5
	all-*cis*-polyacetylene (*cis*-PA)	
	poly-*para*-phenylenevinylene (PPV)	3000 - 5000
	poly(2,5-dialkoxy)-*para*-phenylenevinylene (e.g. MEH-PPV)	n.a.
	polythiophene (PTh)	10 - 1000
	polypyrrole (PPY)	100 - 7500
	polyaniline (PANI)	30 - 200
	poly-*para*-phenylene (PPP)	100 - 1000
	polyethylenedioxythiophene (PEDOT)	300

FIGURE 3.5 Examples of intrinsically conducting polymers.

3.3.1 POLYANILINE

Polyaniline (PANI) is a particularly popular ICP. This popularity may be due to the fact that it was the first material described in the literature [32]; this may be due also to its rich redox and protonation chemistry, as illustrated in a simplified scheme of chemical and electrochemical trans-formations in Figure 3.6. Frequently, its simple synthesis both by chemical and electrochemical oxidation and its stability are invoked as further reasons. The use of aqueous polymerization solutions, including water-based electrolyte solutions for electropolymerization, is a particular advantage of PANI [58]. Most of these advantages are well-established and frequently demonstrated; however, the claimed stability seems to be still somewhat questionable. Both synthesis and further handling and application in aqueous solutions are major advantages.

    The most noticeable effect of the ICP oxidation (and to a lesser extent and less well-studied reduc-tion) starting both in the neutral state (top of Figure 3.6) is the change of electronic conductance; in

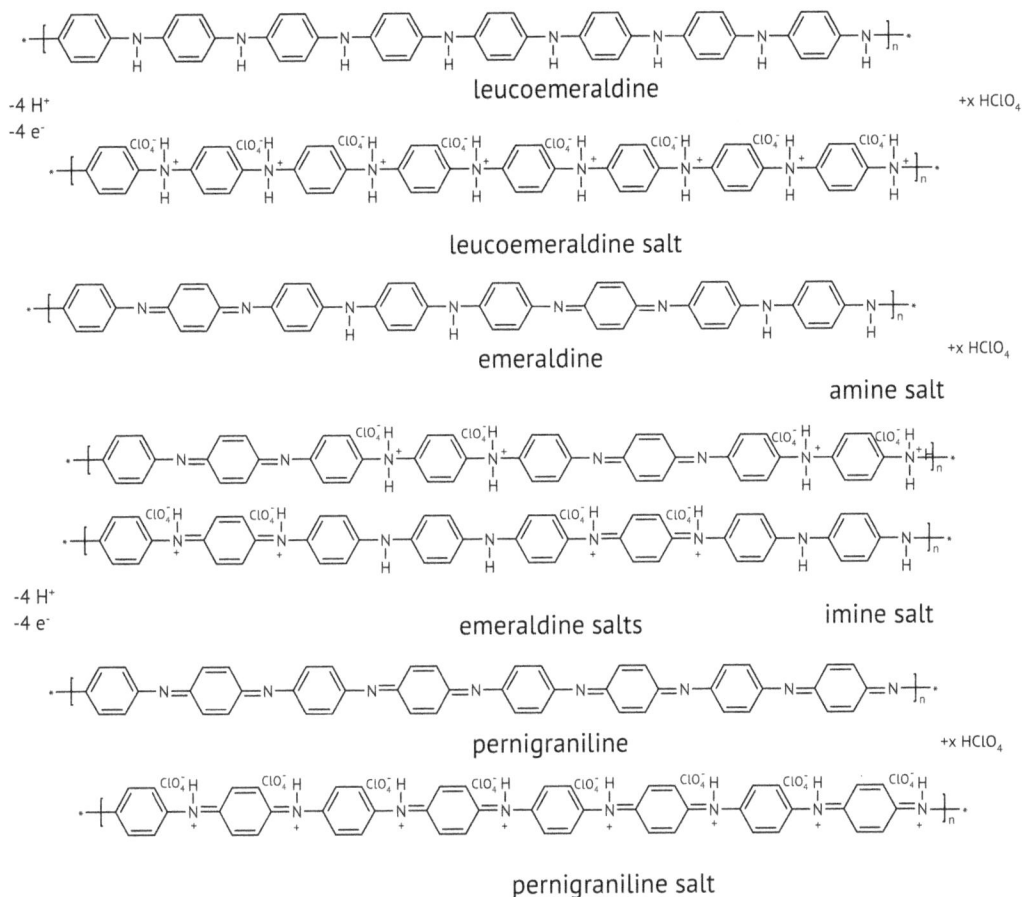

FIGURE 3.6 Possible reactions of PANI in an electrolyte solution.

the case of PANI, protonation/deprotonation may cause similar effects. In case oxidation/reduction is done electrochemically, the process can easily be monitored in a cyclic voltammogram. Current peaks or waves indicate charge transfer, and with respect to the polymer, this is closely related to the creation of radical cations/anions, i.e. mobile charge carriers, participating in electronic conduction. These changes are illustrated in Figure 3.7.

These changes are utilized in e.g. sensors. But they also may cause complications in applications where a constant and, in most cases, high conductance like in a supercapacitor electrode material are needed. The intuitive approach that oxidation provides charge storage (when the ICP is used as the positive electrode mass, this is equivalent to charging the device) and high electronic conductance does not stand the test of practical application: Oxidation, i.e. charging, starts in the poorly conducting state of PANI, providing a rather poor current-carrying capability that only increases as a function of the state of charge. This behavior may be acceptable with very thin film electrodes where the change of conductance moves quickly enough across the thin layer of material, but for a thicker electrode, it is insufficient. Thus, options for mitigating this problem have been discussed; more details of this subject are discussed as follows. Nanostructuring as one attempt to ameliorate some of the problems just addressed has been tried for PANI, and attempts covering e.g. PANI nanofibers have been reported [62]. One-dimensional nanostructuring of PANI has been reviewed [63]; for further aspects of nanostructuring, see below. Further aspects of possible uses of PANI

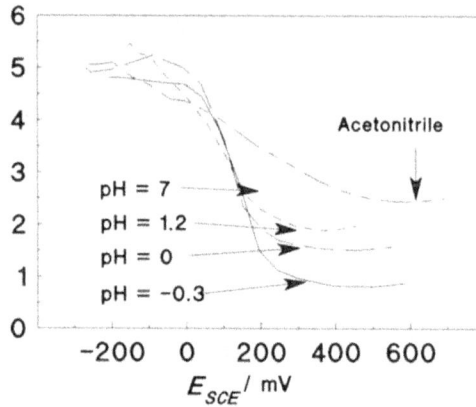

FIGURE 3.7 Change of electronic resistance of PANI as a function of solution pH and electrode potential [59–61].

in electrochemical energy conversion and storage, including supercapacitors, have been inspected elsewhere [64–66]. ICPs prepared from substituted aniline have been the subject of numerous investigations because of the possibilities to modify the molecular and, in turn, polymer properties by suitably selected and positioned substituents [67]. ICPs based e.g. on indole have been suggested as possible electrode materials; for an overview, see [68]. In an attempt to provide an overview of applications of PANI, supercapacitors have been mentioned [69], and perspectives of the application of PANI in both supercapacitors and future batteries are discussed in Ref. [70].

3.3.2 POLYPYRROLE

Polypyrrole (PPy) can be synthesized both in aqueous as well as non-aqueous electrolyte solutions; the influence of the solution composition and further preparation parameters in electrosynthesis has been examined [71, 72]. Traces of water have been found to be essential for establishing satisfactory electrochemical performance. The following scheme (Figure 3.8) depicts the redox process of PPy associated with charge storage and change of electronic conductance.

Because of the advantages of specific structures combining optimized porosity, electronic conductivity, and interfacial area, nanostructured materials have been developed and examined. A "green procedure" for the chemical synthesis of PPy nanospheres and nanofibers has been developed [73]. Bulk synthesis procedures for PPy fibers [74] and PPy nanotubes [75] have been described. A route to PPy nanostructures using reactive templates has been developed [76]. Extended 1D hierarchical PPy nanofibers were obtained via a two-step electropolymerization procedure with the option of keeping the inner part in its highly conducting state during redox switching of the outer

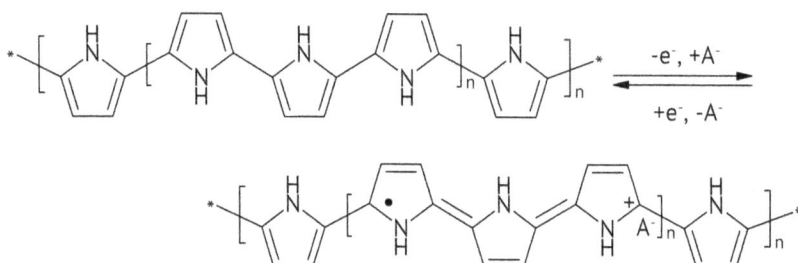

FIGURE 3.8 Redox process of PPy.

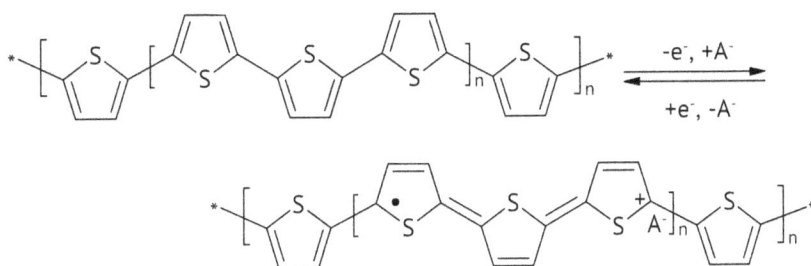

FIGURE 3.9 Redox process of PTh.

part upon charge/discharge [77]. As discussed above, carbon has been suggested as an additive to PPy to keep its electronic conductivity at a constantly sufficiently high value; some aspects of such nanocomposites have been addressed [78].

3.3.3 POLYTHIOPHENE

Polythiophene (PTh) and its substituted polymers can be obtained by chemical or electrochemical oxidation of the respective monomers [79]. The dimers (even more the higher oligomers) are easier to oxidize at less positive electrode potentials. Consequently, they may be used as starting materials, particularly under conditions where electrode potentials required to oxidize the monomers may result in undesirable side reactions. This is due to the larger molecular system available for charge delocalization in the created radical cation. Similar effects can be observed when examining substituted thiophenes [80]. A typical structure and the expected oxidative redox transformation are shown in Figure 3.9.

More frequently investigated and more widely applied in numerous fields of materials science is polyethylenedioxythiophene (PEDOT) (see Figure 3.10). Different from PANI, PPy, and PTh, PEDOT is available as a solution of its oligomer.

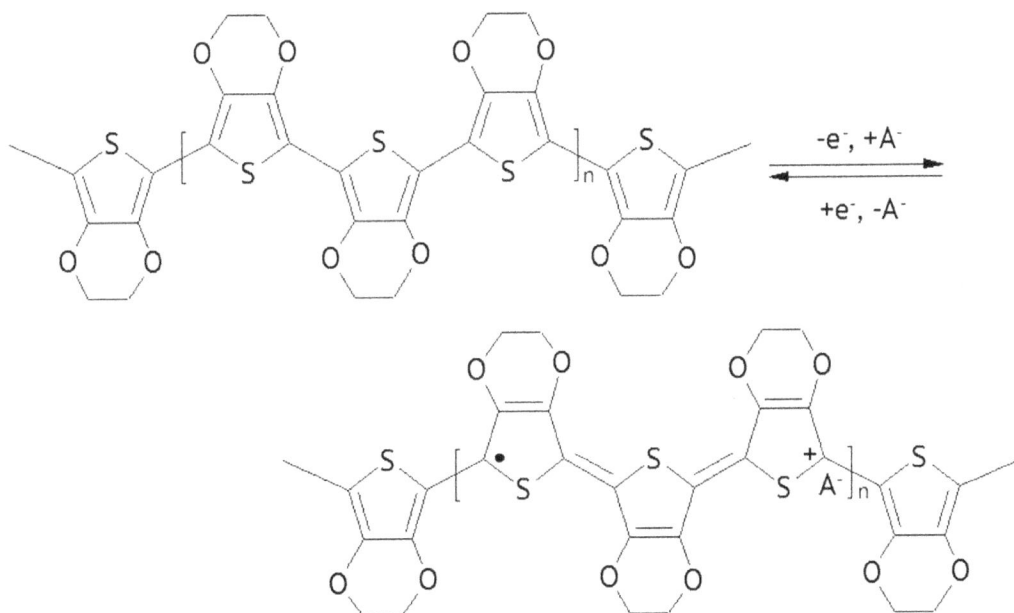

FIGURE 3.10 Redox process of PEDOT.

TABLE 3.1

Electrochemical Data of Selected ICPs (Data from [41], see also [47])

Material	Molecular Weight of Repeat Unit per g	Oxidation Level[a]/-	Theor. Q^b	Measur. $Q/F \cdot g^{-1}$
PANI	93	0.5	750 $F \cdot g^{-1}$	240
PPy	67	0.33	620 $F \cdot g^{-1}$	530
PTh	84	0.33	485 $F \cdot g^{-1}$	–
PEDOT	142	0.33	210 $F \cdot g^{-1}$	92
PbO_2	239	2	807 $As \cdot g^{-1}$	–

[a] Oxidation level, also "dopant level", reports the fraction of oxidized repeat units and also the number of electrons transferred in the electrode reaction.

[b] Gravimetric charge density can be stated with respect to the electrode reaction in units $As \cdot g^{-1}$ or, in the case of a material where no clear electrode reactions can be stated, as the amount of charge stored within a change of electrode potential in units of $As \cdot V^{-1} \cdot g^{-1}$ (i.e. $F \cdot V^{-1} \cdot g^{-1}$).

3.3.4　COMMON ASPECTS

When used as active mass, the ICP can be directly deposited on the support and current collector, preferably during oxidative (electro)polymerization. Because this heterogeneous process' room-time yields tend to be low, the commercial viability of such processes may be poor. In the case of thin films deposited for high material utilization [81], the already-addressed problem of insufficient electronic conductance of the ICP in its neutral state may be less relevant. Unfortunately, the energy density of a cell prepared with such a thin electrode tends to be low. In a second option, higher storage capabilities can be realized with thicker electrodes prepared, preferably by pressing a mixture of the powdery ICP, a binder, and a conducting additive. Both the binder and this conducting additive do not add to the charge storage capability of the material beyond possible, but most likely small, double-layer contributions of the carbon-based conducting additive. In case such conducting additive is definitely needed, materials providing a maximum effect at smallest amounts of added materials are welcome. The addition of specific forms of carbon like graphene [82–84] or carbon nanotubes [52] resulting in composite materials has been suggested. Nanostructuring of such composites, i.e. preparation of nanocomposites, tries to combine the advantages of composite materials with those of suitable nanostructuring of e.g. 1D materials [63, 85].

For further evaluation of ICPs, in particular as active mass in a supercapacitor electrode, the theoretical storage capability Q_{theo} may be of interest. For a battery electrode material, this number can be straightforwardly derived from the electrode reaction equation, always assuming complete material conversion as stated in the equation. In case specific values with respect to weight or volume, further material properties (density) must be known. In the case of supercapacitor materials, in particular, for materials showing a pseudocapacitive response, such a simple calculation is only of uncertain value, as discussed elsewhere [7, 81]. Nevertheless, a rough comparison of the most popular ICPs in terms of possible charge density with specified calculation conditions (change of electrode potential, assumed electrochemical transformation) may be helpful (see Table 3.1). For comparison, values for lead dioxide are included, although Q_{theo} is not directly comparable. Nevertheless, the sometimes-promoted concern that already theoretical data for ICP do not recommend them as storage materials is easily refuted.

3.4　ACTIVE MASS

When used as active mass, an ICP can be the sole constituent of the active mass in the case of direct deposition either chemically or electrochemically upon the application of the support and current collector (metal mesh, metal foil, carbon cloth, etc.). When chemically deposited material obtained

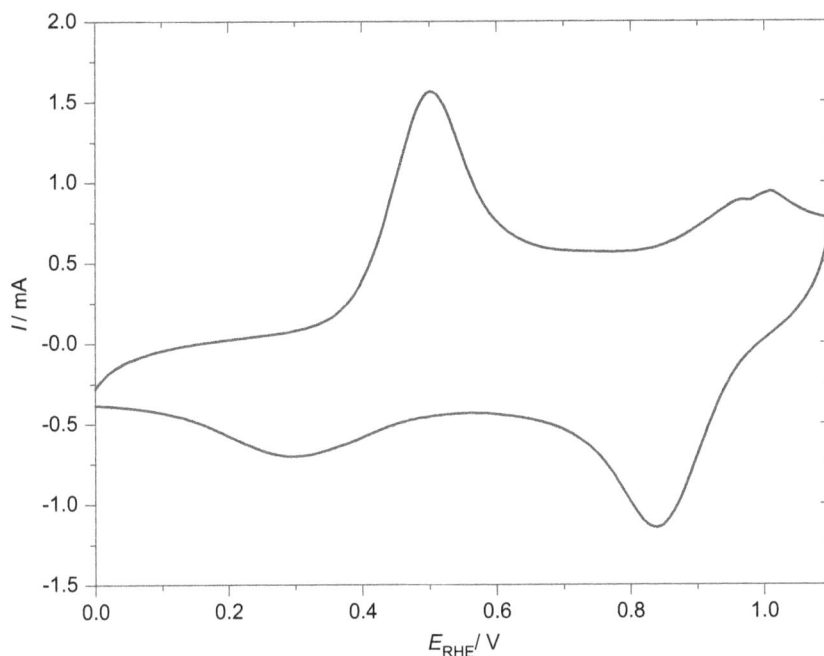

FIGURE 3.11 CV of a PANI-coated (300 electrode potential cycles within the potential range as indicated) stainless steel grid electrode (1 cm^2) in an aqueous electrolyte solution of 0.1 **M** aniline + 1 **M** HClO$_4$, dE/dt = 100 mV·s^{-1}, nitrogen purged.

as a powder is used, it needs an additional binder for manufacturing into an electrode. Because all studied binders (mostly polyvinylidenefluoride, PVDF) are electrochemically inactive and do not contribute to the storage capability of the material, the available charge density will be diminished accordingly. A possible exemption is the use of soluble ICPs as a binder; for details, see below. The already-discussed changes of electronic conductivity that accompany the redox process(es) utilized for charge storage limit the rate capability of the material. Thus, in addition to binder conductive agents, mostly some form of acetylene black, are added. Although these materials provide some charge storage by forming a double layer capacitance, they further limit the possible charge density.

ICPs can store charge in two ways: In the electrochemical double layer (like with an EDLC) and in the redox process. Cyclic voltammetry (CV) and galvanostatic charge/discharge (GCD) can hardly provide a distinction. The CV shown in Figure 3.11 already differs from previously published examples measured with much thinner films.

To avoid overoxidation with associated deterioration of the ICP (see below), potential scans should be limited to electrode potentials only slightly positive to the first oxidation peak. With an even thicker film, the CV shown in Figure 3.12 is obtained.

Following the approach initially suggested by Ardizzone et al. [86], separation of the overall charge stored in the electrode into an outer and an inner fraction is based on spatial distribution and accessibility of reaction sites for ions from the electrolyte solution. The distinction required here does not pertain to location but to charge storage mechanism, i.e. by Faradaic reaction or by double-layer charging. Many ICPs, whether directly deposited onto the current collector or prepared by pressing a mixture of ICP, binder, and conducting carbon onto a support, are highly porous (for example, see Figure 3.13).

Using organic acids as carriers and, additionally, α-alumina as hard templates, different morphologies were obtained by chemical polymerization (Figure 3.13). Because α-alumina is electrochemically inert, it will not cause any current response in the CV, as displayed in Figure 3.14. Instead, the CV shows peaks associated with the redox transformations of PANI, as already discussed above,

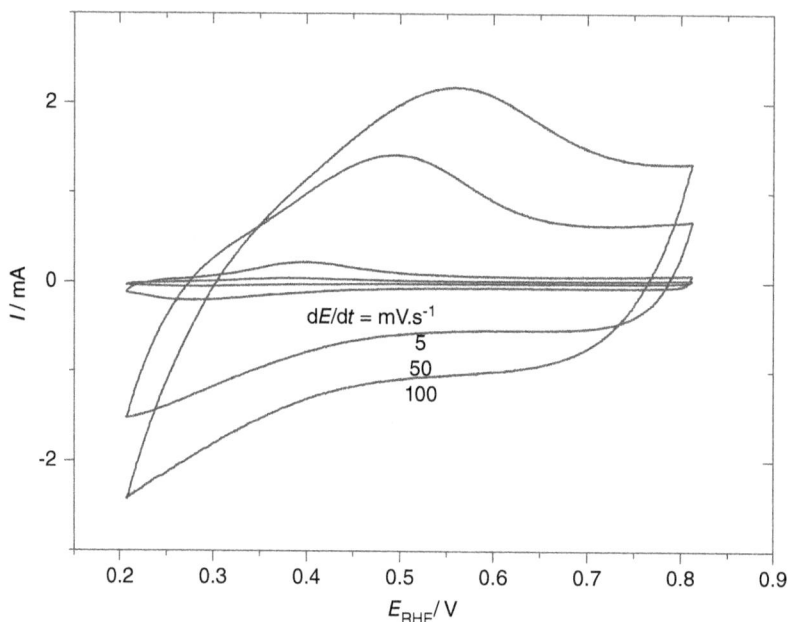

FIGURE 3.12 CVs of a PANI-coated (7.47 mg·cm^{-2}, deposition charge 0.108 C after potentiostatic coating) stainless steel grid electrode (1 cm^2) in contact with an aqueous electrolyte solution of 1 **M** HClO$_4$, at different scan rates, as indicated, nitrogen purged.

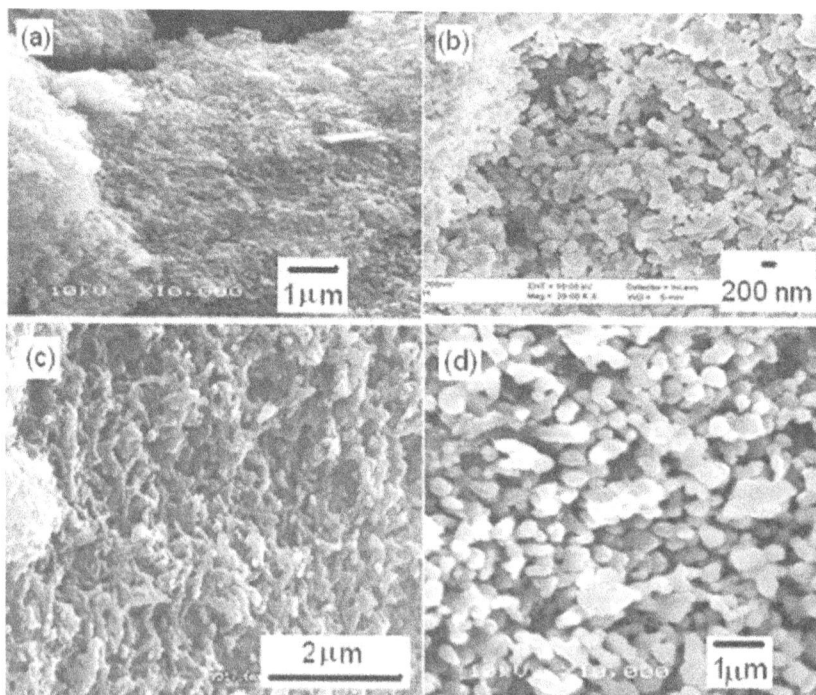

FIGURE 3.13 SEM photographs of the nanostructures (a) PANI + oxalic acid (OA), (b) PANI + OA + α-alumina, (c) PANI + camphoric acid (CA), (d) PANI + CA+α-alumina.

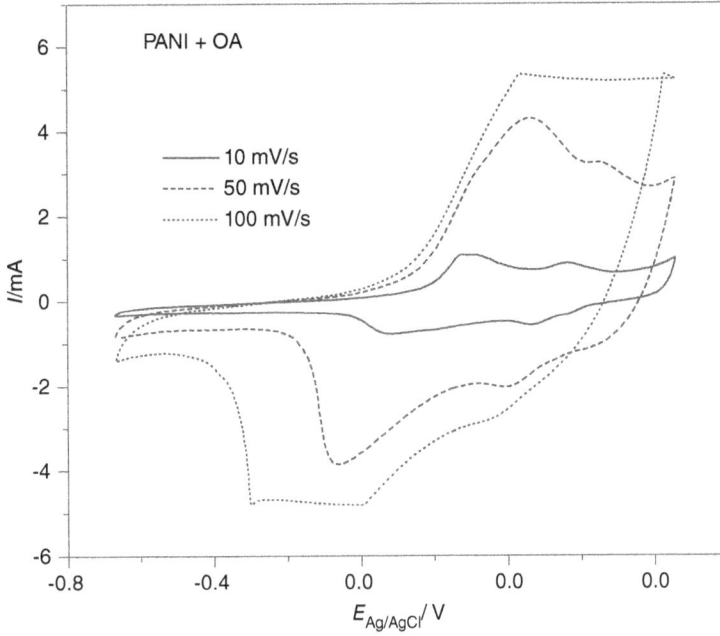

FIGURE 3.14 Cyclic voltammograms of PANI + OA nanostructures obtained at different scan rates in an aqueous electrolyte solution with 2.0 **M** H_2SO_4.

with a substantial background, which may be due both to capacitive contributions from the large surface area of the electrode material in contact with the electrolyte solution as well as Faradaic reactions spread across a wider range of electrode potentials because of interactions between redox sites causing their redox potentials to shift (for a discussion see e.g. [87]).

An option to separate current contributions from interfacial capacitance charging and redox transformations is the use of electrochemical impedance measurements. As discussed elsewhere [88], an equivalent circuit used for elucidating data of the electrode and the electrode reaction contains an element related to the interfacial capacitance C_{DL} and an element related to the redox process of PANI C_{red} (Figure 3.15).

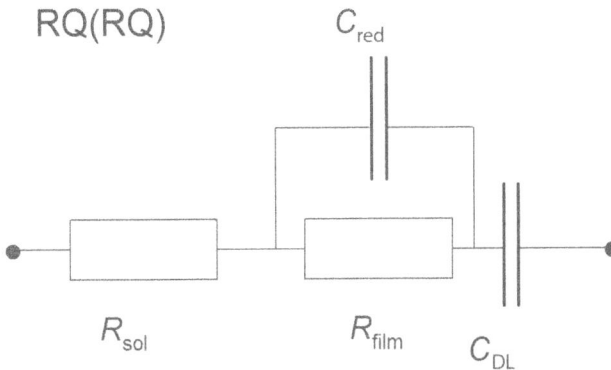

FIGURE 3.15 Modified equivalent circuit used for fitting of electrode impedances.

As expected, the latter is larger in terms of orders of magnitude than the formed one for a typical PANI film. Nevertheless, for practical use, a larger fraction of the double-layer-related fraction is preferable because this mode of charge storage is generally faster than the one associated with a Faradaic process. Because the increase of the interfacial area, i.e. the contact area between the ICP surface and the electrolyte solution, meets practical limits, the utilization of redox storage is needed for increased charge storage. Optimization of ICP morphology with respect to the most suitable porous structure [89] combined with high electronic conductivity of the ICP (either by itself or by added conducting agents) is the most promising approach. The formation of ICPs into hydrogels has been examined for PANI, including possible application in supercapacitors [90].

Three major drawbacks seem to limit the successful application of ICPs in active masses.

3.4.1 SHAPE CHANGE

The shape change includes volume change, shrinking, and swelling during charge/discharge. This is mostly related to the ingress/egress of counterions needed for charge compensation of the (radical) cations formed in the ICP during oxidation. In the case of PANI, the reverse may proceed as an alternative mode of charge balancing; protons may be released/taken up for charge compensation. All ions will be solvated, and their movement with their solvation shell into/out of ICPs causes said volume changes. These changes may result in fragmentation of the polymer, loss of contact between ICP particles, added conducting carbon, and the current collector. Various options to mitigate this shape change have been proposed and examined.

The use of self-doped ICPs, e.g. self-doped PANI, has been suggested in an early review by Malinauskas [91]. In such polymers, the presence of fixed negative charges located on anionic groups changes the mechanism of charge compensation during the redox processes of PANI: instead of ingress of anions during oxidation possibly associated with the problems discussed above, the release of cations (e.g. protons or lithium ions) bound at the anionic sites suffices for charge compensation with less detrimental effects. Such self-doped PANI can be obtained in various ways:

- Sulfonation of PANI by chemical post-treatment.
- Polymerization of suitable substituted monomers.
- Copolymerization of aniline and a second suitably substituted comonomer (e.g. copolymerization of aniline and e.g. N-methylaniline and N(3-sulfopropyl)aniline [92] (see Figure 3.16) or aniline and o-aminobenzene sulfonic acid [93] or m-aminobenzoic acid and aniline [94]).

A copolymer prepared from aniline and N(3-sulfopropyl)aniline (for example, see [94, 95]) has been examined. In the former study with N-methylaniline as the comonomer, the highest electrochemical activity was observed when a comonomer ratio of 1:1 was established in the electropolymerization solution. This agrees quite well with the degree of doping of 0.5 listed above for PANI in Table 3.1, with half of the repeat units participating in the redox reaction. Somewhat surprisingly, this approach has not been exploited in the reported research. In a study employing a copolymer of aniline and m-aminobenzoic acid as the positive electrode, 30% capacitance loss after 1000 cycles for a complete cell with a PPy negative electrode was noticed. Unfortunately, no comparison with a similar system without a self-doped ICP was attempted [96]. The copolymer obtained from aniline and metanilic acid by electropolymerization showed significant electrochemical activity in a

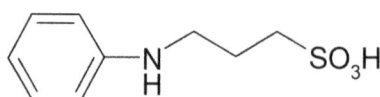

FIGURE 3.16   Structural formula of N(3-sulfopropyl)aniline.

neutral aqueous electrolyte solution of 0.5 **M** Na_2SO_4 with about 50% capacitance retention when assembled into a symmetric supercapacitor [97].

The concept of self-doping has rarely been explored beyond the particularly pH-sensitive PANI. An attempt to prepare an oligomeric bis[3,4-ethylenedioxythiophene]3thiophene butyric acid has been reported [98]; its use in "green energy" application has been proposed but not explored.

Another approach examined more frequently is the formation of micro- and nanostructures, allowing shape and volume change to an extent large enough to keep the operational capability of the ICP at a still-acceptable performance of the material. A comparison of various morphologies (2D: thin film; 3D: microsphere, microtube, microparticles, nanowires networks, nanowires arrays) has been provided [99], and nanowires arrays have been suggested as the most promising approach. A broader overview focused on 1D structures of ICPs has been provided [100].

3.4.2 PEELING OFF

Closely related to and particularly relevant for electrodes where the ICP has been directly deposited onto the current collector (i.e. no binder is used) is peeling off from the current collector, leaving the removed particles lost for charge/discharge. Structuring of the deposited ICPs on a microscale or even nanoscale ameliorating negative effects of shape change of the ICP on mechanical adhesion to the current collector may be an option to reduce the peeling off.

3.4.3 OVEROXIDATION

Overoxidation is closely associated with charge/discharge of an ICP in a supercapacitor electrode. When the electrode potential of an ICP is moved in a positive direction, electrooxidation forms cations on the polymer chain; chemically speaking, they are radicals. These species can be subject to chemical transformation by reaction with electrolyte solution species. When sufficiently positive electrode potentials are reached, in addition, oxidation of water may yield radicalic intermediates, which in turn chemically attack and degrade the ICP. The anions moving into the ICP in most cases for charge compensation may affect this reaction; studies of anion-specific effects show such influences for PANI [101]. Reaction mechanisms and effects of overoxidation vary, and in most studies, decreases in possible electronic conductance, changes of molecular structure, decrease of possible charge storage, and in the extreme case, complete loss of electrochemical redox response are reported (for examples, see [102, 103]). Unfortunately, a wider review on this subject seems to be missing as only the degradation of some forms of PEDOT has been inspected more closely in an overview [104]. At first glance, a simplistic view would suggest an easy solution by limiting the potential excursion with suitable electronic circuitry providing limitation of maximum charge voltage. Unfortunately, practical execution is more complicated. Although the operating voltage limits of electrolytic capacitors and even EDLC supercapacitors have always required appropriate circuit design considerations, the rather low operating voltage of a supercapacitor and the sensitivity toward overoxidation require more precise voltage limitation, making the use of supercapacitors with these materials less attractive. In addition, potential distribution inside of an electrode may significantly increase the exposure of parts of the material to electrode potentials, causing overoxidation. Consequently, other options have been examined (for example, see [105]). The selection of suitable solvents and electrolytes may help to avoid oxidative formation of chemically reactive species (like hydroxyl radicals when water is used as a solvent).

3.5 PART IN COMPOSITES

Composites, sometimes called nanocomposites because of the particular attention focused on the nanostructuring of the obtained material, can be prepared from ICP and a redox-active material (e.g. a metal oxide) or a non-reactive material (e.g. carbon in its many modifications). The latter case has been addressed above when introducing such carbon constituents as electronically conducting additives; for an overview, see e.g. [105]. The former case has been explored intensely; for an exhaustive review, see [53]. Various functions of the ICP and the metal or mixed metal oxide can be envisaged:

- Metal oxide
  - Charge storage material.
  - Template for structuring the ICP morphology.
  - Structural/mechanical stabilizing support for the ICP.
- ICP
  - Charge storage material.
  - Mechanical binder.
  - Mechanical stabilizer compensating for volume changes.
  - Electronic conductance enhancer.
  - Dissolution inhibitor

Unfortunately, in most studies, not even an attempt is made to identify such function for a given constituent. As also discussed below, when evaluating the use of ICPs as binders, embedding a slightly soluble metal oxide into an ICP more or less impermeable for dissolved species of the metal oxide might substantially increase the actual stability of the oxide. This has been illustrated with several composites showing stable cycling performance for more than 10,000 cycles (for further details, see [53]). This applies both to simple composites made in a one-step one-pot process preparing a PANI/TiO_2 composite showing only 16.5% capacitance loss after 30,000 cycles [106] as well as to rather sophisticated materials prepared along a rather complicated procedure involving graphene foam supporting Co_3O_4 nanowires in turn coated with a PEDOT-MnO_2 composite showing negligible capacitance loss after 20,000 cycles [107].

There seems to be evidence of increased reversibility, i.e. greater apparent rate of the redox reaction, when the metal oxide was embedded in an ICP, which may be due to increased utilization of the metal oxide (in particular when it is poorly conducting) because the metal oxide is better accessible for charge transfer when in contact with the ICP.

3.6 PRECURSORS

The enhanced electronic conductivity of carbon in its many forms can be achieved by adding trace amounts (frequently, this is called "doping") of foreign elements. Nitrogen appears to be one of the popular elements. Because almost all ICPs relevant in electrochemistry contain heteroatoms in their monomers, the pyrolysis in an inert atmosphere (carbonization or similar thermal transformations) will result in products containing some or all of these heteroatoms. Consequently, ICPs have been widely used wherein e.g. a metal oxide has been coated with an ICP or composited with an ICP and subsequently thermal treatment resulting in a heteroatom-doped carbon/metal oxide composite. The obtained carbon composite may have higher electronic conductivity because of the incorporated heteroatoms supplied by the ICP [108, 109]. An additional advantage of the use of ICPs is the option of introducing specific morphologies or architectures already addressed above for the "unpyrolyzed" case. For example, PPy nanotubes were pyrolyzed, yielding N-doped carbon nanotubes for the application in negative supercapacitor electrodes [110]. PANI was also used as a starting material to prepare by pyrolysis N-doped carbon nanotubes for use as the negative electrode in a potassium-ion capacitor [109].

3.7 COATINGS

Some electrode materials with basically promising properties for potential use as electrode material, like favorable electrode potential and large charge storage capability, suffer from significant solubility in the electrolyte solution. Attempts to ameliorate this drawback include coating with polymer materials. Such polymers may be electrochemically inert (like some of the binders addressed in the following section), and they will not add to the charge storage capability. The use of an ICP may add beyond the dissolution-inhibiting property further advantages like improved mass utilization and faster electrode kinetics because of the ion as well as electron transport properties of an ICP. These aspects have been reviewed elsewhere, with particular attention also to the use of such materials in secondary batteries [111]. Presumably because of the rather diffuse and hardly defined boundaries between the use of a material as a coating and an embedding material or simply as a constituent in a composite material, as discussed in the previous section, reports on ICP used specifically to coat a supercapacitor electrode material are rare. In a representative study, V_2O_5 nanoribbons have been coated with PPy by chemical polymerization [112]. The coating prevented the dissolution of the vanadate and enhanced charge transport, as evidenced by lifetime cycling stability and rate capability measurements. Long nanowires of V_2O_5 coated with PPy could be manufactured into a flexible, freestanding membrane, showing, as an electrode, improved stability and increased specific capacitance and rate capability [113]. Better stability was attributed to structural improvements: The PPy shell accommodated volume changes of the metal oxide during cycling, whereas the metal oxide nanowires helped in maintaining the structural integrity of the ICP. The inherently low electronic conductivity of MnO_2 could be ameliorated by coating nanowires of MnO_2 with PANI, resulting in a doubled specific capacitance [114]. Similar beneficial effects of PANI-coating yielding coaxial MnO_2/PANI nanowires [115] and coaxial MnO_2/PPy nanotubes [117] in terms of enhanced performance and slightly improved stability were reported.

The low electronic conductivity of CoO nanowires to be used as supercapacitor electrodes could be improved by a coating with PPy yielding high specific capacitance and enhanced rate performance [117]. The claimed improved stability during 5000 cycles is hardly visible in the displayed data. Co_3O_4 nanowires coated uniformly with PPy showed enhanced specific capacitance but practically unchanged rate capability and slightly poorer stability [118]. Coating of hexagonal Co_9S_8 with poly(m-phenylenediamine) resulted in a major increase of storage capability; the stability did not change, presumably because of the already very high stability of the cobalt sulfide [119]. Nanorods of α-Fe_2O_3 were coated with PANI yielding a negative electrode material with significantly increased specific capacitance and stability [121]. Nanobelts of α-MoO_3 coated with PANI demonstrated significantly improved mass utilization but somewhat surprisingly poorer rate capability accompanied with a stability only slightly better than that of plain α-MoO_3 [121]. Nanoflakes of MoS_2 deposited on activated carbon cloth and coated with PANI provided flexible electrodes with increased specific capacitance and enhanced stability at an optimized PANI coating thickness [122].

In a comparative study, the effects of coating either PPy or PEDOT onto $NiCo_2O_4$ were examined [123]. In terms of performance, the PPy coating yielded much better results, but in terms of stability, the plain metal oxide was much more stable. The improved performance was tentatively attributed to enhanced electronic conductivity of the core–shell structured material, and the rather poor stability was discussed in general terms invoking shrinking/swelling of ICPs (see above). In the case of nanotube arrays of $NiCo_2S_4$ coated with PPy beneficial effects regarding all aspects were also clearly visible [124].

A typical case of a material hard to classify is CuS coated/embedded with PPy [125]. The highly porous CuS nanospheres were chemically "filled" (according to the displayed microscope pictures and the reported chemical analysis indicating a presence of 16.7 wt.% of CuS) with PPy. The addition of the ICP doubled the gravimetric charge storage density, but it hardly improved the stability. Synergistic effects between CuS and ICP in the composition might have resulted in a much higher mass utilization, but it must be mentioned that different from other examples, the already-high electronic conductivity of CuS could not be improved with the addition of PPy.

At this point, it appears safe to state that metal oxide electrode materials showing poor electronic conductivity and significant solubility in the electrolyte solution appear to benefit most from a coating with an ICP.

3.8   BINDERS

The poor solubility of most ICPs has limited their application in many cases because of difficult processing. Some oligomers are soluble, even in common solvents. This has made their use e.g. as corrosion protection coatings, possible [126, 127]. Certainly, there are major differences in properties between oligomers and polymers with respect to e.g. electronic conductivity and electrooptical properties. These differences are related in particular to the extension of conjugated systems (conjugation length, not to be confused with effective conjugation length or effective conjugation length coordinate [128]). An aqueous dispersion of PEDOT:polystyrene sulfonate (PSS) has been used as a binder for the carbon used in the active mass of an EDLC capacitor using an organic solvent–based electrolyte solution containing lithium cations [129]. Results suggest that doctor blade–coated electrodes reach higher energy densities when assembled into a full device, whereas spray-coated electrodes support higher power density. This difference was attributed to microstructural differences. For comparison, a binder PSBR100 (presumably modified styrene–butadiene copolymer) was examined. Results do not show a significant contribution of the ICP toward charge storage.

3.9   OUTLOOK AND PERSPECTIVES

The most important challenge is the identification and optimization of long-term stable electrode materials. The sensitivity of some ICPs versus overoxidation (there are apparently no reports on damaging effects of electrode potential excursions into too negative potential ranges) merits attention regarding either careful electrode potential/cell voltage control during charging or development of material combinations, which can suppress the detrimental effects of overoxidation. Approaches identified as being promising for ICPs alone seem to be promising also for material combinations of e.g. ICP and mixed metal oxide. In such studies, the tasks of the constituents should be verified; this may be valuable in optimization procedures. Given the continuously broadening application perspectives of supercapacitors, new and optimized materials enabling the assembly of cells with higher energy density at only slightly diminished power capabilities with increased lifetime expectancies seem to be achievable.

3.10   ACKNOWLEDGMENTS

This overview includes results obtained within a research project at St. Petersburg State University supported by grants № 26455158 and 70037840. Preparation of this communication has been supported in various ways by the Alexander von Humboldt-Foundation, Deutscher Akademischer Austauschdienst, Fonds der Chemischen Industrie, Deutsche Forschungsgemeinschaft, National Basic Research Program of China, and Natural Science Foundation of China (Grant # 51425301 and 52073143) and the National Science Centre, Poland, within OPUS UMO-2017/25/B/ST8/02702.

REFERENCES

1. R. Holze, Y. Wu, Electrochemical Energy Conversion and Storage, VCH-WILEY, Weinheim 2021.
2. F. Beck, K.-J. Euler, Elektrochemische Energiespeicher, VDE-Verlag GmbH, Berlin 1984.
3. P. Kurzweil, O.K. Dietlmeier, Elektrochemische Speicher, Springer Vieweg, Wiesbaden 2015.
4. F.M. Rabiul Islam, K. Al Mamun, M.T.O. Amanullah Eds. Smarter Energy Grid Design for Island Countries, Springer, Cham 2017.
5. B. Drost-Franke, Review of the Need for Storage Capacity Depending on the Share of Renewable Energies in: Electrochemical Energy Storage for Renewable Sources and Grid Balancing (P.T. Moseley, J. Garche Eds.) Elsevier, Amsterdam 2015.
6. R.A. Huggins, Energy Storage, Springer, New York 2010.
7. Y. Ge, X. Xie, J. Roscher, R. Holze, Q. Qu, How to measure and report the capacity of electrochemical double layers, supercapacitors, and their electrode materials, J. Solid State Electr., 24 (2020) 3215–3230.

8. F. Béguin, E. Frackowiak, Carbons for Electrochemical Energy Storage and Conversion Systems, CRC Press, Boca Raton 2010.
9. A. Davies, A. Yu, Material advancements in supercapacitors: From activated carbon to carbon nanotube and graphene, Can. J. Chem. Eng., 89 (2011)1342–1357.
10. A.K. Shukla, S. Sampath, K. Vijayamohanan, Electrochemical supercapacitors: Energy storage beyond batteries, Curr. Sci., 79 (2000) 1656–1661.
11. D.P. Dubal, Y. Wu, R. Holze, Supercapacitors: From the Leyden jar to electric buses, ChemTexts, 2 (2016) 13.
12. B.E. Conway, Electrochemical Supercapacitors: Scientific Fundamentals and Technological Applications, Springer, New York 1999.
13. A. Yu, V. Chabot, J. Zhang, Electrochemical Supercapacitors for Energy Storage and Delivery - Fundamentals and Applications, CRC Press, Boca Raton 2013.
14. Inamuddin, M.F. Ahmer, A.M. Asiri, S. Zaidi Eds. Electrochemical Capacitors: Theory, Materials and Applications, Vol. 26,Materials Research Forum LLC, Millersville 2018.
15. Z. Stevic Ed. Supercapacitor Design and Applications, IntechOpen, London, UK, 2016.
16. F. Béguin, E. Frackowiak, Supercapacitors, Wiley-VCH, Weinheim 2013.
17. L. Kouchachvili, W. Yaici, E. Entchev, Hybrid battery/supercapacitor energy storage system for the electric vehicles, J. Power Sources, 374 (2018) 237–248.
18. J.M. Miller, Ultracapacitor Applications, The Institution of Engineering and Technology, London 2011.
19. F. Wang, X. Wu, X. Yuan, Z. Liu, Y. Zhang, L. Fu, Y. Zhu, Q. Zhou, Y. Wu, W. Huang, Latest advances in supercapacitors: From new electrode materials to novel device designs, Chem. Soc. Rev., 46 (2017) 6816–6854.
20. E.E. Miller, Y. Hua, F.H. Tezel, Materials for energy storage: Review of electrode materials and methods of increasing capacitance for supercapacitors, J. Energy Storage, 20 (2018) 30–40.
21. B.E. Conway, V. Birss, J. Wojtowicz, The role and utilization of pseudocapacitance for energy storage by supercapacitors, J Power Sources, 66 (1997) 1–14.
22. B.E. Conway, E. Gileadi, Kinetic theory of pseudo-capacitance and electrode reactions at appreciable surface coverage, Trans. Faraday Soc., 58 (1962) 2493–2509.
23. B.E. Conway, Transition from "Supercapacitor" to "Battery" behavior in electrochemical energy storage, J. Electrochem. Soc., 138 (1991) 1539–1548.
24. D.P. Dubal, R. Holze, Synthesis, properties, and performance of nanostructured metal oxides for supercapacitors, Pure Appl. Chem., 86 (2014) 611–632.
25. D.P. Dubal, N.R. Chodankar, P. Gomez-Romero, D.H. Kim, Fundamentals of Binary Metal Oxide-Based Supercapacitors in: Metal Oxides in Supercapacitors (D.P. Dubal, P. Gomez-Romero Eds.) Elsevier, Amsterdam 2017, pp. 79–98.
26. C. Arbizzani, M. Mastragostino, L. Meneghello, R. Paraventi, Electronically conducting polymers and activated carbon: Electrode materials in supercapacitor technology, Adv. Mater., 8 (1996) 331–334.
27. C. Arbizzani, M. Mastragostino, L. Meneghello, Polymer-based redox supercapacitors: A comparative study, Electrochim. Acta, 41 (1996) 21–26.
28. R. Holze, Optical and electrochemical band gaps in mono-, oligo-, and polymeric systems: A critical reassessment, Organometallics, 33 (2014) 5033–5042.
29. S.M. Sze, M.K. Lee, Semiconductor Devices - Physics and Technology, 3rd ed., John Wiley & Sons, New York 2012.
30. B.C. Kim, J.-Y. Hong, G.G. Wallace, H.S. Park, Recent progress in flexible electrochemical capacitors: Electrode materials, device configuration, and functions, Adv. Energy Mater., 5 (2015) 1500959.
31. M. Cheng, Y.-N. Meng, Z.-X. Wie, Conducting polymer nanostructures and their derivatives for flexible supercapacitors, Isr. J. Chem., 58 (2018) 1299–1314.
32. R. Holze, J. Stejskal, Recent trends and progress in research into structure and properties of polyaniline and polypyrrole - Topical Issue, Chem. Pap., 67 (2013) 769–770.
33. T.A. Skotheim Ed. Handbook of Conducting Polymers, Marcel Dekker, Inc., New York 1986.
34. H. Shirakawa, E.J. Louis, A.G. MacDiarmid, C.K. Chiang, A.J. Heeger, Synthesis of electrically conducting organic polymers: Halogen derivatives of polyacetylene, $(CH)_x$, Chem. Commun., 1977 (1977) 578–580.
35. J.B. Schlenoff, J.C.W. Chien, Efficiency, stability and self discharge of electrochemically-doped polyacetylene in propylene carbonate electrolyte, Synth. Met., 22 (1988) 349–363.
36. T.F. Otero, I. Cantero, Conducting polymers as positive electrodes in rechargeable lithium-ion batteries, J. Power Sources, 81–82 (1999) 838–841.

37. T. Kita, H. Daifuku, R. Fujio, T. Fuse, T. Kawagoe, Y. Masuda, T. Matsunaga, M. Ogawa, Properties of polyaniline secondary battery, J. Electrochem. Soc., 133 (1986) C291–C297.

38. T. Matsunaga, H. Daifuku, T. Kawagoe, Development of polyaniline-lithium secondary battery, Nippon Kagaku Kaishi, 1990 (1990) 1–11.

39. R.J. Mammone, Use of Electronically Conducting Polymers as Catalytic Electrodes in Aqueous and Inorganic Electrolytes in: Conducting Polymers - Special Applications (L. Alcacer, Ed.), D. Reidel Publishing Company, Dordrecht 1987, pp. 161–172.

40. R. Holze, Y.P. Wu, Intrinsically conducting polymers in electrochemical energy technology: Trends and progress, Electrochim. Acta, 122 (2014) 93–107.

41. C. Daniel, J.O. Besenhard, Handbook of Battery Materials, Wiley-VCH, Weinheim 2011.

42. K. Koga, S. Yamasaki, K. Narimatsu, M. Takayanagi, Electrically conductive composite of polyaniline aramid and its application as a cathode material for secondary battery, Polym. J., 21 (1989) 733–738.

43. Y. Wu, R. Holze, Battery and/or Supercapacitor? – On the Merger of two Electrochemical Storage Families. ChemTexts, submitted.

44. A. Rudge, J. Davey, I. Raistrick, S. Gottesfeld, J.P. Ferraris, Conducting polymers as active materials in electrochemical capacitors, J. Power Sources, 47 (1994) 89–107.

45. S. Sarangapani, B.V. Tilak, C.P. Chen, Materials for electrochemical capacitors - theoretical and experimental constraints, J. Electrochem. Soc., 143 (1996) 3791–3799.

46. G.A. Snook, P. Kao, A.S. Best, Conducting-polymer-based supercapacitor devices and electrodes, J. Power Sources, 196 (2011) 1–12.

47. K. Naoi, M. Morita, Advanced polymers as active materials and electrolytes for electrochemical capacitors and hybrid capacitor systems, Interface, 17(1) (2008) 44–48.

48. P. Novak, K. Müller, K.S.V. Santhanam, O. Haas, Electrochemically active polymers for rechargeable batteries, Chem. Rev., 97 (1997) 207–281.

49. K.D. Fong, T. Wang, S.K. Smoukov, Multidimensional performance optimization of conducting polymer-based supercapacitor electrodes, Sustainable Energy Fuels, 1 (2017) 1857–1874.

50. Q. Meng, K. Cai, Y. Chen, L. Chen, Research progress on conducting polymer based supercapacitor electrode materials, Nano Energy, 36 (2017) 268–285.

51. L. Zhang, W. Du, A. Nautiyal, Z. Liu, X. Zhang, Recent progress on nanostructured conducting polymers and composites: Synthesis, application and future aspects, Sci. China Mater., 61 (2018) 303–352.

52. R. Ramya, R. Sivasubramanian, M.V. Sangaranarayanan, Conducting polymers-based electrochemical supercapacitors-Progress and prospects, Electrochim. Acta, 101 (2013) 109–129.

53. L. Fu, Q. Qu, R. Holze, V.V. Kondratiev, Y. Wu, Composites of metal oxides and intrinsically conducting polymers as supercapacitor electrodes: The best of both worlds? J. Mater. Chem. A, 7 (2019) 14937–14970.

54. S. Bose, T. Kuila, A.K. Mishra, R. Rajasekar, N.H. Kim, J.H. Lee, Carbon-based nanostructured materials and their composites as supercapacitor electrodes, J. Mater. Chem., 22 (2012) 767–784.

55. F. Shen, D. Pankratov, Q. Chi, Graphene-conducting polymer nanocomposites for enhancing electrochemical capacitive energy storage, Curr. Opin. Electrochem., 4 (2017) 133–144.

56. P. Pieta, I. Obraztsov, F. D'Souza, W. Kutner, Composites of conducting polymers and various carbon nanostructures for electrochemical supercapacitors. ECS J. Solid State Sci. Technol., 2 (2013) M3120–M3134.

57. R. Holze, Metal Oxide/Conducting Polymer Hybrids for Application in Supercapacitors in Metal Oxides in Supercapacitors (D.P. Dubal, P. Gomez-Romero, Eds.) Elsevier, Amsterdam 2017, pp. 219–245.

58. H.-M. Zhang, X.-H. Wang, Eco-friendly water-borne conducting polyaniline, Chin. J. Polym. Sci., 31 (2013) 853–869.

59. R. Holze, J. Lippe, A method for electrochemical in situ conductivity measurements of electrochemically synthesized intrinsically conducting polymers, Synth. Met., 38 (1990) 99–105.

60. J. Lippe, R. Holze, Electrochemical in-situ conductivity and polaron concentration measurements at selected conducting polymers, Synth. Met., 41–43 (1991) 2927–2930.

61. R. Holze, Spectroelectrochemistry of Intrinsically Conducting Polymers of Aniline and Substituted Anilines in: Handbook of Advanced Electronic and Photonic Materials, Vol. 2 (H.S. Nalwa Ed.) Gordon and Breach and OPA N.V., Singapore, 2001, p. 171.

62. J.X. Huang, R.B. Kaner, A general chemical route to polyaniline nanofibers, J. Am. Chem. Soc., 126 (2004) 851–855.

63. J. Wang, D. Zhang, One-dimensional nanostructured polyaniline: Syntheses, morphology controlling, formation mechanisms, new features, and applications, Adv. Polym. Technol., 32 (2013) E323–E368.

64. Y. Luo, R. Guo, T. Li, F. Li, Z. Liu, M. Zheng, B. Wang, Z. Yang, H. Luo, Y. Wan, Application of poly-aniline for Li-Ion batteries, lithium-sulfur batteries, and supercapacitors, ChemSusChem, 12 (2019) 1591–1611.

65. A. Eftekhari, L. Li, Y. Yang, Polyaniline supercapacitors, J. Power Sources, 347 (2017) 86–107.

66. Z. Li, L. Gong, Research progress on applications of polyaniline (PANI) for electrochemical energy storage and conversion, Materials, 13 (2020) 548.

67. R. Holze, Spectroelectrochemistry of Intrinsically Conducting Polymers of Aniline and Substituted Anilines in: Advanced Functional Molecules and Polymers, Vol. 2 (H.S. Nalwa Ed.) Gordon and Breach and OPA N.V., Singapore 2001, pp. 171–221.

68. W. Zhou, J. Xu, Progress in conjugated polyindoles: Synthesis, polymerization mechanisms, properties, and applications, Polym. Rev., 57 (2017) 248–275.

69. J. Sebastian, J.M. Samuel, Recent advances in the applications of substituted polyanilines and their blends and composites, Polym. Bull., 77 (2019) 6641–6669.

70. S.K. Simotwo, V. Kalra, Polyaniline-based electrodes: Recent application in supercapacitors and next generation rechargeable batteries, Curr. Opin. Chem. Eng., 13 (2016) 150–160.

71. R. Holze, V. Brandl Der Einfluß der Herstellungsbedingungen auf die Eigenschaften des intrinsisch leit-fähigen Polypyrrols in: GDCh-Monographie, Vol. 14 (J. Russow, G. Sandstede, R. Staab Eds.) GDCh, Frankfurt 1998, pp. 230–238.

72. V. Brandl, R. Holze, Influence of the preparation conditions on the properties of electropolymerised polypyrrole, Ber. Bunsenges. Phys. Chem., 102 (1998) 1032–1038.

73. Z. Liu, Y. Liu, S. Poyraz, X. Zhang, Green-nano approach to nanostructured polypyrrole, Chem. Commun., 47 (2011) 4421–4423.

74. X. Zhang, S.K. Manohar, Bulk synthesis of polypyrrole nanofibers by a seeding approach, J. Am. Chem. Soc., 126 (2004) 12714–12715.

75. X. Zhang, S.K. Manohar, Narrow pore-diameter polypyrrole nanotubes, J. Am. Chem. Soc., 127 (2005) 14156–14157.

76. D.P. Dubal, Z. Caban-Huertas, R. Holze, P. Gomez-Romero, Growth of polypyrrole nanostructures through reactive templates for energy storage applications, Electrochim. Acta, 191 (2016) 346–354.

77. K. Wójcik, M. Grzeszczuk, Polypyrrole nanofibers with extended 1D hierarchical structure, ChemElectroChem, 4 (2017) 2653–2659.

78. A. Sardar, P.S. Gupta, Polypyrrole based nanocomposites for supercapacitor applications: A review, AIP Conf. Proc., 1953 (2018) 030020.

79. G. Inzelt, Conducting Polymers - A New Era in Electrochemistry in: Monographs in Electrochemistry (F. Scholz Ed.) Springer-Verlag, Berlin 2008.

80. J. Arjomandi, F. Alakhras, W. Al-Halasah, R. Holze, Spectroelectrochemical and theoretical tools applied towards an enhanced understanding of structure, energetics and dynamics of molecules and polymers: Polyfuranes, polythiophenes, polypyrroles and their copolymers, Jord. J. Chem., 2009, 4, 279–301.

81. S. Ardizzone, G. Fregonara, S. Trasatti, "Inner" and "outer" active surface of RuO_2 electrodes, Electrochim. Acta, 35 (1990) 263–269.

82. L. Wang, X. Lu, S. Lei, Y. Song, Graphene-based polyaniline nanocomposites: Preparation, properties and applications, J. Mater. Chem. A, 2 (2014) 4491–4509.

83. Z. Huang, L. Li, Y. Wang, C. Zhang, T. Liu, Polyaniline/graphene nanocomposites towards high-performance supercapacitors: A review, Comp. Commun., 8 (2018) 83–91.

84. N.P.S. Chauhan, M. Mozafari, N.S. Chundawat, K. Meghwal, R. Ameta, S.C. Ameta, High-performance supercapacitors based on polyaniline-graphene nanocomposites: Some approaches, challenges and opportunities, J. Ind. Engin. Chem., 36 (2016) 13–29.

85. X. Lu, W. Zhang, C. Wang, T.-C. Wen, Y. Wie, One-dimensional conducting polymer nanocomposites: Synthesis, properties and applications, Prog. Polym. Sci., 36 (2011) 671–712.

86. Y. Ge, Z. Liu, Y. Wu, R. Holze, On the utilization of supercapacitor electrode materials, Electrochim. Acta, 366 (2021) 137390.

87. R. Holze, From current peaks to waves and capacitive currents-on the origins of capacitor-like electrode behavior, J. Solid State Electr., 21 (2017) 2601–2607.

88. M.C.E. Bandeira, R. Holze, Impedance measurements at thin polyaniline films - the influence of film morphology, Microchim. Acta, 156 (2006) 125–131.

89. Z. Liu, X. Yuan, S. Zhang, J. Wang, Q. Huang, N. Yu, Y. Zhu, L. Fu, F. Wang, Y. Chen, Y. Wu, Three-dimensional ordered porous electrode materials for electrochemical energy storage, NPG Asia Mater., 11 (2019) 12.

90. R.D. Pyarasani, T. Jayaramudu, A. John, Polyaniline-based conducting hydrogels, J. Mater. Sci., 54 (2019) 974–996.
91. A. Malinauskas, Self-doped polyanilines, J. Power Sources, 126 (2004) 214–220.
92. A. Malinauskas, R. Holze, Deposition and characterisation of self-doped sulphoalkylated polyanilines, Electrochim. Acta, 43 (1998) 521–531.
93. P.-H. Wang, T.-L. Wang, W.-C. Lin, H.-Y. Lin, M.-H. Lee, C.-H. Yang, Enhanced supercapacitor performance using electropolymerization of self-doped polyaniline on carbon film, Nanomaterials, 8 (2018) 214.
94. H.R. Ghenaatian, M.F. Mousavi, M.S. Rahmanifar, High performance battery-supercapacitor hybrid energy storage system based on self-doped polyaniline nanofibers, Synth. Met., 161 (2011) 2017–2023.
95. A. Malinauskas, M. Bron, R. Holze, Electrochemical and Raman spectroscopic studies of electrosynthesized copolymers and bilayer structures of polyaniline and poly(o-phenylenediamine), Synth. Met., 92 (1998) 127–137.
96. H.R. Ghenaatian, M.F. Mousavi, M.S. Rahmanifar, High performance hybrid supercapacitor based on two nanostructured conducting polymers: Self-doped polyaniline and polypyrrole nanofibers, Electrochim. Acta, 78 (2012) 212–222.
97. L.-J. Bian, F. Luan, S.-S. Liu, X.-X. Liu, Self-doped polyaniline on functionalized carbon cloth as electroactive materials for supercapacitor, Electrochim. Acta, 64 (2012) 17–22.
98. J.F. Franco-Gonzalez, E. Pavlopoulou, E. Stavrinidou, R. Gabrielsson, D.T. Simon, M. Berggren, I.V. Zozoulenko, Morphology of a self-doped conducting oligomer for green energy applications, Nanoscale, 9 (2017) 13717–13724.
99. K. Wang, H. Wu, Y. Meng, Z. Wie, Conducting polymer nanowire arrays for high performance supercapacitors, Small, 10 (2014) 14–31.
100. Z. Yin, Q. Zheng, Controlled synthesis and energy applications of one-dimensional conducting polymer nanostructures: An overview, Adv. Energy Mater., 20 (2011) 1–40.
101. J. Lippe, R. Holze, The anion-specific effect in the overoxidation of polyaniline and polyindoline, J. Electroanal. Chem., 339 (1992) 411–422.
102. R. Goncalves, E.C. Pereira, L.F. Marchesi, The overoxidation of poly(3-hexylthiophene) (P3HT) thin film: CV and EIS measurements, Int. J. Electrochem. Sci., 12 (2017) 1983–1991.
103. M. Bouabdallaoui, Z. Aouzal, S. Ben Jadi, A. El Jaouhari, M. Bazzaoui, G. Lévi, J. Aubard, E.A. Bazzaoui, X-ray photoelectron and in situ and ex situ resonance Raman spectroscopic investigations of polythiophene overoxidation, J. Solid State Electr., 21 (2017) 3519–3532.
104. G.G. Láng, M. Ujvári, S. Vesztergom, V. Kondratiev, J. Gubicza, K.J. Szekeres, The electrochemical degradation of poly(3,4-ethylenedioxythiophene) films electrodeposited from aqueous solutions, Z. Physik. Chem., 230 (2016) 1281–1302.
105. P. Liu, J. Yan, Z. Guang, Y. Huang, X. Li, W. Huang, Recent advancements of polyaniline-based nanocomposites for supercapacitors, J. Power Sources, 424 (2019) 108–130.
106. R. Gottam, P. Srinivasan, One-step oxidation of aniline by peroxotitanium acid to polyaniline-titanium dioxide: A highly stable electrode for a supercapacitor, J. Appl. Polym. Sci., 132 (2015) 41711.
107. X. Xia, D. Chao, Z. Fan, C. Guan, X. Cao, H. Zhang, H.J. Fan, A new type of porous graphite foams and their integrated composites with oxide/polymer core/shell nanowires for supercapacitors: Structural design, fabrication, and full supercapacitor demonstrations, Nano Lett., 14 (2014) 1651–1658.
108. M. Moussa, S.A. Al-Bataineh, D. Losic, D.P. Dubal, Engineering of high-performance potassium-ion capacitors using polyaniline-derived N-doped carbon nanotubes anode and laser scribed graphene oxide cathode, Appl. Mater. Today, 16 (2019) 425–434.
109. Y. He, X. Han, Y. Du, B. Zhang, P. Xu, Heteroatom-doped carbon nanostructures derived from conjugated polymers for energy applications, Polymers, 8 (2016) 366.
110. D.P. Dubal, N.R. Chodankar, Z. Caban-Huertas, F. Wolfart, M. Vidotti, R. Holze, C.D. Lokhande, P. Gomez-Romero, Synthetic approach from polypyrrole nanotubes to nitrogen doped pyrolyzed carbon nanotubes for asymmetric supercapacitors, J. Power Sources, 308 (2016) 158–165.
111. R. Holze, V. Kondratiev, Intrinsically conducting polymers and their combinations with redox-active molecules for rechargeable battery electrodes: an update, Chem. Pap., 75 (2021) 4981–5007. doi: 10.1007/s11696-021-01529-7.
112. Q. Qu, Y. Zhu, X. Gao, Y. Wu, Core-shell structure of polypyrrole grown on V_2O_5 nanoribbon as high performance anode material for supercapacitors, Adv. Energy Mater., 2 (2012) 950–955.
113. J.-G. Wang, H. Liu, H. Liu, W. Hua, M. Shao, Interfacial constructing flexible V_2O_5@Polypyrrole core-shell nanowire membrane with superior supercapacitive performance, ACS Appl. Mater. Interfaces, 10 (2018) 18816–18823.

114. L. Wu, Y. Li, P. Yu, Q. Zhang, Enhanced capacitive properties of manganese dioxide nanowires coating with polyaniline by in situ polymerization, Mater. Res. Soc. Symp. Proc., 1659 (2014) 163–168.

115. A. Sumboja, C.Y. Foo, J. Yan, C. Yan, R.K. Gupta, P.S. Lee, Significant electrochemical stability of manganese dioxide/polyaniline coaxial nanowires by self-terminated double surfactant polymerization for pseudocapacitor electrode, J. Mater. Chem., 22 (2012) 23921–23928.

116. W. Yao, H. Zhou, Y. Lu, Synthesis and property of novel MnO2@polypyrrole coaxial nanotubes as electrode material for supercapacitors, J. Power Sources, 241 (2013) 359–366.

117. C. Yang, H. Chen, C. Guan, Hybrid CoO nanowires coated with uniform polypyrrole nanolayers for high-performance energy storage devices, Nanomaterials, 9 (2019) 586.

118. D. Guo, M. Zhang, Z. Chen, X. Liu, Hierarchical Co_3O_4@PPy core-shell composite nanowires for supercapacitors with enhanced electrochemical performance, Mater. Res. Bull., 96 (2017) 463–470.

119. P. Liu, X. Chang, J. Lin, S. Yan, L. Yao, J. Lian, H. Lin, S. Han, Synthesis of poly(m-phenylenediamine)-coated hexagonal Co_9S_8 for high-performance supercapacitors, J. Mater. Sci., 53 (2018) 759–773.

120. Z. Yang, L. Tang, D. Shi, S. Liu, M. Chen, Hierarchical nanostructured α-Fe_2O_3/polyaniline anodes for high performance supercapacitors, Electrochim. Acta, 269 (2018) 21–29.

121. F. Jiang, W. Li, R. Zou, Q. Liu, K. Xu, L. An, J. Hu, MoO_3/PANI coaxial heterostructure nanobelts by in situ polymerization for high performance supercapacitors, Nano Energy, 7 (2014) 72–79.

122. H. Zhang, G. Qin, Y. Lin, D. Zhang, H. Liao, Z. Li, J. Tian, Q. Wu, A novel flexible electrode with coaxial sandwich structure based polyaniline-coated MoS_2 nanoflakes on activated carbon cloth, Electrochim. Acta, 264 (2018) 91–100.

123. K. Xu, X. Huang, Q. Liu, R. Zou, W. Li, X. Liu, S. Li, J. Yang, J. Hu, Understanding the effect of polypyrrole and poly(3,4-ethylenedioxythiophene) on enhancing the supercapacitor performance of $NiCo_2O_4$ electrodes, J. Mater. Chem. A, 2 (2014) 16731–16739.

124. F.Y. Meng, Y.F. Yuan, S.Y. Guo, Y.X. Xu, $NiCo_2S_4$@PPy core-shell nanotube arrays on Ni foam for high-performance supercapacitors, Mater. Technol., 32 (2017) 815–822.

125. H. Peng, G. Ma, K. Sun, J. Mu, H. Wang, Z. Lei, High-performance supercapacitor based on multi-structural CuS@polypyrrole composites prepared by in situ oxidative polymerization, J. Mater. Chem. A, 2 (2014) 3303–3307.

126. S. Shreepathi, R. Holze, Spectroelectrochemical investigations of soluble polyaniline synthesized via new inverse emulsion pathway, Chem. Mater., 17 (2005) 4078–4085.

127. S. Shreepathi, H. Van Hoang, R. Holze, Corrosion protection performance and spectroscopic investigations of soluble conducting polyaniline- dodecylbenzenesulfonate synthesized via inverse emulsion procedure, J. Electrochem. Soc., 154 (2007) C67–C73.

128. G. Zerbi, M. Veronelli, S. Martina, A.-D. Schlüter, G. Wegner, pi-Electron delocalization in conformationally distorted oligopyrroles and polypyrrole, Adv. Mater., 6 (1994) 385–388.

129. F. Markoulidis, A. Dawe, C. Lekakou, Electrochemical double-layer capacitors with lithium-ion electrolyte and electrode coatings with PEDOT:PSS binder, J. Appl. Electrochem., 51 (2021) 373–385.

4 Supercapacitors Based on Nanocomposites of Conducting Polymers and Metal Oxides

Anil M. Palve[1], Ajay Lathe[1], and Ram K. Gupta[2]
[1]Department of Chemistry, Mahatma Phule ASC College, Panvel, Navi-Mumbai, India

[2]Department of Chemistry, Kansas Polymer Research Center, Pittsburg State University, Pittsburg, Kansas, United States

CONTENTS

4.1 INTRODUCTION

In 1746, the first set-up involving the mechanism of electrostatics was performed in a jar but the patent was filed in 1957 with carbon material [1, 2]. Recently, devices have been developed to deliver better energy and power densities. A transition has been observed toward better energy storage devices due to the increased demand for energy and environmental concerns regarding fossil fuels and other non-renewable resources. This gap was bridged by supercapacitors and gained huge attention mainly due to their longer life and ability to deliver relatively higher power density [3].

DOI: 10.1201/9781003150374-4

TABLE 4.1

Advantages, Disadvantages, and Applications of the Supercapacitors

Advantages of Supercapacitors	Disadvantages of Supercapacitors	Applications of Supercapacitors
Higher charge/discharge rates	Higher self-discharge rate	Electric vehicles (EVs)
Longer cycle life	Lower energy density	On-off system of diesel-engine
Materials with low toxicity	Lower cell voltage	Cordless tools
Functions with a broad range of temperature	Poor voltage-regulation	Safety systems
Low cost for every cycle	High initial cost	

Supercapacitors have been used for different applications based on various advantages and disadvantages (Table 4.1) [3–5].

The nanocomposites based on conducting polymers (CPs) offer various advantages for energy storage applications due to their high conductivity, facile synthesis, and high electrochemical stability. The CPs offer a controlled structural and chemical functionality which made them a good candidate to be used as an electrode material for energy storage devices. The initial inception of CPs was done to investigate the effect as an electrode material on the ongoing research based on metal oxides (MOs). They exhibited good capacity but the cyclic stability and shelf-life of the electrode material were not suitable for commercial applications. The usage of CPs increased after observing the increment in the efficiency along with MOs. Though the incremental efficiency is relatively better than only CPs as the electrode material, the cyclic charge-discharge (CCD) and shelf-life stability were much improved. Nowadays, different derivatives of the CPs have been used as nanocomposites with MOs, which show superior stability and much-improved CCD stability [6]. In the following sections, the charge storage mechanisms in nanocomposites of CPs and MOs are discussed.

4.2 CHARGE STORAGE MECHANISMS

Mainly two types of charge storage mechanisms are observed in nanocomposites of CPs with MOs. Almost all electrodes show an electrochemical double layer formation when electrodes are in contact with any electrolytes. However, the charge storage capacity due to electrochemical double layer formation might be very low for electrode materials having low surface area. Due to the redox-active nature of components in the nanocomposites, they display redox activity, and the energy storage capacity of redox-active materials is much higher than that of an electrochemical double layer. The charge storage capacity in composites is significantly higher due to the combined redox activity of CPs and MOs.

4.2.1 ELECTRICAL DOUBLE-LAYER CAPACITORS

In the electrical double-layer capacitors (EDLCs), the charge storage is due to the electrostatic attraction between the charge at the electrode exposed surface and ions of the electrolyte. As a result, at the electrode/electrolyte boundary, layers of oppositely charged species are created. In 1853, Helmholtz proposed a double-layer capacitance model describing the electrode/electrolyte interface. $C = (\varepsilon_r \varepsilon_0 A)/d$ except for ε_0 all other quantities are ill-defined for the electrode/electrolyte interfaces due to highly disordered materials. In double-layer capacitance, the stored energy is analogous to the conventional capacitor via charge separation. However, an electrode with a larger surface area and optimized pore sizes can store more energy. This mechanism responds rapidly due to the involvement of ions at the electrode surface [5, 7].

Until 2005, the notion amongst researchers was to maximize the double-layer charging to enhance capacitance. After the work of Aurbach and his group, the view changed, and a more efficient mechanistic approach was provided [8]. According to the results, the electrochemical performance can be driven by carbon nanostructure and the pore size (less than the size of ions) and depends not only on the area of the material exposed. The researcher then refined their approach to the charge storage in tiny pores at the interface. The advancement in new techniques like small-angle neutron scattering (SANS), small-angle X-ray scattering (SAXS) along with electrochemical quartz crystal microbalance (EQCM), and nuclear magnetic resonance (NMR) led to a more evidential explanation of electrolyte organization in nano-sized pores under polarization and at the null potential.

4.2.2 REDOX TYPE CAPACITORS

4.2.2.1 Redox Behavior in Battery-Type Materials
The electrochemical energy storage in a battery requires the redox process of electrochemical species. These active species release charges which flow between electrodes kept at different potentials and perform electric work. A battery has a bipolar electrode configuration in a series combination to achieve high voltage. The energy storage is acquired electron transfer, which aims at reversing a redox.

4.2.2.2 Redox Behavior in Pseudocapacitive Materials
In the 1970s, materials with similar electrochemical properties to that of supercapacitors were found to experience redox at or near the respective exterior part of the electrode. Specific materials showing high charge/discharge cycles and energy density exhibit electrochemical property analogous to pseudocapacitive manners. The widely explored pseudocapacitor materials are RuO_2 and MnO_2. The mechanism in redox pseudocapacitance involves redox reactions that occur in the bulk material rather than only the surface of the electrode for charge storage. The mechanism in intercalation pseudocapacitance for storage of charges is not controlled by diffusion, but it is due to redox reactions shown by the fast kinetic response for high power output. Some electrochemical responses linked with pseudocapacitance are (1) redox peaks with small voltage offsets, (2) rate-independent charge storage capacity, and (3) potential has linear dependency on the state of charge. These linkers should be the measuring pointers for considering a material to be pseudocapacitive.

4.2.3 HYBRID SUPERCAPACITORS

Hybrid supercapacitors were designed due to the growing desire for enhancing EDLCs' energy density coupled with pseudocapacitive materials. One of the most advanced hybrid supercapacitor materials is $Li_4Ti_5O_{12}$ (LTO) with a discharge/charge rate of 30 C where material retains a high rate capacity. The major advantages of using LTO are the use of non-aqueous electrolytes, and it supports fast Li^+ ion transport during intercalation-deintercalation steps with zero strain. Initially, the Li^+ capacitor uses a positive capacitive porous-carbon electrode opposite to the graphite electrode. The materials with Li-ion intercalation could significantly increase the high power density while maintaining the current energy density [9, 10].

4.3 METHODS OF CHARACTERIZATION OF SUPERCAPACITORS

The components used in supercapacitors are similar to the batteries. The essential components of the supercapacitor device are electrode materials (anode and cathode), electrolyte, current collectors, binder, and separator. The active electrode materials are socked in an electrolyte and separated using an ion transporting layer (separator). The following techniques are commonly used to characterize a supercapacitor.

4.3.1 CYCLIC VOLTAMMETRY

Cyclic voltammetry (CV) is a fundamental practice for (1) the study of electroactive species, (2) to investigate a material or electrode surface, (3) to observe electrical and redox properties, (4) operating voltage window, (5) capacitance, and (6) cycle life. Since it is so versatile, it finds usefulness in electrochemistry, organic chemistry, inorganic chemistry, biochemistry, etc. CV is the foremost test performed in the electrochemical study of new materials. CV can detect redox behavior over a broad applied potential range, and quickly change those potentials to observe fast redox and chemical reactions. When a potential is applied to a mixture containing a redox-active compound(s) where a redox reaction occurs, the resultant current can be read, amplified, and plotted against the applied potential. The plot of applied voltage versus resultant current is known as cyclic voltammo-gram. The variation of the working electrode is taken against the scan of potential. A CV could be acquired with the controlled potential scan and change in current at the working electrode. A repeti-tive sweep of voltage between two edge potentials occurs in a CV test. Two sweeps in the reverse direction are a cycle that may be run for two-electrode or three-electrode connections.

A three-electrode system can be used to study the electrochemistry of one electrode. The three-electrode system uses a working electrode, a reference electrode, and a counter electrode. All electroana-lytical stations (potentiostats) can run for two-electrode systems by connecting the reference and counter electrodes leading to one side and connecting the working electrode to the other side. A potential when applied between working and reference electrodes in potentiostat then electrolysis occurs at the working electrode. The role of the auxiliary/counter electrode is to provide the current essential for the continual support of electrolysis next to the working electrode. The huge currents that might vary the potential in the measuring arrangement can be prevented by using a reference electrode. Generally, the scan rates for voltage are from 1 mV/s to several hundred mV/s. The downside of the lower scan rate is that it takes a longer time, but allowing the slow processes to occur, which can be studied effectively.

4.3.2 Charge-Discharge

Cyclic charge-discharge (CCD) is the benchmark to check the cycle-life of an electrochemical energy storage device. A recurring loop is known as a cycle. Usually, CCD is performed at a con-stant current awaiting a set voltage. In exercise, the charge is frequently termed as capacity (ampere-hour (Ah); 1 Ah = 3600 C). In general, trade capacitors can be cycled for thousands of successions. Figure 4.1 shows the CV and CD behavior of supercapacitor and battery-type material [11]. As

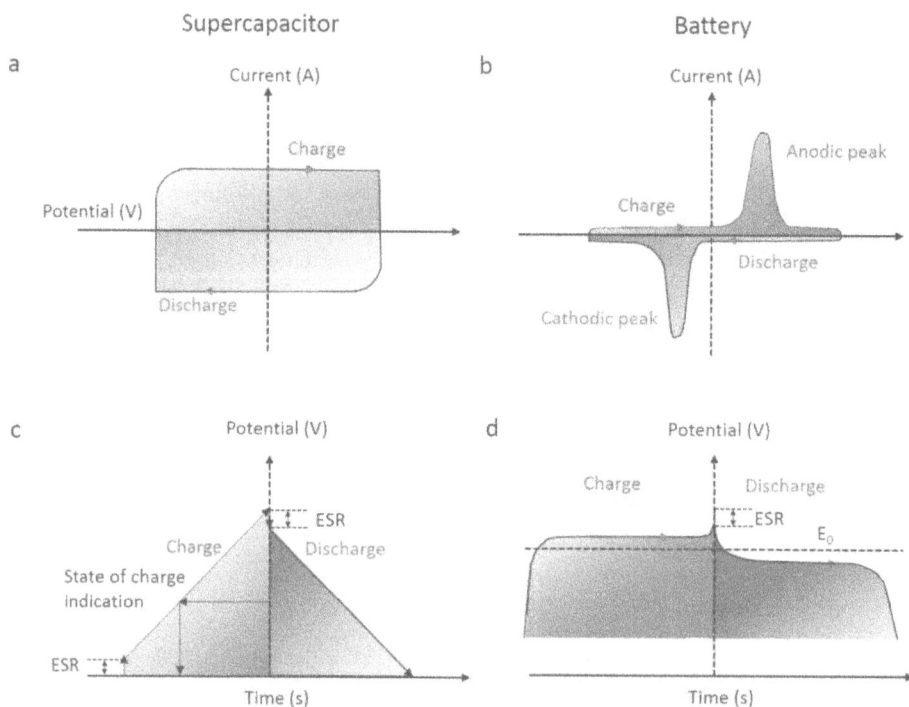

FIGURE 4.1 A schematic of cyclic charge-discharge data showing five cycles. Adapted with permission from reference [11]. Copyright 2018 American Chemical Society.

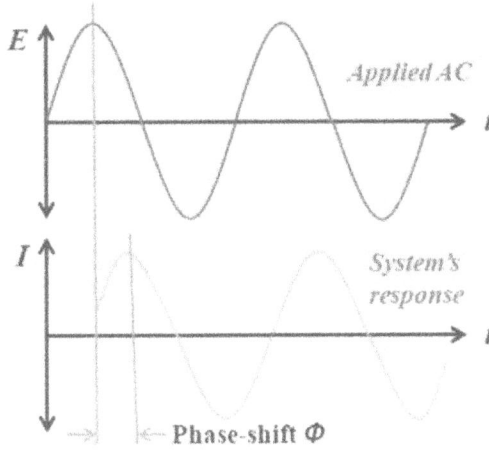

FIGURE 4.2   Representation of relation linking the fed-current and system response.

seen in Figure 4.1, supercapacitor-type materials display a rectangular shape with a linear slope in charge-discharge studies. The battery-type materials show the presence of redox-active peaks in CV curves and non-linear slopes in charge-discharge studies. The presence of plateau in the charge-discharge behavior of battery-type materials is due to the redox process.

4.3.3 ELECTROCHEMICAL IMPEDANCE SPECTROSCOPY

Electrochemical impedance spectroscopy (EIS) is another widely used method to characterize an electrode material. EIS provides many useful information such as solution resistance, charge-transfer resistance, and double-layer capacitance. In EIS, the potentiostat applies a series of low-voltage AC frequencies to the sample, and the impedance of the chemical system (its resistance to current flow) is measured. The following figure shows the relationship linking the fed-current and system response. In Figure 4.2, an applied AC voltage is sent into the chemical system by the potentiostat, and the system's response is shifted with respect to both phase and amplitude.

4.4   MATERIALS FOR SUPERCAPACITORS

An extensive study has been done on carbon-based devices such as supercapacitors, fuel cells, batteries. Nevertheless, each kind of device has different functioning in addition to the type of material used. The carbon-based materials for supercapacitor applications must have (1) good conductivity, (2) high specific surface area, (3) technology for electrode preparation with less loss of energy, and (4) good accessibility toward electrolytes. To obtain good conductivity and accessibility toward electrolytes, the carbon material in the form of fibers, woven clothes, and powders have been used after conditioning the material surface. Due to the existence of micropores, high conductivity cannot be achieved due to reduced effective surface area availability.

High specific surface areas have been achieved by introducing oxygen-based functional groups (ketone, phenol, hydroquinoid) during the activation of carbon. Oxygen-based functional groups result in oxidation or reduction, hence such materials show pseudocapacitance. Activated carbon is affordable as compared to CPs and MOs but has a lesser power density. The observed specific capacitance is about one-fourth the theoretical value based on the surface area of activated carbon reported between 300 and 1000 m^2/g. Various types of CPs, MOs, and their composites are being used to improve the performance of supercapacitors. Recent advancements in CPs, MOs, and their composites for supercapacitor applications are explored in the following sections.

4.4.1 CONDUCTING POLYMERS

A widespread study has been done on CPs for their suitability as an electrode material for supercapacitors. Due to its lower cost and environmental impact, the chemically modified CPs with higher conductivity and storage capacity are used. CPs have better activities in terms of reversible and adjustable redox processes. Commonly used CPs are the derivatives of polythiophene (PTh), polyaniline (PANI), and polypyrrole (PPy).

In general, the main disadvantage of using CPs is the swelling-shrinking during intercalating-deintercalation steps. To overcome this, fabrication of composite and modifications in morphologies have been done. This results in reduced swelling-shrinking and increased specific capacitance with moderately higher power density with a suitable electrolyte. The energy density of CP-based supercapacitors can be optimized by altering the type and concentration of the electrolyte used, the scan rate, current load, and cell configuration.

4.4.2 METAL OXIDES

The MO-based electrode materials are better than CPs in terms of higher energy density and stability. Electrode material and ions involved exhibit electrochemical redox which in turn increases the energy density. A suitable MO should have (1) the metal with variable valency, (2) material to be electrically conducting, (3) protons to get freely intercalated in the lattice during reduction (and deintercalated), allowing inter-conversion of oxide and hydroxide ions, and (4) stable 3-D structure or phase. MOs that are used extensively to date include oxides of vanadium, ruthenium, manganese, cobalt, and nickel [12–15]. Figure 4.3 represents the schematic of production of PPy enveloped iron oxide decorated nanostructured cobalt vanadium oxide hydrate (FeO@CVO) and fabrication of asymmetric supercapacitor device based on electrodes made of PPy enveloped FeO@CVO against graphene nanoplates (GNP).

Ruthenium oxide has been extensively studied due to its highly reversible redox, high proton conductivity, good thermal stability, and wide potential window. The main advantage with RuO_2 is the parallel occurrence of pseudocapacitance and double-layer charging [13, 16, 17]. Despite many advantages of RuO_2, including higher specific capacitance, its affordability is an impediment. Thermally synthesized ruthenium oxide with titanates has $>10^5$ charging cycles up to the potential of 1.4 V [18, 19]. As ruthenium oxide is expensive, it is explored only in the R&D wings of national defenses. For that reason, the trade and economic relevance, other MOs of Co, Ni, and Mn have been investigated and scrutinized. Although these MOs are fairly affordable, their specific capacitance is also in the low range in addition to the conductivity and potential window parameters [18, 19].

4.4.3 NANOCOMPOSITES FOR SUPERCAPACITORS

Conducting polymers are less costly and easily available, but its reduction in cyclic stability encourages the researcher to use the CP with different MOs. Recently various groups were focused on the use of binary nanocomposites or ternary nanocomposites of MOs. Specifically, these materials show advancement in capacitive performances due to synergistic effects.

4.4.3.1 Polyaniline-Based Nanocomposites

Polyaniline displays a higher electro-activity, along with good stability. For an appropriate galvanostatic charge-discharge progression, PANI necessitates the availability of proton in an acidic electrolyte. It has been observed that doping could significantly increase the electrical conductivity of PANI. PANI could provide a capacity in the range of 44–270 mAh/g [4]. The capacity of PANI can be tuned by the quantity and nature of the add-on, chemical binder, fabrication route, and electrode thickness. Boddula et al. introduced PANI to graphene-manganese oxide composite

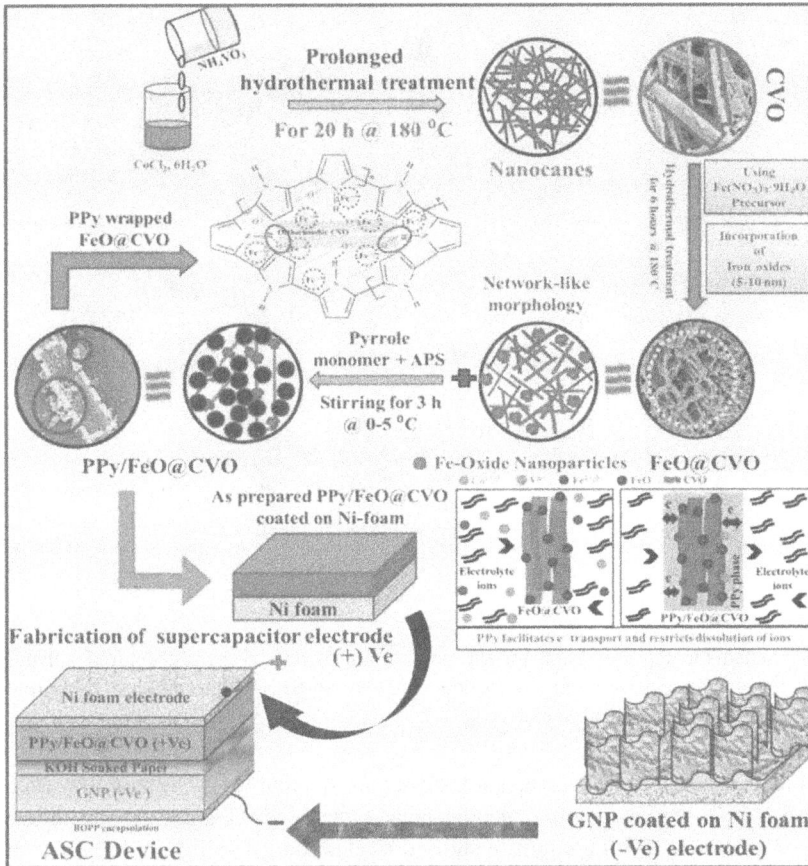

FIGURE 4.3 Schematic of production of PPy enveloped FeO@CVO and fabrication of asymmetric supercapacitor device based on electrodes made of PPy enveloped FeO@CVO against GNP. Adapted with permission from reference [12]. Copyright 2017 American Chemical Society.

by sonochemical polymerization method and used it as a binder-free electrode material [20]. The encapsulation of PANI onto graphene-manganese oxide composite boosted its specific capacitance from 44 to 660 F/g. The capacitance retention was found to be 89% for 4000 charge-discharge cycles. Another group fabricated a functionalized cotton fabric-based electrode using rGO/MnO$_2$ active material [21]. The initial deposition of GO onto cotton fiber was done by the simple method, then reduced GO to rGO. The MnO$_2$ nanoparticles were chemically deposited on rGO/cotton fabric. Then a conductive protecting layer of PANI was coated for restraining cotton fabric from the dissolution of rGO and MnO$_2$ into H$_2$SO$_4$ electrolyte. The specific capacitance found was 888 F/g in 1 M H$_2$SO$_4$ with 70% retention capacity after 3000 charge-discharge cycles at 15 A/g of current density [21]. Yasoda et al. fabricated Mn$_3$O$_4$-GO-PANI via a low-temperature chemical process [22]. The one-dimensional nanostructures with Mn$_3$O$_4$ deposited on the GO surface have an enhanced effect on the mobile ions involved in the electrodes to the electrolyte surface. This hybrid gave 829 F/g specific capacitance at 0.3 A/g current density with 94% retention capability after 1800 cycles of charge-discharge studies [22]. Navale et al. prepared PANI-NiO hybrid thin-film electrodes using the electrodeposition technique [23]. PANI-NiO electrode showed a higher specific capacitance of 936 F/g as weigh against PANI (601 F/g), and NiO (264 F/g). The nanosheets network of PANI-NiO electrodes with a large surface offered a large surface area for rapid transfer of electrons and ions along with enhanced stability. Relekar et al. developed a porous MnO$_2$/PANI composite using

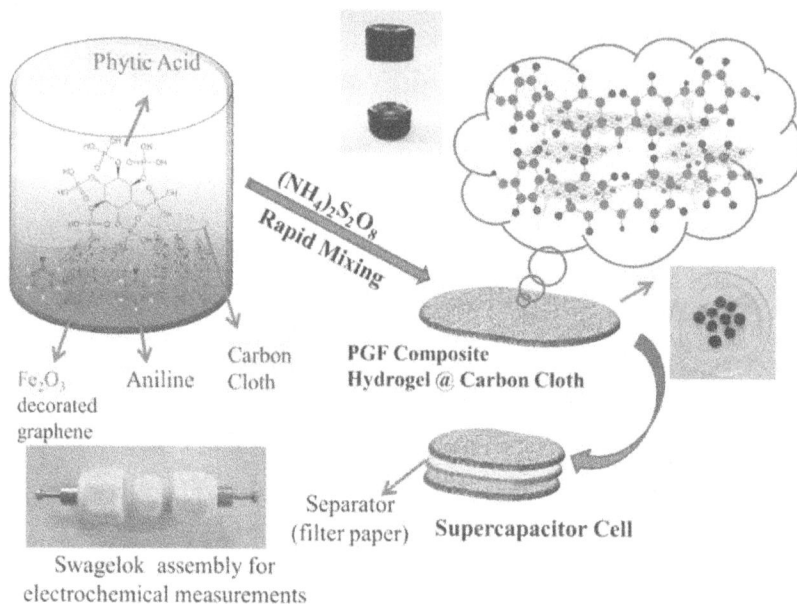

FIGURE 4.4 Schematic representation for the preparation of ternary nanostructured polyaniline/Fe_2O_3-decorated graphene composite coated on carbon cloth. Adapted with permission from reference [24]. Copyright 2020 American Chemical Society.

the electrochemical course for supercapacitor device production [14]. The thin film enclosing 80% PANI and 20% MnO_2 exhibited better specific capacitance of 409 F/g relative to similar composite films. The porous MnO_2/PANI composite showed a retention capability of 92% after 1000 cycles which reduced to 84% after 2000 cycles. Gupta et al. prepared a ternary PANI/Fe_2O_3/graphene composite hydrogel coated on carbon cloth exhibiting 1124 F/g at 0.25 A/g [24]. Schematic representation for the preparation of ternary nanostructured PANI/Fe_2O_3-decorated graphene composite coated on cloth is shown in Figure 4.4.

4.4.3.2 Polypyrrole-Based Nanocomposites

Grover et al. reported the oxide of manganese distribution in PPy medium for electrode performance evaluation [25]. They fabricated a mixed and co-axially multilayered form of MWCNT/PPy:MnO_2. The CV of MWCNT/PPy:MnO_2 gives capacitance of ~365 F/g and the co-axial form gave ~270 F/g, which implies that PPy:MnO_2 medium yields MnO_2 dispersion at molecular point rather better than the formation of multilayered structures. Kulandaivalu et al. fabricated an electrode material consisting of PPy/graphene oxide and PPy/MnO_2 by an electrochemical deposition method [26]. The symmetrical supercapacitor delivered a capacitance of 787 F/g with high cyclic stability up to 1000 cycles. Jose et al. used the electrodeposition method to synthesize a nanocomposite material of palladium oxide-PPy-rGO (PdPGO) and electrodeposited onto stainless steel (SS), which showed a specific capacitance of 595 F/g [27]. Another group prepared nickel oxide integrated polypyrrole (PPy-NiO) composite by chemical polymerization with camphor sulfonic acid (CSA) as surfactant [15]. The PPy-NiO composite exhibits a capacitance of 422 F/g at a scan rate of 10 mV/s with reduced resistance of 9.74 Ω for charge transfer. Sarmah and Kumar prepared MoS_2/rGO/PPyNT in multiple steps, which show 1561 F/g initially, which enhanced up to 1857 F/g after swift heavy ions (SHI) surface modification [28]. This also improved the cyclic stability from 72% to 91% after 10,000 cycles of charge-discharge study. The enhanced overall performance could be due to increased crystallinity of the electroactive material and enhancement in the π–π interaction in the polymer chain. Maitra et al. prepared a mesoporous hybrid composite of PPy enclosed iron oxide decorated cobalt vanadium oxide (PPy/FeO@CVO) using the hydrothermal and oxidative polymerization methods. PPy/FeO@CVO in the electrolyte of 1 M KOH shows ~1202 F/g at 1 A/g in a three-electrode configuration with 95% retention after 5,000 cyclic charge-discharge measurements [12]. Majumder et al. used oxidative polymerization to synthesize PPy/CuO/Eu$_2$O$_3$, which at 1 A/g gave 320 F/g with 62.83% retention capability after 3,000 charge-discharge studies [29]. Scindia et al. used an in situ chemical oxidation route to prepare a core-shell composite of nickel ferrite/polypyrrole (NFO/PPy) and tested its supercapacitive performance [30]. NFO/PPy electrode in 1 N H$_2$SO$_4$ electrolyte solution achieved 722 F/g at 1 A/g of current density with 99.08% efficiency.

4.4.3.3 Polythiophene-Based Nanocomposites

Vijeth et al. constructed a polythiophene/aluminum oxide (PTHA) electrode, which provided a charge storage capacity of 554 F/g at 1 A/g of applied current density [31]. The prepared electrode was used to fabricate an asymmetrical supercapacitor using activated carbon as a cathode. The device showed a specific capacitance of 265 F/g at 2 A/g with 94.61% retention after 2,000 cycles of charge-discharge studies [31]. Hareesh et al. prepared a nanocomposite of poly(3,4-ethylenedioxythiophene) polystyrene sulfonate (PEDOT:PSS)-MnO$_2$-rGO and tested the electrochemical performance in LiClO$_4$ [32]. The fabricated electrode showed an enhanced value of 633 F/g at 0.5 A/g with ~100% stability to nearly 5,000 cyclic tests. Hareesh et al. also prepared a hybrid-supercapacitor with PEDOT:PSS enclosed rGO-Ni$_{0.5}$Co$_{0.5}$Fe$_2$O$_4$, which showed 1286 F/g with 91% retention at 0.5 A/g for over 6,000 cycles of testing [33]. Agnihotri et al. designed PEDOT encapsulated graphene-MnO$_2$ composite using the one-pot solvothermal method and reported a specific capacitance of 213 F/g [34]. Sowmiya and Velraj used the oxidative polymerization method to prepare PPy-PTh-TiO$_2$ nanocomposite, which involves the combination of two conductive polymers, PPy and PTh [35]. The composite electrode showed a specific capacitance of 272 F/g, which is much higher than PTh-TiO$_2$ (110 F/g) and PPy-TiO$_2$ (80 F/g). Amutha et al. prepared rGO/PEDOT:PSS (247 F/g) and rGO/CNF/MnO$_2$ (145 F/g) electrodes with a total potential range of 1.8 V when used in a combined system. The fabricated asymmetric supercapacitor device in 1 M Na$_2$SO$_4$ electrolyte showed 47 F/g of capacitance [36]. Figure 4.5 shows a supercapacitor fabricated using rGO/NiCo$_2$O$_4$@ZrCo$_2$O$_4$

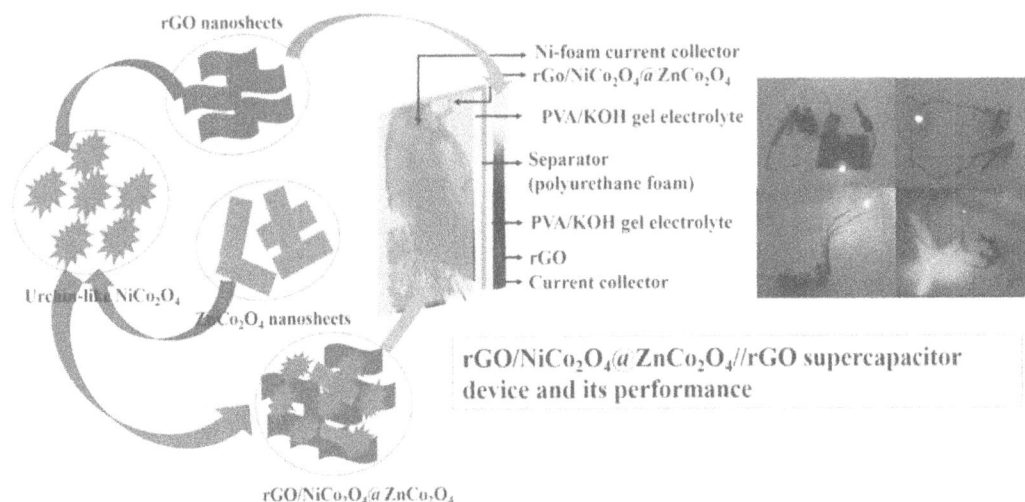

FIGURE 4.5 Schematic representation for the supercapacitor device and its performance using rGO/NiCo$_2$O$_4$@ ZnCo$_2$O$_4$. Adapted with permission from reference [37]. Copyright 2020 American Chemical Society.

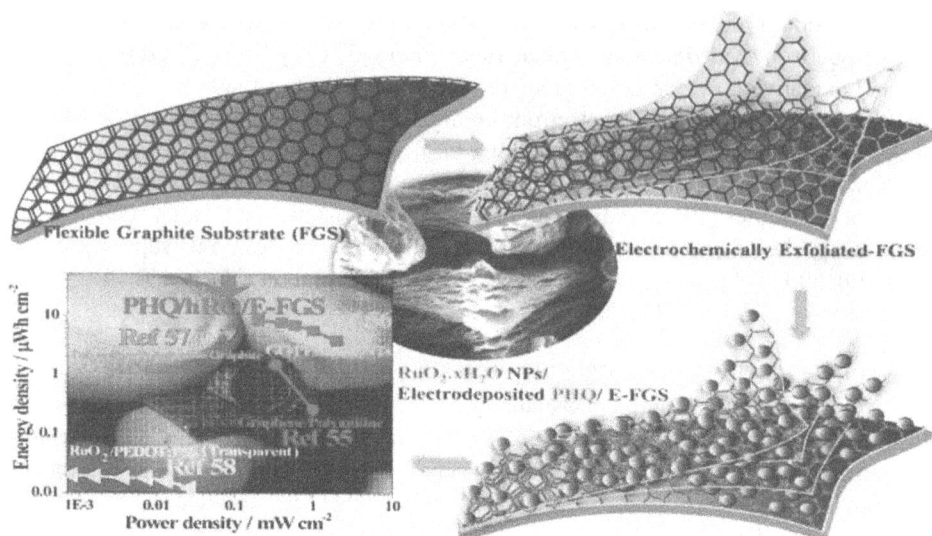

FIGURE 4.6 Schematic showing exfoliation of FGS and electrodeposition of ruthenium oxide hydrate and polyhydroquinone for the supercapacitor application. Adapted with permission from reference [13]. Copyright 2019 American Chemical Society.

(RNZC$_3$) composite, giving 1197 C/g. RNZC$_3$//RNZC$_3$ gave the energy density of 62 Wh/kg and RNZC$_3$//rGO showed 71 Wh/kg [37].

4.5 FLEXIBLE SUPERCAPACITORS BASED ON NANOCOMPOSITES

A flexible asymmetric-supercapacitor based on a nanocomposite of polythiophene wrapped aluminum oxide (PTHA) electrode as anode and charcoal as cathode was constructed which provide a specific capacitance of 265 F/g with 94.61% initial charge retention after 2,000 cycles [31]. Muniraj et al. fabricated exfoliated flexible graphite substrate (E-FGS) based flexible supercapacitor (polyhydroquinone/hydrous RuO$_2$/E-FGS) able to deliver an areal capacitance of 378 mF/cm^2 with retention of 91% after 10,000 cycles (Figure 4.6) [13]. Swapnil and co-workers used γ-Fe$_2$O$_3$ (negative electrode) and MWCNTs/PEDOT: PSS (positive electrode) to yield a wider 1.8 V of voltage window [38]. The flexible supercapacitor device achieved 65 F/g at 2.4 A/g with 80% retention after 5,000 cyclic charge-discharge testings [38]. Mane et al. fabricated a flexible symmetric solid-state supercapacitor device of configuration SS/3%La-MnO$_2$@GO/PVA-Na$_2$SO$_4$/3%La-MnO$_2$@GO/SS operating in the potential window of 1.8 V, which showed a maximum specific capacitance of 140 F/g with 90% retention after 5,000 CV cycles at the scan rate of 100 mV/s [39]. Gupta et al. fabricated a flexible electrode material of PANI/graphene/Fe$_2$O$_3$ composite hydrogel with 1124 F/g capacitance [24].

4.6 CONCLUSION

Supercapacitors are the solution for ideal storage devices. It is useful in renewable energy storage and can be used for many applications such as hybrid electronic devices, new generation electric vehicles, and heavy equipment which require high power. This chapter briefly showcased the progress of the supercapacitors using nanocomposites. It also covered the charge-discharge mechanism such as ECDL, redox, hybrid, etc. The size of capacitors was largely reduced to a micron-scale, and its charge storage capacity was improved from milli-Farad to thousands of Farad. In addition to this, the basic supercapacitor characterization tools such as CV, CCD, and EIS were elucidated. Different kinds of materials that were used to fabricate supercapacitors are surveyed.

REFERENCES

1. Peng C, Zhang S, Jewell D, Chen GZ (2008) Carbon nanotube and conducting polymer composites for supercapacitors. Prog. Nat. Sci. 18:777–788.
2. Becker HJ (1954) United States patent office: low voltage electrolytic capacitor.
3. Pandolfo AG, Hollenkamp AF (2006) Carbon properties and their role in supercapacitors. J. Power Sources. 157:11–27.
4. Snook GA, Kao P, Best AS (2011) Conducting-polymer-based supercapacitor devices and electrodes. J. Power Sources. 196:1–12.
5. Sharma P, Bhatti TS (2010) A review on electrochemical double-layer capacitors. Energy Convers. Manag. 51:2901–2912.
6. Dhibar S, Das P, Mondal S, Rana U, Malik S (2021) Conjugated Polymer Based Nanocomposites as Electrode Materials. In: Conjugated Polymer Nanostructures for Energy Conversion and Storage Applications. Wiley, pp 401–444.
7. Saha S, Samanta P, Murmu NC, Kuila T (2018) A review on the heterostructure nanomaterials for supercapacitor application. J. Energy Storage. 17:181–202.
8. Aurbach D (2005) A review on new solutions, new measurements procedures and new materials for rechargeable Li batteries. J. Power Sources. 146:71–78.
9. González A, Goikolea E, Andoni J, Mysyk R (2016) Review on supercapacitors : Technologies and materials. Renew. Sust. Energ. Rev. 58:1189–1206.
10. Ojha M, Le Houx J, Mukkabla R, Kramer D, Andrew Wills RG, Deepa M (2019) Lithium titanate/ pyrenecarboxylic acid decorated carbon nanotubes hybrid – Alginate gel supercapacitor. Electrochim. Acta. 309:253–263.
11. Shao Y, El-Kady MF, Sun J, Li Y, Zhang Q, Zhu M, Wang H, Dunn B, Kaner RB (2018) Design and mechanisms of asymmetric supercapacitors. Chem. Rev. 118:9233–9280.
12. Maitra A, Das AK, Karan SK, Paria S, Bera R, Khatua BB (2017) A mesoporous high-performance supercapacitor electrode based on polypyrrole wrapped iron oxide decorated nanostructured cobalt vanadium oxide hydrate with enhanced electrochemical capacitance. Ind. Eng. Chem. Res. 56:2444–2457.
13. Muniraj VKA, Dwivedi PK, Tamhane PS, Szunerits S, Boukherroub R, Shelke MV (2019) High-energy flexible supercapacitor – synergistic effects of polyhydroquinone and RuO_2 x H_2O with microsized, few-layered, self-supportive exfoliated-graphite sheets. ACS Appl. Mater. Interfaces. 11:18349–18360.
14. Relekar BP, Fulari A V., Lohar GM, Fulari VJ (2019) Development of porous manganese oxide/ polyaniline composite using electrochemical route for electrochemical supercapacitor. J. Electron. Mater. 48:2449–2455.
15. Vijeth H, Ashokkumar SP, Yesappa L, Vandana M, Devendrappa H (2020) Nickel oxide nanoparticle incorporated polypyrrole nanocomposite for supercapacitor application. In: AIP Conference Proceedings. American Institute of Physics Inc., p 040008.
16. Arunachalam R, Prataap RKV, Pavul Raj R, Mohan S, Vijayakumar J, Péter L, Al Ahmad M (2019) Pulse electrodeposited RuO_2 electrodes for high-performance supercapacitor applications. Surf. Eng. 35:103–109.
17. Muniraj VKA, Kamaja CK, Shelke M V. (2016) RuO_2 NH_2O nanoparticles anchored on carbon nano-onions: An efficient electrode for solid state flexible electrochemical supercapacitor. ACS Sustain. Chem. Eng. 4:2528–2534.
18. Wang YG, Wang ZD, Xia YY (2005) An asymmetric supercapacitor using RuO_2/TiO_2 nanotube composite and activated carbon electrodes. Electrochim Acta. 50:5641–5646.
19. Fugare BY, Thakur A V., Kore RM, Lokhande BJ (2018) Spray pyrolysed Ru:TiO_2 thin film electrodes prepared for electrochemical supercapacitor. In: AIP Conference Proceedings. American Institute of Physics Inc., p 140010.
20. Boddula R, Bolagam R, Srinivasan P (2018) Incorporation of graphene-Mn_3O_4 core into polyaniline shell: Supercapacitor electrode material. Ionics (Kiel). 24:1467–1474.
21. Etana BB, Ramakrishnan S, Dhakshnamoorthy M, Saravanan S, C Ramamurthy P, Demissie TA (2019) Functionalization of textile cotton fabric with reduced graphene oxide/MnO_2/polyaniline based electrode for supercapacitor. Mater. Res. Express. 6:125708.
22. Yasoda KY, Kumar MS, Batabyal SK (2020) Polyaniline decorated manganese oxide nanoflakes coated graphene oxide as a hybrid-supercapacitor for high performance energy storage application. Ionics (Kiel). 26:2493–2500.
23. Navale YH, Navale ST, Dhole IA, Stadler FJ, Patil VB (2018) Specific capacitance, energy and power density coherence in electrochemically synthesized polyaniline-nickel oxide hybrid electrode. Org. Electron. 57:110–117.

24. Gupta A, Sardana S, Dalal J, Lather S, Maan AS, Tripathi R, Punia R, Singh K, Ohlan A (2020) Nanostructured polyaniline/graphene/Fe_2O_3 composites hydrogel as a high-performance flexible supercapacitor electrode material. ACS Appl. Energy Mater. 3:6434–6446.

25. Grover S, Shekhar S, Sharma RK, Singh G (2014) Multiwalled carbon nanotube supported polypyrrole manganese oxide composite supercapacitor electrode: Role of manganese oxide dispersion in performance evolution. Electrochim. Acta. 116:137–145.

26. Kulandaivalu S, Suhaimi N, Sulaiman Y (2019) Unveiling high specific energy supercapacitor from layer-by-layer assembled polypyrrole/graphene oxide|polypyrrole/manganese oxide electrode material. Sci. Rep. 9:4884.

27. Jose J, Jose SP, Prasankumar T, Shaji S, Pillai S, Sreeja PB (2021) Emerging ternary nanocomposite of rGO draped palladium oxide/polypyrrole for high performance supercapacitors. J Alloys Compd. 855:157481.

28. Sarmah D, Kumar A (2019) Ion beam modified molybdenum disulfide-reduced graphene oxide/polypyrrole nanotubes ternary nanocomposite for hybrid supercapacitor electrode. Electrochim. Acta. 312:392–410.

29. Majumder M, Choudhary RB, Thakur AK, Karbhal I (2017) Impact of rare-earth metal oxide (Eu_2O_3) on the electrochemical properties of a polypyrrole/CuO polymeric composite for supercapacitor applications. RSC Adv. 7:20037–20048.

30. Scindia SS, Kamble RB, Kher JA (2019) Nickel ferrite/polypyrrole core-shell composite as an efficient electrode material for high-performance supercapacitor. AIP Adv. 9:055218.

31. Vijeth H, Ashokkumar SP, Yesappa L, Niranjana M, Vandana M, Devendrappa H (2018) Flexible and high energy density solid-state asymmetric supercapacitor based on polythiophene nanocomposites and charcoal. RSC Adv. 31414–31426.

32. Hareesh K, Shateesh B, Joshi RP, Williams JF, Phase DM, Haram SK, Dhole SD (2017) Ultra high stable supercapacitance performance of conducting polymer coated MnO_2 nanorods/rGO nanocomposites. RSC Adv. 7:20027–20036.

33. Hareesh K, Rondiya SR, Dzade NY, Dhole SD, Williams J, Sergey S (2021) An insight from combined experimental and computational approach of PEDOT:PSS wrapped reduced graphene oxide/$Ni_{0.5}Co_{0.5}Fe_2O_4$ nanocomposite as tertiary hybrid supercapacitor. J. Sci. Adv. Mater. Devices. https://doi.org/10.1016/j.jsamd.2021.03.001.

34. Agnihotri N, Sen P, De A, Mukherjee M (2017) Hierarchically designed PEDOT encapsulated graphene-MnO_2 nanocomposite as supercapacitors. Mater. Res. Bull. 88:218–225.

35. Sowmiya G, Velraj G (2020) Designing a ternary composite of PPy-PT/TiO_2 using TiO_2, and multipart-conducting polymers for supercapacitor application. J. Mater. Sci. Mater. Electron. 31:14287–14294.

36. Amutha B, Subramani K, Reddy PN, Sathish M (2017) Graphene-polymer//graphene-manganese oxide nanocomposites-based asymmetric high energy supercapacitor with 1.8 V cell voltage in aqueous solution. ChemistrySelect. 2:10754–10761.

37. Mary AJC, Sathish CI, Vinu A, Bose AC (2020) Electrochemical performance of rGO/$NiCo_2O_4$ @ $ZnCo_2O_4$ ternary composite material and the fabrication of an all-solid-state supercapacitor device. Energy Fuels. 34:10131–10141.

38. Karade SS, Raut SS, Gajare HB, Nikam PR, Sharma R, Sankapal BR (2020) Widening potential window of flexible solid-state supercapacitor through asymmetric configured iron oxide and poly(3,4-ethylenedioxythiophene) polystyrene sulfonate coated multi-walled carbon nanotubes assembly. J. Energy Storage. 31:101622.

39. Mane VJ, Malavekar DB, Ubale SB, Bulakhe RN, In I, Lokhande CD (2020) Binder free lanthanum doped manganese oxide @ graphene oxide composite as high energy density electrode material for flexible symmetric solid state supercapacitor. Electrochim. Acta. 335:135613.

5 Nanocomposites of Conducting Polymers and 2D Materials for Supercapacitors

U.N. Mohamed Shahid[1] and Suddhasatwa Basu[2,3]

[1]Department of Chemical Engineering, School of Engineering, University of Petroleum and Energy Studies, Bidholi, Dehradun, India

[2]Department of Chemical Engineering, Indian Institute of Technology Delhi, Hauz Khas, India

[3]CSIR-Institute of Minerals and Materials Technology, Bhubaneswar, India

CONTENTS

DOI: 10.1201/9781003150374-5

5.1   INTRODUCTION

The energy storage systems designed using renewable sources are given primary importance to address the global demand for energy consumption. Electrochemical energy generation and storage systems are an integral part of clean energy portfolio. There is a need for renewable energy storage devices which provide superior performance and are environment friendly at the same time. Electrochemical capacitors are imminent potential energy storage devices that are used to preserve and control energy consumption. Supercapacitors are more focused and promising energy storage devices compared to other non-conventional energy devices like fuel cells and batteries [1]. Supercapacitors are specifically designed for intermediate energy storage devices, which can have potential applications in electric vehicles to enhance regenerative braking and deliver large acceleration.

A conventional capacitor consists of anode and cathode immersed in a dielectric medium with a small distance of separation between them. The conventional capacitors limit their applicability for potential applications as they are rigid and massive structures [2]. But a practical supercapacitor consists of a cathode and anode immersed into an electrolytic solution with a suitable separator. The practical supercapacitors overcome the drawbacks of conventional supercapacitors as they are lightweight with adaptable features, making them compatible with emerging applications [1, 2]. The performance of a supercapacitor is quantified using parameters like capacitance, energy/power density, and stability. Since the value of capacitance is directly proportional to the quantity of electrode material and the area, the capacitance is usually represented with respect to mass (C_{mp}), area (C_{sp}), or volume (C_{vp}) of the electrode [3], as shown in Equations 5.1–5.3.

$$C_{mp} = \frac{I\ \Delta t}{m\ \Delta V} \tag{5.1}$$

$$C_{sp} = \frac{I\ \Delta t}{s\ \Delta V} \tag{5.2}$$

$$C_{vp} = \frac{I\ \Delta t}{v\ \Delta V} \tag{5.3}$$

where C_{mp}, C_{sp}, and C_{vp} are specific to gravimetric ($F\ g^{-1}$), areal ($F\ cm^{-2}$), and volumetric ($F\ cm^{-3}$) capacitances, respectively, and Δt and ΔV correspond to the time of discharge and electrochemical window of operation of supercapacitors, respectively. In addition [3] to specific capacitance, the energy density (E_{xd}) and power density (P_{xd}) are also represented with respect to mass or area, or volume as shown in Equations 5.4–5.5.

$$E_{xd} = \frac{1}{2} C_{xp} \Delta V^2 \tag{5.4}$$

$$P_{xd} = \frac{E_{xd}}{\Delta t} \tag{5.5}$$

where x represents either mass (m), area (s), or volume (v), respectively. The present supercapacitors still possess lower specific energies as compared to batteries or fuel cells, as shown using Ragone plots [4] for various energy storage devices (Figure 5.1). A deep understanding of the working principle, materials involved in fabrication, and their correlation to specific capacitance in supercapacitors are to be extensively studied to enhance the specific energy for use in potential applications. This chapter discusses in detail the classification of supercapacitors, the current state-of-the-art materials employed, their pros and cons, and their composites which addresses their limitations, thereby improving their performance for high E_{xd}, P_{xd}, and cyclic stability.

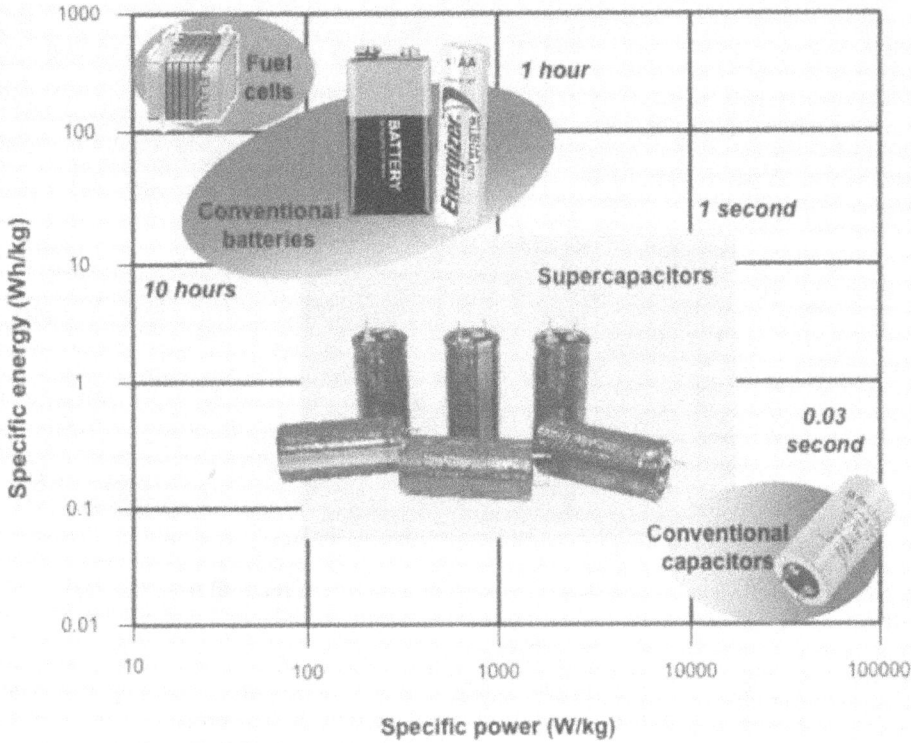

FIGURE 5.1 Ragone plots for various energy storage devices. (**Copyright permission obtained from Ref.** [4].)

5.2 SUPERCAPACITOR CLASSIFICATIONS

Supercapacitors are broadly classified into two types. The energy storage mechanism used to distinguish between the supercapacitors relies on electrochemical reactions taking place over the electrode surface. The supercapacitors are classified as non-Faradaic capacitors and Faradaic capacitors. These two types either independently or cumulatively give rise to a maximum of three subcategories of supercapacitors, as discussed in detail below. Figure 5.2 shows the schematic representation of their classification.

5.2.1 ELECTROCHEMICAL DOUBLE-LAYER CAPACITORS

In the case of electrochemical double-layer capacitors (EDLCs) or non-Faradaic capacitors, charges are stored electrostatically, i.e., charged species are confined to the electrode/electrolyte interface called the electrical double layer. The output capacitance is quantified by the quantity of charge stored in the double layer. Hence, the capacitance value is influenced solely by the properties of the electrode material used in EDLCs [5]. Carbon electrodes are the best examples of electrodes for EDLCs as they possess a high surface area of 1000–2000 m^2 g^{-1} with various multi-dimensional morphologies and optimum pores size distribution. Carbon is used in different forms like activated carbons (ACs), carbon nanospheres, carbon nanotubes (CNTs), graphene. One significant advantage and disadvantage of using EDLC is, their cyclic stability, namely, ≥0.5 million cycles without any degradation of electrodes and low C_{mp}, which leads to low E_{md}, as the charged species are adsorbed/desorbed over the external surface of the electrodes in EDLC [5].

5.2.2 PSEUDOCAPACITORS

Pseudocapacitors are Faradaic capacitors that allow the charges to transfer between the electrode and the electrolyte. The charge storage is through reversible Faradaic reaction at the electrode surface.

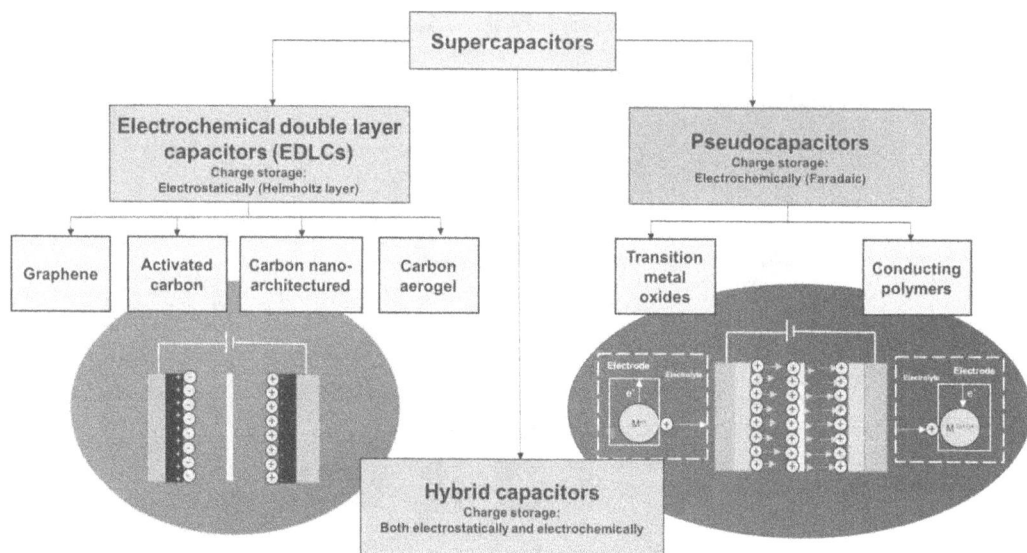

FIGURE 5.2   Classification of supercapacitors. (**Copyright permission received from Ref.** [4].)

The Faradaic process happens through different routes, electro-sorption (or) reversible oxidation and reduction processes (or) reversible intercalation/deintercalation processes, depending on the nature of the active electrode material present in pseudocapacitors. The process of charge transfer is voltage dependent on the redox reaction and intercalation processes [2, 5]. Electrodes made using conducting polymers (CPs) and transition metal compounds (oxides, carbides, and dichalcogenides) constitute pseudocapacitors. CPs have high conductivity ranging from a few S cm^{-1} to 500 S cm^{-1}. Since the reversible intercalation/deintercalation processes take place over the bulk of CPs, they showcase high C_{mp} and E_{md} values. But the cycle life of pseudocapacitors with CPs is short-lived due to reversible intercalation and deintercalation processes accompanied by volume changes causing mechanical failure under prolonged cycling [6]. In the case of TMOs, reversible oxidation and reduction processes occur relatively offering high C_{mp} and E_{md} values. But the major drawback with TMOs is their low conductivity and easy agglomeration tendency at high mass loadings [4], which hinders them from achieving high C_{mp} and E_{md} values.

5.2.3   HYBRID CAPACITORS

Hybrid supercapacitors are combinations of both EDLC and pseudocapacitors. They involve both Faradaic and non-Faradaic processes taking place simultaneously at the electrodes. The presence of pseudocapacitive characteristics increases the C_{mp} thereby increasing the E_{md}, and the presence of EDLC characteristics enables high P_{md} along with long cycle life. Suitable electrode materials for hybrid supercapacitors are composites of carbon and TMOs/CPs. A comparison between all the three sub-types of supercapacitors is shown in Table 5.1.

The electrode material properties like morphology, crystallinity, pore structure, and surface area are dominant physiochemical factors that directly affect the specific capacitance, C_{mp} values. For example, microstructure and morphology are closely related to the surface area and it is correlated to the preparation techniques involved in synthesizing the electrode. In addition, morphologies of zero-dimensional (0D), one-dimensional (1D), and two-dimensional (2D) structures play a vital role in determining the overall performance of the supercapacitors. Similarly, crystallinity has a direct correlation with conductivity and an inverse correlation with electrochemical activity, hence optimum conductivity and electrochemical activity are necessary. Simultaneously, optimum pore structure and redox characteristics are necessary to enhance the reversible adsorption/desorption,

TABLE 5.1
Properties of Electrodes across Various Supercapacitors [2]

Properties	EDLC	Pseudocapacitor	Hybrid Supercapacitor
Electrode composition	Carbon	CPs and TMOs	Carbon with CPs/TMOs
Charge storage	Non-Faradaic	Faradaic	Non-Faradaic and Faradaic
Specific capacitance (C_{mp})	Low C_{mp}	High C_{mp}	High C_{mp}
Specific energy density (E_{md})	Low E_{md}	High E_{md}	High E_{md}
Cyclic stability	Good stability	Low stability	Good stability

intercalation/deintercalation, and overcome the resistances associated with diffusion and migration of ions over the electrode surface.

5.3 MATERIALS FOR SUPERCAPACITOR

The physiochemical characteristics of various electrode materials used in EDLC and pseudocapacitors and their composition, synthesis procedures, and properties are discussed in detail below. In addition to electrode properties, various electrolytes are used in supercapacitors whose properties vary based on their origin. The electrolytes range from aqueous to organic to ionic liquids (ILs). Electrolytes like potassium hydroxide (KOH), sodium sulfate (Na_2SO_4), sulfuric acid (H_2SO_4), and ammonium chloride (NH_4Cl) are aqueous in nature, and they provide high concentration of ions with lower resistance causing higher C_{mp}/P_{md}. The aqueous electrolytes also exhibit high conductivity of ~1 S cm^{-1} [7]. Their major drawback is their narrow electrochemical window of operation due to degradation of the aqueous electrolyte, and their major advantage is the flexibility involved in their synthesis. Organic electrolytes, on the other hand, exhibit less conductivity (10–60 mS cm^{-1}), so lower P_{md}, but have higher E_{md} due to a broad electrochemical window of 2.5–2.7 V.

5.3.1 CARBON ELECTRODES

All nanostructured carbon-based materials ranging from zero to multi-dimensions like AC, carbon nanofibers (CNFs), graphene are employed as electrode material for EDLCs. But a suitable nanoengineered carbon structure with unique morphology and properties is necessary to tune the performance characteristics of EDLCs [8]. The various carbon structures used in EDLCs are shown in Figure 5.3 and discussed along with their advantages and disadvantages.

5.3.1.1 Zero-Dimensional Carbon

Zero-dimensional (0D) carbon refers to particles with an aspect ratio ~1 such that their dimensions in all three dimensions are confined within ~1–100 nm. Examples for 0D carbon are ACs, carbon microspheres, carbon macrospheres, and hollow carbon particles. ACs are synthesized by the conversion of organic substances like wood, nutshell, etc., into carbon by pyrolysis followed by activation. ACs are porous carbon materials that consist of pores ranging from ~2 nm (micropore) to 50 nm (macropore) with a higher specific surface area (SSA). For example, a novel ultra-microporous carbon nanoparticle synthesized through solvothermal polymerization with a particle size of ~30 nm and ~549 m^2 g^{-1} exhibited C_{mp} of 206 F g^{-1} in 6 M KOH aqueous solution. Similarly, a monodisperse carbon microspheres of diameter 250 nm and 456 m^2 g^{-1} SSA synthesized via hydrothermal process exhibited C_{mp} of 245 F g^{-1}. Literature [9] also reports that microspheres of ACs prepared through inverse-microemulsion polymerization with optimum pores abundant with ~2 to 50 nm in addition to pores <2 nm could deliver C_{mp} of 275 F g^{-1} with 94% cycle performance even after 2000 cycles. A hollow carbon nanospheres particle [10] with an outer diameter of 69 nm and a large surface area

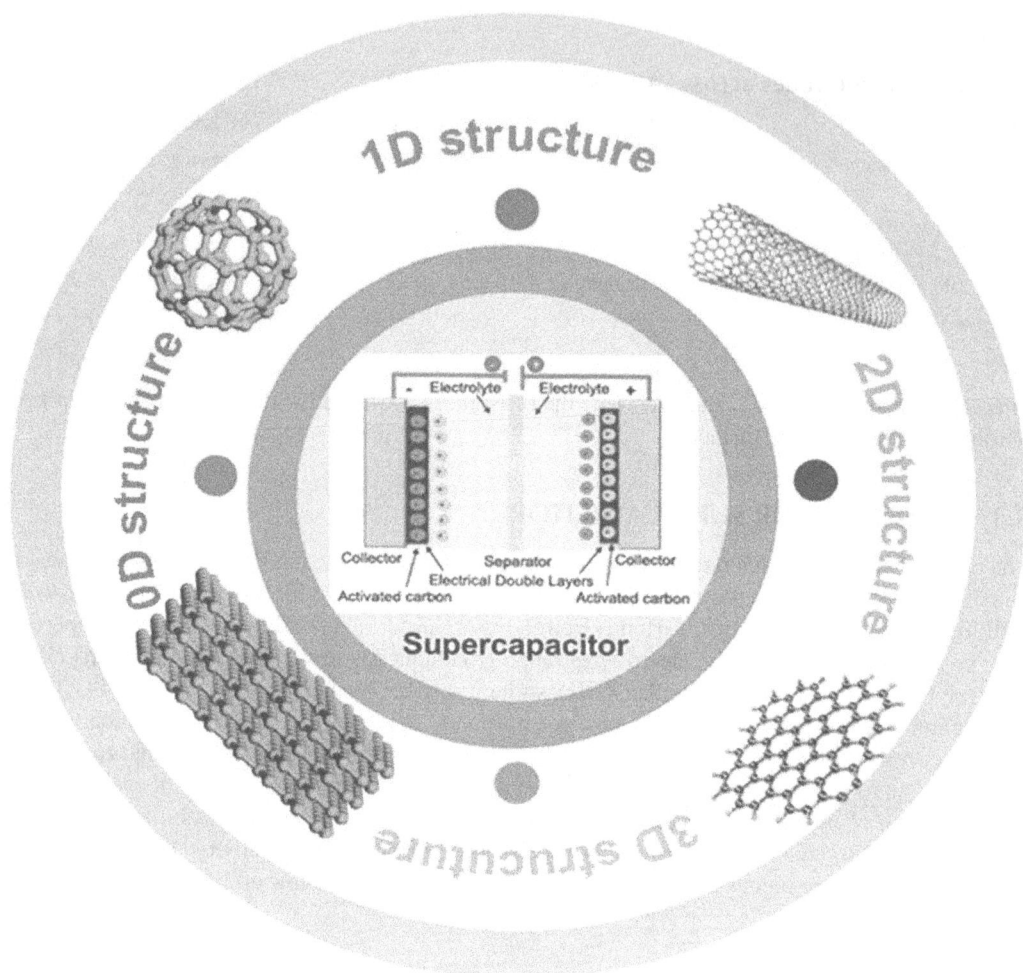

FIGURE 5.3 Schematic representation of various carbon nanostructures for electrochemical double-layer capacitors. (**Copyright permission received from Ref. [12].**)

of 3022 m^2 g^{-1} exhibits C$_{mp}$ of 201 F g^{-1} due to nanopores which reduce the distance for the diffusion of charged species thereby minimizing the resistance.

5.3.1.2 One-Dimensional Carbon

One-dimensional (1D) carbon structures are materials with a large aspect ratio (length/diameter). Carbon nanorods, carbon nanofibers (CNFs), and carbon nanotubes (CNTs) come under 1D carbon materials. The process of charge storage is enhanced by the enhancement in electrical property like conductivity, which in addition to high aspect ratio facilitates the kinetics of the electrochemical reactions. In the case of 1D structures, a continuous network is formed along the longer axis for charge transport. The 1D nanostructures show better charge transportability than 0D nanoparticles by reducing the contact resistance between adjacent particles. Looking into the mechanism of charge transport, the electrons are transferred by two mechanisms, hopping or by diffusion within the extended states in the case of 1D materials [11]. Carbon nanorods, also called carbon nanopillars, are the simplest form of 1D carbon particles with an aspect ratio close to ~10. The 1D nanostructures can be designed with or without templates, as discussed in literature [12]. Literature [13] reports successful synthesis of 1D rod-shaped metal-organic frameworks (MOFs), which are 20–40 nm wide, 100–250 nm long, and 10–25 nm diameter displaying an electrical conductivity of 3.47 S cm^{-1} and C_{mp} of 164 F g^{-1}. Compared to 1D carbon nanorods, 1D carbon nanowires or CNFs possess more aspect ratio resulting in more accessible area. For example, 1D hierarchical porous CNFs [14], which was synthesized using electrospinning technique, employs calcium carbonate acts as hard template (which gets removed through carbonization) thereby providing an area of 679 m^2 g^{-1} through the creation of micropores and mesopores providing 251 F g^{-1} with sustenance in capacitance to 160 F g^{-1}. CNTs are used as electrodes because of high conductivity, strength, and stability, but literature [15] reports the formation of CNT doped with nitrogen shows a C_{sp} of 210 F g^{-1} in 6 M KOH electrolyte.

5.3.1.3 Two-Dimensional Carbon

Materials like graphene, graphene oxide (GO), and reduced graphene oxide (rGO) fall under the category of 2D carbon materials, which are sheet-shaped structures. 2D materials exhibit great mechanical strength, chemical stability, flexibility, transparency, conductivity, and SSA, which are necessary for supercapacitor electrodes [16]. Due to the high-conductivity and aspect ratio of 2D carbon materials, graphene, GO, or rGO are predominantly used for supercapacitors [17]. Although 1D materials are successful in improving the C_{sp} as compared to 0D materials, the synthesis of 1D materials in preferential oriental (longitudinal axis) increases the cost of synthesis. On the other hand, the 2D structures can be easily tailored into various shapes, thereby expanding the horizon of material varieties for energy storage. Graphene is a unique 2D material, with a monolayer of planar carbon atoms (sp^2-hybridized) with delocalized π-electron, which causes ultrahigh conductivity [18]. 2D materials like graphene materials exhibit large SSA (>2600 m^2 g^{-1}) with enhanced mechanical and chemical properties, which make them an ideal candidate for storage applications [19]. It has been reported [20] that graphene's high SSA can cause large C_{mp} of 550 F g^{-1}. However, graphene can easily restack to form a caterpillar-like wrinkled structure during synthesis [36]. This hinders the charge transport at the surface and negatively influences the surface to volume ratio [10] thereby degrading the performance of supercapacitors. N (nitrogen) or B (boron) or S (sulfur) or P(phosphorus) doping is an alternative route to effectively enhance the electrode performance of heterodoped graphenes. The N doped graphenes are more beneficial since the atomic size of N being similar to that of carbon, and the N being highly electronegative and its capacity to contribute the excess free electrons [21]. 3D carbon structures are made using the self-assembly and template-directed methodology of 0D, 1D, and 2D carbon elements. Although the 3D carbon structures provide large surface area, mechanical stability, and well-defined pathway for ion transfer, the preparation of these 3D structures demands a high cost.

5.3.2 Conducting Polymer Electrodes

Fast and reversible redox reactions at the surface and bulk of the electrodes help in storing the charges in CPs like polyaniline (PANi), polypyrrole (PPy), polythiophene (PTh), 3-substituted polythiophene, and poly(3,4-ethylenedioxythiophene) (PEDOT) which exhibit a higher C_{sp} than that of EDLCs [22]. The molecular structures of various CPs used in making electrodes for supercapacitors are shown in Figure 5.4. The C_{mp} of the carbon-based materials is less than 300 F g^{-1}, whereas CPs are able to achieve much higher values. These CPs are either prepared using chemical or electrochemical oxidation of monomers forming conjugated structures [23]. In addition, doping CPs enhance the electrical properties and performance, which help in tailoring their properties for different applications [6]. Doping and de-doping, either create p-type or n-type CPs, as shown below in Equation 5.6–5.7.

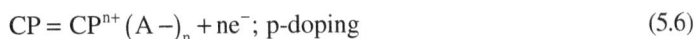

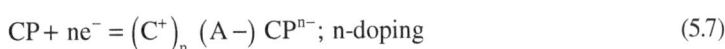

$$CP = CP^{n+}\left(A -\right)_n + ne^-; \text{ p-doping} \tag{5.6}$$

$$CP + ne^- = \left(C^+\right)_n \left(A -\right) CP^{n-}; \text{ n-doping} \tag{5.7}$$

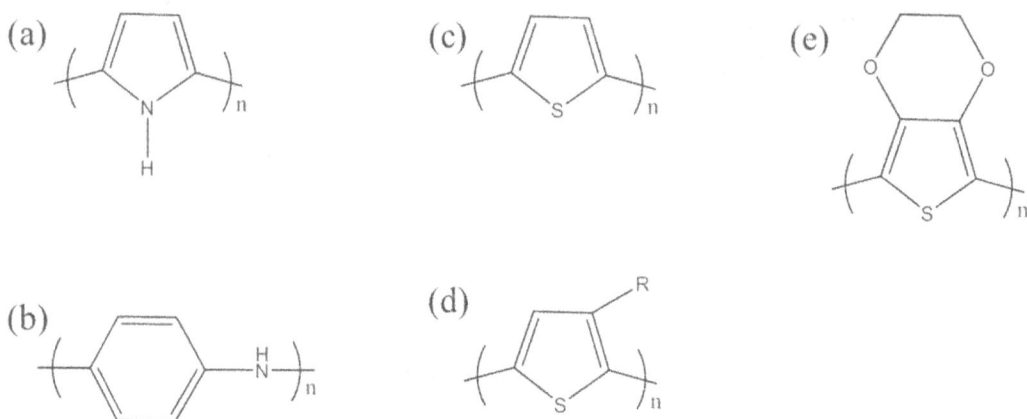

FIGURE 5.4 Molecular structures of various conducting polymers: (a) polypyrrole, (b) polyaniline, (c) polythiophene, (d) 3-substituted polythiophene, and (e) poly(3,4-ethylene-dioxythiophene) used in pseudocapacitive supercapacitors. **(Copyright permission obtained from Ref. [3].)**

The nature of the dopant for CPs depends on the polymer and the voltage required for doping. The CPs like PPy and PANi can only be p-doped, while PTh-based CPs can only be n-doped. The excellent conductivity of CPs is because of delocalized electrons in the polymer chain. Supercapacitor devices made from CPs are used in three different configurations,

1. Type I: These are symmetric supercapacitors where the same p-doped CP is used for both electrodes.
2. Type II: These are asymmetric supercapacitors where two different p-doped CPs with a different range of electroactivity are used as electrodes.
3. Type III: These are symmetric supercapacitors where the same CP is used for both electrodes with the p-doped CP for the positive electrode and the n-doped CP for the negative electrode.

5.3.2.1 Polyaniline

PANi is one of the widely used CPs for supercapacitors due to its good conductivity and a high theoretical capacity in different redox states (leucoemeraldine, emeraldine, and perningraniline) [24] [25]. Aniline monomers are electrodeposited to produce PANi in a low-pH medium like sulfuric acid (H_2SO_4), hydrochloric acid (HCl), phosphoric acid (H_3PO_4), perchloric acid ($HClO_4$), or trifluoroacetic acid (CF_3COOH) [26]. The major difference between PANi and other CPs is that the electrochemical activity of PANi is associated with the proton, so the presence of acidic electrolyte or protic ionic liquid is necessary [27]. PANi, when used as an electrode in a supercapacitor, works normally within a confined electrochemical window to avoid the degradation of PANi. It is also reported that as such synthesized PANi exhibits semiconducting behavior while the doped PANi shows enhanced conductivity upon doping with acidic dopants. The electrical conductivity of the $LiPF_6$-doped PANi improved dramatically to 1.7 S cm^{-1} and the Type I supercapacitor showed an initial C_{mp} of ~100 F g^{-1} and it could retain 70% of it even after 5000 cycles [28].

5.3.2.2 Polypyrrole

PPy is an alternative CP electrode which shows high theoretical C_{mp} of 620 F g^{-1}, improved conductivity with fast cyclic stability [29]. The properties of PPy-based electrode materials are tailored using various preparation methods and the tailored electrode achieve higher C_{vp} of ~500 F cm^3 [30]. PPy can be synthesized in various shapes like porous clusters to nanospheres to nanowires to nanotubes, by using techniques similar to synthesis of PANi. The capacitive performance and stability of the PPy-based supercapacitors is influenced by its level of doping. For example,

SO_4^{2-} doping of PPy, creates physical crosslinks, which enhances the permeable network thereby improving C_{mp} [31]. But the crosslinking could possibly deteriorate the structural integrity of the CP which inevitably degrades the stability of the electrode material. In addition, the PPy materials also exhibits enhanced conductivity of 45 ± 5 S cm^{-1} due to which the supercapacitor could be charged/dis-charged for several thousand cycles between -1.5 and 1.5 V. To further improve the cyclic performance while maintaining or enhancing the capacitance of PPy-based electrode materials designing special micro- or nanostructures for PPy-based electrode materials is preferred.

5.3.2.3 Poly(3,4-ethylenedioxythiophene)

Poly (3,4-ethylenedioxythiophene) (PEDOT) is a sub-group of PThs, which is an alternative CP for pseudocapacitors because of its high-speed kinetics and conductivity [32]. The PEDOT is prepared similar to other CPs, by either chemical routes or by electropolymerization. PEDOT has larger electrochemical window with high conductivity of 400 ± 100 S cm^{-1} and greater stability [33]. The usage of PEDOT nanotubes [34] delivered a P_{md} of 25 kW kg^{-1} with C_{mp} of ~150 F g^{-1}. Although the C_{sp} of PEDOT is relatively lower than PPy and PANi, the PEDOT has very less side-reactions which increases their cycle life. In addition, there are some drawbacks with PEDOT, similar to PANi and PPy, that their processability is poor. But a significant advantage is that the PEDOT can be both n-doped and p-doped. The n-doped PEDOT synthesized at low voltages make them more prone to oxygen and water. It is also observed that the n-doped PEDOT suffers from the charge trapping effect leading to low C_{mp}. The charge trapping effect is where electrically isolated domains are formed in the electrode during cycling thereby leading to the availability of a low quantity of charges during discharge [6].

5.3.3 TRANSITION METAL ELECTRODES

In designing high-performance energy storage systems, major attention is focused on 2D transition metal-containing compounds as plausible pseudocapacitor electrodes. Transition metal oxides (TMOs), transition metal carbonitrides (TMCs), and transition metal dichalcogenides (TMDs) are commonly used 2D transition metal electrodes due to high C_{mp} values. But there also exist some limitations in their usage as discussed below.

5.3.3.1 Transition Metal Oxides

2D nanosheets of TMOs (ruthenium oxide, manganese oxide, vanadium pentoxide, etc.) are a great choice for supercapacitor applications as active electrode materials. Ruthenium oxide (RuO_2) is considered an initial choice for TMO-based electrodes as it has excellent capacitive behavior with high C_{mp} ~2000 F g^{-1} conductivity and rate capability [35]. RuO_2 is regarded as a potential electrode material that shows improved performance compared to carbon and polymer electrodes. But the cost associated with its synthesis and its agglomeration during cycling limits its application. Manganese oxide (MnO_2) is an alternative electrode material to RuO_2 because it has advantages in providing high C_{mp} ~1370 F g^{-1} with reduced synthesis cost in addition to being environment friendly [36]. But its application for commercial applications is limited due to its poor conductivity. Nickel oxide (NiO) is also a plausible TMO because of its high C_{mp} ~2584 F g^{-1}, harmless nature, and reduced cost associated with its synthesis [37]. Lower dimensions of NiO in various morphologies are synthesized using various methods like hydrothermal, solvothermal, and electrodeposition. In recent years, cobalt oxide (Co_3O_4) is emerging as an ideal electrode that can replace the RuO_2 due to its excellent reduction–oxidation characteristics, high C_{mp} ~3560 F g^{-1}, and easy preparatory methods [38]. The characteristics of Co_3O_4 are correlated to their microstructure and to their electronic states of the metal [39].

5.3.3.2 Transition Metal Carbonitrides

2D MXenes are synthesized by carefully etching out the "A" element from the MAX-phase [40], which has a formula of $M_{+1}AX_n$. In the MAX phase, M stands for a transition metal, A stands for group 13 or 14 elements, and X stands for carbon (C) and/or nitrogen (N) with the value of n being 1 or 2 or 3. The high conductivity of MXenes makes them an attractive electrode material for supercapacitors. But their low specific surface degrades the charge storage activity. The C_{vp} of MXenes has been reported close to 300 F cm^{-3} even during the initial stages of operation, which is greater than commercially EDLC electrodes of carbon [41]. This makes MXenes a great potential for energy storage applications. Various etching measurements using HF-etching or LiF-HCl etching make the surface covered with T_x (O, –OH, and –F groups) forming $Ti_3C_2T_x$. For example, titanium carbide (Ti_3C_2), which was produced by etching titanium aluminum carbide (Ti_3AlC_2) in an acid showed sufficient electrochemical activity [42]. With the incorporation of lithium fluoride (LiF_4) during the etching process, the separation between the layers of Ti_3C_2 was maintained, which prevented the restacking of Ti_3C_2 layers thereby enhancing the electrochemically active sites. The variation of Ti oxidation state also influenced the capacitance values [43]. The capacitive behavior of $Ti_3C_2T_x$ in sulfuric acid (H_2SO_4) is due to pseudocapacitance, which arises due to variation in valence state of Ti. The electronic structure is affected by the functionalities which alter the hydrophilicity of the MXenes, influencing electrochemical kinetics and ion movement during the electrochemical process. The superior capacitor performance in H_2SO_4 is additionally due to the involvement of protonation and deprotonation processes.

5.3.3.3   Transition Metal Dichalcogenides

The enhanced diffusion rate of ions in addition to outstanding electrical conductivity is considered an essential qualification of the electrode for providing high performance. The transport of electron is affected by the electronic nature of the electrodes, which spans into three different domains: metals, semiconductors, and insulators. In general, metallic 2D TMDs, show excellent conductivities and offer favorable electron transport properties [44]. Due to the existence of weak Vander Waals forces [45], nanosheets of TMDs in one or two layers are used as active electrode material. The 2D TMDs are an opportunity to bridge, atomic configurations and tune the material structure thereby, tailoring the properties required for the electrodes. For example, an exfoliation technique [46] is used to obtain vanadium disulfide (VS_2) of dimension 150 nm, which exhibits a high C_{mp} ~4760 F g^{-1} and stable performance for 1000 cycles when assembled as an electrode in supercapacitors. In a similar approach, chemical vapor deposition (CVD) was used for the synthesis of VS_2 nanosheets [47] with high crystallinity, conductivity (~3 *10^3 S cm^{-1}), and high C_{mp} ~860 F g^{-1}. In a similar approach, nanosheets of semiconducting tungsten disulfide (WS_2) [48] were used to fabricate stable negative electrodes, demonstrating improved crystallinity and excellent durability. Other types of TMDs, such as MoS_2, $MoSe_2$, $NbSe_2$, and WTe_2 are frequently used as electrode materials for supercapacitor applications.

5.4   NANOCOMPOSITES FOR SUPERCAPACITORS

The inclusion of 2D materials along with CPs is proposed to enhance the stability of the electrodes. The reduced dimensions, 2D structures, of transition metal compounds (TMOs, TMCs, TMDs), increase the redox sites and the 2D in-plane structure promotes the diffusion of the charged species and their movement, thereby improving the performance. Incorporating CPs with 2D materials is considered to improve the conductivity and electrochemical stability in nanocomposite electrodes. Figure 5.5 shows the schematic representation of various 2D materials which can be used with CPs to make nanocomposite electrodes for supercapacitors. The TMO is drawn using VESTA [49] whereas, all other structures are drawn using sketch tool in word. The advantages of using composite materials of CPs with 2D materials are shown in Figure 5.6, where the output parameters like specific capacitance and cyclic stability are studied along with their pure components. The cyclic stability and specific capacitance get enhanced by using composites of 2D materials with CPs. The above composites also find usage as high-performance electrode materials in emerging applications like micro-supercapacitors, compressible supercapacitors, electrochromic supercapacitors, and self-healable supercapacitors, as discussed below.

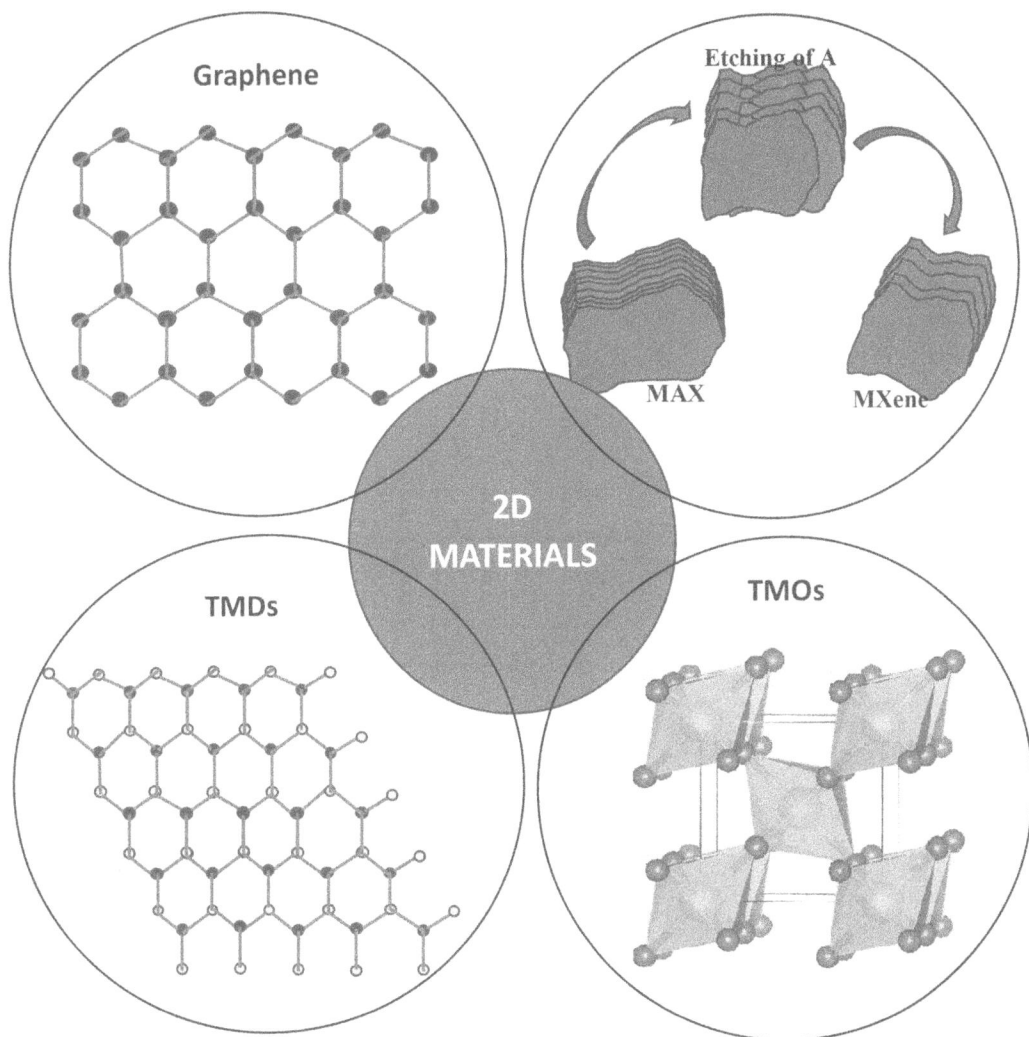

FIGURE 5.5 Various 2D materials, graphene, MXene, TMDs, and TMOs, which are used in composites supercapacitors.

5.4.1 Conducting Polymer—2D Carbon Composites

The 2D carbon materials, graphene or GO or rGO possess limitations for independent use as electrodes due to their low accessible surface area and compact structure. But they are employed as support for polymerization of various monomers like PANi, PPy on their surface and the resultant fiber or film composites show high conductivity and mechanical flexibility thereby providing high C_{sp} and cycling stability. The overall performance is due to the cumulative effect of EDLC (non-Faradaic) and pseudocapacitive (Faradaic) characteristics of carbon and CPs where the former provides a continuous conductive network for rapid charge transfer and also acts as scaffold to stabilize the CPs.

5.4.1.1 Polyaniline-Carbon Composites

PANi-graphene composite films prepared using in situ chemical polymerization using hydroiodic acid as a reducing agent to form PANi-rGO composite films exhibited a C_{sp} of 0.425 F cm^{-2}. The preparation of such composite films over the cellulose paper finds suitable application in flexible solid-state supercapacitors, exhibiting a C_{mp} of ~224 F g^{-1} in polyvinyl alcohol-H_2SO_4 electrolyte

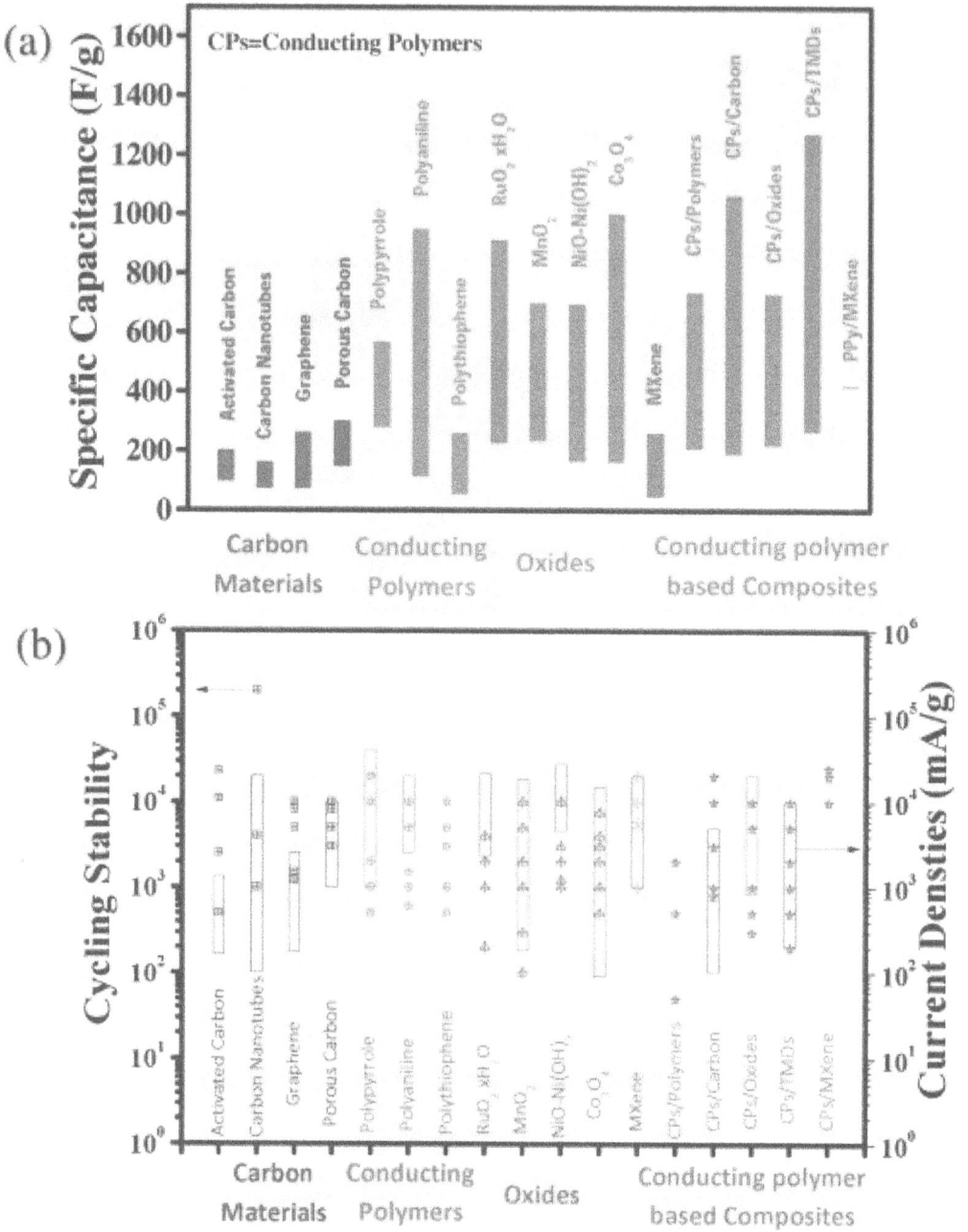

FIGURE 5.6 Comparison of output parameters: (a) specific capacitance and (b) cyclic stability between different types of supercapacitor electrodes. (Copyright permission obtained from Ref. [3].)

and the energy storage properties of the solid-state device remain unchanged on deformation, reflecting its excellent flexibility [50]. PANi depending on its dopant level and energy storage capability finds application in electrochromic supercapacitors. PANi-graphene composites are intensively used for electrochromic applications, and it has been reported [51] that PANi-GO composites exhibit 688 F g^{-1}. But they suffer from cyclic instability, which was overcome by creating hydrogen bonds (Hb). The hydrogen-bonded PANi-HbG exhibited a C_{mp} of 598 F g^{-1} with more excellent capacity retention than that of PANi (217 F g^{-1}), which responded faster to switch complexion under stimulating potentials. Literature [52] also report the formation of PANi-graphene nanocomposites where porous graphene was used to grow 1D PANi which overcomes restacking using the microspherical architecture, thereby facilitating quick diffusion of charged species with a C_{mp} of 338 F g^{-1} and ~87% retention in capacity even after 10,000 cycles.

5.4.1.2 Polypyrrole-Carbon Composites

CP-carbon composite fibers of PPY-rGO were produced using wet spinning process find suitable applications in wearable units or textile structures for smart garments. The diameters of the fibers range from 15 to 80 mm and the resultant PPy-rGO fibers showed high conductivity, mechanical stability, and lightweight characteristics, which provided a C_{sp} of 115 mF cm^{-2}. Paper-based microsupercapacitors [53] were also created using a composite of PPy-graphene foams, which delivered a C_{sp} of 22 mF cm^{-2} and remained unchanged on deformation reflecting its excellent flexibility. A nanoparticle film [54] of composite PPy-rGO electrode prepared using a combination of vacuum filtration and electrochemical reduction showed excellent mechanical and electrochemical properties with C_{sp} of 216 mF cm^{-2} and 87% retention in capacity on operation for 5000 cycles. In addition, in situ oxidative polymerization [55] of PPY along with functionalized graphene sheets produces high-quality electrodes. Similarly, a PPy-GO nanoelectrode with core-shell architecture [56] exhibits C_{mp} up to 370 F g^{-1} with ~91% retention in capacity on operation for 4000 cycles in supercapacitors.

5.4.1.3 Poly(3,4-ethylenedioxythiophene)-Carbon Composites

Literature [57] reports that PEDOT-GO composites exhibited a C_{mp} of 115.15 F g^{-1} at 1.2 V with E_{md} and P_{md} of 13.60 Wh kg^{-1} and 139.09 W kg^{-1} and the obtained value is higher than composites of PEDOT:PSS incorporated with 1D MWCNT. Similarly, a deformable and conductive PEDOT:PSS-rGO film [58] achieved a C_{sp} of 448 mF cm^{-2} without a degradation in performance even after 1000 times of its operation. The PEDOT:PSS-rGO composite fibers also find suitable applications in textile structures [59] by providing a C_{sp} of 304.5 mF cm^{-2} due to the hollow nature of the fibers, which could enhance the capacitance from their interior surfaces. Poly(4-styrene sulfonate), PSS is used along with PEDOT to create a blend with CPs which increases their surface area and provides an uninterrupted network which on compositing with rGO, called PEDOT:PSS-rGO, provides excellent cyclic stability of 96% after 10,000 cycles. Similarly, a flexible supercapacitor [60] using PEDOT:PSS and reduced rGO as nanoelectrode was fabricated using ethylene glycol (EG). Secondary dopant (EG) along with rGO improves the conductivity and the composite film exhibited C_{mp} of 174 F g^{-1} and E_{md} of 810 Wh kg^{-1} with ~90% retention in capacity on operation for 5000 cycles in supercapacitors.

5.4.2 CONDUCTING POLYMER—TRANSITION METAL COMPOSITES

Although the 2D materials such as TMOs, TMCs, and TMDs provide large surface area and pseudocapacitance, they are not suitable for direct use due to their high restacking ability, which degrades the overall performance. But the addition of CPs to TMOs/TMCs/TMDs prevents them from restacking and improves the transport of charges. Similarly, TMOs are added either as fibers or nanoparticles to the CPs to enhance their performance in supercapacitors. The interfacial contact between them is necessary and the transition metal compounds provide strong support, which interconnects the polymer chains of CPs to stabilize the composites and thereby provide high C_{sp} values. The capacitive performance is due to pseudocapacitive (Faradaic) characteristics of 2D materials and CPs.

5.4.2.1 Polyaniline-Transition Metal Composites

A composite of polyaniline-Ti$_3$C$_2$T$_x$ (PANi-MXene) electrode prepared using chemical oxidative polymerization [61] of aniline monomers over Ti$_3$C$_2$T$_x$ surface proved effective in creating a highly conductive open structure which could transfers the ions and electron effectively with increased electrochemical kinetics, thereby exhibiting a maximum C_{mp} of 556.2 F g^{-1} with ~79% rate capability and retention in capacity to ~92% even after 5000 cycles of operation. Similarly, it is also reported in literature [62] that PANi helps increase the spacing between $Ti_3C_2T_x$ layers, which effectively transfers the ions between the electrode/electrolyte interface with improvement in the overall conductivity of the material. Due to these factors, a C_{mp} of 371 F g^{-1} at scan rates of 2 mV s^{-1} was observed over a thick electrode (up to 45 microns) and the output capacitance was retained even after 10,000 cycles, making it useful for portable or self-powered devices. TMDs have different phases due to which the specific capacitance changes. For example, pristine MoS_2 can be present in three different phases, giving rise to various specific capacitance values. A composite electrode [63] prepared through a chemical in situ polymerization technique of PANi with MoS_2 (along with 1D carbon) showed an E_{md} of 0.013 Wh cm^{-3} and a P_{md} of 1.000 W cm^{-3}, thereby providing an unusual volumetric energy density for flexible supercapacitors.

5.4.2.2 Polypyrrole-Transition Metal Composites

A PPy-$Ti_3C_2T_x$ (PPy-MXene) composite fabricated by polymerization [64] of pyrrole monomers on the surface of $Ti_3C_2T_x$ provided a C_{mp} of 416 F g^{-1}. The directional/orientational growth of PPy was due to hydrogen bonds which is caused by the presence of surface-active groups on $Ti_3C_2T_x$ forming an alternating stacked structure. The intercalation of PPy improves the spacing between layers of $Ti_3C_2T_x$ and the oriented structure sustains a high conductivity without any losses. PPY-MoS_2 nanocomposites [65] displayed a C_{mp} of 695 F g^{-1} which is greater than the individual components of the composite and the PPY-MoS_2 nanocomposite exhibited E_{md} and P_{md} of 83.3 Wh kg^{-1} at 3332 W kg^{-1} making it suitable for ultrathin supercapacitors. The enhanced capacitance is due to metallic MoS_2 clubbed with the properties of PPy, which exhibits a large area. Similarly, a ternary hybrid supercapacitor [66], PPy-PANi-MOS$_2$ ternary nanocomposite exhibited a larger C_{mp} of 1273 F g^{-1} with coulombic efficiency above 92%, indicating good electrochemical reversibility.

5.4.2.3 Poly(3,4-ethylenedioxythiophene)-Transition Metal Composites

A high-performance ultra-thin flexible solid-state supercapacitor [67] fabricated using a composite electrode of $Mo_{1.33}$C MXene with PEDOT:PSS delivered a maximum C_{vp} of 568 F cm^{-3} along with an ultrahigh energy density of 33.2 mWh cm^{-3} and a power density of 19470 mW cm^{-3}. The enhanced capacitance and stability are due to the synergistic effect of interlayer spacing between $Mo_{1.33}$C MXene layers with the insertion of PEDOT and redox active sites at the surface of PEDOT and the MXene. Similarly, a composite electrode of PEDOT-MoS_2 [68] provided a C_{mp} of 405 F g^{-1}, which is about greater than that of a pure PEDOT by four times along with excellent improvement in mechanical properties and capacity retention, ~90% after 1000 cycles. Similarly, a nanocomposite material [69] consisting of layered vanadium pentoxide (LVO) and layered manganese oxide (LMO, birnessite-type MnO_2), together with PEDOT, in a sandwich structure of LVO-PEDOT-LMO showed a high E_{md} of 39.2 Wh kg^{-1} with high-rate capability (21.7 Wh kg^{-1} at 2.2 kW kg^{-1}) and good cycle stability (93.5% capacitance retention after 3000 cycles). The usage of PEDOT-MnO_2 electrode in flexible supercapacitor as the positive electrode exhibits a C_{sp} of 60 mF cm^{-2} and a E_{md} of 0.0335 mWh cm^{-2} with good flexibility indicating its applicability in wearable electronics [70].

5.5 CONCLUSIONS

CPs show high electrical conductivity and good charge density with low cost, but the structural destruction (swelling/shrinking) of CPs during the charging/discharging process degrades the cycling stability and rate capability which in turn affects their performance in supercapacitors. The inclusion of 2D materials of carbon—graphene, GO or rGO are proposed to maintain the electrochemical stability of CPs in nanocomposite electrodes. In addition to 2D carbon, reduced dimension of 2D structures like transition metal compounds (TMOs, TMCs, TMDs), increase the redox active sites on the surface of the electrode, and the 2D in-plane structure also promotes the diffusion of the electrolyte ions and charge transport kinetics at the high rates, thereby leading to outstanding electrochemical kinetics. The properties of poly(3,4-ethylenedioxythiophene) CP were greatly enhanced by fabricating composites with 2D materials (rGO, TMO/TMD/TMC) as compared to composites of CPs with PANi or PPy. The improvement in performance and cyclic stability was predominantly seen in composites of poly(3,4-ethylenedioxythiophene)-poly(4-styrene sulfonate), PEDOT:PSS with rGO (with ethylene glycol as secondary dopant) where energy density of 810 Wh kg^{-1} with 90% capacitance retention over 5000 cycles was recorded. The PEDOT composites with layered manganese/vanadium oxides showed a high energy density of 39.2 Wh kg^{-1} with high-rate capability (21.7 Wh kg^{-1} at 2.2 kW kg^{-1}) and good cycle stability (93.5% capacitance retention after 3000 cycles) for flexible solid-state supercapacitors. The properties and performance of PEDOT composites for wearable electronics need to be further improved in terms of cyclic stability and capacitance as observed for composites. Hence, incorporating CPs with 2D materials is necessary to enhance the ionic, electrical conductivity, and electrochemical stability by the synergistic effect of different components, thereby improving the overall performance of the nanocomposite electrodes for application in micro-supercapacitors, compressible supercapacitors, electrochromic supercapacitors, and self-healable supercapacitors.

REFERENCES

1. C. Zhao, X. Jia, K. Shu, C. Yu, G. G. Wallace, C. Wang, Conducting polymer composites for unconventional solid-state supercapacitors, Journal of Materials Chemistry A 8 (9) (2020) 4677–4699.
2. K. Sharma, A. Arora, S. K. Tripathi, et al., Review of supercapacitors: materials and devices, Journal of Energy Storage 21 (2019) 801–825.
3. Z. Wang, M. Zhu, Z. Pei, Q. Xue, H. Li, Y. Huang, C. Zhi, Polymers for supercapacitors: boosting the development of the flexible and wearable energy storage, Materials Science and Engineering: R: Reports 139 (2020) 100520.
4. M. A. A. M. Abdah, N. H. N. Azman, S. Kulandaivalu, Y. Sulaiman, Review of the use of transition-metal-oxide and conducting polymer-based fibres for high-performance supercapacitors, Materials & Design 186 (2020) 108199.
5. S. Arunachalam, B. Kirubasankar, D. Pan, H. Liu, C. Yan, Z. Guo, S. Angaiah, Research progress in rare earths and their composites-based electrode materials for supercapacitors, Green Energy & Environment 5 (3) (2020) 259–273.
6. G. A. Snook, P. Kao, A. S. Best, Conducting-polymer-based supercapacitor devices and electrodes, Journal of Power Sources 196 (1) (2011) 1–12.
7. W. G. Pell, B. E. Conway, Voltammetry at a de levie brush electrode as a model for electrochemical supercapacitor behavior, Journal of Electroanalytical Chemistry 500 (1–2) (2001) 121–133.
8. Y. Zhao, M. Liu, L. Gan, X. Ma, D. Zhu, Z. Xu, L. Chen, Ultra-micro porous carbon nanoparticles for the high-performance electrical double-layer capacitor electrode, Energy & Fuels 28 (2) (2014) 1561–1568.
9. Q. Ruibin, H. Zhongai, Y. Yuying, L. Zhimin, A. Ning, R. Xiaoying, H. Haixiong, W. Hongying, Monodisperse carbon microspheres derived from potato starch for asymmetric supercapacitors, Electrochimica Acta 167 (2015) 303–310.
10. D. Zhang, J. Zhao, C. Feng, R. Zhao, Y. Sun, T. Guan, B. Han, N. Tang, J. Wang, K. Li, et al., Scalable synthesis of hierarchical macropore-rich activated carbon microspheres assembled by carbon nanoparticles for high-rate performance supercapacitors, Journal of Power Sources 342 (2017) 363–370.
11. J. B. Baxter, E. S. Aydil, Nanowire-based dye-sensitized solar cells, Applied physics letters 86 (5) (2005) 053114.
12. J. Jiang, Y. Zhang, P. Nie, G. Xu, M. Shi, J. Wang, Y. Wu, R. Fu, H. Dou, X. Zhang, Progress of nano-structured electrode materials for supercapacitors, Advanced Sustainable Systems 2 (1) (2018) 1700110.
13. P. Pachfule, D. Shinde, M. Majumder, Q. Xu, Fabrication of carbon nanorods and graphene nanoribbons from a metal–organic framework, Nature chemistry 8 (7) (2016) 718–724.
14. L. Zhang, Y. Jiang, L. Wang, C. Zhang, S. Liu, Hierarchical porous carbon nanofibers as binder-free electrode for high-performance supercapacitor, Electrochimica Acta 196 (2016) 189–196.
15. G. Xu, B. Ding, P. Nie, L. Shen, J. Wang, X. Zhang, Porous nitrogen-doped carbon nanotubes derived from tubular polypyrrole for energy-storage applications, Chemistry—A European Journal 19 (37) (2013) 12306–12312.
16. M. Zhi, C. Xiang, J. Li, M. Li, N. Wu, Nanostructured carbon–metal oxide composite electrodes for supercapacitors: A review, Nanoscale 5 (1) (2013) 72–88.

17. G. Xu, Q.-b. Yan, A. Kushima, X. Zhang, J. Pan, J. Li, Conductive graphene oxide-polyacrylic acid (GOPAA) binder for lithium-sulfur battery, Nano Energy 31 (2017) 568–574.
18. C. Tan, X. Cao, X.-J. Wu, Q. He, J. Yang, X. Zhang, J. Chen, W. Zhao, S. Han, G.-H. Nam, et al., Recent advances in ultrathin two-dimensional nanomaterials, Chemical reviews 117 (9) (2017) 6225–6331.
19. Y. Zhang, J. Mei, C. Yan, T. Liao, J. Bell, Z. Sun, Bioinspired 2D nanomaterials for sustainable applications, Advanced Materials 32 (18) (2020) 1902806.
20. J. Xia, F. Chen, J. Li, N. Tao, Measurement of the quantum capacitance of graphene, Nature nanotechnology 4 (8) (2009) 505–509.
21. J. P. Mensing, C. Poochai, S. Kerdpocha, C. Sriprachuabwong, A. Wisitsoraat, A. Tuantranont, Advances in research on 2D and 3D graphene-based supercapacitors, Advances in Natural Sciences: Nanoscience and Nanotechnology 8 (3) (2017) 033001.
22. S. He, W. Chen, 3D graphene nanomaterials for binder-free supercapacitors: Scientific design for enhanced performance, Nanoscale 7 (16) (2015) 6957–6990.
23. K. Lota, V. Khomenko, E. Frackowiak, Capacitance properties of poly (3,4-ethylenedioxythiophene)/carbon nanotubes composites, Journal of Physics and Chemistry of Solids 65 (2–3) (2004) 295–301.
24. X. Cao, H.-Y. Zeng, S. Xu, J. Yuan, J. Han, G.-F. Xiao, Facile fabrication of the polyaniline/layered double hydroxide nanosheet composite for supercapacitors, Applied Clay Science 168 (2019) 175–183.
25. M. Padmini, P. Elumalai, P. Thomas, Symmetric supercapacitor performances of $CaCu_3Ti_4O_{12}$ decorated polyaniline nanocomposite, Electrochimica Acta 292 (2018) 558–567.
26. B. Sari, M. Talu, F. Yildirim, Electrochemical polymerization of aniline at low supporting-electrolyte concentrations and characterization of obtained films, Russian Journal of Electrochemistry 38 (7) (2002) 707–713.
27. M. Wu, G. A. Snook, V. Gupta, M. Shaffer, D. J. Fray, G. Z. Chen, Electrochemical fabrication, and capacitance of composite films of carbon nanotubes and polyaniline, Journal of Materials Chemistry 15 (23) (2005) 2297–2303.
28. K. S. Ryu, K. M. Kim, Y. J. Park, N.-G. Park, M. G. Kang, S. H. Chang, Redox supercapacitor using polyaniline doped with li salt as electrode, Solid State Ionics 152 (2002) 861–866.
29. C. J. Raj, B. C. Kim, W.-J. Cho, W.-g. Lee, S.-D. Jung, Y. H. Kim, S. Y. Park, K. H. Yu, Highly flexible and planar supercapacitors using graphite flakes/polypyrrole in polymer lapping film, ACS Applied Materials & Interfaces 7 (24) (2015) 13405–13414.
30. Y. Huang, H. Li, Z. Wang, M. Zhu, Z. Pei, Q. Xue, Y. Huang, C. Zhi, Nanostructured polypyrrole as a flexible electrode material of supercapacitor, Nano Energy 22 (2016) 422–438.
31. S. Suematsu, Y. Oura, H. Tsujimoto, H. Kanno, K. Naoi, Conducting polymer films of cross-linked structure and their QCM analysis, Electrochimica Acta 45 (22–23) (2000) 3813–3821.
32. K. Zhang, J. Xu, X. Zhu, L. Lu, X. Duan, D. Hu, L. Dong, H. Sun, Y. Gao, Y. Wu, Poly(3,4-ethylenedioxythiophene) nanorods grown on graphene oxide sheets as electrochemical sensing platform for rutin, Journal of Electroanalytical Chemistry 739 (2015) 66–72.
33. G. A. Snook, C. Peng, D. J. Fray, G. Z. Chen, Achieving high electrode specific capacitance with materials of low mass specific capacitance: Potentiostatically grown thick micro-nanoporous PEDOT films, Electrochemistry Communications 9 (1) (2007) 83–88.
34. R. Liu, S. I. Cho, S. B. Lee, Poly (3, 4-ethylenedioxythiophene) nanotubes as electrode materials for a high-powered supercapacitor, Nanotechnology 19 (21) (2008) 215710.
35. C. Tang, Z. Tang, H. Gong, Hierarchically porous Ni-Co oxide for high reversibility asymmetric full-cell supercapacitors, Journal of the Electrochemical Society 159 (5) (2012) A651.
36. J. Qu, L. Shi, C. He, F. Gao, B. Li, Q. Zhou, H. Hu, G. Shao, X. Wang, J. Qiu, Highly efficient synthesis of graphene/MnO_2 hybrids and their application for ultrafast oxidative decomposition of methylene blue, Carbon 66 (2014) 485–492.
37. S. Zhang, Y. Pang, Y. Wang, B. Dong, S. Lu, M. Li, S. Ding, NiO nanosheets anchored on honeycomb porous carbon derived from wheat husk for symmetric supercapacitor with high performance, Journal of Alloys and Compounds 735 (2018) 1722–1729.
38. F. Liu, H. Su, L. Jin, H. Zhang, X. Chu, W. Yang, Facile synthesis of ultrafine cobalt oxide nanoparticles for high-performance supercapacitors, Journal of Colloid and Interface Science 505 (2017) 796–804.
39. N. Iqbal, X. Wang, J. Ge, J. Yu, H.-Y. Kim, S. S. Al-Deyab, M. El-Newehy, B. Ding, Cobalt oxide nanoparticles embedded in flexible carbon nanofibers: Attractive material for supercapacitor electrodes and CO_2 adsorption, RSC Advances 6 (57) (2016) 52171–52179.
40. J. Zhou, X. Zha, F. Y. Chen, Q. Ye, P. Eklund, S. Du, Q. Huang, A two-dimensional zirconium carbide by selective etching of Al_3C_3 from nano laminated $Zr_3Al_3C_5$, Angewandte Chemie International Edition 55 (16) (2016) 5008–5013.

41. B. Anasori, M. R. Lukatskaya, Y. Gogotsi, 2D metal carbides and nitrides (Mxenes) for energy storage, Nature Reviews Materials 2 (2) (2017) 1–17.
42. M. Ghidiu, M. R. Lukatskaya, M.-Q. Zhao, Y. Gogotsi, M. W. Barsoum, Conductive two-dimensional titanium carbide clay with high volumetric capacitance, Nature 516 (7529) (2014) 78–81.
43. M. R. Lukatskaya, S.-M. Bak, X. Yu, X.-Q. Yang, M. W. Barsoum, Y. Gogotsi, Probing the mechanism of high capacitance in 2D titanium carbide using in situ x-ray absorption spectroscopy, Advanced Energy Materials 5 (15) (2015) 1500589.
44. Y. Da, J. Liu, L. Zhou, X. Zhu, X. Chen, L. Fu, Engineering 2D architectures toward high-performance micro-supercapacitors, Advanced Materials 31 (1) (2019) 1802793.
45. Q. Zhang, W. Wang, J. Zhang, X. Zhu, L. Fu, Thermally induced bending of ReS_2 nano walls, Advanced Materials 30 (3) (2018) 1704585.
46. J. Feng, X. Sun, C. Wu, L. Peng, C. Lin, S. Hu, J. Yang, Y. Xie, Metallic few-layered VS_2 ultrathin nanosheets: High two-dimensional conductivity for in-plane supercapacitors, Journal of the American Chemical Society 133 (44) (2011) 17832–17838.
47. Q. Ji, C. Li, J. Wang, J. Niu, Y. Gong, Z. Zhang, Q. Fang, Y. Zhang, J. Shi, L. Liao, et al., Metallic vanadium disulfide nanosheets as a platform material for multifunctional electrode applications, Nano Letters 17 (8) (2017) 4908–4916.
48. S. Liu, Y. Zeng, M. Zhang, S. Xie, Y. Tong, F. Cheng, X. Lu, Binder-free WS_2 nanosheets with enhanced crystallinity as a stable negative electrode for flexible asymmetric supercapacitors, Journal of Materials Chemistry A 5 (40) (2017) 21460–21466.
49. K. Momma, F. Izumi, Vesta 3 for three-dimensional visualization of crystal, volumetric and morphology data, Journal of Applied Crystallography 44 (6) (2011) 1272–1276.
50. L. Liu, Z. Niu, L. Zhang, W. Zhou, X. Chen, S. Xie, Nanostructured graphene composite papers for highly flexible and foldable supercapacitors, Advanced Materials 26 (28) (2014) 4855–4862.
51. W. Xinming, W. Qiguan, Z. Wenzhi, W. Yan, C. Weixing, Enhanced electrochemical performance of hydrogen-bonded graphene/polyaniline for electro-chromo-supercapacitor, Journal of Materials Science 51 (16) (2016) 7731–7741.
52. H. Cao, X. Zhou, Y. Zhang, L. Chen, Z. Liu, Microspherical polyaniline/graphene nanocomposites for high performance supercapacitors, Journal of Power Sources 243 (2013) 715–720.
53. C. Gao, J. Gao, C. Shao, Y. Xiao, Y. Zhao, L. Qu, Versatile origami micro-supercapacitors array as a wind energy harvester, Journal of Materials Chemistry A 6 (40) (2018) 19750–19756.
54. Y. Ge, C. Wang, K. Shu, C. Zhao, X. Jia, S. Gambhir, G. G. Wallace, A facile approach for fabrication of mechanically strong graphene/polypyrrole films with large areal capacitance for supercapacitor applications, RSC Advances 5 (124) (2015) 102643–102651.
55. H. P. De Oliveira, S. A. Sydlik, T. M. Swager, Supercapacitors from free-standing polypyrrole/graphene nanocomposites, The Journal of Physical Chemistry C 117 (20) (2013) 10270–10276.
56. W. Wu, L. Yang, S. Chen, Y. Shao, L. Jing, G. Zhao, H. Wei, Core–shell nano spherical polypyrrole/graphene oxide composites for high performance supercapacitors, RSC Advances 5 (111) (2015) 91645–91653.
57. N. H. N. Azman, H. N. Lim, Y. Sulaiman, Effect of electro-polymerization potential on the preparation of PEDOT/graphene oxide hybrid material for supercapacitor application, Electrochimica Acta 188 (2016) 785–792.
58. Y. Liu, B. Weng, J. M. Razal, Q. Xu, C. Zhao, Y. Hou, S. Seyedin, R. Jalili, G. G. Wallace, J. Chen, High-performance flexible all-solid-state supercapacitor from large free-standing Graphene-PEDOT/PSS films, Scientific Reports 5 (1) (2015) 1–11.
59. G. Qu, J. Cheng, X. Li, D. Yuan, P. Chen, X. Chen, B. Wang, H. Peng, A fiber supercapacitor with high energy density based on hollow graphene/conducting polymer fiber electrode, Advanced Materials 28 (19) (2016) 3646–3652.
60. S. Khasim, A. Pasha, N. Badi, M. Lakshmi, Y. K. Mishra, High performance flexible supercapacitors based on secondary doped PEDOT–PSS–Graphene nanocomposite films for large area solid state devices, RSC Advances 10 (18) (2020) 10526–10539.
61. H. Xu, D. Zheng, F. Liu, W. Li, J. Lin, Synthesis of an MXene/polyaniline composite with excellent electrochemical properties, Journal of Materials Chemistry A 8 (12) (2020) 5853–5858.
62. A. Vahid Mohammadi, J. Moncada, H. Chen, E. Kayali, J. Orangi, C. A. Carrero, M. Beidaghi, Thick and freestanding Mxene/PANI pseudocapacitive electrodes with ultrahigh specific capacitance, Journal of Materials Chemistry A 6 (44) (2018) 22123–22133.
63. I.-W. P. Chen, Y.-C. Chou, P.-Y. Wang, Integration of ultrathin MoS_2/PANI/CNT composite paper in producing all-solid-state flexible supercapacitors with exceptional volumetric energy density, The Journal of Physical Chemistry C 123 (29) (2019) 17864–17872.

64. M. Boota, B. Anasori, C. Voigt, M.-Q. Zhao, M. W. Barsoum, Y. Gogotsi, Pseudocapacitive electrodes produced by oxidant-free polymerization of pyrrole between the layers of 2D titanium carbide (MXene), Advanced Materials 28 (7) (2016) 1517–1522.

65. H. Tang, J. Wang, H. Yin, H. Zhao, D. Wang, Z. Tang, Growth of polypyrrole ultrathin films on MoS_2 monolayers as high-performance supercapacitor electrodes, Advanced Materials 27 (6) (2015) 1117–1123.

66. K. Wang, L. Li, Y. Liu, C. Zhang, T. Liu, Constructing a "pizza-like" MoS_2/polypyrrole/polyaniline ternary architecture with high energy density and superior cycling stability for supercapacitors, Advanced Materials Interfaces 3 (19) (2016) 1600665.

67. L. Qin, Q. Tao, A. El Ghazaly, J. Fernandez-Rodriguez, P. O. Persson, J. Rosen, F. Zhang, High-performance ultrathin flexible solid-state supercapacitors based on solution processable $Mo_{1.33}C$ MXene and PEDOT:PSS, Advanced Functional Materials 28 (2) (2018) 1703808.

68. J. Wang, Z. Wu, H. Yin, W. Li, Y. Jiang, Poly(3,4-ethylenedioxythiophene)/MoS_2 nanocomposites with enhanced electrochemical capacitance performance, RSC Advances 4 (100) (2014) 56926–56932.

69. C. X. Guo, G. Yilmaz, S. Chen, S. Chen, X. Lu, Hierarchical nanocomposite composed of layered V_2O_5/PEDOT/MnO_2 nanosheets for high-performance asymmetric supercapacitors, Nano Energy 12 (2015) 76–87.

70. J. Sun, Y. Huang, C. Fu, Y. Huang, M. Zhu, X. Tao, C. Zhi, H. Hu, A high performance fiber-shaped PEDOT@ MnO_2//C@ Fe_3O_4 asymmetric supercapacitor for wearable electronics, Journal of Materials Chemistry A 4 (38) (2016) 14877–14883.

6 Conducting Polymer-Based Flexible Supercapacitors

Kwadwo Mensah-Darkwa[1], Frank Ofori Agyemang[1], Daniel Nframah Ampong[1], Felipe M. de Souza[2], and Ram K. Gupta[2]

[1]Department of Materials Engineering, College of Engineering, Kwame Nkrumah, and University of Science and Technology, Kumasi, Ghana

[2]Department of Chemistry, Kansas Polymer Research Center, Pittsburg State University, Pittsburg, Kansas, USA

CONTENTS

6.1 INTRODUCTION

Energy is an important element in everyday life; without it, economic activities, health, personal and social well-being cannot take place. Energy provides essential and fundamental services such as heating, cooling, mobility, lighting, cooking, and the operation of appliances in every sector of life

DOI: 10.1201/9781003150374-6

in every country or society. The world's economy, industry, and civilization have been dependent on energy and its infrastructure over the years. Because of that, the increase of the global population pushes technological advancements to improve the quality of life. This can be done by providing sufficient, efficient, reliable, and low-cost energy to improve society's welfare, living standards, and sustainable development. The world reserves of fossil fuels, coal, oil, and gas (conventional sources) are limited and may not be able to power the increasing demand for energy for much longer since they are expected to be depleted within few decades, and hence, requires alternate energy sources. The development of the world's economy is usually related to increasing levels of emissions of greenhouse gases. As a result of industrialization, the growing concern of climate change caused by these emissions poses a new challenge to researchers in reducing CO_2 emissions. To mitigate the effects of climate change, energy efficiency with green energy technology has a key role to play as the energy sector accounts for about 65% of total global greenhouse gas emissions. There is a strong and growing focus on the development and implementation of renewable energy and energy storage technologies. Since most renewable energy sources are intermittent, the ability to store electricity is important, especially in small-scale systems. With energy storage systems, energy can be produced whenever conditions are favorable and then consumed when needed. Energy systems operating with variable renewable sources, in particular, require energy storage devices as an essential component to enable its integration [1].

The quality of the power supply is also a very critical issue considering the fast increase in electric loads, which requires reliable power throughout the year. Hence, the need to develop technologies able to harvest energy that can be sustainable, clean, and be efficiently stored is much needed, particularly with non-conventional energy sources. Batteries, fuel cells, and electrochemical supercapacitors (ESs) are important technologies that can help on overcoming the challenges regarding energy demand. Batteries such as lead-acid and lithium secondary batteries can convert electricity into chemical energy working as energy buffers and are used to meet the energy requirements. However, these are not able to cover fluctuations of fast power without reducing its cycle life. Lead batteries are used in the automobile industries and other industrial applications for utility energy storage for efficient stability in electricity networks; Li batteries are advantageous in terms of energy density and specific energy. Batteries have relatively low charge/discharge efficiencies, shorter life span, and are not eco-friendly. Relatively high energy density storage, fast charging, long charge/discharge cycles of electrochemical supercapacitors, also called supercapacitors have found wide applications as compared to conventional capacitors and batteries, although high manufacturing cost/performance ratio poses a greater challenge [2].

The superior properties of supercapacitors have made them the most important components in energy storage systems. Research geared toward new engineering materials will contribute to reducing manufacturing costs which will enable their use for commercial applications. Supercapacitors consist of two electrodes, a separator, and an electrolyte. The energy is stored between the electrodes and electrolytes by charge transfer. Carbon-based materials such as activated carbon (AC) obtained from biomass are convenient components to build electrochemical capacitors due to their high surface area and energy density. Supercapacitors are usually divided into two categories. The first is the electrical double-layer capacitor (EDLC) which has the charges separated between the interface of the AC electrode's surface and the electrolyte. The second is the redox capacitor which depends on a high reversible redox reaction that occurs either on the electrode surface or inside them [3].

Furthermore, a fuel cell is a technology used for generating power using an electrochemical reaction. Unlike a battery, fuel cells allow for a continuous supply of fuel and electricity. It can also be used for reversible energy storage for later use where excess electricity generated during off-peak hours is used to electrolyze water to produce fuel for the cell reaction. It can be used as energy storage as well as power conversion, which are the features for cost reduction and performance optimization that serve many mobile and stationary applications.

Among the available energy storage technologies, supercapacitors have demonstrated excellent storage properties. Many next-generation portable and wearable consumer electronic devices require lightweight, flexible, transparent, and environment-friendly energy storage systems. Flexible supercapacitors have proven to be a promising storage device for energy that can be used for compact military, medical, and civilian applications due to their excellent power densities. With these features, they can be incorporated into flexible electrode materials with carbon networks which can use an electric double layer or the fast redox reaction mechanism for energy storage. Using flexible carbon networks eliminates the necessity to use additional current collectors, binders, and conductive additives since they are highly conductive, flexible, and binder-free. A major setback for this technology for commercial applications is the high cost of production and low energy density. Hence, research should be directed toward integrating cost-effective active materials used in pseudocapacitors with the EDLCs to obtain high performance in flexible supercapacitors [4].

6.2 MATERIALS AND MECHANISM OF SUPERCAPACITORS

The materials used for the components of the electrochemical supercapacitor are essential to the performance of electrochemical energy devices. Supercapacitor technology has been around for over six decades, and in recent years, the Research and Development (R&D) departments in industries and educational institutes have been more interested in the development of advanced electrode materials and electrolytes. A considerable amount of literature has been published on electrode materials and electrolytes [5]. In a parallel among batteries, fuel cells, and ES, the latter has a low competitive energy density; however, their power density and cycle life are better. For this technology to permeate the R&D of the energy storage device market, researchers must focus on improving energy density. The supercapacitor's energy density (E) is calculated by the expression $E = 1/2\ CV^2$, which means to manipulate energy density, one must investigate the capacitance (C) and the cell voltage (V), which directly affects its electrochemical performances. From the equation, it can be seen that the voltage window is essential to the increase of the energy density. The capacitance and energy density are influenced by the material that composes the electrode. However, increasing the voltage contributes more to the energy density, as it is proportional to the power squared of the cell voltage.

Various attempts have been made to modify the electrochemical properties of supercapacitors. The main routes that have been the focus of research include increasing the specific capacitance by using novel carbon-based materials to increase the specific area; also, the development of electroactive conducting polymer and transition-metal oxides has contributed to the improvement in pseudocapacitive behavior. Finally, the use of improved electrolytes has contributed to improving the voltage window.

6.2.1 Types of Supercapacitors

Supercapacitors are categorized according to the principle of energy storage. There are two mechanisms by which it can store electricity, which depend on the electrode's material as well as its structure. Electrodes can be non-Faradaic or Faradaic. The non-Faradaic is based on an electrostatic process. Hence, there is only the physical adsorption/desorption process that takes place at the electrode's interface. The Faradaic capacitors store energy through the electrochemical redox process. Figure 6.1 shows the various classifications of supercapacitors.

6.2.1.1 Electrical Double-Layer Capacitors

EDLCs are electrochemical storage devices based on electrostatic storage capacity, used in situations that require high power and current, long-cycle times along with fast charge/discharge process.

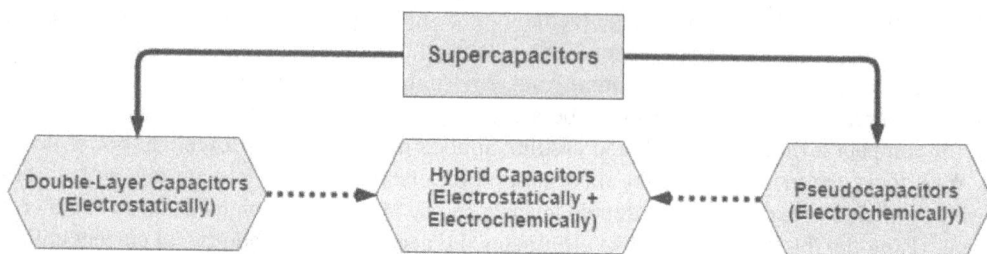

FIGURE 6.1 Classification of supercapacitors.

The charge storage mechanism is associated with the appearance of an electric double-layer (EDL) at the interface between electrode and electrolyte, described by Helmholtz, Gouy-Chapman, and Gouy-Chapman-Stern models. When charged, the charge carriers accumulate on the interface between the electrode and electrolyte, forming the EDL consisting of a compact layer (Helmholtz) and a diffuse layer. The concentration of electrolyte and ion radius influences the double layer thickness. Charge storage in EDLCs is highly dependent on the physical and chemical properties, pore aspects (i.e. distribution and size), and surface area of the electrode material. The chemical stability, high electrical conductivity, and tunable porosity of carbonaceous materials make them the best candidates for EDLCs. Novel carbon structures include AC, CNF, carbon aerogel, graphite, graphene, and carbon nanotubes (CNTs). The morphology and surface characteristics of carbon materials can be modified to increase their specific surface area and accessible surface area, thus, improving their surface chemistry, properties, and efficiency.

6.2.1.2 Pseudocapacitors

Pseudocapacitors are SCs based on Faradaic redox reactions related to the transfer of charges in the electrode/electrolyte interface. The charge transfer is voltage-dependent and storage occurs by redox reactions and intercalation processes. Materials with pseudocapacitive properties include conducting polymers and metal oxides like Ru_2O, MnO_2, and Co_3O_4. The Faradaic process occurs in three possible paths: (a) by the electrolyte adsorbed on the metal surface, (b) by the reversible redox process in the bulk of the transition metal hydroxides or oxides, and (c) by the reversible doping/de-doping process that bulk conducting polymers can perform. Hence, the Faradaic and non-Faradaic processes define the differences between the pseudocapacitance and EDLCs capacitance, respectively. The Faradaic supercapacitor exhibits larger capacitance values and energy density compared to an EDLC.

6.2.1.3 Hybrid Supercapacitors

Despite the respective advancements with EDLCs and pseudocapacitors, they each suffer from drawbacks that limit their electrochemical performance. Pristine transitional metal oxides and conductive polymers suffer from capacitive decay, low conductivity, and structural instability. Furthermore, EDLCs have low energy density (~5 Wh/kg) because their storage mechanism is electrostatic. For these reasons, carbons are combined with pseudocapacitive electrode materials to form hybrid SCs with improved electrochemical performance. Hybrid supercapacitors combine pseudocapacitance with double-layer capacitance resulting in rated voltage, high specific energy as well as capacitance. The conductive carbon backbone supports the pseudocapacitive particles and provides a conductive path for charge transport across the electrode. Additionally, the high specific surface area and porosity of carbons favor electrolytic diffusion, ion transport, and reduce electrolyte/electrode and nanoparticles contact resistance. At the same time, pseudocapacitive reversible redox reactions during the charging/discharging increase the overall capacitance. Exceptional hybrids can be obtained with storage capacities as high as 680 mAh/g.

6.2.2 CHARGE STORAGE MECHANISMS IN SUPERCAPACITORS

A supercapacitor has a higher capacitance value than conventional capacitors with much lower voltage limits, which stores more energy per unit of volume or mass and has much more charge and discharge capabilities. These set-up functional applications have been achieved through efforts in the development of refined devices to store energy that can deal with the overwhelming demand for stable and efficient energy. Supercapacitors combine the mechanisms of both electrostatic double-layer as well as pseudocapacitance, differing from the conventional solid dielectric [6]. Three categories of supercapacitors are known: electrochemical double-layer supercapacitors, pseudo supercapacitors, and hybrid supercapacitors. The first two types are discussed in this chapter.

6.2.2.1 Electrostatic Double-Layer Capacitance

Electrical energy can be stored by using carbon electrodes to separate the charge in the Helmholtz double layer formed at the interface between the electrode and electrolyte, where ions are arranged and the surface of the conducting electrode, where electric charges accumulate. The origin of the separation of charge is static, and charge distance is of the order of 3–8 Å between the double-layer [7]. The electrode materials should have a high superficial area and porous structure sufficiently high for charge accumulation and rapid ion motion, respectively. Pore structure which is randomly connected and ranging from micropores (<2 nm ϕ) to macropores limits the application of the double-layer capacitor even if it is made of conventional carbon with a large active area. Graphitic carbon, having satisfactory electrical conduction, open porosity, and electrochemical stability makes it an interesting electrode material for this application [8]. The specific capacitance has a close relationship with the specific surface area that is available to the electrolyte, density of the electrode material, and the interfacial double-layer capacitance. The charge and discharge cycles of the electrostatic double-layer capacitors are of a higher rate of more than one million lifetime cycles since it is not limited by the kinetics of the electrochemical charge transfer. Figure 6.2 shows schematics of the double-layer capacitor and its functions [9].

Figure 6.2(a) shows the area of two electrode surfaces with a liquid electrolyte in contact with the electrode's metallic surface, which forms a short electric double layer when voltage is applied. This double-layer is separated by polarized solvent molecules with negative ions in the electrode's surface and solvated cations in the liquid electrolyte, held together by electrostatic forces with static

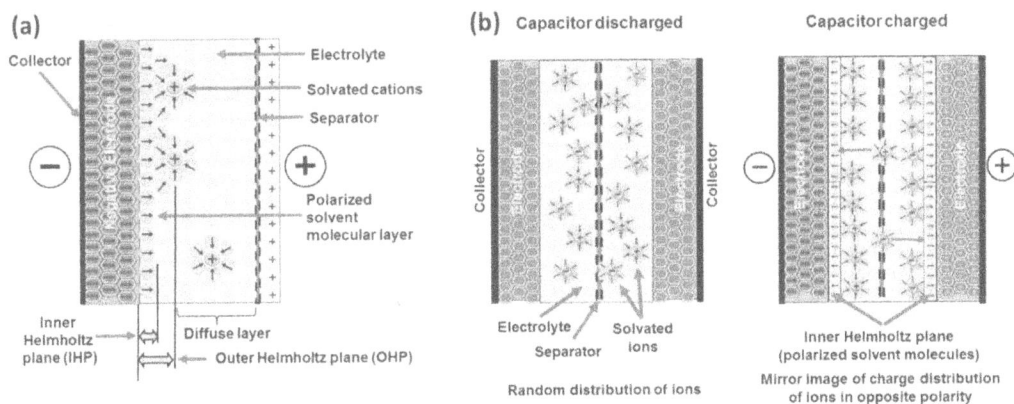

FIGURE 6.2 Schematics of double-layer capacitor: (a) simplified view showing the electrode, liquid electrolyte, and the separator and (b) model of the charge/discharge process of an ideal double-layer capacitor. (Adapted from Ref. [9]. The figures are made available under the Creative Commons CC0 1.0 Universal Public Domain Dedication.)

electric field charge. When a voltage is applied at both electrodes, ions are separated in the liquid electrolyte, forming opposite polarity of mirror charged distribution by Helmholtz double-layer as shown in the charge and discharge capacitor in Figure 6.2(b).

6.2.2.2 Electrochemical Pseudocapacitance

This is a type of supercapacitor with a very high capacitance value due to its several mechanisms to store charge, making it a device that has features of both battery and EDLC. It is composed of two electrodes where chemical reactions occur, which are separated by an electrolyte. The ions from the electrolyte do not interact with the electrode; this condition leads to the energy storage to occur through electrostatic forces using a reversible Faradaic redox reaction. When voltage is applied, the electrodes become polarized and attract the ions from the electrolyte that possess opposite charges compared to the electrodes. These ions have no chemical reaction with the electrode atoms, but there is an electron charge transfer where only one electron pair charge participates, thereby limiting the storage capacity in the available surface. The electrical double layer is formed by the transfer of electrons that come from and to the valence electron states and enter the negative to the positive electrode [10]. Transition-metal oxides such as MnO_2 and conducting polymers such as polyacetylene are materials that exhibit redox behavior, multiple oxidation states, and high specific capacitance, making them good candidates for electrodes. Conducting polymers have high conductivity and are comparatively cheaper than carbon electrodes but are mechanically weak and do not exhibit stability during cycling. Figure 6.3 shows three different pseudocapacitance mechanisms that lead to the electrode's capacitive characteristics [11].

Underpotential deposition is a mechanism in which a single layer of a metal cation is absorbed over the surface of a different metal that is above their Nernst potential or redox potential. An important feature is the thickness of the deposit not exceeding a monolayer, for example, a gold electrode that has lead deposited on its surface [12]. The redox reaction occurs on the electrode's surface as the ions are adsorbed, leading to charge storage. This process accompanied by electron transfer at active redox sites describes the mechanism to accumulate charge. Also, it causes variation in the number of oxidation for the species in the system. For example, ruthenium oxide receiving and releasing protons provided by the electrolyte, which causes a change in the Ru oxidation state. Intercalation occurs when cations are inserted into the electrode's lattice without crystallographic change. These three types of mechanisms happen because of different structures as well as materials used.

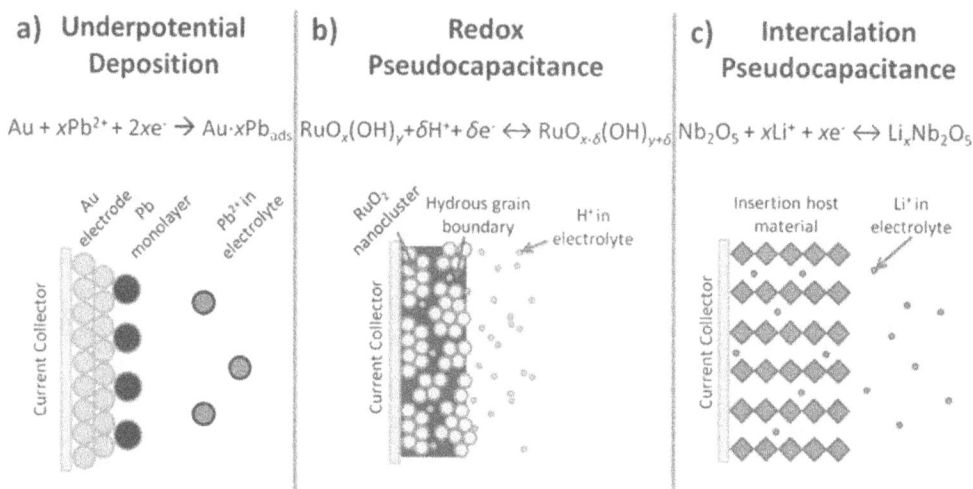

a) Underpotential Deposition

$Au + xPb^{2+} + 2xe^- \rightarrow Au \cdot xPb_{ads}$

b) Redox Pseudocapacitance

$RuO_x(OH)_y + \delta H^+ + \delta e^- \leftrightarrow RuO_{x-\delta}(OH)_{y+\delta}$

c) Intercalation Pseudocapacitance

$Nb_2O_5 + xLi^+ + xe^- \leftrightarrow Li_xNb_2O_5$

FIGURE 6.3 Reversible redox mechanisms that give rise to pseudocapacitance: (a) underpotential deposition, (b) redox pseudocapacitance, and (c) intercalation pseudocapacitance. (Adapted with permission from Ref. [11]. Copyright (2014) Royal Society of Chemistry.)

6.3 SUPERCAPACITOR DEVICES AND TESTING

6.3.1 TYPES OF DEVICE

Coin cells, pouch cells, cylindrical cells are the most predominant supercapacitor designs used for both industrial and commercial purposes. Supercapacitor cell performance is much dependent on its internal resistance and capacitance. The internal resistance of the supercapacitor varies according to the manufacturing condition of the supercapacitor electrode and the assembling of the supercapacitor cells. The conditions associated with supercapacitor cell assembly include the thickness of the electrode, cell pressure, pre-conditioning process, and density of the electrode [13–15]. The preparation of supercapacitor cells comprises the following steps: homogeneous mixture preparation of the supercapacitor materials, formation of an electrode layer on a current collector, assembling and packaging of the supercapacitor cells.

6.3.1.1 Coin Cells

Coin cells are mostly used in the production of supercapacitor cells and batteries, especially for Li and Li-ion batteries. The merits of coin cells in the production of supercapacitors include their capacity or size, the ease of assembling, and the ability to test a small amount of supercapacitor active materials using a large number of coin cells. Characterization of coin cells in full cell arrangement is performed by using a small amount of electrode materials which are mostly packed in the coin cell. These coin cells usually have a limited capacity due to their small magnitude. Coin cell electrodes are designed by spreading a slurry comprising carbon material, active material, additives, liquids, and binders onto a current collector using either a doctor blade or notch bar. After spreading the electrode materials onto the sheet of the current collector, the electrode material is then dried. Before punching the coin-shaped electrodes from the spread on the current collector, the spread is first of roll pressed. The coin cells are assembled and sealed in their casing using a mechanical coin cell crimper (preferable in a controlled environment). The various components, current collector separator, and electrodes are assembled before the top cap of the coin cell are placed on the spring and then with some pressure, the cell is finally assembled as illustrated in Figure 6.4 [16].

6.3.1.2 Cylindrical Cell

Coin cells are mostly employed to test a small amount of electrode material or a supercapacitor cell. But to test a larger capacity cell that ranges between 10 and 1000 F, the electrode materials are mostly packaged in a cylindrical cell. The use of a cylindrical cell is a step pitched toward

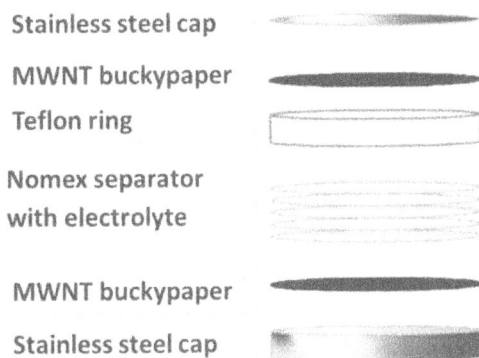

Stainless steel cap

MWNT buckypaper

Teflon ring

Nomex separator
with electrolyte

MWNT buckypaper

Stainless steel cap

FIGURE 6.4 Illustration of coin cell assembly. (Adapted with permission from Ref. [16]. Copyright (2010) American Chemical Society.)

FIGURE 6.5 Illustration of a cylindrical supercapacitor cell design. (Adapted with permission from Ref. [2]. Copyright (2015) John Wiley & Sons, Ltd.)

commercialization. The cylindrical supercapacitor cell design consists of rolled layers of electrode material on a sheet of current collector together with a separator sheet. A cylindrical metal casing is used as the container for the layered sheets in a cylindrical form. To provide a metal connection, the current collect flaps are joined to the rolled sheets. Before the cell is sealed, the electrolyte is injected into the cylindrical cell. A safety outlet is fixed in the cell to ensure the release of any pressure buildup within the cell after sealing and during operation. Figure 6.5 is an illustration of how cylindrical supercapacitor cells are assembled [2].

6.3.1.3 Pouch Cell

Efforts have been made by researchers to scale up the size of a supercapacitor coin cell to an alternative large capacity cell, specifically a pouch cell, aside from the cylindrical supercapacitor cells. Recently reported articles on cylindrical supercapacitor cells usually point out some challenges associated with the preparation of the electrode and processing of the cylindrical cells [15]. The pouch cell configuration appears to be a superior configuration from the research point of view because of its cost-effectiveness, simplicity, industrial compatibility, and ease of modification [17]. Figure 6.6 shows the component and a fabricated pouch cell [17].

6.3.1.4 Flexible Cells

Flexible electronics are the succeeding revolt in the electronics industry. The flexibility and portability of these electronic gadgets improve their esthetic, design, and durability effect thus making them cost-effective commercially comparative to conventional supercapacitors and batteries [19]. Carbon materials used in the production of the flexible cells are mostly shaped macroscopically [20]. These shapes include zero-dimensional (0D), i.e. carbon particles or fullerene, one-dimensional (1D), i.e. carbon fibers or CNT, and two-dimensional (2D), i.e. graphite or graphene nanosheets, which is

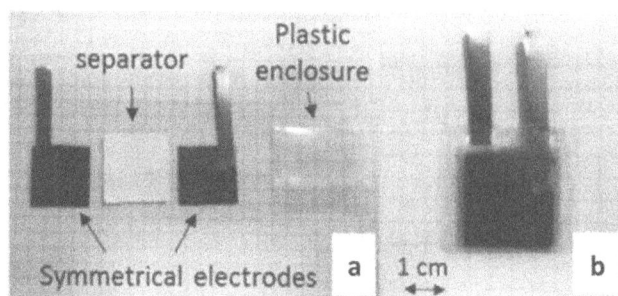

FIGURE 6.6 (a) Pouch cell supercapacitor before and (b) after assembly. (Adapted with permission from Ref. [18]. Copyright (2017) Elsevier.)

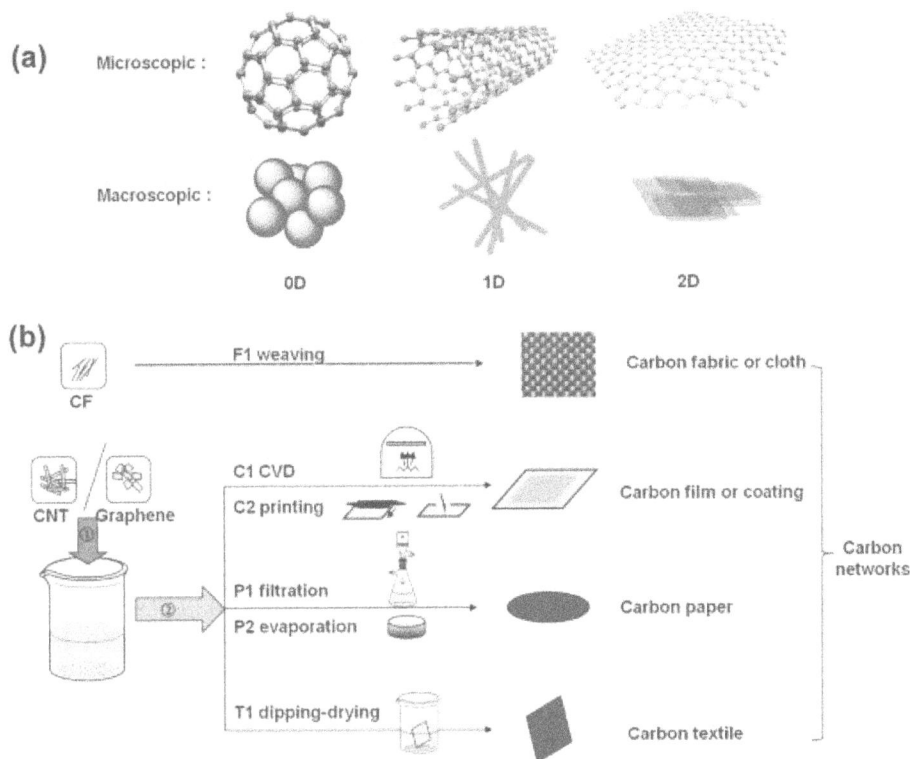

FIGURE 6.7 (a) Schematics for zero-, one-, and two-dimensional carbon-based materials at the micro and macroscopic scale. (b) The process to obtain carbon-based materials with one-dimensional carbon fibers and CNTs or two-dimensional graphene used as starting materials in several approaches. (Adapted with permission from Ref. [21]. Copyright (2013) Elsevier.)

illustrated in Figure 6.7. Both the 1D and 2D are mostly preferred in the production of flexible cells due to their highly conductive and flexible carbon network with outstanding electrochemical performances.

Flexible electrodes are classified into two types, namely single carbon or carbon composite electrodes. On the one hand, single carbon electrodes present only carbon networks in their composition which can be from one or more shapes. On the other hand, carbon composite electrodes have multipurpose carbon networks along with other pseudocapacitor materials [21]. Manufacturing techniques such as chemical vapor deposition, evaporation, printing, filtration, weaving, and dip-drying are used in the fabrication of varied carbon networks from either 1D or 2D carbon particles. According to Masarapu *et al.*, carbon fabric is the most recommended network for flexible electrode fabrication [22]. These fabrics when weaved exhibit high stiffness, excellent flexibility, and good strength. However, it has a low capacity, which limits its application as a supercapacitor. Composite carbon electrodes unlike the single carbon electrodes are formed by incorporating conducting materials or metal oxides into the carbon network as shown in Figure 6.8.

Hence the flexible supercapacitor cells formed from these composite carbon electrodes have a higher energy density according to [21]. Unlike conventional supercapacitors, which comprise a positive and negative electrode, electrolyte, a separator, outer packages, and current collectors, the structure of flexible supercapacitor cells includes a current collector and a binder. Due to the nature of the carbon material, it serves as an active electrode as well as a current collector [23, 24]. Many research groups have technologically advanced carbon and polymeric components to construct flexible supercapacitors cells for electronic gadgets [2, 25].

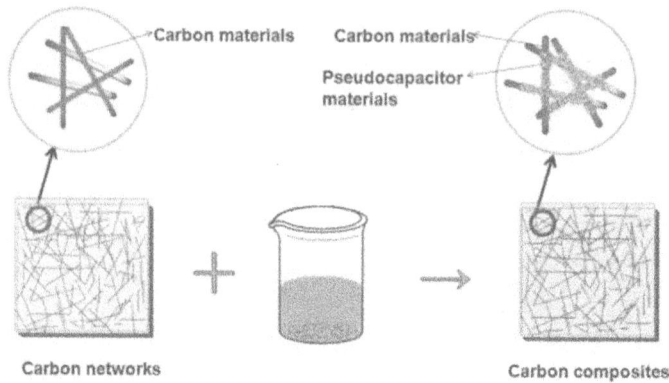

FIGURE 6.8 Use of a method that employs a solution to fabricate carbon composites that are coated with pseudocapacitors materials. (Adapted with permission from [21]. Copyright (2013) Elsevier.)

6.3.2 COMMON METHODS FOR TESTING SUPERCAPACITORS

The testing of supercapacitor devices is an important step for the validation and optimization of the devices. To determine the individual performance of each electrochemical component and the supercapacitor, cyclic voltammetry (CV), galvanostatic charge-discharge curve (GCD), and electro-chemical impedance spectroscopy (EIS) can be used for characterization.

6.3.2.1 Cyclic Voltammetry

The fundamental electroanalysis method provides both quantitative and qualitative information about electrode surface and solution electrochemical processes [26]. Normally, CV analysis is done in a three-electrode system with working, counter, and reference electrodes. For CV testing, the working electrode potential is measured against the potential of the reference electrode by a linear back and forth scanning between the specified upper and lower potential limits as shown in Figure 6.9. The scan rate (v) is obtained from the slope of the linear forward and reverse lines. It is represented by the equation:

$$v = \frac{dE}{dt} \tag{6.1}$$

where E is the potential expressed in V units, while v is expressed in V/s. In CV analysis, the change in current at the working and counter electrodes are recorded during the electrode potential scan.

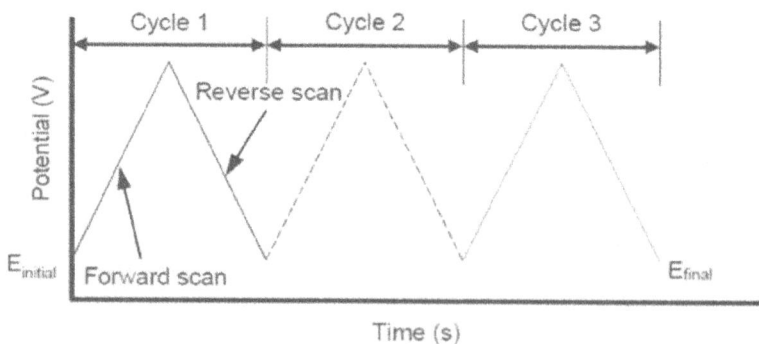

FIGURE 6.9 Potential–time curves for CV with a waveform of triangular potential. (Adapted from Ref. [26]. This article is an open-access article distributed under the terms and conditions of the Creative Commons Attribution (CC BY) license.)

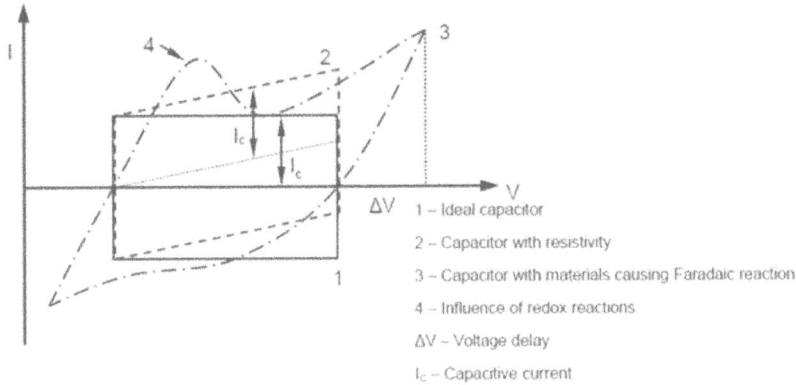

FIGURE 6.10 CV of pure PEO electrolyte on a polyester fabric coated with copper. (Adapted from Ref. [26]. This article is an open-access article distributed under the terms and conditions of the Creative Commons Attribution (CC BY) license.)

As shown in Figure 6.10, a cyclic voltammogram can be obtained when the current variation at the working electrode against the electrode potential is plotted in a graph.

6.3.2.2 Galvanostatic Charge–Discharge Test

GCD analysis is one of the most reliable methods of determining the equivalent circuit, power density, cyclability, and capacitive energy density of a supercapacitor. For GCD analysis, a two-electrode or three-electrode system can be used. In GCD analysis, the electrode is set under constant current, while the voltage of the cell is recorded as a function of time, providing the charging or discharging time. The capacitance is measured as a constant current passes through the cell while the change on electrode potential or voltage is obtained. The curves obtained from the plot of voltage versus time or potential versus time under constant current is named "chronopotentiogram," which is also referred to as the charge/discharge curve for the case two-electrode cell [27].

6.3.2.3 Electrochemical Impedance Spectroscopy

It is a valuable electroanalytical method used to analyze the electrode-electrolyte interface properties of supercapacitors. The specific capacitance and equivalent series resistance of supercapacitors can be obtained from EIS data [28]. EIS analysis can be done in situ and ex situ. For the evaluation of a supercapacitor cell or stack under real practical conditions, the in situ analysis is used. The ex-situ analysis is performed to evaluate single materials used in electrodes and the electrode layers associated with them [29]. The EIS data can be represented as a Nyquist plot that consists of three areas: (a) a high frequency (10^4 Hz) semicircle that represents the interface resistance, (b) an intermediate (from high to medium) frequency area (10^4 to 1 Hz), indicating resistance from charge transfer, and (c) a low-frequency area (<1 Hz) with a vertical line that goes along with the imaginary axis at lower frequencies, suggesting a capacitive behavior [30]. The imaginary impedance $|Z|$ has a relationship with frequency f that is obtained from the EIS analysis [28]. The capacitance (F) can be determined through the equation:

$$C = \frac{1}{2\pi f |Z|} \tag{6.2}$$

A linear plot of $\log|Z|$ vs. $\log f$ gives the Bode plot. From the Bode plot, it can be concluded that capacitance decreases with increasing relative frequency [28].

6.3.3 SUPERCAPACITOR DEVICE EVALUATION

6.3.3.1 Energy and Power Densities

The energy and power density are important measurements to determine the supercapacitor's performance. The energy density is obtained by numerical integration of the discharge curves using the equation [28].

$$E = \int_{t_2}^{t_1} IV\,dt = \frac{1}{2}C(V_1 + V_2)(V_2 - V_1) \tag{6.3}$$

where C is capacitance given in F/g, V_1 and V_2 are end-of-charge and discharge voltages, respectively. $(V_2 - V_1)$ is the specific voltage window. So, when $V_1 = 0$ (0 V), Equation (6.3) becomes

$$E = \frac{1}{2}V_2^2 \tag{6.4}$$

Thus, due to the capacitive behavior exhibited at minimum voltage ($V_1 = 0$ V), Equation (6.4) can be used to calculate the energy density of an EDLC. Nevertheless, for an asymmetric or hybrid supercapacitor, the capacitive behavior exhibited at the minimum voltage (V_1) is greater than 0 V. So, Equation (6.4) would be suitable for calculating the energy density.

Supercapacitor's power density can be calculated by Equation (6.5):

$$P = \frac{E}{t} \tag{6.5}$$

P is power density given in W/kg, E is energy density given in Wh/kg, and t is discharge time given in h. The supercapacitor's maximum power density (P_m) can be calculated using Equation (6.6):

$$P_m = \frac{V^2}{4R_s} \tag{6.6}$$

where V is the maximum voltage, and R_s is the equivalent circuit in series resistance (ESR). It consists of the electrodes, electrolytes, and the diffusion of ions resistance in the pores of the electrode. A plot of energy and power density gives the Ragone plot, which is suitable for the characterization of a supercapacitor's electrochemical profile [31].

6.3.3.2 Cyclic Stability

The most reliable way to determine the degradation of a supercapacitor is to charge and discharge it over many cycles to observe the changes in both specific capacitance and equivalent series resistance [32]. A supercapacitor can have a cycle life of 100,000. Due to the rapid charge-discharge rates, the testing time for cyclic stability of supercapacitors is reduced drastically. Usually, continuous charging–discharging cycling leads to diminishment of capacitance with a corresponding escalation in the equivalent series resistance, leading to decreases in both the energy as well as power densities.

6.4 FLEXIBLE SUPERCAPACITORS FROM CONDUCTING POLYMERS

Flexible supercapacitors became part of the family of energy storage systems in the last few years and have attracted much attention by researchers as satisfactory electrochemical supercapacitors owing to their sensational Faradaic redox reaction reversibility, high power density, cycling life stability, and high rate capability [33]. Conducting polymers, with their high flexibility, superior conductivity, high cycle stability, and great environmental stability have been highlighted as one of the most relevant materials for flexible supercapacitors [34]. Conductive polymers are known to be

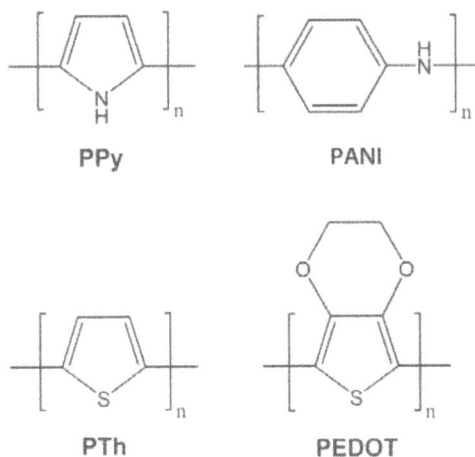

FIGURE 6.11 Structures of some conducting polymers.

a class of polymeric materials with special electrical and optical properties close to metal and semi-conductors. CPs demonstrate their conductive nature via the conjugated π bonds along the polymer backbone spine. CPs are synthesized by either electrochemical oxidation or by chemical means [35]. They can be doped with either n-type or p-type dopants like small salts ions (Br⁻, NO3⁻, Cl⁻). The commonly used conductive polymers for flexible supercapacitors are polyaniline (PANI), polythiophene (PTh), its derivatives such as poly(3-4, ethylene dioxythiophene) (PEDOT), and polypyrrole (PPy) [36] as described in Figure 6.11.

6.4.1 Polyaniline-Based Flexible Supercapacitors

Polyaniline has been researched as one of the most valuable candidates for flexible supercapacitors due to its high theoretical capacitance, electrochemical and chemical stabilities. PANI, when used as an electrode, has high stability in an ionic liquid and long cycling life [37]. Also, the ease of fabrication, inherent flexibility, low cost, good conductivity, high doped capability, variable oxidation levels, and good environmental stability [38] make it a promising flexible material for high-performance flexible supercapacitors. PANI can be synthesized by both chemical or electrochemical means. The type of dopants, structural properties, and morphology greatly influence the electrochemical performance of PANI. Yanilmaz *et al.* [38] fabricated flexible polyaniline-carbon nanofiber (PANI-CNF) to improve the electrode chemical properties for supercapacitor application, by first fabricating flexible carbon nanofiber by sol-gel and electrospinning method followed by PANI coating via chemical polymerization in situ. The electrochemical performance of the composite electrode and the carbon nanofiber was analyzed using two-electrode symmetric cells. Figure 6.12 provides the scanning electron microscopy micrographs and specific capacitance, respectively of flexible CNF and flexible PANI-CNFs grown using different polymerization time. After carbonization, nanofibers with smooth surfaces and uniform PANI coatings were formed. The performance of the composite material was superior to the carbon nanofiber due to the presence of PANI. The capacitance of the flexible PANI-CNF also increased with polymerization time and resulting in a denser and thicker PANI coating deposited onto CNF's surface. The flexible PANI-CNF supercapacitor reached the specific capacitance of 234 F/g at 1 A/g for 12 h of polymerization, whereas the flexible CNF reached 130 F/g. This variation could be associated with the lowest resistance of the flexible PANI-CNF electrode to charge transfer at polymerization that occurred over a longer period. Flexible PANI-CNF electrode experienced a relatively low decrease in capacitance (better C-rate performance) due to the homogeneous distribution of PANI over the porous structure of CNFs arrangement, which also presented high ionic conduction at 12 h polymerization time. Also,

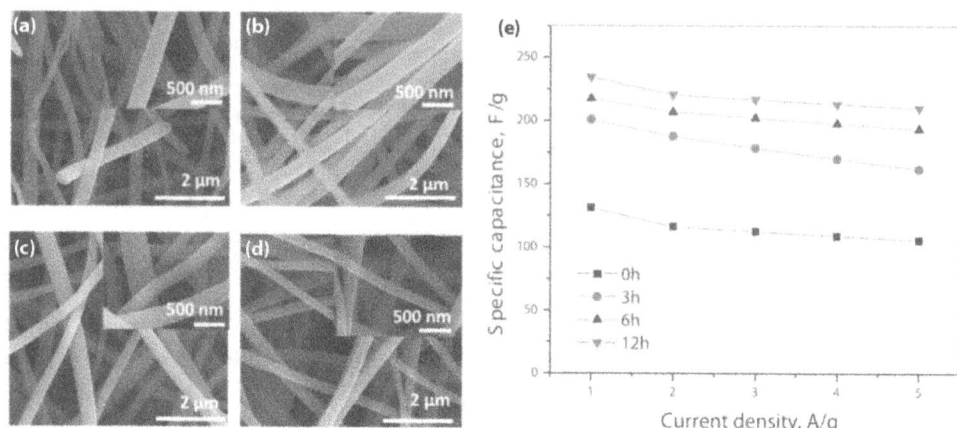

FIGURE 6.12 SEM images of flexible CNF (a) and PANI-CNF electrodes with different polymerization times of 3 h (b), 6 h (c), 12 h (d), and variation of specific capacitance versus current (e). (Adapted with permission from Ref. [38]. Copyright (2019) Elsevier.)

the energy density reached high values of 32 Wh/kg, while mechanical tests demonstrated high flexibility and durability for the PANI-CNF free-standing electrodes at 12 h polymerization.

Oh *et al.* [34] fabricated a high surface area polyaniline composite (Pt_CPPy/PANI:CSA) electrode material for flexible supercapacitor application by using nanoparticles of carboxyl polypyrrole decorated with Pt (Pt_CPPyNPPs) as the nucleating agent for the electrode material (Figure 6.13a).

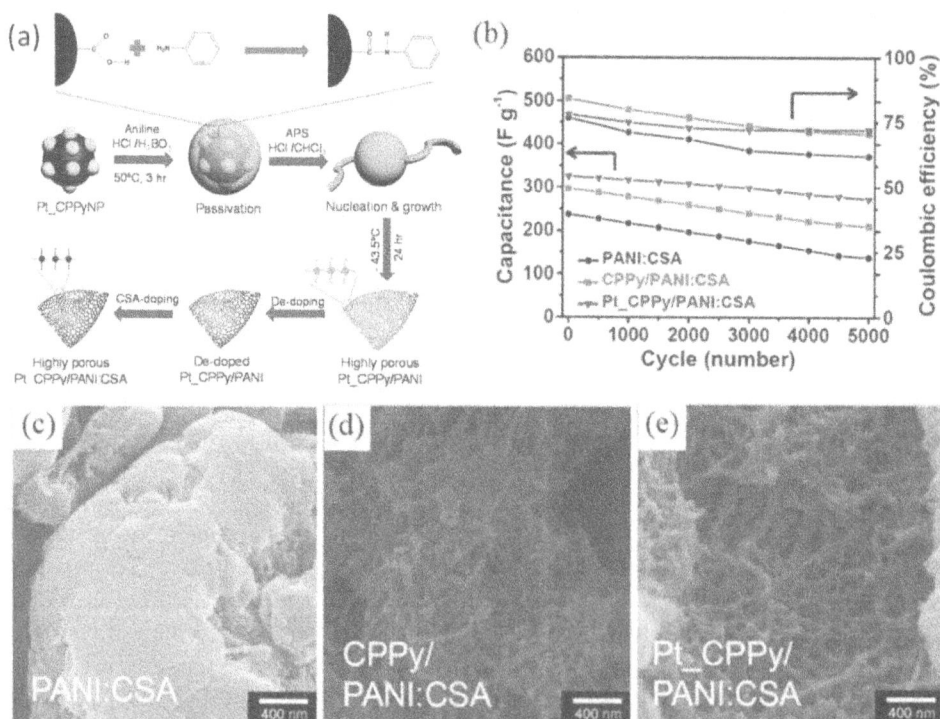

FIGURE 6.13 (a) Synthetical procedure of Pt_CPPyNP embedded PANI:CSA (Pt_CPPy/PANI:CSA) paste. (b) Electrode's cycling performance with charge–discharge numbers at 0.5 A/g of current density. FE-SEM images of (c) PANI:CSA, (d) CPPy/PANI:CSA, and (e) Pt_CPPy/PANI:CSA. (Adapted with permission from Ref. [34]. Copyright (2019) Royal Society of Chemistry.)

They fabricated Pt_CPPy/PANI:CSA composite by first dissolving CPPyNPPs in a distilled water and then injecting $PtCl_4$ solution to which the mixture was ultra-sonicated and filtered to obtain Pt_CPPy. Aniline monomer was then added to the obtained Pt_CPPy for interfacial polymerization to occur. PANI was doped on the sample after the polymerization process to form a composite. The composite material was then dissolved in a CSA solution followed by 24 h of sonication to obtained Pt_CPPy/PANI:CSA composite. They investigated the electrochemical performance of PANI:CSA, CPPy/PANI:CSA, and Pt_CPPy/PANI:CSA composites by spreading the composites paste on a glass substrate to obtain flexible free-standing films. The obtained composite films were then used in a three-electrode system for electrochemical performance. They observed a promising specific capacitance of 325 F/g at 0.5 A/g current density and potential within 0 to 0.8 V for the Pt_CPPy/PANI:CSA film (Figure 6.13b). It was attributed to its highly porous structure composites, which demonstrated satisfactory rate capability and stability over cycling and enhanced the active surface area between PANI and the ions of the electrolytes as shown in Figure 6.13(c–e). The highest specific capacitance as a result of the electrode's lowest resistance accounted for the longest discharging time which helps reduce unwanted power consumption. The polymeric composites presented a crystalline structure due to the addition of Pt_CPPyNP and Pt_CPPy/PANI:CSA, which also exhibits excellent flexibility to be used as supercapacitors.

The specific capacitance for polyvinyl alcohol-carbon nanotube-polyaniline (PVA/CNT/PANI) flexible films have been studied by Ben et al. [39], where a capacitance of 196.5 mF/cm^2 and 71.4% retention after 5000 cycles for PVA/CNT/PANI composite film with high cycling stability was reported. The PANI composite film shows evenly distributed high-density lamellate structures on the surface for the 12-h sample, which could induce different electrochemical behaviors and provides enough spacing for water permeation between active materials and electrolytes as shown in Figure 6.14. They attributed the electrochemical performance of the composite film to the synergic

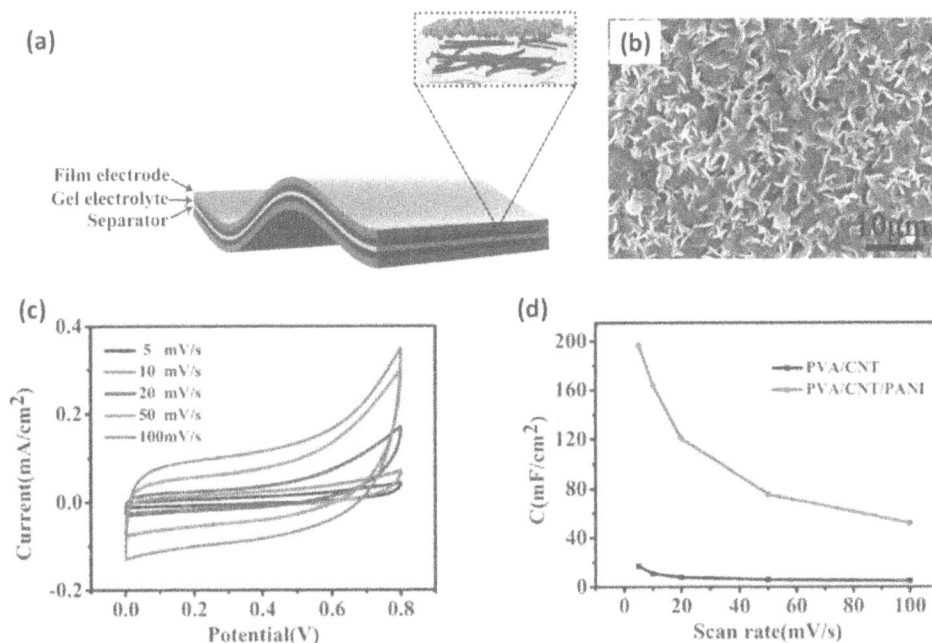

FIGURE 6.14 (a) Scheme for the symmetric supercapacitor composed of PVA/CNT/PANI electrodes and gel electrolyte. (b) SEM image of composite film at 12 h of polymerization time, (c) 12h GCD curves for the 12 h PVA/CNT/PANI with various current densities, and (d) specific capacitance's plot for PVA/CNT and PVA/CNT/PANI composite supercapacitors with increasing scan rate. (Adapted from Ref. [39]. This article is an open-access article distributed under the terms and conditions of the Creative Commons Attribution (CC BY) license.)

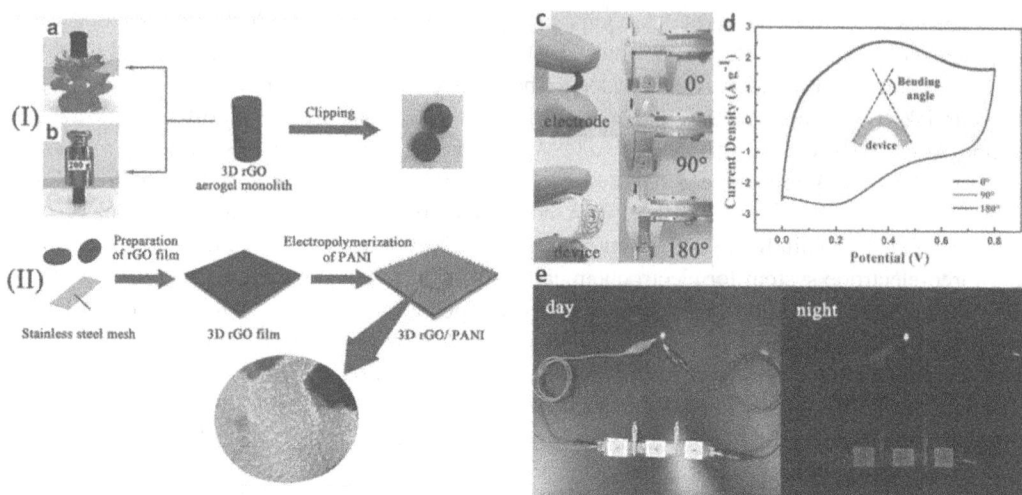

FIGURE 6.15   Scheme for fabrication of (I) 3D rGO aerogel slices ((a) and (b) show the mechanical properties for the aerogel). (II) Hybrid composites are obtained through mechanical pressure and electrodeposition. (c) Photos for the different bending states for the composite electrode, (d) CV curves at various bending angles, and (e) device in work. (Adapted from Ref. [40]. This article is an open-access article distributed under the terms and conditions of the Creative Commons Attribution (CC BY) license.)

effect from PVA's flexibility, CNT's conductivity, and PANI's pseudo-capacity property. This work provides a facile method for preparing flexible electrode materials.

Yang *et al.* [40] have reported on the electrochemical performance of electrodes composed of arrays of PANI coated with graphene aerogel via electrodeposition of PANI on reduced graphene oxide (rGO) aerogel. Figure 6.15 shows the fabrication procedure for aerogel slices of the 3D rGO from the monolith. SEM image for the freeze-dried rGO shows relatively smooth sheets of graphene surface, serving as electrodeposited substrate for PANI arrays with homogenous PANI nanocones as the hybrid composites. High specific capacitance was obtained after analyzing the flexible hybrid composite with the value of 432 F/g at 1 A/g. The stability test was performed by applying 10,000 cycles of charge/discharge into the composite that led to a retention of 85%. Also, satisfactory energy and power densities of 25 Wh/kg and 681 W/kg were achieved, respectively, along with outstanding flexibility. The authors attributed the SC's remarkable performance to the synergy effect between PANI arrays and 3D rGO aerogel. Besides, the SC composite presented remarkable flexibility as well as cyclic stability even after exposure to different bending states. Such electrochemical and mechanical properties give promising applications for this hybrid SC composite as a wearable electronic.

Additionally, Li *et al.* [41] coated PANI on a graphene sheet paper and investigated the electrochemical performance of the composite electrode. SEM images show uniformly dispersed smooth graphene sheets with the presence of wrinkles on the surface of GP and a rough layer of nanorods on the PANi-GP surface which are beneficial to exposing the more electrochemically active surface area and helps transfer electrolyte, respectively. They have reported high areal capacitance in a three-electrode system that reached 176 mF/cm^2 at a current density of 0.2 mA/cm^2 for the flexible composite electrode, which was about ten times higher than that of the pristine graphene paper. They attributed the high capacitance of PANI-GP electrodes to the high electrical conduction of exfoliated graphene sheets and a large accessible surface area of the aligned PANI nanorods that allowed fast ion diffusion, which favors the redox process. The GP samples exhibited a rectangular shape as compared to PANI-GP, which exhibited redox peaks. CV experiments showed that the GP samples had higher rate capability, which matches with the results in the three-electrode system.

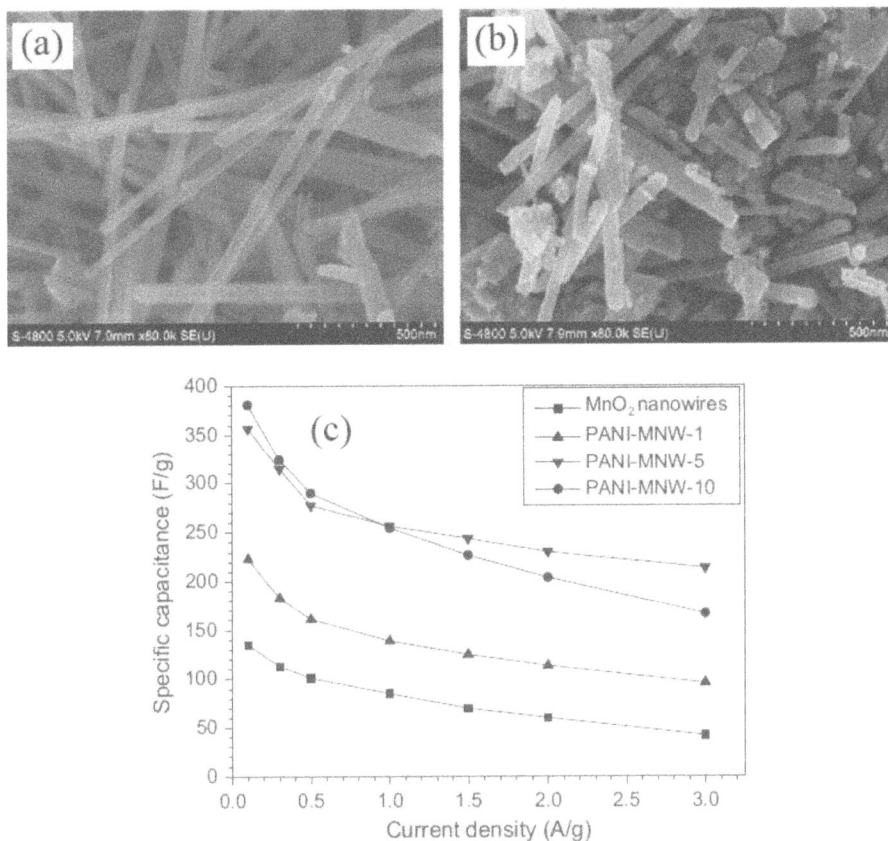

FIGURE 6.16 SEM images of MnO_2 nanowires (a), PANI-MNW-5 (b), and specific capacitance against current density plots (c). (Adapted with permission from Ref. [42]. Copyright (2013) Elsevier.)

Chen *et al.* [42] used in situ polymerization through a chemical oxidative process to prepare PANI–MnO_2 nanowire (PANI–MNW) composites. Figure 6.16(a, b) shows the SEM images of MnO_2 nanowires with crystals of high purity and a highly dispersed porous PANI–MNW–5 composite. The authors observed the highest specific capacitance of 256 F/g at 1A/g of current density for the PANI–MNW–5 composite. Such results showed good cyclic stability and attributed the enhanced properties due to the high porosity of the composite structure Figure 6.16(c).

6.4.2 POLYPYRROLE-BASED FLEXIBLE SUPERCAPACITORS

Polypyrrole with its high flexibility, ductility, and excellent intrinsic characteristics (high electrical conduction and outstanding redox properties) has emerged as a feasible polymer for supercapacitors. Also, PPy with relatively high capacitance (400–500 F/cm³), ease processability, and high cycle stability has gained much attention for supercapacitor applications [43, 44]. Nonetheless, PPy has some shortcomings that limit its applications in high-performance supercapacitors: one is the difference between the predicted theoretical capacitance and the practical value; the other is since it is a p-type conductive polymer, it is only used as a cathode for supercapacitor applications; and lastly, it has poor cycle stability during charge/discharge process [45]. These shortcomings of PPy could be solved by altering the synthesis methods to obtain better morphology and/or structures, by forming composites or hybrid with carbon-based, metal oxides and other conducting polymers, for example, whereas designing a novel configuration of PPy-based supercapacitors is an interesting option [45]. Thus, the specific capacitance of PPy can be enhanced while improving its cycle stability.

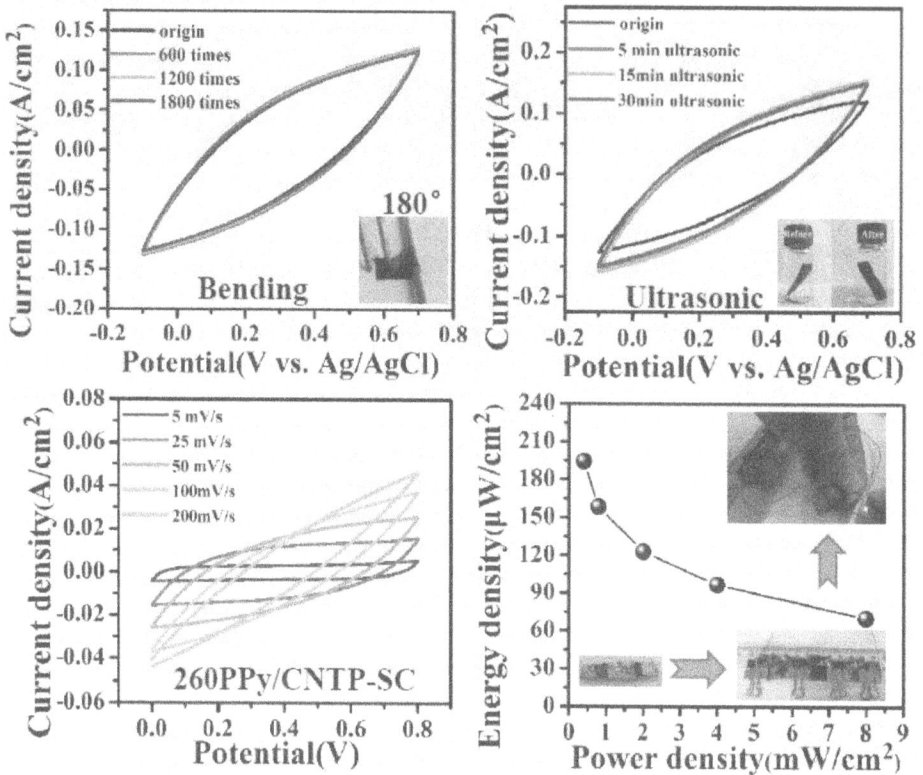

FIGURE 6.17 Electrochemical behavior for the flexible PPy coated with CNT paper electrode fabricated through in situ interfacial polymerization method used for all-solid-state supercapacitors. (Adapted with permission from Ref. [46]. Copyright (2019) Royal Society of Chemistry.)

Tong *et al.* [46] effectively synthesized PPy coated with CNT paper (PPy/CNTP) electrode by in situ polymerization technique and evaluated the electrochemical performance of the composites for supercapacitor applications. Figure 6.17 shows the electrochemical properties of a CNTs interconnected structure and a rough packed nanoporous of the 260-PPy/CNTP electrode, which helps to increase the electrochemical properties of the PPy film. They obtained 8605 mF/cm^2 at 1 mA/cm^2 for the specific capacitance and excellent retention of 107% of the original capacitance after 12,000 charge/discharge cycles. They attributed the high electrochemical performance to the ultrasonic treatment method used and also because of the highly porous structure of the PPy/CNTP electrode. Wang *et al.* [44] reported the synthesis of PPy-based nanostructures based on vertical PPy (VPPy) nanotube array and nano-onions of carbon that were grown on textiles via a one-step polymerization process and investigated the electrochemical characteristics for possible wearable supercapacitors. SEM images displayed smooth fabric of pristine spandex and rough interconnected granules after growing PPyG film which resulted in a highly dense growth in the VPPyNTs fabric surface to increase electrochemical properties (Figure 6.18). The as-fabricated supercapacitor demonstrated high specific gravimetric capacitance of 475 F/g at 0.5 A/g current density and high retention of 99% after 500 cycles and attributed the superior performance to the synergistic stretchable effect of the energy storage device.

Shi *et al.* [47] have also successfully synthesized nanostructured conductive PPy hydrogel electrodes via interfacial polymerization and used them for flexible supercapacitors. SEM images show dominant interconnected spherical hollow structures induced by phytic acid in PPy hydrogel which facilitate electrons and ions transport (Figure 6.19). The as-synthesized nanostructured PPy hydrogel electrode exhibited a satisfactory specific capacitance of ~380 F/g and a high rate capability.

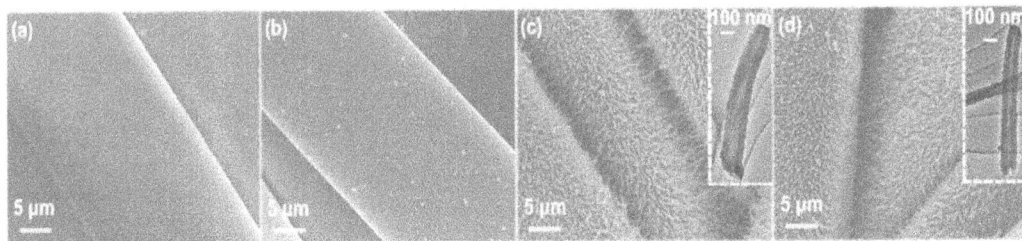

FIGURE 6.18 Images obtained from SEM for (a) neat textile. (b) PPyG functionalized textile. (c) VPPyNTs/PPyG-textile. (d) VPPyNTs/CNOs@PPyG-textile. (Adapted with permission from Ref. [44]. Copyright (2019) Springer Nature.)

They attributed the efficient electrochemical performance combined with the excellent mechanical stability of the nanostructured PPy hydrogel to (a) inherent PPy hydrogel's conductivity, (b) Porous structure that consists of a hierarchical 3D arrangement that allows faster charge transfer during the electrochemical process, (c) hydrogel's mechanical properties, and (d) strong adhesion with substrate and PPy hydrogel.

Additionally, Jyothibasu *et al.* [48] have reported the synthesis of flexible and free-standing PPy/CNT/cellulose composite electrodes to be used as a supercapacitor. The composite made of functionalized CNT (f-CNT) was prepared, and an in situ chemical polymerization of pyrrole was performed to obtain PPy/f-CNT/cellulose composite as shown in the synthesis method in Figure 6.20. The PPy/f-CNT/cellulose composite electrode synthesized served as a buffer and favored the ion's diffusion from the electrolyte as well as transport of electrons during charging/discharge cycles due to its conductive and porous structure. They investigated the electrochemical properties of the as-fabricated composite and obtained a high capacitance value of 2147 mF/cm^2 at 1 mA/cm^2. They ascribed the composite's high performance to the following: (a) good hydrophilic porous support from the regenerated cellulose for homogeneous distribution of the f-CNT, (b) high electrical conductivity of f-CNT for fast charge transport, (c) high structural porosity of the regenerated cellulose matrix that enhanced diffusion and adsorption of electrolyte ions, and (d) increased PPy loading with polymerization time.

6.4.3 Polythiophene and Its Derivatives for Flexible Supercapacitors

Polythiophene and its derivatives materials are compelling candidates for flexible supercapacitor applications and have drawn attention in recent years due to superior electrical conductivity, long-wavelength absorption capabilities, and relatively high environmental stability [43]. PTh and its

FIGURE 6.19 SEM micrographs for (a) PPy hydrogel. (b) A fragment of dehydrated hydrogel (inset figure scale bar: 1 mm). (c) Transmission electron microscopy image for the dehydrated hydrogel (inset figure scale bar: 200 nm). (Adapted with permission from Ref. [47]. Copyright (2014) Royal Society of Chemistry.)

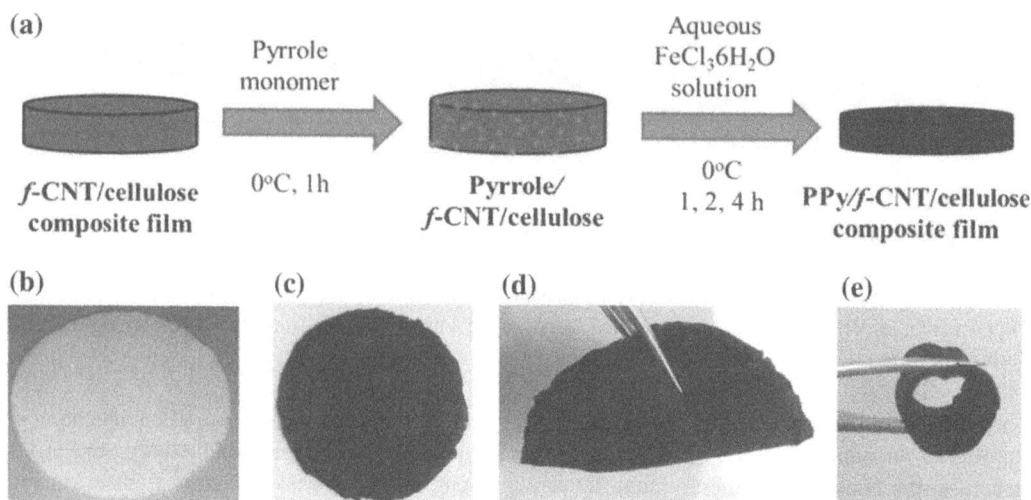

FIGURE 6.20 (a) Fabrication of PPy/f-CNT/cellulose composite film. (b) Photograph of cellulose film. (c) PPy/f-CNT/cellulose composite film. (d, e) Folded PPy/f-CNT/cellulose composite films. (Adapted with permission from Ref. [48]. Copyright (2019) Springer Nature.)

derivatives can be classified as p-doped and n-doped. It is found generally that the conductivity of the n-doped is poor leading to poor mass specific capacitance compared to the p-doped, thus, limiting the use of the n-doped type as anode material for supercapacitor application. The p-doped PTh derivatives are highly stable in air and moist environment [33]. Chen *et al.* [49] have synthesized PTh composites with high strength and good toughness and studied their potential use as flexible electrodes for supercapacitor application. They proposed a new approach of employing an anionic surfactant that functions as a dopant of PTh to bridge the synergy between the rigid PTh chain and the soft polyethylene glycol chain. The dynamic network structure was ascribed as the factor for improved electrochemical performance, along with larger and dense-packed particle sizes formed by the rigid and soft structure of PTh and PEG, respectively. The authors investigated the PTh composite's electrochemical performance and reported a high capacitance of 135 F/g at a current density of 1 A/g.

Wu *et al.* [50] have studied the electrochemical characteristics of flexible PTh by doping Fe^{3+} on PTh using carbon cloth as the substrate through an electrochemical deposition method. The as-synthesized PTh revealed uniform distribution of the iron nanopore in the polymer composite matrix as a result of the doping, which accounted for the improved electrochemical properties. They found out that the specific capacitance of the flexible PTh after Fe^{3+} doping increased to 108 F/g compared to the pristine flexible PTh, which had a 77 F/g of specific capacitance at a current density of 0.5 F/g. Vijeth *et al.* [51] studied the electrochemical behavior of flexible polythiophene/aluminum oxide (PTHA) nanocomposites by constructing an asymmetric supercapacitor employing the PTHA as anode along with charcoal as the cathode. Aggregation in a small amount was seen in the PTHA samples as compared to PTH with uniform particle arrangement under atomic force microscopy (AFM) as shown in Figure 6.21 indicating attractive interaction between the nanoparticles and the polymeric matrix. The specific capacitance obtained was 554 F/g for the PTHA nanocomposite at a current density of 1 A/g with retention of 94.61% after 2000 cycles. They attributed this electrochemical performance to: (a) the pseudocapacitance behavior, which enhances good charge diffusion at the electrode surface, (b) high rate of charge/discharge capabilities of the stable reversible charcoal electrode, and (c) effective transfer of ions and electrons inside the electrode material.

Polythiophene and graphene composite have been investigated by Melo *et al.* [52] for their applications as supercapacitors by employing traditional techniques of chemical polymerization for the

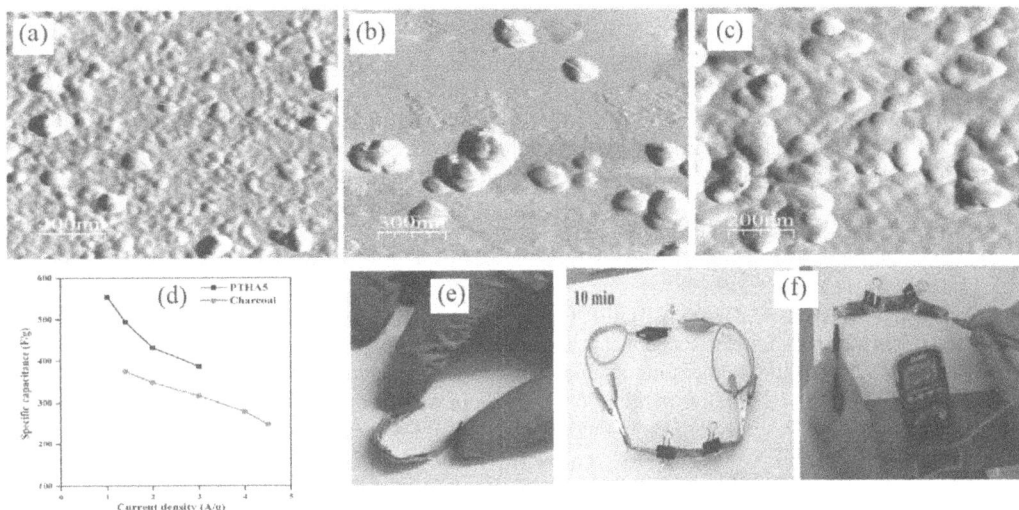

FIGURE 6.21 Atomic force microscopy (AFM) providing the surface images of (a) PTH. (b) PTHA1. (c) PTHA2. (d) Variation of C_{sp} for different CDs. (e) Photocopies of the bent all-solid-state asymmetric SC. (f) ASC device can light a red LED for 10 min under the working voltage of 1.5 V, while showing a voltage up to 4.83 V. (Adapted from Ref. [51]. This article is an open-access article distributed under the terms and conditions of the Creative Commons Attribution (CC BY) license.)

synthesis. Different mass proportions of GR/PTh composites (PTh: GR/PTh-67, –50, and –33) were applied in series, and the electrochemical properties of each nanocomposite were analyzed and compared. Sheets of the multi-layer structure were revealed from GR with repetitive uniform granular structures on PTh, while there was a growth of PTh chains over the GR's surface to form a new GR-PTh composite material with improved properties. They recorded the highest specific capacitance of 365 F/g at 1 A/g for GR/PTh-50, 232 F/g for pure GR, and 92 F/g for pure polymer PTh. The increase in specific capacitance for the composite GR/PTh in comparison to pristine GR and PTh can be due to (a) smaller resistance to electron transfer, (b) short ionic diffusion length caused by good pseudocapacitance of the graphene accompanied to a synergy effect with the PTh, and (c) larger effective surface area which facilitates rapid doping/dedoping of ions.

6.5 CONCLUSION

In recent times, the growing interest in portable, lightweight, and flexible devices such as wearable sensors, smart electronics, touch screens, and electric motors has generated great interest in supercapacitors as a favorable candidate for energy storage devices. Flexible supercapacitors have gained much attention to meet the demands of mankind owing to their high flexibility, lightweight, high power density, high cycle life, and high capacitance performance. Conducting polymers and their composites have been studied as the most attractive candidates for flexible supercapacitors because of their pseudocapacitive nature with high Faradaic redox capacitance. Conducting polymers and composites have shown tremendous improvement in properties of energy storage devices as a result of synergy effects of the composite's components. The performance of various conducting polymers used for flexible supercapacitors such as PANI, PPy, and PTh-based from various reports are discussed in this chapter. It is shown in this chapter that compositing with carbon-based materials such as CNTs, graphene, and nanoparticles does not only improve flexibility but also results in improve electrochemical performance than the bare conducting polymers. Also, the thermal stability of conducting polymers has improved with the compositing of the conductive polymers and as such makes them suitable candidates for energy storage devices.

REFERENCES

1. May GJ, Davidson A, Monahov B (2018) Lead batteries for utility energy storage: A review. J Energy Storage 15:145–157.
2. Kim BK, Sy S, Yu A, Zhang J (2015) Electrochemical supercapacitors for energy storage and conversion. In: Handbook of Clean Energy Systems. John Wiley & Sons. pp 1–25.
3. Hall PJ, Mirzaeian M, Fletcher SI, Sillars FB, Rennie AJR, Shitta-Bey GO, Wilson G, Cruden A, Carter R (2010) Energy storage in electrochemical capacitors: Designing functional materials to improve performance. Energy Environ Sci 3:1238–1251.
4. Palchoudhury S, Ramasamy K, Gupta RK, Gupta A (2019) Flexible supercapacitors: A materials perspective. Front Mater 5:1–9.
5. Mirzaeian M, Abbas Q, Ogwu A, Hall P, Goldin M, Mirzaeian M, Jirandehi HF (2017) Electrode and electrolyte materials for electrochemical capacitors. Int J Hydrogen Energy 42:25565–25587.
6. Bueno PR (2019) Nanoscale origins of super-capacitance phenomena. J Power Sources 414:420–434.
7. Namisnyk A, Zhu J (2003) A survey of electrochemical super-capacitor technology. In: Australian Universities Power Engineering Conference. University of Canterbury, New Zealand.
8. Yushin G, Gogotsi Y (2006) Carbide derived carbon. In: ACS National Meeting Book of Abstracts. CRC Press, Boca Raton, FL, pp 211–254.
9. Wikipedia Supercapacitor. In: Wikipedia. https://en.wikipedia.org/wiki/Supercapacitor. Accessed 4 Nov 2021.
10. Frackowiak E, Béguin F (2001) Carbon materials for the electrochemical storage of energy in capacitors. Carbon N Y 39:937–950.
11. Augustyn V, Simon P, Dunn B (2014) Pseudocapacitive oxide materials for high-rate electrochemical energy storage. Energy Environ Sci 7:1597–1614.
12. Herrero E, Buller LJ, Abruña HD (2001) Underpotential deposition at single crystal surfaces of Au, Pt, Ag and other materials. Chem Rev 101:1897–1930.
13. Liu X, Zhang S, Wen X, Chen X, Wen Y, Shi X, Mijowska E (2020) High yield conversion of biowaste coffee grounds into hierarchical porous carbon for superior capacitive energy storage. Sci Rep 10:1–12.
14. Dsoke S, Tian X, Täubert C, Schlüter S, Wohlfahrt-mehrens M (2013) Strategies to reduce the resistance sources on electrochemical double layer capacitor electrodes. J Power Sources 238:422–429.
15. Bhattacharjya D, Carriazo D, Ajuria J, Villaverde A (2019) Study of electrode processing and cell assembly for the optimized performance of supercapacitor in pouch cell configuration. J Power Sources 439:227106.
16. Hu R, Cola BA, Haram N, Barisci JN, Lee S, Stoughton S, Wallace G, Too C, Thomas M, Gestos A, Cruz ME Dela, Ferraris JP, Zakhidov AA, Baughman RH (2010) Harvesting waste thermal energy using a carbon-nanotube-based thermo-electrochemical cell. Nano Lett 10:838–846.
17. Safitri GA, Nueangnoraj K, Sreearunothai P, Manyam J (2020) Fabrication of activated carbon pouch cell supercapacitor: Effects of calendering and selection of separator-solvent combination. Curr Appl Sci Technol 20:124–135.
18. Scorsone E, Gattout N, Rousseau L, Lissorgues G (2017) Porous diamond pouch cell supercapacitors. Diam Relat Mater 76:31–37.
19. Dubal DP (2018) Advances in flexible supercapacitors for portable and wearable smart gadgets. In: Emerging Materials for Energy Conversion and Storage. Elsevier Inc., pp 209–246.
20. Wang Q, Cao Q, Wang X, Jing B, Kuang H, Zhou L (2013) A high-capacity carbon prepared from renewable chicken feather biopolymer for supercapacitors. J Power Sources 225:101–107.
21. Shi S, Xu C, Yang C, Li J, Du H, Li B, Kang F (2013) Flexible supercapacitors. Particuology 11:371–377.
22. Masarapu C, Wang L, Li X, Wei B (2012) Tailoring electrode/electrolyte interfacial properties in flexible supercapacitors by applying pressure. Adv Energy Mater 2:546–552.
23. Chen X, Paul R, Dai L (2017) Carbon-based supercapacitors for efficient energy storage. Natl Sci Rev 4:453–489.
24. Pushparaj VL, Shaijumon MM, Kumar A, Murugesan S, Ci L, Vajtai R, Linhardt RJ, Nalamasu O, Ajayan PM (2007) Flexible energy storage devices based on nanocomposite paper. Proc Natl Acad Sci 104:13574–13577.
25. El-Kady MF, Strong V, Dubin S, Kaner RB (2012) Laser scribing of high-performance and flexible graphene-based electrochemical capacitors. Science 335:1326–1330.
26. Hui C, Kan C, Mak C, Chau K (2019) Flexible energy storage system – An introductory. Processes 7:922.
27. Inagaki M, Kang F (2016) Materials Science and Engineering of Carbon. Elsevier, Boston, MA.

28. Wang Y, Song Y, Xia Y (2016) Electrochemical capacitors: Mechanism, materials, systems, characterization and applications. Chem Soc Rev 45:5925–5950.

29. Aiping Yu, Chabot V, Zhang J (2013) Electrochemical Supercapacitors for Energy Storage and Delivery-Fundamentals and Applications. CRC Press, London, UK.

30. Sun W, Zheng R, Chen X (2010) Symmetric redox supercapacitor based on micro-fabrication with three-dimensional polypyrrole electrodes. J Power Sources 195:7120–7125.

31. Javed MS, Lei H, Li J, Wang Z, Mai W (2019) Construction of highly dispersed mesoporous bimetallic-sulfide nanoparticles locked in N-doped graphitic carbon nanosheets for high energy density hybrid flexible pseudocapacitors. J Mater Chem A 7:17435–17445.

32. Alipoori S, Mazinani S, Hamed S, Sharif F (2020) Review of PVA-based gel polymer electrolytes in flexible solid-state supercapacitors : Opportunities and challenges. J Energy Storage 27:101072.

33. Snook GA, Kao P, Best AS (2011) Conducting-polymer-based supercapacitor devices and electrodes. J Power Sources 196:1–12.

34. Oh J, Kim YK, Lee JS, Jang J (2019) Highly porous structured polyaniline nanocomposites for scalable and flexible high-performance supercapacitors. Nanoscale 11:6462–6470.

35. Nezakati T, Seifalian A, Tan A, Seifalian AM (2018) Conductive polymers: Opportunities and challenges in biomedical applications. Chem Rev 118:6766–6843.

36. Shi Y, Peng L, Ding Y, Zhao Y, Yu G (2015) Nanostructured conductive polymers for advanced energy storage. Chem Soc Rev 44:6684–6696.

37. Yavuz A, Yilmaz Erdogan P, Zengin H (2020) The use of polyaniline films on flexible tape for supercapacitor applications. Int J Energy Res 44:11941–11955.

38. Yanilmaz M, Dirican M, Asiri AM, Zhang X (2019) Flexible polyaniline-carbon nanofiber supercapacitor electrodes. J Energy Storage 24:100766.

39. Ben J, Song Z, Liu X, Lü W, Li X (2020) Fabrication and electrochemical performance of PVA/CNT/PANI flexible films as electrodes for supercapacitors. Nanoscale Res Lett 15:4–11.

40. Yang Y, Xi Y, Li J, Wei G, Klyui NI, Han W (2017) Flexible supercapacitors based on polyaniline arrays coated graphene aerogel electrodes. Nanoscale Res Lett 12:394.

41. Li K, Liu X, Chen S, Pan W, Zhang J (2019) A flexible solid-state supercapacitor based on graphene/polyaniline paper electrodes. J Energy Chem 32:166–173.

42. Chen L, Song Z, Liu G, Qiu J, Yu C, Qin J, Ma L (2013) Synthesis and electrochemical performance of polyaniline – MnO_2 nanowire composites for supercapacitors. J Phys Chem Solids 74:360–365.

43. Meng Q, Cai K, Chen Y, Chen L (2017) Research progress on conducting polymer based supercapacitor electrode materials. Nano Energy 36:268–285.

44. Wang L, Zhang C, Jiao X, Yuan Z (2019) Polypyrrole-based hybrid nanostructures grown on textile for wearable supercapacitors. Nano Res 12:1129–1137.

45. Huang Y, Li H, Wang Z, Zhu M, Pei Z, Xue Q, Huang Y, Zhi C (2016) Nanostructured Polypyrrole as a flexible electrode material of supercapacitor. Nano Energy 22:422–438.

46. Tong L, Gao M, Jiang C, Cai K (2019) Ultra-high performance and flexible polypyrrole coated CNT paper electrodes for all-solid-state supercapacitors. J Mater Chem A 7:10751–10760.

47. Shi Y, Pan L, Liu B, Wang Y, Cui Y, Bao Z, Yu G (2014) Nanostructured conductive polypyrrole hydrogels as high-performance, flexible supercapacitor electrodes. J Mater Chem A 2:6086–6091.

48. Jyothibasu JP, Kuo DW, Lee RH (2019) Flexible and freestanding electrodes based on polypyrrole/carbon nanotube/cellulose composites for supercapacitor application. Cellulose 26:4495–4513.

49. Chen Q, Wang X, Chen F, Zhang N, Ma M (2019) Extremely strong and tough polythiophene composite for flexible electronics. Chem Eng J 368:933–940.

50. Wu K, Zhao J, Wu R, Ruan B, Liu H, Wu M (2018) The impact of Fe^{3+} doping on the flexible polythiophene electrodes for supercapacitors. J Electroanal Chem 823:527–530.

51. Vijeth H, Ashok Kumar SP, Yesappa L, Niranjana M, Vandana M, Devendrappa H (2018) Flexible and high energy density solid-state asymmetric supercapacitor based on polythiophene nanocomposites and charcoal. RSC Adv 8:31414–31426.

52. Melo JP, Schulz EN, Horswell SL, Camarada MB (2017) Synthesis and characterization of graphene/polythiophene (GR/PT) nanocomposites: Evaluation as high-performance supercapacitor electrodes. Int J Electrochem Sci 12:2933–2948.

7 Nanofibers of Conducting Polymers for Energy Applications

A.R. Ajitha[1,3], H. Akhina[2,3], and P.K. Sandhya[4,5]

[1]Department of Chemistry, Newman College,
Thodupuzha, Kerala, India

[2]Post Graduate Department of Chemistry, MSM College,
Kayamkulam, Kerala, India

[3]International and Interuniversity Centre for Nanoscience and
Nanotechnology, Mahatma Gandhi University, Kottayam,
Kerala, India

[4]School of Chemical Sciences, Mahatma Gandhi University,
Kottayam, Kerala, India

[5]Post Graduate Research Department of Chemistry,
Sree Sankara College, Kalady, Kerala, India

CONTENTS

7.1 INTRODUCTION

Due to the fast development of the current economy, the energy demand has increased. So our current society requires the development of advanced devices for generating and storing energy. Therefore, it is important to design and fabricate novel advanced energy materials with the required properties. In the current era, nanotechnology has great importance since it can provide advanced materials for potential applications, in which the application in the field of sustainable energy and the environment has tremendous value. Nanotechnology can provide solutions to many problems in the future, especially in the field of energy harvesting. Advanced nanomaterials such as carbon nanotubes, nanorods, nanoplates, graphene, nanofibers have great attention in various fields [1–4], in which materials with nanofibrous morphology have great attention to solve numerous energy and environmental issues. Nanofibers of organic and inorganic materials with conducting nature have great applicability in energy devices. In addition to that, the nanofibers of conducting polymers have an inevitable role in the field of energy applications [5] since conducting polymers are good candidates for superior applications in the field of biomedical, environment, electronic, clothing, sensors, energy storage devices, etc. [6].

Nanofibers are one-dimensional (1D) nanomaterials and are good candidates for a wide range of applications in many fields such as in environmental remediation, sensors, electronics, medicine, energy harvesting and storage, tissue engineering, healthcare, etc. due to their unique properties such as high surface area, high aspect ratio, high conductivity, etc.; its diameter is 1000 times smaller than that of human hair, good mechanical properties, etc. Nowadays, nanofiber technology has great importance due to the availability of a wide range of materials that can be used for nanofiber production, like natural and synthetic polymers, metals and metal oxides, carbon-based, and composite nanomaterials. In addition to that, the nanofiber can be modified with various functionalities which are leading to the development of advanced materials for potential applications. Today nanofibers are recognized as electrode and membrane materials for energy devices such as solar cells, capacitors, batteries [1, 4, 7].

From literature studies, it can be clearly said that carbon materials, metal oxides, and conducting polymers can satisfy the requirements for energy devices, in which conducting polymers have great importance since they can offer flexible devices. These individual materials have few drawbacks, such as the high cost and comparatively low conductivity of metal oxides. Also, conducting polymers have a short life cycle. Therefore, current research is focused on the formation of nanostructured materials and composite materials. Conducting polymer nanofibers have a great role in energy applications. Due to the high surface area of the conducting polymer nanofibers they have higher capacitance than the bulk polymers. Therefore, high storage is possible in the case of nanofibers of conducting polymers. In addition to the high surface area, low weight and good environmental stability of conducting polymer nanofibers make them a good candidate for energy applications. Another potential strategy is the incorporation of carbon materials or metal oxides with nanofibers of conducting polymers since it can explore the synergetic properties of individual components. Extensive studies have been done on the combination of nanofibers with carbon materials and metal oxides [4, 7–9].

Polyaniline (PANI) is a well-studied polymer since it possesses a much higher theoretical capacitance than other conducting polymers owing to its multiple redox states. In addition to that PANI has good optical and electrical properties with relatively high conductivity tunable properties by acid/base doping/dedoping chemical stability, environmental stability, ease of synthesis with low-cost monomer [1, 3, 4, 10]. In this chapter, the importance, preparation methods, properties, and various energy applications of nanofibers of conducting polymers are discussed.

7.2 PREPARATION OF NANOFIBERS OF CONDUCTING POLYMERS

Nanofiber fabrication techniques can be done either by top-down approach or bottom-up approach. Top-down techniques involve both chemical and mechanical treatments and these methods are generally used for the fabrication of cellulose nanofibers (CNF). Bottom-up techniques include electrospinning, template synthesis, self-assembly, phase separation, etc., and are widely used for nanofiber fabrication [7]. Out of these techniques, electrospinning is the most widely used method owing to the cost-effectiveness, simplicity, and flexibility to control the morphology and properties of nanofibers. Properties such as aspect ratio, the surface to volume ratio, and pore interconnectivity of nanofibers prepared by electrospinning method are generally higher than that of other 1D materials prepared using other techniques. Hence electrospun nanofibers are more effective for energy conversion and storage [4, 7]. Two main strategies have been used to prepare the nanofibers of conducting polymers, namely template-based synthesis [11] and template-less synthesis. Template-free method includes interfacial approach [12], seeding approach [13, 14], electrospinning, radiolysis, electrochemical nanowire assembly, soft lithography, etc. [15–17].

7.2.1 TEMPLATE-ASSISTED APPROACH FOR THE PREPARATION OF NANOFIBERS OF CONDUCTING POLYMERS

In this approach, a template is used to guide the nanoparticles of conducting polymers to develop in designed shapes and sizes. A nano-sized channel or porous membrane has been used as a template to restrict the deposition/growth for the fabrication of aligned conducting polymer nanowires [18]. Anodic aluminum oxide and particle track-etched membranes have been widely used with good controllability to produce conducting polymer nanowires/fibers/tubes. A fundamental study on poly(3,4-ethylene dioxythiophene) (PEDOT) nanotubes and nanowires in an anodic aluminum oxide template indicate that high monomer concentration and low electrochemical polymerization potential favor the formation of nanowires [19]. Another interesting study demonstrates the preparation of polypyrrole (PPy) nanowires using anodic aluminum oxide as a template and Ti as the substrate. PPy nanowires were produced along with the anodic aluminum oxide pores by an electrochemical polymerization technique, followed by the removal of the template using NaOH solution [20].

7.2.2 TEMPLATE-FREE APPROACHES FOR THE DEVELOPMENT OF NANOFIBERS OF CONDUCTING POLYMERS

7.2.2.1 Interfacial Approach

The interfacial polymerization reaction is a template-less technique. Higher concentrations of both monomer and dopant anions at the liquid–liquid interface are required in this approach to endorse the formation of monomer-anion (or oligomer-anion) aggregates. Here, the monomer of conducting polymers is dissolved in an organic solvent and the ammonium peroxy disulfate is dissolved in an aqueous acidic solution. This approach has been adopted by different research groups for the development of PANI nanofibers [21, 22]. For instance, aniline is dissolved in an organic solvent (chloroform or toluene, insoluble in water), while the oxidant ammonium persulfate is dissolved in an aqueous solution containing strong acid as a dopant for conducting polymers. The oxidation of aniline occurs around the interface of the two immiscible solvents. Within several minutes, PANI nanofibers were formed at the interface [21]. Completion of reaction is indicated by the color change of the organic phase.

7.2.2.2 Seeding Approach

It is a novel template-free method used for the rapid synthesis of large quantities of conducting polymers nanowires/fibers in one step. Nanoscale polymer morphologies can be tailored in this technique. This approach consists of two techniques, namely nanofibers seeding and sacrificial template for the fabrication of nanofibers of conducting polymers [1]. Out of which, the nanofibers seeding method provided precise control of the morphology in mass quantities [13, 23]. A green, one-step approach, consisting of V_2O_5 nanofiber seeds and ferrous chloride ($FeCl_2$), along with the green oxidant H_2O_2 has been suggested to synthesize bulk quantities of PPy [24].

7.2.2.3 Electrospinning Approach

Electrospinning is found to be the most promising template-less technique. It can be considered as an adaptable method as it is simple, cost-effective, highly efficient, and highly reproducible [25]. Electrospinning was patented by Formhals to produce continuous fibers in 1934 [26]. This method has found many applications in the field of fabrication of 1D polymer nanomaterials, inorganic materials, and micro and nanocomposites. Nanofibers with large specific surface areas, a high aspect ratio, and better pore interconnectivity can be prepared using this method [27]. Also, one can tailor the properties of nanofibers by modifying process parameters including applied voltage, feed rate, collector type, tip-to-collector distance, nozzle design, and calcination treatment [28].

7.2.2.3.1 Technical Aspects of Electrospinning

The electrospinning technique involves the use of polymer solutions/melt as the precursor for the fabrication of nanofibers [29]. A typical electrospinning unit consists of three major components: a high voltage power supply, a spinneret, and a grounded collector [30]. The schematic of a typical electrospinning unit is shown in Figure 7.1. The droplet of the spinning solution is discharged from the capillary needle in a continuous flow. Various designs of electrospun nanofibers can be created by the use of different needles. Figure 7.2 shows various types of needles used in electrospinning. A suitable volatile spinning solution with viscosity values of 1–20 poises and surface tension values of 35–55 dynes/cm^2 are required for inducing fiber formation [31]. An infusion syringe pump can be used to adjust the feed rate of the discharged droplet. Then an electric field is applied at the tip of the capillary needle attached to the syringe to induce an electric charge on the surface of the hanging droplet. When the electric field applied reaches a critical value, the repulsive electrical forces overcome the surface tension forces. Ultimately, a charged jet of the solution is ejected from the tip of the Taylor cone and an unstable and rapid whipping of the jet occurs in the space between the capillary tip and collector, which leads to evaporation of the solvent, leaving a polymer behind. [32–34]. When the solution flow rate is increased, the electric field strength should also be enhanced by applying a higher voltage or by decreasing the distance between the syringe tip and the collector [35].

7.2.2.3.2 Electrospin Approach for the Preparation of Conducting Polymer Nanofibers

The preparation of nanofibers using conducting polymers is a tedious process. The low molecular weight, low elasticity, and high aromaticity of the polymers create difficulty for electrospinning.

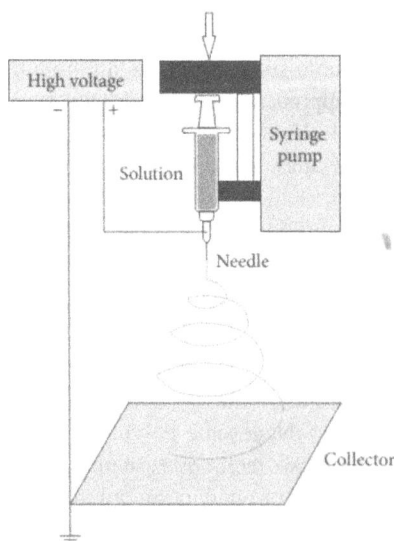

FIGURE 7.1 Schematic representation of an electrospinning setup [36].

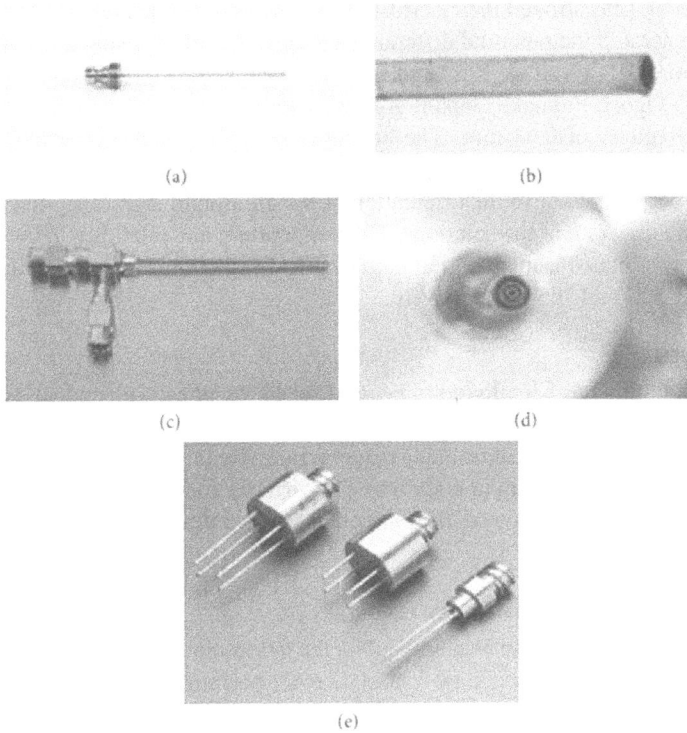

FIGURE 7.2 Images of various needles that can be used in electrospinning techniques: (a) standard needle; (b) inlet of the standard needle; (c) core-sheath needle; (d) inlet of core-sheath needle; and (e) coaxial needle with double or multiple capillary openings [36].

In order to avoid such problems, it is necessary to blend low molecular weight polymers with high molecular weight polymers like poly(ethylene oxide) (PEO), poly(vinyl alcohol) (PVA), polymethyl methacrylate (PMMA), etc. The addition of carrier polymers increases the viscosity of the polymer solution and enhances the ability to spin fibers. Carrier polymer like PEO itself provides no improvement to the electrical properties of the composite fibers and may lead to a reduction in the performance of the fibers as it is an insulating material [25].

PPy is an electrically conducting polymer that is easy to process, possesses good environmental stability, and biocompatibility [37]. Chronakis et al. [38] have developed ultrafine fibers of PPy by blending with poly(ethylene oxide) by electrospinning technique. Higher electrical conductivity has been achieved by increasing the PPy content of the PPy/poly(ethylene oxide) nanofibers. The high initial concentration of poly(ethylene oxide) allows the easy stretching of fibers, which offers higher conductive pathways or charge–carrier mobility of PPy molecules along the fibers [39]. The resulting nanofibers exhibited better electrical conductivity of approximately 3.5×10^{-4} Scm^{-1} for 50 wt% PPy precursors [PPy(SO$_3$H)–DEHS] and 1.1×10^{-4} Scm^{-1} for 37.5 wt% of PPy precursor.

Cong et al. [40] employed electrospinning method to prepare conductive PPy nanofibers with uniform morphology and good mechanical strength. Soluble PPy was manufactured with NaDEHS as a dopant and then applied to electrospinning with or without PEO as a carrier. The PEO contents had a great influence on the morphology and conductivity of the electrospun material. In the typical work, PPy and PEO were dissolved in chloroform with the concentrations 8 wt% and 4 wt%, respectively. PEO and PPy solutions were then mixed with the ratios of 1:10, 1:8, 1:6, 1:4, and 1:2. The 8 wt% PPy and PEO/PPy blends were prepared for electrospinning. Electrospinning was carried out with a 10 kV voltage and 0.8 mL/h feed rate. A ground screen covered by an aluminum sheet was placed 10 cm from the tip of the syringe.

Bhattacharya et al. [25] utilized the electrospinning method for the preparation of PANI nanofibres. They have generated the potential difference for electrospinning using a high-voltage supplier (ES 50P-5W, Gamma High Voltage Research Inc.) and were kept at or below +25 kV. A monoaxial spinneret (MECC, Ogori, Fukuoka, Japan) is fitted with a blunt-tip aluminum needle (23 gauge) having an outer diameter of 0.64 mm. The needle tip-to-collector distance was varied between 20 and 26 cm. Well-aligned conductive fibers having multiple branches to facilitate electron conduction were formed perpendicular to the orientation of the aluminum rods. They have used a syringe pump to control the flow rate of the solutions. The temperature and humidity in the electrospinning box were constantly watched using a digital humidity and temperature monitor (AcuRite) and were maintained at 20 ± 3°C and 16%, respectively.

7.2.2.4 Radiolysis

γ Radiation/UV radiation has been found to be used for the synthesis of conducting polymer nanofibers. [41] This method possesses many advantages like the absence of foreign material, room temperature polymerization, and controllable reaction rate. Radiolytic synthesis of PANI nanofibers (50–100 nm in diameter and 1–3 μm in length) by γ-ray irradiated aqueous solutions of aniline in the presence of initiator has been reported. The nanofibers thus extracted were then doped with acids such as HCl and $HClO_4$ [41].

7.2.2.5 Electrochemical Nanowire Assembly

It is a template-free approach that can be employed for the fabrication of conducting polymer nanofibres/nanowires. PANI nanofibrils were made by controlled multi-potential electropolymerization [42]. They first generated the nucleation sites of PANI on the electrode at a higher potential and obtained nanofibrils finally at a lower potential. Another interesting study demonstrates the fabrication of nanowires of PPy via cathodic electropolymerization using an aqueous solution. They have deposited the as-prepared nanowires directly on the substrate with a nanoporous and interconnected network structure. The effect of time, the concentration of monomer, and the amount of dopant on the morphology of the PPy nanostructures are well elucidated [43].

7.2.2.6 Soft Lithography

Another low-cost, high-resolution, template-free technique used for the fabrication of nanoscale patterns of conducting polymers using micro molds is soft lithography [44–47]. PEDOT: PSS nanowires having a diameter of 278–833 nm were patterned by micromolding in capillaries on glass or a Si wafer. Controlling the height of these grown nanowires was achieved by applying sufficient force on the stamp during micropatterning [48]. The nanoimprint lithography method was used for the direct patterning of functional conducting polymers [49, 50].

7.3 CHARACTERIZATIONS OF NANOFIBERS OF CONDUCTING POLYMERS

Several techniques were employed to analyze the surface morphology and characteristic properties of fabricated nanofibers.

7.3.1 SEM and TEM Analysis

The surface morphologies of the nanofibers can be studied with the help of field emission scanning electron microscope (FESEM) and field emission high-resolution transmission electron microscope (FEHRTEM). The SEM images provide information about the surface morphology of the nanostructures. The SEM images of PPy nanofibers and PPy nanofibers-based composites, including camphor sulfonic acid (CSA), polystyrene sulfonic acid (PTSA), hydrochloric acid (HCA), acetic acid (AA), and diethylene triaminepenta(methylene-phosphonic acid) (DMPA) have 1D helical hollow nanostructures, as shown in Figure 7.3 [51].

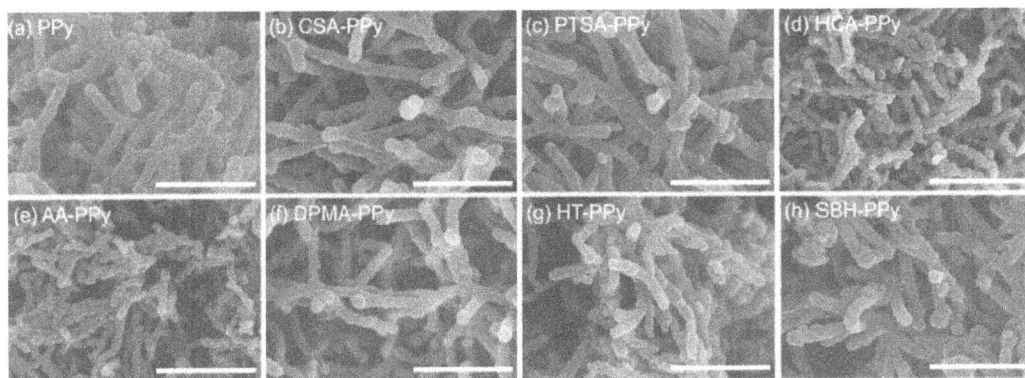

FIGURE 7.3 Scanning electron microscopy images of (a) PPy, (b) CSA-PPy, (c) PTSA-PPy, (d) HCA-PPy, (e) AA-PPy, (f) DMPA-PPy, (g) HT-PPy, and (h) SBH-PPy (1 μm-scale bar) [51].

The electrochemical polymerization of 3,4-ethylene dioxythiophene (EDOT) (monomer) using anodic aluminum oxide (AAO) (template) resulted in the formation of PEDOT nanowires and its SEM images are shown in Figure 7.4. The nanowires showed a 3D arrangement with a diameter of 40 ± 2 nm along the z-direction and the height of the network was about 25–30 mm [52].

FETEM image of nanofibers clearly depicts its structure, diameter, and length. The TEM images also help to identify the alignment of nanorods [52]. If the potential is low (0.65 V), the alignment of PPy nanorods is found to be vertical. With an increase in potential (0.75 V), there is a transformation of polymer nanorods to aligned nanowires occurs. At higher potential, the nanowires are distributed randomly (0.85 V). TEM of the aqueous dispersion of PANI nanofibers shows the uniform fibrous distribution with a diameter of 21–27 nm (Figure 7.5) [53].

Various morphologies and structures of nanofibers that can be prepared by modifying processes can be studied using SEM analysis. Figure 7.6 shows SEM images of various morphologies of nanofibers [8].

7.3.2 FTIR Analysis

The chemical composition of nanofibers can be studied using FTIR spectra. The FTIR spectra of PPy (Figure 7.7a) show characteristic peaks at 1547 cm^{-1} (C=C stretching), 1457 cm^{-1} (C–N stretching), and 1036 cm^{-1} (C–H in-plane deformation mode). The absorption bands observed at 1170 cm^{-1} and 900 cm^{-1} correspond to the doping states of PPy [54]. PANI nanofiber shows peaks at 1562 cm^{-1}

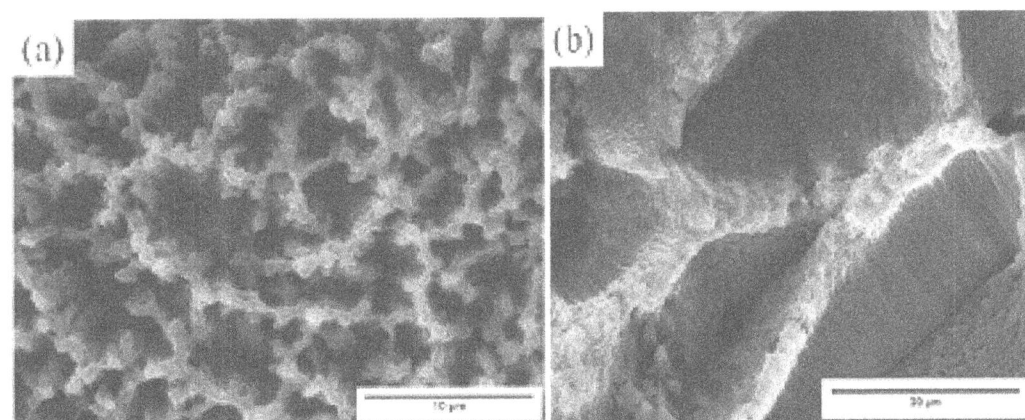

FIGURE 7.4 Scanning electron microscopy images of the 3D arrangement of PEDOT nanowire (a, b) [52].

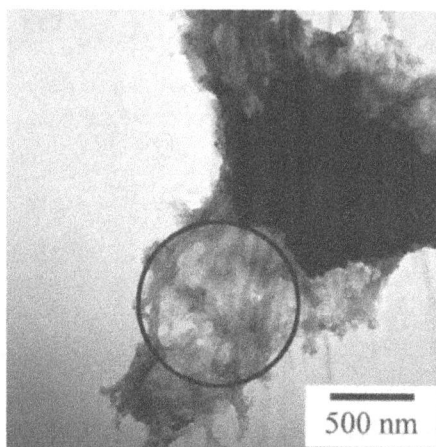

FIGURE 7.5 TEM image of aqueous polyaniline nanofibers dispersion [53].

represent C–C stretching modes in the quinoid while 1493 cm^{-1} assigned to the C–C stretching modes in the benzenoid ring. The C–N, C=N, and C–C stretching vibrations of the benzenoid ring are observed at 1282 cm^{-1}, 1141 cm^{-1}, and 795 cm^{-1}, respectively (Figure 7.7b).

7.3.3 X-Ray Diffraction Analysis

The crystal structure of the nanofibers was determined from X-ray diffraction (XRD) patterns. The orientation and crystallinity of conducting polymers are important for increasing the charge-carrier transport properties, which are obtained by better structural order. Figure 7.8 shows the XRD of

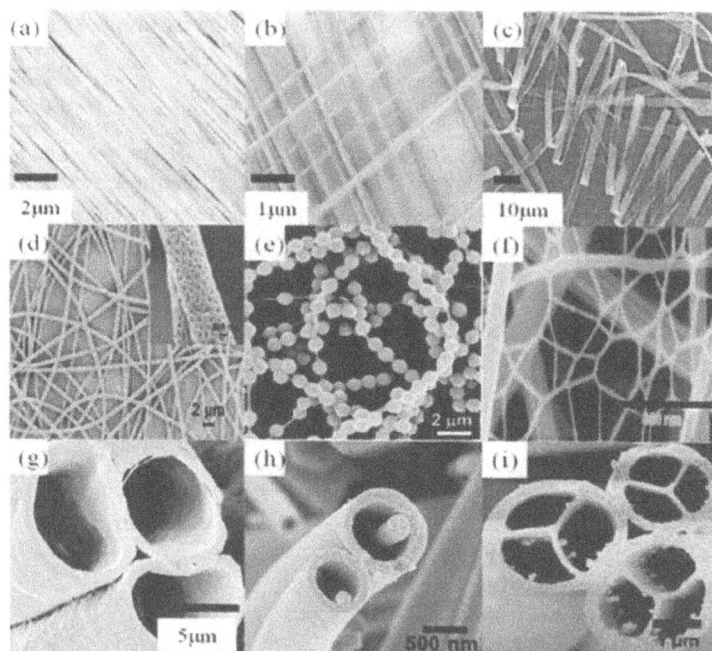

FIGURE 7.6 Different types of nanofiber morphologies: (a) aligned uniaxially, (b) biaxially oriented, (c) ribbon, (d) porous fibers, (e) necklace-like, (f) nanowebs, (g) hollow, (h) nanowire-in-microtube, and (i) multichannel tubular [8].

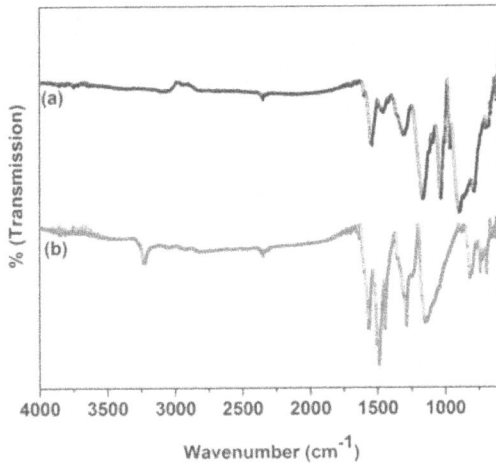

FIGURE 7.7 FTIR spectra results of PPy nanotubes (a) and PANI nanofibers (b) [55].

PPy nanotubes and PANI nanofiber. The partial crystalline structure of PANI nanofibers is represented by the sharp peaks at around 19°, 25°, and 39° in the X-ray diffraction pattern [56]. In the case of PPy, the X-ray diffraction patterns a broad hump in the range of $2\theta = 20°-28°$. The presence of a broad hump represents the amorphous nature of PPy, due to the $\pi-\pi$ interaction of partial PPy chains [55, 57].

7.3.4 ELECTROCHEMICAL CHARACTERIZATION

The cyclic voltammetry characterization of PPy and PANI modified electrode is first measured with a 1 M solution of various electrolytes like H_2SO_4, KCl, Na_2SO_4, and KOH. Then the electrolyte in which the electrode provides superior performance is selected for further studies. The electrochemical performance of PPy and PANI nanofibers were measured at five different scan rates of 5, 10, 20, 50, and 100 mV/s illustrate the capacitive nature of the conducting polymer nanostructures [55]. The galvanostatic charge-discharge behavior of these nanofibers has been examined at various current

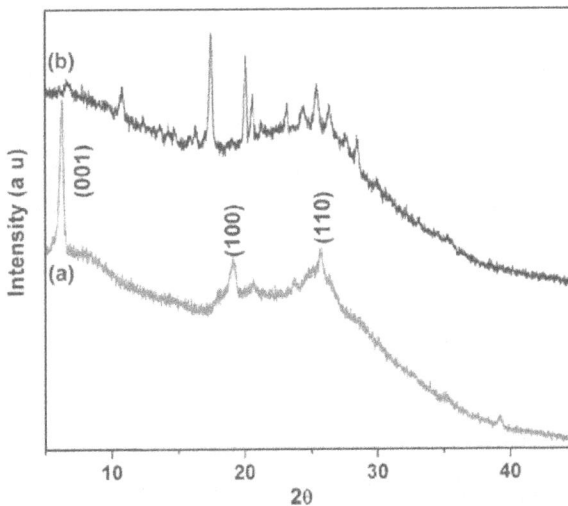

FIGURE 7.8 XRD analysis of PANI nanofiber (a) and PPy nanotubes (b) [55].

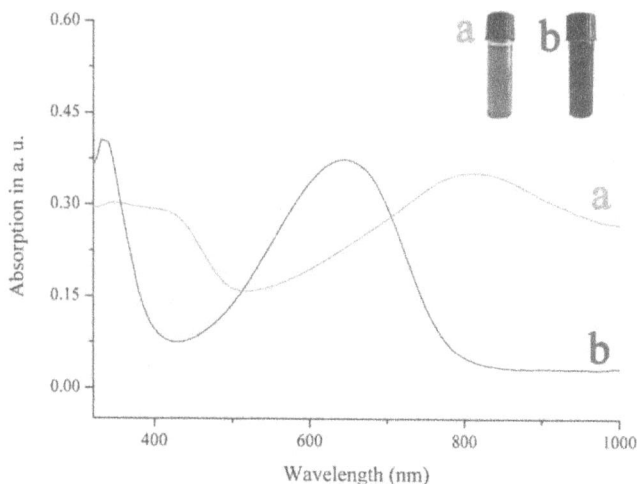

FIGURE 7.9 UV–visible spectra of an aqueous dispersion of polyaniline nanofibers (a) and polyaniline nanofiber emeraldine base in N-methyl-2-pyrrolidone solution (b) [53].

densities such as 0.1, 0.3, 0.5, and 1 A/g. A triangular shape CD curve is observed PPy nanotubes. But PANI nanofiber shows a voltage drop at the initial stage of every charge-discharge curve. The specific capacitance of the electrode is

$$Csp = \frac{I \times \Delta t}{m \times \Delta V}$$

where I represent the discharge current, Δt is discharge time, m is mass of active material and ΔV is the functional potential range [55].

7.3.5 UV–Visible Spectra

UV–visible spectroscopy is another characterization technique for the evaluation of structure. The UV–visible spectra of HCl–doped PANI nanofibers show bands at 330 nm corresponds to π–π* transition and 610 nm represent the π–polaron transition in the benzenoid rings. The band observed at 430 nm is weak and is due to the polaron–π* transition. The presence of this band confirmed the emeraldine salt state of PANI [58]. The doping degree and stability of PANI nanofibers can be improved by the addition of sulfonic acids and it is confirmed by the presence of the strong band at 430 nm. The extension for the conformation of PANI chains is confirmed by when the band at 610 nm is red-shift to above 800 nm [58]. Figure 7.9 represents the UV spectra of PANI nanofibers. The aqueous solution of PANI nanofibers shows a peak at 332 nm. But in emeraldine base (EB), the peak is redshifted to 417 nm, which corresponds to the semiquinoid radical cation (Figure 7.9a). The extended conformation of the PANI chains is confirmed by the strong peak at 815 nm, which indicates the cationic radical polaron band, and at the same time, there is no peak at 646 nm in EB. Figure 7.9b shows the absorption band in EB solution, in which strong absorption peaks at 332 nm and 646 nm were observed due to the transition of the benzenoid structure and the transition of the quinoid structure [53].

7.4 ENERGY APPLICATIONS OF CONDUCTING POLYMER NANOFIBERS

Due to the unique properties of nanofibers of conducting polymers, it can be considered as a good candidate for many energy applications such as in supercapacitors, batteries, solar cells, light-emitting diodes, etc.

7.4.1 SUPERCAPACITORS

Electrochemical capacitors are classified into two types depending on the charge storage method: electric double-layer capacitors and pseudocapacitors. The electric double-layer capacitors are based on carbon electrodes in which energy storage is possible due to the accumulation of electrostatic charges and pseudocapacitors involve Faradaic redox reactions to enable the charge storage. Thus supercapacitors are promising energy storage devices with long life, good cycling stability, simple equipped mechanism, high power density, and can be safely charged and discharged within seconds. Therefore, high electrical conductivity and a large surface area are necessary for the fabrication of high-performance capacitors. Nanofibers of conducting polymers can be satisfied these requirements. In supercapacitors, positive and negative electrodes are separated by an ionic electrolyte (separator). Conducting polymers such as PANI, PPy, PEDOT, and their derivatives have great importance in supercapacitors. Even though conducting polymers have a high capacitance value and create a large total surface charge potential, they suffer from poor cyclic stability due to mechanic instability caused by material swelling and shrinking during the charging and discharging process. Therefore, current research focused on conducting polymer nanostructures having high specific capacitance and good cycling stability in the fabrication of supercapacitors. Fabrication of hybrid materials by incorporation of carbon materials or metal oxides with nanofibers can also lead to the development of high-performance supercapacitors. Even though carbon materials have high conductivity and good cyclic stability, it has low capacitance value. Thereby incorporating with nanofibers, the hybrid materials can provide the synergetic effect of both nanofiber and carbon materials/metal oxides. Hence the hybrid material can perform outstanding conductivity and high capacitance [1, 3, 25, 58].

From the electrical studies of PANI/poly(vinyl alcohol)/ruthenium oxide (RuO_2) composite nanofibers, Chamakh et al. [3] reported that the prepared composite nanofibers can be used as a good material for energy storage applications. They studied electrical properties using impedance spectroscopy tests for the prepared nanofibers as a function of RuO_2 concentration and temperature. They investigated that electrical resistivity and activation energy decreased with increasing RuO_2 concentration for the as-prepared samples, while the annealed samples obtained lower activation energy values of ~0.1 eV with an increase in RuO_2 concentration. Hence they suggested the composite nanofiber for flexible electrochemical double-layer capacitors. In this work, they plotted a graph of measurements of Z′ versus Z″ and obtained a semicircled graph in which the radius of the semicircle indicates the DC resistance of the sample. They observed a decrease in radius with an increase in temperature. This indicates the increased conductivity due to the improved carrier concentrations and mobility of electrons. At high temperatures, ions get enough energy to diffuse toward electrodes [3].

7.4.2 SOLAR CELLS

Even though the solar cells that are being currently used have many advances, it is necessary to improve the solar to the electrical energy conversion efficiency of solar cells. Current research has great attention to it – current strategy is based on the applicability of nanostructured materials. Third-generation solar cells mainly dye-sensitized solar cells and hybrid solar cells still need to improve its conversion efficiency due to the increased energy demand in our day-to-day lives [2, 4, 59].

Dye-sensitized solar cells have great attention other than the conventional solar cells because of their flexibility, low cost, and easy processability. In addition to these properties, dye-sensitized solar cells are stable at low temperature. Dye-sensitized solar cells mainly consist of three parts which include a semiconductor part in which the dye molecule is anchored, counter electrode, and redox electrolyte, in which counter electrodes play a major role to improve the performance of DSSC. Since the electrons are collected at the counter electrode and also the counter electrode act as the catalyzer for the triiodide reduction process. General requirements of a counter electrode mainly include the low resistance and high catalytic efficiency for the triiodide reduction process. Since platinum satisfies these properties, it is used as counter electrodes. But due to its high cost and less availability, platinum cannot be used for large-scale applications. So current research scenario focused on the improvement of the counter electrodes to improve the overall performance of solar cells [2, 59–61].

Therefore, current strategy is based on the usage of nanostructured material. Several studies were also reported based on the applicability of nanostructured conducting polymer to develop advanced solar cells in which nanofibers of conducting polymers can be considered as a good candidate for solar cell applications. Much reported research sheds light on the applicability of nanofibers of conducting polymers in solar cells in which PANI, PEDOT, and PPy have great attention. Also, the incorporation of other conducting materials with conducting polymers has great attention since it can improve the conducting and catalytic properties of hybrid materials, thereby improving the performance of counter electrodes [2, 60, 62]. Very recently, Sheela et al. [59] prepared tungsten diselenide (WSe_2)/PANI composite nanofibers. They investigated the electrochemical studies using cyclic voltammetry.

Yarmohamadi-Vasel et al. reported the applicability of PANI nanofibers/titanium dioxide nanoparticles(PANINFs/TiO_2) composites in solar cells [63]. Sardar et al. developed a light-harvesting material based on poly(diphenyl butadiene) (PDPB) nanofibers and zinc oxide nanoparticles. They reported that the material can be used for the fabrication of solar devices and they reported the development of nanoheterojunctions with high efficiency of light harvesting (Figure 7.10) [2].

Figure 7.10(a, b) represents the SEM images of PDPB nanofibers and PDPB-ZnO light-harvesting nanoheterojunctions and the measured average diameter of nanofiber is 22 nm and length is in the micron range. Figure 7.10(c) represents the TEM image of PDPB nanofiber and the measured average diameter is ~19 nm and a few micrometers in length. This measurement reflects the 1D structure of fabricated nanofibers. Figure 7.10(d) represents the HRTEM of PDPB-ZnO and observed the ZnO NPs embedded in the PDPB nanofibers.

They observed a photocurrent of 7.2 μ A/cm^2 and found that it was 2.5 times higher than pure ZnO NPs. They explained the mechanism as follows: PDPB acts as a p-type, organic semiconductor

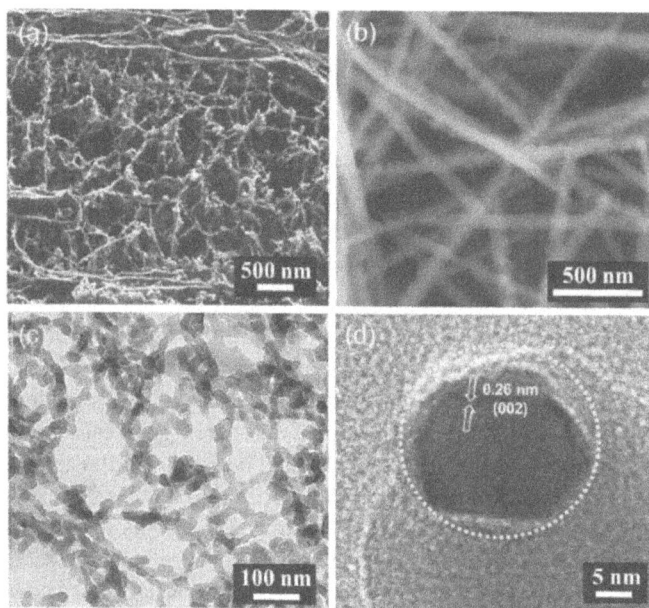

FIGURE 7.10   SEM images of (a) PDPB nanofibers, (b) PDPB-ZnO light-harvesting nanoheterojunction and TEM images of (c) PDPB nanofibers and high-resolution TEM image of (d) PDPB-ZnO nanoheterojunction [2].

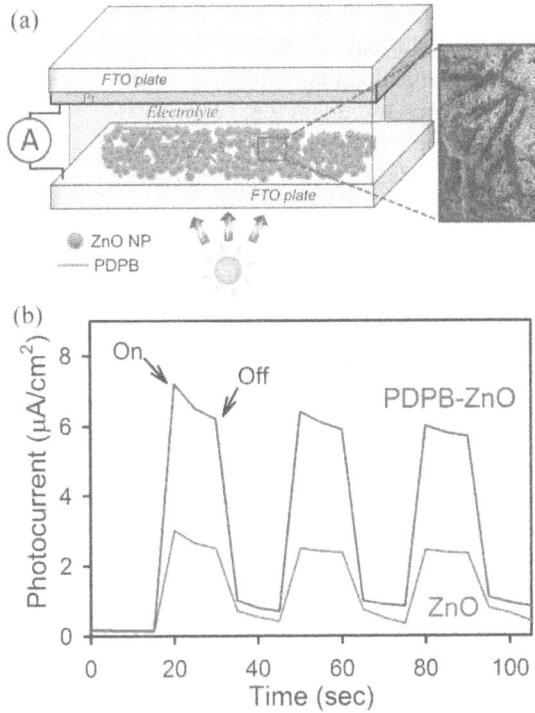

FIGURE 7.11 (a) Schematic picture of photocurrent measurement set up using dye-sensitized solar cell geometry. (b) Photocurrent responses of PDPB-ZnO light-harvesting nanoheterojunction and ZnO NPs [2].

and ZnO acts as an inorganic, n-type semiconductor, therefore a donor–acceptor junction is formed. They considered it as a heterojunction. Under visible light, as a result of the illumination excitation of electrons occurs and it transfers from the highest occupied molecular orbital (HOMO) to the lowest unoccupied molecular orbital (LUMO) of PDPB. Then the excited electrons jump into the conduction band of ZnO (schematic representation given in Figure 7.11a). Hence, they observed enhanced photoresponse and this may be due to the formation of heterojunctions. Figure 7.11(b) shows the photocurrent response of ZnO nanoparticles and PDPB-ZnO light-harvesting nanoheterojunction [2].

Pierini et al. developed fullerene-grafted polythiophene-based single-material organic solar cells with high conversion efficiency. They reported that the fabricated material obtained a high power conversion efficiency of 5.58% [62]. Peng et al. fabricated PANI-polylactic acid composite nanofibers and studied its electrocatalytic performance using electrochemical impedance spectra and cyclic voltammetry measurements. Based on their studies they observed that the fabricated nanofiber achieved a high conversion efficiency of solar energy. And they reported that it can be used as a counter electrode for rigid and flexible dye-sensitized solar cells [60]. Similarly, Lee et al. fabricated a 10–50 nm diameter ranged PEDOT nanofibers for dye-sensitized solar cells as an efficient catalytic counter electrode. They observed low surface resistance and highly porous surfaces with high power conversion efficiency [61].

7.4.3 BATTERIES

One of the most important energy storage applications of nanofibers of conducting polymers involves the usage of cathode materials for batteries. In an electric battery, electricity is produced due to the flow of electrons in a circuit, as part of the chemical reactions taking place in it. Batteries have a positive electrode, a negative electrode, and an electrolyte. The electrolytes can chemically react with electrodes. Lithium-ion (Li-ion) batteries are widely used batteries. In addition, sodium-ion batteries have received much attention than lithium-ion batteries because of their low cost and easy availability of sodium compared to lithium [63–66].

Conventional battery materials have some disadvantages like high cost, low capacity, and low efficiency; therefore, it limits its widespread usage. Metal oxides are generally used as cathodes in rechargeable batteries. But high oxidation potential of metal oxide may cause a decomposition reaction of electrolytes. However, it is very important to develop new advanced electrode materials with high cyclic stability and capacity for batteries. In order to find a solution, the current research scenario focused on nanotechnology. Since nanotechnology can find a solution to develop new electrochemically active alternatives with long shelf life, high efficiency, high capacity, energy density, and low price. Since polymer-based batteries have a great significance towards the fabrication of high energy density batteries. Several research studies were reported based on the applicability of conducting polymers as cathode materials. Due to the π-conjugated bonds in conducting polymers, it shows usual properties than other polymers such as high electron affinity and low energy electronic transitions, and thereby it can be oxidized and reduced easily [9, 61].

Nanostructured/modified conducting polymers receive great attention. Nanofibers of conducting polymers have been recognized as an important material for energy storage applications. The nanofibers of conducting polymers can be used for advanced flexible materials for energy applications. Nanofibers are 1D nanomaterial, therefore it have unique properties such as high surface-to-volume ratio, high aspect ratio, and high conductivity it can offer shorter ion pathways and faster diffusion kinetics for ions in electrodes. In addition to that conducting polymers have high reversible doping/dedoping properties and have good electrical/ionic conducting connectivity. Therefore the 1D nanostructures of conducting polymers with controlled morphology, diameter, and wall sizes have great significance for energy applications. Also, the carbon, metal oxide incorporated nanofibers can make tremendous changes in this field [2, 59, 61, 64–67].

Lu et al. synthesized water dispersed conducting PANI nanofibers. They investigated the applicability of the fabricated material as batteries and found that the fabricated nanofibers can well catalyze the discharge reaction with good cycle stability. Also, they observed that it gives the lithium–oxygen battery a much higher discharge capacity than the PANI cathode with nonfibrous morphology [66]. Han et al. recently prepared polyaniline hollow nanofibers (PANI-HNFs) and studied its applicability as a cathode material. They investigated that it can be used as cathode material for batteries since it exhibited a high reversible capacity and high cycling stability (73.3% capacity retention after 1000 cycles) [64].

7.4.4 MISCELLANEOUS ENERGY APPLICATIONS OF NANOFIBERS OF CONDUCTING POLYMERS

Conducting polymer nanofibers can be used in field-effect transistors. Several studies were reported based on the applicability of conducting polymer nanofibers in field-effect transistors [68–70]. Au/template synthesized PPy nanofiber devices with rectifying behavior were reported by Liu et al.; they reported that these can be used as nano rectifiers [71]. The unique properties of conducting polymer nanofibers shed light on the applicability of these nanostructured materials for light-emitting diodes, fuel cells, field emission, transistors, and electrochromic displays. Hydrogen fuel cells are considered to be the greatest future energy device. Because hydrogen is the most efficient source of renewable energy, it can be used for many energy applications without any pollution problems. Fuel cells convert chemical energy into electrical energy. The by-products of fuel cells are water and heat. Therefore it has great importance in our future life compared to other fossil fuels. In addition to that, the heat produced can be used for other domestic applications. Nanofibers of conducting polymers can be used for fuel cells and have high hydrogen storage capacity [72]. The nanofibers of PANI, PEDOT, and PPy are well studied. The reported works suggested that these nanofibers exhibited stable field-emission behavior with a low threshold voltage. Therefore it can be used for field emission displays as nanotips. Some studies reported that conducting polymer nanostructures can be used as active layers in electrochromic devices due to their ability to change color under an applied potential [73, 74]. By a doping and dedoping process, the electrical and optical properties of nanofibers of conducting polymers can be made reversible. Therefore these nanofibers have a great role in chemical, biological, optical, and gas sensors in which PANI, PEDOT and PPy nanofibers have great exposure as sensors since it shows higher sensitivity and response within a very short time [70, 75–77].

7.5 CONCLUSIONS

Increasing energy demands for current life are looking forward to the fabrication of efficient energy devices. Due to the unique physical and chemical properties of 1D nanostructure such as nanowires, nanotubes, nanorods, nanofibers, etc. have received great attention toward the fabrication of energy devices, in which the nanofibers synthesized by electrospinning have a larger surface area, a high aspect ratio, and have high pore connectivity, which are the major requirements for energy applications. Compared to bulk polymers, nanostructured conducting polymers have specific properties such as high surface area, good conductivity, and optical properties; therefore, nanofibers of conducting polymers are focused on the current area of research and for potential energy applications.

REFERENCES

1. S. Ghosh, T. Maiyalagan, R.N. Basu, Nanostructured conducting polymers for energy applications: towards a sustainable platform, Nanoscale 8(13) (2016) 6921–6947.
2. S. Sardar, P. Kar, H. Remita, B. Liu, P. Lemmens, S.K. Pal, S. Ghosh, Enhanced charge separation and FRET at heterojunctions between semiconductor nanoparticles and conducting polymer nanofibers for efficient solar light harvesting, Scientific Reports 5(1) (2015) 1–14.
3. M. Chamakh, A.I. Ayesh, M.F. Gharaibeh, Fabrication and characterization of flexible ruthenium oxide-loaded polyaniline/poly (vinyl alcohol) nanofibers, Journal of Applied Polymer Science 137(38) (2020) 49125.
4. X. Shi, W. Zhou, D. Ma, Q. Ma, D. Bridges, Y. Ma, A. Hu, Electrospinning of nanofibers and their applications for energy devices, Journal of Nanomaterials 2015 (2015) 140726.
5. V. Thavasi, G. Singh, S. Ramakrishna, Electrospun nanofibers in energy and environmental applications, Energy & Environmental Science 1(2) (2008) 205–221.
6. W. Wang, W. Li, R. Zhang, The preparation of conducting polymer micro-and nanofibers by electrospinning, Journal of Polymer Materials 27(4) (2010) 293.
7. A. Barhoum, K. Pal, H. Rahier, H. Uludag, I.S. Kim, M. Bechelany, Nanofibers as new-generation materials: from spinning and nano-spinning fabrication techniques to emerging applications, Applied Materials Today 17 (2019) 1–35.
8. Z. Su, J. Ding, G. Wei, Electrospinning: A facile technique for fabricating polymeric nanofibers doped with carbon nanotubes and metallic nanoparticles for sensor applications, RSC Advances 4(94) (2014) 52598–52610.
9. K. Ghanbari, M. Mousavi, M. Shamsipur, Preparation of polyaniline nanofibers and their use as a cathode of aqueous rechargeable batteries, Electrochimica Acta 52(4) (2006) 1514–1522.
10. E. Llorens, E. Armelin, M. del Mar Pérez-Madrigal, L.J. Del Valle, C. Alemán, J. Puiggalí, Nanomembranes and nanofibers from biodegradable conducting polymers, Polymers 5(3) (2013) 1115–1157.
11. Y. Liu, J. Goebl, Y. Yin, Templated synthesis of nanostructured materials, Chemical Society Reviews 42(7) (2013) 2610–2653.
12. H.D. Tran, J.M. D'Arcy, Y. Wang, P.J. Beltramo, V.A. Strong, R.B. Kaner, The oxidation of aniline to produce "polyaniline": a process yielding many different nanoscale structures, Journal of Materials Chemistry 21(11) (2011) 3534–3550.
13. X. Zhang, S.K. Manohar, Bulk synthesis of polypyrrole nanofibers by a seeding approach, Journal of the American Chemical Society 126(40) (2004) 12714–12715.
14. X. Zhang, A.G. MacDiarmid, S.K. Manohar, Chemical synthesis of PEDOT nanofibers, Chemical Communications (42) (2005) 5328–5330.

15. Z. Sun, E. Zussman, A.L. Yarin, J.H. Wendorff, A. Greiner, Compound core–shell polymer nanofibers by co-electrospinning, Advanced Materials 15(22) (2003) 1929–1932.
16. Y. Srivastava, I. Loscertales, M. Marquez, T. Thorsen, Electrospinning of hollow and core/sheath nanofibers using a microfluidic manifold, Microfluidics and Nanofluidics 4(3) (2008) 245–250.
17. D. Li, Y. Wang, Y. Xia, Electrospinning of polymeric and ceramic nanofibers as uniaxially aligned arrays, Nano Letters 3(8) (2003) 1167–1171.
18. C.R. Martin, Nanomaterials: a membrane-based synthetic approach, Science 266(5193) (1994) 1961–1966.
19. S.I. Cho, S.B. Lee, Fast electrochemistry of conductive polymer nanotubes: synthesis, mechanism, and application, Accounts of Chemical Research 41(6) (2008) 699–707.
20. S. Cui, Y. Zheng, J. Liang, D. Wang, Conducting polymer PPy nanowire-based triboelectric nanogenerator and its application for self-powered electrochemical cathodic protection, Chemical Science 7(10) (2016) 6477–6483.
21. X. Zhang, R. Chan-Yu-King, A. Jose, S.K. Manohar, Nanofibers of polyaniline synthesized by interfacial polymerization, Synthetic Metals 145(1) (2004) 23–29.
22. X. Zhang, H.S. Kolla, X. Wang, K. Raja, S.K. Manohar, Fibrillar growth in polyaniline, Advanced Functional Materials 16(9) (2006) 1145–1152.
23. X. Zhang, W.J. Goux, S.K. Manohar, Synthesis of polyaniline nanofibers by "nanofiber seeding", Journal of the American Chemical Society 126(14) (2004) 4502–4503.
24. Z. Liu, Y. Liu, S. Poyraz, X. Zhang, Green-nano approach to nanostructured polypyrrole, Chemical Communications 47(15) (2011) 4421–4423.
25. S. Bhattacharya, I. Roy, A. Tice, C. Chapman, R. Udangawa, V. Chakrapani, J.L. Plawsky, R.J. Linhardt, High-conductivity and high-capacitance electrospun fibers for supercapacitor applications, ACS Applied Materials & Interfaces 12(17) (2020) 19369–19376.
26. F. Anton, Process and apparatus for preparing artificial threads, Google Patents, 1934.
27. A. Frenot, I.S. Chronakis, Polymer nanofibers assembled by electrospinning, Current Opinion in Colloid & Interface Science 8(1) (2003) 64–75.
28. Y. Xue, S. Chen, J. Yu, B.R. Bunes, Z. Xue, J. Xu, B. Lu, L. Zang, Nanostructured conducting polymers and their composites: synthesis methodologies, morphologies and applications, Journal of Materials Chemistry C 8(30) (2020) 10136–10159.
29. M. Mirjalili, S. Zohoori, Review for application of electrospinning and electrospun nanofibers technology in textile industry, Journal of Nanostructure in Chemistry 6(3) (2016) 207–213.
30. Y. Zheng, S. Xie, Y. Zeng, Electric field distribution and jet motion in electrospinning process: from needle to hole, Journal of Materials Science 48(19) (2013) 6647–6655.
31. C. Luo, M. Nangrejo, M. Edirisinghe, A novel method of selecting solvents for polymer electrospinning, Polymer 51(7) (2010) 1654–1662.
32. X. Fang, D. Reneker, DNA fibers by electrospinning, Journal of Macromolecular Science, Part B: Physics 36(2) (1997) 169–173.
33. A. Haider, S. Haider, I.-K. Kang, A comprehensive review summarizing the effect of electrospinning parameters and potential applications of nanofibers in biomedical and biotechnology, Arabian Journal of Chemistry 11(8) (2018) 1165–1188.
34. V. Pillay, C. Dott, Y.E. Choonara, C. Tyagi, L. Tomar, P. Kumar, L.C. du Toit, V.M. Ndesendo, A review of the effect of processing variables on the fabrication of electrospun nanofibers for drug delivery applications, Journal of Nanomaterials 2013 (2013) 789289.
35. W.-E. Teo, R. Inai, S. Ramakrishna, Technological advances in electrospinning of nanofibers, Science and Technology of Advanced Materials 12(1) (2011) 013002.
36. S.I. Abd Razak, I.F. Wahab, F. Fadil, F.N. Dahli, A.Z. Md Khudzari, H. Adeli, A review of electrospun conductive polyaniline based nanofiber composites and blends: processing features, applications, and future directions, Advances in Materials Science and Engineering 2015 (2015) 356286.
37. T.S. Kang, S.W. Lee, J. Joo, J.Y. Lee, Electrically conducting polypyrrole fibers spun by electrospinning, Synthetic Metals 153(1–3) (2005) 61–64.
38. I.S. Chronakis, S. Grapenson, A. Jakob, Conductive polypyrrole nanofibers via electrospinning: electrical and morphological properties, Polymer 47(5) (2006) 1597–1603.
39. S. Nair, S. Natarajan, S.H. Kim, Fabrication of electrically conducting polypyrrole-poly (ethylene oxide) composite nanofibers, Macromolecular Rapid Communications 26(20) (2005) 1599–1603.
40. Y. Cong, S. Liu, H. Chen, Fabrication of conductive polypyrrole nanofibers by electrospinning, Journal of Nanomaterials 2013 (2013).
41. S.K. Pillalamarri, F.D. Blum, A.T. Tokuhiro, J.G. Story, M.F. Bertino, Radiolytic synthesis of polyaniline nanofibers: a new templateless pathway, Chemistry of Materials 17(2) (2005) 227–229.

42. X. Yu, Y. Li, K. Kalantar-zadeh, Synthesis and electrochemical properties of template-based polyaniline nanowires and template-free nanofibril arrays: Two potential nanostructures for gas sensors, Sensors and Actuators B: Chemical 136(1) (2009) 1–7.

43. D.-H. Nam, M.-J. Kim, S.-J. Lim, I.-S. Song, H.-S. Kwon, Single-step synthesis of polypyrrole nanowires by cathodic electropolymerization, Journal of Materials Chemistry A 1(27) (2013) 8061–8068.

44. D. Qin, Y. Xia, G.M. Whitesides, Soft lithography for micro-and nanoscale patterning, Nature Protocols 5(3) (2010) 491–502.

45. Z. Nie, E. Kumacheva, Patterning surfaces with functional polymers, Nature Materials 7(4) (2008) 277–290.

46. M. Geissler, Y. Xia, Patterning: principles and some new developments, Advanced Materials 16(15) (2004) 1249–1269.

47. C. Acikgoz, M.A. Hempenius, J. Huskens, G.J. Vancso, Polymers in conventional and alternative lithography for the fabrication of nanostructures, European Polymer Journal 47(11) (2011) 2033–2052.

48. F. Zhang, T. Nyberg, O. Inganäs, Conducting polymer nanowires and nanodots made with soft lithography, Nano Letters 2(12) (2002) 1373–1377.

49. M. Behl, J. Seekamp, S. Zankovych, C.M. Sotomayor Torres, R. Zentel, J. Ahopelto, Towards plastic electronics: patterning semiconducting polymers by nanoimprint lithography, Advanced Materials 14(8) (2002) 588–591.

50. C. Huang, B. Dong, N. Lu, B. Yang, L. Gao, L. Tian, D. Qi, Q. Wu, L. Chi, A strategy for patterning conducting polymers using nanoimprint lithography and isotropic plasma etching, Small 5(5) (2009) 583–586.

51. M. Sun, C. Xu, J. Li, L. Xing, T. Zhou, F. Wu, Y. Shang, A. Xie, Protonic doping brings tuneable dielectric and electromagnetic attenuated properties for polypyrrole nanofibers, Chemical Engineering Journal 381 (2020) 122615.

52. L. Bach-Toledo, B.M. Hryniewicz, L.F. Marchesi, L.H. Dall'Antonia, M. Vidotti, F. Wolfart, Conducting polymers and composites nanowires for energy devices: a brief review, Materials Science for Energy Technologies 3 (2020) 78–90.

53. H. Zhang, Q. Zhao, S. Zhou, N. Liu, X. Wang, J. Li, F. Wang, Aqueous dispersed conducting polyaniline nanofibers: promising high specific capacity electrode materials for supercapacitor, Journal of Power Sources 196(23) (2011) 10484–10489.

54. P. Asen, S. Shahrokhian, One step electrodeposition of V_2O_5/polypyrrole/graphene oxide ternary nanocomposite for preparation of a high performance supercapacitor, International Journal of Hydrogen Energy 42(33) (2017) 21073–21085.

55. J. Upadhyay, T.M. Das, R. Borah, K. Acharjya, Electrochemical performance evaluation of polyaniline nanofibers and polypyrrole nanotubes, Materials Today: Proceedings 32 (2020) 274–279.

56. H.-W. Park, T. Kim, J. Huh, M. Kang, J.E. Lee, H. Yoon, Anisotropic growth control of polyaniline nanostructures and their morphology-dependent electrochemical characteristics, ACS Nano 6(9) (2012) 7624–7633.

57. Y. Ma, S. Jiang, G. Jian, H. Tao, L. Yu, X. Wang, X. Wang, J. Zhu, Z. Hu, Y. Chen, CNx nanofibers converted from polypyrrole nanowires as platinum support for methanol oxidation, Energy & Environmental Science 2(2) (2009) 224–229.

58. Y. Shen, Z. Qin, T. Li, F. Zeng, Y. Chen, N. Liu, Boosting the supercapacitor performance of polyaniline nanofibers through sulfonic acid assisted oligomer assembly during seeding polymerization process, Electrochimica Acta 356 (2020) 136841.

59. S.E. Sheela, V. Murugadoss, R. Sittaramane, S. Angaiah, Development of tungsten diselenide/polyaniline composite nanofibers as an efficient electrocatalytic counter electrode material for dye-sensitized solar cell, Solar Energy 209 (2020) 538–546.

60. S. Peng, P. Zhu, Y. Wu, S.G. Mhaisalkar, S. Ramakrishna, Electrospun conductive polyaniline–polylactic acid composite nanofibers as counter electrodes for rigid and flexible dye-sensitized solar cells, RSC Advances 2(2) (2012) 652–657.

61. T.H. Lee, K. Do, Y.W. Lee, S.S. Jeon, C. Kim, J. Ko, S.S. Im, High-performance dye-sensitized solar cells based on PEDOT nanofibers as an efficient catalytic counter electrode, Journal of Materials Chemistry 22(40) (2012) 21624–21629.

62. F. Pierini, M. Lanzi, P. Nakielski, S. Pawłowska, O. Urbanek, K. Zembrzycki, T.A. Kowalewski, Single-material organic solar cells based on electrospun fullerene-grafted polythiophene nanofibers, Macromolecules 50(13) (2017) 4972–4981.

63. M. Yarmohamadi-Vasel, A.R. Modarresi-Alam, M. Noroozifar, M.S. Hadavi, An investigation into the photovoltaic activity of a new nanocomposite of (polyaniline nanofibers)/(titanium dioxide nanoparticles) with different architectures, Synthetic Metals 252 (2019) 50–61.

64. H. Han, H. Lu, X. Jiang, F. Zhong, X. Ai, H. Yang, Y. Cao, Polyaniline hollow nanofibers prepared by controllable sacrifice-template route as high-performance cathode materials for sodium-ion batteries, Electrochimica Acta 301 (2019) 352–358.

65. F. Cheng, W. Tang, C. Li, J. Chen, H. Liu, P. Shen, S. Dou, Conducting poly (aniline) nanotubes and nanofibers: controlled synthesis and application in lithium/poly (aniline) rechargeable batteries, Chemistry–A European Journal 12(11) (2006) 3082–3088.

66. Q. Lu, Q. Zhao, H. Zhang, J. Li, X. Wang, F. Wang, Water dispersed conducting polyaniline nanofibers for high-capacity rechargeable lithium–oxygen battery, ACS Macro Letters 2(2) (2013) 92–95.

67. F. Yin, X. Liu, Y. Zhang, Y. Zhao, A. Menbayeva, Z. Bakenov, X. Wang, Well-dispersed sulfur anchored on interconnected polypyrrole nanofiber network as high performance cathode for lithium-sulfur batteries, Solid State Sciences 66 (2017) 44–49.

68. S.-Y. Lee, G.-R. Choi, H. Lim, K.-M. Lee, S.-K. Lee, Electronic transport characteristics of electrolyte-gated conducting polyaniline nanowire field-effect transistors, Applied Physics Letters 95(1) (2009) 013113.

69. N. Pinto, A. Johnson Jr, A. MacDiarmid, C. Mueller, N. Theofylaktos, D. Robinson, F. Miranda, Electrospun polyaniline/polyethylene oxide nanofiber field-effect transistor, Applied Physics Letters 83(20) (2003) 4244–4246.

70. Y.-Z. Long, M.-M. Li, C. Gu, M. Wan, J.-L. Duvail, Z. Liu, Z. Fan, Recent advances in synthesis, physical properties and applications of conducting polymer nanotubes and nanofibers, Progress in Polymer Science 36(10) (2011) 1415–1442.

71. L. Liu, Y. Zhao, N. Jia, Q. Zhou, C. Zhao, M. Yan, Z. Jiang, Electrochemical fabrication and electronic behavior of polypyrrole nano-fiber array devices, Thin Solid Films 503(1–2) (2006) 241–245.

72. N. Mahato, H. Jang, A. Dhyani, S. Cho, Recent progress in conducting polymers for hydrogen storage and fuel cell applications, Polymers 12(11) (2020) 2480.

73. B. Kim, M. Kim, K. Park, J. Lee, D. Park, J. Joo, S. Yu, S. Lee, Characteristics and field emission of conducting poly (3,4-ethylenedioxythiophene) nanowires, Applied Physics Letters 83(3) (2003) 539–541.

74. H. Yan, L. Zhang, J. Shen, Z. Chen, G. Shi, B. Zhang, Synthesis, property and field-emission behaviour of amorphous polypyrrole nanowires, Nanotechnology 17(14) (2006) 3446.

75. L. Xia, Z. Wei, M. Wan, Conducting polymer nanostructures and their application in biosensors, Journal of Colloid and Interface Science 341(1) (2010) 1–11.

76. Y. Tao, Y. Zhang, S. Yao, J. Zhu, C. Wang, Y. Liu, J. Yang, Two functional nanofibers of polyaniline with application in supercapacitor and photosensor, Journal of Nanoelectronics and Optoelectronics 15(2) (2020) 291–300.

77. V. Anju, P. Jithesh, S.K. Narayanankutty, A novel humidity and ammonia sensor based on nanofibers/polyaniline/polyvinyl alcohol, Sensors and Actuators A: Physical 285 (2019) 35–44.

8 Conducting Polymers for Organic Solar Cell Applications

Vinod Kumar[1,2], Leta Tesfaye Jule[1,2],
and Krishnaraj Ramaswamy[2,3]

[1]Department of Physics, College of Natural and Computation Science, Dambi Dollo University, Ethiopia

[2]Centre for Excellence in Indigenous Knowledge, Innovative Technology Transfer and Entrepreneurship, Dambi Dollo University, Ethiopia

[3]Department of Mechanical Engineering, Dambi Dollo University, Dembi Dolo, Ethiopia

CONTENTS

8.1 INTRODUCTION

Polymers have been reported widely in the last few decades by the scientific community due to their versatile and adjustable chemical properties [1]. The word "polymer" comes from the Greek words "poly," meaning many, and "meros," meaning parts. Therefore, the meaning of "polymer" is "many parts." Most of the large molecules or macromolecules are polymers. Polymers are composed of macromolecules, which have many repeated subunits called monomers [2, 3]. These repeating units serve as the building blocks of a polymer. The structures of polymers determine if the template is used to fabricate the mesoporous materials or used for a polymeric matrix in the solid electrolyte [4–6]. It has also been used as a counter electrode due to the high catalytic activity for I_3 reduction. The various functional groups of polymers make it possible to control the morphology of perovskite for bulk and interface. Polymers are used as electron and hole transfer materials due to high carrier mobilities. The different functional groups in the polymers determine the use of polymers as an interface layer, which helps in passivation of defects and adjusts the work function of the metal electrode. Therefore an overall improvement in the device's performance is observed [7, 8]. Modifications in structures and the functional group also allow polymers for various optical adsorption properties. Other than this, the variable electron mobility is also used as the photoactive layer or buffer layer in organic solar cells (OSCs) [9, 10]. The processability of polymers also makes it possible to produce polymer-based micro-/nanostructure devices. Different kinds of polymers have already been used in many electronic applications, while conducting polymers have been known for the last few years. Partial polymers with high conductivity are called conductive polymers, and they are broadly used in a range of applications [11–15].

8.2 BASICS OF CONDUCTING POLYMER

Naturally, organic polymers mostly show insulator behavior. However, a few fundamentally conducting polymers exist that have shown alternating single and double bonds along with the polymer backbone or conjugated bonds. These are composed of aromatic rings such as phenylene, pyrrole, anthracene, naphthalene, and thiophene. These are associated with each other by single bonds of carbon–carbon. Conductive polymers are organic polymers that allow the flow of current by the carrier. The theory of orbital hybridization can be explained by the charge transport in the conducting polymer, as in methane, which is a simple hydrocarbon and makes a bond of four hydrogen atoms with one carbon atom. This theory explains that the outer "s" and "p" shell atoms mix with each other to form hybrid orbitals and have the same energy level. One of the two electrons in the 2-s orbital is excited and moves into the last s-p orbital to satisfy Hund's law. This allows the 4 s-p hybrid orbitals to share their electrons with the single electron in the 1-s orbital of hydrogen [16, 17].

Benzene, with an empirical formula of C_6H_6, has alternating double bonds in its structure. It denotes the presence of free, unbonded p orbitals on neighboring carbon atoms. Due to their unbonded nature, the p orbitals of the carbon atoms will start to intermix. This intermixing of the empty p_z orbitals therefore results in the formation of one big delocalized p_z orbital, as shown in Figure 8.1. This delocalization of p_z orbitals allows for the free movement of charge carriers. This means that in the intermixed p orbital, charge carrier transport is possible. This sideways intermixing of p orbitals can occur either constructively or destructively [18]. The constructive intermixing of these orbitals gives rise to a delocalized pi orbital with a lower energy level. Destructive intermixing, on the other hand, gives rise to an antibonding pi state with a higher energy level. This destructive and constructive intermixing splitting of molecular orbitals is what forms the bandgap in such organics. The lower energy state is preferred by electrons and is occupied under normal room temperature conditions. This energy state is entitled as the highest occupied molecular orbital, or HOMO. The unoccupied higher energy state is known as the lowest unoccupied molecular orbital, or LUMO [19]. The HOMO therefore corresponds to the valence band edge in semiconductors, and the LUMO is correspondent to the conduction band edge. This is how organic materials can show semiconductor-like properties. So, the carrier transport in the OSCs occurs through the bonding of unsaturated hydrocarbons. Delocalized pi orbitals and the consequent formation of molecular orbitals are caused by the constructive and destructive overlap of the p_z orbitals. This in turn gives rise to energy states equivalent to the valence and conduction band in inorganic semiconductor materials. This helps us define the bandgap in these materials.

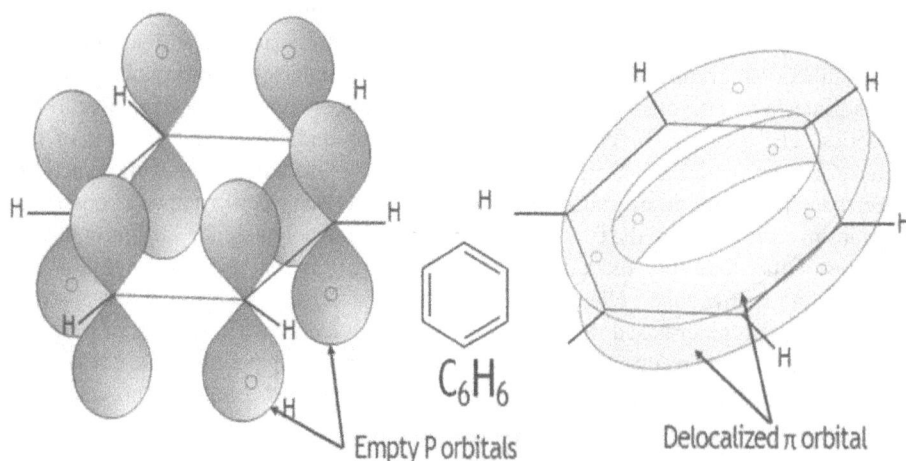

C_6H_6

Empty P orbitals

Delocalized π orbital

FIGURE 8.1 The spatial arrangement of the carbon p orbitals for delocalized p_z orbitals.

8.3 POLYMERS FOR DIFFERENT KIND OF SOLAR CELLS

As a result of the quickly rising population, the world economy has led to a continuously increasing in the energy demand of worldwide [20, 21]. The enormous consumption of traditional fossil energy has caused serious resource exhaustion and pollution of the environment. One of the most urgent solutions is to search for an alternative renewable source. Solar energy is an environmentally friendly resource, free from regional restrictions. The photovoltaic technology is one of the most efficient ways of utilization solar energy. Most research and development is going on silicon-based solar cells. However, the power conversion efficiency (PCE) of silicon-based solar cells on a laboratory scale has already been achieved more than 25% [22]. The fabrication of silicon-based solar cell was limited by the sophisticated manufacturing process and high cost with high energy consumption [23]. In thin-film-based second-generation solar cells such as cadmium telluride (CdTe), gallium arsenide (GaAs), copper indium gallium selenide (CIGS) [24–26] are also inattentive. Although thin-film-based second-generation solar cells have high PCE with good stability, the abundance of gallium and indium in the crust is low, and cadmium is a toxic material. Other than the above two kinds of solar cell, the third-generation solar cells such as dye-sensitized solar cell (DSSC), perovskite solar cell (PSC), and OSC are also in the main field of research on solar cells [27–32]. Due to lightweight and low cost, the OSCs are the emerging solar cells in third-generation solar cells, although fabrication of large-scale OSC is limited. OSCs are needed to be enhanced in the PCE, spectrum response range as well as the device stability [33]. The liquid electrolyte-based DSSCs have always faced a problem of leakage of electrolyte and evaporation of solvent, which challenges the long-term stability of the DSSCs [34]. Solid-based PSC is fabricated using the idea of DSSC, but it overcomes issues of DSSC. Therefore, PSC has achieved more attention from scientists. Although the PCE of PSC has improved rapidly from 3.8% to 23.3%, the stability of the device is still a big challenge that limits the applications of PSC [35–37]. The efficiency curve for different kinds of solar cells is presented in Figure 8.2. Generally, the performances of devices are influenced by each component of the device to perform their respective tasks. Therefore, optimizing the device components is of great significance. Hence, conducting polymer can be used for improving the PCE of different kinds of solar cells.

8.4 POLYMER-BASED ORGANIC SOLAR CELL

OSCs are mainly promising alternatives for a new generation of solar cells because of the abundance of their essential elements/base materials, low cost, and relatively easy method of synthesis. However, the massive production of associated materials has been applied and constitutes an established and robust technology. OSCs are cheaper to produce and more flexible than their counterparts made of crystalline silicon solar cell, but they do not offer the same level of efficiency or stability. OSCs have achieved a large interest in the last few years due to their potential for large-area and low-cost flexible devices. The PCEs of OSCs have increased significantly from the first reports on molecular thin-film devices more than 30 years ago. The progress in the PCE of OSC may be alternative or competitive of inorganic solar cells in the near future. Different concepts for OSC have been published by using either small molecules [38–41], conjugated polymers [42–48], combinations of small molecules, conjugated polymers [49, 50] as well as combinations of inorganic and organic materials [51] used as an active layer. In OSC, the active layer is referred to as a layer in which most of the incident light is absorbed and charges are generated. Usually, the macromolecules with a molecular weight larger than 10,000 amu are called polymers, and lighter molecules are referred to as oligomers or small molecules. OSCs can be defined by their absorber material, which consists of either carbon-based conductive organic polymers or organic molecules. The carbon polymers or molecules can form a cyclic structure, and various carbon structures produce different absorber materials with varying material properties. The OSCs can therefore be produced in a variety of colors. Other advantages of using organic semiconductor materials are the potential low

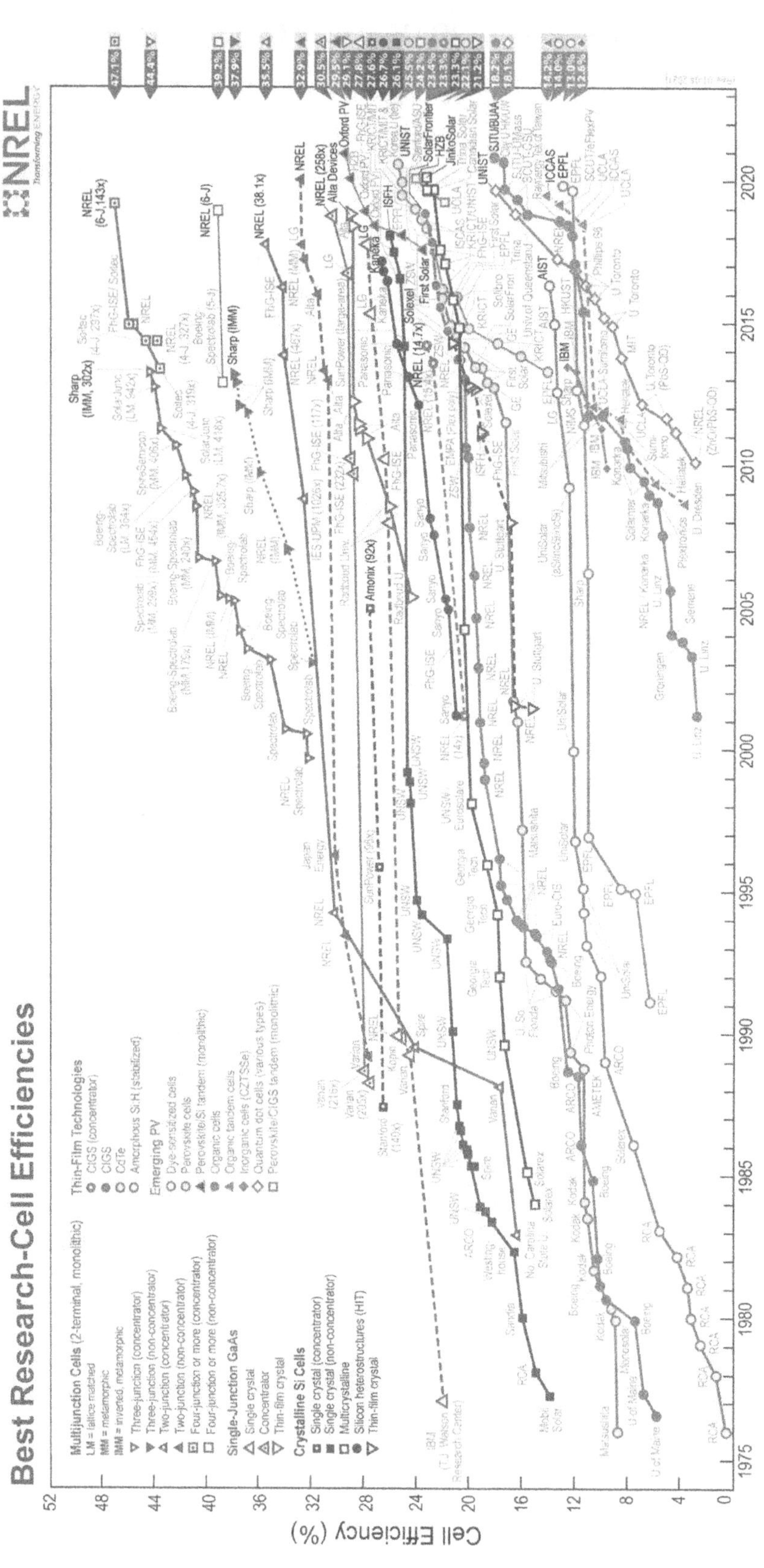

FIGURE 8.2 Efficiencies of different kinds of solar cells chart by National Renewable Energy Laboratory. (https://www.nrel.gov/pv/assets/pdfs/best-research-cell-efficiencies.20200104.pdf)

cost, low-temperature production, and the possibility to deposit on mechanically flexible and other thin or transparent substrates.

In semiconductors, upon being excited by an incident photon, an atom in the lattice absorbs the photon energy and utilizes it to excite an electron from the valence band to the conduction band. The excited electron is left behind a positively charged hole in the lattice. There is an inherent coulombic force of attraction between the electron and hole due to the difference in charge, which is called the binding energy. The force of attraction between these two charges is very minimum and they can easily be split into independent charge carriers in inorganic semiconductors. However, in organic-like polymer, the properties of the material cause an increased coulombic force of attraction between the excited electron and hole. This high binding energy prevents the complete separation of electrons and holes when excited with high-energy photons. As a result, complete charge separation does not occur, but an exciton containing both electron and hole is formed instead. This exciton can be imagined as a sphere and moves around containing the electron and the hole bound by mutual coulombic forces of attraction [52]. Splitting of the electron and hole does not take place unless the exciton is subjected to a strong electric field. An important limitation posed by this exciton is high recombination rates. Due to stronger binding forces between the hole and the electron, the likelihood of electron-hole recombination is very high. This gives rise to very low diffusion lengths. Under normal conditions, the diffusion length of these excitons is 1 to 10 nm [53], which affects the design of organic PV (OPV) cells. Since effective charge carrier separation requires an electric field, in conventional semiconductors, the internal electric field is generated by a PN junction. The PN junction is created by introducing dopant atoms in the semiconductor materials, with either fewer or additional valence electrons. Electrons in a material are naturally bound within the surface of the material and cannot freely diffuse into the free space around it. This is due to a certain energy barrier that prevents the diffusion of electrons into free space. The vacuum level can be visualized as this energy barrier, making it the minimum energy possessed by an electron present outside the material [54]. It can thus be defined as the energy of a free stationary electron that is outside any material. Ionization potential is defined as the energy required for exciting an electron from the HOMO level to the vacuum energy level. Using this definition, we can infer that a material with a low ionization potential requires very little energy to excite an electron from its valence band to the vacuum state. This allows us to refer to such materials as electron donors. Electron affinity is defined as the energy released while moving an electron from just outside the semiconductor to the bottom of the conduction band. A material with a high electron affinity can therefore easily attract an electron into its conduction band from the vacuum state and is therefore called an electron acceptor. So, the materials can be classified as electron donor or electron acceptor materials based on the ionization potential and electron affinity. Doping is not preferred in organic materials since it will lower the diffusion length of charge carriers even more. Therefore intelligently using intrinsic materials, we have to initiate charge separation of an exciton. By bringing together an electron donor on the left and an electron acceptor on the right, the formed interface will resemble a heterojunction. The lack of doping, however, results in the absence of a depletion layer. The size of the depletion layer is inversely proportional to the doping concentration in the materials brought together. When we bring together an electron donor and an electron acceptor, their vacuum levels will align, and misalignment in their HOMO and LUMO levels appears. This produces an electrostatic force at the interface, which can be very useful for the purpose of charge separation. An exciton can be split at this interface if the electrostatic force is high enough [55]. Indirectly, the interface formed will act equivalent to a PN junction that separates charges. The effective bandgap of a heterojunction semiconductor is the difference in energy levels between the LUMO of the electron acceptor and the HOMO of the electron donor. In a normal heterojunction OSC, the organic absorber material has a very high absorption coefficient, which requires an absorber layer with a thickness of about 100 nm to maximize the utilization of the solar spectrum. On the other hand, the inherent nature of excitons to recombine easily gives a diffusion length of around 10 nm. The total thickness of a solar cell absorber layer is limited by its diffusion length, which in this case is much shorter than what is required for a satisfactory absorption of incident light.

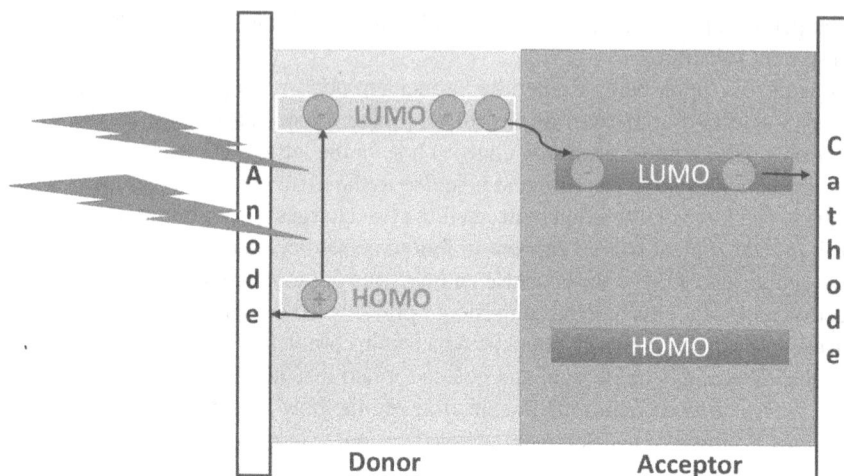

FIGURE 8.3　Schematic diagram of operative sequence of an OSC.

OSC is used for the conversion of the incident solar irradiation into the electrical current, by a process which is shown in Figure 8.3. In this observation, the donor is labeled for the holes transporting material; due to this, it makes contact with the anode. However, the acceptor material is labeled an electron transporting material and makes contact with the cathode electrode. The first step is the absorption of a photon and the formation of an exciton. Then the exciton diffuses inside the material to attain the donor–acceptor interface, where it will be divided. All of these steps are potential targets for the researchers to increase the performances of the OSCs. Generally, the absorption efficiency depends on the absorption spectra of the organic materials. However, the thickness of organic materials in the OSC design plays an important role in absorbing more incident wavelengths. The exciton diffusion length is an important parameter that can impact the efficiency in the second step. So, the longer diffusion length provides more prospect of the exciton extents the donor/acceptor (D/A) interface, raising the holes-electrons production. The morphology of the boundary of D/A also has an impact on the device's efficiency. In fact the acceptor has a possibility to be very close to the donor materials so that the interface of D/A is always the next step to the point where the exciton is created. The exciton split-up into free charges is the third step, which is mainly dependent on the properties of the donor and acceptor materials and the overall device architecture, which has an important role. In the last step of the process, transport of the free charges over the sample and assembly of electrodes for the device is considered. These are the steps involved in designing of an OPV device. This essential function derives from the fact that the structures of the organic film are normally amorphous, so charge recombination is strongly favored in this case.

The probability of absorption of light within a region of 10 nm from the heterojunction interface that leads to effective charge carrier separation is very low. This results in a very low quantum efficiency yield. Bulk heterojunction solar cell is used to reduce the effective path length of the excitons. In the bulk heterojunction solar cell, the electron donor and acceptor material form a sort of blend instead of two distinct layers. This fabrication method of OSC reduces the path length of an exciton before reaching a donor–acceptor interface. Therefore, it effectively reduces the probability of recombination and increases the quantum efficiency of devices [55]. Separated electrons travele through an acceptor layer and are collected at the metal electrode, while the holes in the system move through the donor material. These holes are collected at the front side of the indium tin oxide (ITO) based TCO electrode. In this way, the formation of a heterojunction in organic semiconductors can mimic the working principle of a PN junction. However, a solar cell based on conjugated polymers shows a low efficiency with a low lifetime compared to the first-generation silicon solar cells. The reason behind a low PCE of these kinds of bulk heterojunction solar cells is short exciton

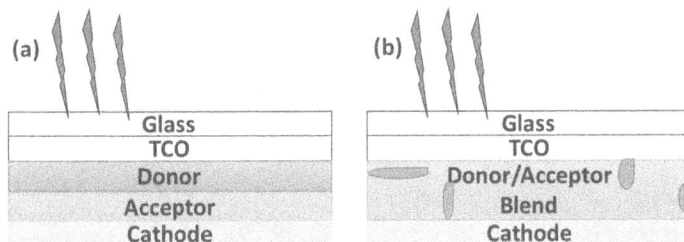

FIGURE 8.4 Schematic diagrams of different approaches in (a) bi-layer and (b) bulk heterojunctions.

diffusion length, poor carrier mobility, charge trapping, and mismatching in the absorption curve of photoactive layer with the solar spectrum [50–59]. The different approaches for OSCs are shown in Figure 8.4. Researchers have reported different OSC based on conducting polymer using these approaches [60–72].

In the light-harvesting layer of OSCs, donor and acceptor materials are merged together to form a so-called bulk heterojunction. This layer absorbs solar photons and converts these photons into free charge carriers for devices, which is called the active layer. It is shown that the development of organic polymer-based light-harvesting materials has played an important role in the enhancement of the PCE from 1% to the present efficiency [60–75]. Fullerene derivatives such as [6]-phenyl-C61-butyric acid methyl ester (PCBM) have been widely used as electron acceptors because of their high electron affinity and excellent electron-transport capability [76, 77]. An enormous amount of polymeric or small molecular donor materials have been considered and manufactured to make a good matching of absorption spectra and molecular energy levels with the PCBM. Fullerene-based OSCs have been constructed using low bandgap polymer donors and have achieved a high value of PCEs, i.e., ~12% [78]. However, researchers are devoted to develop non-fullerene acceptors because of the considerable limitations of fullerene acceptors such as restricted tunability of the electronic and optical properties as well as large device energy losses. A variation in the organic polymer materials has been prepared and then used to fabricate the OSCs. In 2015, a milestone molecule named ITIC was reported in 2015. It lead to a remarkable progress in the PCE of OSC via the use of a PBDB-T polymer as a donor in devices [79, 80]. Recently, Y6 and its derivatives have shown high PCEs (15% to 18%) by using suitable donor materials in OSCs because it is reduced the energy losses and increased wider absorption spectra [81–83]. The PCEs for OSCs have been observed to have increased very quickly over the last few years. The efficiency of large-area and ecocompatible processed devices has been found to be very low than the efficiency observed for small-area OSCs devices, as shown in Figure 8.5. So, the research community needs to incorporate a lot of work and effort to decrease this big gap in the near future.

The efficiency of OSC is enhanced by different approaches such as plasmonic, nanocomposite, photon management, and Forster resonance energy transfer (FRET) [84–90], and has recorded a 20%–50% enhancement in efficiency with respect to the baseline devices' efficiency. The development of a $ZnO:Eu^{3+}$ bifunctional layer for use in solar cells was reported by Wu et al. [91]. The $ZnO:Eu^{3+}$ layer was applied in OSCs as an electron buffer layer with the configuration (ITO/$ZnO:Eu^{3+}$/P3HT:PC61BM/MoO$_3$/Al) for understanding the impact of Eu^{3+} doping on the photoactive performance. The short circuit current density (J_{sc}) of the OSC is varied from 8.94 to 9.11 mA/cm^2 using ZnO and $ZnO:Eu^{3+}$ as buffer layers. This change in the J_{sc} is due to a spectral down conversion process by using photon management in the OSC.

Gao et al. reported small molecules with cooperative plasmon enhanced based OSCs and thermal coevaporates of Au and Ag nanoparticles (NPs). The improvement is mainly attributed to the increase in the J_{sc} because of the enhancement in light harvesting due to localized surface plasmon resonance of Au:Ag NPs. In Figure 8.6, the J-V curves of the OSCs for optimized MoO$_3$ layers are shown. For the reference OSC without any metal NPs layer, J_{sc}, open-circuit voltage (V_{oc}), and fill

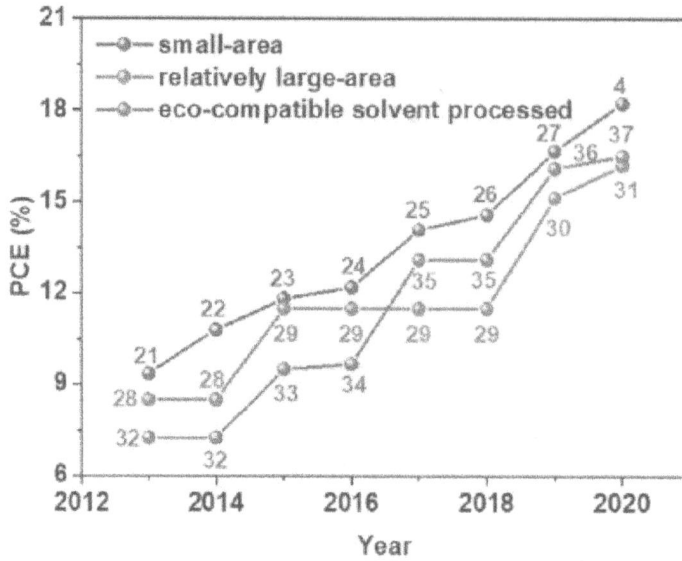

FIGURE 8.5 PCE trend for small-area devices compared with large-area and ecocompatible solvent-processed OSCs. (Reproduced from Ref. [84] with permission.)

factor (FF) are recorded to be 5.53 mA/cm^2, 0.84 V, and 0.64, respectively. The corresponding PCE of this device is $\sim$ 2.98%. However, a maximum J_{sc} of 6.39 mA/cm^2 has been observed in the OSC with plasmonic Ag NPs with an FF of 0.58. These OSCs using Ag NPs exhibit a PCE of 3.19%. However, the OSC with Au NPs is recorded to have the highest FF of 0.65 with a minimum J_{sc} of 5.89 mA/cm^2, and the PCE of the devices is recorded to be 3.29%. Meanwhile, OSCs based on the mixture of Au:Ag NPs have been recorded to have moderate J_{sc} and FF values. OSCs with a molar ratio of 3:1 for Au:Ag have shown the highest performance. The J_{sc}, V_{oc}, FF, and PCE values of these devices are recorded to be 6.11 mA/cm^2, 0.86 V, 0.63, and 3.32%, respectively. This indicates that the

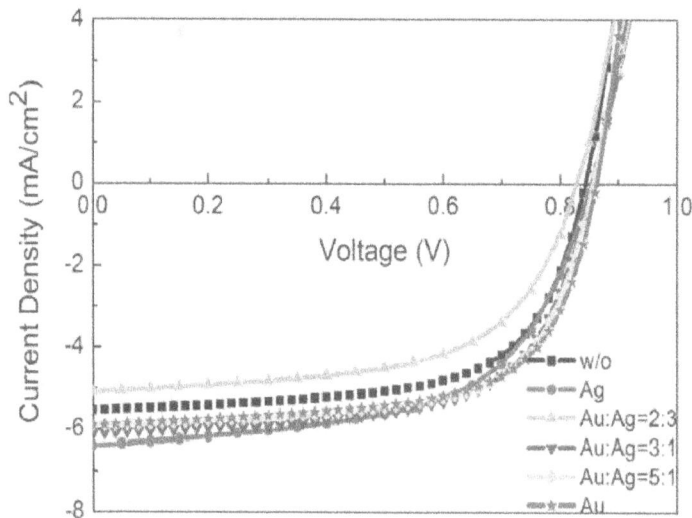

FIGURE 8.6 Effect of plasmonic NPs molar ratio on the J-V curves of the OSCs. (Reproduced from Ref. [92] with permission.)

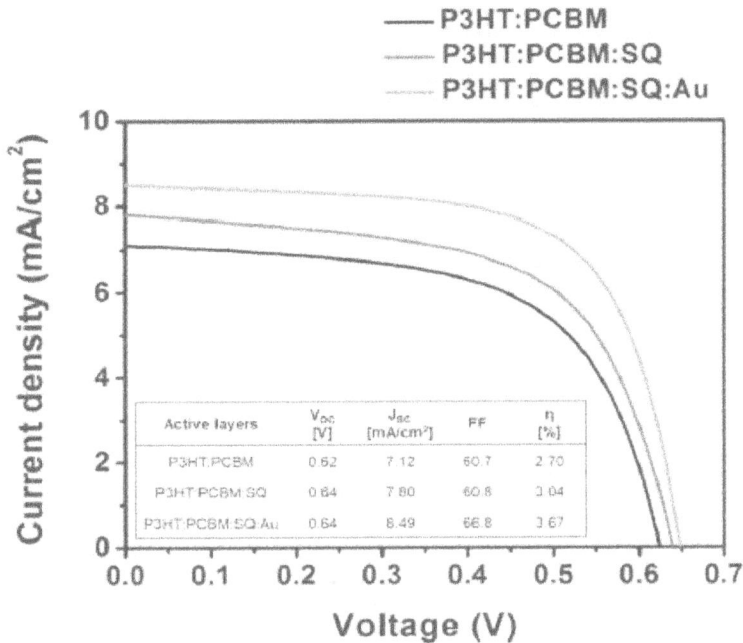

FIGURE 8.7 The J-V curves of different OSCs with photovoltaic parameters of devices shown as insert table in the figure. (Reproduced from Ref. [93] with permission.)

optimized OSCs show a PCE of 3.32% with an optimized molar ratio of Au:Ag NPs. The optimized PCE of these devices is observed to be 22.5% higher than that of the reference OSC.

Jang et al. reported two different strategies for enhancing the efficiency of OSCs such as FRET and the incorporation of plasmonic NPs. The photovoltaic performance of OSCs is studied by introducing Au NPs and squaraine (SQ) with a binary mixture of poly(3-hexylthiophene):phenyl-C61-butyric acid methyl ester. The J-V curve of the device and details of the photovoltaic parameters table are shown in Figure 8.7. These results are observed from OSC devices with a binary (P3HT:PCBM) or ternary (P3HT:PCBM:SQ) active thin layer or P3HT:PCBM:SQ:Au. Ternary blend–based OSCs have been observed to have a 12.6% higher PCE value compared to binary OSCs due to the enhancement of J_{sc}. The introduction of Au NPs into P3HT:PCBM:SQ has been shown to significantly enhance J_{sc} and FF values. This evidently exhibits that the optical gain in P3HT by using plasmonic NPs contributes to the number of charge carriers which is then converted into PCE of OSC devices.

8.5 SUMMARY

OSC technologies have substantially improved due to extensive research and development activities. Research on the technology of OPV are attractive to the research community due to an increase in the efficiency of OSCs. So, the output comes in the form of cost reduction, which makes the use of OPVs as a replacement for inorganic solar cells. According to the demand of the market, a PCE of 10% or higher is necessary for commercial viability. So, research efforts on OPVs are continuing earnestly. The main advantage of OSCs is high efficiency per unit cost, simple fabrication/processing technique, and mechanical flexibility. These advantages have led to extensive research to overcome the shortcomings of OPVs. The short diffusion length and low absorption of the active layer are actually some of the limitations of OSCs devices in terms of PCE. Using the FRET phenomenon in OSCs incorporated with plasmons and photon-conversion particles can increase the absorption range of solar energy. Different kinds of nanocomposites-based OSCs are probably at the forefront of the next generation of materials for solar cells. There are other possible structures/arrangements for devices that will help OSCs to be economically competitive with larger area as compared to conventional inorganic solar cells.

REFERENCES

1. S. Cichosz, A. Masek, M. Zaborski, Polymer-based sensors: A review, Polym. Test., 67 (2018) 342–348.
2. X. Yang, T. Dai, Z. Zhu, Y. Lu, Electrochemical synthesis of functional polypyrrole nanotubes via a self-assembly process, Polymer, 48 (2007) 4021–4027.
3. C.I. Awuzie, Conducting polymer, Materials Today: Proceedings, 4 (2017) 5721–5726.
4. M.M. Lee, J. Teeuscher, T. Miyasaka, T.N. Murakami, H.J. Snaith, Efficient hybrid solar cells based on meso-superstructured organometal halide perovskites, Science, 338 (2012) 643–647.
5. J. Wei, H. Li, Y.C. Zhao, W.K. Zhou, R. Fu, Y.L. Wang, D.P. Yu, Q. Zhao, Suppressed hysteresis and improved stability in perovskite solar cells with conductive organic network, Nano Energy, 26 (2016) 139–147.
6. A.M. Stephan, Review on gel polymer electrolytes for lithium batteries, Eur. Polym. J., 42 (2006) 21–42.
7. C.L. Gao, H.Z. Dong, X.C. Bao, Y.C. Zhang, A. Saparbaev, L.Y. Yu, S.G. Wen, R.Q. Yang, L.F. Dong, Additive engineering to improve efficiency and stability of inverted planar perovskite solar cells, J. Mater. Chem. C, 6 (2018) 8234–8241.
8. Q.F. Xue, Z.C. Hu, J. Liu, J.H. Lin, C. Sun, Z.M. Chen, C.H. Duan, J. Wang, C. Liao, W.M. Lau, Highly efficient fullerene/perovskite planar heterojunction solar cells via cathode modification with an amino-functionalized polymer interlayer, J. Mater. Chem. A, 2 (2014) 19598–19603.
9. C.M. Chang, W.B Li, X. Guo, B. Guo, C.N. Ye, W.Y. Su, Q.P. Fan, M.J. Zhang, A narrow bandgap donor polymer for highly efficient as-cast non-fullerene polymer solar cells with a high open circuit voltage, Organ. Electron., 58 (2018) 82–87.
10. M.C. Scharber, N.S. Sariciftci, Efficiency of bulk-heterojunction organic solar cells, Prog. Polym. Sci., 38 (2013) 1929–1940.
11. J. Gao, Y. Yang, Z. Zhang, J.Y. Yan, Z.H. Lin, X.Y. Guo, Bifacial quasi-solid-state dye-sensitized solar cells with poly (vinyl pyrrolidone)/polyaniline transparent counter electrode, Nano Energy, 26 (2016) 123–130.
12. C.P. Lee, C.A. Lin, T.C. Wei, M.L. Tsai, Y. Meng, C.T. Li, K.C. Huo, C.I. Wu, S.P. Lau, J.H. He, Economical low-light photovoltaics by using the Pt-free dye-sensitized solar cell with graphene dot/PEDOT:PSS counter electrodes, Nano Energy, 18 (2015) 109–117.
13. S.S. Jeon, C. Kim, T.H. Lee, Y.W. Lee, K. Do, J. Ko, S.S. Im, Camphorsulfonic acid-doped polyaniline transparent counter electrode for dye-sensitized solar cells, J. Phys. Chem. C, 116 (2012) 22743–22748.
14. T.K. Das, S. Prusty, Review on conducting polymers and their applications, Polym. Plast. Techn. Eng., 51 (2012) 1487–1500.
15. A.O. Paul, A.J. Heeger, F. Wudl, Optical properties of conducting polymers, Chem. Rev., 88 (1988) 183–200.
16. L. Lu, T. Zheng, Q. Wu, A.M. Schneider, D. Zhao, L. Yu, Recent advances in bulk heterojunction polymer solar cells, Chem. Rev., 115 (2015) 12666–12731.
17. M.M. Alam, S.A. Jenekhe, Efficient solar cells from layered nanostructures of donor and acceptor conjugated polymers, Chem. Mater., 16 (2004) 4647–4656.
18. W. Zhao, S. Li, H. Yao, S. Zhang, Y. Zhang, B. Yang, J. Hou, Molecular optimization enables over 13% efficiency in organic solar cells, J. Am. Chem. Soc., 139 (2017) 7148–7151.
19. K.W.J. Barnham, A new approach to high efficiency multi-band-gap solar cells, J. Appl. Phys., 67 (1990) 3490–3493.
20. B.K. Sovacool, National context derives concerns, Nat. Energy, 3 (2018) 820–821.
21. S. Quirin, T. Je, S. Tony, W. Alexandra, M. Oliver, Energy alternatives: Electricity without carbon, Nature, 454 (2008) 816–823.
22. M.A. Green, The path to 25% silicon solar cell efficiency: History of silicon cell evolution, Prog. Photovolt. Res. Appl., 17 (2009) 183–189.
23. T.M. Razykov, C.S. Ferekides, D. Morel, E. Stefanakos, H.S. Ullal, H.M. Upadhyaya, Solar photovoltaic electricity: Current status and future prospects, Sol. Energy, 85 (2011) 1580–1608.
24. S.M. Hubbard, C.D. Cress, C.G. Bailey, R.P. Raaelle, S.G. Bailey, D.M. Wilt, Effect of strain compensation on quantum dot enhanced GaAs solar cells, Appl. Phys. Lett., 92 (2008) 123512.

25. A. Chirila, S. Buecheler, F. Pianezzi, P. Bloesch, C. Gretener, A.R. Uh, C. Fella, L Kranz, J. Perrenoud, S. Seyrling, Highly efficient Cu(In,Ga)Se$_2$ solar cells grown on flexible polymer films, Nat. Mater., 10 (2011) 857–861.

26. W. Li, R.L. Yang, D.L. Wang, CdTe solar cell performance under high-intensity light irradiance, Sol. Energy Mater. Sol. Cells, 123 (2014) 249–254.

27. T.R. Andersen, H.F. Dam, M. Hosel, M. Helgesen, J.E. Carle, T.T. Larsen-Olsen, S.A. Gevorgyan, J.W. Andreasen, J. Adams, N. Li, Scalable, ambient atmosphere roll-to-roll manufacture of encapsulated large area, flexible organic tandem solar cell modules, Energy Environ. Sci., 7 (2014) 2925–2933.

28. M. Freitag, Q. Daniel, M. Pazoki, K. Sveinbjornsson, J.B. Zhang, L.C. Sun, A. Hagfeldt, G. Boschloo, High efficiency dye-sensitized solar cells with molecular copper phenanthroline as solid hole conductor, Energy Environ. Sci., 8 (2015) 2634–2637.

29. Y.Z. Wu, F.X. Xie, H. Chen, X.D. Yang, H.M. Su, M.L. Cai, Z.M. Zhou, T. Noda, L.Y. Han, Thermally stable MAPbI$_3$ perovskite solar cells with efficiency of 19.19% and area over 1 cm^2 achieved by additive engineering, Adv. Mater., 29 (2017) 1701073.

30. M.E. Ragoussi, T. Torres, New generation solar cells: Concepts, trends and perspectives, Chem. Commun., 51 (2015) 3957–3972.

31. Y.M. Xiao, J.H. Wu, J.Y Lin, G.T. Yue, J.M. Lin, M.L. Huang, Z. Lan, L.Q. Fan, A dual function of high performance counter-electrode for stable quasi-solid-state dye-sensitized solar cells, J. Power Sources, 241 (2013) 373–378.

32. V. Kumar, S.K. Swami, A. Kumar, O.M. Ntwaeaborwa, V. Dutta, H.C. Swart, Eu^{3+} doped down shifting TiO$_2$ layer for efficient dye-sensitized solar cells, J. Colloid Interface Sci., 484 (2016) 24–32.

33. D. Wohrle, D. Meissner, Organic solar cells, Adv. Mater., 3(3) (1991).

34. I. Mesquita, L. Andrade, A. Mendes, Perovskite solar cells: Materials, configurations and stability, Renew. Sust. Energ. Rev., 82(3) (2018) 2471–2489.

35. A. Kojima, K. Teshima, Y. Shirai, T. Miyasaka, Organometal halide perovskites as visible-light sensitizers for photovoltaic cells, J. Am. Chem. Soc., 131 (2009) 6050–6051.

36. F.X. Xie, C.C. Chen, Y.Z. Wu, X. Li, M.L. Cai, X. Liu, X.D. Yang, L.Y. Han, Vertical recrystallization for highly efficient and stable formamidinium based inverted structure perovskite solar cells, Energy Environ. Sci., 10 (2017) 1942–1949.

37. Research Cell Efficiency Records. Available online https://www.nrel.gov/pv/assets/pdfs/best-research-cell-efficiencies.20200104.pdf

38. C.W. Tang, A.C. Albrecht, Photovoltaic effects of metal-chlorophyll-a-metal sandwich cells, J. Chem. Phys., 62(6) (1975) 2139–2149.

39. C.W. Tang, Two layer organic photovoltaic cell, Appl. Phys. Lett., 48(2) (1986) 183–185.

40. J. Xue, S. Uchida, B.P. Rand, S.R. Forrest, 4.2% efficient organic photovoltaic cells with low series resistances, Appl. Phys. Lett., 84(16) (2004) 3013–3015.

41. J. Xue, S. Uchida, B.P. Rand, S.R. Forrest, Asymmetric tandem organic photovoltaic cells with hybrid planar mixed molecular heterojunctions, Appl. Phys. Lett., 85(23) (2004) 5757–5759.

42. K. Takahashi, N. Kuraya, T. Yamaguchi, T. Komura, K. Murata, Three layer organic solar cell with high power conversion efficiency of 3.5%, Sol. Energy Mater. Sol. Cells, 61(4) (2000) 403–416.

43. G. Yu, A.J. Heeger, Charge separation and photovoltaic conversion in polymer composites with internal donor/acceptor heterojunction, J. Appl. Phys., 78(7) (1995) 4510–4515.

44. M. Granstrom, K. Petritsch, A.C. Arias, A. Lux, M.R. Andersson, R.H. Friend, Laminated fabrication of polymeric photovoltaic diodes, Nature, 395 (1998) 257–260.

45. S.A. Jenekhe, S. Yi, Efficient photovoltaic cells from semiconducting polymer heterojunctions, Appl. Phys. Lett., 77(17) (2000) 2635–2637.

46. A.J. Breeze, Z. Schlesinger, S.A. Carter, H. Tillmann, H.-H. Horhold, Improving power efficiencies in polymer polymer blend photovoltaics, Sol. Energy Mater. Sol. Cells, 83(2–3) (2004) 263–271.

47. C.J. Brabec, N.S. Sariciftci, J.C. Hummelen, Plastic solar cells, Adv. Funct. Mater., 11(1) (2001) 15–26.

48. C. Winder, N.S. Sariciftci, Low bandgap polymers for photon harvesting in bulk heterojunction solar cells, J. Mater. Chem., 14(7) (2004) 1077–1086.

49. A.J. Breeze, A. Salomon, D.S. Ginley, B.A. Gregg, H. Tillmann, H.-H. Hörhold, Polymer perylene diimide heterojunction solar cells, Appl. Phys. Lett., 81(16) (2002) 3085–3087.

50. J.I. Nakamura, C. Yokoe, K. Murata, K. Takahashi, Efficient organic solar cells by penetration of conjugated polymers into perylene pigments, J. Appl. Phys., 96(11) (2004) 6878–6883.

51. A.J. Breeze, Z. Schlesinger, S.A. Carter, P.J. Brock, Charge transport in TiO$_2$/MEH-PPV polymer photovoltaics, Phys. Rev. B, 64(12) (2001) 125205.

52. C.J. Brabec, S. Gowrisanker, J.J.M. Halls, D. Laird, S. Jia, S.P. Williams, Adv. Mater., 22 (2010) 3839–3856.

53. S.R. Scully, M.D. McGehee, Effects of optical interference and energy transfer on exciton diffusion length measurements in organic semiconductors, J. Appl. Phys., 100 (2006) 34907.

54. K. Nakano, K. Tajima, Organic planar heterojunctions: From models for interfaces in bulk heterojunctions to high performance solar cells, Adv. Mater., 29 (2017) 1603269.

55. G. Yu, J. Gao, J.C. Hummelen, F. Wudl, A.J. Heeger, Polymer photovoltaic cells: Enhanced efficiencies via a network of internal donor-acceptor heterojunctions, Science, 270 (1995) 1789–1791.

56. Z. Xie, L.W.S. Shao, F. Liu, High-efficiency hybrid polymer solar cells with inorganic P- and N-type semiconductor nanocrystals to collect photogenerated charges, J. Phys. Chem., 114 (2010) 9161.

57. S. Rafique, S. Mah Abdullah, K. Sulaiman, M. Iwamoto, Fundamentals of bulk heterojunction organic solar cells: An overview of stability/degradation issues and strategies for improvement, Renew. Sust. Energ. Rev., 84 (2018) 43–53.

58. M. Koppe, H.J. Egelhaaf, G. Dennler, M.C. Scharber, C.J. Brabec, P. Schilinsky, C.N. Hoth, Near IR sensitization of organic bulk heterojunction solar cells: Towards optimization of the spectral response of organic solar cells, Adv. Funct. Mater., 20 (2010) 338–346.

59. M.M. Wienk, M.G.R. Turbiez, M.P. Struijk, Low-band gap poly(di-2-thienylthienopyrazine):fullerene solar cells, J. Appl. Phys. Lett., 88 (2006) 153511.

60. T. Kietzke, Recent advances in organic solar cells, Adv. OptoElectron., Special Issue (2007) 40285.

61. S. Yang, X. Zhang, L. Ding, 18% efficiency organic solar cells, Sci. Bull., 65(4) (2020) 272–275.

62. S.H. Liao, H.J. Jhuo, Y.S. Cheng, S.A. Chen, Fullerene derivative doped zinc oxide nanofilm as the cathode of inverted polymer solar cells with low-bandgap polymer (PTB7-Th) for high performance, Adv. Mater., 25(34) (2013) 4766–4771.

63. W. Zhao, S. Li, S. Zhang, X. Liu, and J. Hou, Ternary polymer solar cells based on two acceptors and one donor for achieving 12.2% efficiency, Adv. Mater., 29(2) (2017) 1604059.

64. Z. Xiao, Z. Jia, L. Ding, Ternary organic solar cells offer 14% power conversion efficiency, Sci. Bull., 62(23) (2017) 1562–1564.

65. T. Yan, W. Song, J. Huang, R. Peng, L. Huang, Z. Ge, 16.67% rigid and 14.06% flexible organic solar cells enabled by ternary heterojunction strategy, Adv. Mater., 31(39) (2019) 1902210.

66. B. Fan, Z. Zeng, W. Zhong, L. Ying, D. Zhang, M. Li, F. Peng, N. Li, F. Huang, Y. Cao, Optimizing microstructure morphology and reducing electronic losses in 1 cm² polymer solar cells to achieve efficiency over 15%, ACS Energy Lett., 4(10) (2019) 2466–2472.

67. Y. Cui, H. Yao, J. Zhang, K. Xian, T. Zhang, L. Hong, Y. Wang, Y. Xu, K. Ma, C. An, C. He, Z. Wei, F. Gao, J. Hou, Single-junction organic photovoltaic cells with approaching 18% efficiency, Adv. Mater., 32(19) (2020) 1908205.

68. C.C. Chueh, K. Yao, H.L. Yip, C.Y. Chang, Y.X. Xu, K.S. Chen, C.Z. Li, P. Liu, F. Huang, Y. Chen, W.C. Chen, A.K.Y. Jen, Non-halogenated solvents for environmentally friendly processing of high-performance bulk heterojunction polymer solar cells, Energy Environ. Sci., 6(11) (2013) 3241.

69. C. Sprau, F. Buss, M. Wagner, D. Landerer, M. Koppitz, A. Schulz, D. Bahro, W. Schabel, P. Scharfer, A. Colsmann, Highly efficient polymer solar cells cast from non-halogenated xylene/anisaldehyde solution, Energy Environ. Sci., 8(9) (2015) 2744–2752.

70. H. Zhang, H. Yao, W. Zhao, L. Ye, J. Hou, High-efficiency polymer solar cells enabled by environment-friendly single-solvent processing, Adv. Energy Mater., 6(6) (2016) 1502177.

71. W. Zhao, S. Zhang, Y. Zhang, S. Li, X. Liu, C. He, Z. Zheng, J. Hou, Environmentally friendly solvent processed organic solar cells that are highly efficient and adaptable for the blade-coating method, Adv. Mater., 30(4) (2018) 1704837.

72. L. Hong, H. Yao, Z. Wu, Y. Cui, T. Zhang, Y. Xu, R. Yu, Q. Liao, B. Gao, K. Xian, H.Y. Woo, Z. Ge, J. Hou, Eco-compatible solvent-processed organic photovoltaic cells with over 16% efficiency, Adv. Mater., 31(39) (2019) e1903441.

73. R. Sun, T. Wang, Z. Luo, Z. Hu, F. Huang, C. Yang, J. Min, Achieving eco-compatible organic solar cells with efficiency >16.5% based on an iridium complex-incorporated polymer donor, Sol. RRL, 4 (2020) 2000156.

74. Y. Huo, H.-L. Zhang, X. Zhan, Nonfullerene all small molecule organic solar cells, ACS Energy Lett., 4(6) (2019) 1241–1250.

75. Y. Cui, H. Yao, L. Hong, T. Zhang, Y. Tang, B. Lin, K. Xian, B. Gao, C. An, P. Bi, W. Ma, J. Hou, Organic photovoltaic cell with 17% efficiency organic photovoltaic cell with superior processability, Natl. Sci. Rev., 7 (2020) 1239–1246.

76. E. M. Speller, The significance of fullerene electron acceptors in organic solar cell photo-oxidation, Mater. Sci. Technol. 33(8) (2017) 924–933.

77. G. Yu and A.J. Heeger, Charge separation and photovoltaic conversion in polymer composites with internal donor/acceptor heterojunctions, J. Appl. Phys., 78(7) (1995) 4510–4515.

78. J. Zhao, Y. Li, G. Yang, K. Jiang, H. Lin, H. Ade, W. Ma, H. Yan, Efficient organic solar cells processed from hydrocarbon solvents, Nat. Energy, 1(2) (2016) 15027.

79. W. Zhao, D. Qian, S. Zhang, S. Li, O. Inganäs, F. Gao, J. Hou, Fullerene free polymer solar cells with over 11% efficiency and excellent thermal stability, Adv. Mater., 28(23) (2016) 4734–4739.

80. Y. Lin, J. Wang, Z.G. Zhang, H. Bai, Y. Li, D. Zhu, X. Zhan, An electron acceptor challenging fullerenes for efficient polymer solar cells, Adv. Mater., 27(7) (2015) 1170–1174.

81. J. Yuan, Y. Zhang, L. Zhou, G. Zhang, H.-L. Yip, T.-K. Lau, X. Lu, C. Zhu, H.H. Peng, P.A. Johnson, M. Leclerc, Y. Cao, J. Ulanski, Y. Li, Y. Zou, Single junction organic solar cell with over 15% efficiency using fused-ring acceptor with electron-deficient core, Joule, 3(4) (2019) 1140–1151.

82. Y. Cui, H. Yao, J. Zhang, T. Zhang, Y. Wang, L. Hong, K. Xian, B. Xu, S. Zhang, J. Peng, Z. Wei, F. Gao, J. Hou, Over 16% efficiency organic photovoltaic cells enabled by a chlorinated acceptor with increased open-circuit voltages, Nat. Commun., 10(1) (2019) 2515.

83. S. Liu, J. Yuan, W. Deng, M. Luo, Y. Xie, Q. Liang, Y. Zou, Z. He, H. Wu, Y. Cao, High-efficiency organic solar cells with low non-radiative recombination loss and low energetic disorder, Nat. Photonics, 14(5) (2020) 300–305.

84. L. Hong, H. Yao, Y. Cui, Z. Ge, J. Hou, Recent advances in high-efficiency organic solar cells fabricated by eco-compatible solvents at relatively large-area scale, APL Mater., 8 (2020) 120901.

85. J. Chuan, L. Tianze, H. Luan, Z. Xia, Research on the characteristics of organic solar cells, J. Phys.: Conf. Ser., 276 (2011) 012169.

86. B. Zacher, J.L. Gantz, R.E. Richards, N.R. Armstrong, Organic solar cells at the interface, J. Phys. Chem. Lett., 4 (11) (2013) 1949–1952.

87. J. Wu, X. Che, H.C. Hu, H. Xu, B. Li, Y. Liu, J. Li, Y. Ni, X. Zhang, X. Ouyang, Organic solar cells based on cellulose nanopaper from agroforestry residues with an efficiency of over 16% and effectively wide-angle light capturing, J. Mater. Chem. A, 8 (2020) 5442–5448.

88. Y. Cui, P. Zhu, X. Liao, Y. Chen, Recent advances of computational chemistry in organic solar cell research, J. Mater. Chem. C, 8 (2020) 15920–15939.

89. B.R. Gautam, R. Younts, J. Carpenter, H. Ade, K. Gundogdu, The Role of FRET in non-fullerene organic solar cells: Implications for Molecular Design, J. Phys. Chem. A, 122(15) (2018) 3764–3771.

90. Y.J. Cheng, S.H. Yang, C.S. Hsu, Synthesis of conjugated polymers for organic solar cell applications, Chem. Rev., 109(11) (2009) 5868–5923.

91. N. Wu, Q. Luo, X. Qiao, C.Q. Ma, The preparation of a Eu^{3+} doped ZnO Bi-functional layer and its application in organic photovoltaics, Mater. Res. Express, 2 (2015) 125901.

92. Y. Gao, F. Jin, Z. Su, H. Zhao, Y. Luo, B. Chu, W. Li, Cooperative plasmon enhanced organic solar cells with thermal coevaporated Au and Ag nanoparticles, Organic Electronics, 48 (2017) 336–341.

93. Y.J. Jang, D. Kawaguchi, S. Yamaguchi, S. Lee, J.W. Lim, H. Kim, K. Tanaka, D.H. Kim, Enhancing the organic solar cell efficiency by combining plasmonic and Forster Resonance Energy Transfer (FRET) effects, J. Power Sourc., 438 (2019) 227031.

9 Hybrid Conducting Polymers for High-Performance Solar Cells

Kelsey Thompson[1], Muhammad Rizwan Sulaiman[1,2], and Ram K. Gupta[1]
[1]Department of Chemistry, Kansas Polymer Research Center, Pittsburg State University, Pittsburg, Kansas, USA
[2]Plastic Engineering, Kansas Polymer Research Center, Pittsburg State University, Pittsburg, Kansas, USA

CONTENTS

9.1   INTRODUCTION

In today's modern society, many people believe that reliable access to electricity is a necessity of life. While the actual human necessities have always been food, water, and shelter, but there are other amenities that many people in the developed world have come to believe they cannot do without them. This view has become widespread even though there have been past civilizations that thrived without electricity. However, access to energy certainly makes it easier to acquire and maintain necessary things in life. A reliable supply of electrical energy allows us to safely store food, clean water, and heat homes in the winter. Access to energy has enabled us to save time in menial activities, such as laundry, so that time may be better spent learning and developing new technologies to further enhance the quality of life. From heating and lighting our homes and places of business to modernizing the industries and traveling, energy has undoubtedly changed our perception of the living standard. As we have developed new technologies, the way of utilizing energy has also changed. For example, burning coal in a stove to heat a room transformed into burning coal in power plants to produce electricity for distribution in the grid. Therefore, the forms that energy storage has taken over the years need to change as well. It is not practical to carry around electrical generators that run on fossil fuels to check the time or light our path, so the batteries were developed. As new devices become smaller and more compact, the ways to power them must change as well. Green forms of energy generation and storage are more suitable for these applications than fossil fuels. Once fossil fuels are burned, they can no longer provide any sort of energy generation and storage, but devices like batteries, fuel cells, and solar cells have recycling capabilities, and the elemental components can be used again in new devices. The recyclability of the components of green energy devices will be of paramount importance in the future, which would decrease the need to mine new raw materials for making devices. In contrast, fossil fuels cannot be recycled and have finite resources, and the search for land for fossils fuel reservoirs would never end.

Among the current forms of green energy, solar cells are of chief importance. The harvesting of solar energy has a minor impact on the environment compared to fossil fuels, wind energy, and hydroelectric energy from dams and tidal stations. This would be true in the future too because the raw materials needed for green energy generation may be acquired by recycling old technologies or from elements that are readily available and require little mining, like carbon. Solar energy is also readily available in many parts of the world, unlike wind and hydroelectric energy. Solar energy may also be generated on-site and does not have to be transferred from where it is mined to a refinement facility and then to the location where it will be used, unlike fossil fuels. Together, these things mean that solar energy has the opportunity to positively impact the lives of more people. The impact of solar cells on our lives has already been demonstrated in several ways. Even though the first known observation of the photovoltaic effect was made in 1839 by Alexandre Edmond Becquerel [1], the earliest solar cells had not been developed as an energy generation technology for use in the space program until the 1950s by Chapin from Bell Labs [2]. These solar cells were made of silicon and are still in use today in satellites, probes, and even the International Space Station, and the same technology was also put to use here on earth. However, silicon solar cells only have a lifespan of about 25 years, based on their degradation rate [3]. They can also be challenging to manufacture since single crystal wafers require a clean-room condition at high temperatures for production. Even minor defects in their crystalline structure can rob their efficiency. Therefore, new solar materials have been developed that can absorb solar radiation and in return release electrons.

Perovskites are one such class of materials that have been researched for use in solar cells. These solar cells are based on a special class of materials that take their name due to their spinel crystalline structure. Figure 9.1 shows an example of the perovskite structure and how perovskite solar cells (PeSCs) have changed over time [4]. Perovskite is a crystalline structure based on the naturally occurring mineral calcium titanium oxide ($CaTiO_3$) and compounds with similar crystalline structures. They typically have a chemical structure of ABX_3, where A and B represent cations and the X is an anion, which is usually a halide, but it can be an oxide as well. Some of the earliest perovskite materials researched for solar cell applications contained lead as one of the cations. Due to the crystal's high solubility in water, this posed a health hazard. Therefore, other materials are currently being developed to eliminate lead. The further development of polymer solar cells with other two-dimensional materials is of great interest as the bandgap is tuneable based on the thickness of the material, such as black phosphorus, with a bandgap range of approximately 0.3–2 eV [5]. Another type of solar cell being researched as a possible replacement for silicon-based solar cells are dye-sensitized solar cells (DSSCs). DSSCs may be a potential replacement for silicon solar cells because of the low material cost

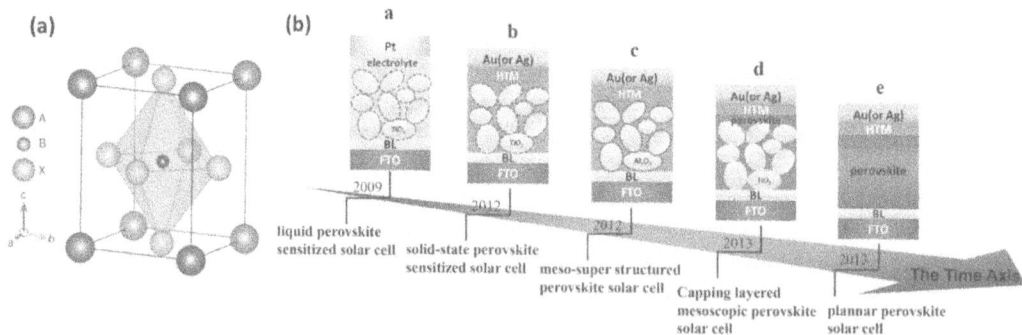

FIGURE 9.1   (a) Perovskite structure and (b) changes in PeSCs over time. (Adapted with permission from Ref. [4]. Copyright (2019) American Chemical Society.)

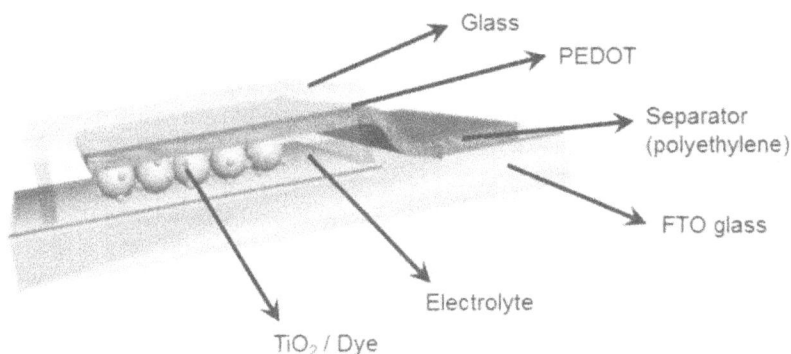

FIGURE 9.2 The layered structure of a DSSC. (Adapted with permission from Ref. [6]. Copyright (2013) American Chemical Society.)

and the possibility for roll to roll processing. DSSCs are different from other types of solar cells and contain electrolytes with a dissolved redox couple, which is located between the photoanode made up of a semiconductor oxide covered with a photochemically active dye and a counter electrode (CE). The dye releases the electrons to other layers upon being struck by sunlight. Figure 9.2 shows an example of a DSSC that incorporates a polymer between the dye and CE [6].

Conducting carbon materials like graphene and carbon nanotubes (CNTs) have also been researched for solar cell applications. These carbons, are naturally conducting and can be structured on thin films with a high degree of transparency and flexibility. The carbon structures can also be oxidized to increase their conductivity and be used as a base material for hole transport instead of other commonly used materials, such 2,2′,7,7′-tetrakis-9,9′-spirobifluorene (Spiro-OMeTAD) [7, 8]. In addition to other carbon materials, many polymers are also used for solar cells in a wide range of applications. The research for the application of polymers in solar cells has become popular due to the wide range of properties polymers can exhibit and the availability of precursor material that allows large-scale manufacturing [9]. The applications of polymers in the different aspects of solar cell technology will be discussed in detail in subsequent sections.

Silicon solar cells were originally made with a single p-n junction, which was achieved by doping the silicon with phosphorus for the n-type layer and boron for the p-type layer. The doping allows the electrons in the semiconducting silicon to move from the valence band to the conduction band more easily. One of the advances of polymer solar cells made in 1995 was the implementation of bulk heterojunction, which allows several junction areas in the solar cell to operate at the nanoscale [9]. Multijunction cells have also been studied as tandem solar cells. These multijunction cells have the potential to be more tuneable in absorbing solar radiation since the materials with different absorbance wavelengths can be layered in very thin sheets, and a wider range of solar radiation can therefore be absorbed. Some polymeric materials with carbon backbones are also of interest in various parts of a solar cell. Certain polymers can function well as an absorber, while others could function as charge carriers. Polymers and carbon-based materials are also potentially useful in solar cells because they have a much higher degree of flexibility and may have a high degree of transparency than thin films. Flexibility and the capacity to absorb a shock without breaking are important parameters for materials. By using these types of materials, the life cycle of solar panels can also be lengthened, which would make them much more cost-effective in the long run.

9.2 POLYMERS IN SOLAR CELLS

Solar cells consist of different parts, and each part can be produced by using different materials to optimize its performance. The versatility of polymeric materials opens many doors for their use in solar cells. The polymers used in solar cells are chosen based on desired functionality. Much focus has been given to replacing inorganic photon absorber materials with organic materials that are less toxic and easier to manufacture. However, there are possibilities for using polymeric materials in every part of a solar cell, from the substrate to the coating material on the electrode.

Currently, both manufactured solar cells and ongoing research are focused on employing polymers in two ways: an encapsulant and substrate or back-sheet material. Encapsulant polymeric materials are used to protect solar cells and entire panels from water damage. Since silicon wafers are not water soluble, it is believed that the sudden decrease in efficiency upon their installment is due to an oxidation reaction between the boron and oxygen in the p-type material [10]. Sealing the cell or panel in a thin, clear material can protect them from an increase in surface oxygen content without interfering with their performance. Other water-soluble absorber materials, such as inorganic perovskites, do not survive long in damp conditions. The absorber material will leech out from the cell into the environment [11]. This raised significant concerns since some commercial perovskites use lead or other harmful elements [5, 7]. Ethylene vinyl acetate (EVA) is a commonly used encapsulant that is one of the cheapest and has been used for over 20 years in manufacturing. Since EVA turns brownish and thereby decreases the power output of the solar panel, an effort to find a better encapsulant has been pursued. Like EVA, polyaniline (PANI) is another encapsulant or coating material known to inhibit the corrosion of materials while being weather resistant itself [12].

Similarly, many electrode materials need a substrate to adhere upon, which needs to be removed afterward and, that can result in damaging the electrode. For example, when graphene is transferred from transition metal substrates [8]. Therefore, having some sort of substrate that the electrode can be formed on without removing it later is very beneficial. Also, various types of glass have been used as substrates, but these glass-like materials are not flexible. Different kinds of polymers have been employed to address this issue, including some Tedlar (PVF)-based materials, like Tedlar/PET/Tedlar (TPT), Tedlar/PET/EVA (TPE), Tedlar/PET/Al foil/Tedlar (TPAT), and Tedlar/PET/Oxide/Tedlar (TPOT). Since Tedlar is a relatively expensive, polyethylene terephthalate (PET), and polyethylene naphthalate (PEN)-based materials have also been researched, examples of which include Protekt, Teijin Teonex, and barium sulfate–filled PET. Polymeric materials can be used whether the substrate needs to be flexible or rigid. Some examples of polymer substrates include PET and PEN on indium tin oxide (ITO) [13]. Certain substrate polymer materials can even be used to complete an entirely transparent solar cell [13]. Using polymers as an encapsulant material or as a substrate can provide a tough and cheap replacement for glass and other materials that have been used in the past.

Another avenue for polymer use in solar cells can be found as the electrode material. In commercial solar cells, silver is often used as a top electrode due to its conductivity and corrosion resistance. However, some drawbacks, such as rareness, expensiveness, and rigidity, make silver a poor option for flexible solar cells. Since polymers are not as conductive as metals, the carbon backbone or substrates engineering with double bonds or other charge carriers allows materials scientists to tune the conductivity of polymer-based electrodes. Examples of electrodes made entirely of polymeric materials include but are not limited to poly(3,4-ethylene dioxythiophene):poly(styrenesulfonate) (PEDOT:PSS) [14], polypyrrole (PPy) [13], PANI [15], and poly(3-hexylthiophene) (P3HT) [15]. Some conducting polymers can also be employed as a binder in electrodes. Some electrode materials do not bind closely with various types of transport layers. Poly(methyl methacrylate) (PMMA) has been used to bind graphene as a top electrode without disrupting the movement of electrons or holes [8]. PEDOT was used as a conductive polymeric binder for transition metal phosphates to make a Pt-free CE in a DSSC [16].

Recently, tremendous attention has been given to multijunction cells to convert as much sunlight as possible to a usable energy form. Thin films can be used to level up the bandgap difference between various layers of multijunction cells by providing hole transport, electron transport, or buffer/interface layers (ETLs). Fullerene derivatives are an effective interface layer to facilitate charge transport as an ETL [9]. Some polymers are also used to enhance ETLs, such as polyethylene oxide (PEO) with ZnO [17] or PANI and Au [18]. One common addition to the solar cells is Spiro-OMeTAD, which is commonly used as a hole transfer layer (HTL) [7]. Jeon et al. found that poly[bis(4-phenyl)(2,4,6-trimethyl phenyl) amine] (PTAA) can also act as an effective HTL [19]. These polymeric materials have an advantage over other transport layers by being added onto the cell as thin films and do not impact the photons absorption ability of the photoactive layer. A second advantage of using polymeric materials in multijunction cells is that they can allow the cell to be flexible, whereas other highly crystalline layers do not.

Polymers may be also employed in solar cells as photon absorbers. Since these PCSs have not yet achieved the same efficiency as many commercially available solar cells, they have made tremendous gains in recent years. Examples of polymers that have been used as the photoactive layer include polymers that have been employed in other parts of solar cells, such as PEDOT:PSS, P3HT, and PANI. And there are several examples of heterojunction materials that make use of [6]-phenyl-C61-butyric acid methyl ester (PCBM) [9]. Early on in organic solar cell research, poly[2-methoxy,5-(20-ethyl-hexoxy)-p-phenylenevinylene] (MEH-PPV) and poly[2-methoxy-5-(30,70-dimethyloctyloxy)-1,4-phenylenevinylene] (MDMO-PPV) were found to yield acceptable power conversion efficiency (PCE) of around 3% when used as absorber materials [9]. More recently, Xu et al. fabricated a polymer solar cell that could reach a PCE of 9.45% [20], and Kim et al. used polydimethylsiloxane (PDMS) to improve the PCE of a PeSC up to a value of 19% [21]. In reality, there are almost limitless options for polymer-based absorber materials in solar cells. The challenge is to find a material that can be manufactured on a large scale through a simple process while remaining highly efficient and cost effective.

9.3 HYBRID CONDUCTING POLYMERS IN SOLAR CELLS

The application of conducting polymers in cells displayed lower efficiencies, but when used with other materials like metal oxides, they can help increase the performance of the cell [18]. In addition to improving the cell's photovoltaic performance, conducting polymers provide other positive features, such as straightforward processing, superior flexibility, and tunable conductivity [15]. These features make the continued research and development of hybrid polymer and metal oxide solar cells a worthwhile endeavor. Examples of hybrid conducting polymers in DSSCs are widespread and addressed briefly in this section, emphasizing PANI and its hybrid materials. PANI, as a homopolymer, can conduct electric charge due to the existence of conjugated pi-bonds along the length of the polymer chain and the presence of nitrogen atoms between each ring. The nitrogen atoms along the polymer chain can have varying oxidation states depending on whether or not the extra hydrogen atoms are introduced, which alters the conductivity of the polymer. The structure allows PANI to fall into the category of "synthetic metals," which have alternating single and double bonds along the polymer backbone [15]. Several compounds could fall into this category, but PANI is of special interest for several reasons. Firstly, it has been widely studied since 1862, when Henry Letheby first reported its synthesis [12]. Second, it is easy to synthesize through oxidative polymerization and modify through copolymerization or doping [12, 15]. And finally, it is inexpensive to manufacture, has a good shelf life, and, as mentioned above, is conductive [15]. Thus, there have been several studies made on PANI.

One such study was performed by Ameen et al. that focused on nanorod composites of PANI and cadmium sulfide (CdS), which focused on the impedance of the fabricated nanorods [15]. It was found that the impedance of the material decreased when the CdS amount was increased during the production process. The impedance of the pure PANI nanorod electrode decreased from ~17kΩ to 8.4 kΩ with a slight addition of 0.01 M CdS. After that, the decreases in resistance were less drastic but still notable, going from 8.4 to 5.7 to 4.1 kΩ with the addition of 0.01M, 0.05 M, and 0.10 M CdS, respectively. The observed reduction in impedance was directly attributed to the increased metal sulfide coating on the PANI nanorods, thereby increasing the conductivity of the sample. Another option for producing PANI hybrids involves graphene. A 1.54% increase in the PCE was seen in

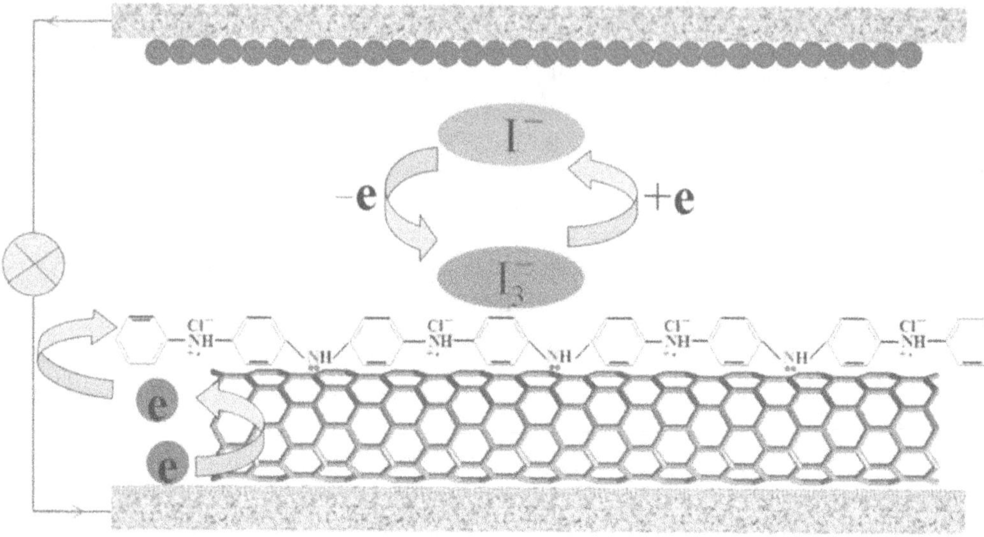

FIGURE 9.3   Graphene PANI complex as a DSSC redox agent. (Adapted with permission from Ref. [23]. Copyright (2015) Elsevier.)

PANI – 8 wt% graphene complex, compared to a pure PANI CE [22]. Similarly, PANI–multiwalled carbon nanotubes (MWCNTs) complexes have also been prepared via a reflux approach [23]. The synthesized material was used as a CE in bifacial DSSCs and observed a superior PCE of 9.24% than the 8.08% with a pure PANI CE. The researchers attributed the improved PCE to the ease with which the electrons could flow from the N atoms in the PANI to the C atoms of CNTs via the covalent bond made when the complex was formed (Figure 9.3).

Furthermore, CdS has also been utilized in other hybrid material solar cell research as early as 1996 [24]. This study determined that different polymeric materials need to be used since the applied electrical field used in the study was higher than the typical working voltage of solar cells of that time. For this reason, Sun et al. and Wang et al. both combined CdS and CdSe with MEH-PPV instead [24, 25]. It was found that eliminating the ligands used in manufacturing quantum dots (QD) achieved an improved cell efficiency of 0.2%. The earlier work was built upon by other groups, and as techniques progressed, nanorods and then later tetrapods were synthesized [26]. Eventually, the CdSe tetrapod reached a PCE of 3.13% in 2010 by incorporating a smaller bandgap polymer than the original MEH-PPV.

Several metal oxides have also been investigated for improved results in DSSCs. For this reason, the conjunction of TiO_2 with different materials, including PANI, is discussed below. Since TiO_2 is a durable, nontoxic, and widely available material, it is an excellent material to be employed as a photoanode for DSSCs. In a review of several different works, Hou et al. concluded that the porosity of the TiO_2 photoanodes and, therefore, the overall efficiency of the cell could be affected by the type and amount of polymer used during the synthesis of the photoanode [13]. TiO_2 nanoparticles have also been employed in DSSCs as nanocomposite electrolytes. TiO_2 NPs were included in a three-part polymer electrolyte by Kang et al. that resulted in a 7.2% PCE [27]. Later on, Katsaros et al. fabricated a four-part polymer electrolyte using PEO that reached a PCE of 4.2% [28], and Stergiopoulos' group synthesized a polyethylene glycol methyl ester and titanium oxide nanoparticle (PEGME-TiO_2) composite to test heat treatment on the effect of PCE [29]. Also, TiO_2 has been modified with PANI to form a hybrid photoanode material. Rutile TiO_2 was synthesized and layered with PANI by Roy et al. in 2019, which resulted in increased PCE by 1.95% compared to the nanorods without the coating [30].

Likewise, ZnO can also be hybridized with PANI for utilization as a photoanode in a DSSC. Zhu et al. synthesized a ZnO-based nanograss structure and later layered it with PANI by chemisorption

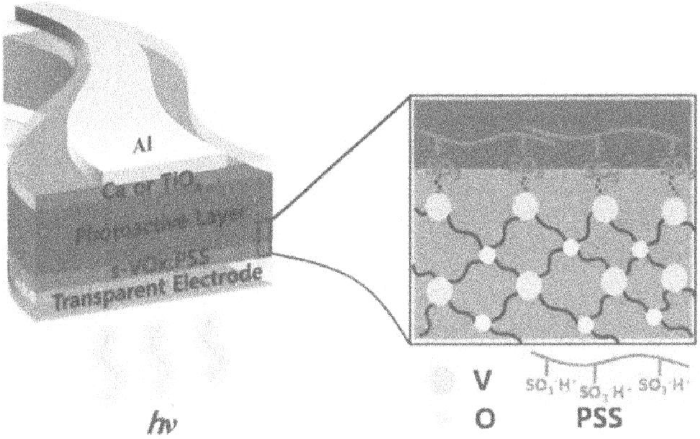

FIGURE 9.4 The device structure of PSCs incorporating s-VO$_x$:PSS composite as a hole transport layer. (Adapted with permission from Ref. [34]. Copyright (2014) American Chemical Society.)

[31]. The result of their work was a 60% improvement in the PCE due to the hybridization. The improved PCE and EIS data were ascribed to the charge separation mechanism in the hybrid PANI/ZnO nanograss. Similarly, Ameen et al. made a thin film electrode out of PANI/Ga-ZnO and compared it with an electrode of PANI/ZnO [15]. The author determined that doping the ZnO with Ga atoms led to better performance, which was postulated to be caused by the higher amount of minority charge carriers. These charge carriers could be generated and more swiftly transferred through the junction between the PANI and the doped ZnO. Another interesting PANI and zinc hybrid was developed by Mitra et al. using aluminum to dope the ZnO instead of Ga and displayed improved conductivity, which was attributed to the increased carrier mobility [32]. Other examples of metal oxides used in conjunction with polymers to form hybrid materials are tin oxide (SnO_2) and vanadium oxide (VO). Kudo et al. synthesized SnO_2 nanoparticles, which had been modified with the fullerene derivative $C_{60}C(COOH)_2$ and then hybridized it with MDMO-PPV (poly(2-methoxy-5-(3′,7′-dimethyl octyl oxy)-1,4-phenylenevinylene)) [33]. They determined that the boundary between the organic MDMO-PPV and inorganic SnO_2 formed a beneficial electronic junction. Kim et al. synthesized an s-VO$_x$/PSS composite and tested it as a HTL [34]. Compared with the pristine s-VO$_x$, the composite showed superior performance, which, according to the group, was due to enhanced stability and decreased leakage currents. Figure 9.4 shows the schematic for the s-VO$_x$/PSS as a hybrid hole transport layer in a polymer-based solar cell.

9.4 POLYMERS AS HOLE TRANSPORT LAYERS

The efficiency of the PSCs needs to be improved to make them more competitive in the market. The PCE performances of OSCs, organic tandem cells, and DSSCs (all of which make wide use of polymers) are still much lower than other multijunction SCs. Researchers have sought to improve the PCE of many types of solar cells to reduce recombination in the cell by adding hole transport layers. Since indirect recombination in cells has been calculated to happen approximately 100 times faster than direct recombination and occurs more quickly in HTL/perovskite junctions than in ETL/perovskite junctions [13], ways to smooth the carrier transition from the absorber material to the electrode have been undertaken. Resistance losses are also common at the cell/electrode interface, and adding a hole transport layer can help reduce these losses. Moreover, HTL materials are also used to passivate defects between layers due to their wide functional group and structural possibilities [13]. Another way HTLs help improve PSCs is by smoothing the bandgap transition from the absorber material to the electrode material because of many functional groups. HTL materials should also have a high charge carrier mobility to reduce recombination, and conducting polymers have demonstrated this ability [5, 7, 13, 35]. The improvement in the hole mobility of polymeric HTL materials has also been demonstrated by synthesizing it as a composite and increasing its crystalline character [36]. Another property of conducting polymers that makes them an excellent candidate for HTLs is that they can be both chemically and thermally stable, which is very important when it comes to being used in PeSCs, which can be unstable. This makes conducting polymers a better option than the commonly used Spiro-OMeTAD, which is not stable in the presence of air and is also more expensive [7]. The high degree of transparency of many conducting polymers is also a significant property since the incident light needs to reach the photoactive material for the solar cell to function [5, 35].

One material that has been widely utilized for hole transportation is PEDOT:PSS because of its high conductivity and excellent hole mobility. For example, one group has reported hole mobility of 2.0 S/cm for PEDOT:PSS [35]. Another positive aspect of using PEDOT:PSS in solar cells is its high work function, which enables the materials to function as an effective electrode and as an HTL at the same time [14]. However, the PSS component is acidic and hydrophobic and may cause the degradation of some types of solar cells [13]. Therefore, many efforts to modify the PEDOT:PSS material have been undertaken to increase its positive attributes and enhance overall cell life as well. Ibanez et al. reported in a review that PEDOT:PSS has been used for HTL materials in DSSCs because of the ability of PEDOT to oxidize the dye used as the photo absorber to regenerate the dye while at the same time serving as the charge carrier layer [35]. However, most of the reported applications of PEDOT:PSS have been in PeSCs. Because of the acidity of the PSS component of PEDOT:PSS, many groups have attempted to modify the PEDOT:PSS with various forms of graphene so that the PCE of various PeSCs would be more stable. Yu et al. used a fluorinated graphene composite and suggested the improvement was due to "the interfacial dipoles at the PEDOT/rGO interface. These interfacial dipoles promoted the electron-blocking and higher built-in potential in the devices" [5]. The fluorinated composite showed a 4.3% improvement in PCE over the solely PEDOT HTL device.

Other polymers that have been employed as HTLs include but are not limited to polytriarylamine (PTAA), P3HT, PANI, and PPy. Several researchers used PTAA and obtained varying results. Focus has been given to research on P3HT and its modifications because it offers more stability to PeSCs because P3HT is hydrophobic [5]. P3HT is also preferred in DSSCs over other HTL polymers because of "its film-forming ability, good hole transport, and small band gap" [35]. Some researchers have attempted to address this stability issue by modifying P3HT to make it more efficient as an HTL. Ahmad et al. reported work with P3HT and MoS_2 functionalized nano squares composited with P3HT by differing weights [36]. It was found that by adding 3% wt of MoS_2 to the P3HT, the crystallization of the polymer was improved, which increased the hole mobility of the overall device. A third focus has been placed on PANI as an HTL in addition to its use as a conducting polymer electrode. The advantages to PANI as an HTL material are the large band gap and the ability to deposit it as a thin film on a wide range of substrates [15]. Finally, PPy and its derivatives have been used as HTL materials. One such study used PPy in conjunction with polyacrylic acid and polyethylene glycol and found that the PCE of the device with the HTL material was higher than the non-HTL device, which points out to the potential application of such HTL material [35]. More research needs to be done to further improve HTL materials in solar cells, and polymers provide a large pool of options to do this.

9.5 POLYMERS AS ELECTRON TRANSPORT LAYERS

Another technique to optimize the efficiency of a solar cell is to improve the charge transfer by reducing electron recombination at the interface of the photoactive layer and the electrode contact. One method to accomplish this is introducing a layer between the two layers that can block hole transport, has high electron mobility, is stable, transparent, and can match the band gaps of the photoanode and electrode layers. Having a transparent ETL is essential since the sunlight has to

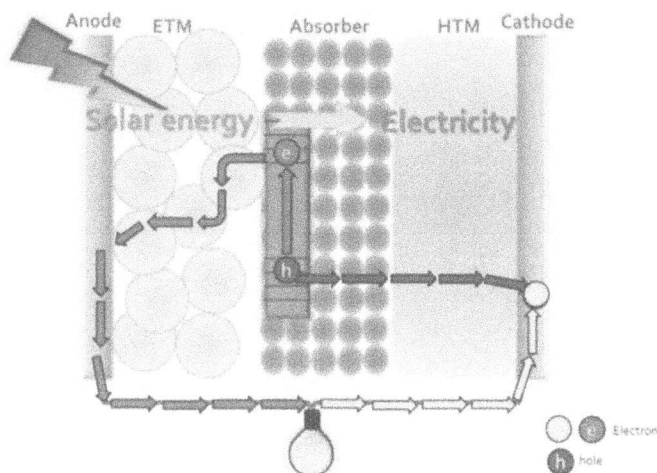

FIGURE 9.5 Schematic of photovoltaic conversion in DSSCs. (Adapted with permission from Ref. [35]. Copyright (2018) American Chemical Society.)

pass through it to reach the absorber layer, as shown in Figure 9.5. Polymers have been explored as ETL materials because of several characteristics. Notably, they have a tunable band gap, satisfactory electron mobilities, the ability to form thin films, and cheap manufacturing capability. The myriad of modifications that can be made to various polymers can help modify the electron mobility of the fabricated material and enhance their electron transport properties, which in turn improves the overall performance of the solar cell [37]. Compared to other ETL materials, polymers tend to have lower electron mobilities, which causes decreased photocurrents and fill factors. Two approaches can be followed to improve the electron mobility of polymer ETLs, optimize the polymer material with certain functional groups, or use polymers to enhance other ETL materials, such as graphene or metal oxides [5, 26, 35].

The first method uses conjugated polymers with high electron mobilities due to a network of π–π bonds and tunable energy levels due to chemical doping [5]. One widely studied polymers family for electron transport are the diimides, specifically naphthalene diimide (NDI) and perylene diimide (PDI) [13, 37]. Three main characteristics of NDIs highlighted by Choi and the group are: "(1) the high electron affinity of the NDI core caused by two strong electron-withdrawing diimide groups, (2) their highly extended π-conjugated structure for producing strong π–π intermolecular interaction, and (3) the controllable physicochemical properties by introducing a variety of alkyl chains on the N-position in diimide group" [37]. NDI copolymers have also exhibited excellent electron-hole mobilities, with one example having a value of 1.8 cm^2/V·s [37]. The structures of the NDIs used as well as their relative electron mobilities are shown in Figure 9.6.

Two other reasons that NDIs and PDIs are attributed to increasing device performance are the ability to reduce the work function of electrodes and passivate surface traps specifically for perovskites [13]. It was determined by Sun et al. that by incorporating PF-2TNDI into the amine groups of the NDI, a PCE of 16.7% can be obtained [38]. Wang et al. synthesized NDI-based ETLs named N2200 and PNDI2OD-TT and found that they helped produce PeSCs PCEs of 8.15% and 6.11%, respectively [39]. Kim et al. also synthesized a novel NDI ETL copolymer of NDI and dicyano-terthiophene, which had an impressive PCE of 17.0% compared to a similar device with a PCBM ETL that had a PCE of 14.3% [40]. They also tested the mechanical properties of the dicyanothiophene [P(NDI2DT-TTCN)] ETL-based cell and observed superior stability to the PCBM-based cell [40]. On a slightly different tack, Guo et al. tested several different PDI polymers as ETLs copolymers with vinylene, thiophene, selenophene, dibensosilole, and cyclopentadithiophene with the best PCE out of the group reaching 10.14% [41].

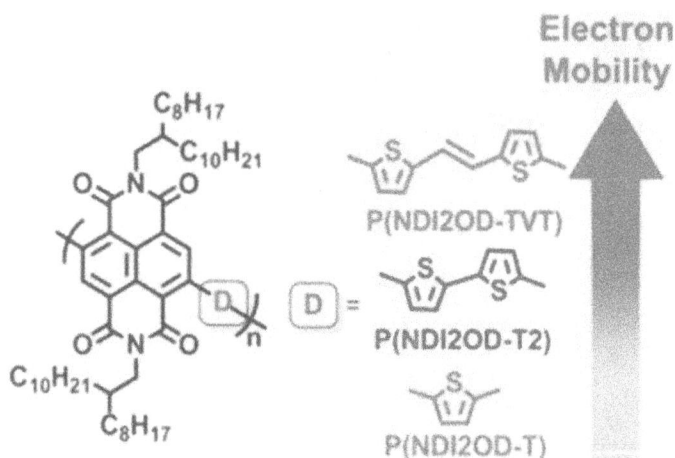

FIGURE 9.6 Chemical structure of various NDI polymers and their relative electron mobilities. (Adapted with permission from Ref. [37]. Copyright (2015) American Chemical Society.)

Polymers have also been used to modify other ETL materials to improve their performance through improving electron mobility or by passivating surfaces [13]. One common method for achieving enhanced performance is to modify graphene or graphene derivatives like reduced graphene oxide with various polymers [5]. Tong et al. reported the employment of PANI to modify graphene and make an ETL [42]. The synthesized layer completely covered a perovskite absorber layer, increased the cell performance from 9.3% to 13.8%, and protected the perovskite layer from moisture and temperature effects. Various researchers also investigated graphite, polyvinylpyrrolidone, and N-methyl pyrrolidone to functionalize the graphite [43]. Graphene oxide has also been composited with polymers like PEO, polyvinyl alcohol, polystyrene, polyurethane, and PMMA [43].

Another way to use polymers in ETLs is in conjunction with metals or metal oxides. One group prepared a nanocomposite of PANI and Au and determined that the Au particles improved the electron transfer of the PANI [18]. While other groups have focused their attention on CdSe composites with P3HT, MDMO-PPV, and poly[2,7-(9,9-dioctyl-fluorene)-alt5,5-(40,70-di-2-thienyl-20,10,30-benzothiadiazole)] (APFO-3) [24–26, 44]. Zhou et al. synthesized CdSe nanocrystals with P3HT and saw better electron transport [44]. CdSe tetrapods and MDMO-PPV were used to improve the efficiency of CdSe NRs and P3HT [25, 26]. The obtained positive result was suggested to be due to the shape of the tetrapods to increase electron transport. Wang et al. further improved the performance of CdSe-based materials by using APFO-3 instead of P3HT or MDMO-PPV because of its added ability to absorb part of the red spectrum of sunlight. Zinc and titanium oxides are often used as ETLs [45]. ZnO is easy to synthesize and has a low T_c, making it an attractive material to use as an ETL. However, there can be a large number of surface traps [17], which can be passivated by combining them with PEO and PEI (polyethyleneimine) [17, 46]. Xu et al. reported a water-/alcohol-soluble conjugated polymer, specifically poly[(9,9-bis(6′-((N,N-diethyl)-N-ethylammonium)-hexyl)-2,7-fluorene)-alt-1,4-diphenylsulfide]dibromide)] (PF6NPSBR) and Cs_2CO_3 material, as an ETL that was not sensitive to the layer thickness [20]. These results also show the flexibility available when employing polymers as electron transport enhancing materials and point to the need for further research and study.

9.6 POLYMERS AS COUNTER ELECTRODES

DSSCs consist of two electrodes that sandwich a photoactive layer. The CE is an important part of the DSSC since it drives the electrocatalytic reduction of the electrolyte and the CE is also responsible for hole collection from the cell [15]. Pt is often used as a CE since it can catalyze the reduction of I_3^-, which is often used in the electrolyte [13]. However, using Pt as a CE poses some issues as it is rare and expensive, and though usually considered stable, has been noted to affect the lifetime of DSSCs because it could break down in the triiodide/iodide electrolyte. For these reasons, attempts to replace the CE materials have been undertaken by several groups, and polymeric substances have been among those options. Polymers have been of particular interest because of their various properties, including cost, availability, translucence, and catalytic activity [13]. Three main polymers, namely PANI, PPy, and PEDOT, are often used as counter electrodes.

PANI has been used as a possible counter electrode material because of its low cost, ease of synthesis, and properties, such as oxidation state variation with color, ability to be doped with acids or bases, good conductivity, and stability. Researchers demonstrated the advantageous use of PANI by demonstrating that the thin PANI layer–based counter electrode can "show an increment on the active reaction interface and a decrease in the interfacial charge transfer resistance. The potential peak separation of I_3^-/I^- (I_2/I^-) for a PANI CE was comparable to that for a Pt CE, indicating that the PANI film was electrochemically active for the triiodide redox reaction" [35, 47]. Several groups have used various oxidizing agents to synthesize PANI CEs in multiple morphologies with varying results yielding increases in PCE from 0.19% all the way up to 1.46% improvement [13, 47–54]. Additionally, PANI hybrids have also been studied with variations based on carbon compounds and different doping types. This is because "GR–PANI complex allows the combination of efficient electron transport and transfer from graphene to PANI by covalent bonds between C atoms (graphene) and N atoms (aniline), enhancing the redox performance of PANI" [22, 35]. PANI-graphene hybrid CEs via electrodeposition were reported by The et al., and Wang et al. synthesized a thin film, both of which had acceptable results [55, 56]. Another trend that has been used is MWCNTs with PANI to enhance the CE's conductivity with the carbon structure properties [23, 57, 58]. As another example of PANI modification, Li et al. studied the morphological changes of PANI CEs due to doping with various ions (SO_4^{2-}, ClO-, BF-, Cl-, and p-toluenesulfonate (TsO-)), with the best results obtained from the sulfate doped PANI [59].

Polypyrrole is another polymer that has been widely tested as a CE in DSSCs. The properties that make PPy a possible CE replacement for Pt are that it can be easily synthesized, environmentally stable, and shows good catalytic activity [13]. The oxidation polymerization of PPy can be achieved via many different methods with many options available for oxidants. Peng et al. used $FeCl_3 \cdot 6H_2O$ to form a flexible PPy nanotube membrane with a slightly lower PCE than the Pt reference, as shown in Figure 9.7 [60]. Instead of a metal halide, Wu et al. used iodine as an oxidizing agent and then coated the resulting PPy on an FTO substrate with a 0.76% better PCE than the Pt CE [61]. Bu et al. used APS to formulate a transparent PPy CE for a bifacial DSSC with a resultant efficiency of

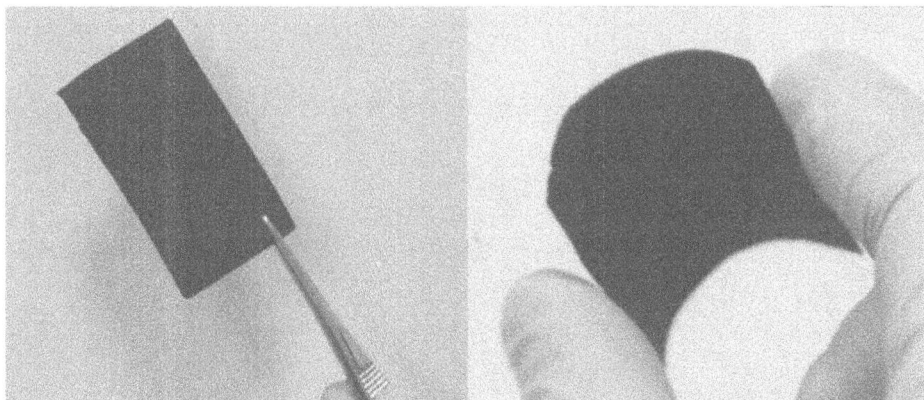

FIGURE 9.7   Flexible PPy paper-like membrane. (Adapted with permission from Ref. [60]. Copyright (2014) American Chemical Society.)

5.74% [62]. Wang et al. used V_2O_5 and $FeCl_3$ to synthesize a nanostructured PPy CE that reached an efficiency similar to a Pt reference electrode [63]. Veerender et al. synthesized a substrate-free PPy CE with limited efficiency results, but that showed good catalytic activity of the triiodide electrolyte [64]. Jeon et al. synthesized PPy nanoparticles that were drop coated onto an FTO substrate as a CE with a 7.73% PCE [65]. PPy hybrids can also be formed and have been tested as CEs in DSSCs by using them in conjunction with carbon-containing materials and with various doping types. One example of this was a PPy coating on cotton fabric that was in turn coated on Ni foam, which gave rise to good catalytic activity but only achieved a PCE of 3.83% [66]. Like PANI, PPy has also been composited with MWCNTs, and Yue et al. demonstrated that the composite performed better (3.78% PCE) than either the PPy (2.68% PCE) or the MWCNT (0.72% PCE) CEs [67]. Zhang et al. proposed that anion-doped PPy could replace Pt-based CEs by testing a dodecylbenzenesulfate-doped PPy CE because of its high catalytic activity [68]. Thus PPy also presents many possibilities for future research as well.

PEDOT has also been tested as a pure and hybrid CE. It shows great promise as a CE material due to its superior catalytic activity and conductivity, which is higher than that of both PANI and PPy, as well as its thermal and chemical stabilities [13, 69]. A bifacial honeycomb PEDOT CE was developed by Li et al. through electrodeposition that yielded a 1.07% improvement in efficiency, compared to a flat PEDOT CE [70]. Another structured PEDOT CE made via electrodeposition formed nanotube arrays and exhibited a PCE of 8.3% [71]. Apart from changing the morphology of the PEDOT to obtain different results, PEDOT can also be doped in a variety of ways to test its ability as a hole collector [13]. Xu et al. synthesized a TiN-doped PEDOT:PSS CE (Figure 9.8) with favorable results due to the TiN's positive interactions with the FTO substrate [72]. Di et al. made transition metal phosphate composite CEs and compared them to a plain PEDOT CE (PCE 5.44%) and saw some positive results with the Ni_3PO_4 (6.41%) and Co_3PO_4 (6.11%) composites [16].

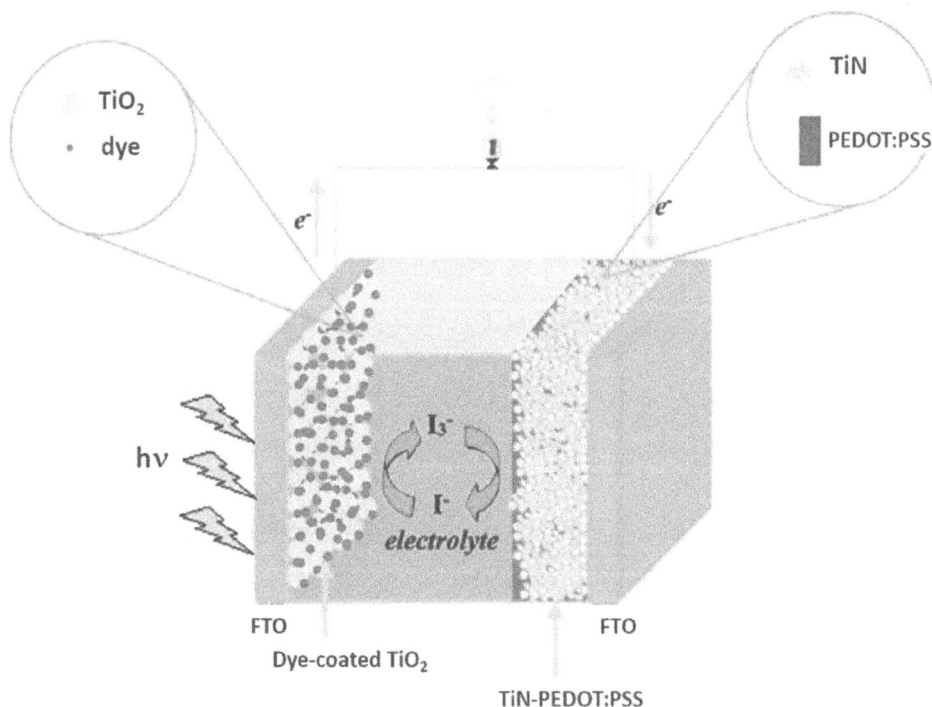

FIGURE 9.8 DSSC device structure with TiN-doped PEDOT:PSS counter electrode. (Adapted with permission from Ref. [72]. Copyright (2012) American Chemical Society.)

In addition to ion-doped PEDOT CEs, other polymers can be added into the mix, and polymer/polymer hybrids can be formed. One such example is PEDOT:PSS/PPy, which was compared to CEs of PEDOT:PSS, PPy, and Pt and resulted in PCEs of 7.60%, 6.31%, 5.23%, and 7.73%, respectively [73]. These are just a few of the many examples of PEDOT applications as a CE in DSSCs, showing the myriad of ways PEDOT can be employed to improve the properties of DSSCs.

9.7 POLYMERS AS AN INTERLAYER

Interfacial defects are responsible for efficiency loss in all types of solar cells due to recombination [13], and as solar cells become thinner and thinner, the interfacial properties of the device become increasingly important [74]. For electron and hole transfer to occur efficiently, each layer needs to be seamlessly bonded to the next layer in the cell. If the lattice structure of one layer is very different from another layer, then the flow of electrons becomes obstructed. Figure 9.9 shows an example of how the transition from a perovskite active layer to an inorganic HTL can be smoothed using PDMS to improve the function of the cell.

Interlayers in PeSCs are essential because they can help prevent the degradation of the active layer while also passivating these surface defects, which creates better contact between the active layer and the electrode, thus facilitating charge transport [21]. Interlayers are also important in OCSs for improving the charge transport function of the electrodes [75]. Some features that lead to the desired interlayer benefits involve neutral pH, solubility for easy processing in a solvent that will not degrade the active layer, good conductivity, appropriate band gap that matches the adjoining layers, and weak optical absorption. There are two main approaches to using polymers as interlayer materials in solar cells, which are to serve as an anode interface layer (AIL) or a cathode interface layer (CIL) [76]. The basic structure for the placement of interlayer materials in both types of solar cell structures is shown in Figure 9.10.

Cui et al. stated that conjugated polymers are well suited to be employed as AILs due to the wide variety of approaches they can be modified and tuned for their electrical, chemical, and optical properties [76]. For this reason, PCP-Na, PCF-Na, and PFS-Na (which are all thiophene and benzene derivatives) as AILs were used between an ITO glass substrate and a polymer-based heterojunction active layer. Compared to the fabricated polymer solar cell without an interfacial layer, the PCP-Na saw the highest increase in PCE at 9.89% compared to 5.18%, and the PFS-Na obtained the lower increase in PCE to a value of 8.16%. Some other examples of AILs in polymer solar cells include

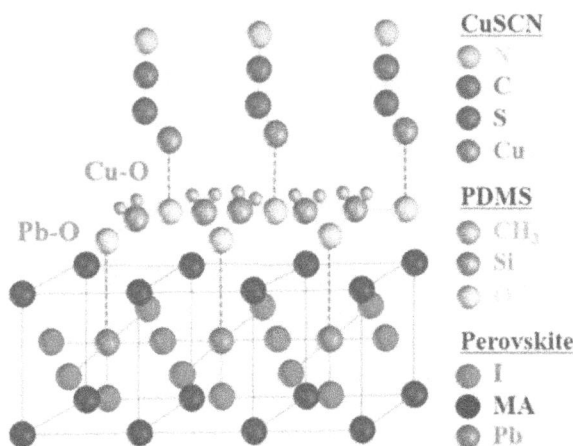

FIGURE 9.9 Schematic of the passivation of defects via PDMS interlayer. (Adapted with permission from Ref. [21]. Copyright (2019) American Chemical Society.)

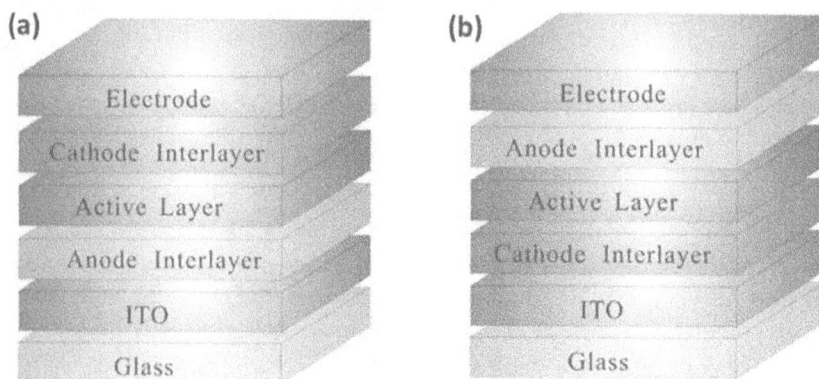

FIGURE 9.10 (a) Conventional and (b) inverted structures of organic solar cells. (Adapted with permission from Ref. [74]. Copyright (2020) American Chemical Society.)

PEDOT:PSS and polyfluorene thiophene derivatives [77–81]. AIL materials are also useful in PeSCs, as demonstrated by Malinkiewicz et al. with a poly(N,N′-bis(4-butylphenyl)benzidine) layer [82], Lin et al. with the best results from a poly[(9,9-bis(6′-((N,N-diethyl)-N-ethylammonium)-hexyl)-2,7-fluorene)-alt-1,4-diphenylsulfide] dibromide interlayer [83], and Wen et al. with a polystyrene (PS) layer [84]. These researchers found that their chosen AIL materials were able to increase the hole transfer through the device while blocking electron transfer. These results support the above statement that interlayers can increase device efficiencies by decreasing the chance for recombination.

CIL materials can range from simple polymer structures, such as polystyrene, to more complex copolymers to ions-doped polymers. Simple polymers may be used as interlayer material like PS and PEO [85–87], but complex substances, such as carboxylic acid functionalized hyperbranched polyether ketone (CHBPEK), sulfonic acid functionalized hyperbranched polyether sulfone (SHBPES) [88], and perylene diimide polyelectrolytes (PPDI-X) [89], have also been tested. Some approaches even use insulating polymers as interlayer materials; for example, PS, Teflon, and polyvinylidene-trifluoroethylene (PVDF-TrFE) were used and then compared as electron tunneling materials to improve overall device performance [90]. The PS device outperformed the control device by 3.4% in its PCE. Another approach in using CILs is to use the layer to reduce the work function of the cathode. Zhang et al. used PEI and poly[3-(6-trimethylammoniumhexyl)thiophene] (P3TMAHT) to form a dipole and thereby decrease the Ag cathode work function, which improved the PCE of the devices [91]. Some have used modified polythiophenes as CILs to improve device performance [92–94], while others chose to test modified polyfluorenes [95–97]. A specific example of a synthesized polyfluorene CIL that doubles as ETL was Cs_2CO_3-doped PF6NPSBr (poly[(9,9-bis(6′-((N,N-diethyl)-N-ethyl ammonium)-hexyl)-2,7-fluorene)-alt-1,4-diphenylsulfide]dibromide) [20]. While some interlayer materials' performance is dependent on the thickness of the layer, this was not the case in this investigation which determined that the prepared Cs_2CO_3-doped PF6NPSBr CIL was thickness insensitive. The researchers also discovered that doping the polymer changed the surface morphology of the polymer enough to improve the contact to the cathode. The best result was obtained by a device prepared with the Cs_2CO_3-doped PF6NPSBr CIL/ETL and achieved a PCE of 10.78%.

Many investigations involve the testing of the target materials in one capacity, while some studies have been performed to determine if the synthesized interlayer material can function as dual or bifunctional interlayers. For example, PS [84, 90] can be employed as CIL or AIL with positive results, but both polyfluorenes and polycarbonates can also be used in each capacity. A further study of the interlayer material poly[9,9-bis(1-sulfopropane-3-yl)-fluorene-2,7-diyl-alt-(2,2′-bithiophene-5,5′)diyl] (PFS) was explicitly synthesized to test it as a CIL and an AIL and to improve PCE while simultaneously simplifying the manufacturing process [98]. The researchers synthesized PFS in a simple two-step process and used it in several devices to test its AIL, CIL, and dual-use efficiencies. The fabricated device achieved a PCE of 9.60%, 9.67%, and 9.48% as an AIL, CIL, and dual-use, respectively. These results showed that device fabrication could be simplified without a significant loss in device efficiency if the right material is used. Another application of polymer interlayer materials was explored by Liu et al. to smooth the PET substrate surface for a flexible PeSC [99]. A layer of ZEOCOAT™ was laid down between the PET substrate and graphene electrode and thereby increased the overall device efficiency.

9.8 POLYMERS AS ELECTROLYTES

Dye-sensitized solar cells cannot function without a mechanism to regenerate the dye after being oxidized by light absorption. The electrolyte between the active layer dye and the counter electrode facilitates charge transfer and regenerates the redox couple to activate the dye [13, 100]. Since these are kinetic processes that require the reactants to make contact with each other, early electrolytes were all some sort of liquid phase [100]. The liquid electrolyte accomplishes these charge transfer and kinetic processes through diffusion from the counter electrode to the active dye material. It means that several factors need to be taken into account when choosing an electrolyte material. They are viscosity, concentration, distance inside the cell, and composition of the components [13]. The principles by which a DSSC operates are shown in Figure 9.11. The electrolyte has an impact on the device performance by influencing the values of current density (J_{sc}), open-circuit voltage (V_{oc}), and fill factor (FF). Therefore, it is imperative to use materials that remain stable and have a high charge transfer rate [13]. Although liquid phase electrolytes allow for somewhat efficient charge transfer, they also pose a set of problems that need to be overcome for the practical application of DSSCs. One of the main issues with liquid electrolytes is inadequate encapsulation because the electrolyte tends to evaporate or leak out of the cell, impacting device stability. Another problem is that the electrolyte may encourage cell degradation by reacting with cell components [69]. The approaches taken to solve these problems fall into two main categories: improving ions conduction

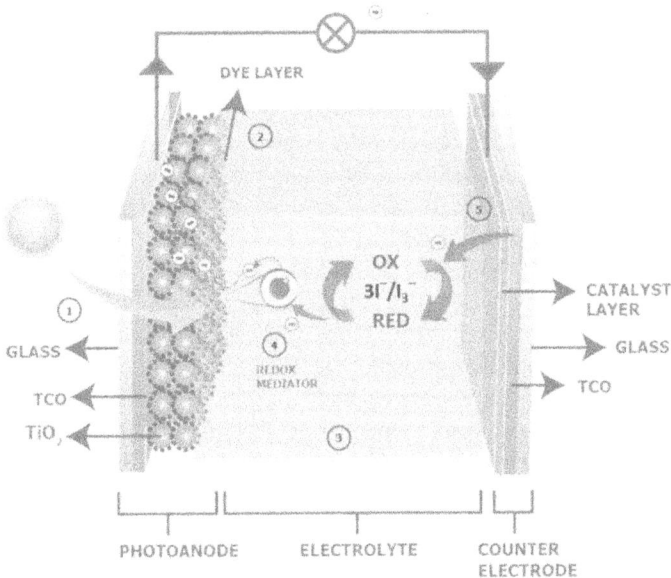

FIGURE 9.11 Schematic and operating principle of a DSSC [101]. (Adapted from Ref. [101]. This article was published as an open-access article which was distributed under the terms and conditions of the Creative Commons Attribution (CC BY).)

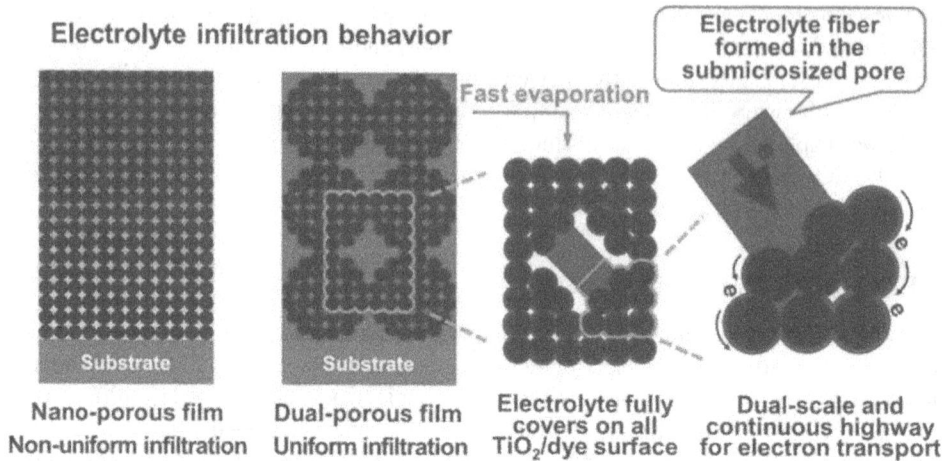

FIGURE 9.12 Polymer electrolyte infiltration of TiO₂ film in a DSSC. (Adapted with permission from Ref. [110]. Copyright (2016) American Chemical Society.)

in the electrolyte or improving contact between the electrolyte and the dye. Some researchers have chosen to address one or both of these issues by employing solid or gel electrolytes, while other groups focus on the type of polymer used (thermoplastics, thermosets, or composites).

As stated above, in liquid electrolytes, the conductivity occurs through diffusion, but in solid polymer electrolytes, the rapid segmental motion of the polymer and the interaction between ions and the atoms of the polymer are responsible for the conductivity of the electrolyte [69]. Several researchers have set about to improve the conductibility of the electrolyte and, therefore, the efficiency of the device by using polymeric ionic liquids or adding fillers or plasticizers to less conductive polymers like PEO [102]. For this reason, plasticizers like ethylene carbonate and propylene carbonate were added to swell PEO and PAN (polyacrylonitrile) by Ren and Cao, respectively [103, 104]. Several researchers also attempted another conduction improving approach in which a solvent with lower compatibility was added to the polymer [105–109]. This approach creates a "binary phase of coexisting liquid and solid phases," and the improved conductivity is attributed to the liquid phase [102]. However, another method for improving device performance is to help the polymer electrolyte penetrate the dye-coated active layer, which improves contact between the electrolyte and the photoanode by filling in the mesopore TiO₂ structure. It can be performed by tuning the mesopore size of the photoanode or by controlling the polymerization process of the electrolyte. An example of a pore-filling strategy is shown in Figure 9.12, which involves modifying TiO₂ film shape to allow for more infiltration of the polymer and brought about a 140% increase in the device efficiency [110].

Polymers have also been used as the matrix material in all-solid-state electrolytes because they can form 3D structures that can fill in the mesopores of the dye-sensitized material. The all-solid-state DSSCs face challenges with charge mobility as well as contact between the electrolyte and the dye. For this reason, researchers have switched their focus from solid polymer electrolytes to polymer gels instead. In its most basic form, a polymer gel is a mixture of a polymer and some sort of solvent, which may or may not be water. Gel electrolyte devices can improve both liquid electrolyte and all-solid-states DSSCs by avoiding most of the issues associated with leakage and by providing better wetting, conductivity, and filling properties, respectively. One of the challenges in developing polymer gel electrolytes is finding the right polymer to solvent/electrolyte ratio so that the polymers are not so abundant as to impede conduction and prevent electrolyte solvent from causing stability issues. There are different approaches to achieve this balance, as various researches have attempted to find successful thermoplastic, thermoset, and even composite polymer electrolytes to improve the function of DSSCs.

A thermoplastic polymer mainly exhibits entanglement among the polymer chains and interactions between its substituents groups to produce its physical properties. Thermoplastic polymers may be reprocessed by melting them to reuse them later. Thermoplastics are added to the DSSC as oligomers (small molecule forms of the polymer), which are thoroughly incorporated throughout the device framework and the solvent, thus initiating the polymerization process [13, 111]. The small oligomers are initially added to the framework, which allows the electrolyte material to fill the mesopores of the active material in the DSSC more effectively, giving better contact between the electrolyte and the active material of the cell [102]. Some examples of thermoplastics employed as polymer electrolyte materials include PAN with I_2/NaI [104], a more complex matrix of ethylene carbonate (EC)/propylene carbonate(PC)/N-methyl pyridine iodide, and poly(acrylonitrile-co-styrene) [112], a PC/PEG (poly-ethylene glycol) matrix with KI/I_2 [113], and PS dissolved in a liquid electrolyte [114]. A different approach involves utilizing thermoset polymers in the gel electrolyte as the matrix material because they offer more device stability at a varying temperatures [13]. A thermoset polymer undergoes cross-linking in the matrix, trapping the liquid electrolyte inside the polymer matrix [115]. Examples of thermoset polymers' employment as electrolytes in DSSCs include the utilization of PEG/PEGDA (poly(ethylene glycol) diacrylate) monomer solutions to permeate the TiO_2 mesopores followed by UV initiation of cross-linking [116] or taking polymer precursors in solution with I_2/I^- and heating them to form a polypyridyl-pendant poly(amidoamine) dendritic derivative/I_2/I^-/N-methylbenzimidazol/tert-butylpyridine layer [117] and using PEG and poly(acrylic acid) (PAA) to absorb a high amount of liquid electrolyte [118]. The third approach for producing polymer gel electrolytes, based on the type of material used in the electrolyte, is to make a polymer composite. These polymer composite electrolytes can be made by including inorganic nanoparticles into the polymer solution so that these "inorganic nanoparticles serve as the gelatins to solidify the liquid electrolyte and to enlarge the amorphous phase in the electrolyte" [13, 29]. The inorganic nanoparticles are included to address the issues with device conductivity and stability. Several researchers have incorporated titania as the inorganic nanoparticle, while others have tested SiO_2. Some have tested zinc oxide or alumina [111, 119, 120] and mineral-based nanoparticles as well [121, 122].

9.9 CONCLUSION

Solar energy will continue to grow in importance in the coming years; therefore, better materials must continuously to be developed as demand increases. With the ever-increasing production of panels and arrays, a more sustainable and affordable photovoltaic device is the goal of solar energy developers and companies. Polymeric materials are in a unique position to meet this need because of their low cost and the many different ways they can be synthesized and then processed. It makes them not only versatile in their application but much cheaper than current alternative materials, such as Pt or Pd in counter electrodes. As devices become smaller and have begun to transform into wearables, polymers continue to become an even more attractive material for use in solar devices. These flexible materials can fit various needs while retaining their photovoltaic properties, unlike silicon or other single-crystal materials. The high transparency of polymeric materials while absorbing solar radiation is another positive aspect for future wearable devices that could run on solar energy instead of relying on batteries or the constant need for charging. While the PCE of all-polymer solar cells is currently lower than other types of collector or multijunction cells, the above-mentioned favorable properties of polymers make them an ideal candidate for further research and improvement now.

REFERENCES

1. Parkinson B (1983) An overview of the progress in photoelectrochemical energy conversion. J Chem Educ 60:338–340.
2. Chapin DM, Fuller CS, Pearson GL (1954) A new silicon p-n junction photocell for converting solar radiation into electrical power. J Appl Phys 25:676–677.

3. Jordan DC, Kurtz SR (2013) Photovoltaic degradation rates – An analytical review. Prog Photovoltaics Res Appl 21:12–29.
4. Zhen C, Wu T, Chen R, Wang L, Liu G, Cheng HM (2019) Strategies for modifying TiO_2 based electron transport layers to boost perovskite solar cells. ACS Sustain Chem Eng 7:4586–4618.
5. You P, Tang G, Yan F (2019) Two-dimensional materials in perovskite solar cells. Mater Today Energy 11:128–158.
6. Kwon J, Park NG, Lee JY, Ko MJ, Park JH (2013) Highly efficient monolithic dye-sensitized solar cells. ACS Appl Mater Interfaces 5:2070–2074.
7. Das S, Pandey D, Thomas J, Roy T (2019) The role of graphene and other 2D materials in solar photovoltaics. Adv Mater 31:1–35.
8. Liu Z, Lau SP, Yan F (2015) Functionalized graphene and other two-dimensional materials for photovoltaic devices: Device design and processing. Chem Soc Rev 44:5638–5679.
9. Salikhov RB, Biglova YN, Mustafin AG (2018) New Organic Polymers for Solar Cells. In: Ameen S, Shin H-S, Akhtar MS (eds) Emerging Solar Energy Materials. IntechOpen, London, United Kingdom.
10. Vaqueiro-Contreras M, Markevich VP, Coutinho J, Santos P, Crowe IF, Halsall MP, Hawkins I, Lastovskii SB, Murin LI, Peaker AR (2019) Identification of the mechanism responsible for the boron oxygen light induced degradation in silicon photovoltaic cells. J Appl Phys 125:185704.
11. Idígoras J, Aparicio FJ, Contreras-Bernal L, Ramos-Terrón S, Alcaire M, Sánchez-Valencia JR, Borras A, Barranco Á, Anta JA (2018) Enhancing moisture and water resistance in perovskite solar cells by encapsulation with ultrathin plasma polymers. ACS Appl Mater Interfaces 10:11587–11594.
12. Bhandari S (2018) Polyaniline: Structure and Properties Relationship. In: Visakh PM, Falletta E, Pina C Della (eds) Polyaniline Blends, Composites, and Nanocomposites. Elsevier, Cambridge, United States, pp 23–60.
13. Hou W, Xiao Y, Han G, Lin JY (2019) The applications of polymers in solar cells: A review. Polymers (Basel) 11:1–46.
14. Hu L, Song J, Yin X, Su Z, Li Z (2020) Research progress on polymer solar cells based on PEDOT:PSS electrodes. Polymers (Basel) 12:145.
15. Ameen S, Shaheer M, Song M, Shik H (2013) Metal Oxide Nanomaterials, Conducting Polymers and Their Nanocomposites for Solar Energy. IntechOpen, London, United Kingdom.
16. Di Y, Xiao Z, Liu Z, Chen B, Feng J (2018) Hybrid films of PEDOT containing transition metal phosphates as high effective Pt-free counter electrodes for dye sensitized solar cells. Org Electron 57:171–177.
17. Shao S, Zheng K, Pullerits T, Zhang F (2013) Enhanced performance of inverted polymer solar cells by using poly(ethylene oxide)-modified ZnO as an electron transport layer. ACS Appl Mater Interfaces 5:380–385.
18. Iqbal S, Ahmad S (2018) Recent development in hybrid conducting polymers: Synthesis, applications and future prospects. J Ind Eng Chem 60:53–84.
19. Jeon NJ, Noh JH, Kim YC, Yang WS, Ryu S, Seok S Il (2014) Solvent engineering for high-performance inorganic-organic hybrid perovskite solar cells. Nat Mater 13:897–903.
20. Xu R, Zhang K, Liu X, Jin Y, Jiang XF, Xu QH, Huang F, Cao Y (2018) Alkali salt-doped highly transparent and thickness-insensitive electron-transport layer for high-performance polymer solar cell. ACS Appl Mater Interfaces 10:1939–1947.
21. Kim J, Lee Y, Yun AJ, Gil B, Park B (2019) Interfacial modification and defect passivation by the cross-linking interlayer for efficient and stable CuSCN-based perovskite solar cells. ACS Appl Mater Interfaces 11:46818–46824.
22. He B, Tang Q, Wang M, Chen H, Yuan S (2014) Robust polyaniline-graphene complex counter electrodes for efficient dye-sensitized solar cells. ACS Appl Mater Interfaces 6:8230–8236.
23. Zhang H, He B, Tang Q, Yu L (2014) Bifacial dye-sensitized solar cells from covalent-bonded polyaniline-multiwalled carbon nanotube complex counter electrodes. J Power Sources 275:489–497.
24. Wang P, Abrusci A, Wong HMP, Svensson M, Andersson MR, Greenham NC (2006) Photoinduced charge transfer and efficient solar energy conversion in a blend of a red polyfluorene copolymer with CdSe nanoparticles. Nano Lett 6:1789–1793.
25. Sun B, Marx E, Greenham NC (2003) Photovoltaic devices using blends of branched CdSe nanoparticles and conjugated polymers. Nano Lett 3:961–963.
26. Martínez-Ferrero E, Albero J, Palomares E (2010) Materials, nanomorphology, and interfacial charge transfer reactions in quantum dot/polymer solar cell devices. J Phys Chem Lett 1:3039–3045.
27. Kang MS, Ahn KS, Lee JW (2008) Quasi-solid-state dye-sensitized solar cells employing ternary component polymer-gel electrolytes. J Power Sources 180:896–901.

28. Katsaros G, Stergiopoulos T, Arabatzis IM, Papadokostaki KG, Falaras P (2002) A solvent-free composite polymer/inorganic oxide electrolyte for high efficiency solid-state dye-sensitized solar cells. J Photochem Photobiol A Chem 149:191–198.
29. Stergiopoulos T, Arabatzis IM, Katsaros G, Falaras P (2002) Binary polyethylene oxide/titania solid-state redox electrolyte for highly efficient nanocrystalline TiO_2 photoelectrochemical cells. Nano Lett 2:1259–1261.
30. Roy A, Mukhopadhyay S, Devi PS, Sundaram S (2019) Polyaniline-layered rutile TiO_2 nanorods as alternative photoanode in dye-sensitized solar cells. ACS Omega 4:1130–1138.
31. Zhu S, Wei W, Chen X, Jiang M, Zhou Z (2012) Hybrid structure of polyaniline/ZnO nanograss and its application in dye-sensitized solar cell with performance improvement. J Solid State Chem 190:174–179.
32. Mitra M, Kargupta K, Ganguly S, Goswami S, Banerjee D (2017) Facile synthesis and thermoelectric properties of aluminum doped zinc oxide/polyaniline (AZO/PANI) hybrid. Synth Met 228:25–31.
33. Kudo N, Shimazaki Y, Ohkita H, Ohoka M, Ito S (2007) Organic-inorganic hybrid solar cells based on conducting polymer and SnO_2 nanoparticles chemically modified with a fullerene derivative. Sol Energy Mater Sol Cells 91:1243–1247.
34. Kim J, Kim H, Kim G, Back H, Lee K (2014) Soluble transition metal oxide/polymeric acid composites for efficient hole-transport layers in polymer solar cells. ACS Appl Mater Interfaces 6:951–957.
35. Ibanez JG, Rincón ME, Gutierrez-Granados S, Chahma M, Jaramillo-Quintero OA, Frontana-Uribe BA (2018) Conducting polymers in the fields of energy, environmental remediation, and chemical-chiral sensors. Chem Rev 118:4731–4816.
36. Ahmad R, Srivastava R, Yadav S, Chand S, Sapra S (2017) Functionalized 2D-MoS_2-incorporated polymer ternary solar cells: Role of nanosheet-induced long-range ordering of polymer chains on charge transport. ACS Appl Mater Interfaces 9:34111–34121.
37. Choi J, Kim KH, Yu H, Lee C, Kang H, Song I, Kim Y, Oh JH, Kim BJ (2015) Importance of electron transport ability in naphthalene diimide-based polymer acceptors for high-performance, additive-free. All-Polymer Solar Cells Chem Mater 27:5230–5237.
38. Sun C, Wu Z, Yip HL, Zhang H, Jiang XF, Xue Q, Hu Z, Hu Z, Shen Y, Wang M, Huang F, Cao Y (2016) Amino-functionalized conjugated polymer as an efficient electron transport layer for high-performance planar-heterojunction perovskite solar cells. Adv Energy Mater 6:1501534.
39. Wang W, Yuan J, Shi G, Zhu X, Shi S, Liu Z, Han L, Wang HQ, Ma W (2015) Inverted planar heterojunction perovskite solar cells employing polymer as the electron conductor. ACS Appl Mater Interfaces 7:3994–3999.
40. Kim H Il, Kim M-J, Choi K, Lim C, Kim Y-H, Kwon S-K, Park T (2018) Improving the performance and stability of inverted planar flexible perovskite solar cells employing a novel NDI-based polymer as the electron transport layer. Adv Energy Mater 8:1702872.
41. Guo Q, Xu Y, Xiao B, Zhang B, Zhou E, Wang F, Bai Y, Hayat T, Alsaedi A, Tan Z (2017) Effect of energy alignment, electron mobility, and film morphology of perylene diimide based polymers as electron transport layer on the performance of perovskite solar cells. ACS Appl Mater Interfaces 9:10983–10991.
42. Tong SW, Balapanuru J, Fu D, Loh KP (2016) Thermally stable mesoporous perovskite solar cells incorporating low-temperature processed graphene/polymer electron transporting layer. ACS Appl Mater Interfaces 8:29496–29503.
43. Kim H, Abdala AA, MacOsko CW (2010) Graphene/polymer nanocomposites. Macromolecules 43:6515–6530.
44. Zhou Y, Riehle FS, Yuan Y, Schleiermacher HF, Niggemann M, Urban GA, Krüger M (2010) Improved efficiency of hybrid solar cells based on non-ligand-exchanged CdSe quantum dots and poly(3-hexylthiophene). Appl Phys Lett 96:013304.
45. Hau SK, Yip HL, Baek NS, Zou J, O'Malley K, Jen AKY (2008) Air-stable inverted flexible polymer solar cells using zinc oxide nanoparticles as an electron selective layer. Appl Phys Lett 92:253301.
46. Chen HC, Lin SW, Jiang JM, Su YW, Wei KH (2015) Solution-processed zinc oxide/polyethylenimine nanocomposites as tunable electron transport layers for highly efficient bulk heterojunction polymer solar cells. ACS Appl Mater Interfaces 7:6273–6281.
47. Zhang J, Hreid T, Li X, Guo W, Wang L, Shi X, Su H, Yuan Z (2010) Nanostructured polyaniline counter electrode for dye-sensitised solar cells: Fabrication and investigation of its electrochemical formation mechanism. Electrochim Acta 55:3664–3668.
48. Li Q, Wu J, Tang Q, Lan Z, Li P, Lin J, Fan L (2008) Application of microporous polyaniline counter electrode for dye-sensitized solar cells. Electrochem Commun 10:1299–1302.

49. Hou W, Xiao Y, Han G, Fu D, Wu R (2016) Serrated, flexible and ultrathin polyaniline nanoribbons: An efficient counter electrode for the dye-sensitized solar cell. J Power Sources 322:155–162.

50. Wang H, Feng Q, Gong F, Li Y, Zhou G, Wang ZS (2013) In situ growth of oriented polyaniline nanowires array for efficient cathode of Co(III)/Co(II) mediated dye-sensitized solar cell. J Mater Chem A 1:97–104.

51. Xiao Y, Lin JY, Wang WY, Tai SY, Yue G, Wu J (2013) Enhanced performance of low-cost dye-sensitized solar cells with pulse-electropolymerized polyaniline counter electrodes. Electrochim Acta 90:468–474.

52. Tai Q, Chen B, Guo F, Xu S, Hu H, Sebo B, Zhao XZ (2011) In situ prepared transparent polyaniline electrode and its application in bifacial dye-sensitized solar cells. ACS Nano 5:3795–3799.

53. Xiao Y, Han G, Li Y, Li M, Chang Y (2014) High performance of Pt-free dye-sensitized solar cells based on two-step electropolymerized polyaniline counter electrodes. J Mater Chem A 2:3452–3460.

54. Lin JY, Wang WY, Lin YT (2013) Characterization of polyaniline counter electrodes for dye-sensitized solar cells. Surf Coatings Technol 231:171–175.

55. He B, Tang Q, Wang M, Ma C, Yuan S (2014) Complexation of polyaniline and graphene for efficient counter electrodes in dye-sensitized solar cells: Enhanced charge transfer ability. J Power Sources 256:8–13.

56. Wang G, Xing W, Zhuo S (2012) The production of polyaniline/graphene hybrids for use as a counter electrode in dye-sensitized solar cells. Electrochim Acta 66:151–157.

57. Niu H, Qin S, Mao X, Zhang S, Wang R, Wan L, Xu J, Miao S (2014) Axle-sleeve structured MWCNTs/polyaniline composite film as cost-effective counter-electrodes for high efficient dye-sensitized solar cells. Electrochim Acta 121:285–293.

58. Xiao Y, Lin JY, Wu J, Tai SY, Yue G, Lin TW (2013) Dye-sensitized solar cells with high-performance polyaniline/multi-wall carbon nanotube counter electrodes electropolymerized by a pulse potentiostatic technique. J Power Sources 233:320–325.

59. Li Z, Ye B, Hu X, Ma X, Zhang X, Deng Y (2009) Facile electropolymerized-PANI as counter electrode for low cost dye-sensitized solar cell. Electrochem Commun 11:1768–1771.

60. Peng T, Sun W, Huang C, Yu W, Sebo B, Dai Z, Guo S, Zhao XZ (2014) Self-assembled free-standing polypyrrole nanotube membrane as an efficient FTO- and Pt-free counter electrode for dye-sensitized solar cells. ACS Appl Mater Interfaces 6:14–17.

61. Wu J, Li Q, Fan L, Lan Z, Li P, Lin J, Hao S (2008) High-performance polypyrrole nanoparticles counter electrode for dye-sensitized solar cells. J Power Sources 181:172–176.

62. Bu C, Tai Q, Liu Y, Guo S, Zhao X (2013) A transparent and stable polypyrrole counter electrode for dye-sensitized solar cell. J Power Sources 221:78–83.

63. Wang G, Dong W, Yan C, Hou S, Zhang W (2018) Facile synthesis of hierarchical nanostructured polypyrrole and its application in the counter electrode of dye-sensitized solar cells. Mater Lett 214:158–161.

64. Veerender P, Saxena V, Jha P, Koiry SP, Gusain A, Samanta S, Chauhan AK, Aswal DK, Gupta SK (2012) Free-standing polypyrrole films as substrate-free and Pt-free counter electrodes for quasi-solid dye-sensitized solar cells. Org Electron 13:3032–3039.

65. Jeon SS, Kim C, Ko J, Im SS (2011) Spherical polypyrrole nanoparticles as a highly efficient counter electrode for dye-sensitized solar cells. J Mater Chem 21:8146–8151.

66. Xu J, Li M, Wu L, Sun Y, Zhu L, Gu S, Liu L, Bai Z, Fang D, Xu W (2014) A flexible polypyrrole-coated fabric counter electrode for dye-sensitized solar cells. J Power Sources 257:230–236.

67. Yue G, Wang L, Zhang X, Wu J, Jiang Q, Zhang W, Huang M, Lin J (2014) Fabrication of high performance multi-walled carbon nanotubes/polypyrrole counter electrode for dye-sensitized solar cells. Energy 67:460–467.

68. Zhang X, Wang S, Lu S, Su J, He T (2014) Influence of doping anions on structure and properties of electro-polymerized polypyrrole counter electrodes for use in dye-sensitized solar cells. J Power Sources 246:491–498.

69. Theerthagiri J, Senthil AR, Madhavan J, Maiyalagan T (2015) Recent progress in non-platinum counter electrode materials for dye-sensitized solar cells. ChemElectroChem 2:928–945.

70. Li H, Xiao Y, Han G, Hou W (2017) Honeycomb-like poly(3,4-ethylenedioxythiophene) as an effective and transparent counter electrode in bifacial dye-sensitized solar cells. J Power Sources 342:709–716.

71. Trevisan R, Döbbelin M, Boix PP, Barea EM, Tena-Zaera R, Mora-Seró I, Bisquert J (2011) PEDOT nanotube arrays as high performing counter electrodes for dye sensitized solar cells. Study of the interactions among electrolytes and counter electrodes. Adv Energy Mater 1:781–784.

72. Xu H, Zhang X, Zhang C, Liu Z, Zhou X, Pang S, Chen X, Dong S, Zhang Z, Zhang L, Han P, Wang X, Cui G (2012) Nanostructured titanium nitride/PEDOT:PSS composite films as counter electrodes of dye-sensitized solar cells. ACS Appl Mater Interfaces 4:1087–1092.

73. Yue G, Wu J, Xiao Y, Lin J, Huang M, Lan Z (2012) Application of poly(3,4-ethylenedioxythiophene): polystyrenesulfonate/polypyrrole counter electrode for dye-sensitized solar cells. J Phys Chem C 116:18057–18063.
74. Li T, Chen Z, Wang Y, Tu J, Deng X, Li Q, Li Z (2020) Materials for interfaces in organic solar cells and photodetectors. ACS Appl Mater Interfaces 12:3301–3326.
75. Chen Y, Liu H, Jiang B, Zhao Y, Meng X, Ma T (2020) Hierarchical porous architectures derived from low-cost biomass *Equisetum arvense* as a promising anode material for lithium-ion batteries. J Mol Struct 1221:128794.
76. Cui Y, Xu B, Yang B, Yao H, Li S, Hou J (2016) A novel pH neutral self-doped polymer for anode interfacial layer in efficient polymer solar cells. Macromolecules 49:8126–8133.
77. Rider DA, Harris KD, Wang D, Bruce J, Fleischauer MD, Tucker RT, Brett MJ, Buriak JM (2009) Thienylsilane-modified indium tin oxide as an anodic interface in polymer/fullerene solar cells. ACS Appl Mater Interfaces 1:279–288.
78. Lee EJ, Choi MH, Han YW, Moon DK (2017) Effect on electrode work function by changing molecular geometry of conjugated polymer electrolytes and application for hole-transporting layer of organic optoelectronic devices. ACS Appl Mater Interfaces 9:44060–44069.
79. Jo JW, Jung JW, Bae S, Ko MJ, Kim H, Jo WH, Jen AKY, Son HJ (2016) Development of self-doped conjugated polyelectrolytes with controlled work functions and application to hole transport layer materials for high-performance organic solar cells. Adv Mater Interfaces 3:1500703.
80. Jo JW, Yun JH, Bae S, Ko MJ, Son HJ (2017) Development of a conjugated donor-acceptor polyelectrolyte with high work function and conductivity for organic solar cells. Org Electron 50:1–6.
81. Tang CG, Ang MCY, Choo KK, Keerthi V, Tan JK, Syafiqah MN, Kugler T, Burroughes JH, Png RQ, Chua LL, Ho PKH (2016) Doped polymer semiconductors with ultrahigh and ultralow work functions for ohmic contacts. Nature 539:536–540.
82. Malinkiewicz O, Yella A, Lee YH, Espallargas GM, Graetzel M, Nazeeruddin MK, Bolink HJ (2014) Perovskite solar cells employing organic charge-transport layers. Nat Photonics 8:128–132.
83. Lin Q, Armin A, Nagiri RCR, Burn PL, Meredith P (2015) Electro-optics of perovskite solar cells. Nat Photonics 9:106–112.
84. Wen X, Wu J, Ye M, Gao D, Lin C (2016) Interface engineering: Via an insulating polymer for highly efficient and environmentally stable perovskite solar cells. Chem Commun 52:11355–11358.
85. Zhang F, Ceder M, Inganäs O (2007) Enhancing the photovoltage of polymer solar cells by using a modified cathode. Adv Mater 19:1835–1838.
86. Zhou Y, Li F, Barrau S, Tian W, Inganäs O, Zhang F (2009) Inverted and transparent polymer solar cells prepared with vacuum-free processing. Sol Energy Mater Sol Cells 93:497–500.
87. Nie R, Zhao Z, Deng X (2017) Roles of electrode interface on the performance of organic photodetectors. Synth Met 227:163–169.
88. Ai L, Ouyang X, Liu Z, Peng R, Mi D, Kakimoto M, Ge Z (2016) Multi-channel interface dipole of hyperbranched polymers with quasi-immovable hydrion to modification of cathode interface for high-efficiency polymer solar cells. Prog Photovoltaics Res Appl 24:1044–1054.
89. Hu Z, Xu R, Dong S, Lin K, Liu J, Huang F, Cao Y (2017) Quaternisation-polymerized N-type polyelectrolytes: Synthesis, characterisation and application in high-performance polymer solar cells. Mater Horizons 4:88–97.
90. Wang Q, Dong Q, Li T, Gruverman A, Huang J (2016) Thin insulating tunneling contacts for efficient and water-resistant perovskite solar cells. Adv Mater 28:6734–6739.
91. Zhang H, Azimi H, Hou Y, Ameri T, Przybilla T, Spiecker E, Kraft M, Scherf U, Brabec CJ (2014) Improved high-efficiency perovskite planar heterojunction solar cells via incorporation of a polyelectrolyte interlayer. Chem Mater 26:5190–5193.
92. Zilberberg K, Behrendt A, Kraft M, Scherf U, Riedl T (2013) Ultrathin interlayers of a conjugated polyelectrolyte for low work-function cathodes in efficient inverted organic solar cells. Org Electron 14:951–957.
93. Seo JH, Gutacker A, Sun Y, Wu H, Huang F, Cao Y, Scherf U, Heeger AJ, Bazan GC (2011) Improved high-efficiency organic solar cells via incorporation of a conjugated polyelectrolyte interlayer. J Am Chem Soc 133:8416–8419.
94. Hu L, Wu F, Li C, Hu A, Hu X, Zhang Y, Chen L, Chen Y (2015) Alcohol-soluble n-type conjugated polyelectrolyte as electron transport layer for polymer solar cells. Macromolecules 48:5578–5586.
95. Sun X, Li C, Ni J, Huang L, Xu R, Li Z, Cai H, Li J, Zhang Y, Zhang J (2017) A facile two-step interface engineering strategy to boost the efficiency of inverted ternary-blend polymer solar cells over 10%. ACS Sustain Chem Eng 5:8997–9005.

96. Sun C, Wu Z, Hu Z, Xiao J, Zhao W, Li HW, Li QY, Tsang SW, Xu YX, Zhang K, Yip HL, Hou J, Huang F, Cao Y (2017) Interface design for high-efficiency non-fullerene polymer solar cells. Energy Environ Sci 10:1784–1791.

97. Yang T, Wang M, Duan C, Hu X, Huang L, Peng J, Huang F, Gong X (2012) Inverted polymer solar cells with 8.4% efficiency by conjugated polyelectrolyte. Energy Environ Sci 5:8208–8214.

98. Xu B, Zheng Z, Zhao K, Hou J (2016) A bifunctional interlayer material for modifying both the anode and cathode in highly efficient polymer solar cells. Adv Mater 28:434–439.

99. Liu Z, You P, Xie C, Tang G, Yan F (2016) Ultrathin and flexible perovskite solar cells with graphene transparent electrodes. Nano Energy 28:151–157.

100. Goncalves LM, De Zea Bermudez V, Ribeiro HA, Mendes AM (2008) Dye-sensitized solar cells: A safe bet for the future. Energy Environ Sci 1:655–667.

101. Iftikhar H, Sonai GG, Hashmi SG, Nogueira AF, Lund PD (2019) Progress on electrolytes development in dye-sensitized solar cells. Materials (Basel) 12:1998.

102. Song D, Cho W, Lee JH, Kang YS (2014) Toward higher energy conversion efficiency for solid polymer electrolyte dye-sensitized solar cells: Ionic conductivity and TiO_2 pore-filling. J Phys Chem Lett 5:1249–1258.

103. Ren Y, Zhang Z, Gao E, Fang S, Cai S (2001) A dye-sensitized nanoporous TiO_2 photoelectrochemical cell with novel gel network polymer electrolyte. J Appl Electrochem 31:445–447.

104. Cao F, Oskam G, Searson PC (1995) Solid state dye sensitized photoelectrochemical cell. J Phys Chem 99:17071–17073.

105. Wang P, Zakeeruddin SM, Moser JE, Nazeeruddin MK, Sekiguchi T, Grätzel M (2003) A stable quasi-solid-state dye-sensitized solar cell with an amphiphilic ruthenium sensitizer and polymer gel electrolyte. Nat Mater 2:402–407.

106. Wang P, Zakeeruddin SM, Exnar I, Grätzel M (2002) High efficiency dye-sensitized nanocrystalline solar cells based on ionic liquid polymer gel electrolyte. Chem Commun 2972–2973.

107. Xiang W, Huang W, Bach U, Spiccia L (2013) Stable high efficiency dye-sensitized solar cells based on a cobalt polymer gel electrolyte. Chem Commun 49:8997–8999.

108. Choi H, Paek S, Lim K, Kim C, Kang MS, Song K, Ko J (2013) Molecular engineering of organic sensitizers for highly efficient gel-state dye-sensitized solar cells. J Mater Chem A 1:8226–8233.

109. Chen CL, Chang TW, Teng H, Wu CG, Chen CY, Yang YM, Lee YL (2013) Highly efficient gel-state dye-sensitized solar cells prepared using poly(acrylonitrile-co-vinyl acetate) based polymer electrolytes. Phys Chem Chem Phys 15:3640–3645.

110. Li Y, He XL, Lian CX, Yao HL, Yang GJ, Li CX, Li CJ, Fang B (2016) In situ formation of continuous charge transfer pathways for highly efficient, solvent-free, polymer electrolyte-based dye-sensitized solar cells. ACS Sustain Chem Eng 4:4013–4020.

111. Lan Z, Wu J, Wang D, Hao S, Lin J, Huang Y (2007) Quasi-solid-state dye-sensitized solar cells based on a sol-gel organic-inorganic composite electrolyte containing an organic iodide salt. Sol Energy 81:117–122.

112. Wu J, Lan Z, Wang D, Hao S, Lin J, Huang Y, Yin S, Sato T (2006) Gel polymer electrolyte based on poly(acrylonitrile-co-styrene) and a novel organic iodide salt for quasi-solid state dye-sensitized solar cell. Electrochim Acta 51:4243–4249.

113. Wu JH, Hao SC, Lan Z, Lin JM, Huang ML, Huang YF, Fang LQ, Yin S, Sato T (2007) A thermoplastic gel electrolyte for stable quasi-solid-state dye-sensitized solar cells. Adv Funct Mater 17:2645–2652.

114. Lee KS, Jun Y, Park JH (2012) Controlled dissolution of polystyrene nanobeads: Transition from liquid electrolyte to gel electrolyte. Nano Lett 12:2233–2237.

115. Su JY, Tsai CH, Wang SA, Huang TW, Wu CC, Wong KT (2012) Functionalizing organic dye with cross-linked electrolyte-blocking shell as a new strategy for improving DSSC efficiency. RSC Adv 2:3722–3728.

116. Parvez MK, In I, Park JM, Lee SH, Kim SR (2011) Long-term stable dye-sensitized solar cells based on UV photo-crosslinkable poly(ethylene glycol) and poly(ethylene glycol) diacrylate based electrolytes. Sol Energy Mater Sol Cells 95:318–322.

117. Wang L, Fang S, Lin Y, Zhou X, Li M (2005) A 7.72% efficient dye sensitized solar cell based on novel necklace-like polymer gel electrolyte containing latent chemically cross-linked gel electrolyte precursors. Chem Commun 5687–5689.

118. Wu JH, Lan Z, Lin JM, Huang ML, Hao SC, Sato T, Yin S (2007) A novel thermosetting gel electrolyte for stable quasi-solid-state dye-sensitized solar cells. Adv Mater 19:4006–4011.

119. Nazmutdinova G, Sensfuss S, Schrödner M, Hinsch A, Sastrawan R, Gerhard D, Himmler S, Wasserscheid P (2006) Quasi-solid state polymer electrolytes for dye-sensitized solar cells: Effect of the electrolyte components variation on the triiodide ion diffusion properties and charge-transfer resistance at platinum electrode. Solid State Ionics 177:3141–3146.

120. Zhang J, Han H, Wu S, Xu S, Yang Y, Zhou C, Zhao X (2007) Conductive carbon nanoparticles hybrid PEO/P(VDF-HFP)/SiO$_2$ nanocomposite polymer electrolyte type dye sensitized solar cells. Solid State Ionics 178:1595–1601.
121. Tu C-W, Liu K-Y, Chien A-T, Yen M-H, Weng TH, Ho K-C, Lin K-F (2008) Enhancement of photocurrent of polymer-gelled dye-sensitized solar cell by incorporation of exfoliated montmorillonite nanoplatelets. J Polym Sci Part A Polym Chem 46:47–53.
122. Ito BI, De Freitas JN, De Paoli MA, Nogueira AF (2008) Application of a composite polymer electrolyte based on montmorillonite in dye-sensitized solar cells. J Braz Chem Soc 19:688–696.

10 Nanocomposites Based on Conducting Polymers and Metal Sulfides for Solar Cell Applications

Shikha Chander[1,2], Kalpana Madgula[3], Venkata Sreenivas Puli[4,5], and Meenu Mangal[6]

[1]Department of Chemistry, University of Technology, Jaipur, Rajasthan, India

[2]St. Francis Degree and P.G. College for Women, Begumpet, Hyderabad, Telangana, India

[3]SAS Nanotechnologies LLC, Wilmington, Delaware, USA

[4]Smart Nanomaterials Solutions LLC, Orlando, USA

[5]Materials and Manufacturing Directorate, Wright Patterson Air Force Base, Ohio, USA

[6]Department of Chemistry, Poddar International College, Jaipur, Rajasthan, India

CONTENTS

10.1 INTRODUCTION

Conductive polymers have hybrid qualities with traditional methods, which often add unique electrical characteristics of metals or semiconductors. Research on nanostructured polymers has turned into a buzz in the past few years because of their remarkable unique properties of bulk materials, energy transfer, sensors, and actuators. Several methods are being investigated for providing multiple conductive polymer nanostructures and high-performance devices based on these nanostructure conducting polymers (CPs). The applications covered in this chapter include capacitors, electrodes, and practical materials for next-generation solar cells and nanostructured conductive polymers for an active electrode in electrochemical capacitors and high energy-density batteries. A great deal of basic

DOI: 10.1201/9781003150374-10

information about solar cell construction and design mechanisms is provided, beginning with the first few pages. These first segments feature approaches in which authors incorporated more complex manufacturing or product techniques from the literature to enhance electrochemical performance.

The world's population and economy are increasingly growing, necessitating the development of alternative renewable energy sources. Solar energy stands out among them even though it is eco-sustainable and with no geographic restrictions. The harvesting by solar (or photovoltaic, PV) cell is one of the most effective ways of harnessing solar energy. The main emphasis of photovoltaic solar research and innovation is on [1] mature silicon-based solar cells, as typical silicon-based solar cells are restricted in power consumption and cost. Gallium arsenide (GaAs), copper indium gallium selenide (CIGS), and cadmium telluride (CdTe) are usually thin-film based and all efficient solar cells [2]. Despite the high efficiency and stability of the mentioned thin-film solar cells, the accumulation of gallium and indium in the crust is limited, and cadmium is hazardous [3]. Organic photovoltaic (OPV), dye-sensitized solar cells (DSSCs), and perovskite solar cells (PSCs) are three other types of emerging photovoltaic cells. The benefits of these types of solar cells are they are lighter in weight with relatively inexpensive [3, 4] to produce them. OPV cells need enhanced performance, and electrolyte-based DSSCs face challenges such as oxidative damage and solvent evaporation that need to be resolved. With an all PSC, the conversion efficiency has fluctuated between 3.8% and 23.3% [5–7]. Each portion of the system influences its output by performing its respective tasks, as shown in Figure 10.1 [2, 8].

Polymers, in general, have received a lot of attention in recent years due to their flexibility and functionality. The three-dimensional polymer network structures determine whether they can be used as a template to fabricate mesoporous materials or as a polymeric matrix [9] in solid electrolytes owing to high catalytic reducing activity limitation and is a powerful counter electrochemical (or electrode) in a cell. Polymers with complex functional groups can control perovskite microstructure from both mass and surface viewpoints; polymers with high (charge) carrier mobilities can behave as electron/hole transition materials; and polymers with organic compounds could be utilized for bridging layers to passivate defects, to adjust the metal electrode activation energy, and to better the morphology of perovskites. Polymers that are used as the photoactive layer or buffer layer in OPV have a variety of optoelectronic characteristics and electrical conductivity due to their different structures and functional modifications [10]. The photocatalytic efficiency and excellent stability of conductive polymers make them common additives in a wide range of industries [11–13].

10.2 DESCRIPTION OF SOLAR CELLS

Solar energy conversion has now proven to be a feasible way to meet a significant portion of the world's energy needs while lowering carbon dioxide emissions, generating employment, and understanding how to deal with instabilities derived from fossil fuel global politics. There is an increasing recognition that energy sources are unsustainable, that their sector is rapidly depleting, and that they would be either biologically or economically unviable [5–6]. There is a need for the exponential growth in power plants based on renewable energy to be mounted for self-sustainability by utilizing infinite natural resources, sun, and wind as a continuously increasing chain of supply and demand and complete understanding of the basic properties required for both economical and efficient solar energy converters.

Hereafter, the following sections (10.2.1–10.2.3) are dedicated to explaining the functioning of various solar cells.

10.2.1 Photovoltaic (PV) Solar Cell

When a solar cell or the P-N junction comes in contact with the solar rays, light photons easily penetrate through a very thin P-type layer, as represented in Figure 10.2 [7]. The PV solar cells are

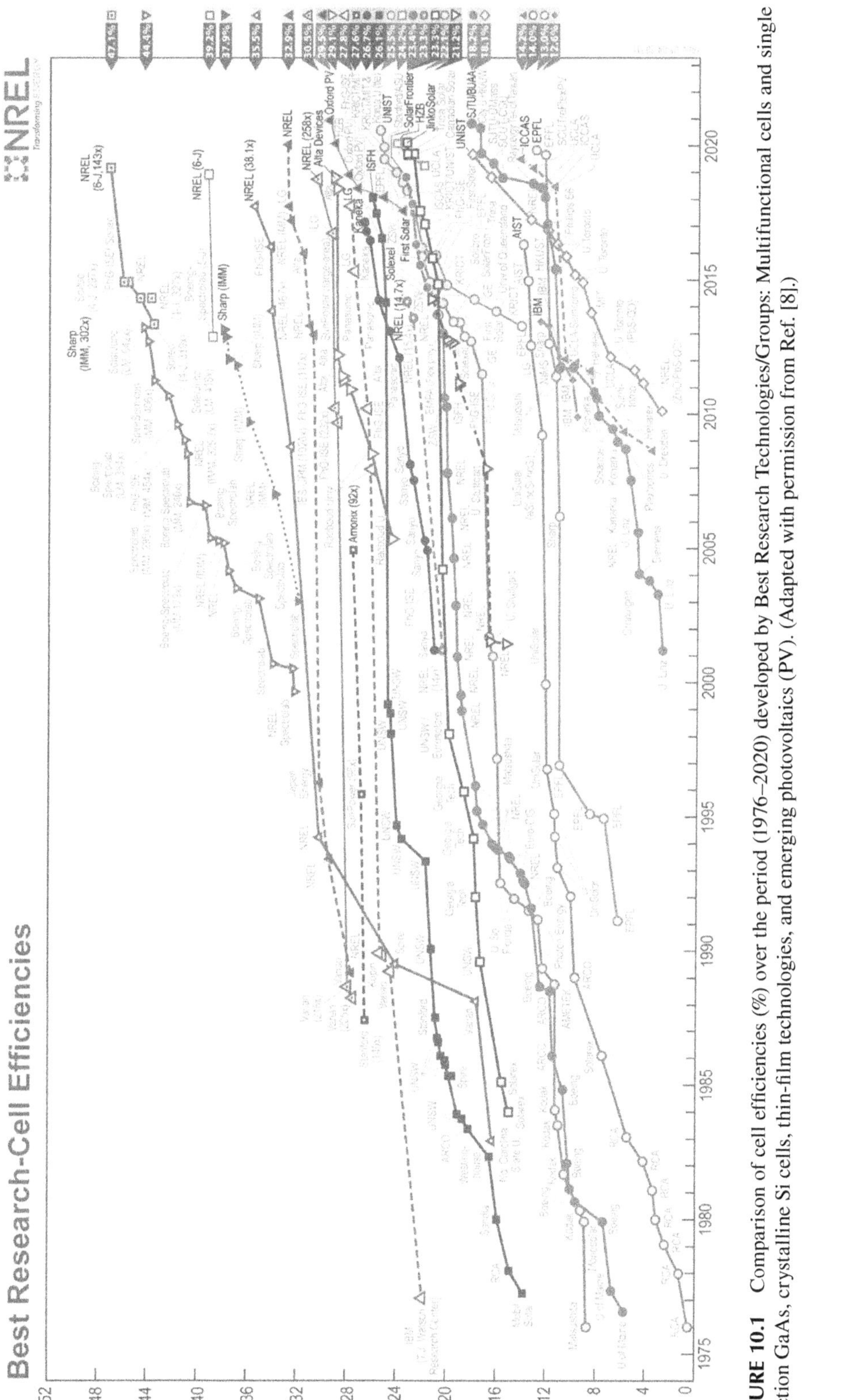

FIGURE 10.1 Comparison of cell efficiencies (%) over the period (1976–2020) developed by Best Research Technologies/Groups: Multifunctional cells and single junction GaAs, crystalline Si cells, thin-film technologies, and emerging photovoltaics (PV). (Adapted with permission from Ref. [8].)

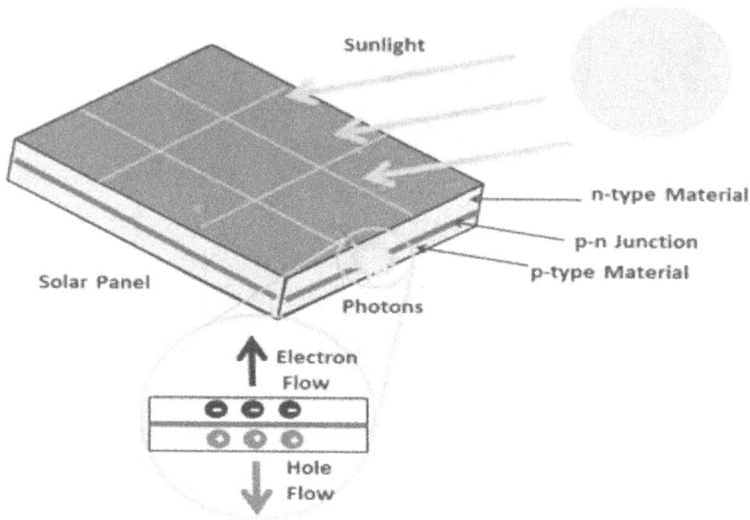

FIGURE 10.2 Working of a photovoltaic cell; created internally by a member of the Energy Education team. (Copyright 2015. Adapted with permission from Ref. [7]. Ecogreen Electrical and https://images.app.goo.gl/ u2cBkfHMfBGjYZ8w8.)

further classified into monocrystalline, polycrystalline, or non-crystalline (used in thin films or along with other materials) silicon solar cells. The photons supplied by light energy are sufficient enough to the junction to produce a number electron-hole pairs (or as valence band to the conduction band) and this energy is further transported to the load via link. Ninety percent of the world's photovoltaic cells are dependent on differences in silicon purity, expense, and efficiency. Crystalline silicon solar cells are most commonly used, conventional/traditional or as wafers in solar panels. Its energy conversions are over 25%. They are also of two types, namely monocrystalline silicon (Si) solar cells, which are single-crystal silicon cells or mono-Si or mono silicon, and polycrystalline silicon (Si) solar cells. The efficiency of mono-Si solar cells is 21% now; it has improved to 26.7% due to technology development. The mono-Si solar cells (heterojunction with intrinsic layer, HIT) are constructed as shown below in Figure 10.3. It is made from single, ultra-thin layer silicon

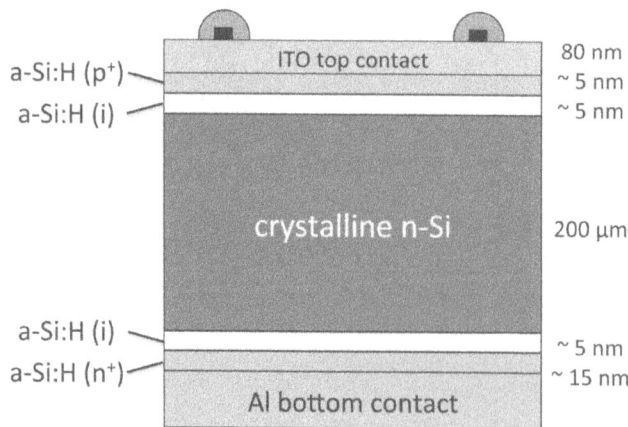

FIGURE 10.3 Graphic representation of a heterojunction with intrinsic thin layer (HIT). (Adapted with permission from CC BY-SA 4.0; https://upload.wikimedia.org/wikipedia/commons/thumb/1/16/HIT_cell. jpg/220px-HIT_cell.jpg.)

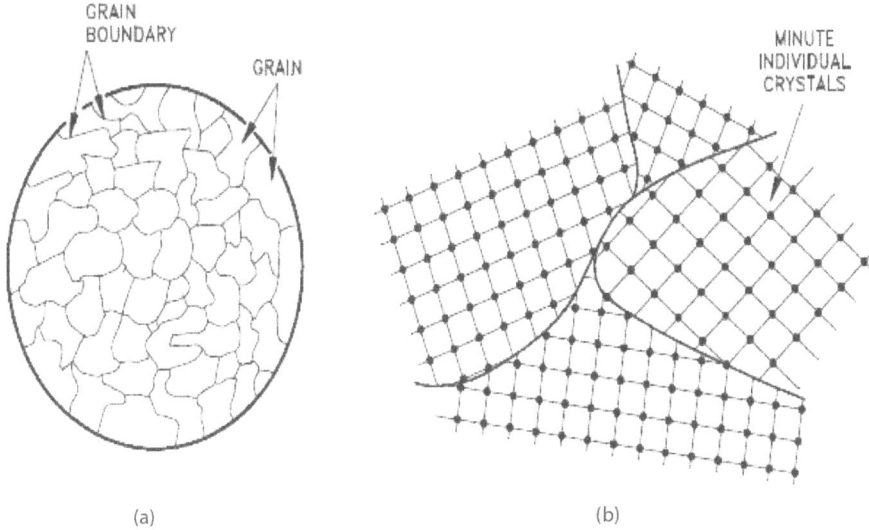

FIGURE 10.4 Representation of grains and boundaries and polycrystalline nature of Si semiconductors (a) and (b). (Adapted from U.S. Department of Energy, Material Science. DOE Fundamentals Handbook, Volume 1 and 2. January 1993. https://material-properties.org/wp-content/uploads/2020/07/Grains-Grain-Boundaries.png.)

surrounded by a much thicker amorphous silicon wafer. HIT is the better option because it has many benefits at a lower cost owing to their passivation and emitting effectiveness of intrinsic Si layer, deposition of these layers at low temperatures with low coefficients of temperature as compared to traditional Si solar cells.

Polycrystalline silicon (polycrystalline Si) solar cells (Figure 10.4a and b) [14] are simple and cheaper to manufacture and are made from raw silicon which is melted and poured into a square mold. Polycrystalline Si is less efficient than monocrystalline ones, and also these materials are of different compositions with distinct sizes and shapes including materials like ceramic, rock, and ice. The efficiency is in the range of 13%–16% because of the low quality of silicon material used. CdTe is the most commonly used and the least costly of the ones. Whereas, CIGS is comparable to crystalline solar cells in terms of performance and price.

10.2.2 Organic Solar Cells

Organic solar cells (OSCs) have attracted researchers due to their benefits over Si-based devices, such as their lighter weight, suppleness, less expensive to process, and less impact on the environment. Moreover, spin coating techniques or printing from the solutions are used to deposit active layers, allowing massive fabrication of devices at reduced temperatures allowing the final reduced cost of the device consequently. OSCs are essentially the combination of electron donor and acceptor materials. The donor absorbs the photons (i.e. from the solar radiation), in which excited states are created and therefore exciton is generated.

The exciton can be divided into electron-pair components (exchanges) by strong electric fields. Once the electron is removed the dissociated from the cavity, the acceptor will take the missing electrons. More attention has been paid to OSCs over the last three decades, as they are smaller, versatile, and less damaging to the atmosphere because of manufacturing. Also, functional layers can be fabricated using interfacial polymerization methods, which allow for low-temperature large-scale fabrication and thus reducing the costs of the devices [15]. Several semiconductor materials and ligands are implicated in OSCs. The radiation from the source causes solar explosions, which

a) Polymer Solar cell

SUNLIGHT

| BACK ELECTRODE |
| HOLE TRANSPORT LAYER |
| ACTIVE LAYER |
| ELECTRON TRANSPORT LAYER |
| TRANSPARENT CONDUCTIVE ELECTRODE |

| TRANSPARENT CONDUCTIVE ELECTRODE |
| HOLE TRANSPORT LAYER |
| ACTIVE LAYER |
| ELECTRON TRANSPORT LAYER |
| BACK ELECTRODE |

SUNLIGHT

b) Inverted Polymer Solar cell

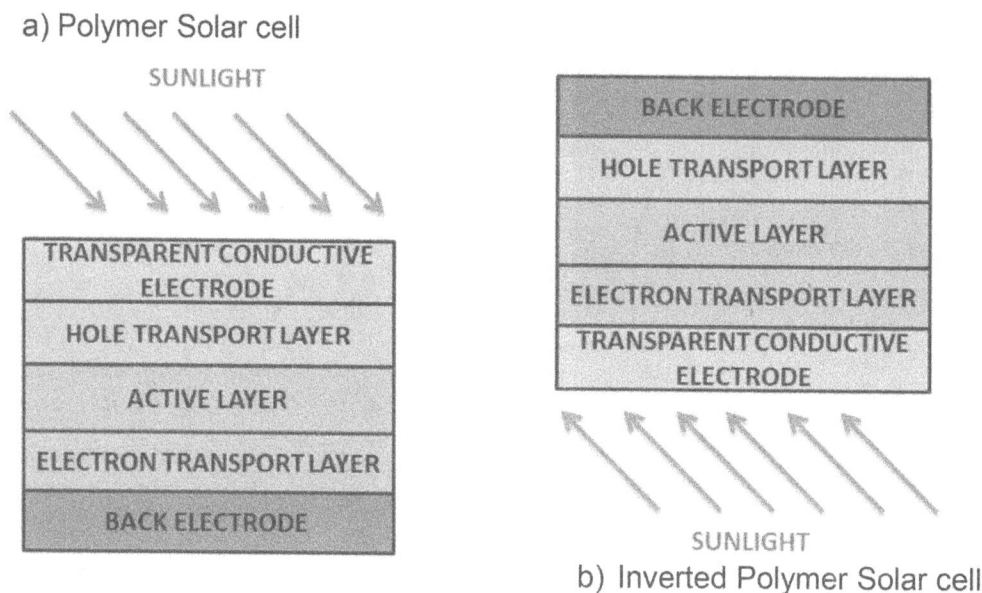

FIGURE 10.5 Typical polymer solar cells (PSCs) (a) and polymer and inverted polymer solar cells made from graphene-polymer nanocomposites (b). (Copyright 2018, MDPI. Adapted with permission from Ref. [17].)

produces excitons, i.e. electron-hole pair. According to the composition of the materials used in the active layer, namely, PSCs and small molecules, P-type solar cells can be further divided into polymer solar cells (PSC) and small-molecule solar cells [16]. Due to their transparency feature, a typical PSC and an inverted PSC made from graphene polymer nanocomposite [17] have PSCs used in glass, modular electronics but drawbacks being less efficient converting the power and degrading photochemically. Many materials such as indium tin oxide (ITO), conductive glass plate, lithium fluoride (LiF), gold, silver, platinum are used in many devices as cathode, a hole transport layer, an electron transport layer buffer layer, and anode transparent conductor [18–20].

PSCs (Figure 10.5a and b) [17] are composed of ITO superconducting glass, a hole-transporting layer of poly 3,4-ethylenedioxythiophene and polystyrene sulfonate (PEDOT:PSS), and an active layer of polymer, donor and recipient materials, an electrochemical sensor made of LiF, and a relatively low function metal electrode made of aluminum. Zinc oxide (ZnO) is often used in an inverted system ITO with a buffer layer (made from oxides of titanium and zinc). The inverted system has a typical cathode made of ITO with buffer layer made from oxides of titanium or zinc, and anode can be large work function materials, i.e. silver, Ag, or gold, Au with a hole transporting layer such as PEDOT:PSS [18–20]. The performance of the PSC method is determined by the qualities and layer thicknesses of the metal, as well as the nature of the metal used as electrode. The most popular PSC design is made up of a bulk electron-rich conjugated polymer (donor) and an electron-deficient fullerene (acceptor) as the combination.

A solar cell typically comes in three varieties: (a) the open-circuit voltage (V_{oc}); (b) the short-circuit density (J_{sc}), which shows how close the current-voltage characteristics of the PSC are to that of ideal; (c) the fill factor (FF), which indicates how close the current-voltage characteristics of the PSC are to that of perfect. The range between 0.75 and 1.0 is of FF values considered outstanding in practice. The overall output or power conversion efficiency (PCE) can be determined based on these parameters and the incident solar power (Pin) [20, 21]. This is equivalent to the calculation of the top occupied molecular orbital (HOMO) and the bottom unoccupied molecular orbital (LUMO) of V_{oc} dyes using product density of the photo charge carrier and mobility of the molecule [22–24]. A significant quantity of bulk heterojunction polymer and pyrrole conjugates have been produced over the last few years. Aspects of both electronic and performed with the straightforwardness of traditional semiconductors are remarkably simple. Conjugated polymers are presented in the graphical

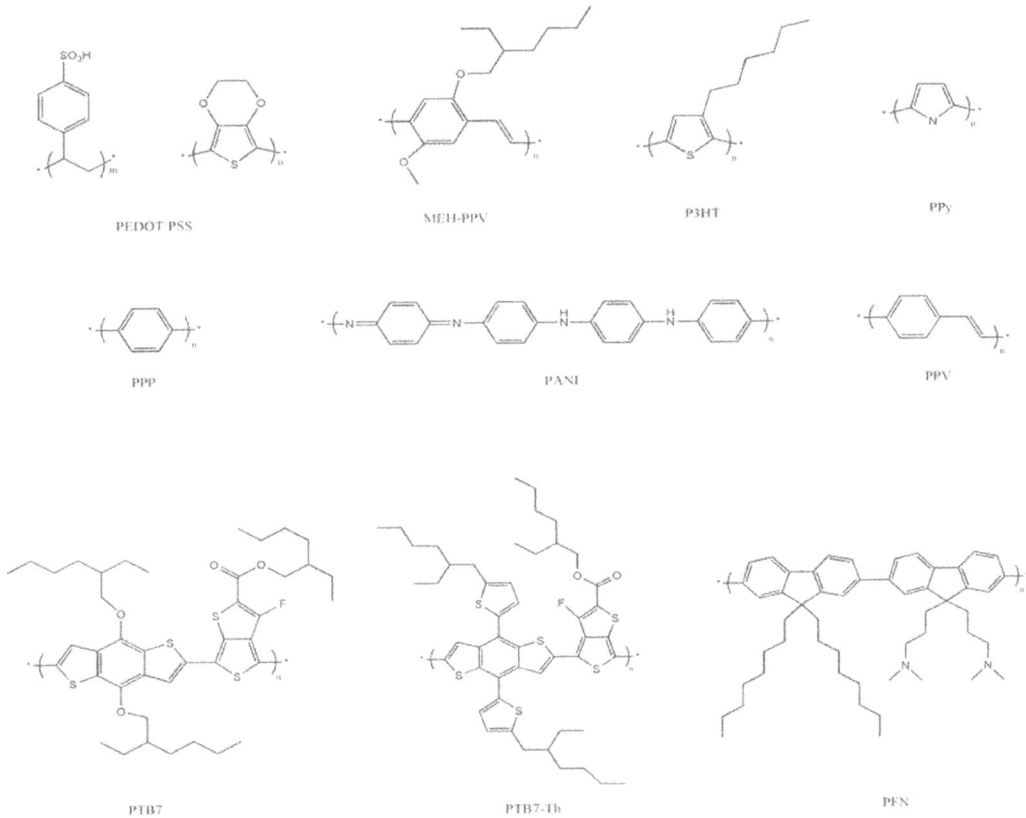

FIGURE 10.6 Chemical structures of conjugated polymers typically used in PSCs. (Copyright 2018 MDPI. Adapted with permission from Ref. [27].)

form in Figure 10.6. Conjugated polymers in the PSCs of BHJs are depicted in their chemical structure. All the ring structures are designed with the alternating carbon–carbon double bond, which causes their low energy levels, optical and chemical affinity.

Since synthesized conjugated polymers are insulators, they must be oxidized (p-doping) or reduced (n-doping) to become conductive [25]. Examples of typical CP applications in solar applications (PSCs) are given [2] in Figure 10.6. PEDOT is made up of monomers called ethylene dioxythiophene (EDOT). It has exceptional properties, including good electrochemical properties as compared to other polythiophenes [26]. Since it reacts with oxygen rapidly in air, it is insoluble in many different solvents and unstable in its neutral state. In most cases, polyelectrolyte solution – PSS added results in an aqueous dispersion of PEDOT in its oxidized form and PSS functions as a counterion, indicating that the PEDOT chains are dispersed in the aqueous solution with increase in ease of processing and stable/strengthened thermal property. PEDOT:PSS has a number of attractive features. However, PSS has many disadvantages, including high acidity, hygroscopicity, and inhomogeneous electrical properties, all of which contribute to low durability.

The outcome of the polymer test will be attained by doing several checks on the solar cells [25] as described in the following PV material examples. Poly (3-hexylthiophene) (P3HT) is a common PV material owing to superior electronic properties, self-assembling capability, and ease of solubility in most of the organic solvents. Poly(p-phenylene) (PPV) is a water-insoluble polymer. The addition of side groups to the conjugated backbone increases the polymer's solubility substantially. PPV and its derivatives, on the other hand, have low absorption and photodegradation. Similarly, polypyrrole's (Ppy) solubility is also limited due to crosslinking of the polymer or polymer chains copolymerization. Even though neutral Ppy is insoluble, it can expand or swell in contact with certain solvents. Ppy can be doped in acid or basic media and dissolved in a range of solvents.

Only organic solvents dissolve poly(p-phenylene) (PPP), which is useful because it is very stable (particularly when undoped) and can be used as an electrode. Among CP's, polyaniline (PANI) is particularly interesting because it is a relatively low-cost way to synthesize it. PANI with its ease of control in oxidation and protonation states and also its response toward acid base doping can be advantageous to apply in broad range of electrical properties [25, 27].

Recent advancements in graphene (G)/polymer nanocomposite techniques for PSC applications have found importance as G has emerged as a nanomaterial useful for a variety of components, such as transparent conductive electrodes (TCEs), active layers (ALs), and interfacial layers (IFLs), based on specific as well as flexible applications. Graphene, its forms in the reduced state or oxidized state and along with derivatives or composites, preparation methods thereof and integrating these nanocomposites to derive photovoltaic properties of PSCs, are all raised the great interest to researchers. Even though a broad range of reviews talk about PSCs, the ones that deal best with solar cell applications are limited and do not focus on nanocomposites [28]. Supercapacitors and applications in solar cells have also been investigated using graphene and related composites [29].

10.3 CONDUCTING POLYMER NANOCOMPOSITE APPLICATIONS

10.3.1 GRAPHENE AND ITS DERIVATIVES – SYNTHESIS AND PROPERTIES

Graphene (G) is graphite's fundamental structure, consisting of a single sheet of carbon atoms. The graphene, graphene oxide (GO), and rGO (reduced graphene oxide) structures are depicted in Figure 10.7. The oxidation of graphite produces GO via a number of reduction methods, reducing agents minimize GO to form rGO. It is well known that G is composed of a single sheet of sp^2 carbon atoms

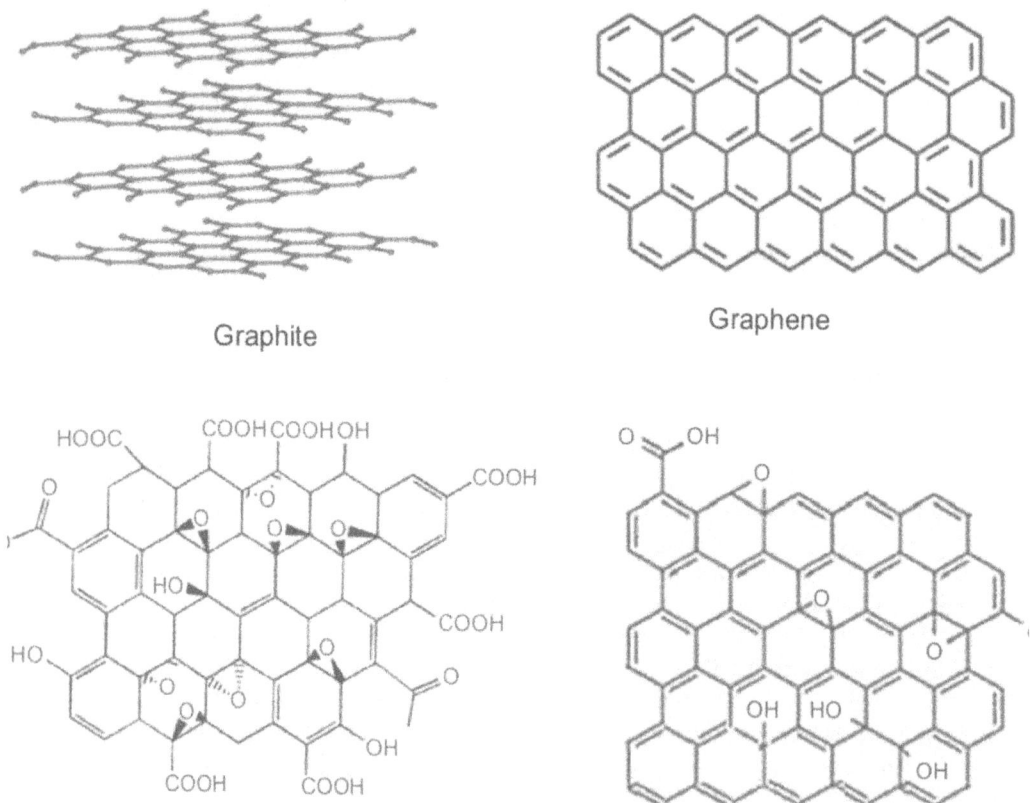

Graphite

Graphene

FIGURE 10.7 Structure of graphite, graphene (G), graphene oxide (GO), and reduced graphene oxide (rGO). (Copyright 2019, Dove Press. Adapted with permission from Ref. [30].)

that is smooth and atomically dense, forming a honeycomb shape. According to its two-dimensional existence, the electronic band arrangement exhibits a linear relationship between energy and momentum at the Fermi stage [29–30]. G possesses exceptional physical, thermal, and mechanical properties, including extremely high electron mobility (15,000 cm^2/Vs) and a wide surface area (2630 m^2/g). It has a thermal conductivity of 3000–6000 W/m K and a conductivity of 9.6×107 S/m [31]. Additionally, G is one of the hardest materials known to man, possessing an elastic modulus equal to 1 T Pa [32–34], an ultimate strength of 130 Gpa, and a breaking strength of approximately 40 N/m. G is an outstanding material for photovoltaic cell applications due to the combination of these extraordinary properties. Nonetheless, many problems remain unresolved, most notably, its sheet resistance is extremely high in comparison to materials commonly used as electrodes, such as ITO [35–37], making it critical to improve electronic transport while losing optical clarity or other properties. Apart from that, G sheets' flexibility and chemical inert nature, they can be used as both an electrode and a barrier plate. G can be synthesized through a variety of processes, including exfoliation, epitaxial growth, chemical vapor deposition (CVD), and chemical and electrochemical methods. The structure of G, GO, and rGO, as well as their synthesis, are described below, along with their feasibility for use in solar energy devices. Numerous articles discuss the properties and applications of G and G-based composite materials in detail [38–39]. Numerous methods for preparing G/polymer nanocomposites have been published, including covalent and non-covalent approaches [40].

10.3.1.1 Graphene/Polymer Nanocomposites in Solar Cells

Graphene/polymeric nanocomposites may be synthesized using various means, e.g., covalent or non-covalent approaches. G has also been incorporated into PSCs, such as TCS, AL, and IFL nanocomposites. Due to the volume of literature on this topic, only the most representative examples involving polymer and G nanocomposites are discussed [41]. Advanced electrodes with a high energy capacity and high power density are desperately required for high-performance energy storage systems, such as lithium-ion batteries (LIBs) and supercapacitors (SCs), to meet the needs of future electrochemical power supplies for applications such as electric hybrid cars. Metal sulfides are being investigated as attractive supercapacitors due to their unusual physical and chemical properties, as well as their high specific capacity/capacitance, which is usually several times that of carbon/graphite-based materials. Nanostructures benefit from high surface-to-volume ratios, favorable transport properties, and a high degree of volume shift freedom during ion insertion reactions – a chance to develop next-generation LIBs and SCs. Thus, the advancement of innovative concepts in materials science in order to create novel nanostructures paves the way for enhanced electrochemical efficiency.

We review the latest progress in nanostructured metal sulfides (MSx), including iron sulfides, copper sulfides, nickel sulfides, manganese sulfides, molybdenum sulfides, and tin sulfides, for LIB and SC applications. Additionally, the idea of combining conductive matrices, especially graphene, with metal sulfide nanomaterials will be discussed. Finally, some problems and opportunities for the production of LIB and SC devices based on metal sulfide are discussed [41].

10.3.2 Conducting Polymer-Metal Sulfide Nanocomposites

Low-cost conductive polymer composites are attractive due to their good mechanical flexibility. As well, CPs are promising semiconducting materials due to their outstanding electrical properties and relative ease of processing. Well studied conductive polymers for solar cell applications include Ppy, PANI, and PEDOT etc. In particular, these materials are frequently used in DSSC as hole transport materials. Recently research attention increased toward nanostructured conductive polymer materials due to their unique combination of the large surface area of nanomaterials and electronic properties of conductive polymers [42–43].

Even though metal oxide-based conductive polymer nanocomposites have been extensively studied for hybrid solar cell applications with high photovoltaic efficiencies, further research is being conducted in such devices using metal sulfide or selenide nanoparticles in combination with conductive polymers to achieve high charge separation yields and efficient charge transport [44]. Various strategies are currently being discussed to design effective hybrid solar cells with high yield and that can also support long-standing charge separation at the polymer-nanoparticles interface [45–48]. Metal sulfide–polymer nanocomposite thin films for photovoltaic applications were reported by Eugen Maier et al. [49]. A direct forming route was used to make polymer nanocomposites with poly(3-(ethyl-4-butanoate) thiophene) (P3EBT) and a sulfur-based semiconductor (i.e., CdS, PbS, or ZnS) thin films. Several other groups published optical properties, such as metal-sulfide-conjugated polymers (e.g. (poly 3-hexylthiophene)-ZnS using Na_2S as a sulfur source) using a direct synthesis route [50].

In addition, Vidya et al. [51] reported MeS (Me = Zn, Cd, Cu)-poly(3-octylthiophene) polymer composites prepared using Langmuir–Blodgett-techniques and gaseous H_2S as a desulfurization reagent. Liao et al. [52] Reported CdS-poly(3-hexylthiophene) [CdS/PVP] nanocomposite films for solar cells applications. Alternatively, thioacetamide was used as a sulfur source, in the preparation of CdS-polyvinylpyrrolidone (PVP) [CdS/PVP] blends using *in situ* synthesis for optical applications. CdS/PVP films have shown promising optical properties such as non-linear optical refractive index and an absorption coefficient [53].

Some of the researchers [53–55] have designed solar cells aimed at ligand-free *in situ* approaches that avoided separate steps involving nanoparticle (NP) synthesis (i.e. ZnO-PNC) or binary solar cells are made from solutions of CP, a source of sulfur and metal salts, where polymer can control growth by capping the NPs. In other words, MSx like CdS, ZnS, CdS, ZnS, and $CuInS_2$ [56] are successfully utilized to form SCs by thermal annealing step (below the thermal stable temperature of a polymer involved) from the solutions of their respective metal salts and thiourea. But these SCs have low or moderate PCEs due to the formation of non-volatile by-products. Recently, new methods [57] are explored to convert metal xanthates directly to metal sulfides, with the removal of few volatile compounds to increase the power conversion efficiency. For example, polymer/cadmium sulfide (CdS) SCs approach has extended to polymer/CIS nanocomposites that have the benefit of lower bandgap (1.5 eV) as compared to other metal sulfides or selenides of cadmium (1.74 eV in CdSe or 2.4 eV in CdS) [58, 59], homogeneous and stable coating solutions of copper xanthates, indium xanthates, and the conjugated polymer poly[(2,7-silafluorene)-alt-(4,7-di-2-thienyl-2,1,3-benzothiadiazole)] (PSiF-DBT). By spin coating or doctor blading, homogeneous layers are prepared and subsequently converted into nanocomposite layers by thermal annealing step below 300°C (or 200°C), where they are thermally stable, as shown in Figure 10.8.

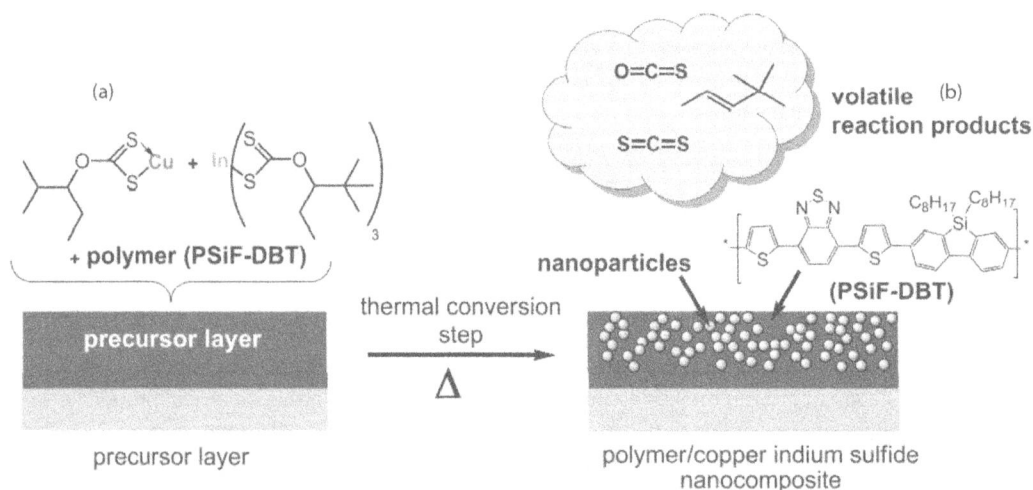

FIGURE 10.8 Scheme for the preparation of polymer/copper indium sulfide (CIS) nanocomposite solar cells – (a) homogeneous and stable coating solutions of copper xanthates, indium xanthates, and the conjugated polymer poly[(2,7-silafluorene)-alt-(4,7-di-2-thienyl-2,1,3-benzothiadiazole)] (PSiF-DBT) by spin coating or doctor blading; (b) homogenous layers are subsequently converted into nanocomposite layer by thermal annealing step below 300°C (≈ 200°C) with the release of few volatile products. (Copyright 2011. Adapted with permission from Ref. [59].)

10.4 CONCLUSIONS

A solar cell or a photovoltaic cell is a device based on the changes in physical or chemical aspects (of materials) that could convert the light (sun) energy into electrical energy based on photovoltaic effect. In contrast to other power sources, solar power is the most sought environmentally friendly and clean energy form. To take full advantage of such sustainable energy forms, the development and utilization of novel solar cells/technologies that harvest the sun's energy efficiently and for long term are continuously in need.

ACKNOWLEDGMENTS

One of the authors, KM acknowledges the support of her supervisor, Sumedh P. Surwade, CEO and Founder, SAS Nanotechnologies LLC, for working toward polymer nanocomposite self-healing materials, and VSP acknowledges Young Talent Fellowship (File Number: 88887.468130/2019-00 funding by CAPES-PRINT Program – Call no. 41/2017 (PRINT-Programa Institucional De Internacionalização), Brazil.

REFERENCES

1. Ye L.P., Ding Y., Zhao Y. Yu G. Nanostructured conductive polymers for advanced energy storage. Chem. Soc. Rev. 2015; 44(19) (May):6684–6696. http://hdl.handle.net/2152/41175.
2. Hou W., Xiao Y., Han G., Lin J-Y. The applications of polymers in solar cells: A review. Polymers 2019; 11(1):143. https://doi.org/10.3390/polym11010143.
3. Andersen T.R., Dam H.F., Hosel M., Helgesen M., Carle J.E., Larsen-Olsen T.T., Gevorgyan S.A., Andreasen J.W., Adams J., Li N. et al. Scalable, ambient atmosphere roll-to-roll manufacture of encapsulated large area, flexible organic tandem solar cell modules. Energy Environ. Sci. 2014; 7:2925–2933. doi: 10.1039/C4EE01223B.
4. Ragoussi M.E., Torres T. New generation solar cells: Concepts, trends and perspectives. Chem. Commun. 2015; 51:3957–3972. doi: 10.1039/C4CC09888A.
5. Luque A., Hegedus S. (Eds.), Handbook of Photovoltaic Science and Engineering, second edition, John Wiley & Sons Ltd., Chichester, UK, 2011.
6. Praveena U. et al., Effect of synthesis on properties of Gd doped LaBi5Fe2Ti3O18. Mater. Today 2019; 11:1041–1048.
7. Donev J.M.K.C. et al., Energy Education – Photovoltaic effect, 2015 –https://energyeducation.ca/encyclopedia/Photovoltaic_effect#cite_note-3, Eco-green Electrical and https://images.app.goo.gl/u2cBkfHMfBGjYZ8w8.
8. Research Cell Efficiency Records (accessed on 10 January 2019). Available online: https://www.nrel.gov/pv/assets/images/efficiency-chart-20180716.jpg.
9. Gao C.L., Dong H.Z., Bao X.C., Zhang Y.C., Saparbaev A., Yu L.Y., Wen S.G., Yang R.Q., Dong L.F. Additive engineering to improve efficiency and stability of inverted planar perovskite solar cells. J. Mater. Chem. C. 2018; 6:8234–8241. doi: 10.1039/C8TC02507J.
10. Xue Q.F., Hu Z.C., Liu J., Lin J.H., Sun C., Chen Z.M., Duan C.H., Wang J., Liao C., Lau W.M. et al. Highly efficient fullerene/perovskite planar heterojunction solar cells via cathode modification with an amino-functionalized polymer interlayer. J. Mater. Chem. A. 2014; 2:19598–19603. doi: 10.1039/C4TA05352D.
11. Chang C.M., Li W.B., Guo X., Guo B., Ye C.N., Su W.Y., Fan Q.P., Zhang M.J. A narrow-bandgap donor polymer for highly efficient as-cast non-fullerene polymer solar cells with a high open circuit voltage. Organ. Electron. 2018; 58:82–87. doi: 10.1016/j.orgel.2018.04.001.
12. Scharber M.C., Sariciftci N.S. Efficiency of bulk-heterojunction organic solar cells. Prog. Polym. Sci. 2013; 38:1929–1940. doi: 10.1016/j.progpolymsci.2013.05.001.
13. Solar Cells: Materials Solar Cells: Materials, Manufacture and Operation, McEvoy A., Castañer L., Markvart T. (Eds), second edition, Elsevier, Oxford, UK, 2012.
14. Grains and Boundaries. U.S. Department of Energy, Material Science. DOE Fundamentals Handbook, Volume 1 and 2. January 1993. Available online: https://www.standards.doe.gov/standards-documents/1000/1017-BHdbk-1993-v1/@@images/file.
15. Brabec C.J., Sariciftci N.S., Hummelen J.C. Plastic solar cells. Adv. Funct. Mater. 2001; 11:15–26.

16. Cheng Y.-J., Yang S.-H., Hsu C.-S. Synthesis of conjugated polymers for organic solar cell applications. Chem. Rev. 2009; 109:5868–5923.

17. Pascual A.M.D., Luceño J.A., Peña R., García-Díaz P. Recent developments in graphene/polymer nanocomposites for application in polymer solar cells. Polymers 2018; 10(2):217. doi: 10.3390/polym10020217.

18. Das S., Keum J.K., Browning J.F., Gu G., Yang B., Dyck O., Chen W., Chen J., Ivanov I.N., Hong K. et al. Correlating high power conversion efficiency of PTB7:PC71BM inverted organic solar cells with nanoscale structures. Nanoscale 2015; 14:15576–15583.

19. Jørgensen M., Norrman K., Krebs F.C. Stability/degradation of polymer solar cells. Sol. Energy Mater. Sol. Cells 2008; 92:686–714; Thompson B.C.; Fréchet J.M. Polymer-Fullerene Composite Solar Cells. Angew. Chem. Int. Ed. Engl. 2007; 47:58–77.

20. Chen H.Y., Hou J., Zhang S., Liang Y., Yang G., Yang Y., Yu L., Wu Y., Li G. Polymer solar cells with enhanced open-circuit voltage and efficiency. Nat. Photonics 2009; 3:649–653.

21. Alam S. et al. Thermally induced degradation of PBDTTT-CT:PCBM based polymer solar cells, IOP science. J. Phys. D: Appl. Phys. 2019; 52:475501. https://www.ossila.com/pages/organic-photovoltaics-introduction.

22. Green M.A. Solar cell fill factors: General graph and empirical expressions. Solid State Electron. 1981; 24:788–789.

23. Luo J., Wu H., He C., Li A., Yang W., Cao Y. Enhanced open-circuit voltage in polymer solar cells. Appl. Phys. Lett. 2009; 95:043301.

24. De Kok M.M., Buechel M., Vulto S.I.E., van de Weijer P., Meulenkamp E.A., de Winter S.H.P.M., Mank A.J.G., Vorstenbosch H.J.M., Weijtens C.H.L., van Elsbergen V. Modification of PEDOT:PSS as hole injection layer in polymer LEDs. Phys. Status Solidi A Appl. Res. 2004; 201:1342–1359.

25. Ganesamoorthy R., Sathiyan G., Sakthivel P. Review: Fullerene based acceptors for efficient bulk heterojunction organic solar cell applications. Sol. Energy Mater. Sol. Cells 2017; 161:102–148.

26. Yun J.-M., Yeo J.-S., Kim J., Jeong H.-G., Kim D.-Y., Noh Y.-J., Kim S.S., Ku B.-C., Na S.I. Solution-processable reduced graphene oxide as a novel alternative to PEDOT:PSS hole transport layers for highly efficient and stable polymer solar cells. Adv. Mater. 2011; 23:4923–4928.

27. Taleghani H.G., Aleahmad M., Eisazadeh H. Preparation and characterization of polyaniline nanoparticles using various solutions. World Appl. Sci. J. 2009; 6:1607–1611; Recent Developments in graphene/polymer nanocomposites for application in polymer solar cells. Polymers 2018; 10(2):217. doi: 10.3390/polym10020217.

28. Walker B., Kim C., Nguyen T.Q. Small molecule solution-processed bulk heterojunction solar cells. Chem. Mater. 2011; 23:470–482.

29. Gupta A., Sardana S., Dalal J., Lather S., Maan A.S., Tripathi R., Punia R., Singh K., Ohlan A. Nanostructured polyaniline/graphene/Fe 2 O 3 composites hydrogel as a high-performance flexible supercapacitor electrode material. ACS Appl. Energy Mater. 2020; 3: 6434–6446.

30. Bai R.G., Muthoosamy K., Manickam S., Hilal-Alnaqbi A. Graphene-based 3D scaffolds in tissue engineering: fabrication, applications, and future scope in liver tissue engineering. Int. J. Nanomedicine 2019; 14:5753–5783. doi: 10.2147/IJN.S192779.

31. You J., Dou L., Yoshimura K., Kato T., Ohya K., Moriarty T., Emery K., Chen C.-C., Gao J., Li G., Yang Y.A. Polymer tandem solar cell with 10.6% power conversion efficiency. Nat. Commun. 2013; 4:1446.

32. Lin X.-F., Zhang Z.-Y., Yuan Z.-K., Li J., Xiao X.-F., Hong W., Chen X.-D., Yu D.-S. Graphene-based materials for polymer solar cells. Chin. Chem. Lett. 2016; 27:1259–1270.

33. Sun Y., Zhang W., Chi H., Liu Y., Hou C.L., Fang D. Recent development of graphene materials applied in polymer solar cells. Renew. Sustain. Energy Rev. 2015; 43:973–980.

34. Lee C., Wei X., Kysar J.W., Hone J. Measurement of the elastic properties and intrinsic strength of monolayer graphene. Science 2008; 321:385–388.

35. Diez-Pascual A.M., Gomez-Fatou M.A., Ania F., Flores A. Nanoindentation in polymer nanocomposites. Prog. Mater. Sci. 2015; 67:1–94.

36. Eda G., Chhowalla M. Chemically derived graphene oxide: towards large-area thin-film electronics and optoelectronics. Adv. Mater. 2010; 22:2392–2415.

37. Pei S., Chen H.-M. The reduction of graphene oxide. Carbon 2012; 50:3210–3228.

38. Cote L.J., Cruz-Silva R., Huang J. Flash reduction and patterning of graphene oxide and its polymer composite. J. Am. Chem. Soc. 2009; 131:11027–11032.

39. Periasamy M., Thirumalaikumar M. Methods of enhancement of reactivity and selectivity of sodium borohydride for applications in organic chemistry. J. Organomet. Chem. 2000; 609:137–151.

40. Bai H., Li C., Shi G. Functional composite materials based on chemically converted graphene. Adv. Mater. 2011; 23:1089–1115.

41. Das S., Sudhagar P., Kang Y., Choi W. Graphene synthesis and application for solar cells. J. Mater. Res. 2014; 29(3):299–319. doi:10.1557/jmr.2013.297.
42. Rui X., Tan H., Yan Q. Nanostructured metal sulfides for energy storage. Nanoscale 2014; 6:9889–992. https://doi.org/10.1039/c4nr0305.
43. Lu X., Zhang W., Wang C., Wen T.-C., Wei Y. One-dimensional conducting polymer nanocomposites: Synthesis, properties and applications. Prog. Polym. Sci. 2011; 36:671–712.
44. Dowland S., Lutz T., Ward A., King S.P., Sudlow A., Hill M.S., Molloy K.C., Haque S.A. Direct growth of metal sulfide nanoparticle networks in solid-state polymer films for hybrid inorganic–organic solar cells. Adv. Mater. 2011; 23:2739–2744.
45. Wong W.Y., Wang X., Zhang H.-L., Cheung K.-Y., Fung M.-K., Djurisic A.B., Chan W.K. Synthesis, characterization and photovoltaic properties of a low-bandgap platinum(II) polyyne functionalized with a 3,4-ethylenedioxythiophene-benzothiadiazole hybrid spacer. J. Organomet. Chem. 2008; 693:3603–3612.
46. Lokteva I., Radychev N., Witt F., Borchert H., Parisi J., Kolny-Olesiak J. Surface treatment of CdSe nanoparticles for application in hybrid solar cells: The effect of multiple ligand exchange with pyridine. J. Phys. Chem. C. 2010; 114:12784–12791.
47. Zhang Q., Russell T.P., Emrick T. Synthesis and characterization of CdSe nanorods functionalized with regioregular poly(3-hexylthiophene). Chem. Mater. 2007; 19:3712–3716.
48. Seo J., Kim W.J., Kim S.J., Lee K.S., Cartwright A.N., Prasad P.N. Polymer nanocomposite photovoltaics utilizing CdSe nanocrystals capped with a thermally cleavable solubilizing ligand. Appl. Phys. Lett. 2009; 94:133302.
49. Maier E., Fischereder A., Haas W., Mauthner G., Albering J., Rath T., Hofer F., List E.J.W., Trimmel G. Metal sulfide–polymer nanocomposite thin films prepared by a direct formation route for photovoltaic applications. Thin Solid Films 2011; 519:4201–4206.
50. Dong Y., Lu J., Yan F., Xu Q. Optical properties of poly(3-hexylthiophene)/ZnS nanocomposites. High Perform. Polym. 2009; 21:48.
51. Vidya V., Ambily S., Narang S.N., Major S., Talwar S.S. Development of metal sulfide-poly (3-octylthiophene) composite LB multilayers. Colloids Surf. A 2002; 198–200:383.
52. Liao H., Chen S., Liu D. In-situ growing CdS single-crystal nanorods via P3HT polymer as a soft template for enhancing photovoltaic performance. Macromolecules 2009; 42:6558.
53. Jing C., Xu X., Zhang X., Liu Z., Chu J. In situ synthesis and third-order nonlinear optical properties of CdS/PVP nanocomposite films. J. Phys. D 2009; 42:075402.
54. Oosterhout S.D., Wienk M.M., van Bavel S.S., Thiedmann R., Koster L.J.A., Gilot J., Loos J., Schmidt V., Janssen R.A.J. The effect of three-dimensional morphology on the efficiency of hybrid polymer solar cells. Nat. Mater. 2009; 8:818.
55. Dayal S., Kopidakis N., Olson D.C., Ginley D.S., Rumbles G. Direct synthesis of CdSe nanoparticles in poly(3-hexylthiophene). J. Am. Chem. Soc. 2009; 131:17726.
56. Maier E., Rath T., Haas W., Werzer O., Saf R., Hofer F., Meissner D., Volobujeva O., Bereznev S., Mellikov E., Amenitsch H., Resel R., Trimmel G. CuInS2–Poly(3-(ethyl-4-butanoate)thiophene) nanocomposite solar cells: Preparation by an in situ formation route, performance and stability issues. Sol. Energy Mater. Sol. Cells 2011; 95:1354.
57. Leventis H.C., King S.P., Sudlow A., Hill M.S., Molloy K.C., Haque S.A. Nanostructured hybrid polymer–inorganic solar cell active layers formed by controllable in situ growth of semiconducting sulfide networks. Nano Lett. 2010; 10:1253.
58. Yue W., Han S., Peng R., Shen W., Geng H., Wu F., Tao S., Wang M. CuInS2 quantum dots synthesized by a solvothermal route and their application as effective electron acceptors for hybrid solar cells. J. Mater. Chem. 2010; 20:7570.
59. Rath T., Edler M., Haas W., Fischereder A., Moscher S., Schenk A., Trattnig R., Sezen M., Mauthner G., Pein A., Meischler D., Bartl K., Saf R., Bansal N., Haque S.A., Hofer F., List E.J.W. and Trimmel G. A direct route towards polymer/copper indium sulfide nanocomposite solar cells. Adv. Energy Mater. 2011; 1:1046–1050.

11 Thin Films of Conducting Polymers for Photovoltaics

S. Rijith[1], S. Sarika[1], and V.S. Sumi[2]
[1]Postgraduate and Research Department of Chemistry,
DST-FIST Supported Department, Sree Narayana
College, Kollam, Kerala, India

[2]Department of Chemistry, Government College, Attingal,
Thiruvananthapuram, Kerala, India

CONTENTS

11.1 INTRODUCTION

Energy crisis is the concern that the global energy exigency increases date to date due to the exponential population rise and industrialization. The demand is predicted to be increasing in the forthcoming years [1] and has been increased far more than the production; an imbalance between demand and supply has been created. This increasing demand is coupled with the speculations of depletion of carbon-centered non-renewable fuel economy. Thus along with the invention of newer technologies, an alternate source that explores newer and safer ways of clean and sustainable energy is the foremost goal. From this point of view, energy captured from daylight by organic photovoltaic technologies is a suitable green approach to subdue growing energy demand [2], and a large variety of conducting polymers (CPs) and their derivatives used light-harvesting materials in organic photovoltaic devices.

The Nobel prize-winning revolutionary discovery of Alan J Heeger et al. in 1977 [3–6] proved that the normal organic polymers are not conventional insulators but can act as metal and semiconductors. This leads to a new field of research that induced the progression to understand fundamental chemistry and physics. They fathom out that the oxidative doping of polyacetylene with iodine engender conducting behavior in it [3–5], and this drew interest among worldwide researchers to an exciting domain of CPs which are now considered as fourth-generation polymeric materials with fascinating electrical transport properties and macromolecular characters [7]. A large variety of conducting polymers and their derivatives used light-harvesting materials in organic photovoltaic (OPV) devices [8–9].

DOI: 10.1201/9781003150374-11

Thus an opportunity to fabricating innovative CPs with enhanced and anticipated properties gain the attention of scientists and a wide range of CPs with modified functionalities were synthesized. The unique characteristics of CPs lead to rapid growth of its development in a short period [10]. Currently, CPs are chief important materials which contribute countless benefits in a wide range of sectors such as organic photovoltaic (OPV), organic light emitting diodes (OLED), organic field effect transmitters (OFET), dye-sensitized solar cells (DSSC), and organic-inorganic hybrid cells (OIH) [11–14]. Admittedly, due to their promising practical use in everyday life, the world accelerated the commercialization of this material.

This chapter focuses on up-to-date development in fine-tuning the physicochemical properties of conduct CPs and their real-time application toward photovoltaics.

11.2 UNIQUE PROPERTIES AND CLASSIFICATION OF CONDUCTING POLYMERS

Until the mid of the last century, polymers are regarded as insulators used to coat electric wires to prevent short circuits. But after the discovery of A. Heeger on polyacetylene, a special class of organic polymers has a fascinating electrical conductivity those of the metals due to its conjugated π electrons backbone chain and are termed as CPs. CP can achieve metal-like conductivity due to its extraordinary beneficial properties, which depend on any factor influencing the electron delocalization along the main chain, such as the composition of monomer, the functional group attached to the polymer and its position, molecular weight, and conformation of polymer. Thus the side chains and functional groups attached to the backbone of CP chain can enhance their physicochemical properties such as the energy of optical transition, ionization energy, and electron affinity to a favorable range and in fact by choosing the specific functional groups, CPs can be more easily customized and tailored by desired properties for any particular application [3, 15]. CP can be classified as their chemical composition, such as nitrogen containing CP (polpyrrole, polyaniline), sulfur-containing CP (polythiophene, poly-3-alkyl thiophene, polyisothianapthene, polyethylenedioxythiophene), etc. The chemical structures of some of the CPs are depicted in Figure 11.1.

Furthermore, in the case of CPs, the conjugated backbone chain alone is not sufficient for conductivity. CP can easily undergo oxidation and reduction than normal polymers by doping with a dopant, which modifies its backbone band structure by inducing carriers, and their delocalization enables an extended 3D electrical conductivity by hole-electron pair mobility and interchain electron transfer. Thus by doping, an insulating or semi conducting polymer can be converted into CP with the conductivity typically in the range of ca. 1 to 10^4 S cm^{-1}. But for measuring the conductivity, it requires that the CPs be made into thin films. As mentioned, suitable features of CPs turn them out to be an interesting applicant in the photovoltaic field. Thus the exploration of their conduction mechanism is of great significance.

11.3 THE CONDUCTION MECHANISM IN CP

11.3.1 THE ELECTRONIC STRUCTURE OF CP

As we already mentioned, CPs are different from conventional saturated polymers like polyethylene, in which all the four valence electrons in each C atom are utilized for four sigma bond formation via sp^3 hybridization, and there is no more free electron contribution for conducting electricity throughout the polymer chain (Figure 11.2). Whereas in CP, only three valance electrons are utilized for three sigma bond formation. That is, there forms a conjugated backbone chain of sp^2 hybridized C, where each C is bonded to two adjacent C and one hydrogen atom [16]. Here each C can contribute remaining one electron from their unhybridized 2p$_z$ orbital which is perpendicular to the molecular plane of trigonal planar geometry for electrical conduction.

For planar CPs, the overlapping of neighboring 2p$_z$ orbitals causes alternate conjugated π bond formation along the polymer chain leads to the formation of a 1-dimensional electronic band [3] and

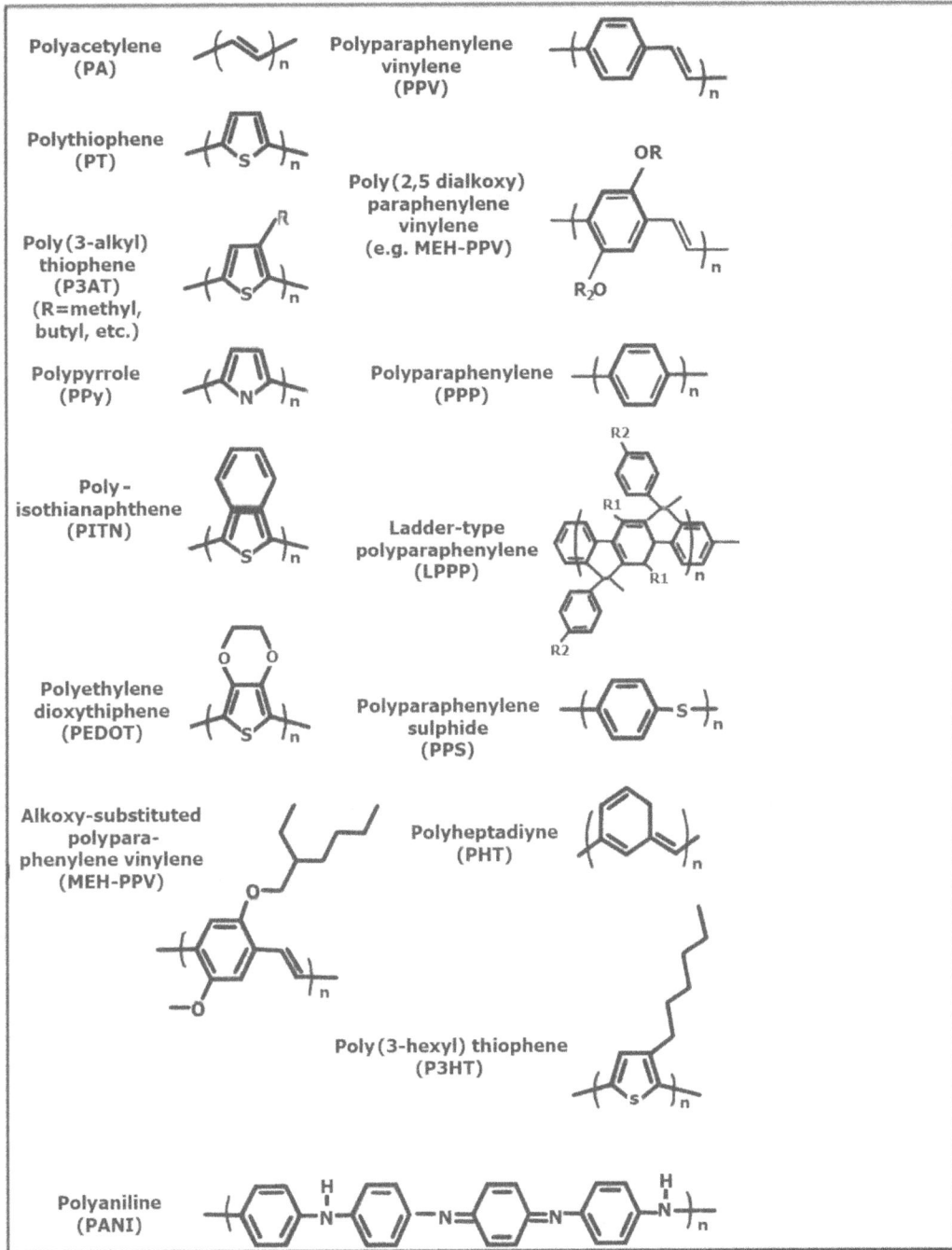

FIGURE 11.1 Chemical structures of some commonly used CPs. (Reproduced with permission from Ref. [16].)

partially emptying it by dopants leads to the introduction of delocalized π system having free charge carriers and resulting in carrier mobility within the band. As a result, the carriers gain high charge mobility and are supposed to move along the polymer backbone chain and bring about electrical conductivity [3, 4, 17].

The electrical conductivity is greatly influenced by the geometry of CP. For instance, in the case of polyacetylene, its trans-polymer has conductivity in the range of that of a semiconductor

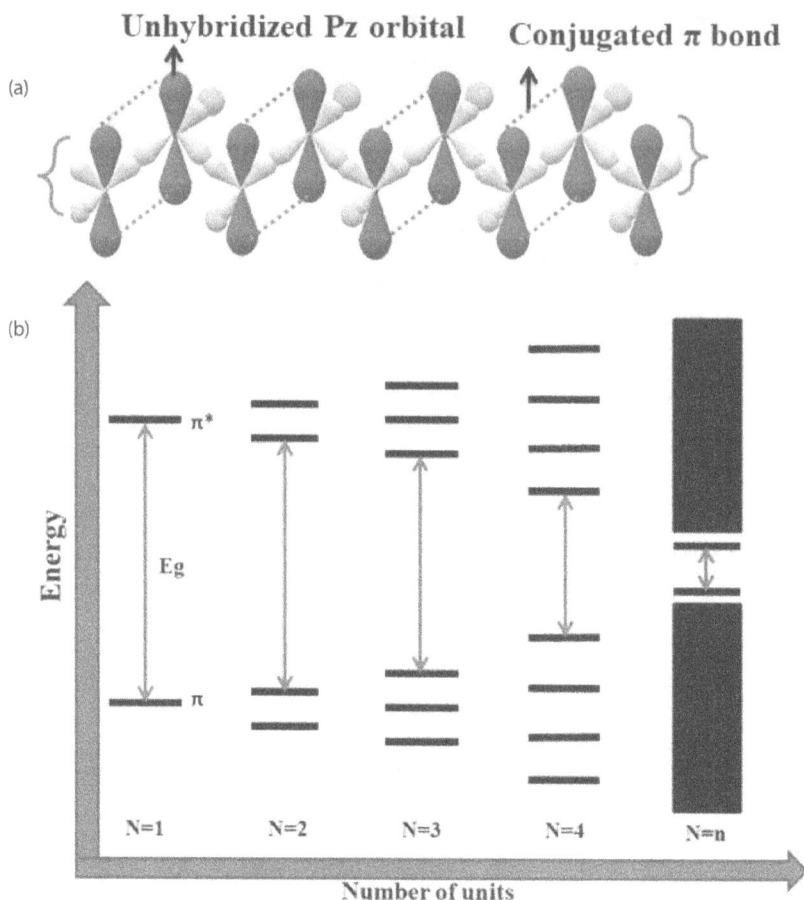

FIGURE 11.2 (a) Delocalized Pz orbital in the conjugated carbon backbone chain. (b) Effect of extended conjugation in CPs.

(4.4×10^{-5} S·cm^{-1}), but its cis-polymer is regarded as an insulator (1.7×10^{-9} S·cm^{-1}) [18]. Extended conjugation also has an effect on electrical conductivity. As the number of conjugation along the polymer chain increases, the distance of transition between conduction molecular orbital, π and valence molecular orbit, π* decreases and leads to a lowering in bandgap energy, and ultimately the band gap falls in the semiconductor or metallic range.

11.3.2 DOPING IN CP

The electrical conductivity, δ is given by the Equation (11.1), where n is the concentration of the carrier and μ is the mobility of the carrier. The number of charge carriers can be enhanced with the process of doping, which can be achieved by several methods like acid-base doping, redox doping, charge injection, and photo-doping [19]. While doping, the δ increases due to the increases in the number of carriers by oxidation or reduction with electron acceptors or electron donors respectively and fastens the free charge carrier transfer.

$$\delta = n\mu e \qquad\qquad (11.1)$$

According to Hall effect, thermopower and junction measurement, CP oxidized with an electron acceptor causes the removal of electrons and leads to the formation of a polaron with a positive charge hole site, which moves along the polymer chain to contribute conductivity, whereas reduction of CP with an electron donor creates a polaron with a negative charge which is responsible for electric conduction along the chain. The initial accepter doped polymer is called p-type and later one is called n-type.

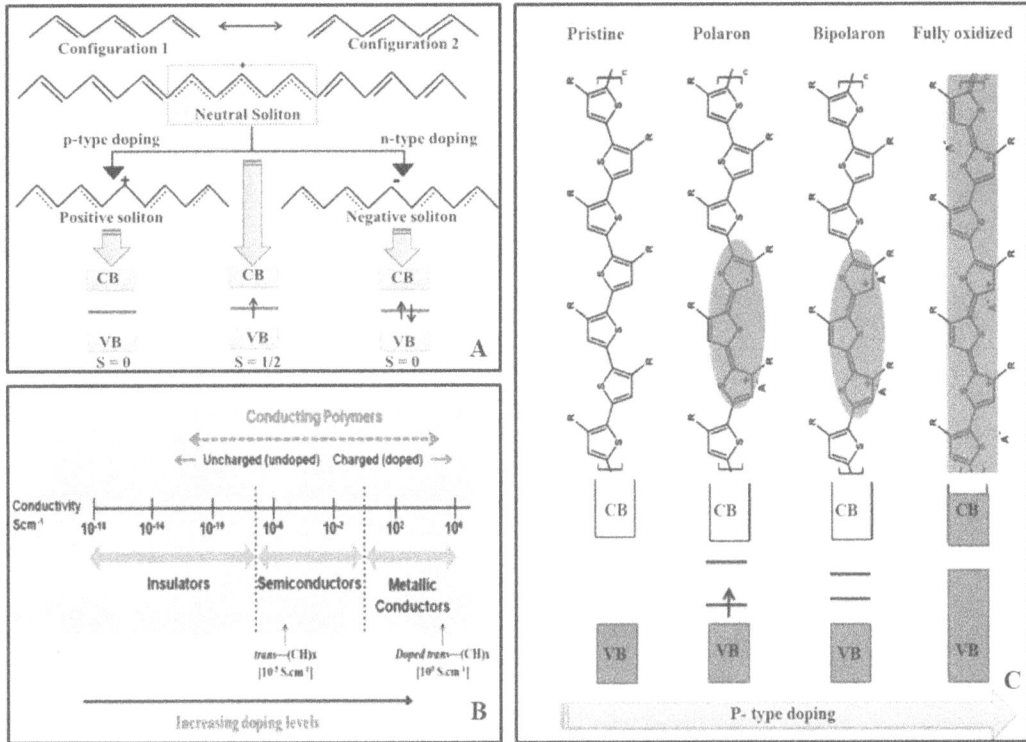

FIGURE 11.3 (a) Formation of soliton in trans polyacetylene; (b) conductivity of CP (reprinted with permission from Ref. [23]; (c) formation of polarons and bipolarons. (Reprinted with permission from Ref. [24].)

Doping produces two types of CPs, one with degenerate ground states of configurations with the same total energy and the other with non-degenerate ground states of configurations with different total energies. In the first category, configurations are different only in terms of their position of single and double bonds [20]. For instance, trans polyacetylene (t-PA) possesses two configurations as shown in Figure 11.3(a), where C centers in each polymer backbone chain offer an unpaired electron. These two configurations are separated by a domain wall boundary called "Soliton" and a new localized energy level forms at the middle region in its energy band structure. A neutral soliton forms in polymer chain having odd number of C centers [21], where the middle gap is half occupied. The doping induces the formation of charged solitons. The n-type doping creates a negatively charged soliton with two electrons in the middle gap, whereas p-type produces a positively charge soliton with an empty middle gap [22].

The second category, where the configurations having non-degenerate ground states differ in their resonance structures, is called aromatic form with lower energy and quinoid form with higher energy. Here doping creates radical ions called polarons with two new energy levels in its electronic band structure. The n-type and p-type doping form negatively and positively charge polarons, respectively. In some species, a positive bipolaron is formed by the attraction of two positively charged polarons that are so close to each other that they can share the same distortions by lowering their energy as shown in Figure 11.3(c). The formation of bipolaron from polaron is a thermodynamically feasible reaction [25]. Its band structure is slightly different from polaron because in bipolarons, the lower energy level is more shifted from the valence band and the upper level is shifted more from conduction band and energy levels are unoccupied. Thus doping generates polarons and bipolaron band levels within the electronic band of CP and as the doping level increases, the bandgap decreases and charge mobility increases [24]. As the dopant concentration approaches 100% (theoretically), the bipolaron band merge with conduction and valence band. i.e., the bandgap vanishes, which is in consistent with the results of Skotheim [26]. Thus doping distinguishes CP over all other polymers and favors the

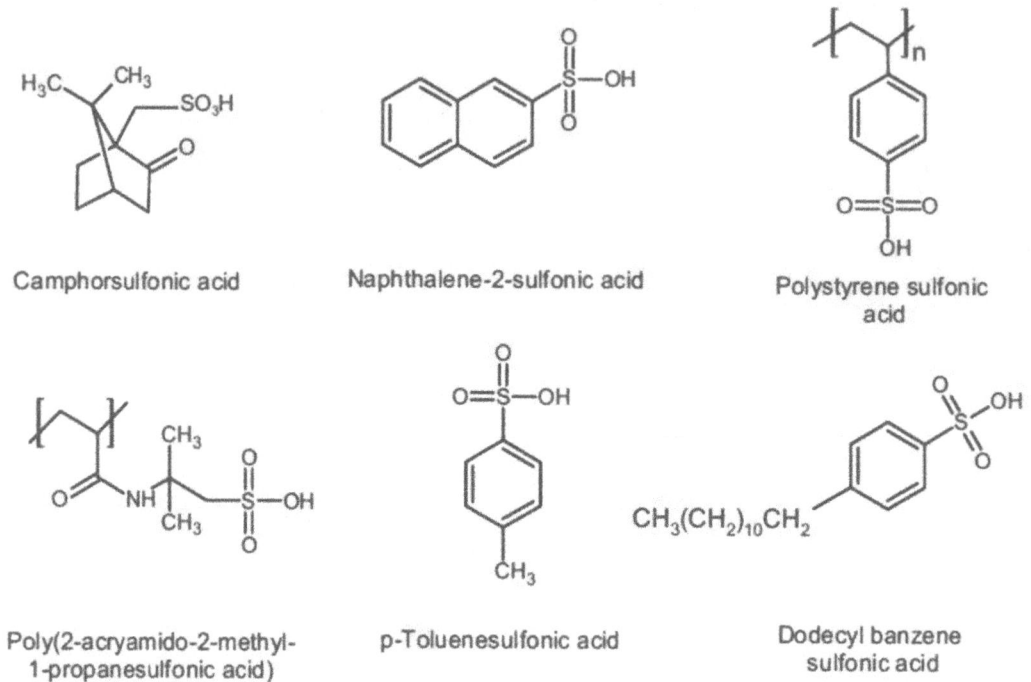

FIGURE 11.4 Chemical structures of some commonly used dopants.

electrical conduction by altering the electronic structure of CP [27]. The effect of doping on the conductivity of CP is depicted in Figure 11.3(b). However, the main issues faced by CPs are their limited solubility, which can be rectified by the usage of long-chain dopants whose alkyl chain has an effect in decreasing the interaction between CPs, thus can acts as a better solubilizing agent. The chemical structures of some commonly used dopants [28] are shown in Figure 11.4.

11.4 METHODS FOR SYNTHESIS OF CP

The major line approaches for the synthesis of CPs are direct route, in which polymers are synthesized by conventional condensation or addition polymerization reactions which has some disadvantage of practical difficulties, and the indirect route approaches deal with the advantage of enhanced synthesis flexibility. A large variety of synthesis routes are available for the synthesis of CPs and can be broadly classified into three main categories [19]: condensation polymerization reaction such as Gilch method; oxidative method such as electrochemical and chemical oxidation; and metal cross-coupling reaction methods such as Kumada-corriu, Negishi, Stille, Suzuki, direct hetero arylation and Yamamoto cross-coupling reactions. The initial step to obtaining CP is the synthesis of monomer and then a polymerization steps is conducted. In the oxidative method, the CP is obtained by the coupling of radical cations in monomer units. In pyrrole-based CPs, the coupling is selective due to the reaction of the radical cation at second and fifth α-carbons. The electrochemical parameters such as applied potential, current, duration of plating, the chemical environment has a vital role in the synthesis of CP. Also, the chemical parameters like solvent, electrolyte, and monomer concentration influence the synthesis. The chemical method is more controllable for polymerization reaction to achieve an enhanced solubility and conductivity, which lead the reaction to a solid–liquid interface; it is more difficult to obtain a thin film from such interface. Thus, the synthesis of new CPs and engineering the existing CPs structure to obtain altered CPs with beneficial properties is the goal of the research field since the 1980s. Some of the general synthesis methods of CPs are described in Table 11.1.

TABLE 11.1
Schematic Representation of Some General Synthesis Procedure of CPs

Synthesis Method	Schematic Representation of Reaction	Reference
Electrochemical oxidative reaction	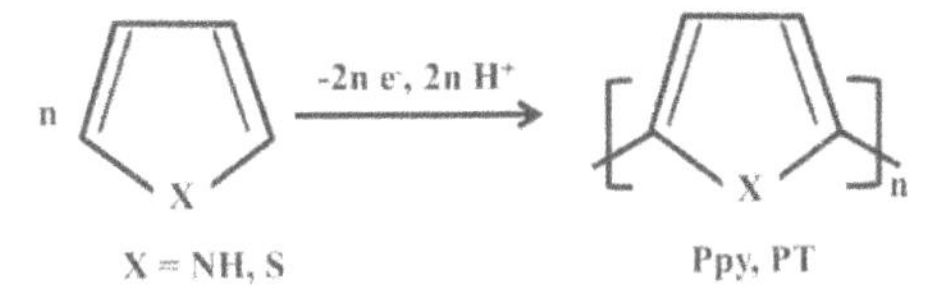 -2n e⁻, 2n H⁺ X = NH, S Ppy, PT	[29]
Chemical oxidative reaction	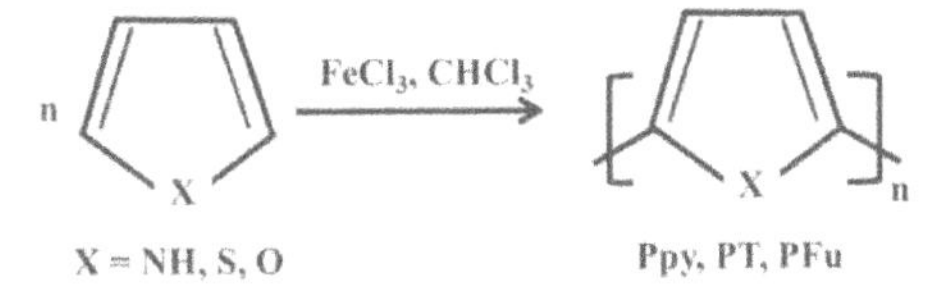 FeCl3, CHCl3 X = NH, S, O Ppy, PT, PFu	[30]
Kumada-corriu coupling reactions	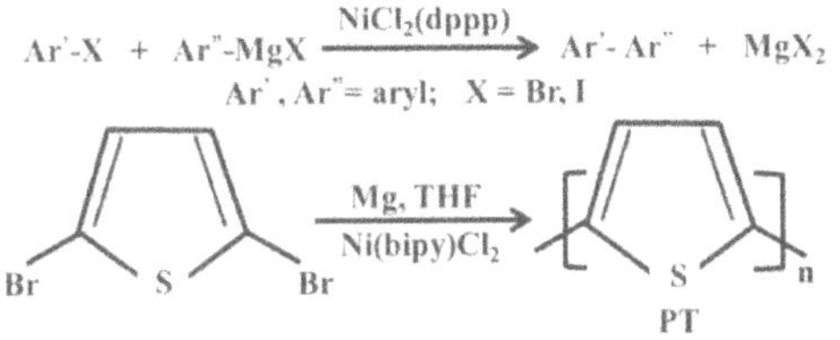 Ar'-X + Ar''-MgX NiCl2(dppp) Ar'-Ar'' + MgX2 Ar', Ar''= aryl; X = Br, I Mg, THF Ni(bipy)Cl2 Br S Br PT	[31]
Negishi coupling reactions	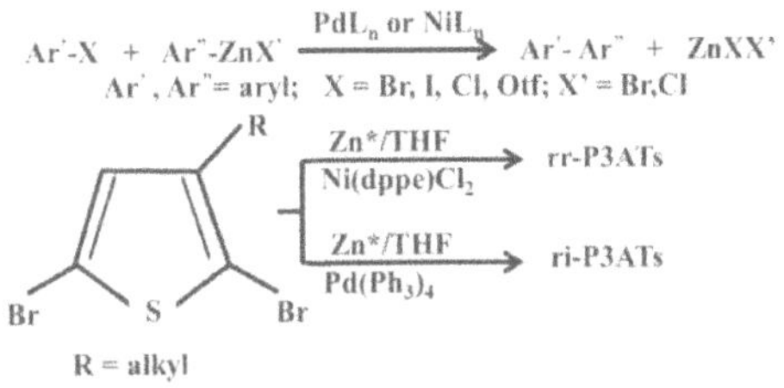 Ar'-X + Ar''-ZnX' PdLn or NiLn Ar'-Ar'' + ZnXX' Ar', Ar''= aryl; X = Br, I, Cl, Otf; X' = Br,Cl Zn*/THF Ni(dppe)Cl2 rr-P3ATs Zn*/THF Pd(Ph3)4 ri-P3ATs R = alkyl	[32]
Stille coupling reactions	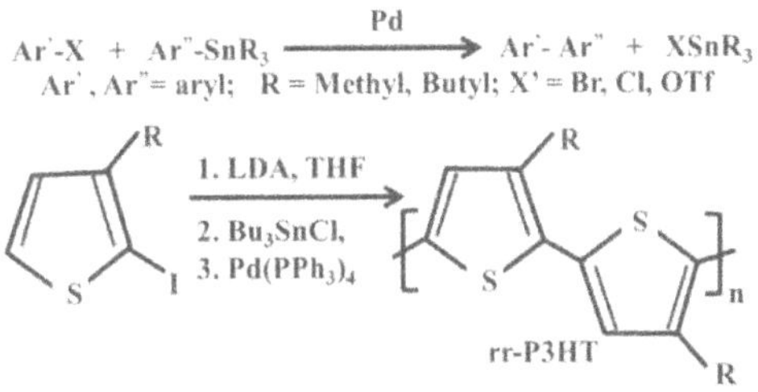 Ar'-X + Ar''-SnR3 Pd Ar'-Ar'' + XSnR3 Ar', Ar''= aryl; R = Methyl, Butyl; X' = Br, Cl, OTf 1. LDA, THF 2. Bu3SnCl, 3. Pd(PPh3)4 rr-P3HT	[33]
Suzuki coupling reactions	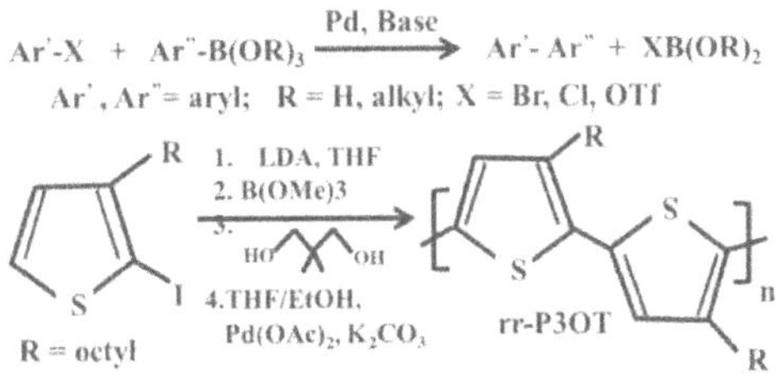 Ar'-X + Ar''-B(OR)3 Pd, Base Ar'-Ar'' + XB(OR)2 Ar', Ar''= aryl; R = H, alkyl; X = Br, Cl, OTf 1. LDA, THF 2. B(OMe)3 3. 4.THF/EtOH, Pd(OAc)2, K2CO3 R = octyl rr-P3OT	[34]
Direct hetero arylation cross coupling reactions	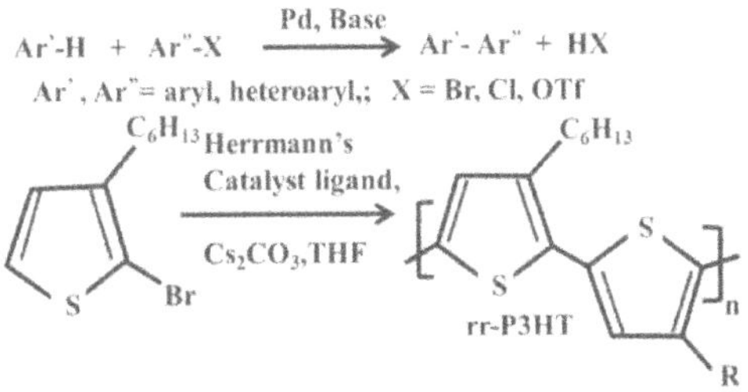 Ar'-H + Ar''-X Pd, Base Ar'-Ar'' + HX Ar', Ar''= aryl, heteroaryl,; X = Br, Cl, OTf C6H13 Herrmann's Catalyst ligand, Cs2CO3,THF rr-P3HT	[35]

(Continued)

TABLE 11.1   (Continued)
Schematic Representation of Some General Synthesis Procedure of CPs

Synthesis Method	Schematic Representation of Reaction	Reference
Yamamoto coupling reactions	R= alkyl Ni(COD)$_2$, COD / 2,2'-bipyridyl, DMF	[36]
Condensation polymerization reaction or Gilch method	X = Cl, Br; R = 2-ethylhexyl t-BuOK, THF	[37]

11.5   ROLE OF CP IN PHOTOVOLTAICS

The solar power that reaches the earth in an hour is sufficient to afford the global energy demand per year [38]. Thus, a daylight harvesting device that converts solar energy to electricity is a good choice for the upcoming energy crisis. CPs-based organic photovoltaic (OPV) devices are a promising candidate for the renewable energy field due to their flexibility, low-cost nature, and ease of commercialization [39]. The attracting feature of adjustability of electrical conductivity varied from the range of insulators to metals and excellent optical properties, brought CP to the research field toward the energy applications not only in organic photovoltaics but also in optoelectronics, organic field emission transistors, supercapacitors, batteries, sensors, and data storage.

11.5.1   Types of CP-Based OPV Solar Cell Devices

The CP solar cells can be classified into three categories based on their architecture. The simplest CP solar cell is single-layer solar cells [40], which consists of a single organic layer sandwiched between cathodic electrodes such as Al, Mg, or Ca and anodic electrode like indium tin oxide (ITO) coated on a glass substrate. One of the major disadvantages of these type OCP devices are their low external quantum efficiency (EQE), due to the fact that in CP the exciton has a higher order of binding energy value (> 0.5 eV) [41] and the electric field produced in the single-layer solar cell device is insufficient to separate the holes and electrons in the exciton. This in turn decreases the carrier charge mobility and affect the EQE as well as PCE.

The second category, the bilayer planar heterojunction OPV devices contain a sandwich of two materials with layered electron donor and electron acceptor nature between cathode and anode electrodes. These types of devices were first developed by Tang [42], in which perylene tetracarboxylic acid derivative and copper phthalocyanine were used as electron acceptor and electron donor species, and the device was found to have an enhanced PCE than single-layered solar cells [43] due to the exciton dissociation at donor–acceptor interface. But this type of solar cell still has a low PCE because of the diffusion length of exciton which is in the range of 4–20 nm [44]. Thus, the dissociation occurs only for excitons near donor–acceptor interface and since most of the excitons are far away from the interface, and they undergo rapid hole–electron recombination and adversely affect the performance of CP solar cell devices [45]. This can be rectified by replacing n-type species

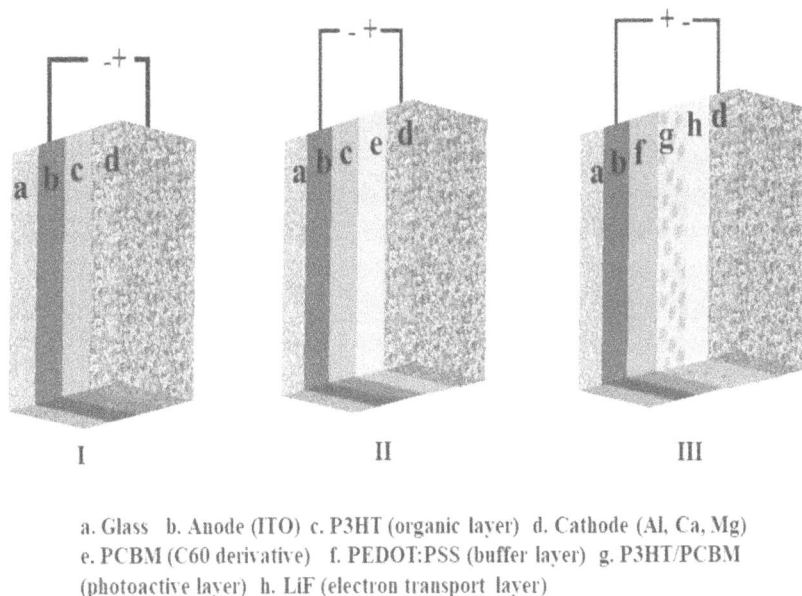

a. Glass   b. Anode (ITO)   c. P3HT (organic layer)   d. Cathode (Al, Ca, Mg)
e. PCBM (C60 derivative)   f. PEDOT:PSS (buffer layer)   g. P3HT/PCBM
(photoactive layer)   h. LiF (electron transport layer)

FIGURE 11.5   Schematic representation of different types of CP solar cells. I. Single layer CP solar cell. II. Bilayer heterojunction CP solar cells. III. Bulk heterojunction CP solar cells.

in OPV devices with buckminsterfullerene (C_{60}) and their derivatives like $PC_{61}BM$ and $PC_{71}BM$. Various beneficial properties of C_{60}, such as fastest photon-induced electron transfer rate in the range of femtoseconds, high electron affinities, and triply degenerated LUMO level make them a standard acceptor for OPV devices.

The third category, namely bulk heterojunction solar cells contain an active layer of blended mixture system of CP and C_{60} derivatives [46], which can reduce the distance of exciton migration and can induce its rapid dissociation into free electrons and holes more efficiently as compared to a single layer and bilayer planar heterojunction CP solar cells and leads to an improvement in EQE. Here electron donor and acceptor species are not in direct contact with anode and cathode. Thus electrons and holes are transports to respective electrodes via two separate channels. These solar cells consist of an ITO-glass anode and negatively charged electrodes like Al, Ca, or Mg cathode. A buffer layer of PEDOT:PSS is placed between the anode and an active layer of CP with C_{60} [47] in order for the hole extraction [48]. A layer of LiF is placed in between the photoactive layer and cathode for electron transport [49]. Different types of PC-based solar cells are depicted in Figure 11.5.

11.5.2 PRINCIPLE OF CP SOLAR CELLS

Harvesting daylight and converting into electricity resulting from a numerous charge transfer chain process inside the organic photovoltaics device. The active layer in the photovoltaic device absorbs incident photons and creates the photogenerated electron-holes species termed as excitons, which further undergoes dissociation to produce free charge carriers and goes to respective electrodes and in turn result in power output, and the entire phenomenon is called photovoltaic process [50].

The working principle of CP solar cells consists of four processes as depicted in Figure 11.6. The initial step is the electronic excitation of donor species from HOMO to LUMO via photon absorption leaving behind the hole in HOMO for the generation of excitons [51]. The next step is the diffusion step, which involves the diffusion of excitons to the interface of donor–acceptor [52]. This step follows the transfer of electrons from donor LUMO to acceptor LUMO. But still, the electrons

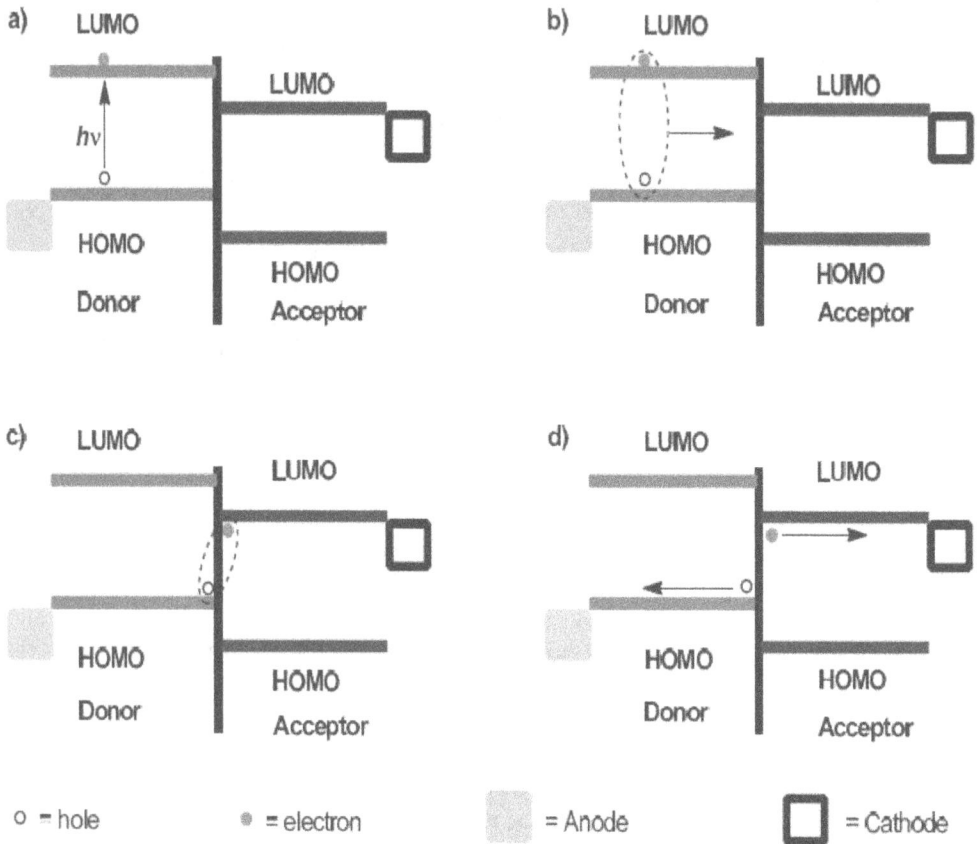

FIGURE 11.6   Process involved in the working principle of CP solar cells. (Reprinted with permission from Ref. [55].)

and holes in acceptor and donor species, respectively are paired as geminate pairs due to strong coulombic interactions [53]. Finally, the electrons and holes in the geminate pair overcome the coulombic interaction and the geminate pair dissociates into holes and electrons, which are collected on respective electrodes [54].

11.5.3   CHARACTERIZATION PARAMETERS FOR CP-BASED OPV DEVICES

Various techniques and parameters were opted for the characterization of organic photovoltaic devices such as open-circuit voltage, short circuit current, incident photon to current efficiency, fill factor, excitation diffusion length [50].

The open-circuit voltage (V_{oc}), which is normally the difference between E_{fermi} levels of n-doped and p-doped semiconductors. It has a linear dependence on the donor's HOMO level and acceptor's LUMO level [56]. It is measured when the external current passing through the system is zero. The energy levels of different components of the photovoltaic device are shown in Figure 11.7(a).

The second factor is the short circuit current (I_{sc}), which is the measure of current in the cell when the applied voltage is zero. I_{sc} has a linear dependence on charge carrier mobility (μ), carrier density (η), and electric field (E) as represented in Equation (11.2). The incident photon to current efficiency (IPCE) can be calculated from the I_{sc} value. IPCE is denoted as the ratio of electrons collected in a short circuit to the incident photons (P_{in}) as given in Equation (11.3), where λ is the wavelength of incident radiation. The next parameter is called fill factor (FF), which is the ratio of

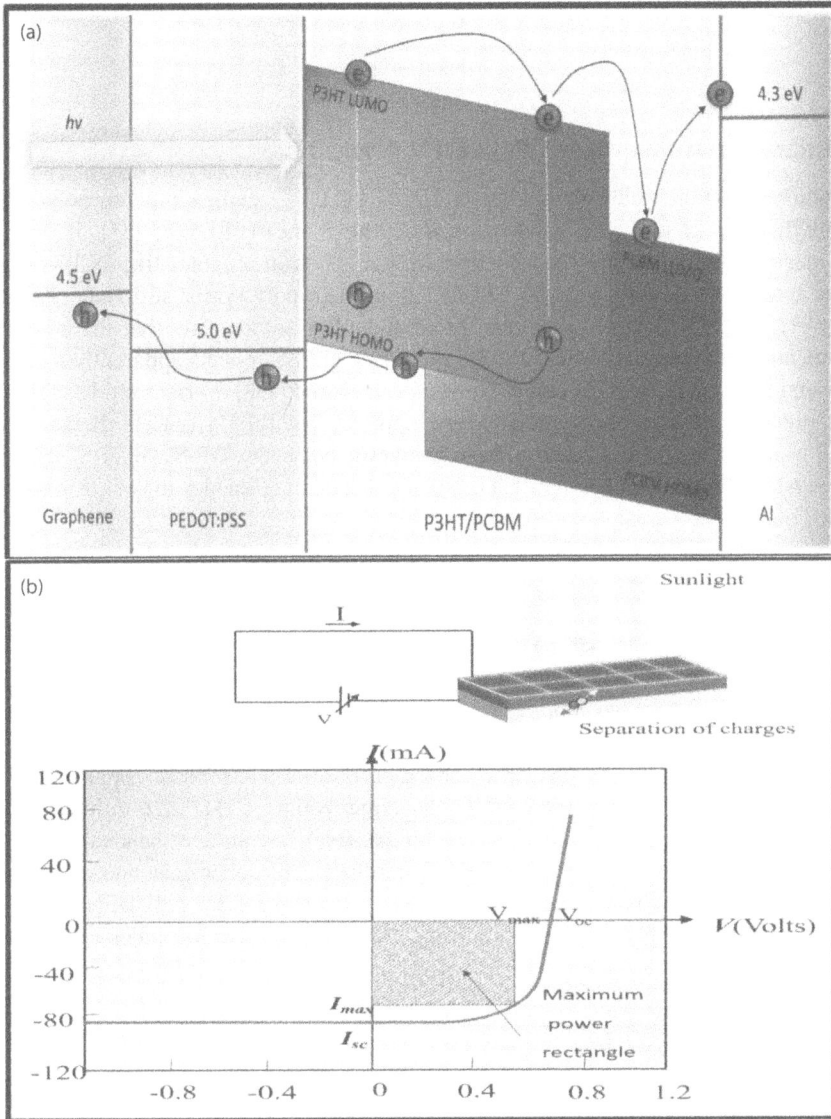

FIGURE 11.7 (a) The energy levels of different components of photovoltaic device; (b) largest rectangular area within the I-V curve of OPV device. (Reprinted with permission from Ref. [65].)

highest possible output power from OPV device to the $V_{oc} \times I_{sc}$, Equation (11.4). The FF depends on the largest rectangular area within the I-V curve of OPV device as depicted in Figure 11.7(b).

The distance (d) covered by free carriers also influences the performance of OPV devices. The d value can be measured by the Equation (11.5), where r is the life time of carrier and μ denotes its mobility. The other parameter includes exciton diffusion length, which should be around 100 nm for an OPV species. The dipolar coupling interactions between the molecules in OPV devices has a great influence in determining exciton diffusion length. For a pure species, the order of defects is considerably small due to trap free environment thus the value of exciton diffusion length is large.

$$I_{sc} = \eta\mu eE \qquad (11.2)$$

$$IPCE = 1240 \ I_{sc}/\lambda P_{in} \qquad (11.3)$$

$$FF = P_{o/p\ max} / V_{oc}\ I_{sc} \tag{11.4}$$

$$d = \mu\ r\ E \tag{11.5}$$

11.5.4 Recent Developments of CPs in OPV Applications

The two major CPs with well-known characteristics for OPV applications are poly(3-hexylthiophene-2,5-diyl) (P3HT) and poly(3,4-ethylenedioxythiophene) (PEDOT) [57], in which P3HT acts as an electron donor and are incorporated with an electron acceptor like PCBM ([6]-phenyl-C61-butyric acid methyl ester), whereas PEDOT along with polystyrene sulfonate (PSS) acts as a hole transporter in OPV devices. Previous research studies authenticate that the mixed blend of the above-mentioned systems of P3HT:PCBM and PEDOT:PSS reflects apparently a high value of power conversion efficiency (PCE) and enhanced performance [58]. Various studies related to thin layer CP applications toward photovoltaics were reported. The thin layer CP-fullerene [59], mixed CP-CP [60] were developed and found to have good efficiency. The CP-Si (amorphous) system has a solar conversion efficiency of 2.0% [61]. The solar conversion efficiency (η) of OPV is represented in equation 11.6.

$$\eta = (FF \times V_{oc} \times J_{sc}) / P_{in} \tag{11.6}$$

It is evident that the η value has a strong dependence on the thickness of CP film in OPV devices [62]. The η versus thickness of POT-PTSA/n-Si is depicted in Figure 11.8, which clearly states that thin-film CP with small thickness has more suitable surface morphology for light trapping than smaller or larger CP films.

A variety of thin layer CP was developed for the purpose of photovoltaic applications. A donor–acceptor CP, PDTPDTBT were synthesized by copolymerization of DTP (2,6-dithieno[3,2-b:2′,3′-d] pyrrole) and DTBT(4,7-di(2′-thienyl)-2,1,3-benzothiadiazole) having low bandgap and soluble in most of the common organic solvents, like THF, chloroform, chlorobenzene, and toluene due to the presence of the bulky side chain. A CP solar cell of PDTPDTBT:PCBM bulk heterojunction gives a PCE of 2.18%, indicating that DTP is a promising competitor for photovoltaics applications [P12].

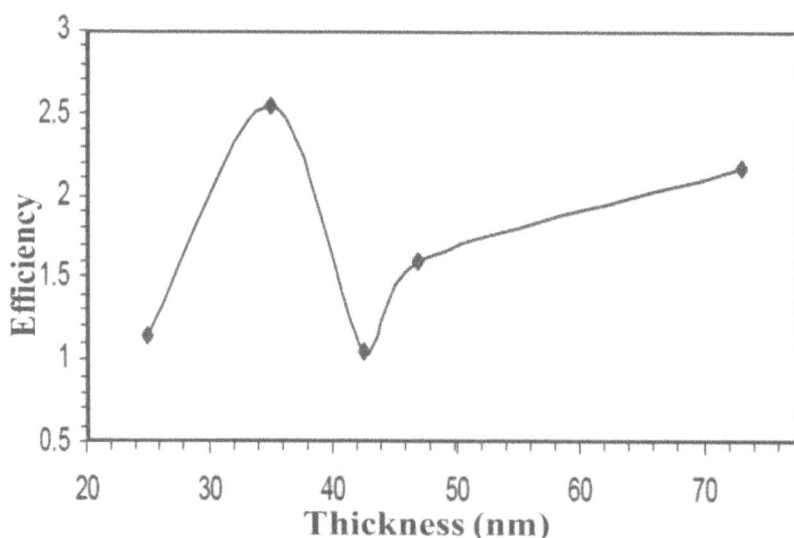

FIGURE 11.8 Solar conversion efficiency, η versus thickness of POT-PTSA/n-Si. (Reprinted with permission from Ref. [62].)

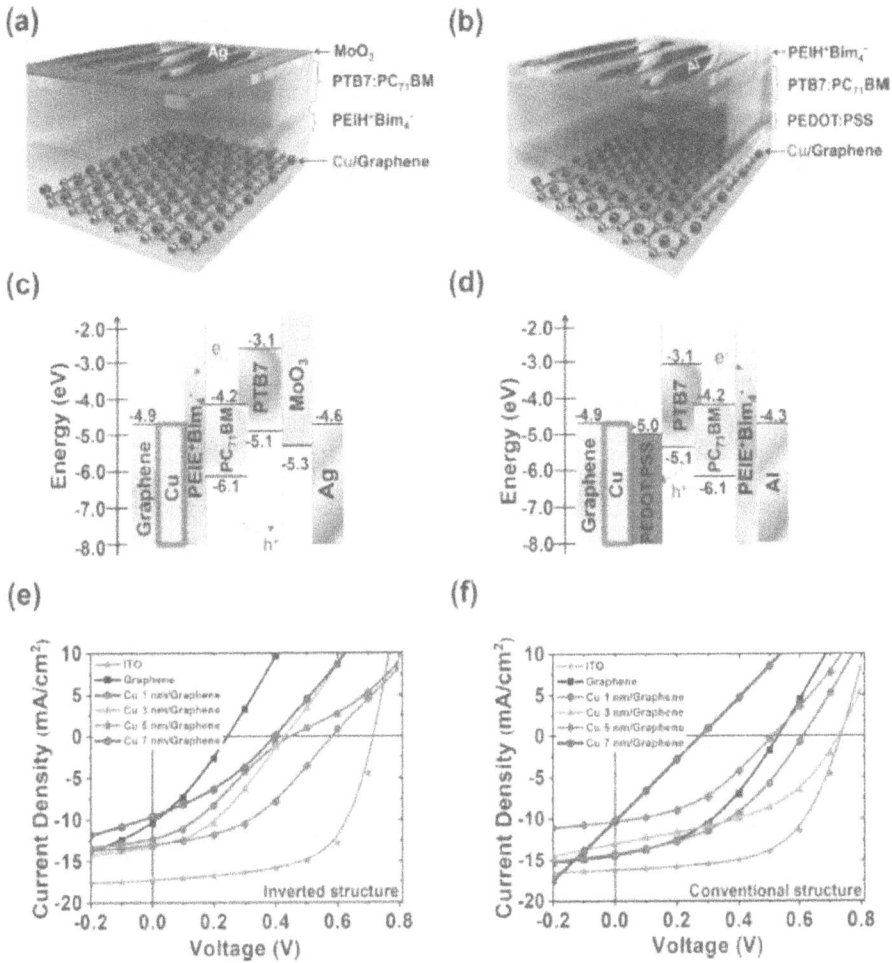

FIGURE 11.9 (a) Inverted metal-graphene-based PTB7:PC71BM CP solar cells; (b) conventional metal-graphene based PTB7:PC71BM CP solar cells; (c) energy level diagram of inverted CP solar cells; (d) energy level diagram of conventional CP solar cells; (e) and (f) are their respective J-V curves. (Reprinted with permission from Ref. [64].)

A visibly transparent CP OPV device with maximum transparency of 66% was designed by mild solution process with Ag nanowires-based thin layer composite (ITO/PEDOT:PSS/PBDTT-DPP:PCBM/TiO$_2$/AgNW) composite electrode which can harvest daylight from IR region and has a PCE of about 4%. [63].

The latest work reveals the enhanced electronic characteristics of CP-OPV devices by the incorporation of metal-graphene hybrids into CP solar cells [64]. The schematic representation of the metal-graphene hybrid conventional and inverted CP solar cells along with their energy diagram and J-V curves are depicted in Figure 11.9. The conventional metal-graphene hybrid CP solar cells achieved a PCE of 4.38%, and the inverted one has a PCE of 2.67%. The studies of Tiwari et al. [65] also suggest graphene as a significant electrode in various solar cells.

The chlorination can increase the photovoltaic performance of CPs by generating a large number of excitons and enhances the charge carriers' separation and hence facilitate exciton dissociation by reducing its binding energy. This can be authenticated by the work of Hwan-Il Je et.al. [66], that the incorporation of Cl in BDT-Th unit in PBDBTS (PBDB-TS-4Cl) increases the transportation of charge, probability of exciton dissociation as well as open-circuit voltage and hence enhances the overall photovoltaic activity of CP. Zou et al. designed a CP solar cell of Y6 with PBDB-T-2F shows a PCE of 15.7% [67], which has a low energy loss below the bandgap in the range of 0.5–0.6 V due to a small value of energy offset in the system causing almost negligible radiative energy loss and a lower non-radiative energy loss of about 0.25 V [67].

11.6　CONCLUSION AND FUTURE ASPECTS

For the forthcoming OPV devices, CPs are considered as potential materials because of their unique beneficial properties. Developing CPs via various methods such as condensation, oxidative, and coupling methods can be employed in several forms of applications in CP solar cells. As mentioned in this chapter, the thin film CP is used as innovative active conductive material whose conductivity ranges from semiconductor to metallic. Moreover, CP has substantial optoelectrical properties and can easily be customized and tailored by desired properties for any particular application by adjusting molecular designs. Also, since CP has a high absorption coefficient, only a thin film of thickness 100–200 nm is necessary for adequate daylight absorption. The photovoltaic performance of CP relies on certain parameters as discussed in the chapter. However, there is some constraint which weakens the rapid growth in developing CP-based OPV devices, which includes their PCE and life time compared to inorganic OPV devices.

Several innovative technologies and progressive CP and derivatives material species are yet to be figured out in order to change our viewpoint toward the promising future of organic photovoltaics devices.

ACKNOWLEDGMENTS

The authors would like to thank Principal, Sree Narayana College, Kollam, Kerala, India, and Professor and Head, Department of Chemistry, Sree Narayana College, Kollam, Kerala, India for providing the laboratory facilities. One of the authors, Sarika. S is grateful to the University of Kerala, Trivandrum, Kerala, India, for providing assistance in the form of a research fellowship.

REFERENCES

1. Omer A M, Energy, environment and sustainable development, Renew. Sust. Energ. Rev., 2008, 12, 2265–2300.
2. Cheng Y-J, Yang S H, Hsu C S, Synthesis of conjugated polymers for organic solar cell applications. Chem. Rev., 2009, 109, 5868–5923.
3. Heeger A J, Semiconducting and metallic polymers: The fourth generation of polymeric materials, J. Phys. Chem. B, 2001, 105, 8475–8491.
4. MacDiarmid A G, "Synthetic Metals": A novel role for organic polymers (Nobel Lecture), Angewandte Chemie International Edition, 2001, 40 (14), 2581–2590.
5. Shirakawa H, The discovery of polyacetylene film: The dawning of an era of conducting polymers (Nobel Lecture), Angewandte Chemie International Edition, 2001, 40, 2574–2580.
6. Shirakawa H, Nobel lecture: The discovery of polyacetylene film—the dawning of an era of conducting polymers. Rev. Mod. Phys., 2001, 73, 713.
7. Kim Y H, Sachse C, Machala M L, May C, Müller-Meskamp L, Leo K, Highly conductive PEDOT:PSS electrode with optimized solvent and thermal post-treatment for ITO-Free organic solar cells, Adv. Funct. Mater., 2011, 21, 1076–1081.
8. Dou L T, You J B, Yang J, Chen C C, He Y J, Murase S, Moriarty T, Emery K., Li G, Yang Y, Tandem polymer solar cells featuring a spectrally matched low-bandgap polymer, Nat. Photonics, 2012, 6, 180–185.
9. Chen C C, Dou L, Zhu R, Chung C H, Song T B, Zheng Y B, Hawks S, Li G, Weiss P S, Yang Y. Visibly transparent polymer solar cells produced by solution processing, ACS Nano., 2012, 6 (8):7185–90.
10. Zhou Q, Shi G, Conducting polymer-based catalysts, J. Am. Chem. Soc., 2016, 138, 2868–2876.

11. Ostroverkhova O, Organic optoelectronic materials: Mechanisms and applications, Chem. Rev., 2016, 116, 13279–13412.
12. Dou, L, Liu Y, Hong Z, Li G, Yang Y, Low-bandgap near-IR conjugated polymers/molecules for organic electronics, Chem. Rev., 2015, 115, 12633–12665.
13. Wang, C, Dong H, Hu W, Liu Y, Zhu D, Semiconducting π-conjugated systems in field-effect transistors: A material odyssey of organic electronics, Chem. Rev., 2012, 112, 2208–2267.
14. Liu M, Gao Y, Zhang Y, Liu Z, Zhao L, Quinoxaline-based conjugated polymers for polymer solar cells, Polym. Chem., 2017, 8, 4613–4636.
15. Grimsdale A C, Chan K L, Martin R E., Jokisz P G, Holmes A B, Synthesis of light-emitting conjugated polymers for applications in electroluminescent devices, Chem. Rev., 2009, 109, 897–1091.
16. Muller P, Glossary of terms used in physical organic chemistry (IUPAC Recommendations 1994), Pure Appl. Chem., 1994, 66, 1077.
17. Wan, M., Conducting Polymers with Micro or Nanometer Structure, Springer-Verlag, Berlin Heidelberg, 2008.
18. Shirakawa H, Louis E J, MacDiarmid A G, Chiang C K, Heeger A J, Synthesis of electrically conducting organic polymers: Halogen derivatives of polyacetylene, (CH), J. Chem. Soc., Chem. Commun., 1977, 16, 578–580.
19. Murad R A, Iraqi A, Aziz S B, N Abdullah S, Brza M A, Conducting polymers for optoelectronic devices and organic solar cells: A review, Polymers, 2020, 12, 2627.
20. Su W P, Schrieffer J, Heeger A J, Solitons in polyacetylene, Phys. Rev. Lett., 1979, 42, 1698.
21. Bredas J, Themans B, Andre J, Chance R, Silbey R, The role of mobile organic radicals and ions (solitons, polarons and bipolarons) in the transport properties of doped conjugated polymers, Synth. Met., 1984, 9, 265–274.
22. Heeger A J, Kivelson S A, Schrieffer, J R, Su W-P, Solitons in conducting polymers, Rev. Mod. Phys., 1988, 60, 781–850.
23. Wang X X, Yu G F, Zhang J, Yu M, Ramakrishna S, Long Y Z, Conductive polymer ultrafine fibers via electrospinning: Preparation, physical properties and applications, Prog. Mater. Sci., 2020, 100704.
24. Cobet C, Gasiorowski, J, Farka D, Stadler P, Polarons in conjugated polymer, Ellipsometry of Functional Organic Surfaces and Films, Springer, Cam, 2018, 355–387.
25. Kiess G H, Conjugated Conducting Polymers, Springer-Verlag, Berlin Heidelberg, 1992.
26. Skotheim T, A Handbook of Polymer Synthesis Part B, Mercel Dekker, N Y, 1986.
27. Heeger A J, Semiconducting and metallic polymers: the fourth generation of polymeric materials. Curr. Appl. Phys, 2001, 1, 247.
28. Khokhar D, Jadoun S, Arif R, Jabin S, Functionalization of conducting polymers and their applications in optoelectronics, Polym-Plast. Tech. Mat., 2021, 60, 463–85.
29. Heinze J, Frontana-Uribe B A, Ludwigs S, Electrochemistry of conducting polymers—persistent models and new concepts, Chem. Rev., 2010, 110, 4724–4771.
30. Yoshino K, Hayashi S, Sugimoto R-I, Preparation and properties of conducting heterocyclic polymer films by chemical method, Jpn. J. Appl. Phys., 1984, 23, L899–L900.
31. Kumada M, Nickel and palladium complex catalyzed cross-coupling reactions of organometallic reagents with organic halides, Pure Appl. Chem., 1980, 52, 669–679.
32. Chen T A, Rieke, R D, The first regioregular head-to-tail poly(3-hexylthiophene-2,5-diyl) and a regiorandom isopolymer: Nickel versus palladium catalysis of 2(5)-bromo-5(2)-(bromozincio)-3-hexylthiophene polymerization, J. Am. Chem. Soc., 1992, 114, 10087–10088.
33. Stille J K, The palladium-catalyzed cross-coupling reactions of organotin reagents with organic electrophiles, Angew. Chem. Int. Ed., 1986, 25, 508–524.
34. Guillerez S, Bidan G, New convenient synthesis of highly regioregular poly(3-octylthiophene) based on the Suzuki coupling reaction, Synth. Met., 1998, 93, 123–126.
35. Wang Q, Takita R, Kikuzaki Y, Ozawa F, Palladium-catalyzed dehydrohalogenative polycondensation of 2-bromo-3-hexylthiophene: An efficient approach to head-to-tail poly(3-hexylthiophene), J. Am. Chem. Soc., 2010, 132, 11420–11421.
36. Zhang Z-B, Fujiki M, Tang H-Z, Motonaga M, Torimitsu K, The first high molecular weight poly(N-alkyl-3,6-carbazole)s, Macromolecules, 2002, 35, 1988–1990.
37. Akcelrud L, Electroluminescent polymers, Prog. Polym. Sci., 2003, 28, 875–962.
38. Morton O, Solar energy: A new day dawning? Silicon Valley sunrise, Nature, 2006, 443, 19–22.
39. Krebs F C, Polymer solar cell modules prepared using roll-to-roll methods: Knife-over-edge coating, slot-die coating and screen printing, Sol. Energy Mater. Sol. Cells, 2009, 93, 465–475.
40. Nelson J, Organic photovoltaic films, Mater. Today, 2002, 5, 20–27.

41. Brabec C J, Sariciftci N S, Hummelen J C, Plastic solar cells, Adv. Funct. Mater., 2001, 11, 15–26.
42. Tang C W, Two-layer organic photovoltaic cell, Appl. Phys. Lett., 1986, 48, 183–185.
43. Blom P W M, Mihailetchi V D, Koster L J A, Markov D E, Device physics of polymer: Fullerene bulk heterojunction solar cells, Adv. Mater., 2007, 19, 1551–1566.
44. Markov D E, Amsterdam E, Blom P W M, Sieval A B, Hummelen J C, Accurate measurement of the exciton diffusion length in a conjugated polymer using a heterostructure with a side-chain cross-linked fullerene layer, J. Phys. Chem. A, 2005, 109, 5266–5274.
45. Günes S, Neugebauer H, Sariciftci N S, Conjugated polymer-based organic solar cells, Chem. Rev., 2007, 107, 1324–1338.
46. Son H J, Carsten B, Jung I H, Yu L, Overcoming efficiency challenges in organic solar cells: Rational development of conjugated polymers, Energy Environ. Sci., 2012, 5, 8158–8170.
47. Groenendaal L, Jonas F, Freitag D, Pielartzik H, Reynolds J R, Poly(3,4 ethylenedioxythiophene) and its derivatives: Past, present, and future, Adv. Mater., 2000, 12, 481–494.
48. Kirchmeyer S, Reuter K, Scientific importance, properties and growing applications of poly(3,4-ethylenedioxythiophene), J. Mater. Chem., 2005, 15, 2077–2088.
49. Benanti T L, Venkataraman D, Organic solar cells: An overview focusing on active layer morphology, Photosynth. Res., 2006, 87, 73–81.
50. Chamberlain G A, Organic solar cells: A review, Sol. Cells., 1983, 8, 47–83.
51. Son H J, He F, Carsten B, Yu L, Are we there yet? Design of better conjugated polymers for polymer solar cells, J. Mater. Chem., 2011, 21, 18934–18945.
52. Brédas J-L, Norton J E, Cornil J, Coropceanu V, Molecular understanding of organic solar cells: The challenges, Acc. Chem. Res., 2009, 42, 1691–1699.
53. Thompson B C, Fréchet J M J, Polymer–fullerene composite solar cells, Angew. Chem. Int. Ed., 2008, 47, 58–77.
54. Yeh N, Yeh P, Organic solar cells: Their developments and potentials, Renew. Sustain. Energy Rev., 2013, 21, 421–431.
55. R Murad A, Iraqi A, Aziz S B, N Abdullah S, Brza M A, Conducting polymers for optoelectronic devices and organic solar cells: A review, Polymers, 2020, 12 (11), 2627.
56. Scharber M C, Mühlbacher D, Koppe M, Denk P, Waldauf C, Heeger A J, Brabec C J, Design rules for donors in bulk-heterojunction solar cells—towards 10% energy-conversion efficiency, Adv. Mater., 2006, 18, 789–794. https://doi.org/10.1002/adma.200501717.
57. Lee S J, Pil Kim H, Mohd Yusoff A R b, Jang J, Organic photovoltaic with PEDOT:PSS and V_2O_5 mixture as hole transport layer, Sol. Energy Mater. Sol. Cells, 2014, 120, 238–243.
58. Liang C, Wang Y, Li D, Ji X, Zhang F, He Z, Modeling and simulation of bulk heterojunction polymer solar cells, Sol. Energy Mater. Sol. Cells, 2014, 127, 67–86.
59. Gao J, Yu G, Hummelen J C, Wudl F, Heeger A J, Polymer photovoltaic cells: Enhanced efficiencies via a network of internal donor-acceptor heterojunctions, Science, 1995, 270, 1789–1791.
60. Granström M, Petritsch K, Arias A C, Lux A, Andersson M R, Friend R H, Laminated fabrication of polymeric photovoltaic diodes, Nature, 1998, 395, 257–260.
61. Williams E L, Jabbour G E, Wang Q, Shaheen S E, Ginley D S, Schiff E A, Conducting polymer and hydrogenated amorphous silicon hybrid solar cell, Appl. Phys. Lett., 2005, 87, 13.
62. De Souza Gomes A, New polymers for special applications, Conducting Polymers Application, 2012 (Chapter 1). https://doi.org/10.5772/48316.
63. Chen C C, Dou L, Zhu R, Chung C H, Song T B, Zheng Y B, Hawks S, Li G, Weiss P S, Yang Y, Visibly transparent polymer solar cells produced by solution processing, ACS Nano., 2012, 6, 7185–90.
64. Kang J H, Choi S, Park Y J, Park J S, Cho N S, Cho S, Walker B, Choi D S, Shin J W, Seo J H, Cu/graphene hybrid transparent conducting electrodes for organic photovoltaic devices, Carbon, 2021, 171, 341–349.
65. Tiwari S, Purabgola A, Kandasubramanian B, Functionalised graphene as flexible electrodes for polymer photovoltaics, J. Alloys Compd., 2020, 825, 153954.
66. Je H I, Shin E Y, Lee K J, Ahn H, Park S, Im S H, Kim Y H, Son H J, Kwon S K, Understanding the performance of organic photovoltaics under indoor and outdoor conditions: Effects of chlorination of donor polymers, ACS Appl. Mater. Interfaces, 2020, 12, 23181–9.
67. Yuan J, Zhang Y, Zhou L, Zhang G, Yip H-L, Lau T-K, Lu X, Zhu C, Peng, H, Johnson P A, Leclerc M, Cao Y, Ulanski J, Li Y, Zou Y, Single-junction organic solar cell with over 15% efficiency using fused-ring acceptor with electron deficient core, Joule, 2019, 3, 1140–1151.

LIST OF ABBREVIATIONS

C_{60}	Buckminsterfullerene
CB	Conduction band
CP	Conducting polymer
DSSC	Dye-sensitized solar cells
DTBT	4,7-di(2'-thienyl)-2,1,3-benzothiadiazole
DTP	2,6-dithieno[3,2-b:2',3'-d]pyrrole
E	Electric field
EQE	External quantum efficiency
FF	Fill factor
HOMO	Highest occupied molecular orbitals
IPCE	Incident photon to current efficiency
I_{sc}	Short circuit current
ITO	Indium tin oxide
LPPP	Ladder-type polyparaphenylene
LUMO	Lowest unoccupied molecular orbitals
MEH-PPV	Alkoxy-substituted poly para-phenylene vinylene
η	Carrier density charge
OFET	Organic field effect transmitters
OIH	organic-inorganic hybrid cells
OLED	Organic light emitting diodes
OPV	Organic photovoltaic
P3AT	Poly (3-alkyl) thiophene
P3HT	Poly (3-hexyl) thiophene
P3OT	Poly (3-octylthiophene)
PA	Polyacetylene
PANI	Polyaniline
$PC_{61}BM$	[6]-phenyl-C61-butyric acid methyl ester
$PC_{71}BM$	[6]-phenyl-C71-butyric acid methyl ester
PEDOT	Polyethylenedioxythiophene
PHT	Polyheptadiyne
P_{in}	Incident photons
PITN	Polyisothianaphthene
POT	Poly octyl thiophene
PPP	Polyparaphenylene
PPS	Polyparaphenylene sulfide
PPV	Polyparaphenylene vinylene
PPy	Polypyrrole
PT	Polythiophene
PTSA	Para-toluenesulfonic acid
t-PA	Trans-polyacetylene
VB	Valance band
V_{oc}	Open circuit Voltage
μ	Carrier mobility

12 Application of 2D Materials in Conducting Polymers for High Capacity Batteries

Muhammad Rizwan Sulaiman[1,2], Shrestha Tyagi[3], Manohar Singh[3], Beer Pal Singh[3], R.K. Soni[4], Rahul Singhal[5,6], and Ram K. Gupta[1]

[1]Department of Chemistry, Kansas Polymer Research Center, Pittsburg State University, Pittsburg, Kansas, USA

[2]Department of Plastic Engineering, Pittsburg State University, Pittsburg, Kansas, USA

[3]Department of Physics, Chaudhary Charan Singh University, Meerut, India

[4]Department of Chemistry, Chaudhary Charan Singh University, Meerut, India

[5]Department of Physics and Engineering Physics, Central Connecticut State University, New Britain, Connecticut, USA

[6]Shivaji College, University of Delhi, Delhi, India

CONTENTS

DOI: 10.1201/9781003150374-12

12.1 INTRODUCTION

In the 21st century, energy is an essential factor for economic competitiveness due to the continual increase in the global population and energy needs. Rising energy demands and the increase of carbon dioxide emissions in the environment are significant global concerns that garnered international attention toward renewable energy technologies. The energy obtained from natural sources and the environment has huge potential for an alternative of energy generated from fossil fuels [1–3]. The primary energy consumption throughout the world was estimated at nearly 149,634 and 157,064 Terawatt-hours (TWh) in the year 2015 and 2018, respectively [4]. In 2018, the Asia Pacific and North America were the largest energy consumers, whereas Africa utilized the minimum energy [4]. Energy storage is a necessity in order to balance the demand and supply of energy. According to studies, energy storage is a form of battery for stationary applications that is predicted to grow 17-folds by 2030 [5]. Various advanced and modernized energy storage devices and increasing production are required to fulfill the increasing energy demand. The efficiency of conventionally used energy storage techniques and various devices such as magnets, batteries, dry cells, and capacitors has been steadily improving.

12.1.1 Types of Energy Storage Devices

Energy storage is the process of transforming energy from different hard-to-store sources in easy or economically storable forms. Energy storage devices accumulate energy in many forms, such as electrochemical, potential, electromagnetic, chemical, and thermal, via various energy storage devices, including rechargeable batteries [6–8], fuel cells [9, 10], capacitors [11], supercapacitors [12, 13], flywheels [14], super magnets [15], and pumped hydro energy [16, 17]. Several commonly used energy storages are as follows:

12.1.1.1 Capacitors

A capacitor is a passive two-terminal device that stores energy electrostatically and releases it as per the requirement of the circuit. It is also used as a backup or in other types of energy storage systems [18]. It contains two parallel conducting plates with an insulator between them. In the uncharged state, the charge on either of the parallel conductors in the capacitor is zero. An electrical charge of +Q and –Q builds up on the capacitor plates, respectively, when DC voltage V is applied to a capacitor electrode. Current flows continuously until the potential difference between anode and cathode equals the supply voltage in a fully charged capacitor. The ratio of charge Q and voltage V is defined as the capacitance as given in Equation (12.1).

$$C = \frac{Q}{V}$$

(12.1)

The unit of capacitance (C) is Farad (F). Basically, energy is stored in capacitors due to an electric field develop between their plates. When a battery is connected to a capacitor for the required period of time, no current will pass through it. Capacitors are widely employed in electronic devices to retain power supply when batteries are being replaced.

12.1.1.2 Supercapacitors

The field of a supercapacitor is relatively new as compared to other electrochemical devices, such as electrochemical batteries and fuel cells [13]. Because of its extremely high capacitance, a supercapacitor differs from a regular capacitor. Supercapacitors can be charged much faster than a battery and store much more charge per volume unit than an electrolytic capacitor. As a consequence, it is considered between a battery and electrolytic capacitor. Supercapacitors have a high energy density in contrast to capacitors. The productivity of supercapacitors related to specific capacitance, energy density, and power density are significantly affected by numerous factors, such as electrode material, type of electrolyte, and operating-potential windows. Among these factors, the selection of electrode material is the key component that remarkably affects the development of supercapacitor technology [19, 20]. Various materials, like carbon-based materials [21, 22], metal oxides [23, 24], conducting polymers (CPs), and their composites [25, 26] are used for electrodes in supercapacitors. Usually, supercapacitors are categorized into two types, i.e., electric double-layer capacitors (EDLCs) and pseudocapacitors. The technique for energy storage in EDLCs is strictly electrostatic, while in pseudocapacitors, the energy storage mechanism is a revisable Faradic redox reaction [27]. The calculation of the specific capacitance of the supercapacitor is performed by using Equation (12.2).

$$C = \varepsilon_o \varepsilon_r \frac{A}{d} \tag{12.2}$$

where ε_o stands for the dielectric constant, ε_r is the relative dielectric constant, A is the specific surface area, and d represents the distance between the supercapacitor electrode and electrolyte ion. The energy density of the supercapacitor can be calculated by using Equation (12.3).

$$E = \frac{1}{2} C (\Delta V)^2 \tag{12.3}$$

Here, C is the specific capacitance, and V represents the operating potential. Also, the power density of the supercapacitor is given by Equation (12.4).

$$P = \frac{1}{4} \frac{(\Delta V^2)}{R} \tag{12.4}$$

where R is symbolized for equivalent series resistance. Supercapacitors offer high-peak-power output, which makes them able to be charged and discharged millions of times without any damage. However, supercapacitors also have some limitations related to the lower energy density in comparison to other energy storage devices like batteries.

12.1.1.3 Batteries

A battery is an electrochemical component that produces electrical energy by undergoing a chemical reaction. Batteries are composed of two or more cells joint in series or parallel and transform chemical energy into electrical energy via electrochemical reactions that take place at their electrodes [28]. Batteries contain three active components, including two-conductor electrodes and an electrolyte. If the reactions that occur at electrodes are irreversible, the battery is known as the primary battery. Whereas, if the reactions are reversible, the battery may be recharged again and termed as secondary batteries [28]. A lot of research has been conducted on lead-acid batteries over the past years. Recently, researchers focused on new battery technologies such as nickel-cadmium [29, 30], sodium-sulfur [31, 32], lithium-ion batteries (LIBs) [33, 34], and lithium-air batteries [35] for high-power consuming electronics. There are several challenges with battery technology, such as low power density, self-discharge, shelf life, and cyclability.

12.2 IMPORTANCE OF BATTERIES

It is well known that compact electronic devices, such as smartphones, tablets, and wearable sensors, and also smart cards with integrated circuits have become an indispensable part of human existence. Also, uninterruptible power supply, reliable and lightweight marine performance, surveillance or alarm systems in sites, hybrid/full electric vehicles, etc., have been playing a significant role in everyday life. The activity of all these devices depends on the energy supplier, which is presently prevailed by high-capacity batteries due to their superior energy density, long lifetime, and broad working temperature in comparison to other energy storage systems. There are mainly five factors that decide any battery capacity, i.e., size, temperature, cut-off voltage, discharge rate, and past usage. The improvement in the battery technology can be achieved by the development of new materials for electrodes that can improve the cycling stability, rate capability, and energy density of the batteries. There is a wide range of materials that are being used as electrode materials for rechargeable battery applications, such as transition metals oxides [36], LiFePO$_4$ and their derivatives [37, 38], graphene [39], and carbon nanotubes [40, 41]. The major concerns while using these materials are environmental, cost-effectiveness, and limited resources. However, electroactive organic materials having intrinsic redox properties are the potential candidates in comparison to normally used inorganic intercalation materials. Some organic electrode materials are CPs, organic carbonyl compounds, organosulfur compounds, and organic free radical compounds. Other than that, CPs like polyaniline (PANI), polypyrrole (PPy), poly(3,4-ethylene dioxythiophene) (PEDOT), polythiophene (PT), polyacetylene (PA), polyparaphenylene (PPP), and others are the most promising for sustainable batteries owing to their exclusive electronic and mechanical properties [42–44]. Moreover, their unique features such as diversifying structures, flexibility, processability, ease of synthesis, tunable physical properties, and facile chemistry, and good electrochemical performance make them suitable for various battery systems, including lithium-ion, lithium-sulfur, sodium ion, and all-polymer batteries [45–47]. However, the mechanical strengths and conductivities of CPs still need improvements. Also, the inferior charge/discharge, rate capability, and cyclic durability of CPs limit their applications in high-capacity batteries. The inclusion of two-dimensional (2D) materials can enhance the properties of CP-based electrodes by creating different polymer-fillers interfacial geometries. Various 2D materials such as graphene, MoS$_2$, and hBN were used for enhancing the performance of CP-based electrodes for high-performance battery application.

12.3 CHARACTERISTICS AND TYPES OF BATTERIES

12.3.1 Characteristics of Batteries

The conceptual working principle of batteries can be better interpreted by understanding their fundamental characteristics and parameters. This section will define the basic terminologies, which are very frequently used in batteries and fuel cells, such as the anode, cathode, the theoretical potential of the battery, the theoretical capacity of the battery, specific capacity, theoretical energy, specific energy, energy density, coulombic efficiency, and C-rate.

12.3.1.1 Anode and Cathode

The anode, also called the negative electrode, is the electrode that undergoes an oxidation reaction during the discharging process of an electrochemical cell or battery. The cathode, also known as a positive electrode, works as the electrode in a cell or battery, where the reduction reaction occurs during the discharging process.

12.3.1.2 Theoretical Voltage

The voltage of the battery can be theoretically described on the basis of the electrode's redox potentials and the activity of the electrodes' material. It is also known as standard cell voltage and can be measured using the standard voltage of the individual electrodes employed in the fuel cell, as shown in Equation (12.5). Factors that influence the cell voltage include concentration, temperature, etc.

Cathode standard potential − Anode standard potential = standard cell potential (12.5).

12.3.1.3 Theoretical and Specific Capacity

The theoretical capacity (C_{th}) can be defined as the cell's maximum capability to store electric charges by utilizing the overall quantity of the electrode's active materials. The unit of capacity is Coulomb (1 Coulomb = 1 Ampere-hour). The specific capacity (C_{sp}) is defined as the fuel cell's capability to store charge per unit mass of the electrodes' active material. It can be determined by using the following equations.

$$c_{sp} = \frac{nxF}{M_w} = \frac{nx\,96485 \left[\frac{As}{mol}\right]}{M_w \left[\frac{g}{mol}\right]} \tag{12.6}$$

$$C_{sp} = \frac{nx26801}{M_w} \left[\frac{mAh}{g}\right] \tag{12.7}$$

In the above equation, n refers to the certain electrons transmitted in a reaction, F refers to the Faraday constant, and M_w is the molar weight of active material used in the electrode. In other cases, the specific capacity of the cell can also be measured by using the specific capacity of the particular electrodes, as shown in Equation (12.8).

$$Csp_{cell}^{-1} = Csp_{anode}^{-1} + Csp_{cathode}^{-1} \tag{12.8}$$

12.3.1.4 Theoretical and Specific Energy

Another important parameter of any cell is theoretical and specific energy. Theoretical energy (E_{th}) is the highest energy that can be produced by the cell with a theoretical capacitance and voltage, as shown in Equation (12.9). Its unit is Watt-hour.

$$E_{th} = C_{th} \times V \tag{12.9}$$

Similarly, specific energy (E_{sp}) is the highest energy of the cell that can be produced by the unit mass of the active material (Equation 12.10). Its unit is Wh/g.

$$E_{sp} = \frac{E}{m} \tag{12.10}$$

Energy density is another important term frequently used to describe cell characteristics. The energy density (E_{den}) is the energy produced per volume (Vol) of active material. The equation for calculating E_{den} is displayed in Equation (12.11).

$$E_{den} = \frac{E}{Vol} \tag{12.11}$$

There is a huge variation between the theoretically obtained value of the discussed parameters and the actual practical values. These variations come from the addition of the materials, which do not contribute to the reaction and are added for other purposes, such as current collectors, binders, and conductive additives. In addition to these fabrication losses, losses due to incomplete charging of the active material, polarization effects, and depth of discharge also play a critical part in minimizing the E_{den} of the cell. In practice, only 25% to 30% of the total theoretical energy can be generated from the cell if it runs at its optimum level [48].

12.3.1.5 Coulombic Efficiency, C-Rate, and Current Density

The coulombic efficiency, also known as Faradaic efficiency, is the ratio of the overall charge derived from the battery to the overall charge transferred into the battery during a complete cycle. In other words, it is the ratio discharge and charge capacities of the cell.

$$\eta_C = \frac{C_{discharge}}{C_{charge}} \tag{12.12}$$

The C-rate is defined as the ratio of current required for charging to current obtained at discharging of the cell. It can be described by the equation below.

$$C = \frac{i_{applied}}{i_{1h}} \qquad (12.13)$$

Here, i_{1h} refers to the current required to fully charge the cell within 1 hour based on the theoretical capacity. Thus, 1C C-rate represents the complete charge/discharge in 1 hour. The current density is one of the most frequently used parameters in any electrochemical cell. It can be explained as the current that flows per cross-sectional area of the electrode. Generally, materials with large surface areas are employed in batteries and electrochemical cells to lower current density out of the provided overall operating current.

12.3.2 TYPES OF BATTERIES

The fundamental operating principle of any battery and the electrochemical cells is almost the same: converting chemical energy into electrical energy. The reaction that governs this transformation of energies is called a redox reaction. A redox reaction is a combination of reduction and oxidation reactions carried out at separate redox-active electrodes, positioned in an electrolytic solution, and connected through the external circuit. One electrode (anode) undergoes an oxidation reaction that produces electrons and ions. These electrons are forced to move through the external circuit to the electrode (cathode), where the reduction reaction occurs. The ions are moved via the electrolyte to achieve charge equilibrium. Therefore, high conductivity of ions is required for the faster completion of the reaction. Consequently, electronic movement through the electrolyte must be restricted to prevent a short circuit internally.

The operation function in any electrochemical cell remains the same. For example, the galvanic cell is the simplest known electrochemical cell comprising two electrodes and a corresponding electrolyte with the same functions, as described above. The galvanic cell employs a salt bridge or a semipermeable membrane for the ions transfer, as shown in Figure 12.1. Generally, three basic elements are always present in an electrochemical cell: the anode, cathode, and electrolyte. In addition to this, some of the cells also employ a porous membrane for the separation of two electrodes. The electrochemical cells can be categorized based on their redox reaction reversibility. Cells that cannot be recharged are known as primary cells. The redox reaction in the primary cells cannot be reversible. Therefore, they

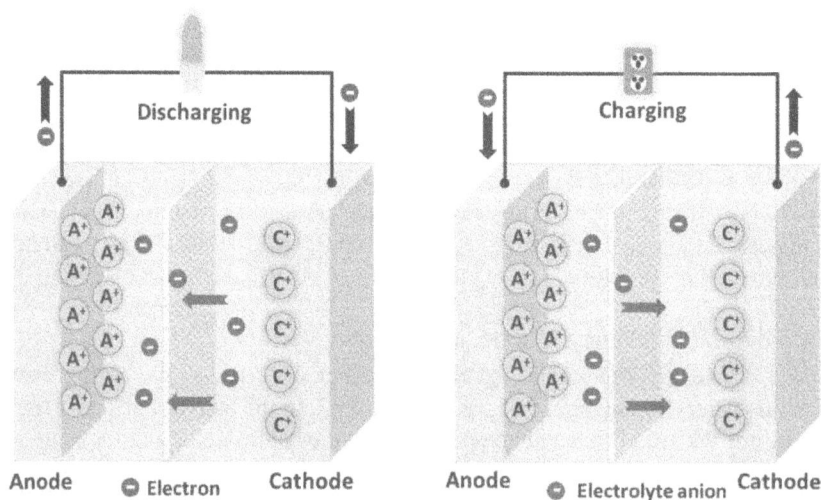

FIGURE 12.1   Schematic illustration of discharging and charging of a metal-based secondary battery.

are one-time-use cells only and are discarded after discharge. In contrast, secondary cells hold redox reversibility, enabling them to be recharged repeatedly after discharge. For recharging the cell, each electrode must be meet redox reversibility criteria. Recharging can be done by plug the external power source in the opposite direction to push the reaction backward and achieve its charged state. The number of charge/discharge cycles depends on the stability of electrodes and the reversibility efficiency of the cell. Several types of secondary cells have been studied. In this section, types of secondary cells based on their material and housing methods will be discussed.

12.3.2.1 Cells Based on Different Materials

12.3.2.1.1 Metal-Based Secondary Cell

Metal-containing secondary cells consist of two metal electrodes made up of different metals such as copper and silver. The electrodes are dipped into the electrolyte, which contains respective metal ions. The electrodes are generally separated by a porous, permeable membrane that allows ions to transfer and restrict electrons transfer through the electrolyte. When the circuit is open, charges are developed at the electrodes' surface to achieve the electrode-solution equilibrium. When the circuit is closed, the potential difference between the electrode acts as a driving force that moves electrons from the anode via the external circuit to the cathode. The electrons movement disturbs the equilibrium, which equilibrates through a redox reaction, creating electrons and A+ ions that relocate to the cathode. These electrons moved to the cathode, resulting in increased electron density, which is balanced by C+ ions' reduction at the cathode. A steady flow of anions and cations continues to their respective electrode, maintaining charge equilibrium throughout the process. The process continues until the anode is depleted, or the cathode uses all the C+ ions. The ion transfer plays a vital role in cell efficiency; therefore, a high ionic conductive electrolyte is required. For recharging, the cell is connected in the opposite way, reducing the anode, oxidizing the cathode, and restoring its original state.

12.3.2.1.2 Li-Ion Battery

LIBs possess high energy density, enabling them to be used in various applications, such as mobile phones, watches, laptops, and electric vehicles; it follows the same working principle as described above. The anode is composed of graphite-containing Li atoms. In comparison, the cathode is comprised of Li-ion-based material, like lithium cobalt oxide. The anode gets oxidized and produces Li-ions and electrons. The electrons transfer through the external circuit, while the Li-ions move through the electrolyte toward the cathode, which eventually converts into $LiCoC_2$, as displayed in Figure 12.2. The charging process remains the same as described in Section 12.3.2.1.1. The Li intercalation technique provides a major benefit in the development of LIBs. It restricts the creation of metallic Li and dendrites because of less mobility and enhanced separation of Li-ions in the graphitic structure. The main drawback of LIBs is the formation of some layered structures at the electrode during the intercalation of Li-ions, which results in sluggish reactions and heat emission. This drawback hinders the application of LIBs in a high-performance application.

12.3.2.1.3 Organic Batteries

Unlike LIBs, organic batteries provide better redox reactions, prevent any morphological variation, and offer an enhanced reaction rate. Like other batteries, organic batteries consist of two oppositely redox-active electrodes connected through an external circuit and dipped in the electrolyte, as shown in Figure 12.3. Organic materials that possess reversible redox characteristics can be applied as the electrode in organic batteries. For organic materials, the redox reaction takes place on the basis of the alteration in the charge state of the organic molecules. So, organic materials are classified into three main categories based on their reaction pathway, such as p-type, n-type, and b-type. The p-type materials undergo oxidation reactions that produce cations and are applied as anode; the n-type materials are reduced to produce anions and are employed as a cathode. Both oxidation and

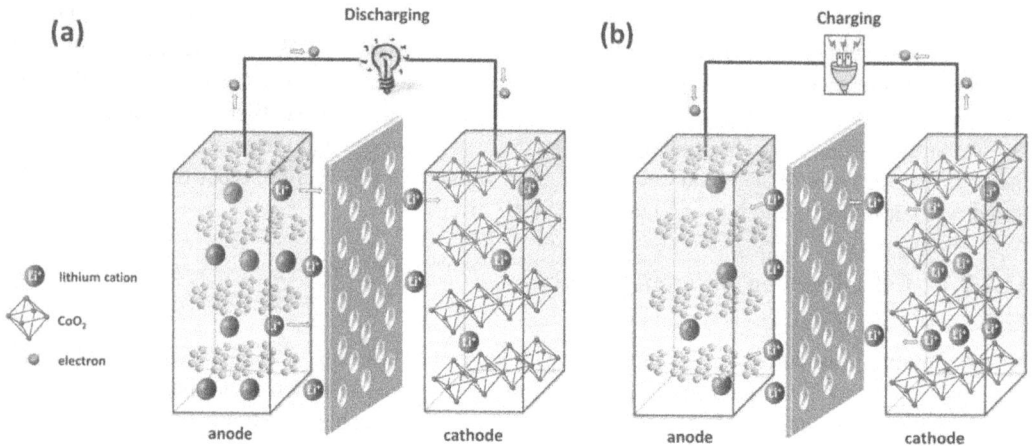

FIGURE 12.2 Schematic illustration of (a) discharging and (b) charging of Li-ion battery. (Adapted with permission from Ref. [49]. Copyright (2016) American Chemical Society.)

reduction reactions can be carried out with b-type materials and used as both anode and cathode. A permeable membrane is used as an electrode separator.

Firstly, the cell is charged by oxidizing the cathode and reducing the anode with the help of applied current. The created charges are balanced by the ions present in the electrolyte. The charging limitation includes the cathode's full oxidation, full reduction of the anode, or complete salt consumption in the electrolyte. The electrodes are connected to the outer circuit at discharging, allowing the electrons to move from anode to cathode. Similarly, the ions transfer from the polymer into the electrolyte. The reaction rate in organic batteries is restricted due to the electrolyte ions transfer and standard rate constant for the electrons transfer (k^0) [50]. Normally, the value of k^0 for the application of electrodes is in the range of 10^{-1} cm/s [51, 52]. Higher N^0 values are good for organic batteries because of their fast charge/discharge characteristics. The ionic conductivity and the presence of counterions can be improved by increasing the salt concentration. But, excessive concentration can also result in low ion movement due to increased viscosity. Additionally, the organic electrode's amorphous morphology possesses fast ionic movement, especially when

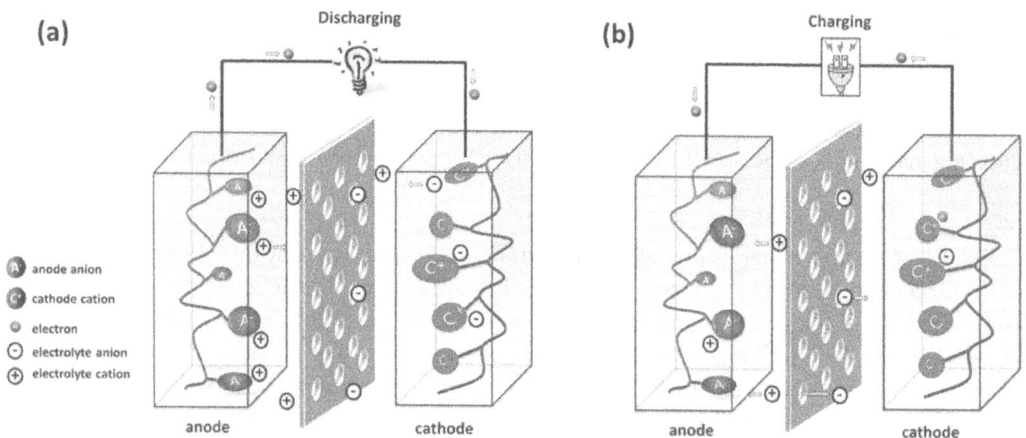

FIGURE 12.3 Diagrammatic illustration of (a) discharging and (b) charging of the all-organic battery. (Adapted with permission from Ref. [49]. Copyright (2016) American Chemical Society.)

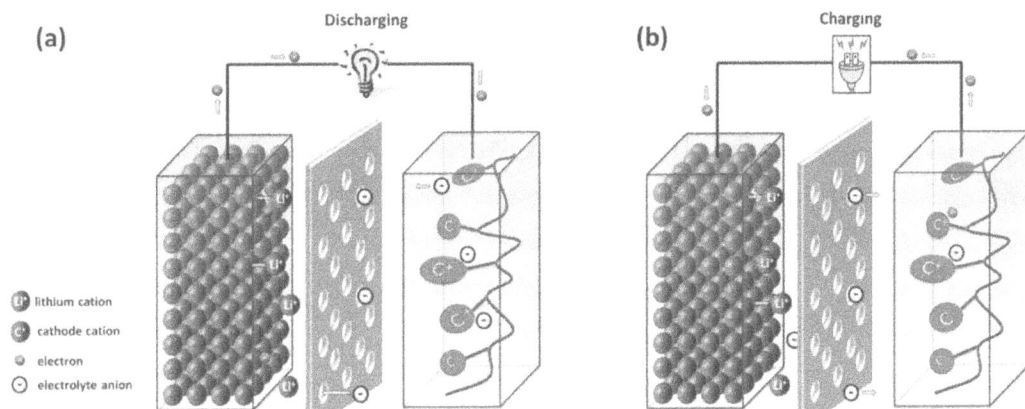

FIGURE 12.4 Diagrammatic illustration of (a) discharging and (b) charging of the metal-organic battery. (Adapted with permission from Ref. [49]. Copyright (2016) American Chemical Society.)

discharging and charging. During this process, excellent apparent diffusion co-efficient values are obtained, which is about 10^{-8} cm^2/s.

12.3.2.1.4 Metal-Organic Battery

When it comes to comparative testing of the organic electrodes, an all-organic electrode battery is not applied. Instead, the metal-organic system is used to test the organic electrode's performance, as shown in Figure 12.4. A metal electrode, such as lithium, is used as the anode, and an organic cathode, such as n-type and p-type, is employed in the battery. The electrolyte in such batteries must contain cations of the corresponding metal electrode. Li is used as the anode due to two reasons. First, Li has a small redox potential, which allows the organic cathode to utilize most of the potential. Second, in the case of n-type material, a small Li$^+$ radius helps in the fast migration of ions to the cathode. During the charging of p-type material, the cations are reduced and deposited at the metal anode. At the same time, the p-type material is oxidized. The discharging is the reverse of the process described above. When the n-type organic material is employed, the basic process remains the same. However, in this case, the battery stays charged after the fabrication, and the discharging takes place by reducing the cathode and oxidizing the anode.

12.3.2.2 Cells Based on Housing

Various housing techniques are used to fabricate the cell. The cells for testing the electrode materials are usually done with a small quantity of materials. Different housing allows the best setting to test a particular material. Some types of housing are discussed below.

12.3.2.2.1 Coin Cell

Coin cell is a common type of test cell housing. It requires a small amount of electrode materials. The porous membrane is used between the electrodes, and the electrodes are pushed against each other with springs, as shown in Figure 12.5(a). This simple method enables fast cell fabrication and testing.

12.3.2.2.2 Swagelok Cell

This type of cell housing is similar to the coin cell and has few advantages. Unlike coin cells, Swagelok cells can be reused and offer easy dismantling. This feature helps scientists to carry out morphological studies before and after the charge/discharge cycles. Furthermore, reusability provides environmental and economic benefits. Swagelok cells can be modified to enable more testing

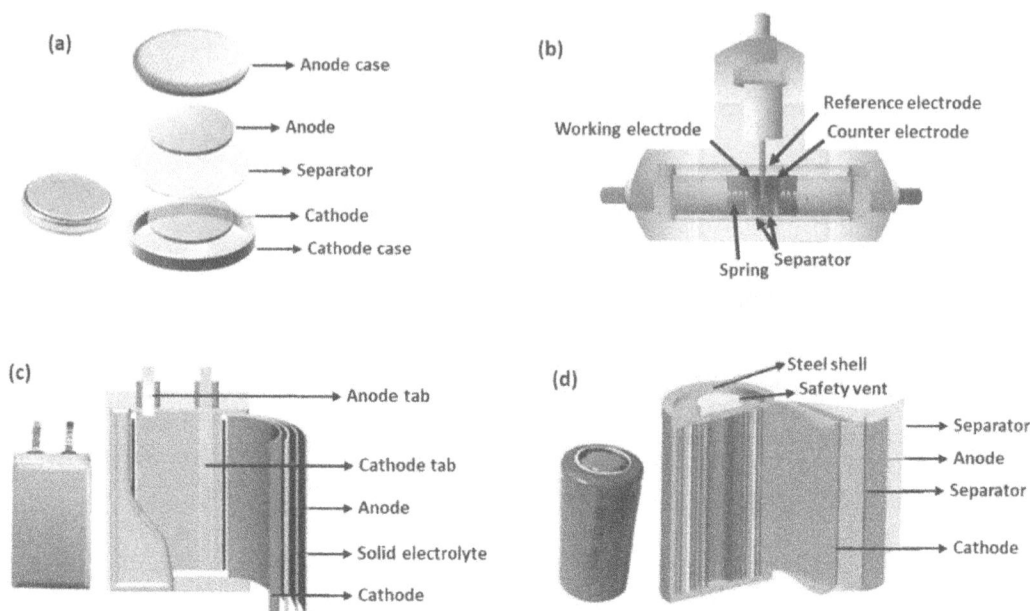

FIGURE 12.5 Schematic illustration of (a) coin cell, (b) Swagelok cell, (c) pouch cell, and (d) cylindrical cell. ((a), (c), and (d) are adapted from Ref. [54]. The article was published with CC-BY license; (b) is adapted with permission from Ref. [49]. Copyright (2016) American Chemical Society.)

advantages, such as Swagelok T-cell. Swagelok T-cell is a three-electrode cell used to carry out more complex studies (Figure 12.5b).

12.3.2.2.3 Pouch Cell

The simple pouch cell was introduced in 1995 and offered 90% efficient packing [53]. In pouch cells, several layers of cathode and anodes are packed together with foils in between like a sandwich (Figure 12.5c). The foils are used to avoid direct electrode contact. The current from each electrode is collected by attaching metal collectors, usually copper and aluminum, for the anode and the cathode, respectively. The piles of electrodes are packed in an aluminized plastic box.

12.3.2.2.4 Cylindrical Cell

Cylindrical cells are mostly used in commercial applications because of their good mechanical stability. The electrodes are separated by a porous membrane, rounded, and packed in a cylinder, as shown in Figure 12.5(d).

12.4 ELECTROCHEMICAL METHODS FOR BATTERY TESTING

The electrochemical technique of testing batteries involves the potentiostat, which can be connected in two ways: two-electrode cell or three-electrode cell. These three electrodes include the reference electrode (RE), counter electrode (CE), and the working electrode (WE). The potentiostat is an electronic device that evaluates and regulates the potential difference between the WE and RE and evaluates the current flow between the WE and CE. The schematic illustration of two and three-electrode cell connections of the potentiostat are displayed in Figure 12.6. The two basic approaches to test the battery are cyclic voltammetry (CV) and galvanostatic charge-discharge (GCD). The CV proceeds in the controlled voltage, while the GCD proceeds in the controlled current. Both CV and GC are essential in the battery characterization, but the GCD is more preferred due to its ability to show crucial capacities

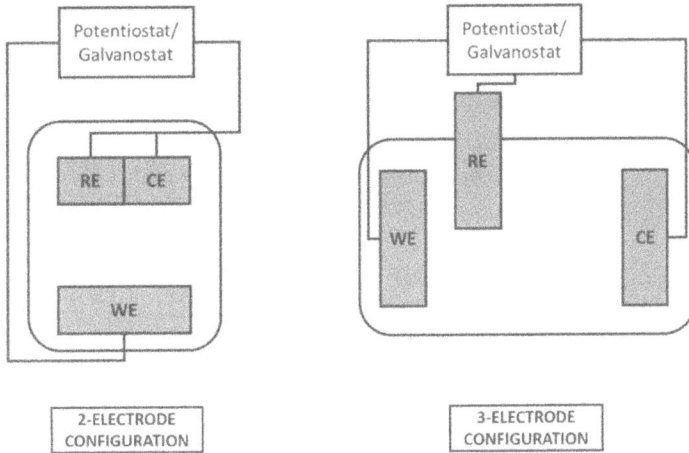

FIGURE 12.6 Schematic illustration of a potentiostat connection in (a) two-electrode cell and (b) a three-electrode cell configuration. (Adapted from Ref. [55]. The article was published with CC-BY license.)

within the small potential window. Furthermore, GCD has superior preciseness in calculating the coulombic efficiency and capacity. Mostly, in a three-electrode arrangement, the RE voltage stays constant throughout cycling, which makes both the GCD and CV approaches reliable.

Another important parameter is the ohmic drop or most commonly known as IR drop. The IR drop in the electrolyte occurs between the WE and RE, which emerges because of the three variables. These variables include electrolyte ionic conductivity, the gap between the WE and RE, and current passing through WE. The first couple of variables remain constant for any electrochemical setting, but the current changes in the CV approach and remain constant in the GC approach. As a result, the IR drop only moves the entire GC curve and significantly affects the CV curve structure. Therefore, a high unbalanced IR drop can result in a misunderstanding of the electrochemical procedure.

Electrochemical impedance spectroscopy (EIS) is a characterization approach that has extensively been utilized for testing batteries. It involves the application of current or voltage disturbance of different frequencies to the cell. The disturbance should be small enough for the cell to respond linearly, which will be the sinusoidal wave at a similar frequency but with specific phase shifts. As the result of this phase change, the current and voltage ratios become a complex impedance, which is stated as Bode or Nyquist plot. Impedance spectroscopy is an essential characterization that is used to figure out various processes happening in the cell, like mass transport in the electrolyte or through solid-electrolyte interphase, charge transport, or diffusivity. Thus, they significantly participate in the electric conductivity according to its frequency. There are two types of EIS approaches that can be employed to obtain the correct impedance of the battery, which include potentiostatic EIS (PEIS) and galvanostatic (GEIS). Mostly, PEIS is considered more suitable because it provides a wide range of current and shifts itself to the most suitable one to carry out the measurement. While for the GEIS, some idea about the range of impedance is necessary to choose the adequate current. There are some severe cases, such as high and low impedance, where PEIS and GEIS are preferred, respectively. Except for these cases, both methods, such as PEIS and GEIS, must be equivalent.

12.5 MATERIALS FOR BATTERIES

In this era of technology, rechargeable batteries are considered an effective technique for energy storage and conversion that is gaining great attention. Much work has been poured into designing improved batteries with higher energy and power capacity, large capacity, long cyclic stability, quick response, and reduced price. The functionality of rechargeable batteries relies on the active materials used in cathode and anode and on the battery components' integration technologies [56].

The demands of advanced batteries have inspired the scientific community to develop rechargeable battery systems and improve existing electrode materials and electrode reactions. Enhancing electrical materials focuses primarily on (a) large battery capacity, improved energy/power density, (b) the high reactiveness, reversibility, and framework durability during the charging/discharging cycles, (c) fast diffusion of ions and electron transfer at elevated charging/discharging; and (d) price-effectiveness, high protection, and environmental friendliness [57]. Followings are some of the commonly used materials for energy devices.

12.5.1 Conducting Polymers

The polymeric materials are comprised of lengthy repeating units of molecules and possess exceptional characteristics, depending on the kind of molecules and the way these molecules are bonded. CPs are materials that conduct electricity due to the delocalization of π-electrons and are thus also known as conjugated polymers [58]. These materials are either metallic or semiconducting. Like insulating polymers, CPs also exhibit superior electrical conductance but have different mechanical characteristics than commercial polymers. Some CPs, including polythiophene, polypyrrole, polyaniline, polyacetylene, and poly(3,4-ethylene di oxyethylene), have been produced in recent years [59]. These CPs have specific characteristics, including high optical properties, high electrical conductivity, good magnetic properties, structural diversity, mechanical flexibility, often lightweight and efficient microwave absorbing properties [60]. CPs are grouped into three categories based on method and process: ex situ fabrication, in situ fabrication, and one-pot fabrication. During ex situ fabrication, the formation of CPs is done separately, and by blending two or more different components, the composite is obtained. Furthermore, during in situ fabrication, one ingredient of the material is prepared in attendance of another one. The essential benefit of this technique is the interfacial control obtained at the molecular level, and it helps in achieving the synergistic effect. In the one-pot method of synthesis, the monomer and other components contact each other and make a composite in a single step. CPs are frequently used as cathodes in advanced LIB systems and have benefits like good processability, cost-effectiveness, and convenience in modifying their chemical framework [61]. Except for these advantages, CPs cathodes may have some disadvantages, including inferior cyclic stability and poor conductance in the reduced state. To resolve these deficits, inorganic metal oxides, carbon materials, and various 2D materials are incorporated with CPs and form nanostructured hybrid electrode materials [58]. CPs interact gradually with inorganic materials and provided a conducting backbone, electrical conductivity, and plastic property.

The nanocomposites based on CPs and 2D materials have characteristics of both materials having a synergistic effect owing to the smooth charge transport through them. Furthermore, CPs not only help in charge transmission between 2D nanosheets but they also refrain those 2D nanosheets from being stacked again. Due to the high surface area that exposes an enormous amount of atoms at the surface, 2D materials interact with the matrix, and very few amounts of them cause a remarkable increase in the performance of CPs. 2D materials like graphene, graphene oxide (GO), reduced graphene oxide (rGO), transition metal dichalcogenides (TMDCs), and hexagonal boron nitride act like an appropriate filler for polymeric composites for different application purposes [62].

12.5.2 Graphene and Composites

Graphene and graphene-CPs composites significantly improve the electrochemical properties of CPs, having a major effect on the morphologies, electrical characteristics, and structural stability of CPs. In situ polymerization and melt-intercalation are the two main techniques for synthesizing polymer composites with graphene. For the synthesis of graphene-containing polymer composites, the most commonly used method is in situ polymerization, which produced various nanocomposites such as epoxy/graphene, poly(vinylidene fluoride) (PVDF)/graphene, and poly(vinyl alcohol) (PVA)/graphene. [63]. Furthermore, the melt intercalation technique employs elevated temperature and high shear force to mix graphene and polymer material in the liquefied form. Two significant factors that affect this technique are the molecular weight of the polymeric material and the interaction pattern of the 2D materials and the polymeric material. In situ oxidative polymerization was used to create graphene/PPy composites. At a scan rate of 10 mV/s, these composites had the highest capacitance and energy density of 409 F/g and 227 Wh/Kg, respectively. However, at a scan rate of 200 mV/s, the maximum power density obtained was 4617 W/kg [64]. GO comprises oxygenated groups like hydroxyl groups, epoxides, and carboxyl functional groups on its basal planes and edges and has excellent compatibility with polymers. It has been commonly used for the construction of CP-based composites due to their extraordinary morphology and strong hydrophilicity. A hybrid composite of RGO oxide/polypyrrole and Ni-Co layered double hydroxides (RGO/PPy/NiCo-LDH) was synthesized easily [65]. The polypyrrole embedded in the 2D nanosheets enhanced the surface area and the conductance of the nanocomposite. The electrode showed an elevated specific capacitance of 2534 F/g at 1 A/g with excellent cyclic efficiency of 78% after 5000 cycles.

Similarly, highly active cathode materials for biocompatible zinc/polymer batteries have also been produced by employing biofluids as electrolytes [66]. The PPy fiber/rGO cathode and Zn anode battery exhibited a superior energy density of 264 mWh/g in 0.1 M phosphate buffer saline. Furthermore, silicon/graphene/polyaniline (Si/G/PANI)-based composite material has been prepared via in situ polymerization. The PANI sheet is inextricably attached to the graphene created an enhanced conjugated π linkage among PANI and graphene-attached Si particles [67]. The schematics of the synthesis process and π linkage among PANI and graphene-attached Si particles are shown in Figure 12.7. A 3D composite material containing graphene/polyaniline/polyoxometalates

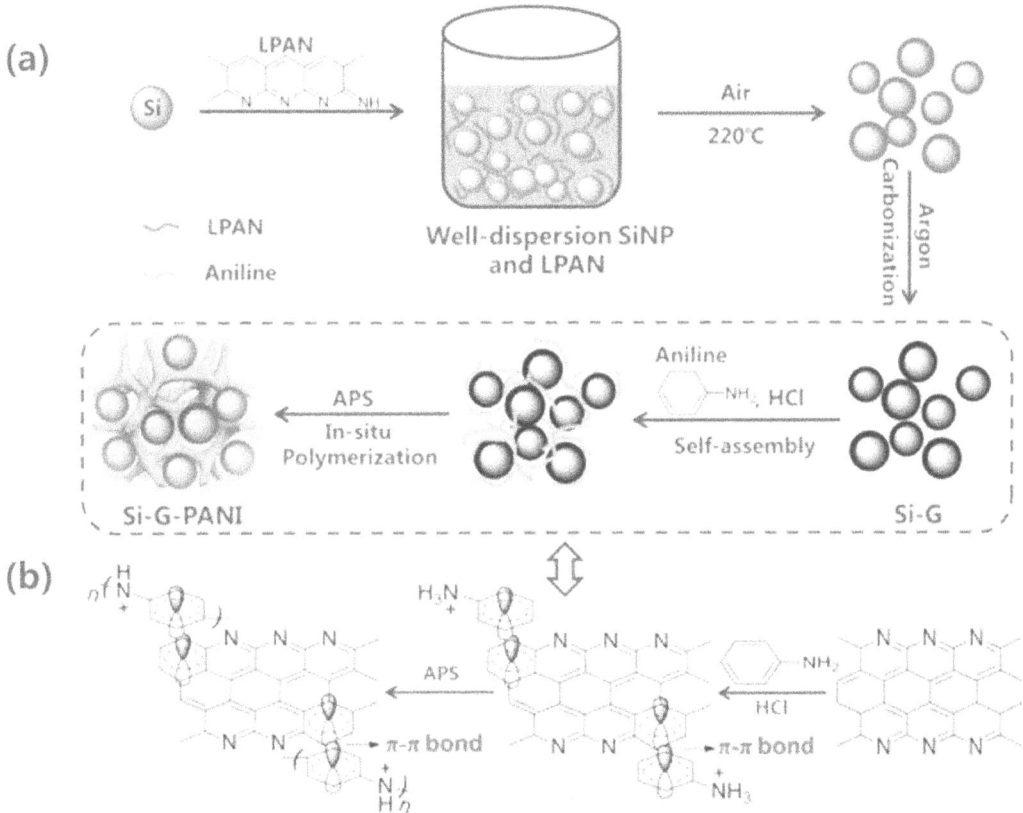

FIGURE 12.7 The schematics of the synthesis process and π linkage among PANI and graphene-attached Si particles. (Adapted with permission from Ref. [67]. Copyright (2016) Elsevier.)

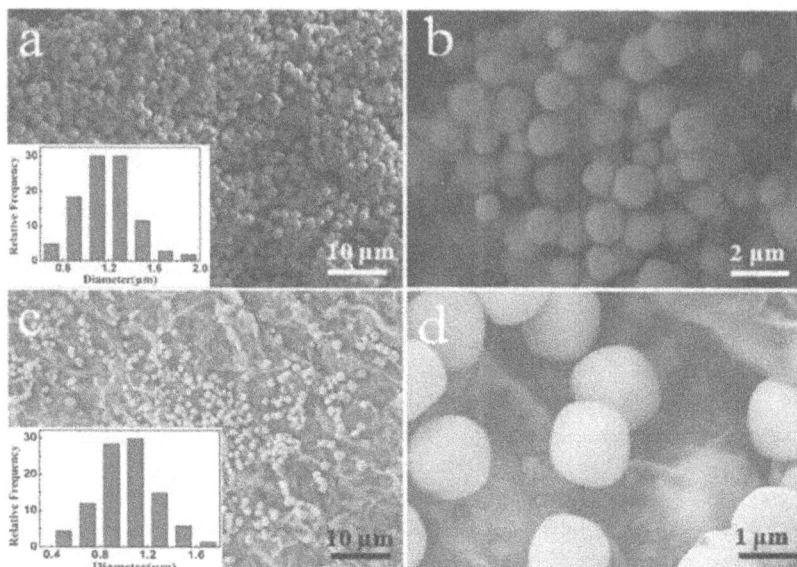

FIGURE 12.8 SEM micrographs of (a–b) PANI/PW$_{12}$ nanospheres, (c–d) rGO@PANI/PW$_{12}$ hybrids. (Inset: Statistical study revealing the size of the spheres.) (Adapted with permission from Ref. [68]. Copyright (2017) Elsevier.)

(rGO@PANI/PW$_{12}$) homogeneous nanospheres, having a mean diameter of 1.1–1.3 μm, has been synthesized to be utilized as a cathode in LIBs through a single-step hydrothermal method. Figure 12.8 displays the surface structure of the rGO@PANI/PW$_{12}$ [68]. The as-synthesized rGO@PANI/PW$_{12}$ composite in half-cell exhibited excellent electrochemical activity with a superior specific capacitance of about 285 mAh/g at 50 mA/g. It also showed an exceptional rate capability of 140 mAh/g at 2 A/g. A good cyclability with the capacity deterioration rate of 0.0280% per cycle even after 1000 cycles at 2 A/g were also observed during its characterization. In Table 12.1, some of the electrochemical properties of graphene-based materials are compared.

12.5.3 MoS$_2$/Polymer Composites

The 2D molybdenum disulfide (MoS$_2$) is a type of transition metal dichalcogenide, having a layered structure analogous to graphene. The S–Mo–S layers in MoS$_2$ are attached due to a weak van der Waals interaction. Being an alternative to graphene, MoS$_2$ can be used to compensate for the shortcomings of gapless graphene. It has various applications in energy storage, energy conversion, powering nanodevices, and flexible and stretchable electronics [74]. It has been reported that MoS$_2$/PANI composite with hierarchical morphology similar to tremella was prepared using a simple polymerization technique [75]. PANI increases MoS$_2$ conductivity while also rebuilding a hierarchically porous Framework, resulting in faster Li-ions transfer during the Li-ions insertion/extraction process. At 0.1 A/g, the MoS$_2$/PANI composite displayed a superior preliminary reversible capacity of 910 mAh/g and a preliminary coulombic efficiency of 80%. Furthermore, after 200 cycles at 1 A/g, the synthesized material demonstrated an excellent capacity of 915 mAh/g, demonstrating its possible applications in LIBs. Figure 12.9 showed the CV diagrams of the MoS$_2$/PANI composite [75].

Furthermore, the synthesis of MoS$_2$@polypyrrole core-shell microspheres is accomplished in two steps. The MoS$_2$ microspheres were wrapped in conductive polypyrrole coating and demonstrated an excellent specific capacity of 1012 mAh/g after 200 cycles at 200 mA/g. Even at 4000 mA/g, it provided a high-rate output of 600 mAh g^{-1}. An in situ polymerization technique was

TABLE 12.1

Different Graphene and Graphene Oxide-Based Composite Materials and Their Performances

Composition	Method	Electrolyte	Energy Density (Wh/Kg)	Scan Rate (mV/s)	Stability	References
Graphene-polypyrrole	In situ oxidative polymerization	1 M KCl	227.2	10	–	[64]
RGO/Ag/PPy nanocomposites	In situ polymerization	1 M H$_2$SO$_4$	36.92	4	66% after 1000 cycles	[69]
RGO/PPy/ NiCo-LDH	Modified Hummers	6 M KOH	41.9	10	78% after 5000 cycles	[65]
Polypyrrole/ reduced graphene oxide (RGO)	One-step chemical polymerization	0.2 M Phosphate-buffered saline (PBS)	264	5	–	[66]
ZnS/reduced graphene oxide/ polypyrrole	In situ polymerization	1.0 M LiPF$_6$	80.0	–	151% after 5000 cycles	[70]
Si-G-PANI	In situ polymerization	1 M LiPF$_6$	1200	0.1		[67]
Graphene/ polyaniline/ polyoxometalate	In situ anodic electropolymerization	1.0 M LiPF$_6$	–	–	0.028% decay per cycle	[68]
Polyaniline/ graphene/ZrO$_2$	Facile polymerization	1 M H$_2$SO$_4$	104.76	2	93% after 1000 cycles	[71]
Polystyrene/ polyaniline/ graphene	Emulsifier-free emulsion polymerization	1 M H$_2$SO$_4$	49	0.5	74% retention after 5000 cycles	[72]
Zn-PANi/G	Electrochemical polymerization	1.0 M ZnCl$_2$ and 0.5 M NH$_4$Cl	162.3	5	Capacity loss 0.15% per cycle	[73]

used to prepare poly(3,4-ethylene dioxythiophene)/molybdenum disulfide (PEDOT/MoS$_2$) intercalated composites. With increasing MoS$_2$ fraction, PEDOT/MoS$_2$ composites showed improvement in thermal stability, conductivity, and capacitance. The specific capacitance obtained was 405 F/g, which is approximately four-folds of a PEDOT electrode, with a capacity retention of approximately 90% even after 1000 cycles [76]. Hierarchical MoS$_2$/PANI nanoflowers having a diameter between 300 and 700 nm were successfully produced via a simple hydrothermal technique and polymerization [77]. The schematics of the sample preparation and SEM images are shown in Figure 12.10. MoS$_2$/PANI sample exhibited remarkable cyclability and high reversible specific capacitance of 888 mAh/g after 50 cycles. These MoS$_2$/PANI nanoflowers have a high specific area and large empty rooms beside each adjacent nanosized plate resulting in better electrochemical performance [77]. Some of the electrochemical properties of MoS$_2$-based composite are compared in Table 12.2.

12.5.4 h-BN and Its Composites

Two-dimensional h-BN contains a graphene-like morphology with high thermal conductivity, mechanical robustness, low dielectric constant, and poor electrical conductivity. Dangling bonds and surface charge traps are also lacking in h-BN. It is a potential candidate for applications that required

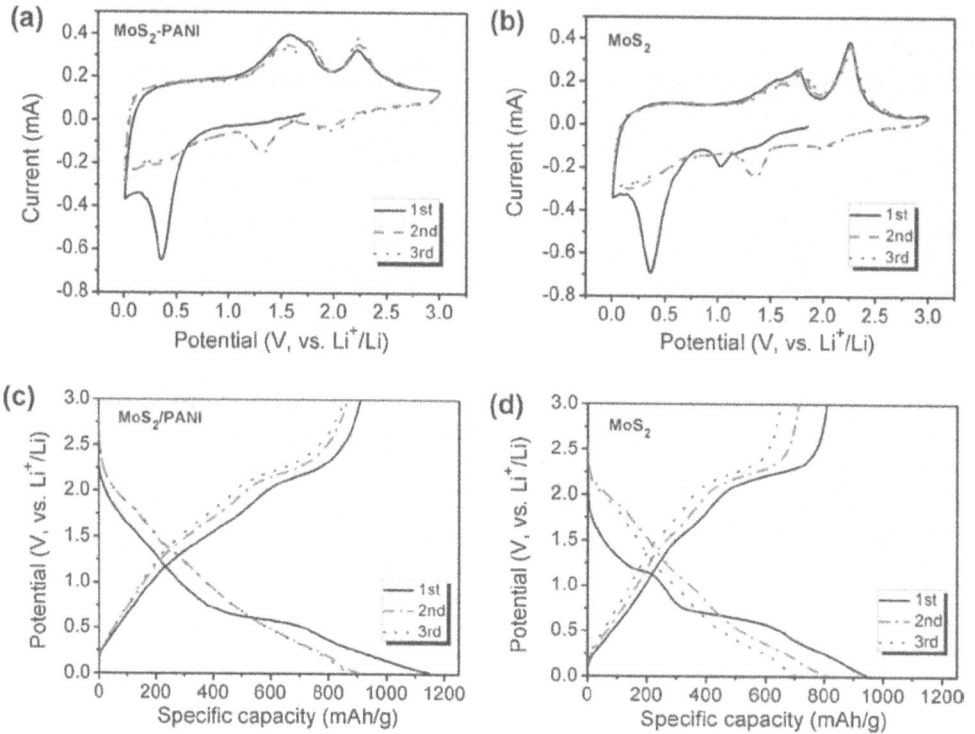

FIGURE 12.9 (a) Cyclic voltammetry diagrams of the MoS_2/PANI composite; (b) initial three cycles estimated at a scan rate of 0.1 mV/s for pure MoS_2; (c) initial three voltage–capacity diagrams for MoS_2/PANI composite; (d) first three voltage–capacity diagrams for pure MoS_2 at a current density of 0.10 A/g. (Adapted with permission from Ref. [75]. Copyright (2015) Elsevier.)

high thermal conductivity and good insulating properties. Moreover, in a study, h-BN was introduced in polymer matrices like PEs and poly(vinylidene fluoride)hexafluoropropylene for the formation of composite separators (Figure 12.11). The produced material displayed enhanced thermal homogeneity that promoted regular nucleation and development of Li metal while improving electrochemical efficiency [80]. The nanosize h-BN allowed more uniformity in composites with less thickness while still offering a large surface area for solvent/polymer system interaction. Electrospinning was used to create new h-BN/polyacrylonitrile composite nanofibers having an average fiber diameter of approximately 220 nm to apply in rechargeable LIBs. At ambient temperature, the electrolyte-loaded nanofiber exhibited superior ionic conductance of 10^{-3} S/cm, an elevated discharge capacity of 144 mAh/g, and 92.4% capacitance retention. The h-BN/PAN nanofiber electrolytes are proven to be a promising separator material for LIBs due to their high safety and cost-effectiveness [81]. h-BN is also used as a separator and is an important component in the secure operation of rechargeable batteries. Furthermore, traditional polymer-containing separators have reduced thermal durability, resulting in catastrophic thermal runaway failure modes that afflicted LIBs. Phase-inversion composite separators based on carbon-coated hBN nanosheets and PVDF polymers have demonstrated superior porosity, excellent electrolyte wettability, and good thermal durability. The carbon-layered h-BN separators resulted in excellent wettability for a broad variety of liquid electrolytes [80].

12.5.5 METAL-ORGANIC FRAMEWORK-BASED COMPOSITES

Metal-organic frameworks (MOFs) are organic–inorganic hybrid porous materials with a unique crystalline morphology comprised of positively charged metal ions surrounded by organic "linker molecules." MOFs are typically classified into four categories, namely clusters (0D), chains (1D),

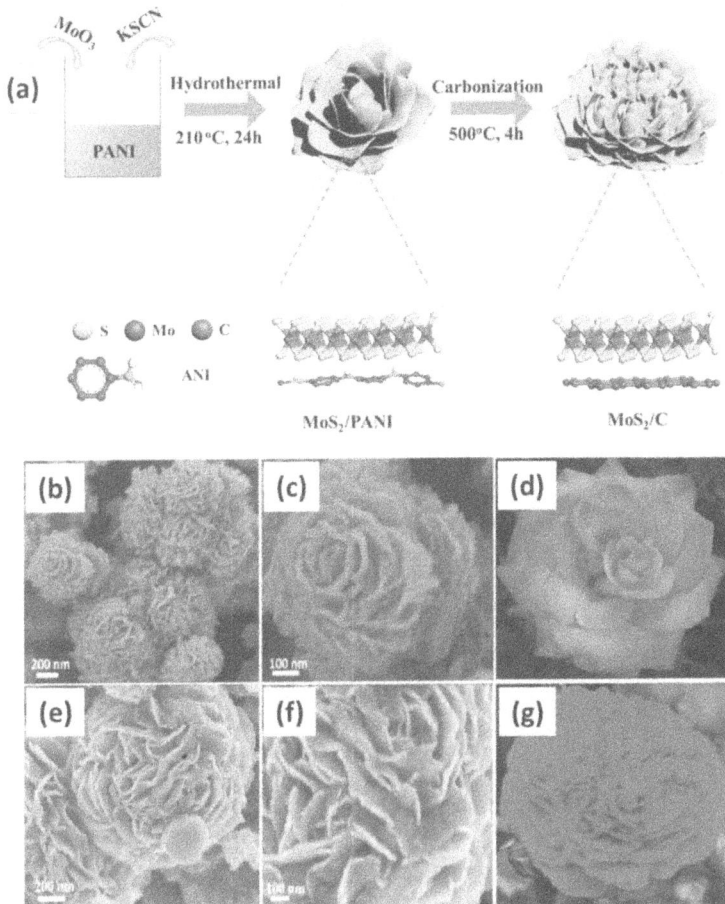

FIGURE 12.10 (a) Schematics of synthesis of 3D hierarchical MoS_2/PANI and MoS_2/C nanoflowers. Morphology of MoS_2/PANI (b, c) and MoS_2/C (e, f) nanoflowers. Pictures of two types of Chinese roses (d, g). (Adapted with permission from Ref. [77]. Copyright (2014) American Chemical Society.)

TABLE 12.2
Various MoS_2-Based Composite and Their Performances

Composition	Method	Electrolyte	Current Densities (A/g)	Reversible Capacity (mAh/g)	Scan Rate (mV)	Stability	References
MoS_2/PANI	Facile polymerization	1 M LiPF$_6$	1	915	0.1	80% after 200 cycles	[75]
MoS_2/PANI	Chemical polymerization	1 M LiPF$_6$	0.1	888.1	0.1	–	[77]
Hierarchical MoS_2/PANI nanowires	In situ polymerization	1 M LiPF$_6$	0.1	952.6	0.1	90% after 50 cycles	[78]
MoS_2@ Polypyrrole	Hydrothermal/ polymerization	1 M LiPF$_6$	0.2	1012	–	–	[79]
PEDOT/ MoS_2	In situ polymerization	1 M H$_2$SO$_4$	–	405 F/g	–	90% after 1000 cycles	[76]

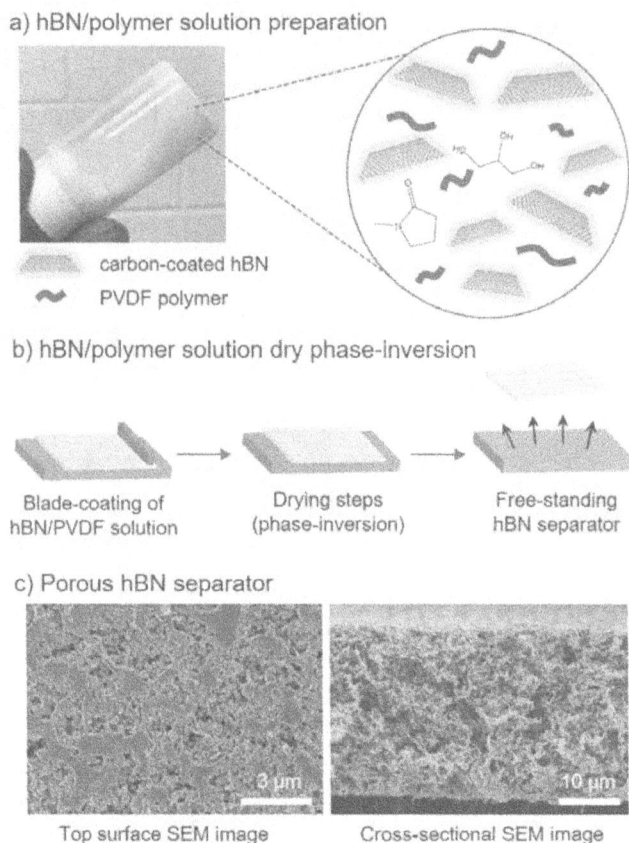

a) hBN/polymer solution preparation

carbon-coated hBN
PVDF polymer

b) hBN/polymer solution dry phase-inversion

Blade-coating of
hBN/PVDF solution

Drying steps
(phase-inversion)

Free-standing
hBN separator

c) Porous hBN separator

3 μm

10 μm

Top surface SEM image

Cross-sectional SEM image

FIGURE 12.11 Preparation of h-BN separator. (a) The hBN/PVDF solution was prepared by dispersing the previously annealed EC-exfoliated hBN nanosheet powder and PVDF polymer in a 95:5 ratio of NMP and glycerol. (b) The hBN/PVDF solution was blade-coated on a glass substrate followed by drying. (c) Top-view and cross-sectional SEM images of a separator. (Adapted with permission from Ref. [80]. Copyright (2020) American Chemical Society.)

layers (2D), and frameworks (3D). 2D MOF is the ultrathin nanosheets that have numerous cavities exposed on their surface, and their thickness is typically lower than 10 nm. MOFs provide good structural diversity in comparison to various other porous materials. Their unique features include uniformity in pore shape, atomic-level structural uniformity, and modifiable porosity, allowing for precise control over framework topology, porosity, and functionality [82]. Polypyrrole/MOF (PPy/MOF) composites were prepared using the polypyrrole solution-based polymerization method. Three MOFs, having different pore shapes of distinct 1D channels (MIL-53), 3D hierarchical nano-cages (MIL-101), and cross-linked pore and tunnels (PCN-224), were prepared. The PPy-MOF composites were employed in lithium-sulfur (Li-S) batteries to ensure proper sulfur containment. The PPy-S-in-MOF electrodes perform better than their MOF and PPy counterparts, particularly at an increased charge/discharge rate. Among them, a PPy-S-in-PCN-224 electrode having cross-linked pores and tunnels showed good capacity around 670 and 440 mAh/g at 10 C after 200 and 1000 cycles, respectively [83]. PBA along with sodium iron cyanide ($Na_2Fe[Fe(CN)_6]$) was used for the confinement of polysulfides through Lewis acid-base bonding along with pore confining effects in high-capacity lithium-sulfur batteries. The large interstitial sites in the as-synthesized PBA structure can host sulfur molecules and polysulfides successfully. The $S@Na_2Fe[Fe(CN)_6]@PEDOT$

composite electrode exhibited a superior specific ability of 1291 mAh/g at 0.1 C current rate and excellent cycling efficiency after the addition of PEDOT [84].

12.5.6 ELECTROLYTES

A traditional LIB contains anode and cathode along with an electrolyte system. The electrolyte works as a medium that offers the ion transport mechanism during charging and discharging processes between the anode and cathode. Therefore, the electrolyte must be in stable form within the duration of battery operation. Electrolytes usually perform two functions; the first is the efficient conduction of ions, and the second is the electrode separation in order to avoid short-circuiting. Electrolytes are typically supposed to be liquids like water with dissolved salts and acids, which are essential for efficient ionic conduction. Also, batteries along with the traditional (AA/AAA/D) batteries, comprised solid electrolytes (SEs), which act like ionic conductors even at room temperature. However, leakage, poor compatibility, and safety threats are few drawbacks associated with the liquid electrolyte [85]. The polymer electrolyte (PE) replaced the liquid electrolyte that prevails in commercialized batteries in the present time. These electrolytes are also vital for the advancement in currently used electronic devices, including superior energy density batteries, fuel cells, and electrochromic displays. Polymer electrolytes act as a separator as well as an electrolyte in solid-state configuration. Polymer electrolytes of the first generation that used the poly(ethylene oxide) (PEO)/LiX system only delivered a quite low ionic conductance of the order 10^{-8} S/cm at ambient temperature. But for practical applications, the ion conductivity must be more than 10^{-3} S/cm [86]. For the last three decades, the main focus of the research and advancement in polymer electrolytes was to improve ion conductivity. The polymer electrolytes are primarily categorized into three main types, namely liquid polymer electrolytes (LPEs), gel polymer electrolytes (GPEs), and solid polymer electrolytes (SPEs).

Liquid electrolytes comprise lithium salt that is dissolved in an organic solvent and a separator to avoid electrode short-circuiting. Typically used lithium salts are $LiClO_4$, $LiPF_6$, and $LiBF_4$. Additionally, liquid electrolytes are further classified into two types based on the solvent used: non-aqueous liquid (NALE) and aqueous liquid electrolyte (ALE). A novel type of liquid electrolyte is ionic liquids (ILs) which are pure organic salts similar to materials and occur at room temperature in liquid form owing to weak coordination of ions or low packing of atoms [85]. GPEs, on the other hand, belong to the third generation of PEs and comprise an inactive polymer structure having dissolved low molecular weight compounds and a lithium salt. According to their preparation method, gels have been categorized into two types, i.e., physical gels and chemical gels. GPEs provide various advantages in terms of shape availability, increased charge-discharge rate, lightweight, low cost, and high power density. SPEs are formed by dissolving salt in a polymer matrix containing an electron donor group and have several benefits such as design flexibility, miniaturization feasibility, safety, ease in device fabrication, and so on. Polymer electrolytes, which are made by dissolving salt in a solvent, serve as ions transfer media (organic or inorganic water). Polar solvents like water, amide, and others are adopted for improved solvation because they have a high dielectric constant. PVDF/graphene gel polymer electrolyte was successfully produced via the non-solvent induced phase separation (NIPS) method. The findings revealed that graphene disperses uniformly in PVDF, increasing its porosity while decreasing its crystallinity. With 0.002 wt.% graphene applied to the PVDF/graphene polymer electrolyte, the ionic conductivity enhanced dramatically from 1.850 mS/cm in pristine PVDF to 3.61 mS/cm. PVDF/graphene gel polymer electrolyte is high performing material for LIBs [87]. The solution cast method was used to create ionic liquid-based polymer electrolytes from polymer polyethylene oxide, salt lithium bis(fluoromethylsulfonyl) imide, and IL N-propyl-N-methyl pyrrolidinium-bis (fluorosulfonyl) imide (PYR13FSI). For investigating the effect of graphene oxide layering on the electrochemical activity of the lithium cell, a $LiFePO_4$ (LFP) cathode and a graphene oxide-layer enclosed LFP cathode (GO-LFP) was formed. Concerning the LFP cathode, GO-LFP has a high basic discharge power of 163 mAh/g at 0.1 C [88].

12.6 CONCLUSION

The employment of redox-active natural materials, such as polymers, has gained considerable attention recently. This book chapter provided an extensive overview of the 2D CP-based high capacitance batteries. Different battery types and their packing arrangement are also described in detail. Also, various important electrochemical testing for the battery is also summarized to understand the discussed results better. Various composite 2D materials are presented along with their battery characterization to show their suitability for their application in the battery. Furthermore, superior material content and performance also remained a challenge. However, the employment of redox-active 2D material is just at its start, and yet no system has been commercialized. The growing market demand for high performance and sustainable battery systems keeps pushing the development of advanced materials toward reality.

REFERENCES

1. Fic K, Platek A, Piwek J, Frackowiak E (2018) Sustainable materials for electrochemical capacitors. Mater. Today 21:437–454.
2. Wei L, Zhao W, Yushin G (2019) Carbons from Biomass for Electrochemical Capacitors. Springer, Singapore, pp 153–184.
3. Mirzaeian M, Abbas Q, Ogwu A, Hall P, Goldin M, Mirzaeian M, Jirandehi HF (2017) Electrode and electrolyte materials for electrochemical capacitors. Int. J. Hydrogen Energy 42:25565–25587.
4. Primary energy consumption. In: Our World Data. https://ourworldindata.org/explorers/energy?tab =chart&country=USA~GBR~CHN~OWID_WRL~IND~BRA~ZAF&Total+or+Breakdown=Total& Energy+or+Electricity=Primary+energy&Metric=Annual+consumption
5. International Renewable Energy Agency (2017) Electricity storage and renewables: Costs and markets to 2030.
6. Deng J, Luo W-B, Chou S-L, Liu H-K, Dou S-X (2018) Sodium-ion batteries: From academic research to practical commercialization. Adv. Energy Mater. 8:1701428.
7. Ru Y, Zheng S, Xue H, Pang H (2019) Different positive electrode materials in organic and aqueous systems for aluminium ion batteries. J. Mater. Chem. A 7:14391–14418.
8. Nitta N, Wu F, Lee JT, Yushin G (2015) Li-ion battery materials: Present and future. Mater. Today 18:252–264.
9. Das V, Padmanaban S, Venkitusamy K, Selvamuthukumaran R, Blaabjerg F, Siano P (2017) Recent advances and challenges of fuel cell based power system architectures and control – a review. Renew. Sustain. Energy Rev. 73:10–18.
10. Rezk H, Sayed ET, Al-Dhaifallah M, Obaid M, El-Sayed AHM, Abdelkareem MA, Olabi AG (2019) Fuel cell as an effective energy storage in reverse osmosis desalination plant powered by photovoltaic system. Energy 175:423–433.
11. Gunawardane K (2015) Capacitors as energy storage devices-simple basics to current commercial families. In: Energy Storage Devices for Electronic Systems: Rechargeable Batteries and Supercapacitors. Elsevier Inc., San Diego, CA, pp 137–148.
12. Wang F, Wu X, Yuan X, Liu Z, Zhang Y, Fu L, Zhu Y, Zhou Q, Wu Y, Huang W (2017) Latest advances in supercapacitors: From new electrode materials to novel device designs. Chem. Soc. Rev. 46:6816–6854.
13. Muzaffar A, Ahamed MB, Deshmukh K, Thirumalai J (2019) A review on recent advances in hybrid supercapacitors: Design, fabrication and applications. Renew. Sustain. Energy Rev. 101:123–145.
14. Mousavi GSM, Faraji F, Majazi A, Al-Haddad K (2017) A comprehensive review of Flywheel Energy Storage System technology. Renew. Sustain. Energy Rev. 67:477–490.
15. Lewis LH, Jiménez-Villacorta F (2013) Perspectives on permanent magnetic materials for energy conversion and power generation. Metall. Mater. Trans. A 44:2–20.
16. Beevers D, Branchini L, Orlandini V, De Pascale A, Perez-Blanco H (2015) Pumped hydro storage plants with improved operational flexibility using constant speed Francis runners. Appl. Energy 137:629–637.
17. Zeng M, Zhang K, Liu D (2013) Overall review of pumped-hydro energy storage in China: Status quo, operation mechanism and policy barriers. Renew. Sustain. Energy Rev. 17:35–43.
18. Tahalyani J, Akhtar MJ, Cherusseri J, Kar KK (2020) Characteristics of Capacitor: Fundamental Aspects. In: Springer Series in Materials Science. Springer, Gewerbestrasse, Switzerland, pp 1–51.

19. Bonaccorso F, Colombo L, Yu G, Stoller M, Tozzini V, Ferrari AC, Ruoff RS, Pellegrini V (2015) Graphene, related two-dimensional crystals, and hybrid systems for energy conversion and storage. Science 347:1246501.
20. Lukatskaya MR, Dunn B, Gogotsi Y (2016) Multidimensional materials and device architectures for future hybrid energy storage. Nat. Commun. 7:12647.
21. Miao L, Song Z, Zhu D, Li L, Gan L, Liu M (2020) Recent advances in carbon-based supercapacitors. Mater. Adv. 1:945–966.
22. Han Y, Lai Z, Wang Z, Yu M, Tong Y, Lu X (2018) Designing carbon based supercapacitors with high energy density: A summary of recent progress. Chem. – A Eur. J. 24:7312–7329.
23. An C, Zhang Y, Guo H, Wang Y (2019) Metal oxide-based supercapacitors: Progress and prospectives. Nanoscale Adv. 1:4644–4658.
24. Guan BY, Kushima A, Yu L, Li S, Li J, Lou XWD (2017) Coordination polymers derived general synthesis of multishelled mixed metal-oxide particles for hybrid supercapacitors. Adv. Mater. 29:1605902.
25. Yang Z, Ma J, Bai B, Qiu A, Losic D, Shi D, Chen M (2019) Free-standing PEDOT/polyaniline conductive polymer hydrogel for flexible solid-state supercapacitors. Electrochim. Acta 322:134769.
26. Basnayaka PA, Ram MK (2017) A Review of Supercapacitor Energy Storage Using Nanohybrid Conducting Polymers and Carbon Electrode Materials. Springer, Cham, pp 165–192.
27. Saha S, Samanta P, Murmu NC, Kuila T (2018) A review on the heterostructure nanomaterials for supercapacitor application. J. Energy Storage 17:181–202.
28. Torabi F, Ahmadi P (2020) Fundamentals of batteries. In: Simulation of Battery Systems. Elsevier, pp 55–81.
29. Pourabdollah K (2017) Development of electrolyte inhibitors in nickel cadmium batteries. Chem. Eng. Sci. 160:304–312.
30. Petrovic S, Petrovic S (2021) Nickel–cadmium batteries. In: Battery Technology Crash Course. Springer International Publishing, pp 73–88.
31. Syali MS, Kumar D, Mishra K, Kanchan DK (2020) Recent advances in electrolytes for room-temperature sodium-sulfur batteries: A review. Energy Storage Mater. 31:352–372.
32. Wang Y, Zhou D, Palomares V, Shanmukaraj D, Sun B, Tang X, Wang C, Armand M, Rojo T, Wang G (2020) Revitalising sodium-sulfur batteries for non-high-temperature operation: A crucial review. Energy Environ. Sci. 13:3848–3879.
33. Mohamed N, Allam NK (2020) Recent advances in the design of cathode materials for Li-ion batteries. RSC Adv. 10:21662–21685.
34. Kim HJ, Krishna TNV, Zeb K, Rajangam V, Muralee Gopi CVV, Sambasivam S, Raghavendra KVG, Obaidat IM (2020) A comprehensive review of li-ion battery materials and their recycling techniques. Electron. 9:1–44.
35. Yang CS, Gao KN, Zhang XP, Sun Z, Zhang T (2018) Rechargeable solid-state Li-air batteries: A status report. Rare Met. 37:459–472.
36. Su H, Jaffer S, Yu H (2016) Transition metal oxides for sodium-ion batteries. Energy Storage Mater. 5:116–131.
37. Park S, Oh J, Kim JM, Guccini V, Hwang T, Jeon Y, Salazar-Alvarez G, Piao Y (2020) Facile preparation of cellulose nanofiber derived carbon and reduced graphene oxide co-supported LiFePO$_4$ nanocomposite as enhanced cathode material for lithium-ion battery. Electrochim. Acta 354:136707.
38. Yuan LX, Wang ZH, Zhang WX, Hu XL, Chen JT, Huang YH, Goodenough JB (2011) Development and challenges of LiFePO$_4$ cathode material for lithium-ion batteries. Energy Environ. Sci. 4:269–284.
39. Zhang Y, Gao Z, Song N, He J, Li X (2018) Graphene and its derivatives in lithium–sulfur batteries. Mater. Today Energy 9:319–335.
40. Kang C, Patel M, Rangasamy B, Jung KN, Xia C, Shi S, Choi W (2015) Three-dimensional carbon nanotubes for high capacity lithium-ion batteries. J Power Sources 299:465–471.
41. Varzi A, Täubert C, Wohlfahrt-Mehrens M, Kreis M, Schütz W (2011) Study of multi-walled carbon nanotubes for lithium-ion battery electrodes. J. Power Sources 196:3303–3309.
42. Nguyen VA, Kuss C (2020) Review – conducting polymer-based binders for Lithium-ion batteries and beyond. J. Electrochem. Soc. 167:065501.
43. Amanchukwu CV, Gauthier M, Batcho TP, Symister C, Shao-Horn Y, D'Arcy JM, Hammond PT (2016) Evaluation and stability of PEDOT polymer electrodes for Li-O$_2$ batteries. J. Phys. Chem. Lett. 7:3770–3775.
44. Geng P, Cao S, Guo X, Ding J, Zhang S, Zheng M, Pang H (2019) Polypyrrole coated hollow metal-organic framework composites for lithium-sulfur batteries. J. Mater. Chem. A 7:19465–19470.

45. Yan H, Zhang G, Li Y (2017) Synthesis and characterization of advanced Li 3 V 2 (PO 4) 3 nanocrystals@conducting polymer PEDOT for high energy lithium-ion batteries. Appl. Surf. Sci. 393:30–36.
46. Yu X, Manthiram A (2019) Sodium-sulfur batteries with a polymer-coated NASICON-type sodium-ion solid electrolyte. Matter 1:439–451.
47. Katz HE, Searson PC, Poehler TO (2010) Batteries and charge storage devices based on electronically conducting polymers. J. Mater. Res. 25:1561–1574.
48. Linden D (1995) Handbook of Batteries. McGraw-Hill, New York.
49. Muench S, Wild A, Friebe C, Häupler B, Janoschka T, Schubert US (2016) Polymer-based organic batteries. Chem. Rev. 116:9438–9484.
50. Nishide H, Koshika K, Oyaizu K (2009) Environmentally benign batteries based on organic radical polymers. Pure Appl. Chem. 81:1961–1970.
51. Nicholson RS (1965) Theory and application of cyclic voltammetry for measurement of electrode reaction kinetics. Anal Chem. 37:1351–1355.
52. Novák P, Müller K, Santhanam KSV, Haas O (1997) Electrochemically active polymers for rechargeable batteries. Chem. Rev. 97:207–281.
53. Aiken CP, Self J, Petibon R, Xia X, Paulsen JM, Dahn JR (2015) A survey of in situ gas evolution during high voltage formation in Li-Ion pouch cells. J. Electrochem. Soc. 162:A760–A767.
54. Liang Y, Zhao C-Z, Yuan H, Chen Y, Zhang W, Huang J-Q, Yu D, Liu Y, Titirici M-M, Chueh Y-L, Yu H, Zhang Q (2019) A review of rechargeable batteries for portable electronic devices. InfoMat 1:6–32.
55. Girimonte A, Stefani A, Innocenti M, Fontanesi C, Giovanardi R (2021) Influence of magnetic field on the electrodeposition and capacitive Performances of MnO2. Magnetochemistry 7:19.
56. Cheng F, Liang J, Tao Z, Chen J (2011) Functional materials for rechargeable batteries. Adv. Mater. 23:1695–1715.
57. Liu J, Xu C, Chen Z, Ni S, Shen ZX (2018) Progress in aqueous rechargeable batteries. Green Energy Environ. 3:20–41.
58. Le TH, Kim Y, Yoon H (2017) Electrical and electrochemical properties of conducting polymers. Polymers (Basel) 9:150.
59. Kaur G, Adhikari R, Cass P, Bown M, Gunatillake P (2015) Electrically conductive polymers and composites for biomedical applications. RSC Adv. 5:37553–37567.
60. Namsheer K, Rout CS (2021) Conducting polymers: A comprehensive review on recent advances in synthesis, properties and applications. RSC Adv. 11:5659–5697.
61. Yang J, Liu Y, Liu S, Li L, Zhang C, Liu T (2017) Conducting polymer composites: Material synthesis and applications in electrochemical capacitive energy storage. Mater. Chem. Front. 1:251–268.
62. Liu W, Ullah B, Kuo C-C, Cai X (2019) Two-dimensional nanomaterials-based polymer composites: Fabrication and energy storage applications. Adv. Polym. Technol. 2019:1–15.
63. Potts JR, Dreyer DR, Bielawski CW, Ruoff RS (2011) Graphene-based polymer nanocomposites. Polymer (Guildf). 52:5–25.
64. Sahoo S, Karthikeyan G, Nayak GC, Das CK (2011) Electrochemical characterization of in situ polypyrrole coated graphene nanocomposites. Synth. Met. 161:1713–1719.
65. Liang J, Xiang C, Zou Y, Hu X, Chu H, Qiu S, Xu F, Sun L (2020) Spacing graphene and Ni-Co layered double hydroxides with polypyrrole for high-performance supercapacitors. J. Mater. Sci. Technol. 55:190–197.
66. Li S, Shu K, Zhao C, Wang C, Guo Z, Wallace G, Liu HK (2014) One-step synthesis of graphene/polypyrrole nanofiber composites as cathode material for a biocompatible zinc/polymer battery. ACS Appl. Mater. Interfaces 6:16679–16686.
67. Mi H, Li F, He C, Chai X, Zhang Q, Li C, Li Y, Liu J (2016) Three-dimensional network structure of silicon-graphene-polyaniline composites as high performance anodes for Lithium-ion batteries. Electrochim. Acta 190:1032–1040.
68. Ni L, Yang G, Sun C, Niu G, Wu Z, Chen C, Gong X, Zhou C, Zhao G, Gu J, Ji W, Huo X, Chen M, Diao G (2017) Self-assembled three-dimensional graphene/polyaniline/polyoxometalate hybrid as cathode for improved rechargeable lithium ion batteries. Mater. Today Energy 6:53–64.
69. Ates M, Caliskan S, Ozten E (2018) Synthesis of ternary polypyrrole/Ag nanoparticle/graphene nanocomposites for symmetric supercapacitor devices. J. Solid State Electrochem. 22:773–784.
70. Xu Z, Zhang Z, Li M, Yin H, Lin H, Zhou J, Zhuo S (2019) Three-dimensional ZnS/reduced graphene oxide/polypyrrole composite for high-performance supercapacitors and lithium-ion battery electrode material. J. Solid State Electrochem. 23:3419–3428.
71. Giri S, Ghosh D, Das CK (2014) Growth of vertically aligned tunable polyaniline on graphene/ZrO2 nanocomposites for supercapacitor energy-storage application. Adv. Funct. Mater. 24:1312–1324.

72. Chen J, Liu Y, Li W, Wu C, Xu L, Yang H (2015) Nanostructured polystyrene/polyaniline/graphene hybrid materials for electrochemical supercapacitor and Na-ion battery applications. J. Mater. Sci. 50:5466–5474.

73. Ghanbari K, Mousavi MF, Shamsipur M, Karami H (2007) Synthesis of polyaniline/graphite composite as a cathode of Zn-polyaniline rechargeable battery. J. Power Sources 170:513–519.

74. Li X, Zhu H (2015) Two-dimensional MoS_2: Properties, preparation, and applications. J. Mater. 1:33–44.

75. Liu H, Zhang F, Li W, Zhang X, Lee CS, Wang W, Tang Y (2015) Porous tremella-like MoS_2/polyaniline hybrid composite with enhanced performance for lithium-ion battery anodes. Electrochim. Acta 167:132–138.

76. Wang J, Wu Z, Yin H, Li W, Jiang Y (2014) Poly(3,4-ethylenedioxythiophene)/MoS_2 nanocomposites with enhanced electrochemical capacitance performance. RSC Adv. 4:56926–56932.

77. Hu L, Ren Y, Yang H, Xu Q (2014) Fabrication of 3D hierarchical MoS_2/polyaniline and MoS_2/C architectures for lithium-ion battery applications. ACS Appl. Mater. Interfaces 6:14644–14652.

78. Yang L, Wang S, Mao J, Deng J, Gao Q, Tang Y, Schmidt OG (2013) Hierarchical MoS_2/polyaniline nanowires with excellent electrochemical performance for lithium-ion batteries. Adv. Mater. 25:1180–1184.

79. Xie D, Zhang M, Cheng F, Fan H, Xie S, Liu P, Tu J (2018) Hierarchical MoS_2@Polypyrrole core-shell microspheres with enhanced electrochemical performances for lithium storage. Electrochim. Acta 269:632–639.

80. De Moraes ACM, Hyun WJ, Luu NS, Lim JM, Park KY, Hersam MC (2020) Phase-inversion polymer composite separators based on hexagonal boron nitride nanosheets for high-temperature lithium-ion batteries. ACS Appl. Mater. Interfaces 12:8107–8114.

81. Aydın H, Çelik SÜ, Bozkurt A (2017) Electrolyte loaded hexagonal boron nitride/polyacrylonitrile nanofibers for lithium ion battery application. Solid State Ionics 309:71–76.

82. Bradshaw D, Garai A, Huo J (2012) Metal–organic framework growth at functional interfaces: Thin films and composites for diverse applications. Chem. Soc. Rev. 41:2344–2381.

83. Jiang H, Liu X-C, Wu Y, Shu Y, Gong X, Ke F-S, Deng H (2018) Metal-Organic Frameworks for High Charge-Discharge Rates in Lithium-Sulfur Batteries. Angew. Chemie 130:3916–3921.

84. Su D, Cortie M, Fan H, Wang G (2017) Prussian blue nanocubes with an open framework structure coated with PEDOT as high-capacity cathodes for lithium–sulfur batteries. Adv. Mater. 29:1700587.

85. Arya A, Sharma AL (2017) Polymer electrolytes for lithium–ion batteries: A critical study. Ionics (Kiel). 23:497–540.

86. Zhao S, Wu Q, Ma W, Yang L (2020) Polyethylene oxide-based composites as solid-state polymer electrolytes for lithium metal batteries: A mini review. Front Chem. 8:640.

87. Liu J, Wu X, He J, Li J, Lai Y (2017) Preparation and performance of a novel gel polymer electrolyte based on poly(vinylidene fluoride)/graphene separator for lithium ion battery. Electrochim. Acta 235:500–507.

88. Gupta H, Kataria S, Balo L, Singh VK, Singh SK, Tripathi AK, Verma YL, Singh RK (2018) Electrochemical study of Ionic Liquid based polymer electrolyte with graphene oxide coated $LiFePO_4$ cathode for Li battery. Solid State Ionics 320:186–192.

13 Conducting Polymers in Batteries

Arunima Reghunadhan[1], Jiji Abraham[2], Malavika S.[3], P. Hema[3], Sreedha Sambhudevan[3], and Sabu Thomas[1]

[1]School of Energy Materials, International and Inter University Center for Nanoscience and Nanotechnology (IIUCNN) and School of Chemical Sciences, Mahatma Gandhi University, Kottayam, Kerala, India

[2]Department of Chemistry, Vimala College, Ramavarmapuram, Thrissur, Kerala, India

[3]Department of Chemistry, Amrita Viswavidyapeedom, Amrithapuri Campus, Kollam, Kerala, India

CONTENTS

13.1   INTRODUCTION

Energy storage and conversion are some of the most discussed topics in research. The prerequisite for advanced electronic materials and their usage is tremendously increasing. As different devices demand differences in the properties of energy storage materials, the development of highly efficient materials is always a requirement. We are passing through a variety of energy devices in our day-to-day life. Most of these materials are associated with batteries. Batteries are one of the unavoidable materials in the current era of electronics. So, we can undoubtfully state that "batteries" is a term frequently coming in daily life. The strong interest in energy devices demands their further development in terms of materials and efficiency. In most of the fields, nowadays, polymers are finding their places, which is not different for the batteries also. Conducting polymers along with their composites are being employed in a variety of ways in batteries.

13.2   CONDUCTING POLYMERS

Polymers are by definition macromolecular, randomly coiled, amorphous materials. They have been in most cases employed in insulating applications. But there are quite a good number of polymeric materials which are excluded from insulation purposes due to their conductivity. Such polymers, which have the ability to conduct electricity are often termed as "conducting polymers." They have many applications in rechargeable batteries, memory cells, coatings, solar cells, barriers, molecular electronic circuits, and intelligent sensor systems (1). The most significant advantage is processability, which can be achieved by dispersion. They are not thermoplastics, which means they cannot be thermoformed. They are, however, organic materials, like insulating polymers. They have high electrical conductivity, but mechanical properties are not comparable to those of other commercially available polymers. Examples are polyaniline (PANI), polyacetylene (PA), polythiophene (PT), etc.

Conducting polymers have got a wide range of attention over the last two decades, owing to their wide range of possible applications. A simple application of these materials is the development of conductive plastics, which combine electrical conductivity with good mechanical properties (2). These electrically conductive composites with a thermoplastic matrix are regarded as a significant category of low-cost materials for special applications. Mixtures of conductive additives, such as fine metal powders, conductive polymers, are used to achieve conductivity in a strongly insulating matrix. Carbon-based fillers such as carbon black, graphite, carbon fibers are the most used conductive fillers due to their high conductivity, low cost, and excellent final and processing properties of the composites. For the design of electroconductive thermoplastic composites, a proper balance of electrical conductivity, mechanical properties, and processing characteristics is essential (3). It is well established that a small increase in conductivity is observed with rising filler content when the conductive filler is loaded low. A sudden increase in conductivity occurs in a relatively small concentration range around the percolation threshold. A conductive network is built within the insulating phase at this concentration, resulting in a drastic increase in conductivity. The existence of these fillers in a thermoplastic matrix has a major impact on the material's behavior, especially in terms of the mechanical properties and processing characteristics (4).

Polymers are being abandoned in their conventional position as electrical insulators in favor of assuming the position of conductors with a variety of new applications. Scientists from various fields are integrating their knowledge to investigate organic solids with unusual conduction properties. Charge transfer complexes/ion radical salts, organometallic species, and conjugated organic polymers are three classes of organic compounds that effectively transport charge. Intrinsically conducting polymers, also known as electroactive conjugated polymers, are a new form of polymer that has recently appeared (5). Electrical conductivity, low energy optical transitions, low ionization

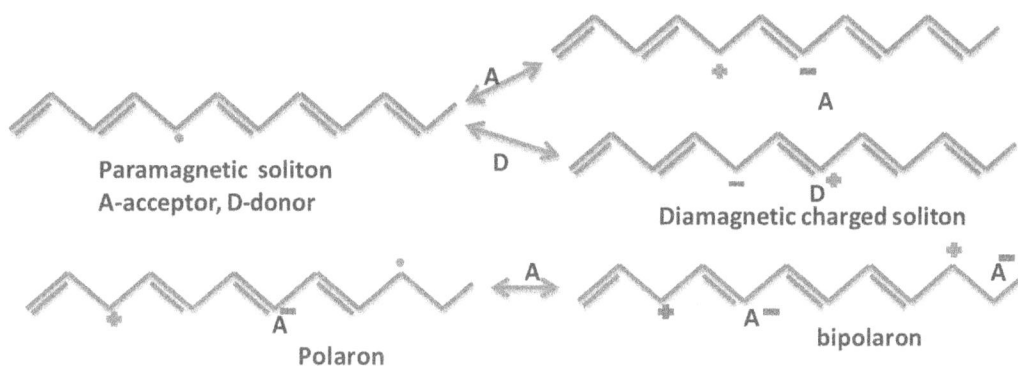

FIGURE 13.1   Mechanism behind the formation of polarons and bipolarons in polyacetylene.

potential, and high electron affinity are all electronic properties of conducting polymers, which are due to their π-electron backbone. The word "synthetic metals" was coined to describe organic polymers with higher electrical conductivity values. Conducting polymers have a wide range of uses, including analytical chemistry and biosensing systems. Doping increases the electrical conductivity of conducting polymers by many orders of magnitude. To describe the electronic phenomena in these systems, the concepts of solitons, polarons, and bipolarons have been used (6).

13.3   WHY DO SOME POLYMERS CONDUCT?

Well, the answer should be elaborated. Primarily, conductivity is the mobility of ions or charged species. In the case of polymers, the conductivity arises from their structure. Mostly, the polymers that conduct electricity are said to possess multiple bonds (conjugated) in their skeleton. Through the rearrangement of the multiple bonds (accurately the pi bonds), the conducting nature is possible. One of the simplest conducting polymers, PA, is the polymerization product of acetylene and the structure is rich in conjugated bonds. In the PA, when it undergoes reduction or oxidation (addition or removal of electrons), the pi bonds are disturbed and free electrons are produced. The electrons with added or removed skeletons are called doped polymers, n-doped or p-doped, respectively. These electrons result in a species called polarons and these polarons are responsible for the movement of electrons through the polymeric chain. In the case of some other polymers, there exist bipolarons. This can be schematically represented as in Figure 13.1.

13.4   APPLICATIONS OF CONDUCTING POLYMERS

Conducting polymers are being exploited for a good number of electronic applications these days. The most common conducting polymers are PANI, poly(3,4-ethylenedioxythiophene (PEDOT), polypyrrole (PPy), PT, PA, etc. These polymers in combination with each other and also their composites by adding suitable fillers such as graphene and carbon nanotubes have been employed in various applications in the biomedical field (7–10). The applications such as tissue engineering, drug delivery, and bioactuators use PEDOT-PPy, PEDOT-nanotubes, PANI grafted to biopolymers, conducting polymer-based hydrogels, and a lot more (11–17). The properties such as biocompatibility, hydrophobicity, redox stability, and the possibility of surface modification make them suitable for biomedical and biotechnological applications (9).

Conducting polymers and their composites (nanocomposites) are used in the supercapacitors, as electrode materials, in the light-emitting diodes, in flexible displays, in energy harvesting devices like solar cells, in different type of sensors, and so on. One of the most important and interesting applications of conducting polymers is in batteries. They have been employed as electrodes, electrolytes, battery separators, etc. (18–21).

Some other utilization of conducting polymers is:

1. Biosensors in health care
   a. Glucose biosensors
   b. Urea biosensors
   c. Lactate biosensors
   d. Cholesterol biosensors
2. Immunosensors
3. Biosensors such as
   a. DNA biosensors
   b. for environmental monitoring
   c. for food industry

Conducting polymers have several advantages over metals, as a result, metals are constantly replaced by conducting polymers:

1. Weight is lighter
2. Processing is simple and requires less energy
3. Corrosion safety
4. Blends well with other plastics
5. It compacts the material
6. It is used in smart materials (22)

Intrinsically conducting polymers are the organic polymers that contain repeated units of oxidized or reduced monomers, as first described by Chiang et al. ICPs, also known as organic semiconductors, are polymers with conjugated or delocalized π-electrons and an intrinsically large bandgap that determines their affinity for electrons (23).

13.5 BATTERIES

Benjamin Franklin first used the concept "battery" to describe related capacitors in 1749. By definition, a battery is a collection of cells that are used to convert chemical energy to electrical energy. They ensure the flow of electrons through them. All the batteries are self-possessed with three main components – two electrodes (positive and negative) and an electrolyte. In batteries, in addition to these components, a separator is inserted to prevent the contact of anode and cathode. The separators are usually porous materials. Conventional batteries typically generate electricity chemically from one or more electrochemical cells.

13.6 ELECTROCHEMISTRY OF BATTERIES

In order to work, batteries usually necessitate a number of chemical reactions. One or more reactions occur in the anode, and at least one reaction occurs in or near the cathode. The anode reaction produces excess electrons by the process of oxidation, while the cathode reaction uses the extra electrons by the process of reduction. The basic electrochemistry of batteries is that of an electrochemical cell which is depicted in Figure 13.2.

The cathode is one of the electrodes, and the anode is the other. The cathode side of the cell is reduced, which means it gains $\bar{e}$ and serves as an oxidizing agent for the anode. The anode side of the cell is where oxidation occurs, which means it lacks $\bar{e}$s and serves as a reducing agent for the cathode. Each electrode is immersed in an electrolyte, which is an ion-containing compound. This electrolyte serves as a concentration gradient on both sides of the half reaction, allowing electron transfer through the wire. This movement of electrons generates electricity, which is then used to fuel the battery (24).

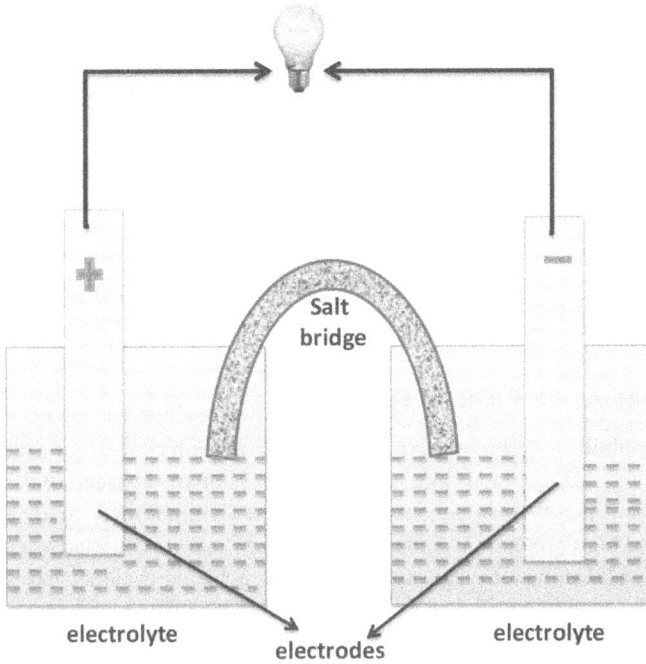

FIGURE 13.2   Schematic representation of an electrochemical cell.

13.7   TYPES OF BATTERIES

Batteries are divided into two divisions based on their charging profiles and usage: primary and secondary.

13.8   PRIMARY BATTERIES/CELLS

By the term, primary cells or primary batteries mean that these are single-use materials (Figure 13.3). They can be charged once and become dead when the chemicals complete the reactions. These cells are also called dry cells. They are preferred when long-term charging is required.

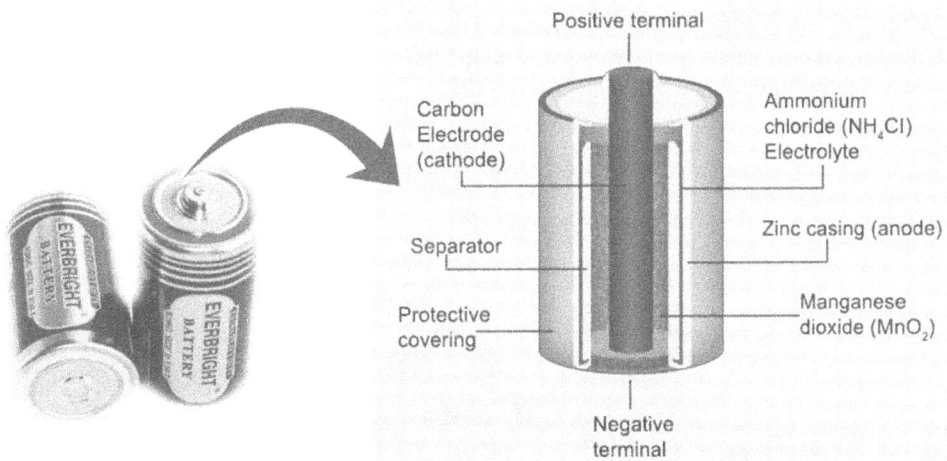

FIGURE 13.3   Schematic representation of the structure of a primary cell.

13.9 STRUCTURE OF PRIMARY BATTERIES

Primary batteries are much similar to the normal electrochemical cell. It contains an electrode at the center having positive and negative terminals. It is surrounded by a layer of electrolytes.

13.10 RECHARGEABLE BATTERIES/SECONDARY BATTERIES

Secondary cells are those which can be charged many times and can be reused. In this type of battery, the electrochemical reactions are reversible. They have been employed in personal digital assistance such as laptops, notebooks, and smart devices. The most common examples are lead-acid batteries, nickel-cadmium batteries, lithium-based batteries such as Li-ion batteries, Li-S batteries, and Li-Mn batteries.

13.11 POLYMERS AS ELECTROLYTES

Because of their low boiling point, organic liquid electrolytes have certain drawbacks when used in batteries, including leaking, smoking, and even fires. Polymer-based solid electrolytes are an alternative option because of their benefits, such as low flammability, good flexibility, excellent thermal stability, and high protection. Conducting polymers are widely used as electrolytes for various batteries. Both solid polymer electrolyte (SPE) and gel-type polymer electrolyte can be used for this purpose. The poor ionic conductivity of SPE can be further improved by the addition of numerous new types of lithium salt. It is also found that mechanical strength as well as transport properties and electrochemical properties can also be enhanced by the addition of inorganic salts (25). Figure 13.4 represents the schematic diagram of lithium polymer battery where conducting polymer is using as an electrolyte.

Studies on electrolytes based on polymers started in 1970 with the discovery of ionic conductivity in polyethylene oxide (PEO) (26). Even though PEO has so many advantages, such as increased solubility for salts deprived of phase separation, it experiences some drawbacks, expressly crystallization at temperatures below 60°C. In order to overcome the restrictions of PEO, many other electrolytes have been introduced. This comprises polycarbonates, polynitriles, polyamines, etc. (27). There are different categories of polymer-based electrolytes like gels, block copolymers, and polymer composites. Gel polymer electrolytes (GPEs) offer the mandatory room temperature ionic conductivity for commercial battery applications. Dry polymer electrolyte is not appropriate for battery applications at room temperature. Figure 13.5 represents the prevailing accomplishments in the area of PEs. Among these, gel-based polymer electrolytes have adequate ionic conductivity. Polymer composite electrolytes, exclusively the ones preoccupied with ceramic nanowires, are approaching the mandatory boundary of conductivity and are fortuitous to be employed in solid-state batteries. Dry polymer electrolytes with single lithium-ion conductivity still endure markedly below the essential limit at ambient temperatures (28).

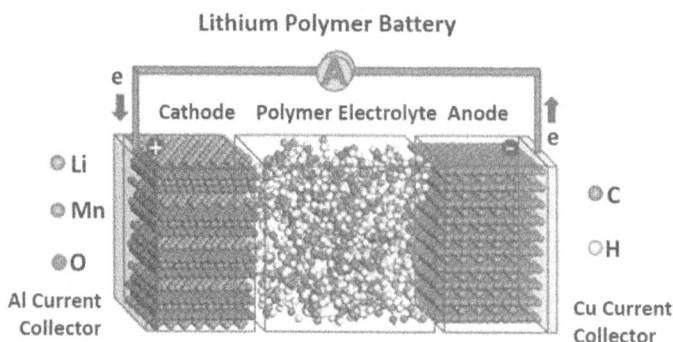

FIGURE 13.4   Schematic diagram of lithium polymer battery.

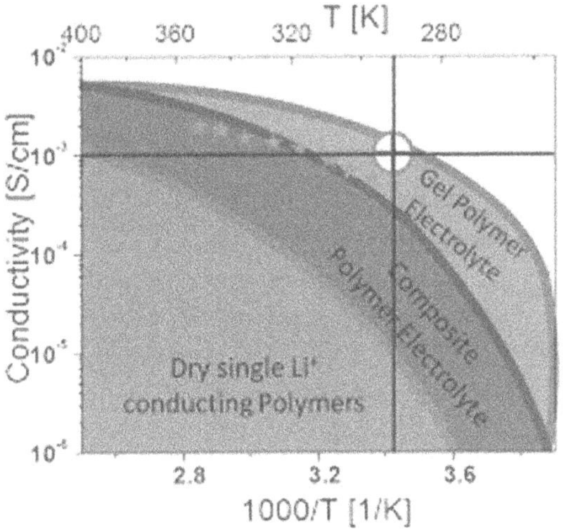

FIGURE 13.5 Representation of the development of diverse categories of polymer electrolytes (circle marks the value for conductivity at ambient temperature essential for real-world application) (Reprinted (adapted) with permission from Ref. (28). Copyright (2020) American Chemical Society.)

SPEs are light weight, flexible, and non-flammable, making them a viable solution to lithium-ion battery safety concerns. Ionic conductivity is weak in SPEs. The addition of lithium salts and inorganic fillers will increase ionic conductivity even further. Lithium salts are dissolved in a high-molecular-weight polymer matrix to make SPEs. In the motion of polymer segments, the polymer serves as a host for the transport of lithium ions. Polymer-based solid electrolytes have a number of advantages. This includes the ability to compensate for electrode volume variations by elastic and plastic deformation (29). This is one of the best candidates for electrolytes for all-solid-state Li metal, Li-ion, Li-sulfur, and Li-O$_2$ batteries (30). Figure 13.6 categories of the existing polymer-based composite solid electrolytes.

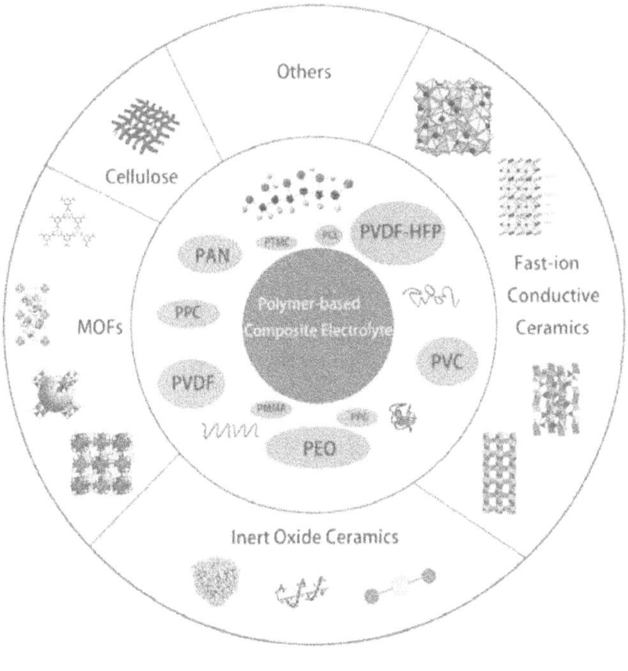

FIGURE 13.6 Different types of polymer-based composite solid electrolytes existing in the usage. (Reproduced with permission from Ref. (31).)

Nanocomposite polymer electrolytes can be used in lithium batteries. Poly Ethylene Oxide-Li$_x$ conductivity grasps virtually useful values above 60°C (32).

13.12   POLYMERS AS ELECTRODE MATERIALS

Polymers having extended backbone limits their solubility in the electrolyte solution. This also improves the cyclic performance (33). Conducting polymers have a wide range of electromechanical properties. Among various applications of conducting polymers in various sectors, the battery sector is a promising one. This conductive polymer can be used as a positive electrode along with other negative electrodes. This is because of the inherent electronic and ionic conductivity of conducting polymers (34).

13.13   CONDUCTING POLYMER ELECTRODE

Several studies have been reported on the use of conducting polymers as electrode material. Conducting polymers such as PEDOT (35), PPy (36), polythiophene (37), poly-3-methylthiophene (38), polyacetylene (39), and poly(aniline) have been widely explored as electrodes. Among these studies, based on PANI have got special attention. This is because of its stability against aerial oxidation and moisture, moreover its electrochemical response is reversible. Based on the above properties, several PANI batteries have been prepared both in aqueous and non-aqueous solvents (40). This polymer is used as electrodes for several batteries like magnesium reserve battery (41), lithium battery (42), lithium-sulfur battery (43), and zinc battery (44). In n/p-type supercapacitors, thiophene-based conducting polymers were also used as positive and negative electrodes (45). Amanchukwu et al. appraised the steadiness of PEDOT polymer electrodes for Li-O$_2$ batteries. The establishment of sulfone functionalities on the exterior of PEDOT and cleavage of the polymer replication impairs ē conductivity and leads to poor cycling. So authors proposed that new Li-O$_2$ electrodes should be discovered for better performance (46). Poly(m-amino thiophenol) crosslinked with sulfur is used as cathode materials for high rate, ultralong-life lithium-sulfur batteries (47).

13.14   DOPED POLYMERS AS ELECTRODES

Shi et al. reported the practice of self-doped PANI in place of cathode material for aqueous zinc batteries (44). Chloroaluminate-doped conducting polymers can serve as anodes in rechargeable aluminum batteries (48). When PPy was doped with chloroaluminate ionic liquid electrolytes and the cycling of the conducting polymer electrodes befallen via the electrochemical insertion and removal of these anions. Zhang et al. reported the use of PPy with different dopants as catalysts in non-aqueous Li-O$_2$ batteries for the first time. PPy doped with Cl$^-$ and ClO$_4$$^-$ were synthesized and applied as the cathode catalyst in Li-O$_2$ batteries (49).

13.15   MIXED POLYMERS AS ELECTRODES FOR BATTERIES

An inter-tangled network of redox-active and conducting polymers as a cathode for ultrafast rechargeable batteries was developed by Kim et al. Here, PEDOT: PSS (poly(3,4-ethylenedioxythiophene) doped with poly (styrene sulfonate)) and tangled with another 1D polymer poly(vinyl carbazole) (PVK) is used as anode for lithium-ion battery. PEDOT:PSS plays the role of both binder and conductive agent (50). Electron transfer is from PEDOT:PSS to PVK because of the greater positive reduction potential of PVK. The TEM images of PSS-PEDOT electrodes are given in Figure 13.7.

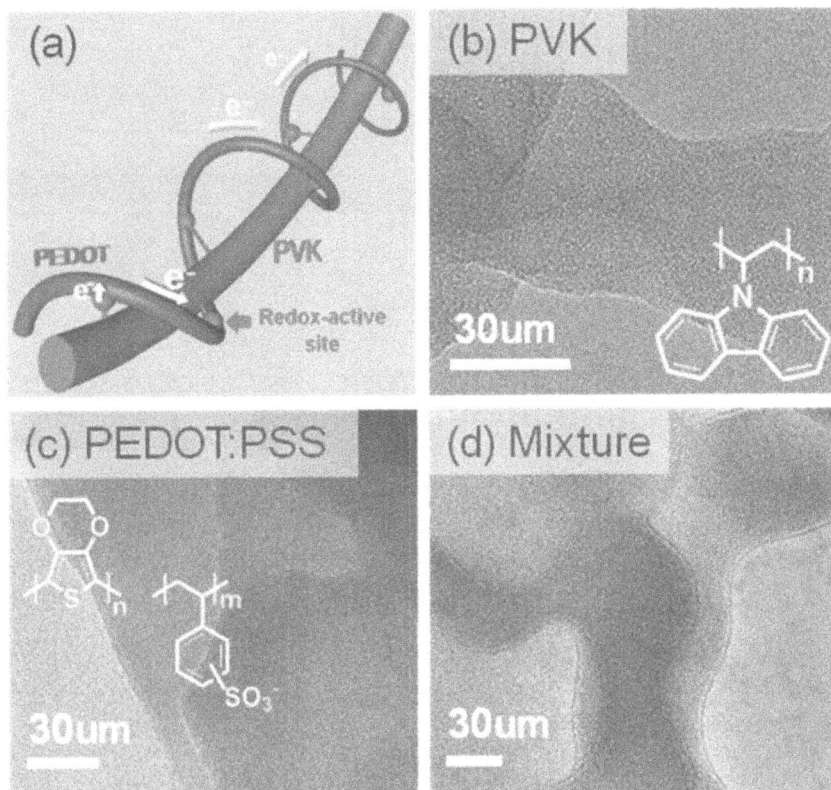

FIGURE 13.7 Schematic picture and the corresponding TEM images of PSS-PEDOT electrodes. (Reproduced with permission from Ref. (50), RSC Publishing.)

13.16 CONDUCTING POLYMER/CARBON-BASED MATERIAL AS ELECTRODES

As an electrode medium for batteries, high charge density conducting polymer/graphite fiber may be used. The graphite surface has a large surface area and a lightweight structure in this application. Over 50 cycles, more than 70% of the available charge was removed from the cell with no loss in cell efficiency (51). Lee et al. experimented with the use of poly(diallyl dimethylammonium chloride) (PDDA)/graphene oxide-sulfur composites as electrode material for rechargeable Li-S batteries (52). It was described that chopped carbon-filament (CCF) reinforced PPy–nanocellulose network can be used as cathode material in a lithium-ion battery. This was a binder-free and free-standing paper alike cathode material. The CCF PPy@nanocellulose LIB cell unveiled the highest area-based dimensions and lowermost self-discharge rate hitherto tested for a conducting polymer-based LIB cathode material (53). In a study was reported by Cao et al., to improve the performance of Li-O$_2$ batteries, carbon material can be modified with organic polymers such as (poly(2,3-dihydrothieno-1,4-dioxin), PPy, and PANI can be inserted. Among these materials, PEDOT exhibited better stability and cell performance (54). Interconnected 3D CNT–CP hydrogel network for high-performance flexible battery electrodes was explored by Chen et al. The CNT-conducting polymer hydrogel's interpenetrating network exhibits improved mechanical properties, high conductivity, and easy ion transfer, resulting in simple electrode kinetics and high strain resistance during electrode volume transition (55). Aligned carbon nanotube/PEDOT conducting polymer can be used as electrodes for lithium-ion batteries. The problem with this electrode is its inability to form free-standing films. This can be overcome by casting the second film of polyvinylidene fluoride (PVDF) over to this (56) (Figure 13.8).

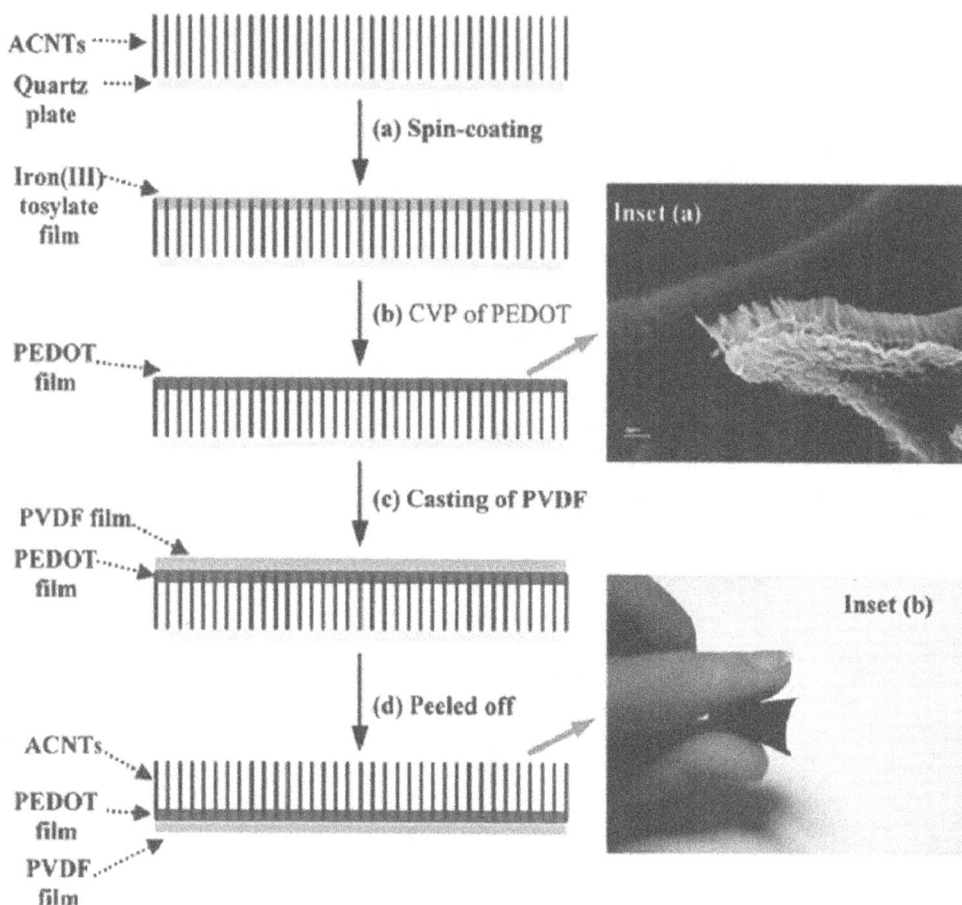

FIGURE 13.8 Schematic representation of the procedures for the fabrication of a free-standing and highly conductive ACNT/PEDOT/PVDF membrane electrode. (Reproduced (adapted) with permission from Ref. (56). Copyright (2007) American Chemical Society.)

13.17   CONDUCTING POLYMER/METAL OXIDE COMPOSITES FOR ELECTRODES IN BATTERIES

Even though inorganic compounds have high lithium storage properties, their commercial application in the battery is limited because of a lack of conductivity and cyclability. It can be overcome by its combination with conductive polymers. The synergistic combination of conducting polymer and inorganic material showed improvements in the electrode lifetime, rates, and voltage, mechanical and thermal stability (57). Hybrid material composed of PANI and titanium dioxide (TiO_2) was used as cathode material in a rechargeable battery. It was found that this hybrid material showed better charge-discharge characteristics up to 50 cycles than PANI alone and also exhibited better cyclability (58). MnOx/PPy is explored as a cathode material in aqueous zinc-ion batteries by Zhang et al. PPy is epitaxially polymerized around the 2D MnOx nanosheets. MnOx nanosheets are integrated into PPy via the interaction between Mn and N atoms and forms material with 2D morphology. Various factors like its 2D structure, high flexibility of polymer chains, proper interaction between Mn and N, and direct surface contact of MnOx nanosheets with Zn^{2+} improve ultra-discharge capacity, structural stability, capacity retention, etc. (59).

13.18 HYBRID BIOPOLYMER ELECTRODES

PEDOT/lignin composite material is using as cathode material for sodium-ion and lithium-ion batteries. The use of this hybrid material as an electrode improved the storage capacity of both batteries (60).

13.19 CONDUCTING POLYMERS AS BINDER FOR BATTERIES

Polymers are using as binders for the fabrication of electrodes. Binders should have the properties such as distribution/dispersion and steadiness of the active material, electronic and ionic conductivity of the composite electrode, and grip with the current collector. The cycling capacity of electrodes can be improved with the use of many polymer binders like poly(methyl methacrylate) (PMMA) and carboxymethyl chitosan mixed with styrene-butadiene rubber, PVDF (61). Figure 13.9 represents the chemical structures of various polymer binders. Binder is a critical factor that accounts for 10% to 20% of the overall electrode weight. Furthermore, rather than being used in bulk, the polymer binder is often found on the surface of conductive particles (such as carbon). As a result, the polymer binder's volatility may have a big impact on the $Li-O_2$ battery's output (62).

Biopolymers also can be used as a binder in batteries. Prasanna et al. demonstrated the use of chitosan as a binder and compared its performance with traditional binder PVDF. Figure 13.10 represents the Field Emission Scanning Electron Microscopy (FE-SEM) images of $LiFePO_4$ electrodes coated with two polymer binders. Here a uniform surface coverage which enables the strong network between electrode and binder. In comparison to the electrode with the PVDF binder, which had a discharge capacity of 127.9 mAh/g and a capacity retention ratio of 85.13%, the electrode with the chitosan binder had a higher discharge capacity of 159.4 mAh/g and an outstanding capacity retention ratio of 98.38% (64). Carvalho et al. investigated the use of guar gum-based biopolymers as binders in batteries. These polymers are electrochemically stable within the operating voltage of LIBs and do not show evidence of thermal decomposition up to 200°C (65). Linear and cross-linked ionic liquid polymers as binders in lithium-sulfur batteries were studied. The cell capacity,

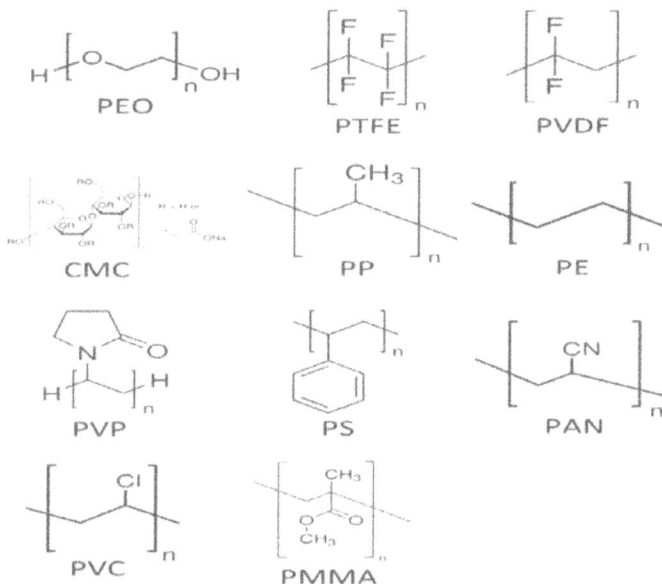

FIGURE 13.9 Structures of polymers commonly used as binders. (Reproduced with permission from Ref. (63), Elsevier (2013).)

FIGURE 13.10   The FE-SEM images of LiFePO$_4$ electrodes with PVDF and chitosan binder. Schematic representation on the right shows the electrodes prepared with (c) PVDF and (f) chitosan binders. (Reproduced (adapted) with permission from Ref. (64). Copyright (2015) American Chemical Society.)

cyclability, and morphology of the cathode itself were found to be influenced by the PIL molecular structure, polymer backbone, and polymer architecture (66).

13.20   CONDUCTING POLYMERS AS SEPARATOR IN BATTERY

Good battery separators are characterized by electrolyte wettability, heat tolerance, and electrochemical properties. Many polymers like polyethylene, PVDF, polyamide, cellulose nanofibers, polylactic acid, polyacrylonitrile, polypropylene, PMMA, are used as battery separators (67).

13.21   PROPERTIES AND CHARACTERIZATION OF POLYMERIC BATTERY MATERIALS

One of the big scientific issues right now is the establishment of renewable energy sources. The most important things are energy conversion and energy storage. As a result, developing reliable, high-capacity, high-current rechargeable battery systems are critical, and polymer materials are noteworthy parts of these gadgets. As the market for high-energy-density gadgets upsurges, contemporary materials that depend on fundamental comprehension of physical phenomena and structure–property relationships will be expected to empower high-capacity next-generation battery chemistry (68).

A polymeric battery is a type of rechargeable battery that uses a polymer-based electrolyte rather than a liquid-type electrolyte. These electrolytes are made up of semisolid (gel) polymers with high conductivity. These batteries have superior specific energy than other lithium battery types, and they are used in applications such as handheld devices, radio-controlled airplanes, and some electric vehicles where weight is a concern (69). A polymer-based battery is made up of organic materials rather than bulk metals. Metal-based batteries face extraordinary difficulties because of the restricted framework, harmful ecological impacts, and moving toward growth cap. Along with their engineered accessibility, high performance, versatility, lightweight, ease, and low toxicity, redox dynamic polymers are appealing choices for electrodes in cells. Recent research has looked at how to improve performance and reduce obstacles in order to bring polymeric active materials closer to battery practicality. Conductive, non-conductive, and radical polymers are among the polymers being investigated (70).

13.22 PROPERTIES OF POLYMERIC BATTERY MATERIALS

A polymer-based battery, like metal-based batteries, has a reaction that occurs between a positive and negative electrode with varying redox potentials. Between these electrodes, an electrolyte transports charge (71). An item should have the option to take part in a synthetically and thermodynamically reversible redox response to be an appropriate battery dynamic substance. The redox mechanism of polymer-based batteries is based on a change in the state of charge in the organic material, as opposed to metal-based batteries, whose redox cycle depends on the valence charge of metals. The electrode terminals should have comparable specific energies to the number of charges for a high energy density (72). A battery is characterized by the properties like operating voltage, which is the battery's available charging/working voltage, C-rate. The C-rate is a measurement of a battery's loading current, and capacity, which is the total electric charge, measured in Amp hours, that can be stored inside the battery, specific capacity, which is defined as the capacity per mass of active material used for making the battery, its theoretical capacity, which is defined as the entire amount of used active material contributed, coulometric efficiency, which is known as the ratio of the received discharge to the respective charge capacity, the shelf life, which refers to the consistency of usable discharge power as a function of storage time of the charged battery, and the cycle life, which refers to the consistency of available discharge capacity discharge cycles (73).

Because of the popularity of lithium-based battery packs, there is still an increasing need for new technology. There are a variety of current drivers for battery science and the advancement of new battery technology. Here safety is the prime concern. Solid electrolytes like polymer electrolytes or self-detecting separators are included in this sense. The required/desired energy density of the batteries is the next critical consideration (74). Mobile systems, in particular, and, in particular, electric vehicles, necessitate high energy densities. Lithium–sulphur and lithium–air batteries are currently being researched extensively in this sense. In comparison to metal-based batteries, polymer-based batteries have a number of advantages. The electrochemical reactions are easier, and the structural diversity of polymers and the polymerization process allow for greater application flexibility. New forms of inorganic materials are a challenge to come by but recently discovered organic polymers are much easier to create. Another benefit is that, although polymer electrode materials have smaller redox potentials than inorganic materials, they have a greater energy density value (75). Organics also have greater value of power density and rate efficiency than inorganics due to their higher redox reaction kinetics. Polymeric electrodes can be cast, vapor-deposited, and printed due to the inherent durability and lightweight of organic materials compared to inorganic materials, allowing them to be used in thinner and more lightweight products. Furthermore, the majority of polymers can be synthesized at a reduced rate, derived from biomass, and even recycled, while inorganic metals can be limited and hazardous to the atmosphere (76).

13.23 CHARACTERIZATION OF POLYMERIC BATTERY MATERIALS

Characterization is becoming increasingly necessary to recognize manufacturing and design defects as battery producers and end-users demand better performance and improved protection. Batteries must be tested using both in situ and destructive testing to assess the best materials and methods for improving efficiency and protection while keeping battery costs down. The selection method of characterization technique to use is determined by the information required, the needed level of precision, and the expenditure available for qualification and testing (77). Non-destructive evaluation and testing have the advantage of not having the battery to be dismantled but the amount of information that can be analyzed is restricted. On the contrary, the properties of the active substances themselves, and on the other, the critical variables of the respective system assemblies. Significant methods for electrochemical and electrooptical characterization are introduced to researchers with a more synthesis-focused perspective (78). Competency tests of organic and polymeric materials proposed to be used in energy-storage devices are conducted using electrochemical and spectroscopic techniques. The presented series of characterization techniques enable researchers to conduct a thorough investigation of redox-active materials, allowing them to recognize promising prospects for use in batteries, laying the groundwork for potential renewable energy conversion and storage technology (79).

13.24   ELECTROCHEMICAL METHODS

13.24.1   VOLTAMMETRIC METHODS

To determine redox potentials and basic details regarding electrochemical stability and reversibility, voltammetric methods are used. Cyclic voltammetry (CV) is a technique for studying the redox properties of substances that are electrochemically active and, as a result, determining the redox potential, which is essential for energy storage systems (Figure 13.11). A three-electrode device consisting of a plane-disk working electrode, a counter electrode, and a reference electrode is typically used. The plane-disk working electrode is made of platinum, graphite, or glassy carbon and should be similar to the material used in the device (80). The counter electrode and the electrolyte device in use are the same. Because of their higher peak resolution and sensitivity, differential pulse and supporting methods such as square wave voltammetry are extremely useful. CV offers important data about the reversibility and reliability of examined systems in addition to possible values. A 1:1 ratio of anodic and cathodic peak current defines reversibility. The study of the scan

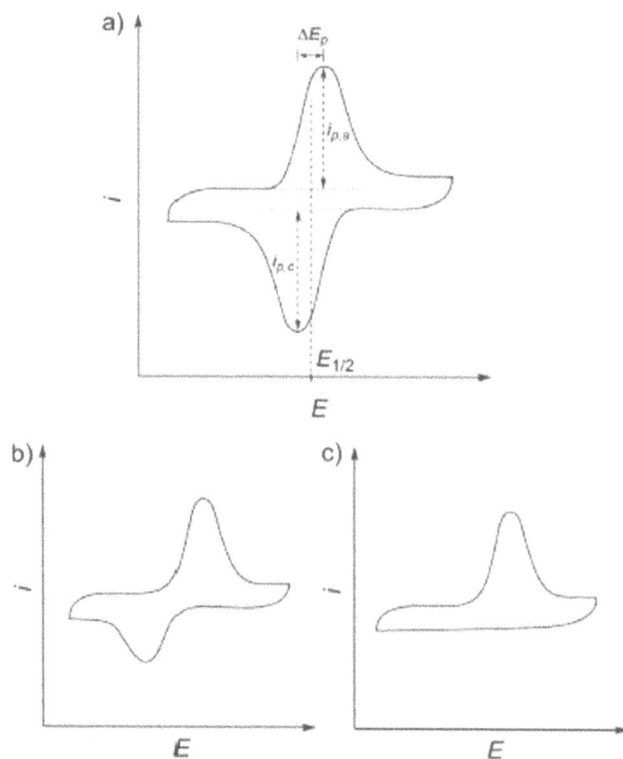

FIGURE 13.11   Common types of CV. (Copyright © 2016, American Chemical Society.)

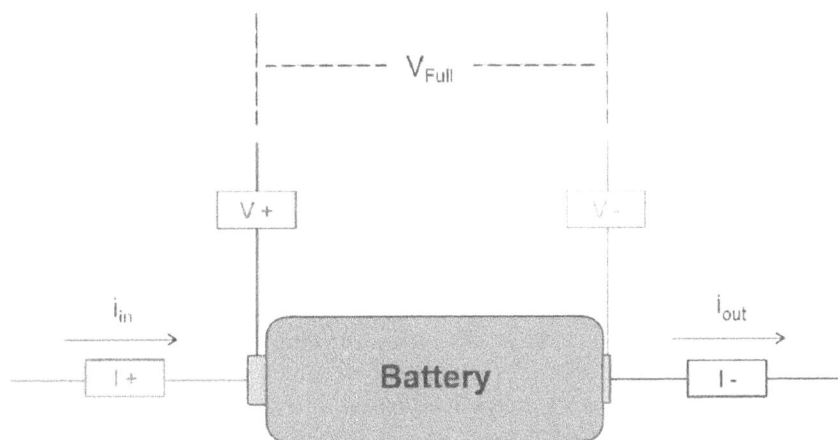

FIGURE 13.12 Experimental setup for impedance measurement of a battery. (Republished with permission. Copyright © 2020 Published by Elsevier Ltd.)

rate-dependent behavior of the cyclic voltammogram can be used to investigate electrode transfer kinetics, which is especially interesting for energy storage devices (81).

13.24.2 ELECTROCHEMICAL IMPEDANCE SPECTROSCOPY

Electrochemical impedance spectroscopy (EIS) is a measurement software that provides all of the tools needed to investigate impedance in surface engineering, electrochemical power generation, corrosion analysis, catalysis, and even basic electrochemical research. It is an in situ technique that gives kinetic information on the cells. EIS has many advantages over DC electrochemical techniques. EIS can be used to monitor changes in battery properties under various consumption or storage conditions, in addition to providing detailed kinetic information (82). EIS is a highly sensitive technique that provides a wealth of knowledge about battery systems, including state of charge analysis, reaction mechanism studies, change of active surface area during activity, separator assessment, passivating film behavior, electrode kinetics separation and comparison, and identification of potential electrode corrosion processes. To probe the impedance characteristics of a cell, EIS uses a small amplitude alternating current (AC) signal. The AC signal is sampled over a wide range of frequencies to generate an impedance distribution for the electrochemical cell under evaluation. Researchers can use EIS to investigate capacitive, inductive, and diffusion processes in the electrochemical cell, which is different from DC techniques (83). The data can be used to determine corrosion rate, electrochemical mechanisms and reaction kinetics, identification of localized corrosion, battery life, and cell efficiency. EIS is usually used in a three-electrode mode. Experimental setup for impedance measurement of a battery is shown in Figure 13.12.

The EIS technique is described by the application of an AC signal to a conductive electrode, resulting in a characteristic response from the device interface. The resultant polarization of the electrode is in a linear potential region because the perturbing ac signal is very weak. As a result, the electrode is not damaged, and EIS can be used to determine the time relationship of device parameters. EIS is now commonly used in the analysis of batteries and fuel cells due to its numerous advantages (84).

13.25 SPECTROSCOPIC AND SPECTRO-ELECTROCHEMICAL METHODS

UV-vis, IR, EPR, and Raman spectroscopic methods are commonly used in a spectro-electrochemical approach to analyses critical redox reactions by tracking spectral characteristics as potential changes. The in situ UV-vis calculation technique is a fast and simple way to determine how redox shuttle overcharge safety additives work in secondary lithium batteries. It allows one to distinguish redox-active species, calculate the number of electrons transported by the shuttle, and evaluate the diffusivity and stability of both the oxidized and reduced forms. We can also monitor the accumulation of inactive degradation products by analyzing spectra obtained under various potential and current density conditions (85).

Electron spin resonance (ESR) is also called as electron paramagnetic resonance spectroscopy (EPR). EPR is especially important in redox-active species study. As at least one of the classes involved in the underlying oxidation-reduction reactions is usually a radical, this is of special interest for research on redox-active compounds (86). The number of present radicals as well as their position in the molecule are both determined by ESR spectroscopy. This technique has been commonly used in electrochemistry to obtain quantifiable, structural, and active knowledge of materials due to its sensitivity to analyses surface defects and the electronic structure of materials, as well as their selectivity against paramagnetic species. It also gives details about the number of radicals that are currently present in the ground state and in the oxidized/reduced molecule (87).

The different layers in a battery are imaged using microscopy techniques. Larger cracks can be examined with optical microscopy, but to investigate layer depth and observe fluctuations in the microstructure, such as microscopic holes (voids) and defects, scanning electron microscopy (SEM) or transmission electron microscopy (TEM) is needed (88).

X-ray photoelectron spectroscopy (XPS) is a useful method for determining the assembly and arrangement of different layers in a battery, as well as illustrating lithium and another element migration within the cathode, anode, and separator. XPS offers comprehensive quantitative data that can be used to assist in failure analysis or to demonstrate how improvements in materials or design affect solid electrolyte interface (SEI) creation (89).

The Fourier transform infrared (FTIR) spectrometer and Raman spectroscopy are the two other spectroscopic techniques used in battery analysis. The benefits of FTIR include a short data collection period, a minor sample dimension requirement, and a qualitative data collection process. Surface-sensitive characterization can also be thought of as FTIR spectra (90). Raman spectra can be applicable to exterior layers which are thick enough or when augmentation of the response is obtained (termed surface-enhanced Raman spectra) owing to the distinct morphology of the sample surface. Specific element recognition provides additional useful methods for tracking electrode processes (91). The analysis of transmission and intercalation processes involving NMR-active elements is possible with solid-state NMR technique. X-ray diffraction technique and neutron diffraction analysis allow for precise element identification and position inside an electrode or electrode coating in batteries (92).

Glow discharge mass spectrometry is a technique for sensing trace amounts of elements. This procedure can be used in battery applications to detect froths and pollutants that may have a negative impact on battery output. One caution with this method is that excessive battery dismantlement can introduce pollutants into the battery that were not previously present (93).

Inductively coupled plasma optical emission spectrometry (ICP-OES) detects the existence of trace metals by applying heat and ionizing the species in a sample. Laboratories established a method to discrete the cathode from a battery cell in different charging states or after a fixed amount of charge/discharge cycles. The investigators then used ICP-OES indeed to quantify lithium and manganese levels in the cathode, detecting slight variations in lithium content that are linked to poor battery efficiency (94).

For battery applications, gas chromatography–mass spectrometry (GCMS) necessitates siphoning gases from a hole penetrated into a battery cell in order to analyze gases emitted throughout electrolyte breakdown. Since the failure mechanisms in batteries are linked to premature electrolyte breakdown, this technique is particularly useful in cases where the battery has swollen or undergone thermal runaway (95). The use of these practices under specific electrochemical environments allows for monitoring the respective elements as the future insights, revealing new information about relevant processes (96).

13.26   OTHER ADVANCED TECHNIQUES

Scanning probe microscopy, specifically atomic-force (AFM) and scanning-tunneling microscopy (STM), is commonly used to gain accurate info about the syllable structure of the electrode surface, whether for energy storage or inside photovoltaic devices (97).

Aside from pure morphological investigations including smoothness, the presence of defects that contribute to electrode material destruction or passivation, and the distribution of components of heterogeneous, phase-separated surface assemblies, (redox) processes may also be tracked, for example, by the volume shift of the studied film during the reaction. Researchers can use the electrochemical quartz crystal microbalance (EQCM) to see how coatings change mass during electrochemical treatment. This is accomplished by coating thin quartz crystals with a controlled oscillation frequency (98). It is possible to investigate deposition and adsorption/desorption processes in response to potential modulation.

13.27   DEVICE CHARACTERIZATION METHODS

The final devices are described by output tests, such as charging/discharging trials in the case of electrical storage and energy conservation studies in batteries. Important parameters such as power, coulometric efficiency, and cycle lifetime can be easily determined by repeatedly charging and discharging the battery under controlled conditions (72). Overall, only collective features and constraints that characterize the entire system are attained, with the exception of special techniques that can offer statistics on single procedures and components, such as impedance spectroscopy. As a result, comparative experiments with reference systems are needed to determine the actual impact and efficiency of the practically active materials.

13.27.1   CHARGING/DISCHARGING CHARACTERISTICS

For better performance and to resolve safety issues, proper battery characterization is essential. The measurement of charge/discharge plots, i.e. measuring the battery voltage at the time of galvanostatic charging or discharging of the battery, is an important experiment. The applied current, as well as the beginning and end (termination) voltages, dictate the duration of the experiment. Although, the voltage window must be wide enough to include the majority of the device's power, whereas, on the other hand, it must be sufficient in size enough to prevent unfavorable (irreversible) redox reactions (99). The so-called C-rate is used to express the applied current. The examination of the resulting plots yields a few distinguishing values. The form of the curve, for starters, indicates the operational stability of the battery. Multiple plateaus result from the existence of multiple underlying redox steps and contribute to a variety of system operating voltages. Furthermore, if the curve slopes, the redox system is changing dramatically during the charge/discharge phase, preventing a constant operating voltage. Knowing the applied current and the loading time obtained, one may calculate the transferred charge or the capacity of the battery, which is typically expressed in Ah. At a lower charge/discharge current, the performance is of the system is higher. Temperature is another significant experimental parameter (100). At reduced temperatures, the resistance inside the system rises, resulting in a lower voltage and hence a lower power and stored energy. At higher temperatures, however, the chemical reactions of the redox-active species elevate, leading to increased self-discharge. The lifespan of an energy-storage system is an important property that can be investigated. The shelf life and the cycle life are two distinct characteristics. The former refers to the variation in discharge capacity as a function of the quantity of time the charged device is kept between charging and discharging. This property is primarily affected by self-discharging and, as a result, the stability of charged species; it is particularly important for long-term storage requirements. The latter is calculated primarily by the reversibility of the electrochemical and related processes, and is expressed by the development of capacity and coulometric efficacy over many

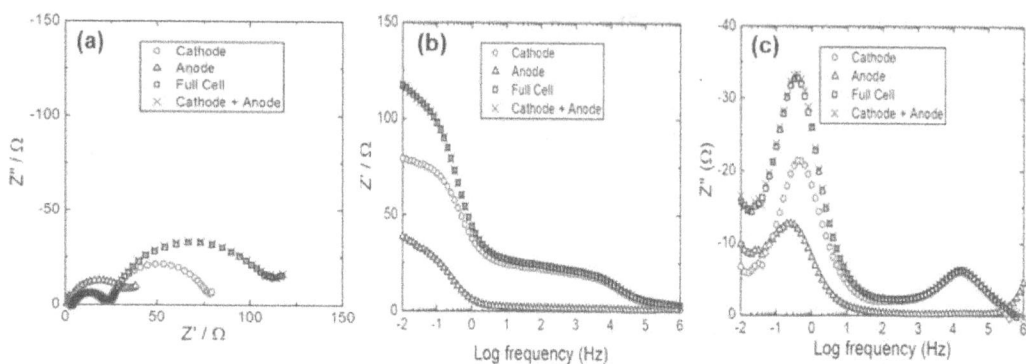

FIGURE 13.13 Some three-electrode EIS data for an in-house constructed sodium-ion battery. (Reproduced with permission from Ref. (101), 2020 Published by Elsevier Ltd.)

charge/discharges cycles. The number of cycles is normally specified until the capacity reaches 80% of its initial value. This property appears to be more relevant in applications that require a lot of charging and discharging in a short period of time (25).

13.27.2 ELECTROCHEMICAL IMPEDANCE SPECTROSCOPY

EIS is an anti-destructive technique that produces a significant amount of data in a very limited amount of time while maintaining the quality of battery packs. This technique allows us to understand the details of the layers and interfaces inside the battery. Here a relatively low-frequency AC signal is applied over a wide range of frequencies and the response is calculated (Figure 13.13). Calibration of the received response provides a distinct characteristic value for various processes inside the battery, which enables the detection of error causes and rate-limiting measures. Monitoring the impedance inside the battery pack clearly states the changes at various levels during battery cycling. Significant examination of impedance values can yield considerably more information than simple plane plots. A three-electrode setup is needed to maximize the effectiveness of the process since the impedance inputs at each electrode must be isolated and analyzed (101).

13.27.3 SPECTROSCOPIC METHODS

Different spectroscopic techniques are excellent for learning about the various processes that occur during electrochemical reactions. Despite the fact that most electrochemically active organic and polymeric species have distinct optical features, spectroscopic experiments on complete devices remain uncommon due to the greater percentage of substances needed to achieve acceptable energy densities. As a result, only sample cells with a reduced content concentration or elaborate models can be used even though nuclear magnetic resonance studies on battery materials are used to understand various chemical species and the interfaces of electrochemical cells in accordance with the signal shifts obtained (102).

13.28 CONCLUSIONS

Conducting polymers are much employed in energy applications such as batteries. Batteries contain conducting polymers as electrolytes, binders, separators, etc. The common conducting polymers are PANI, polyacetylene, their copolymers, biodegradable polymers such as PVDF, etc. When considering batteries, they are characterized by the properties like operating voltage, which is the battery's available charging/working voltage, C-rate, etc. The chapter discussed in detail about these properties and the different characterization techniques to measure the characteristics of batteries that use the conducting polymers. The role of conducting polymers as different components is also detailed. The chapter also discussed the general application of batteries and conducting polymers.

REFERENCES

1. Mahitha PM, Sneha K, Advaith PS, Rashid Sultan K, Sajith M, Jose B, et al. Conducting polyaniline based rubber nanocomposites – Synthesis and characterization studies. Mater Today Proc. 2020;33:1429–33.
2. Hema S, Sambhudevan S, Mahitha PM, Sneha K, Advaith PS, Rashid Sultan K, et al. Effect of conducting fillers in natural rubber nanocomposites as effective EMI shielding materials. Mater Today Proc. 2020;25:274–7.
3. Bhadra S, Chattopadhyay S, Singha NK, Khastgir D. Improvement of conductivity of electrochemically synthesized polyaniline. J Appl Polym Sci. 2008;108(1):57–64.
4. Rajaboopathi S, Thambidurai S. Heterostructure of CdO-ZnO nanoparticles intercalated on PANI matrix for better thermal and electrochemical performance. Mater Sci Semicond Process. 2017;59:56–67.
5. Chandrasekhar P. Basics of Conducting Polymers (CPs). In: Conducting Polymers, Fundamentals and Applications. Springer Science+Business Media, New York. 1999. p. 3–22.
6. S H, Sambhudevan S. Ferrite-based polymer nanocomposites as shielding materials: A review. Chem Pap. 2021. Available from: https://doi.org/10.1007/s11696-021-01664-1
7. Park Y, Jung J, Chang M. Research progress on conducting polymer-based biomedical applications. Appl Sci (Switzerland). 2019;9(6):1070.
8. Guo B, Glavas L, Albertsson AC. Biodegradable and electrically conducting polymers for biomedical applications. Prog Polym Sci. 2013;38:1263–86.
9. Ravichandran R, Sundarrajan S, Venugopal JR, Mukherjee S, Ramakrishna S. Applications of conducting polymers and their issues in biomedical engineering. J R Soc Interface. 2010;7:S559–79.
10. Dubey N, Kushwaha CS, Shukla SK. A review on electrically conducting polymer bionanocomposites for biomedical and other applications. Int J Polym Mater Polym Biomater. 2020;69(11):709–27.
11. Oh WK, Kwon OS, Jang J. Conducting polymer nanomaterials for biomedical applications: Cellular interfacing and biosensing. Polym Rev. 2013;53(3):407–42.
12. Joy N, Gopalan GP, Eldho J, Francis R. Conducting Polymers: Biomedical Applications. In: Biomedical Applications of Polymeric Materials and Composites. Wiley-VCH, Weinheim, Germany. 2016. pp. 37–89.
13. Shahadat M, Khan MZ, Rupani PF, Embrandiri A, Sultana S, Ahammad SZ, et al. A critical review on the prospect of polyaniline-grafted biodegradable nanocomposite. Adv Colloid Interface Sci. 2017;249:2–16.
14. Stejskal J, Hajná M, Kašpárková V, Humpolíček P, Zhigunov A, Trchová M. Purification of a conducting polymer, polyaniline, for biomedical applications. Synth Met. 2014;195:286–93.
15. Kaur G, Adhikari R, Cass P, Bown M, Gunatillake P. Electrically conductive polymers and composites for biomedical applications. RSC Adv. 2015;5:37553–67.
16. Zare EN, Makvandi P, Ashtari B, Rossi F, Motahari A, Perale G. Progress in conductive polyaniline-based nanocomposites for biomedical applications: A review. J Med Chem. 2020;63:1–22.
17. da Silva AC, Córdoba de Torresi SI. Advances in conducting, biodegradable and biocompatible copolymers for biomedical applications. Front Mater. 2019;6:98.
18. Mishra AK. Conducting polymers: Concepts and applications. J At Mol Condens Nano Phys. 2018;5(2):159–93.
19. Vernitskaya TV, Efimov ON. Polypyrrole: A conducting polymer (synthesis, properties, and applications). Uspekhi Khimii. 1997;66:502–5.
20. Kumar R, Singh S, Yadav BC. Conducting polymers: Synthesis, properties and applications conducting polymers: Synthesis, properties and applications. Int Adv Res J Sci Eng Technol. 2016;2(11):110–24.
21. Almeida LCP. Conducting Polymers: Synthesis, Properties and Applications. In: Conducting Polymers: Synthesis, Properties and Applications. Nova Science Publisher, UK. 2013. pp. 1–358.
22. Saxena V, Malhotra BD. Prospects of conducting polymers in molecular electronics. Curr Appl Phys. 2003;3(2–3):293–305.
23. Holze R, Wu YP. Intrinsically conducting polymers in electrochemical energy technology: Trends and progress. Electrochim Acta. 2014;122:93–107.
24. Schmidt-Rohr K. How batteries store and release energy: Explaining basic electrochemistry. J Chem Educ. 2018;95(10):1801–10.
25. Long L, Wang S, Xiao M, Meng Y. Polymer electrolytes for lithium polymer batteries. J Mater Chem A. 2016;4:10038–9.

26. Fenton DE, Parker JM, Wright P V. Complexes of alkali metal ions with poly(ethylene oxide). Polymer (Guildf). 1973;14(11):589.

27. Tan S-J, Zeng X-X, Ma Q, Wu X-W, Guo Y-G. Recent advancements in polymer-based composite electrolytes for rechargeable lithium batteries. Electrochem Energy Rev. 2018;1(2):113–38.

28. Bocharova V, Sokolov AP. Perspectives for polymer electrolytes: A view from fundamentals of ionic conductivity. Macromolecules. 2020;53:4141–57.

29. Murata K, Izuchi S, Yoshihisa Y. Overview of the research and development of solid polymer electrolyte batteries. Electrochim Acta. 2000;45(8):1501–8.

30. Kim JG, Son B, Mukherjee S, Schuppert N, Bates A, Kwon O, et al. A review of lithium and non-lithium based solid state batteries. J. Power Sources. 2015;282:299–322.

31. Yao P, Yu H, Ding Z, Liu Y, Lu J, Lavorgna M, et al. Review on polymer-based composite electrolytes for lithium batteries. Front Chem. 2019;7:522.

32. Croce F, Appetecchi GB, Persi L, Scrosati B. Nanocomposite polymer electrolytes for lithium batteries. Nature. 1998;394(6692):456–8.

33. Zhao Q, Whittaker AK, Zhao XS. Polymer electrode materials for sodium-ion batteries. Materials. 2018;11(12):2567.

34. Novák P, Müller K, Santhanam KS V, Haas O. Electrochemically active polymers for rechargeable batteries. Chem Rev. 1997;97(1):207–82. Available from: https://doi.org/10.1021/cr941181o

35. Yao Y, Liu N, McDowell MT, Pasta M, Cui Y. Improving the cycling stability of silicon nanowire anodes with conducting polymer coatings. Energy Environ Sci. 2012;5(7):7927–30.

36. Kuwabata S, Masui S, Yoneyama H. Charge-discharge properties of composites of $LiMn_2O_4$ and polypyrrole as positive electrode materials for 4 V class of rechargeable Li batteries. Electrochim Acta. 1999;44(25):4593–600.

37. Liu L, Tian F, Wang X, Yang Z, Zhou M, Wang X. Porous polythiophene as a cathode material for lithium batteries with high capacity and good cycling stability. React Funct Polym. 2012;72(1):45–9.

38. Otero TF, Cantero I. Conducting polymers as positive electrodes in rechargeable lithium-ion batteries. J Power Sources. 1999;81–82:838–41.

39. Park KS, Schougaard SB, Goodenough JB. Conducting-polymer/iron-redox-couple composite cathodes for lithium secondary batteries. Adv Mater. 2007;19(6):848–51.

40. Mirmohseni A, Solhjo R. Preparation and characterization of aqueous polyaniline battery using a modified polyaniline electrode. Eur Polym J. 2003;39(2):219–23.

41. Kumar G, Sivashanmugam A, Muniyandi N, Dhawan SK, Trivedi DC. Polyaniline as an electrode material for magnesium reserve battery. Synth Met. 1996;80(3):279–82.

42. Matsunaga T, Daifuku H, Nakajima T, Kawagoe T. Development of polyaniline–lithium secondary battery. Polym Adv Technol. 1990;1(1):33–9.

43. Xiao L, Cao Y, Xiao J, Schwenzer B, Engelhard MH, Saraf L V., et al. A soft approach to encapsulate sulfur: Polyaniline nanotubes for lithium-sulfur batteries with long cycle life. Adv Mater. 2012;24(9):1176–81.

44. Shi HY, Ye YJ, Liu K, Song Y, Sun X. A long-cycle-life self-doped polyaniline cathode for rechargeable aqueous zinc batteries. Angew Chemie – Int Ed. 2018;57(50):16359–63.

45. Banerjee S, Kar KK. Conducting Polymers as Electrode Materials for Supercapacitors. In: Springer Series in Materials Science. Springer, Switzerland. 2020. pp. 333–52.

46. Amanchukwu C V., Gauthier M, Batcho TP, Symister C, Shao-Horn Y, D'Arcy JM, et al. Evaluation and stability of PEDOT polymer electrodes for $Li-O_2$ batteries. J Phys Chem Lett. 2016;7(19):3770–5.

47. Zeng S, Li L, Xie L, Zhao D, Wang N, Chen S. Conducting polymers crosslinked with sulfur as cathode materials for high-rate, ultralong-life lithium–sulfur batteries. ChemSusChem. 2017;10(17):3378–86.

48. Hudak NS. Chloroaluminate-doped conducting polymers as positive electrodes in rechargeable aluminum batteries. J Phys Chem C. 2014;118(10):5203–15.

49. Zhang J, Sun B, Ahn HJ, Wang C, Wang G. Conducting polymer-doped polypyrrole as an effective cathode catalyst for $Li-O_2$ batteries. Mater Res Bull. 2013;48(12):4979–83.

50. Kim J, Park HS, Kim TH, Yeol Kim S, Song HK. An inter-tangled network of redox-active and conducting polymers as a cathode for ultrafast rechargeable batteries. Phys Chem Chem Phys. 2014;16(11):5295–300.

51. Coffey B, Madsen PV, Poehler TO, Searson PC. High charge density conducting polymer/graphite fiber composite electrodes for battery applications. J Electrochem Soc. 1995;142(2):321–5.

52. Lee HY, Jung Y, Kim S. Conducting polymer coated graphene oxide electrode for rechargeable lithium-sulfur batteries. J Nanosci Nanotechnol. 2016;16(3):2692–5.

53. Wang Z, Xu C, Tammela P, Zhang P, Edström K, Gustafsson T, et al. Conducting polymer paper-based cathodes for high-areal-capacity lithium-organic batteries. Energy Technol. 2015;3(6):563–9.

54. Cao D, Shen X, Wang Y, Liu J, Shi H, Gao X, et al. Conductive polymer coated cathodes in Li-O$_2$ batteries. ACS Appl Energy Mater. 2020;3(1):951–6.

55. Chen Z, To JWF, Wang C, Lu Z, Liu N, Chortos A, et al. A three-dimensionally interconnected carbon nanotube-conducting polymer hydrogel network for high-performance flexible battery electrodes. Adv Energy Mater. 2014;4(12):1400207.

56. Chen J, Liu Y, Minett AI, Lynam C, Wang J, Wallace GG. Flexible, aligned carbon nanotube/conducting polymer electrodes for a lithium-ion battery. Chem Mater. 2007;19(15):3595–7.

57. Sengodu P, Deshmukh AD. Conducting polymers and their inorganic composites for advanced Li-ion batteries: A review. RSC Adv. 2015;5(52):42109–30.

58. Gurunathan K, Amalnerkar DP, Trivedi DC. Synthesis and characterization of conducting polymer composite (PAn/TiO$_2$) for cathode material in rechargeable battery. Mater Lett. 2003;57(9–10):1642–8.

59. Zang X, Wang X, Liu H, Ma X, Wang W, Ji J, et al. Enhanced ion conduction via epitaxially polymerized two-dimensional conducting polymer for high-performance cathode in zinc-ion batteries. ACS Appl Mater Interfaces. 2020;12(8):9347–54.

60. Navarro-Suárez AM, Carretero-González J, Casado N, Mecerreyes D, Rojo T, Castillo-Martínez E. Hybrid biopolymer electrodes for lithium- and sodium-ion batteries in organic electrolytes. Sustain Energy Fuels. 2018;2(4):836–42.

61. Nguyen VH, Wang WL, Jin EM, Gu H-B. Impacts of different polymer binders on electrochemical properties of LiFePO$_4$ cathode. Appl Surf Sci. 2013;282:444–9. Available from: https://www.sciencedirect.com/science/article/pii/S0169433213010891

62. Nasybulin E, Xu W, Engelhard MH, Nie Z, Li XS, Zhang J-G. Stability of polymer binders in Li–O$_2$ batteries. J Power Sources. 2013;243:899–907. Available from: https://www.sciencedirect.com/science/article/pii/S037877531301104X

63. Nasybulin E, Xu W, Engelhard MH, Nie Z, Li XS, Zhang JG. Stability of polymer binders in Li-O$_2$ batteries. J Power Sources. 2013;243:899–907.

64. Prasanna K, Subburaj T, Jo YN, Lee WJ, Lee CW. Environment-friendly cathodes using biopolymer chitosan with enhanced electrochemical behavior for use in lithium ion batteries. ACS Appl Mater Interfaces. 2015;7(15):7884–90.

65. Carvalho DV, Loeffler N, Hekmatfar M, Moretti A, Kim GT, Passerini S. Evaluation of guar gum-based biopolymers as binders for lithium-ion batteries electrodes. Electrochim Acta. 2018;265:89–97.

66. Vizintin A, Guterman R, Schmidt J, Antonietti M, Dominko R. Linear and cross-linked ionic liquid polymers as binders in lithium-sulfur batteries. Chem Mater. 2018;30(15):5444–50.

67. Deimede V, Elmasides C. Separators for lithium-ion batteries: A review on the production processes and recent Developments. Energy Technol. 2015;3(5):453–68.

68. Reddy MV, Mauger A, Julien CM, Paolella A, Zaghib K. Brief history of early lithium-battery development. Materials (Basel). 2020;13(8):1884.

69. Armand M, Tarascon JM. Building better batteries. Nature. 2008;451:652–7.

70. Larcher D, Tarascon JM. Towards greener and more sustainable batteries for electrical energy storage. Nat Chem. 2015;7:19–29.

71. Snook GA, Kao P, Best AS. Conducting-polymer-based supercapacitor devices and electrodes. J Power Sources. 2011;196:1–12.

72. Muench S, Wild A, Friebe C, Häupler B, Janoschka T, Schubert US. Polymer-based organic batteries. Chem Rev. 2016;116:9438–84.

73. Nishide H, Oyaizu K. Toward flexible batteries. Science. 2008;319:737–8.

74. Kim US, Shin CB, Kim CS. Modeling for the scale-up of a lithium-ion polymer battery. J Power Sources. 2009;189(1):841–6.

75. Wang G, Zhang L, Zhang J. A review of electrode materials for electrochemical supercapacitors. Chem Soc Rev. 2012;41(2):797–828.

76. Miroshnikov M, Divya KP, Babu G, Meiyazhagan A, Reddy Arava LM, Ajayan PM, et al. Power from nature: Designing green battery materials from electroactive quinone derivatives and organic polymers. J Mater Chem A. 2016;4:12370–86.

77. Kermani G, Sahraei E. Review: Characterization and modeling of the mechanical properties of lithium-ion batteries. Energies. 2017;10:1730.

78. Manthiram A. A reflection on lithium-ion battery cathode chemistry. Nat Commun. 2020;11:1550.

79. Ditch B, Yee G, Chaos M. Estimating the Time-of-Involvement of Bulk Packed Lithium-Ion Batteries in a Warehouse Storage Fire. In: Fire Safety Science. International Association for Fire Safety Science, University of Canterbury, Christchurch, New Zealand. 2014. pp. 1024–34.

80. Gharbi O, Tran MTT, Tribollet B, Turmine M, Vivier V. Revisiting cyclic voltammetry and electrochemical impedance spectroscopy analysis for capacitance measurements. Electrochim Acta. 2020;343:136109.

81. Kim T, Choi W, Shin HC, Choi JY, Kim JM, Park MS, et al. Applications of voltammetry in lithium ion battery research. J Electrochem Sci Technol. 2020;11:14–25.

82. Benshatti A, Islam SMR, Link T, Park SY, Park S. Design and Control of AC Current Injector for Battery EIS Measurement. In: Conference Proceedings – IEEE Applied Power Electronics Conference and Exposition – APEC. 2020. p. 3452–5.

83. Gharbi O, Tran MTT, Tribollet B, Turmine M, Vivier V. CPE analysis from cyclic voltammetry and electrochemical impedance spectroscopy. ECS Meet Abstr. 2020;MA2020-02(20):1573–1573.

84. Osaka T, Nara H, Mukoyama D, Yokoshima T. New analysis of electrochemical impedance spectroscopy for lithium-ion batteries. J Electrochem Sci Technol. 2013;4(4):157–62.

85. Patel MUM, Demir-Cakan R, Morcrette M, Tarascon JM, Gaberscek M, Dominko R. Li-S battery analyzed by UV/vis in operando mode. ChemSusChem. 2013;6(7):1177–81.

86. Lozeman JJA, Führer P, Olthuis W, Odijk M. Spectroelectrochemistry, the future of visualizing electrode processes by hyphenating electrochemistry with spectroscopic techniques. Analyst. 2020;145:2482–509.

87. Bartl A, Doege HG, Froehner J, Domschke G. In-situ EPR and charge storage in conjugated organic compounds. Synth Met. 1991;43(1–2):2881–4.

88. Andersson AM, Abraham DP, Haasch R, MacLaren S, Liu J, Amine K. Surface characterization of electrodes from high power lithium-ion batteries. J Electrochem Soc. 2002;149(10):A1358.

89. Shutthanandan V, Nandasiri M, Zheng J, Engelhard MH, Xu W, Thevuthasan S, et al. Applications of XPS in the characterization of battery materials. J Electron Spectros Relat Phenomena. 2019;231:2–10.

90. de Vries J, Ditch BD. Multi-Spectral Infrared Analysis of Lithium-Ion Battery Bulk-Storage Fire Tests. Research Publishing, Singapore. 2013. pp. 271–80.

91. Novák P, Panitz JC, Joho F, Lanz M, Imhof R, Coluccia M. Advanced in situ methods for the characterization of practical electrodes in lithium-ion batteries. J Power Sources. 2000;90(1):52–8.

92. Liu X, Liang Z, Xiang Y, Lin M, Li Q, Liu Z, et al. Solid-state NMR and MRI spectroscopy for Li/Na batteries: Materials, interface, and in situ characterization. Adv Mater. 2021 (in press: https://doi.org/10.1002/adma.202005878).

93. Gamez G, Finch K. Recent advances in surface elemental mapping via glow discharge atomic spectrometry. Spectrochimica Acta – Part B Atomic Spectroscopy. 2018;148:129–36.

94. Schwieters T, Evertz M, Mense M, Winter M, Nowak S. Lithium loss in the solid electrolyte interphase: Lithium quantification of aged lithium ion battery graphite electrodes by means of laser ablation inductively coupled plasma mass spectrometry and inductively coupled plasma optical emission spectroscopy. J Power Sources. 2017;356:47–55.

95. Kahr J, Rezqita A, Antoniadou M, Rosenberg E, Fontana D, Jahn M. Fast operando GCMS gas analysis for monitoring electrolyte decomposition in lithium ion batteries. ECS Meet Abstr. 2020;MA2020-01(1):108–108.

96. Lagadec MF, Zahn R, Wood V. Characterization and performance evaluation of lithium-ion battery separators. Nature Energy. 2019;4:16–25.

97. Hiesgen R, Sörgel S, Costa R, Carlé L, Galm I, Cañas N, et al. AFM as an analysis tool for high-capacity sulfur cathodes for Li-S batteries. Beilstein J Nanotechnol. 2013;4(1):611–24.

98. Yang H, Kwon K, Devine TM, Evans JW. Aluminum corrosion in lithium batteries an investigation using the electrochemical quartz crystal microbalance. J Electrochem Soc. 2000;147(12):4399.

99. Zheng Q, Xing F, Li X, Liu T, Lai Q, Ning G, et al. Investigation on the performance evaluation method of flow batteries. J Power Sources. 2014;266:145–9.

100. Scrosati B. Recent advances in lithium ion battery materials. Electrochim Acta. 2000;45(15–16):2461–6.

101. Middlemiss LA, Rennie AJR, Sayers R, West AR. Characterisation of batteries by electrochemical impedance spectroscopy. Energy Reports. 2020;6:232–41.

102. Pecher O, Carretero-Gonzalez J, Griffith KJ, Grey CP. Materials' methods: NMR in battery research. Chem Mater. 2017;29:213–42.

14 The Role of Chalcogenide in Conducting Polymers for Enhanced Battery Performance

Nobel Tomar[1], *Deepti Rawat*[2], *Parmod Kumar*[1],
and Rahul Singhal[3]

[1]J.C. Bose University of Science and Technology, Faridabad, India
[2]Miranda House College, University of Delhi, India
[3]Shivaji College, University of Delhi, India

CONTENTS

14.1 INTRODUCTION

Demand for the fabrication of high-power density energy storage devices is continuously increasing with the advent of portable electronic devices. Thus, for sustainable supply of energy, an efficient energy storage system is required, which majorly is dominated by alkali metal ion batteries like lithium-ion batteries (LIBs); having extraordinary energy density, extensive shelf life and variable temperature range unlike other energy storage systems [1–3]. LIBs work on the principle of converting chemical energy into electrical energy by using electrodes and electrolytes. Nowadays, there is an increased demand for flexible wearable electronics due to our hectic lifestyle. So for flexibility, polymers can be used but they are generally insulators that are known to show conductivity with a conjugated system of extended pi electrons and large carrier concentration. Thus the solution was found in conducting polymers (CPs) which are organic molecules having a conjugated carbon system [4]. They were used since 1980 in Li-ion batteries, which later found application in various battery systems [5–7]. They have shown great electrochemical performance in various battery systems and are also known to show magnetic, electrical and optical properties which resemble with metals and semiconductors. The uniqueness about conducting polymers is that it combines the properties of plastic as well as metals. Thus can work as synthetic metals.

DOI: 10.1201/9781003150374-14

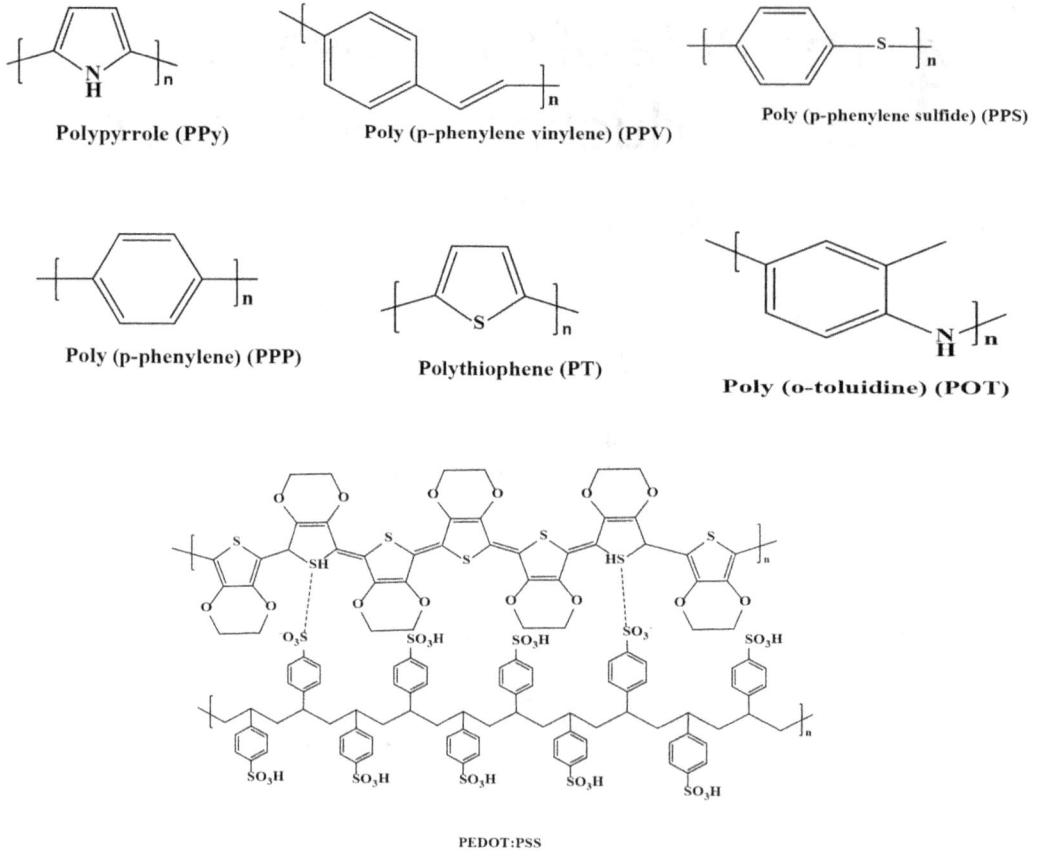

FIGURE 14.1   Examples of some common conducting polymers.

Recent research deals with improving the intrinsic properties of CPs and tailoring them as most of them are known to be extrinsically conducting. The most commonly used CPs for battery applications are polyheterocycles, such as polypyrrole (PPy), polyaniline (PANI), polythiophene (PTh), poly(3,4-ethylenedioxythiophene) (PEDOT) and their composites as shown in Figure 14.1 having 0.01–500 S cm^{-1} as the electronic conductivity range [8] (Table 14.1). The conducting polymers are known to show conductivity due to the conjugated double bonds present. The electrolyte ions get extracted and inserted during the charging and discharging of the conducting polymers, which work between the two known states p-type doped and n-type doped. A conducting polymer can be synthesized in an undoped state (pristine) or in a doped state. For the pristine CPs, doping can be

TABLE 14.1

Conductivity Values of a Few of Conducting Polymers (Reproduced from Refs. 29–31)

Conducting Polymer	Conductivity (Scm^{-1})	Type of Doping
Polyaniline (PANI)	30–200	n, p
Polypyrrole(PPy)	10–7500	p
Poly(3,4-ethylenedioxythiophene)	0.4–400	n, p
Polyacetylene	200–1000	n, p
Polyparaphenylene (PPP)	500	n, p

done either chemically or electrochemically and the dopants used can be anions (ClO_4^-, BF_4^-) or cations (Na^+, H_3O^+). The doping is measured by the number of dopant ions per monomer unit and generally depends on the extent of oxidation/reduction of the polymer. Oxidation of the conducting polymer results in the generation of a positive charge and an associated anion, thus the charges are strongly delocalized on several monomer units, which results in relaxation of the geometry of the polymer to a more energetically favoured conformation. Reduction can be observed by donation of charge to the conduction band of conducting polymer thus generating a negatively charged conducting polymer and an associated cation. Thus, addition results in the insertion of holes into the HOMO (p-type doping) or electrons into the LUMO (n-type doping) to create free-charge carriers, thus showing high electrical conductivity [9]. So, during the electrochemical oxidation and reduction, the electroneutrality of electrodes is maintained by accepting or giving off the ions thus causing doping and undoping of polymer. Thus conducting polymers act as both electronic and ionic conductors as charge compensating ions can move within the polymers. In the case of polymers, it is difficult to attain the doping level of 1:1, some polymers such as poly(pyrrole) show maximum doping of 33% for ClO_4^- and poly(thiophene) of 34% by ClO_4^- [10].

In 1996, John Hopkins University researchers showed that battery electrodes and electrolytes can be made up of polymers that conduct polymers unlike the previous designs that used CPs only as cathode. The design gains on high plasticity and lightweight along with good battery performance [11–13]. CPs can be moulded to various shapes so they can be easily integrated into electronic paper, structural panels etc. [14, 15]. A lot of work has been reported in the past, which majorly focuses on the usage of conducting polymers in battery technology [16–18]. For battery performance, the polymer conductivity depends upon the depth of the charge-discharge cycle, which is determined by the polymer's doping level. Chalcogens and metal chalcogenides are well-known potential candidates for anodes in batteries, especially alkali metal ion ones, due to the large range of species, abundance, cheap, exceptional physical and chemical properties, and high theoretical capacities [19–31]. These semiconductor chalcogenides are profoundly investigated for renewable energy conversion devices both as electrodes as well as dopants in electrolytes along with conducting polymers. In the current chapter, we want to elaborate on the role of chalcogenide in improving the properties of conducting polymers in batteries, their various ways of synthesis, how it affects their electrochemical performance, the challenges and future outcomes.

14.2   LITHIUM-ION BATTERIES

For an uninterrupted supply of energy, the system has to build which can store energy thus bridging the gap between energy generated through renewable power sources and high energy demand. Nowadays, mostly electronic devices come in handy, such as smartphones, computers, integrated circuit smart cards, and sensors [32, 33]. For these gadgets to work at their maximum capacity, they need a continuous energy supply which is dominated by Li-ion batteries. The performance of the energy storage device is controlled by ion/electron transport inside the battery. It depends on three most important parameters.

1. Energy density: Energy density is mainly determined by the specific capacity and loading of active material.
2. Power density: Power density is proportional to the current that LIB generates during the discharge process.
3. Cyclic stability: Controlled by factors such as structural/phase stability of active material, solid electrolyte-electrode interphase, contact stability for electron transport and electrolyte stability for ion conduction.

Lithium-ion batteries have high energy densities, long shell life, competitive operating voltage and low-cost fabrication making them ideal and commercially available energy storage systems. The growing demand of the market lies in improving power and energy densities in the electrode material. Thus tailoring the battery designs to improve their efficiency is the major task which includes tuning the electrode thickness, particle size of the electrolytes being used, cell geometry and the speed of diffusion of the ions through the solid phase [34, 35]. LIB comprises of solid inorganic particles as active material (cathode and anode particles), organic liquids or inorganic/ polymeric solids as electrolytes, polymeric materials as separator or binder [36], carbon nanomaterials conductive agents and metals as a current collector. In working of LIB, the electrode serves as a reacting warehouse where storing/releasing of Li^+ accompanied by electron transfer in active material takes place. Electrolytes and separators work in conjunction to block electron transport but allow only the transport of ions across the electrode.

The working current of the device during charge/discharge is collectively contributed by the following factors: electron transport along with Li^+ ion transport and electrochemical reactions in both cathode and anode. Since the external circuit is in series with ion/electron transport inside, the maximum current available for LIB is determined by slowest one among of all these factors.

During the past few decades, considerable efforts have been made to modify inside ion-electron transport systems for LIB. Due to high energy density and lightweight, LIB has captured the global market, but growing concern regarding sustainability and cost of Li, researchers are exploring towards other promising areas of battery research. Replacement of Li with more abundant materials, researchers have also focused their attention towards other alkali metal ion like potassium and sodium, which can be used in batteries. But as we move down the periodic table, the increase in ionic radii results in further decreased mobility of metal ions which is already a limitation in the case of LIBs, thus leading to sluggish movement in the host materials. This leads to faster degradation of the batteries due to enhanced ionic interaction of host material with the negative charges resulting in less energy holding capacity than lithium-ion batteries. Thus the efficiency of sodium and lithium-ion batteries falls way below the expectation for energy storage devices.

14.3 Li-S BATTERIES

We have seen and understood that lithium-ion batteries are one of the major devices used for storage. They have already reached the energy density limit so more research aims at developing high energy density materials with low cost and more stability. Chalcogenide family can provide potential candidates due to low cost, stability and high theoretical capacity. Thus transition metal chalcogenide (TMD) has attracted the attention for using them commercially to enhance the performance of Li-ion batteries. The electrode-specific energy of Li-S is found to be 2600 Wh/kg and was used for the first time in 1962. The major limitation faced by these products is low sulphur utilization, poor conductivity of sulphur and discharge products like Li_2S. With all these limitations, a great deal of work is going on in improving the energy capacity of Li-S batteries as S, which is used as a cathode material has a high energy capacity of 1675 mAhg^{-1}. The driving force for these batteries are direct chemical reactions:

At anode: $2Li \rightarrow Li^+ + 2e^-$
At cathode: $S + 2e^- \rightarrow S^{2-}$

The combined reaction can be written as $2Li + S \rightarrow Li_2S$.

The formation of dilithium sulphide as the final reduction product takes place through the formation of series of intermediates like S_6^{2-}, S_3^{2-}, S_3^- (Figure 14.2). Although these batteries have a promising future but practical difficulties like shuttle effect of polysulphides [37–39] make them less efficient. This effect rapidly decreases the battery capacity by significant sulphur migration, high self-discharge etc. A lot of work is underway to reduce the sulphur dissolution and increase the electronic conductivity of these cathodes. This use of the right kind of separator reduces this self-discharge problem and tuning the correct composition of the electrolyte can help in reducing the

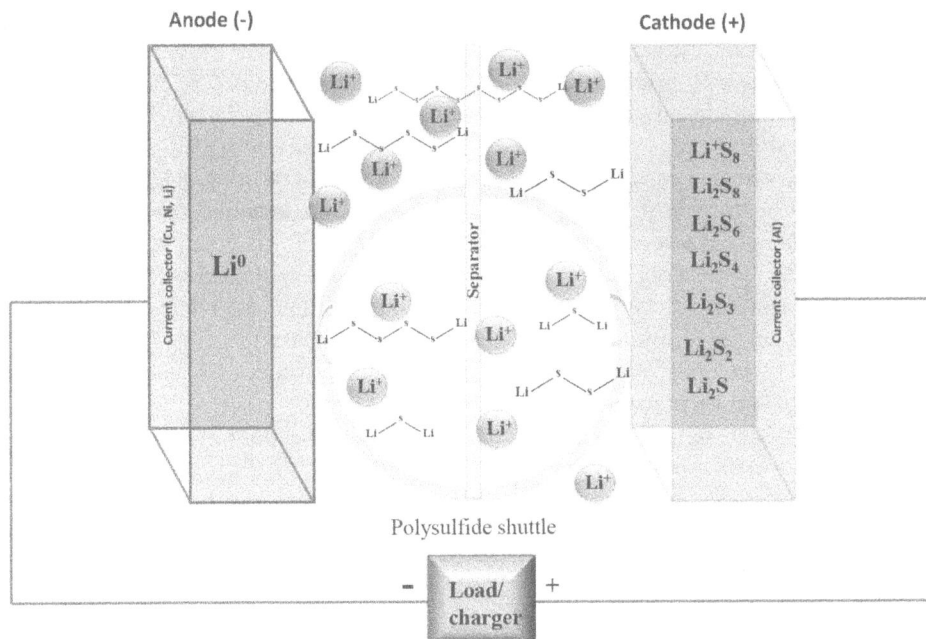

FIGURE 14.2 Working of Li-sulphur batteries.

polysulphide issue, thus enhancing the battery capacity [40]. The liquid electrolytes used generally are of majorly two types: (i) ethers such as tertraethylene glycol and diglyme; (ii) organic carbonates such as diethyl carbonate and propylene carbonate.

Literature study [41, 42] showed the reports that the tetra (ethylene glycol) dimethyl ether (TEGDME) and 1,3-dioxolane (DOL) were mixed in the ratio of 1:2 by volume to give the best discharge capacity. It was proved by Choi et al. that DOL contributed to improved compatibility with lithium and reduced viscosity thus giving increased cycling stability for Li-S cells. Zhang et al. support the usage of Li-S batteries by promoting the polysulphide dissolution and managing the interaction of electrodes [43, 44]. After a detailed study by Boros et al. [45] on the dissolution of sulphides in ionic liquids it was concluded that they could also be used in the mixture of electrolytes can reduce the polysulphide dissolution, but at this stage, no combination has yet been proved to be significantly reducing the steady decrease in discharge capacity of the batteries.

Various limitations shown by Li-S batteries are due to the liquid electrolyte solutions i.e. diffusion, migration, dissolution thus the development of solid electrolytes became the new area of research. Solid electrolytes can be used both as polymer or non-polymer electrolytes.

But the best performance to date is found when the materials and the configuration of the cell are chosen in such a way that the limitations are controlled by more than one means like tuning the sulphur concentration by suppressing the solubility of these species in electrolytes, designing cathodes having high sulphur affinity like that made up of reduced graphene oxide (rGO), etc. [46].

14.4 APPLICATIONS OF CONDUCTING POLYMERS IN BATTERY

The conducting polymers are known to be classified into three different categories on the basis of type of conduction: proton conducting, electron conducting and ion conducting. All three types have applications in the electronic industry. Polyacetylene, polypyrrole, PEDOT, PANI are the most commonly used conducting polymers used in battery application. They can be employed both as cathodic and anodic materials but generally used as cathodes in Li-ion batteries. The conduction by electrons in CPs can be tuned by varying the chemical structure of the conjugated polymeric backbone. This can be done by doping using either electron donors (i.e. alkali metals, etc.) or electron acceptors (i.e. halides, etc.).

Polyacetylene has widely been used as anode and cathode electrodes in batteries. It exhibits excellent battery performance. Literature reports that n-(CH)/LiClO$_4$/p-(CH) (in 1.0 LiClO$_4$ in propylene carbonate) exhibited open circuit potential of 3.5 V and a current density of 28 mA/cm^2 [47]. Experimentally, the energy density has been calculated to be 424 W h/kg [47, 48]. Polyacetylene battery showed outstanding results with each cycle of 10-minute discharge at a constant resistance of 50 kΩ and a 10-minute charge at 0.1 mA even after 2000 successive charge/discharge cycles. This configuration claims high energy density, flexibility, lightweight, high power and portability. The use of polyacetylene in the battery greatly enhanced the surface characteristics such as the uniform movement of Li ions in the film and weaving of (CH) network and the prevention of precipitation of Li-ion on the cathode surface. The use of dehydrated LiClO$_4$, propylene carbonate in the polyacetylene aids the battery to achieve long-term stability.

PANI is another interesting CP that is extensively studied for the improvement of battery performance. Among its various forms, PANI-EB emeraldine base which is the half oxidized state of PANI finds application in batteries, supercapacitors and fuel cells due to its stability in air and humidity. The other two forms which are formed by complete oxidation and reduction respectively, are not stable in air and are easily degradable [49]. PANI exhibits good electrochemical sensitivity, that is, as. It is synthesized from aniline as a monomer by chemical or electrochemical oxidation [50]. The electrochemical method of polymerization is generally preferred because it is faster and environmentally benign as it is free of oxidant or the addition of any additives. Owing to their simple facile synthesis, cycling stability and energy density, they are widely explored in batteries. It has been reported that polyaniline-based CPs have a theoretical capacity of 294 mA h/g because of their high doping level. But the experimental capacity of PANI is less than the theoretical capacity. Practically, it is difficult to exceed a capacity greater than 150 mA h/g [51].

It has also been used as an anode material by using Li-doped aluminium as cathode material [52]. Polyaniline/graphene composite paper or thin film [53–56] has also been prepared, which generates high electrochemical performance and high flexibility [57]. Polyaniline with Li–Al alloy is now used in liquid crystal display (LCD) and memory backup of power source of random access memory (RAM) [58]. Cheng and co-workers have shown the highest discharge capacity after 80 cycles using doped PANI nanotubes [59].

Poly (p-phenylene) (PPP) batteries are electroconducting polymers that have a high redox potential. PPP is used both as anode and cathode materials in the battery due to its extraordinary stability and processability. It can be selectively used along with other metals. PPP, when employed in Li batteries, exhibits an open circuit potential of 4.4 V, which is higher than polyacetylene/Li battery due to elevated ionization potential of pristine PPP (5.6 eV) than polyacetylene (4.7 eV). The major drawbacks for PPP are lower specific capacity, instability due to irreversible oxidation reaction at elevated temperatures and in conventional non-aqueous solvents like propylene carbonate [60], and lower efficiency due to self-oxidation reaction. These limitations hinder its practical utilization in rechargeable Li batteries [57]. Heterocyclic conducting polymers have been employed in multivalent metal-ion batteries. Polythiophene is one of the interesting heterocyclic classes of electroactive material and can be used both as anode and cathode for battery application. In some cases, it has been proven that polythiophene is used for the solubility of the redox-active species [61]. A polypyrrole-based battery fabricated showed a charge capacity of 82 mAhg^{-1} [62, 63]. It has been reported that a polythiophene-based battery offers a power density and an energy density of 89 W/kg and 93 W h/kg, respectively [64, 65]. The major advantage of the heterocyclic polymer battery is its stability at room temperature and that it can be easily modified to fabricate a suitable electrochemical system so, further research should be focused on the heterocyclic CP with respect to surface characteristics, long-term stability, flexibility, and so on, so that the future energy-related problems might be solved. Table 14.2 summarizes the comparison of the performance of conducting polymers with Li-ion battery materials.

TABLE 14.2

Performance Characteristics of Some Conducting Polymers Compared to Current Inorganic Li-Ion Battery Materials

Electrode	Capacity (mAhg⁻¹) (Theoretical	Capacity (Experimental)	Energy Density (mWhg⁻¹)	Power Density (mWhg⁻¹)	Reference
Graphite	372	360	372–744	-	[66]
LiCoO₂	274	135	30–144	100–10000	[67]
LiFePO₄	170	109	90–100	23–70	[68]
Polythiophene	326	82	93	89	[64, 65]
Polypyrrole	412	82	60	10	[62, 63]
PEDOT	191	30–70	1–4	35–2500	[67]
Polyaniline	294	100–147	300	100	[51]

14.5 STRATEGIES TO FABRICATE CP/CHALCOGENIDE NANOCOMPOSITES

Conducting polymers properties can be tuned when metal chalcogenides, especially TMC are added in nanometre range to the polymers to form a composite. These nanocomposites show enhanced properties since the metal chalcogenides added work as fillers thus symbiotically enhancing the stability and properties of the material unlike when polymer or chalcogenides are used independently for the energy storage application. These MCs do not achieve the theoretical rates of charging/discharging practically. But the nanoscale chalcogenides show enhanced diffusivity based on Fick's law due to a decrease in diffusion length thus enhancing the cycling stability and rate capability. The kinetics can be enhanced by designing the morphologies of composites in various forms like core-shell, porous, hierarchical, etc. Nanocomposites comprise of unique category of composite designed by combination of two or more components in which one of dispersed phase is in nanoscale range which possess high surface volume ratio thus contributing to new behaviours and functionalities. The properties of the polymeric backbone can be tuned and the interfacial interaction of metal-C bond can be enhanced by the presence of heteroatom which suppresses the dissolution of chalcogenides as well lead to change in band structure thus giving better rate capability and high electronic conductivity. The characteristics of conducting polymers can further be enhanced by forming composites with metal oxides, chalcogenides, clays, etc. In this section, we describe three strategies for fabrication of conducting polymer/metal chalcogenide synthesis: in situ, ex situ and one pot.

14.5.1 IN SITU SYNTHESIS

This is the most widely used method to synthesize metal chalcogenide/CP nanocomposites with varied morphologies. In this method, the metal chalcogenides are prepared in nanometre range by tuning their size along with the polymerization process taking place in the presence of an oxidant. The nano range building blocks of metal chalcogenides can control the geometry of resultant nanotubes/wires/particles formed. The resultant composite formed can be tailored and tuned during synthesis by controlling the ratio of oxidant to monomer added, temperature and time for which polymerization is done as well controlling the amount of dopant loaded on the polymer. This process results in efficient attachment of organic moieties to the inorganic metal chalcogenides thus showing a better interfacial interaction between the components.

Polyaniline is the most commonly used conducting polymer to incorporate nanostructures of TMD [69–87]. Poly(3,4-ethylenedioxythiophene) (PEDOT) [88, 89], polythiophene (PTh) [90]

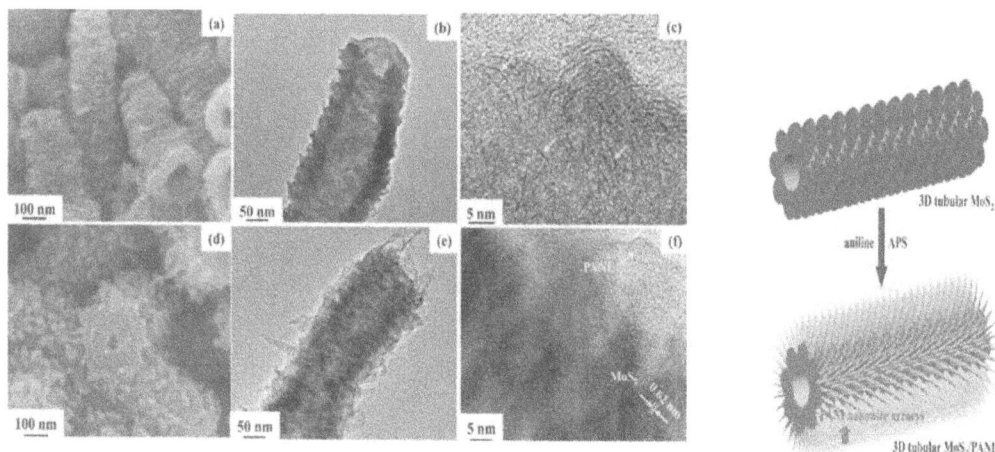

FIGURE 14.3 FESEM, TEM, and HRTEM images of MoS$_2$ (a–c) and MoS$_2$/PANI-60 (d–f). Schematic illustration of the 3D tubular MoS$_2$/PANI. (Reproduced with permission. Copyright ACS, American Chemical Society.)

and polypyrrole (Ppy) [91, 92] have also been used to synthesize CP/MC hybrid have also been employed to construct nanohybrid composites.

Chen et al. [59] had synthesized a stable organic–inorganic hybrid composite of Mo disulphide loaded on an organic polymeric backbone of PANI by the same polymerization technique. The MoS$_2$ was dispersed in 1:1 mixture of water and ethanol and ultrasonicated to produce well-dispersed colloidal suspension. The monomer aniline was added dropwise to the colloidal suspension of MoS$_2$. The mixture solution was cooled in an ice bath followed by the addition of oxidant ammonium persulphate to initiate the polymerization of aniline. Similar methodology and oxidants were used by Liu and co-workers [64] synthesized 3D tubular morphology of MoS$_2$/PANI hybrid composite. The tubular structure is the result of the unsaturated dangling bonds of Mo and S along the sheets. This results in energetically unfavourable interlayer interaction thus forcing the sheets to bend and resulting in nanostructured tubes. This strategy has also been applied by some other groups to prepare CP/MC nanocomposites. This methodology results in variable morphology and which can be tuned the way nanowires are growing around the polymeric chains thus improving the transport and storage performance. The conducting polymer can also intercalate between the MoS$_2$ nanosheets besides growing on the surface.

The doped PANI was prepared using dodecyl benzene sulphonic acid (DBSA) as a doping agent. The reaction took place in the presence of a strong oxidant which results in initiating the reaction as well. The usage of dopant also enhances the organic–inorganic hybrid interaction of MoS$_2$ layers with the polymeric chain by bringing the polarity of both close to each other. This method results in better intercalation of the MCs into the polymeric chain.

Transmission electron microscopy (TEM) image, as shown in Figure 14.3 clearly depicts the exfoliated morphology with a large surface area. The enhanced conductivity of layered MoS$_2$/polyaniline nanocomposites is shown in Figure 14.4 [59–63].

Fabrication of hierarchical hybrid MC/CP nanocomposites is achieved by constructing 2D transition metal dichalcogenide nanosheets in the presence of a polymer matrix. Yang et al. [57] for demonstrated fabrication of through MoS$_2$/PANI nanowires through a bottom up approach. MoSx/PANI nanowires involved polymerization of anilinium trimolybdate, which on treatment with thiourea converted to MoS$_2$/PANI nanowires via hydrothermal process. High-resolution transmission electron microscopy (HRTEM) technique revealed that MoS$_2$ nanosheets were embedded into the polymer in a particular direction thus enhancing its anodic properties making it a prominent material for lithium-ion batteries. The polyaniline hybrid and hierarchical structure presents outstanding performance for Li storage presenting new opportunities for developing electrode nanomaterial.

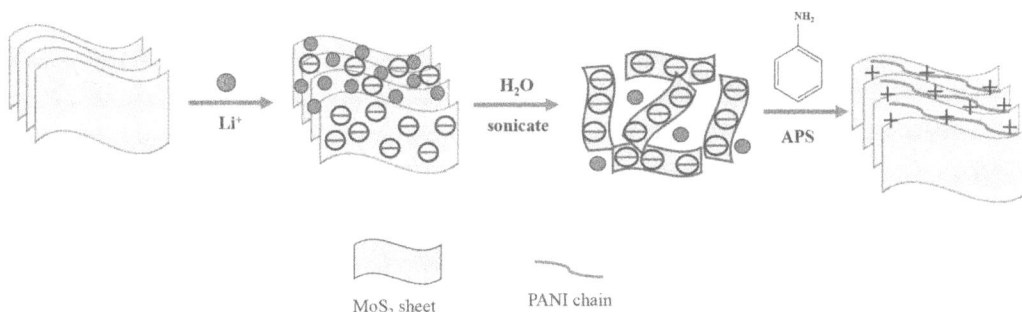

FIGURE 14.4 Diagrammatic representation of PANI and MoS$_2$ nanosheets.

14.5.2 Ex Situ Synthesis

This method involves the synthesis of the conducting polymer and the nano chalcogenide by a standard method separately and finally blending both of them using various known procedures. Two major approaches to synthesize metal chalcogenides are top-down (exfoliation) and bottom-up (wet chemical synthesis) for layered MCs and non-layered structures, respectively. This wet chemical synthesis at low temperature mostly forms amorphous structures with abundant defects that may affect the band structure and finally the performance. The ex situ method may involve either controlled addition of one into another through solution or layer by layer deposition of one component on another using electrochemical process. The methodology decides the properties as the nanofillers used for loading the polymer have high interfacial tension, which does not allow a strong interaction between the organic–inorganic moieties. The structure of monomer, size and morphology are defined previously.

14.5.2.1 Solution Mixing Method

As the name suggests, the polymer and inorganic metal chalcogenides are dispersed separately in different solutions and then mixed homogeneously either by mechanical stirring or using the technology of sonication. This is the direct and simplest method for synthesizing MC/CP nanocomposites [68–71]. TMD and conducting polymers nanosheets are synthesized separately and dispersed in the solvent and sonicated, followed by mixing the two components.

Films of disulphides and diselenides of molybdenum combine with conducting polymer (PEDOT) finds application in dye-sensitized solar cells as counter electrodes. They are cheaper than the conventional platinum electrodes already in use. To enhance the interactions and electron transfer from nanosheets of MCs to the polymeric matrix, the film of the mixture of conducting polymer can be used i.e. PEDOT:PSS [63, 64].

The same conductive matrix was used as a dispersing medium for synthesis of MoS$_2$/poly(3,4-ethoxythiophene):poly(styrenesulphonate) nanocomposite by using a dip coating method in an ultrasonic cell [40]. This is the facile dip-coating method synthesis that can be easily applied to other electrodes. Kim et al. [67] incorporated SnSe nanosheets into the matrix of conducting polymer, PEDOT:PSS, by solution process fabrication method. SnSe nanosheets were prepared from layered structure of synthesized bulk SnSe ingot via Li intercalation method followed by exfoliation process. The exfoliated SnSe exhibited 2D sheet structure was distributed uniformly over PEDOT:PSS matrix by dip-coating method.

Jiang et al. [66] have also reported the synthesis of organic solvent-assisted exfoliated MoS$_2$ nanosheets using vacuum filtration.

14.5.2.2 Electrophoretic Deposition

Electrophoretic deposition is a low-cost technique that can be used easily to scale up for bulk production of nanomaterial films of desired shape and size on a conducting substrate. The charged particles dispersed in a colloidal solution move under the effect of the electric field. The suspension

TABLE 14.3

Highlights of Advantages and Disadvantages Associated with the Different Methodology

Methodology	Advantages	Disadvantages	Examples
In situ	Prevents inorganic particle agglomeration while retaining good spatial distribution in polymer matrix	Complexity may be attained due to involvement of number of parameters	In situ polymerization, electrodeposition, in situ reduction
Ex situ	Simple solution – processable	Limited applications, poor control of interfacial contact between each component	Mechanical mixing, layer by layer deposition
One pot	Simple method, short processing time	Limited control over structure and morphology of the product	Redox reaction, co-precipitation

is prepared using an ultra sonicator in a preferably common solvent. Jun et al. [72] have reported the deposition of layered MoS_2 sheets prepared by exfoliation method on stainless sheet substrate using an electrode made up of graphite. Nanowires of polyaniline grown in one dimension using oxidative polymerization were used during this method. In this case, both exfoliated PANI and MoS_2 nanosheets were suspended in acetonitrile and sonicated for homogeneous dispersion. In contrast to the chemically synthesized method, this process enables precise control over film deposition, thickness with uniformity material to improve the electrochemical performance.

14.5.3 ONE POT SYNTHESIS

This methodology uses a unique approach where the reactants simultaneously react in a single reactor or pot. The polymerization process, as well as the formation of nano range metal chalcogenides takes place simultaneously, resulting in combining all of the reactants [72]. Wang et al. [75] synthesized for the first time algal-like MoS_2/PPy nanocomposites in one pot reaction. The composite was prepared through a redox reaction between ammonium tetra thiomolybdate $[(NH_4)_2MoS_4]$ and chemical oxidative polymerization of pyrrole using the hydrothermal route. The pressure in the hydrothermal vessel helps in the reduction of ammonium molybdate to molybdenum disulphide. Structural morphology revealed that nanosheets of MoS_2 were grown and covered by a sheet of Ppy. The self-assembled structure is obtained with an algal-like morphology which was further confirmed by TEM. Nanohybrids obtained by in situ methods have several advantages over ex situ method because organic–inorganic interface can be controlled at the molecular level. Highlights of advantages and disadvantages associated with the different methodology adopted are shown in Table 14.3.

14.6 APPLICATIONS OF CONDUCTING POLYMER/METAL CHALCOGENIDES IN BATTERY PERFORMANCE

Metal chalcogenides such as MoS_2, NiS, SnS and CoS are extensively used in high energy density batteries as cathodic materials due to their structural architecture and outstanding theoretical capacitance. TMDs like MoSe, TeS_2 and WTe_2 are promising materials for energy storage devices like in battery applications. They possess good electric double-layer properties and possess excellent mechanical and electrochemical properties. The nanotubes/wires/particles of MX, MX_2 and ternary type surface-functionalized metal chalcogenides can be built and tailored to show unique physical properties. Nanosheets of metal dichalcogenides have been extensively employed as

TABLE 14.4

Various Chalcogenide/Conducting Polymer Nanocomposites Used as Electrodes in Battery and Their Performance

Electrodes	Electrolyte	Battery Performance	Cycle Performance	Reference
MoS_2-Py-rGO	1 M $LiPF_6$ in mixture of ethylene carbonate/ dimethyl carbonate (1:1)	Specific capacity 1085 mAh g^{-1} at 0.2 A g^{-1}	1070 mAh g^{-1} is retained after 400 cycles at 0.2 A g^{-1}	[75]
MoS_2/PANI	1 M $LiPF_6$ in mixture of ethylene carbonate/ dimethyl carbonate (1:1)	Specific capacity 893 mAh g^{-1} at 0.1 A g^{-1}	801.2 mAh g^{-1} is retained after 50 cycles at 0.1 A g^{-1}	[64]
SnS/PPy	1 M $LiPF_6$ in mixture of ethylene carbonate/ dimethyl carbonate (1:1)	Specific capacity 1000 mAh g^{-1} at 0.1 C	703 mAh g^{-1} is retained after 500 cycles at 1 C	[62]
MoS_2/PEDOT:PSS	1 M $LiPF_6$ in ethyl carbonate/dimethyl carbonate (1:1)	Specific capacity of 712 mAh g^{-1} at 0.05 A g^{-1}	439 mAh g^{-1} retained 81% after 100 cycles at 0.05 A g^{-1}	[89]
SnS_2/PANI	1M $LiPF_6$ in EC-DEC-EMC(1:1:1 vol.%)	Specific capacity of 356.1 mAh g^{-1} at 5000 mAh g^{-1}	730.8 mAh g^{-1} after 80 cycles	[53]
Sb_2Se_3/PPy		Specific capacity of 630 mAh g^{-1} at 0.05 Ah g^{-1}	No obvious capacity fading over 80 cycles at 300 and 500 mA g^{-1}	[81]
Bi_2S_3/PPy	For Li-S battery: DOL and DME (1:1 v/v) with 1 mol L^{-1} lithium bis(trifluoromethane sulphonyl)imide (LiTFSI) and 2% $LiNO_3$	Reversible capacity of 729 mAh g^{-1} after 500 cycles at 1 C	90% capacity retention of second cycle after 5000 cycles at 1000 mA g^{-1}	[82]

electrode material due to ultrathin structures, tunable properties, and high surface-to-volume ratio. They are the better metallic conductors and they also undergo faradic redox reaction among various reduction/oxidation states of the metal ion, but they suffer from short cycling stability, in which the capacitance fades rapidly. The construction of composite structure utilizing metal chalcogenide nanosheets and CP as a component is considered a promising approach paving the way for its application in energy storage devices. To circumvent these shortcomings, integration of chalcogenides into the conducting polymers creates hybrid nanocomposites combining the valuable properties of both. It results in enhanced conductivity of the electrode and accelerates electron transfer and electrochemical performance due to the synergistic effect involved between metal sulphide and conducting polymer. Conjugated π-building blocks of polymer chain allow easy permeability of electrolyte thus facilitating the ion transfer in the electrode. Table 14.4 lists various chalcogenide/ conducting polymer nanocomposites used as electrodes in battery with their performance.

Reduced graphene oxide is a well-known material used as electrodes in alkali metal-ion batteries. rGO has oxygenated functional groups that provide better anchoring to the chalcogenides as well as show adsorption of the polysulphides formed, thus preventing the loss of material. The porous network also prevents the aggregation of chalcogenides which is the common problem in Li-S batteries. Tu et al. [76] reported the fabrication of MoS_2-Ppy-rGO composite by solution method. The nano rGO sheets are intercalated with MoS_2-polypyrrole composite to form the electrode, which exhibits a superior discharge rate capacity of 1070 mAh g^{-1} after several hundred cycles, which is way better than obtained with MoS_2-PPy, MoS-GO, MoS_2 composite.

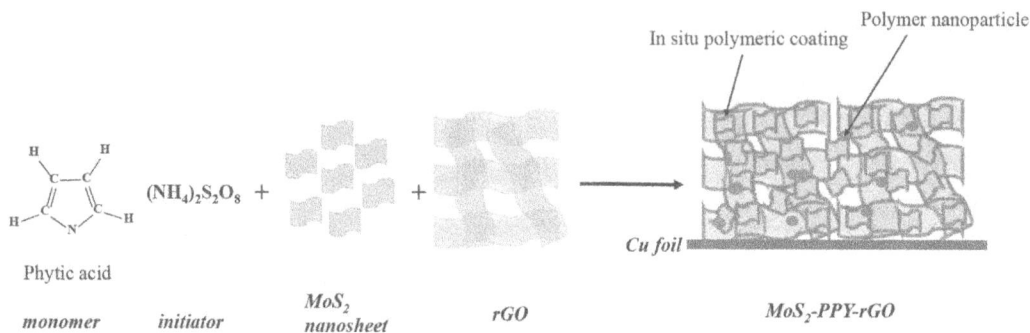

FIGURE 14.5 Applications of conducting polymer-metal chalcogenide in battery performance.

Nanoflowers of MoS_2 was grown with polyaniline polymer using the hydrothermal process as reported by Xu et al. [77]. This 3D-hierarchial structure was prepared hydrothermally by annealing at 500°C. A high charge capacity of 893 mAh g^{-1} was obtained for the prepared composite. The interlayer spacing, as revealed by structural characterization of the nanoflower, effectively increases the surface area and allows diffusion of lithium ion thus enhancing the electrochemical performance. The polymeric network further improves structural stability.

Similarly, the SnS-Ppy framework prepared by Zhu et al. [78] also shows better diffusion of an alkali metal with good mechanical strength. The hydroxide of zinc and tin [$(ZnSn(OH)_6$] was used as a single-source precursor as the tin source. The composite performed very well as an electrode showing around 500 life cycles with retention of around 700 mAh/g at 1 C. The hydrothermally obtained SnS nanosheet enhances the metal ion diffusion and the polymeric backbone shows fast electronic transport.

Wang and the group [79] effectively designed PEDOT/MoS_2 nanocomposite via in situ polymerization method utilizing ammonium persulphate. The resultant intercalated structure exhibited excellent conductivity, thermal stability and specific capacitance dependent on the amount of MoS_2 employed. In fact, with 45% of MoS_2, the specific capacitance was quadrupled in comparison to PEDOT alone. Nanocomposite retained 90% capacity even after 1000 cycles. Enhanced performance was also shown by a new Mo composite prepared by mixing two different conducting polymers as reported in the previous section by Zhang et al. [80]. The designed composite showed high current density, which was retained up to 81% even after 100 cycles. Other metal chalcogenides like SnS_2, $SnSe_2$, and Bi_2S_3 also showed enhanced cyclic stability and performance [81–84].

CdSe in nanometre range was reported to form a composite with polythiophene conducting polymer and the impedance studies showed it to have low charge transfer resistance and high capacitive behaviour, making it another possible material to be considered for battery [85].

Pure selenium doped polymer like PANI also has been reported after understanding the wide applications of Se in the electronics industry. The green emeraldine form of PANI doped with different concentrations of Se was studied [86]. The conductivity values tripled to that reported for undoped polymer samples due to a shift in Fermi levels of doped samples. This increase in conductivity is shown maximum for 15% doped sample by hopping mechanism followed by the further decrease with increase in concentration due to saturation of charges. Various applications are schematically described in Figure 14.5.

14.7 COMPUTATIONAL STUDIES OF CHALCOGENIDES/CONDUCTING POLYMERS AS ENERGY MATERIALS

Computational calculations help us in reducing the number of practical trials by providing powerful algorithms based on theoretical principles. This can provide a breakthrough for experimental verification by providing fundamental understanding. The theoretical calculations provide us the energetics of the ongoing electrochemical reactions, bonding information and charge distribution on the chalcogenides loaded in the polymeric backbone. As we have mentioned before, the practical performance of sulphides and other chalcogenides in the batteries is way below the theoretical expected efficiency. To understand and bridge this gap, theoretical studies using density functional calculations are used, which gives an insight in rationally designing the host material for S and other chalcogenides, which can restrain the formation of unnecessary side products like lithium polysulphides in the case of Li-S batteries thus reducing the shuttle effect. These studies can further help in understanding the diffusion barrier and the binding strength of alkali metals to the chalcogenide layers, thus help in enhancing the capacity.

The studies in Li-S batteries have shown that carbon materials like graphene, when used do not suppress the shuttle effect [86, 87]. This can be addressed by introducing polarity in the layer by doping heteroatoms [88]. The N-doped graphene shows increase adsorption of polysulphides due to dipole–dipole electrostatic interaction formed, thus improving the electrochemical performance. Similarly, various other atoms are chosen and modelled (B, O, P, S, Cl) based on their periodic properties, almost similar ionic radius like lithium-ion were studied [89].

The catalytic conversion of lithium polysulphides was further studied on the surface of other metal sulphides and selenides both experimentally and theoretically. The cobalt and tungsten sulphide DFT (density functional theory) calculations show a strong anchoring effect, thus convert catalytically the polysulphides by adsorption. During the discharge process, the adsorption energy of around 1.4 eV was still shown by WS_2 even after combining with the extracted Li^+ ion. Similarly, CoSe is known to show strong adsorption behaviour for the sulphides, which was proved by theoretical calculations and thus accelerating their conversion [88].

Due to the weak interactions of alkali metals with traditional anodic materials, layered calculations were extended to metal chalcogenides like 2D MoX_2 (where X = S, Se, Te) [90] and ternary layered transition metals chalcogenides like Ti_2PX_2 (X = S, Se, Te) [91, 92]. The results showed the binary chalcogenides to be potential candidates for lithium-ion batteries doped in conducting polymers. The nanosheets of $MoSe_2$ have increased specific surface area, resulting in contact area being more between electrode and electrolyte, thus providing decreased diffusion length for lithium-ion. Similarly, the computation of electronic band structure and density of states of ternary materials showed metallic character, high sodium ion storage and low metal ion diffusion barrier, thus making them highly desirable anode materials for alkali ion batteries. The DFT calculations have further proven the bonding strength of chalcogenides and other lithium polysulphides [88] to be higher with the linear conducting polymer like polyethyleneimine than for PVDF. The increased cathodic performance and slow capacity decay are the effects of substitution of heteroatoms and the pi conjugation effect present in the polymer. Even after so many studies are available, still, more systematic theoretical and experimental efforts are required to bring these materials to commercial space.

14.8 CONCLUSION AND FUTURE PROSPECTS

The demand for Li-ion batteries in portable devices is increasing tremendously. But the high cost, low stability, relatively low energy and low density obstruct its commercial application. This chapter features current trends in synthesis, properties and applications of conducting polymer composites for advanced electrode material for battery devices. CP has widely been used in batteries electrode materials because of their conductivity, flexibility processability and the easy fabrication and they deliver energy rapidly. Besides the above-mentioned advantages, the main drawbacks associated with conducting polymer electrodes involve the low amount of stability after various cycles, which can be bypassed by effective combination with materials (inorganic metal oxides, metal sulphides hydroxides) to form composite electrode material. The hybrid nanocomposite-based battery electrodes ensure composite electrode boosts capacity good and long-term cyclic stability, low cost and simple reproducible fabrication methods. During the past few years, considerable efforts have been made to develop hybrid conducting polymer nanocomposites for realization in battery performance. There is plenty of room/challenges that need to be addressed to understand basic in rational fabrication and improvement in capacitance of CP/MC nanocomposite. The main challenges include a wide gap between the electrode and electrolyte and system level for optimization of ion/electron transport and optimizing the relationship between component and system. Cost-effective, scalable nanotechnology is the need of the hour for efficient and stable ion/electron transport in practical energy storage remains a challenge. The potential for regulating ion transport of these hybrid CP/chalcogenides has been realized but on a small scale. Efforts in the direction should be laid on how to realize efficient, stable and uniform ion/electron transport. The enhanced electrochemical performance of the hybrid nanocomposite in batteries provides long-term solution for further improvement that will make them viable for commercialization.

REFERENCES

1. J. Xie and Q. Zhang, Recent progress in rechargeable lithium batteries with organic materials as promising electrodes, J. Mater. Chem. A, 2016, 4 (19), 7091–7106.
2. X.-B. Cheng, R. Zhang, C.-Z. Zhao and Q. Zhang, Toward safe lithium metal anode in rechargeable batteries: A review, Chem. Rev., 2017, 117 (15), 10403–10473.
3. C. P. Grey and J. M. Tarascon, Sustainability and in situ monitoring in battery development, Nat. Mater., 2016, 16, 45–56.
4. Shumaila, G. B. V. S. Lakshmi, M. Alam, A. M. Siddiqui, M. Zulfequar and M. Husain, Synthesis and characterization of Se doped polyaniline, Curr. Appl. Phys., 2011, 11, 217–222.
5. F. Goto, K. Abe, K. Ikabayashi, T. Yoshida and H. Morimoto, The polyaniline/lithium battery, J. Power Sources, 1987, 20 (3), 243–248.
6. T. Matsunaga, H. Daifuku, T. Nakajima and T. Kawagoe, Development of polyaniline–lithium secondary battery, Polym. Adv. Technol., 1990, 1 (1), 33–39.
7. Q. Meng, K. Cai, Y. Chen and L. Chen, Research progress on conducting polymer based supercapacitor electrode materials, Nano Energy, 2017, 36, 268–285.
8. A. Rudge, J. Davey, I. Raistrick, S. Gottesfeld and J. P. Ferraris, Conducting polymers as active materials in electrochemical capacitors, J. Power Sources, 1994, 47 (1), 89–107.
9. A. J. Heeger, S. Kivelson, J. R. Schrieffer and W. P. Su, Solitons in conducting polymers, Rev. Mod. Phys., 1988, 60, 781–850.
10. P. Chandrasekhar, Conducting Polymers, Fundamentals and Applications: A Practical Approach, Springer Science Business Media, New York, 1999, 14 (chapter 1).
11. J. A. Irvin, D. J. Irvin and J. D. Stenger-Smith, Handbook of Conducting Polymers: Conjugated Polymers Processing and Applications, Eds. Skotheim, T. and Reynolds, J.R. (3rd edition), CRC Press, Boca Raton, 2007, 9–1.
12. J. G. Killian, B. M. Coffey, F. Gao, T. O. Poehler and P. C. Searson, J. Electrochem. Soc., 1996, 354, 1555.
13. C. Y. Wang, V. Mottaghitalab, C. O. Too, G. M. Spinks and G. G. Wallace, J. Power Sources, 2007, 163, 1105.
14. P. Sengodua and A. Deshmukh, Conducting polymers and their inorganic composites for advanced Li-ion batteries: a review, RSC Adv., 2015, 5, 42109–42130.
15. D. Wei, D. Cotton and T. Ryhanen, All-solid-state textile batteries made from nano-emulsion conducting polymer inks for wearable electronics, Nanomaterials, 2012, 2, 268.
16. R. Jalili, J. M. Razal, P. C. Innis and G. G. Wallace, One-step wet-spinning process of poly(3,4-ethylenedioxythiophene):poly(styrenesulfonate) fibers and the origin of higher electrical conductivity, Adv. Funct. Mater., 2011, 21 (17), 3363–3370.
17. P. Novák, K. Müller, K. S. V. Santhanam and O. Haas, Electrochemically active polymers for rechargeable batteries, Chem. Rev., 1997, 97 (1), 207–282.
18. J. F. Mike and J. L. Lutkenhaus, Electrochemically active polymers for electrochemical energy storage: Opportunities and challenges, ACS Macro Lett., 2013, 2 (9), 839–844.
19. G. Kaur, R. Adhikari, P. Cass, et al. Electrically conductive polymers and composites for biomedical applications, RSC Adv., 2015, 5, 37553–37567.
20. S. Chu, Y. Cui, N. Liu, The path towards sustainable energy, Nat. Mater., 16,16–22,2017.
21. D. O. Akinyele and R. K. Rayudu, Review of energy storage technologies for sustainable power networks, Sustainable Energy Technol. Assess., 2014, 8, 74–91.
22. Y. Liang, Z. Tao and J. Chen, Organic electrode materials for rechargeable lithium batteries. Adv. Energy Mater., 2012, 2 (7), 742–769.

23. W. Xu, J. Wang, F. Ding, X. Chen, E. Nasybulin, Y. Zhang, Ji-G. Zhang, Lithium metal anodes for rechargable batteries, J. Energy Environ. Sci., 2014, 7, 513–537.
24. S. E. Cheon, K. S. Ko, J. H. Cho, S. W. Kim, E. Y. Chin and H. T. Kim, Rechargable sulphur battery I. Structural change of sulphur cathode during discharge and charge, J. Electrochem. Soc., 2003, 150, A796–A799.
25. S. E. Cheon, K. S. Ko, J. H. Cho, S. W. Kim, E. Y. Chin and H. T. Kim, Rechargable sulphur battery II. Rate capability and cycle characteristics, J. Electrochem. Soc., 2003, 150, A800–A805.
26. J. Gao, M. A. Lowe, Y. Kiya and H. D. Abruña, Effects of liquid electrolytes on the charge–discharge performance of rechargeable lithium/sulfur batteries: Electrochemical and in-situ X-ray absorption spectroscopic studies. J. Phys. Chem. C, 2011, 115 (50), 25132–25137.
27. J.-W. Choi, J.-K. Kim, G. Cheruvally, et al., Rechargeable lithium/sulfur battery with suitable mixed liquid electrolytes. Electrochim. Acta, 2007, 52 (5), 2075–2082.
28. S. S. Zhang, Liquid electrolyte lithium/sulfur battery: Fundamental chemistry, problems, and solutions, J. Power Sources, 2013, 231, 153–162.
29. S. S. Zhang, New insight into liquid electrolyte of rechargeable lithium/sulphur battery, Electrochim. Acta, 2013, 97, 226–230.
30. E. Boros, M. J. Earle, M. A. Gilea, et al., On the dissolution of non-metallic solid elements (sulfur, selenium, tellurium and phosphorus) in ionic liquids. Chem. Commun., 2010, 46 (5), 716–718.
31. L. Ji, M. Rao, H. Zheng, et al., Graphene oxide as a sulfur immobilizer in high performance lithium/sulfur cells, J. Am. Chem. Soc., 2011, 133 (46), 18522–18525.
32. T. Nakatomo, T. Homma, C. Yamamoto, K. Negishi and O. Omoto, A long-lasting polyacetylene battery with high energy density, Jpn. J. Appl. Physics., 1983, 22, 275.
33. G. Louran, M. L. Apkowski, S. Quillard, A. Pron, J. P. Buisson and S. Lefrant, Vibrational properties of polyaniline-isotopic effects, J. Phy. Chem., 1996, 100, 69998–7006.
34. D. W. Hatchett, M. Josowi, J. Janata, Comparison of chemically and electrochemically synthesized polyaniline films, J. Electrochem. Soc, 1999, 146, 4535–4538.
35. L. Shao, J. W. Jeon and J. L. Lutkenhaus, Polyaniline/vanadium pentoxide layer-by-layer electrodes for energy storage, Chem. Mater., 2012, 24, 181.
36. E. M. Genies, A. A. Syed and C. Tsintavis, Electrochemical study of polyaniline in aqueous and organic medium. Redox and kinetic properties, Mol. Cryst. Liq. Cryst., 1985, 121, 181–186.
37. D.-W. Wang, F. Li, J. Zhao, W. Ren, Z.-G. Chen, J. Tan, Z.-S. Wu, I. Gentle, G. Q. Lu and H.-M. Cheng, Fabrication of graphene/polyaniline composite paper via in situ anodic electropolymerization for high-performance flexible electrode, ACS Nano, 2009, 3, 1745.
38. H.-P. Cong, X.-C. Ren, P. Wang and S.-H. Yu, Flexible graphene-polyaniline composite paper for high-performance supercapacitor, Energy Environ. Sci., 2013, 6, 1185–1191.
39. Z. Song and H. Zhou, Towards sustainable and versatile energy storage devices: An overview of organic electrode materials, Energy Environ. Sci., 2013, 6, 2280–230.
40. T. Kita, M. Ogawa, Y. Masuda, T. Fuse, H. Daifuka, R. Fujio, T. Kawagoe and T. Matsunaga, Extended abstracts of the electrochemical society (USA), Electrochem. Soc., 1986, 86, 37.
41. W. Deng, X. Liang, X. Wu, J. Qian, Y. Cao, X. Ai and H. Yang, A low cost, all-organic Na-ion battery based on polymeric cathode and anode, Sci. Rep., 2013, 3, 2671.
42. M. Morita, K. Komaguchi, H. Tsutsumi and Y. Matsuda, Electrosynthesis of poly(p-phenylene) films and their application to the electrodes of rechargeable batteries, Electrochem. Acta., 1992, 37, 1093–1099.
43. S. H. Oh, C. W. Lee, D. H. Chun, J. D. Jeon, J. Shim, K. H. Shin and J. H. Yang, A metal-free and all-organic redox flow battery with polythiophene as the electroactive species, J. Mater. Chem. A., 2014, 2, 19994–19998.
44. K. Kaneto, K. Yoshino and Y. Inuishi, Characteristics of polythiophene battery, Jpn. J. Appl. Phys., 1983, 22, L567.
45. C. Arbizzani, M. Mastragostino and M. Rossi, Preparation and electrochemical characterization of a polymer $Li_{1.03}Mn_{1.97}O_4$/pEDOT composite electrode Electrochem. Commun., 2002, 4, 545–549.
46. J. Chen, J. Wang, C. Wang, C. O. Too and G. G. Wallace, Lithium-polymer battery based on polybithiophene as cathode material, J. Power Sources, 2006, 159, 708–711.
47. B. Scrosati and J. Garche, Lithium batteries: Status, prospects and future, J. Power Sources, 2010, 195, 2419–2430.
48. D. A. Pasquier, I. Plitz, S. Menocal and G. Amatucci, A comparative study of Li-ion battery, supercapacitor and non-aqueous asymmetric hybrid devices for automotive applications, J. Power Sources, 2003, 115, 171–178.
49. M. Dubarry and B. Y. Liaw, Identify capacity fading mechanism in a commercial $LiFePO_4$ cell, J. Power Sources, 2009, 194, 541–549.

50. J. C. Carlberg and O. Inganas, Poly(3,4-ethylenedioxythiophene) as electrode material in electrochemical capacitors, J. Electrochem. Soc., 1997, 144, L61.

51. J. Zhu, W. Sun, D. Yang, Y. Zhang, H. H. Hoon, H. Zhang and Q. Yan, Multifunctional architectures constructing of PANI nanoneedle arrays on MoS$_2$ thin nanosheets for high energy supercapacitors, Small, 2015, 11, 4123–4129.

52. T. Yang, H. Chen, T. Ge, J. Wang, W. Li and K. Jiao, Highly sensitive determination of chloramphenicol based on thin-layered MoS$_2$/polyaniline nanocomposites, Talanta, 2015, 144, 1324–1328.

53. G. Wang, J. Peng, L. Zhang, J. Zhang, B. Dai, M. Zhu, L. Xia and F. Yu, Two-dimensional SnS$_2$@PANI nanoplates with high capacity and excellent stability for lithium-ion batteries, J. Mater. Chem. A, 2015, 3, 3659–3666.

54. K. Gopalakrishnan, S. Sultan, A. Govindaraj and C. N. R. Rao, Supercapacitors based on composites of PANI with nanosheets of nitrogen-doped RGO, BC$_{1.5}$N, MoS$_2$ and WS$_2$, Nano Energy, 2015, 12, 52–58.

55. S. Liu, P. Gordiichuk, Z.-S. Wu, Z. Liu, W. Wei, M. Wagner, N. Mohamed-Noriega, D. Wu, Y. Mai, A. Herrmann, K. Müllen and X. Feng, Patterning two-dimensional free-standing surfaces with mesoporous conducting polymers, Nat. Commun., 2015, 6, 8817.

56. M. Kim, Y. K. Kim, J. Kim, S. Cho, G. Lee and J. Jang, Fabrication of a polyaniline/MoS$_2$ nanocomposite using self-stabilized dispersion polymerization for supercapacitors with high energy density, RSC Adv., 2016, 6, 27460–27465.

57. X. Li, C. Zhang, S. Xin, Z. Yang, Y. Li, D. Zhang and P. Yao, Facile synthesis of MoS2/reduced graphene oxide @polyaniline for high-performance supercapacitors, ACS Appl. Mater. Interfaces, 2016, 8, 21373–21380.

58. J. Wang, Z. Wu, H. Yin, W. Li and Y. Jiang, Poly(3,4-ethylenedioxythiphene)/MoS2 nanocomposites with enhanced electrochemical capacitance performance, RSC Adv., 2014, 4, 56926–56932.

59. B.-Z. Lin, C. Ding, B.-H. Xu, Z.-J. Chen and Y.-L. Chen, Preparation and characterization of polythiophene/molybdenum disulphide intercalation material, Mater. Res. Bull., 2009, 44, 719–723.

60. B.-H. Xu, B.-Z. Lin, Z.-J. Chen, X.-L. Li and Q.-Q. Wang, Preparation and electrical conductivity of polypyrrole/WS$_2$ layered nanocomposites, J. Colloid Interface Sci., 2009, 330, 220–226.

61. H. Tang, J. Wang, H. Yin, H. Zhao, D. Wang and Z. Tang, Growth of polypyrrole ultrathin films on MoS$_2$ monolayers as high-performance supercapacitor electrodes, Adv. Mater., 2015, 27, 1117–1123.

62. J. Liu, M. Gu, L. Ouyang, H. Wang, L. Yang and M. Zhu, Sandwich-like SnS/polypyrrole ultrathin nanosheets as high-performance anode materials for Li-ion batteries, ACS Appl. Mater. Interfaces, 2016, 8, 8502–8510.

63. Y. Gao, C. Chen, X. Tan, H. Xu and K. Zhu, Polyaniline-modified 3D-flower-like molybdenum disulfide composite for efficient adsorption/photocatalytic reduction of Cr(VI), J. Colloid Interface Sci., 2016, 476, 62–70.

64. L. Ren, G. Zhang, Z. Yan, L. Kang, H. Xu, F. Shi, Z. Lei and Z.-H. Liu, Three-dimensional tubular MoS$_2$/PANI hybrid electrode for high rate performance supercapacitor, ACS Appl. Mater. Interfaces, 2015, 7, 28294–28302.

65. J. Wang, Z. Wu, K. Hu, X. Chen and H. Yin, High conductivity graphene-like MoS$_2$/polyaniline nanocomposites and its application in supercapacitor, J. Alloys Compd., 2015, 619, 38–43.

66. L. Yang, S. Wang, J. Mao, J. Deng, Q. Gao, Y. Tang and O. G. Schmidt, Hierarchical MoS$_2$/polyaniline nanowires with excellent electrochemical performance for lithium-ion batteries, Adv. Mater., 2013, 25, 1180–1184.

67. H. Ju and J. Kim, Chemically exfoliated SnSe nanosheets and their SnSe/poly(3,4-ethylenedioxythio phene):poly(stryrenesulphonate) composite films for polymer based thermoelectric applications, ACS Nano, 2016, 10, 5730–5739.

68. X. Zhao, Y. Mai, H. Luo, D. Tang, B. Lee, C. Huang and L. Zhang, Nano-MoS$_2$/poly poly(3,4-ethylenedioxythiophene):poly(stryrenesulphonate) composite prepared by a facile dip-coating process for Li-ion battery anode, Appl. Surf. Sci., 2014, 288, 736–741.

69. Y.-J. Huang, M.-S. Fan, C.-T. Li, C.-P. Lee, T.-Y. Chen, R. Vittal and K.-C. Ho, MoSe/poly(3,4-ethylenedioxythiophene):poly(stryrenesulphonate)composite film as a Pt-free counter electrode for dye-sensitized solar cells, Electrochim. Acta, 2016, 211, 794–803.

70. D. Song, M. Li, Y. Jiang, Z. Chen, F. Bai, Y. Li and B. Jiang, Facile fabrication of MoS$_2$/PEDOT–PSS composites as low-cost and efficient counter electrodes for dye-sensitized solar cells, J. Photochem. Photobiol. A: Chemistry, 2014, 279, 47–51.

71. F. Jiang, J. Xiong, W. Zhou, C. Liu, L. Wang, F. Zhao, H. Liu and J. Xu, Organic solvent assisted exfoliated MoS$_2$ for the optimized thermoelectric performance of flexible PEDOT:PSS thin-film, J. Mater. Chem. A, 2016, 4, 5265–5273.

72. M. S. Nam, U. Patil, B. Park, H. B. Sim and S. C. Jun, A binder free synthesis of 1D PANI and 2D MoS2 nanostructured hybrid composite electrodes by the electrophoretic deposition (EPD) method for supercapacitor application, RSC Adv., 2016, 6, 101592–101601.

73. A. Bahuguna, S. Kumar, V. Sharma, et al., Nanocomposite of MoS_2-RGO as facile, heterogeneous, recyclable, and highly efficient green catalyst for one-pot synthesis of indole alkaloids, ACS Sustain. Chem. Eng., 2017, 5, 8551–8567.

74. X. Lei, X. Lu, G. Nie, Z. Jiang and C. Wang, One-pot synthesis of algae-like MoS_2/PPy nanocomposite: A synergistic catalyst with superior peroxidase-like catalytic activity for H_2O_2 detection part. Part. Syst. Charact., 2015, 32, 886–892.

75. D. Xie, D. H. Wang, W. J. Tang, X. H. Xia, Y. J. Zhang, X. L. Wang, C. D. Gu and J. P. Tu, Binder-free network-enabled MoS_2-PPy-rGO ternary electrode for high capacity and excellent stability of lithium storage, J. Power Sources, 2016, 307, 510–518.

76. L. Hu, Y. Ren, H. Yang and Q. Xu, Fabrication of 3D hierarchical MoS_2/polyaniline and MoS_2/C architectures for lithium-ion battery application, ACS Appl. Mater. Interfaces, 2014, 6, 14644–14652.

77. J. Liu, M. Gu, L. Ouyang, H. Wang, L. Yang and M. Zhu, Sandwich-like SnS/polypyrrole ultrathin nanosheets as high-performance anode materials for Li-ion batteries, ACS Appl. Mater. Interfaces, 2016, 8, 8502–8510.

78. J. Wang, Z. Wu, H. Yin, W. Li, and Y. Jiang, Poly (3, 4-ethylenedioxythiophene)/MoS_2 nanocomposites with enhanced electrochemical capacitance performance. RSC Adv., 2014, 4 (100), 56926–56932.

79. X. Zhao, Y. Mai, H. Luo, D. Tang, B. Lee, C. Huang and L. Zhang, Nano-MoS_2/poly (3, 4-ethylenedioxythiophene): Poly (styrenesulfonate) composite prepared by a facial dip-coating process for Li-ion battery anode. Appl. Surf. Sci., 2014, 288, 736–741.

80. G. Wang, J. Peng, L. Zhang, J. Zhang, B. Dai, M. Zhu, … and, F. Yu, Two-dimensional SnS2@ PANI nanoplates with high capacity and excellent stability for lithium-ion batteries, J. Mater. Chem. A, 2015, 3 (7), 3659–3666.

81. Y. Fang, X. Y. Yu and X. W. Lou, Formation of polypyrrole-coated Sb_2Se_3 microclips with enhanced sodium-storage properties, Angewandte Chemie, 2018, 130 (31), 10007–10011.

82. B. Long, Z. Qiao, J. Zhang, S. Zhang, M. S. Balogun, J. Lu and Y. Tong, Polypyrrole-encapsulated amorphous Bi_2S_3 hollow sphere for long life sodium ion batteries and lithium–sulfur batteries, J. Mater. Chem. A, 2019, 7 (18), 11370–11378.

83. R. Singh, A. K. Bajpai and A. K. Shrivastava, CdSe nanorod-reinforced poly(thiophene) composites in designing energy storage devices: Study of morphology and dielectric behaviour, Polym. Bull., 2020, 78, 6861–6877.

84. M. Alam, A.M. Siddiqui and M. Husain, Synthesis, characterization and properties of Se nanowires intercalated polyaniline/Se nanocomposites, Express Polym. Lett., 2013,7 (9), 723–732.

85. Y. Lu, S. Gu, J. Guo, K. Rui, C. H. Chen, S. P. Zhang, J. Jin, J. H. Yang and Z. Y. Wen, Sulfonic groups originated dual-functional interlayer for high performance lithium-sulfur batteries, ACS Appl. Mater. Interfaces, 2017, 9, 14878–14888.

86. J. Li, Y. Qu, C. Chen, X. Zhang and M. Shao, Theoretical investigation on lithium polysulfides adsorption and conversion for high-performance Li–S batteries, Nanoscale, 2020. DOI: 10.1039/D0NR06732F.

87. T.-Z. Hou, X. Chen, H.-J. Peng, J.-Q. Huang, B.-Q. Li, Q. Zhang and B. Li, Design principles for heteroatom-doped nanocarbon to achieve strong anchoring of polysulphides for lithium-sulfur batteries, Small, 2016, 12, 3283.

88. Y. M. Chen, X. Y. Yu, Z. Li, U. Paik and X. W. D. Lou, Hierarchical MoS_2 tubular structures internally wired by carbon nanotubes as a highly stable anode material for lithium-ion batteries, Sci. Adv., 2016, 2, e1600021.

89. M. R. Panda, R. Gangwar, D. Muthuraj, S. Sau, D. Pandey, A. Banerjee, A. Chakarbarti, A. Sagdeo, M. Weyland, M. Majumder, Q. Bao and S. Mitra, High performance lithium-ion batteries using layered 2H-$MoTe_2$ as anode, Small, 2020, 2002669.

90. Z. Luo, J. Zhou, L, Wang, G. Fang, A. Pan and S. Liang, Two-dimensional hybrid nanosheets of few layered $MoSe_2$ on reduced graphene oxide as anode for long-cycle-life lithium-ion batteries, J. Mater. Chem. A, 2016. DOI: 10.1039/C6TA04390A.

91. J. Liu, M. Qiao, X. Zhu, Y. Jing and Y. Li, Ti_2PTe_2 monolayer: A promising two-dimensional anode material for sodium-ion batteries, RSC Adv., 2019, 9, 15536–15541.

92. B. Ge, B. Chen and L. Li, Ternary transition metal chalcogenides Ti_2PX_2 (X = S, Se, Te) anodes for high performance metal-ion batteries: A DFT study, Appl. Surf. Sci., 2021, 550, 149177.

LIST OF ABBREVIATIONS

Lithium ion batteries	LIB
Polypyrrole	PPy
Polyaniline	PANI
Polythiophene	PTh
poly(3,4-ethylenedioxythiophene)	PEDOT
the tetra (ethylene glycol) dimethyl ether	TEGDME
1,3-dioxolane	DOL
Poly (p-phenylene)	PPP
Metal chalcogenides	MC
dodecyl benzene sulphonic acid	DBSA

15 Conducting Polymers for Flexible Devices

Amir Ershad Langroudi and Hamidreza Parsimehr
Color and Surface Coating Group, Polymer Processing Department,
Iran Polymer and Petrochemical Institute (IPPI), Tehran, Iran

CONTENTS

15.1   INTRODUCTION

In 1977, Shirakawa, MacDiarmid, and Heeger introduced conducting silvery polyacetylene as a conductive polymer. The findings led to the Nobel Prize in Chemistry for them. The electrical conductivity of CPs can be tunable in the conductors or semiconductors range and increased by multi-million times on exposure to halogens vapors. CPs find essential applications in various industrial sectors because of high conductivity, optical properties, reasonable price, and high mechanical flexibility. The conductive polymers have structurally double (or triple) covalent bonds in their backbone to increase their electrical conductivities during the doping process. Unlike inorganic semiconductors and metals, π bond in the chemical structure of the polymer and a doping process is responsible for electron transfer in a conductive polymer. Their conductivity can be adjusted by modifying molecular structures, doping levels, and molecular order. By doping ions, electron acceptor, or donor materials, and oxidation or reduction reactions, conjugated polymers' conductivity can be increased by creating chemical stability. The doping ions can penetrate the weak bonds of polymer chains. Some useful dopers are Br_2, I_2, and $FeCl_3$ for electron acceptors and Li, Na, and K for electron donors. Polymer duplication causes charge transfer by oxidation or doping of type p and reduction or duplication of type n. Among the properties in conductive polymers, the following properties can be mentioned: electrical, magnetic, optical, wettability, mechanical, and electromagnetic wave absorption properties[1].

Conducting polymers' methods involved the electrically conductive coating on a substrate polymer or adding the electrically conductive material to the polymer mixture with similar electrical, electronic, or magnetic properties of common metals while retaining the flexibility and other general properties of polymers[2]. Among the applications of conductive polymers are the following: electronic nanocouples; transistors and light-emitting diodes; sensors: gas and chemical sensors, optical sensors, and biosensors; catalyst: optical and chemical catalysts and electrodes, fuel cells,

DOI: 10.1201/9781003150374-15

microwave absorption, and shielding of electromagnetic frequency interference, electrheological fluids; biomedical applications: protein delivery and purification, tissue engineering, neural interfaces, and activators[2].

One of the advantages of conductive polymers is that they have good mechanical properties and flexibility similar to conventional polymers. Furthermore, they can be made through various processes and in a variety of structures. Conductive polymer gels have several advantages in converting energy storage. They provide an integrated electrical conductive framework to enhance electron transfer, ion or molecule diffusion by having micro and mesopores in polymer matrices. They expand the effective surface area between the polymeric chains and the electrolyte electrochemical reactions. The porous structure of conductive polymer gels can help localize induced strains by changing volume during electrochemical reactions. Moreover, good mechanical properties give them the opportunities to be used to build lightweight and flexible devices.

15.2 CONVENTIONAL CONDUCTIVE POLYMERS

The most common conductive polymers here are polymers in which the resonance of π electrons leads to conductivity. Polypyrrole (PPY), polyaniline (PANI), poly(p-phenylene vinylene) (PPV), poly(p-phenylene) (PPP), polythiophene (PTh), and poly(3,4-ethylene dioxythiophene) (PEDOT) are among the popular conductive polymers (CPs). Figure 15.1 shows their chemical structures.

Polyaniline (PANI) and its derivatives, such as poly orthoaminophenol (POAP), are used to boost ion transfer by reducing penetration routes in energy storage apparatus. Recently, environment-friendly electrodes have been synthesized using chitosan/polyorthoaminophenol biopolymer (POAP/CTS). POAP/CTS nanocomposite exhibits higher electrochemical activity than POAP. The specific capacity for POAP/CTS was 345 F/g higher than 91 F/g for pure POAP.

Cyclic stability with continuous charge/discharge measurement significantly increases by 92%, maintaining capacity after 1000 cycles due to the layered microstructures limiting the polymer backbone's frequent stretching during the discharge cycle. Hence, chitosan compositions limit the easy degradation of the polymer. Based on the results, the chitosan framework can be classified

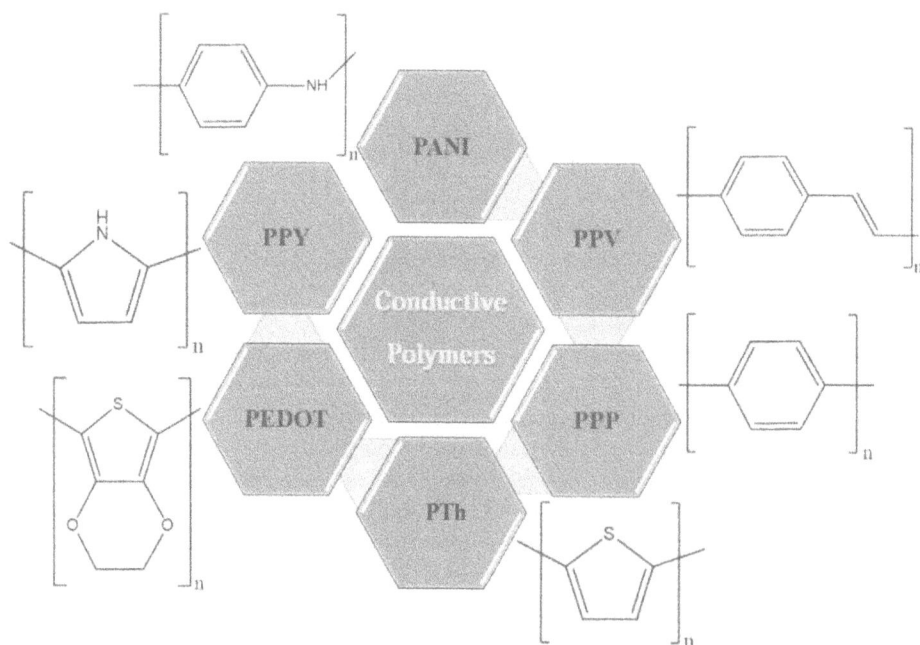

FIGURE 15.1 Chemical structure of common CPs.

into n-type (electron-donor) and p-type (electron-acceptor) parts, and therefore parallel and vertical intramolecular parts with different electronic behaviors/electrochemical properties in the molecule[3].

The effect of nanostructured VO-acetyl acetone coordination system in films made with POAP/VO(acac) 2L nanocomposite in comparison with POAP pure electrodes on the capacitive pseudo-capacitance performance of p-type conductive polymer by various electrochemical methods was investigated.

Based on the obtained results, the nanocomposite electrode showed better electrochemical performance compared to the performance of POAP electrodes. This improvement can be attributed to the Faradic contribution of VO(acac) 2L and its synergistic effect. The nanocomposite electrode showed limit maintenance of 89 after 1000 charge/release cycles at 100 mA cm^{-2} current density in addition to excellent long cycle life. This indicates that the POAP/VO(acac) 2L nanocomposite electrode is one of the most promising options for SC electrochemical applications. Moreover, the POAP/VO(acac)2L nanocomposite electrode indicated a capacitance retention value of 89%, at the end of 1000 cycles at 100 mA·cm^{-2} current density[4].

However, research in conductive polymers is mainly for energy production and storage, which is the largest part of research on conductive polymers. One of the best ways to store energy is as an electrochemical reaction. There are three major types of electrochemical energy storage (EES) apparatus: batteries, capacitors, and hybrid frameworks that combine capacitors and batteries. In general, EES systems consist of four primary parts, including an electrode, electrolyte, binder, and separator or membrane. Essential parts of batteries and capacitors (especially electrodes and binders) can be made using conductive polymers. Biochar can also be used as a battery and capacitor electrode.

15.3 OPTOELECTRONIC DEVICES

Recently, the application of conductive polymer nanocomposites and the technology resulting from advances in this field have been reviewed. Preparation strategies mainly include ex situ mixing and in situ (in situ polymerization or nanostructure synthesis) methods. Electronic and optical instruments that can use or emit light (in the spectrum of electromagnetic radiation) are one of these materials' applications[5]. Table 15.1 shows several classes of light-emitting polymers (LEPs). Combining LEPs with some nanofillers gives unique properties to conductive polymer nanocomposites that can be used as different OLED fabrication components such as emitting layer (EML), electrode, and hole transfer layer (HTL)[6].

Another application of conductive polymer nanocomposites is in photovoltaic devices. Organic solar cells (or organic photovoltaic devices (OPVs)) convert solar energy, such as visible light and ultraviolet light, into electrical energy using photovoltaics. Light currents are generated through active organic photo materials[8]. Photovoltaic cells made of conductive polymer nanocomposites are light, flexible, and thin that can be produced in the industry at low cost and have high light absorption coefficients[9]. They can also be divided into several categories such as monolayer OPV cells, bilayer OPV cells, uncoordinated OPV cells, based on their structures and types of connections[10,11].

Nanocomposites made of conductive polymer get a lot of applications in electronic devices with flexibility. In the field of flexible/wearable electronics, conductive nanomaterials have made significant progress. Stretch electronics can be used for applications such as robotics and electronic shells due to their flexible elastomers. Conductive polymers have high specific energy capacity and high specific power; EES equipment such as lithium-ion supercapacitor is an excellent choice. The battery life is one of the main limiting factors. Recently reported a stable lithium-ion battery anode in which silicon nanoparticles incorporated into a conductive polymer hydrogel network[12]. Silicon is considered an ideal rechargeable battery anode component due to its immense strength power. However, cycling instability has limited its applications. Electrically conductive polyaniline hydrogel showed more than 90% capacity maintenance after 5000 charge and discharge cycles. One of the advantages of this unique hierarchical three-dimensional material is the continuous conduction

TABLE 15.1

Representative Classes of Light-Emitting Polymers[7]

Poly(p-phenylene vinylene) (PPV)

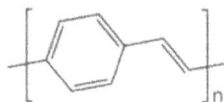

Poly[2-methoxy-5-(2-ethylhexyloxy)-
  1,4-phenylene vinylene] (MEH-PPV)

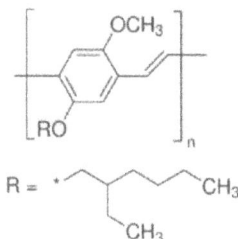

Poly(9,9-dioctyl fluorene) (PFO)

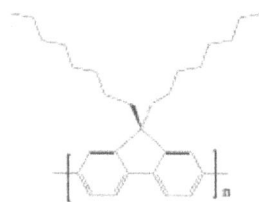

poly(3-octylthiophene) (P3OT)

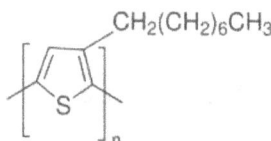

Polyquinoline

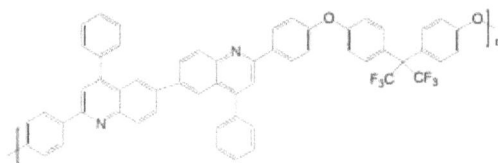

path, the porous space for expanding silicon volume. These materials can prolong the life of the lithium-ion battery. Conductive polymers have been shown to improve the performance of other electrodes such as Co_3O_4. A conductive polymer layer (polyaniline) was constructed on Co_3O_4 to fabricate organic/inorganic crust nanostructures[13].

Since graphene and graphene oxide are present in the matrix of certain polymers, they have stable electrical conductivity and mechanical flexibility. Under the condition of conductivity, the reduced graphene oxide was shown to form zinc oxide nanoclusters in polyaniline and polypyrrole, leading to better solar cell performance[14]. Combining conductive fillers such as dextran sulfate or polystyrene sulfonate and copolymerization with processable polymers improve its performance. Polymer-based electronic devices such as flexible photovoltaics have been made on a large scale at a low cost[15].

15.4   CONDUCTIVE POLYMERS IN ENERGY STORAGE

The evolution of electrochemical capacitors (ECs) is toward high power density and a long life cycle. In ECs, ionic and electronic transport processes occur at the electrode's surface and the electrodes–electrolyte solution interfaces. The electrode is the main component of the capacitor, and the final performance depends on it. In general, for the high performance of electrodes, it is necessary to simultaneously minimize initial resistances in the electrochemical charge–discharge reaction, such as ion transfer in the electrolyte and the electrode, electrochemical reaction, and charge transfer in the electrode and current collector[16].

In general, swelling and shrinkage in conductive polymers' structures lead to instability during the charging and discharging processes. Also, bulk films of conductive polymers have compact structures that prevent electrolyte ions' penetration and reduce their performance. Therefore, some strategies need to be considered to improve their performance.

The porous structure of PPy composites on a flexible cellulosic substrate can rapidly transfer the mass of ions required during oxidation and reduction of PPy[17,18].

Conductive polymer nanowire arrays such as PPy provide the desired surface area and ion diffusion pathway to prepare the electrode structure for high-performance supercapacitors so that PPy nanowires can maintain about 70% of their original capacity in the charge/discharge cycles[19].

Another strategy for fabricating high-performance ECs is to use conductive polymer hydrogels to stretch the structure derived at the biphasic organic/aqueous interface[20]. This strategy performed well for samples such as PPy and PANI hydrogels with adjustable porous nanostructures. Because of the rapid ion transfer, the three-dimensional nanostructure will result in low capacitance loss and good cyclic stability when the current density is increased by a factor of ten.

One strategy for swelling and shrinkage of conductive polymers that occurs in the course of the electrochemical cycle is to use composite materials by combining the inherent electrochemical properties of each component as well as their synergistic effect. For example, CCG-PANI-NF composite has more advantages over its components, including its stand-up properties and high flexibility and conductivity without the need for additives[16,21].

Another solution is to use vertical nanowire array electrodes on graphene oxides as electrodes with several advantages: First, all nanowires' electrical connection to a conductive substrate increases storage capacity. Second, nanowire arrays result in higher speed performance with better electrical charge carrying and shorter ion transport lengths. Finally, the space between directional nanowires can cause volumetric changes without breaking the bulk material[16,22,23].

The use of pseudo-transition metal oxides such as ruthenium(IV) oxide, cobalt(II, III) oxide, vanadium oxide, manganese dioxide, and nickel(II) oxide due to high energy density, high purity capacity, and higher electrochemical activities is another strategy used in the fabrication of hybrids of conductive polymers[24,25].

The capacitance of quasi-capacitors is mainly due to the extremely reversible redox reactions of electrodes contained in metal oxides or conductive polymers. Since the redox reaction process's velocity involves charge transport and ion diffusion are the phase for evaluating velocity for the potential rate of pseudo-capacitors. As a result, quasi-capacitors can store 10 to 100 times the energy of double-layer capacitors per unit mass or volume[26].

Recently, the structure of two-dimensional materials, predominantly two-dimensional black phosphorus, has received considerable attention for the construction of flexible supercapacitors (FSCs). Black phosphorus fabrication processes are fully described in various dimensions, such as multilayer and quantum dots. Besides, the design approaches of black phosphorus-based FSCs, including conventional structures such as sandwiches, on-screen micro-configuration, are reviewed with potential challenges and opportunities. Based on a review of BP-based FSCs, they are believed to meet high electrochemical performance and excellent flexibility for storage devices based on the latest energy requirements[27].

Hollow or porous nanostructures can be used to store energy, convert and produce technology, increase the specific surface area of electrodes, lithium batteries, and capacitors, and provide numerous electrically active locations for electrochemical reactions. Furthermore, such a microstructure can enable a fast channel for electron and ion exchange because of the high fluid absorption capacity[28].

Recently, the arrangement of hollow nanostructures for power-related technologies, like the battery, supercapacitor, color-sensitive solar cell, and photoelectrochemical cell, has been studied[28,29].

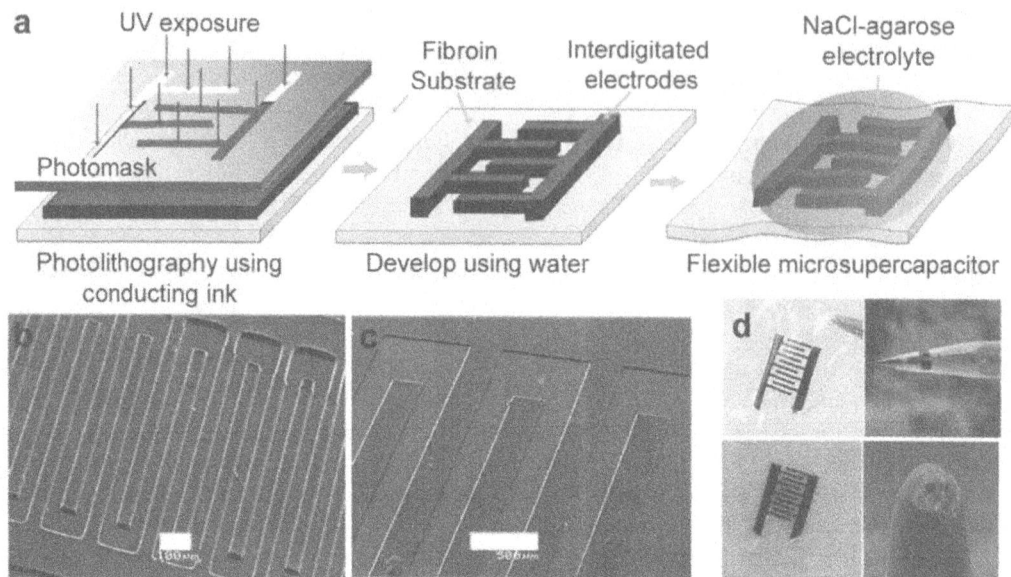

FIGURE 15.2 Photolithography steps for making compact micro capacitors. (a) Water and ultraviolet radiation are used to coat conductive ink on a fibrin substrate. (b) SEM photographs of electrodes at a scale of 100 μm and (c) at a magnification of 500 μm. (c) A flexible monitor that can be placed on the skin or folded over a pencil[30]. (Adapted with permission from Ref. [30]. Copyright (2018) American Chemical Society.)

Implantable bioelectronic systems use flexible and thin-film devices. Silk protein has recently been used to produce low-volume, biodegradable, flexible, biocompatible, and biodegradable micro-capacitors. A protein carrier with the conductive polymer PEDOT:PSS and a graphene oxide reducing agent were used to create a biocomposite ink used in light through the photolithographic process. With an agarose-NaCl gel electrolyte, these electrodes are printed on flexible protein sheets to create biodegradable organic devices. This device can provide light-emitting diodes and has the strength and stability to cycle more than 500 times. Another benefit is their flexibility, which can endure more than 450 cycles of rotational mechanical stress. They keep their capacitive properties for several days in the liquid before being lost after one month. As a result, they may provide suitable temporary options for bioelectronic energy supply that is flexible and implantable. Figure 15.2 depicts the photolithographic construction of a degradable capacitor[30].

Although carbon materials seem to be the most suitable material for making high capacitors, each material's limitations in composites need to be studied from energy density and longevity. Recent advances in composite materials with carbon components have focused on constructing supercapacitor electrodes, focusing on the limitations of the material alone and the possibility of combining it with other components to build a composite material with advanced performance. Also, several essential compounds and the main pathways of synthesis and composites' performance in this field have been studied[31,32].

15.5 FUEL CELL

In addition to EES, conductive polymers are also used in the production of electrochemical energy. The most important means of generating electrochemical energy are fuel cells.

According to polyether sulfonate and polysulfonate ether ketone, new organic and inorganic composite hybrid membranes were prepared for use in methanol fuel cells. The inorganic alteration, such as silica, titania, or zirconia, reduces membranes' permeability to water and methanol. The use of zirconia resulted in a 60-fold decrease in methanol flux. The use of organosilane resulted in a 40-fold reduction in water permeability without affecting proton conductance significantly. The combination of zirconia and zirconium phosphate can significantly reduce water and methanol permeability with a slight decrease in high conductivity[33].

Solution casting was used to make proton-conducting polymer electrolytes based on polyvinyl alcohol, ammonium acetate, and 1-butyl 3-methyl imidazolium chloride. The ionic conductivity achieved about 5.74 mS cm^{-1} in containing 50% by weight of butyl methyl imidazolium chloride. Ionic liquid doping made proton-conducting polymer electrolyte stable up to 250°C. At room temperature, the electrochemical fuel cell with the polymer electrolyte membrane had a power density of 18 mW cm^{-2}[34].

The proton exchange membrane's performance is affected by the carbon support's degradation, especially at the cathode. A catalytic layer containing an electrically conductive polymer-PPy nanowire has been developed as catalyst support to overcome this problem. In the longitudinal direction of PPy nanowires, a thin layer catalyst of platinum and palladium is formed. The resulting arrays are compressed on both sides of the Nafion membrane and form a membrane electrode assembly. A thin layer of catalyst arranged about 1.1 μm thick is applied to a single cell as anode and cathode without additional ionomers. Single-cell with low load Pt indicated a high performance of 762.1 mW cm^{-2}. The catalyst layer shows higher mass transfer at high-level current densities than the platinum carbon-based electrode. Because the prepared electrodes' activity is more than the targets in 2017, they can be used in fuel cells[35].

15.6 SOLAR CELL

Besides, conductive polymers can be used to make solar cells. Organic solar cells were fabricated on flexible polymer and glass layers with comparable performance to indium tin oxide (ITO)-containing samples. Furthermore, in the flexibility test, cells without ITO showed higher mechanical strength on flexible substrates than cells based on ITO. Recently, a combination of PEDOT:PSS has been used as a polymer anode in a small molecule-based organic emitting diode. These cells performed almost identically to ITO-based devices built on polymer and glass layers. Moreover, while ITO-based cells on flexible substrates after 75 flexural cycles during the flexibility test showed a power conversion efficiency of 0%, the efficiency without ITO with PEDOT anodes (power conversion efficiency 2.73%) after more than 300 bending cycles remained unchanged. It shows superior mechanical strength compared to ITO-based cells[36].

A combination of Si nanoparticles and conductive polymer PEDOT:PSS has been investigated for fabricating affordable photovoltaic systems. The optimal nanocone structure allows the polymer surface to be matched according to the rotation while providing excellent anti-reflectance and light-trapping properties. Consistent inconsistency over nanocone with increased light absorption leads to a power conversion yield of more than ten percent. Optimal nanostructures can achieve short-circuit current densities very close to the theoretical level. The current-voltage density of the following Si microstructures is shown in Figure 15.3(a): planar and nanocone, with or without Au finger grid.

The results of Figure 15.3(b) show that changing the nanostructure from a flat plate to a nanocone can increase the short-circuit current density from about 50% to more than 80% in the 400 to 950 nm wavelength.

Consequently, a silicone nanocone combined PEDOT:PSS has a suitable microstructure for polymer coating and solar cells[37].

Roll-to-roll technology is used in solar cell production due to its low cost, lightweight, flexible layer compatibility, high power processing, and large-scale solar cell production. The reverse roll-to-roll process can be applied by thin-film adhesion to polyethylene terephthalate (PET) layers. The adhesion energy of adjacent layers is measured very simply and independently of the mechanical properties and thickness by quantitatively analyzing different processing parameters and microstructural variables.

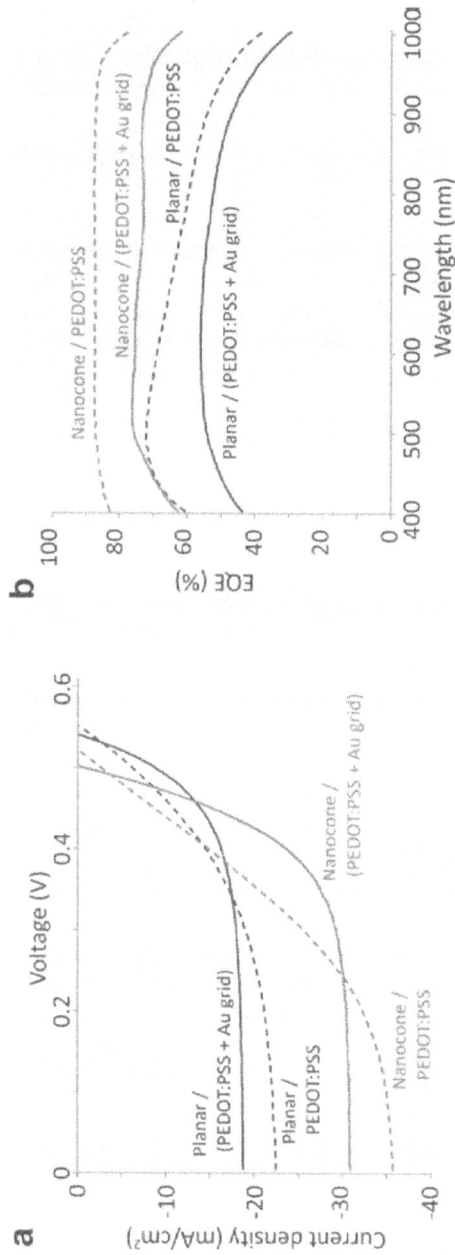

FIGURE 15.3 Properties of four hybrid samples of Si with planar and nanocones morphologies coated with PEDOT:PSS and with Au grid, (a) current density–voltage and (b) EQE spectra of the four systems[37]. (Adapted with permission from Ref. [37]. Copyright (2012) American Chemical Society.)

Besides, it enables the analysis of a combination of interlayer interfaces. In one study, adhesion was measured between the photoactive zone of poly 3-hexyl thiophene (P3HT):phenyl-C61-butyric acid methyl ester (PCBM) and the hole transfer layer, poly 3.4-ethylene dioxythiophene (PEDOT):polystyrene sulfonate (PSS). The adhesion failure energy was measured at the interface, and it varied greatly depending on the layer structure (P3HT:PCBM). Annealing operations improve interface adhesion, and the fracture energy of the adhesive increases with the annealing time and temperature. Also, layering in roll-to-roll reverse flexible polymer without PEDOT has been investigated. The conductive polymer PEDOT:PSS was changed by a metal oxide such as vanadium oxide, which plays a hole transfer layer. Interlayer adhesion in P3HT:PCBM inverted layer has been investigated and quantified[38].

ITO is a suitable candidate for transparent conductors for many electronic applications. However, due to resource scarcity, high market demand, and price fluctuations, efforts are underway to replace it for polymer solar cells (PSCs). Although there are several options for ITO, not most of them bring significant benefits to PSC. In this article, a number of reduced manufacturing solutions suitable for PSC use are discussed. These choices are divided into four categories: polymers, polymer-metal composites, ultra-thin metal films, and metallic nanowires, and carbon and graphene nanotubes. This article discusses the success of using these choices in PSC as well as the obstacles that it will face in the future[39].

Using an active layer material, the impact of ITO on the production of transparent electrodes from flexible conductive polymer-based on PEDOT:PSS printed on a combination of silver meshes and a bulk incompatibility based on PCDTBT:PC70BM was investigated. The results showed that the PEDOT:PSS stamped silver mesh performs similarly to ITO-coated PET surfaces, with more advantages[40].

Transparent silver electrodes based on silver nanogrids are an alternative to ITO electrodes due to the low-cost assembly and good flexibility. To smooth the surface and reduce layer resistance, a thin film of conductive polymers, such as HC-PEDOT:PSS, is often used for the silver mesh electrode. Recently, a solar cell device based on the inverted flexible polymer with silver mesh, PEDOT:PSS, zinc oxide, reaction layer, molybdenum oxide, and aluminum structure was made by eliminating light soaking treatment. Based on this discovery, WF adjustments were made to PEDOT:PSS layers doped with polyethylene and ammonia, and a flexible reverse polymer solar cell without light soaking with an energy conversion efficiency of about 7% was built[41].

Although energy is the most critical application of conductive polymers, they have other applications such as medical applications, sensors, and remediation, which are mentioned below.

15.7 MEDICAL APPLICATIONS

Microbial activity and aggregation in polymeric thermal conductors in electronic medical equipment impair these devices' performance and longevity. Recently, ternary thermal conductive composites with good antimicrobial properties have been developed using epoxy and boron hexagonal nano-sheets to solve this problem. The results of the tests showed that this composite was resistant to *Staphylococcus aureus* and *Escherichia coli*. Furthermore, using these composites will increase the thermal conductivity of pure epoxy by more than 11 times, indicating that they have a lot of potential in the production of electronic medical devices[42].

Redox polymers are electrically active macromolecules that contain components that can be oxidized or reduced, respectively, by losing or gaining electrons. Redox properties describe the polymer applications. In addition to their application in energy sectors, redox oxidation polymers are studied in different branches of medical technologies, like drug delivery. The redox potential values obtained in a specific cyclic voltammetric test can be varied by adjusting experimental parameters including the electrolyte. Cathodes in batteries are typically made of high-redox-potential polymers, while anodes are made of low-redox-potential polymers. Polymers with rapid and extensive redox reactions, leading by the plateau zone, are necessary for large capacitors with energy density capacities. Furthermore, polymers are used as oxidants with redox enzymes and reaction solutions for applications like biosensors as well as biofuel cells[43].

Neurological diseases are increasingly being treated with microelectrodes inserted in the head. The long-term and consistent use of neural microelectrodes, on the other hand, stays a problem. The application of conductive polymeric nanotubes of PEDOT was studied on neural microelectrodes. The results indicated an improvement compared to metallic electrodes. Coating PEDOT nanotubes made the neural microelectrodes, and their long-term performances were evaluated. The relationship between the impedance mechanism and tissue responses is consistent with the brain's reaction to nerve implants and the time course of the findings.

During this analysis, PEDOT nanotubes had a slightly lower impedance at lower noise, allowing for the detection of more neural units. Furthermore, they capture neural signals more effectively in a chronic environment, creating an opportunity for chronic recording and drug delivery than metal electrodes. PEDOT nanotubes can also be used as biosensors to transmit an inflammatory reaction to a chronic reply, as they minimize low-frequency impedance in severe environments. The ability to use these nanotubes to make high-quality neural recordings is also a plus[44].

Creating an optimized substrate using conductive polymers is essential for supporting and surviving stem cells in repairing heart function in many cases. Both endothelial and cardiac cells have found sufficient protection in conductive polymeric materials. Furthermore, roughness and surface energy have been found to affect both precursor cell types. Identifying the relationship between cardiac stem cells and the surface of conductive polymers may be a crucial step in optimizing these substances for treating heart tissue regeneration in this regard[45].

In recent years, conductive hydrogels bind to vital organs, such as heart transplants, can improve the cardiovascular regeneration process. In this regard, it can mention the construction of a pigmented hydrogel, which is based on Fe^{3+} triggered and simultaneous polymerization of pyrrole and dopamine in super branched chains, in which the conducting polyprolol formed in situ also creates crosslinking in the network. This conductive and adhesive hydrogel can be easily painted in pieces on the heart's surface without harmful liquid leakage. The available patch, whose conduction is equivalent to normal myocardium, attaches tightly to the beating heart within four weeks, effectively boosting electrophysiological signal transmission. Finally, the remodeling of myocardial infarction and cardiovascular function has improved dramatically. A suture-free strategy can be very promising in addressing human clinical challenges in cardiac tissue engineering[46].

Polymer microelectrodes have optical and physical properties that can be used to create a new generation of in vivo probes with less invasive, soft, and versatile features, such as possible cortical field capture, brain stem implantation, stents, and middle ear sensor arrays. Conductive microchannels were fabricated in polydimethylsiloxane as a traceable insulating plastic carrier using a replication technique. The microchannels were then packed with conductive polymer films like PEDOT:PSS or graphite-polydimethylsiloxane composite. All-polymer microelectrodes that were flexible and biocompatible enabled researchers to calculate potentials from acute myocardial preparation of cells and complete retinal assembly[47].

Flexible biosensors in the wearable and implantable fields for human health are receiving more and more attention. Flexible sensors include support zones compatible with sensing areas that can maintain their function in transmitting signals, even when the system is moving or physically deformed. Flexible devices are utilized to work on various surfaces, including irregular, soft, and non-static biological surfaces. It is essential to develop strategies for detecting proteins and biomarkers to expand flexible biosensors' applications in health care programs. Recently, fully organic, biodegradable biosensors have been developed for a vital biomarker. This sensor is based on a conductive ink made of a photoreactive fiber sequence and a conductive polymer called PEDOT:PSS. These active electrodes are mounted on adjustable fibroin surfaces with a controlled thickness and can stand alone or be used on smooth ground. Antibodies to vascular endothelial growth factors, which were conjugated in the conduction system, were used to detect the proteins.

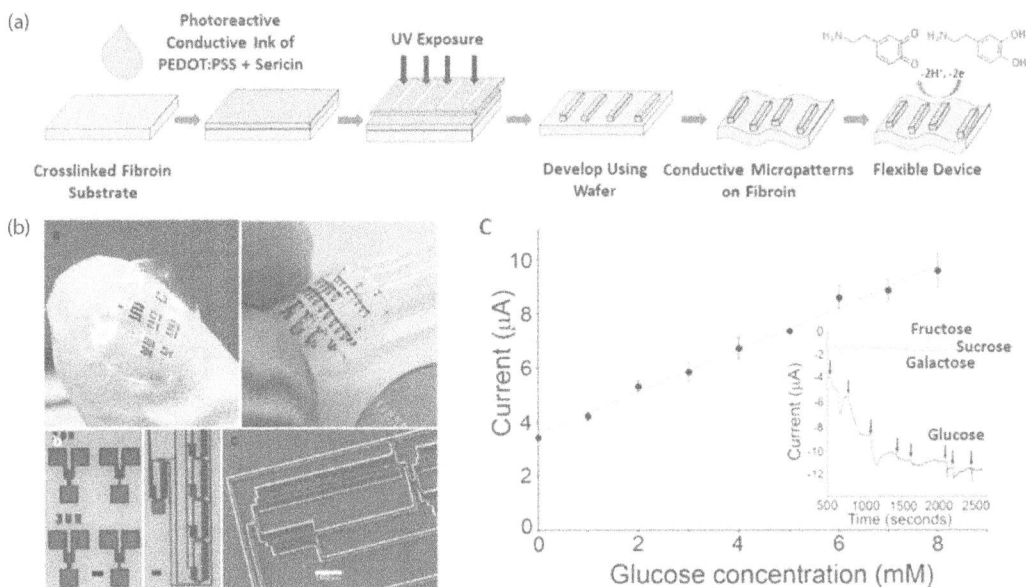

FIGURE 15.4 CP in biosensing application. (A) Schematic illustrating the fabrication process of CD (PEDOT:PSS) on a flexible substrate. (B) CD on different substrates: (a) on flexible silk fibroin sheets and (b–c) glass, and (c) SEM image, and (C) the current variation – glucose concentration curve in the biosensor[49]. (Adapted with permission from Ref. [49]. Copyright (2018) American Chemical Society.)

Unlike rigid biosensors, where antibodies covalently bind to surfaces and limit their use. Antibodies are distributed in the conductive matrix, which protects them and enables the detectors to operate in a soft environment without interfering with protein biomarker measurement. Biosensors must be adaptive and reproducible in a wide variety of biological liquids and act quite carefully and selectively to the protein structure. Flexible biosensors are used to monitor tissue repair and timely diagnosis[48].

A biocompatible and biodegradable biosensor was created on a silk fibroin substrate using a water-based conductive photoreactive ink and PEDOT:PSS containing a light-resistant sericin protein (SPP) compound. The steps involved in its production are depicted schematically in Figure 15.4A). It entails coating conductive ink on a fibrin protein-coated substrate and crosslinking under ultraviolet light (Figure 15.4B). The decreased average resistivity reduces from 11% to 50% of the resistance from 39 to 1 Ω-cm as the ink content increases. After four weeks in an enzymatically active protease solution at 37°C, the sensor is fully degraded. This is due to SPP components in the PEDOT:PSS structure.

Figure 15.4C depicts the sensor's selective output against the enzyme glucose oxidase (GOx) for detecting glucose concentrations against other sugars such as fructose, galactose, and sucrose, which is completely selective at tested concentrations and can be worn as a biosensor[49].

15.8 SENSOR AND ACTUATOR

Conductive polymers are classified into different types based on charge transfer (e.g., double bond and ions). Active electrical polymers, including polymer-ionic metal composites and other nanomaterial elastomers, react to pressure, humidity, and temperatures, for example. As a result, they can experience changes in their thermal, mechanical, and chemical properties. Changes in material properties can make calibration more difficult and have a negative impact on the sensor output. Another major challenge in designing electrochemical sensors is adhesion conductive polymers to the electrode for effective signal transmission. In addition to this challenge, conventional methods of fabricating sensors, such as lithographic methods, may not apply to these polymers and may require the use of other fabrication methods, such as printing or electrochemical deposition. Studies on the possibility of CP's chemical composition with the substrate surface should be investigated to overcome these limitations.

It is also necessary to study the conduction failure mechanism of polymers due to various factors such as heterogeneous distribution of thermal, mechanical, and microstructures before using them for commercial applications[50].

A high-performance, high-sensitivity pressure sensor was fabricated in low-pressure areas by varying the contact resistance between the coated gold micro-column assembly on a PPy layer. The pressure sensor's detection can be adjusted from 0.03 to 17 kPa[-1] by changing the column diameter and detection limit to 2 Pa. The simulation behavior of the sensor was done by understanding the basic principles of sensor response generation.

Conductive polymer-based pressure sensors can detect small-scale objects due to their high sensitivity to low-pressure regimes. Finally, this mechanism predicts contact-based measurement that can be used for touchscreen devices[51].

Nanostructured conjugated organic thin films from an array of 15 nm conductive PSS conductive polymer array using the etching and topography method. The linear resistance response of the nanowire array is higher than that of a non-patterned layer of the same thickness. Conducive conjugated polymers may also be converted into electronic components such as laser diodes, memory, or transistor arrays using self-assembled nanolithographic methods[52].

A combination of a repairable conductive polymer, highly elastic and soluble in solution for measuring pressure, and ultra-pressure creating electronic shells by simulating human skin is still a significant challenge in designing and synthesizing soft, conductive electronic materials that are stretchable and repairable. In this regard, a polymer composite consisting of polyaniline, phytic acid, and polyacrylic acid showed good self-healing properties and high tensile strength. The electrical conductivity of optimal composition was 0.12 S cm[-1]. During a rupture, active hydrogen bonding and electrostatic forces will recover electrical and mechanical properties with nearly 99% in less than 24 hours. Besides, this composite is pressure sensitive and can be used to build pressure sensors and detect various types of mechanical deformation with extremely high sensitivity. A soluble casting method can also be used to make this form of composite, which could lead to the development of massive, low-cost electronic shells[53].

Today, intelligent, flexible strain biosensors based on conductive electrical composites have been considered due to their lightweight, flexibility, traction, and easy fabrication method. They have immense potential for applications such as artificial muscles and skin, movement detection, human-machine interaction, and wearable devices. Essential factors in the preparation of these pressure sensors are the polymer matrix, the type of conductive filler, and the morphological design of the polymer background and filler phase, which affects the obtained properties.

In general, elastomers with good flexibility and elongation, like thermoplastic polyurethane, rubber, and PDMS, are polymers that have been widely selected for the production of flexible strain sensors. For the preparation of composites with excellent electrical conductivity, carbon fillers (such as graphene, carbon black, and carbon nanotubes), inherently conductive polymers (such as PPy, PANI, and PEDOT:PSS), and electrical conductive nanomaterials (such as silver or copper nanowires) are used.

In general, the conductivity mechanism in this type of polymer composites is attributed to the charge of the conductive fillers that can be divided into two categories depending on the filler content, one including the tunneling effect with fewer loading fillers, and the other ohmic conductivity with increasing fillers and creating a percolation effect.

Several review articles recently examined the development of electrical conductive polymer in intelligent, versatile strain detectors. The conductor network structures of these sensors, such as segmented, porous, microcrack, and wrinkled, are also examined[54].

15.9   EES

Due to metal corrosion and the formation of heterogeneous deposits during charging, metal anodes, especially zinc anodes, become unstable. A simultaneous electrospraying and electrospinning process (Figure 15.5a) has recently been developed to create a new class of versatile Zn-air batteries, in which three-dimensional continuous ion/electron transport channels are created at the electrodes. Zinc-air cells (including an anode, cathode, and PVA/PAA GPE separator film) were made using the specific structure (Figure 15.5b). Under the galvanostatic cycling test of zinc cell discharge/charge, the cycling behavior showed stability (Figure 15.5c). In fact, different deformation modes (such as bending and twisting) and in situ analysis of their galvanostatic discharge/charge profiles

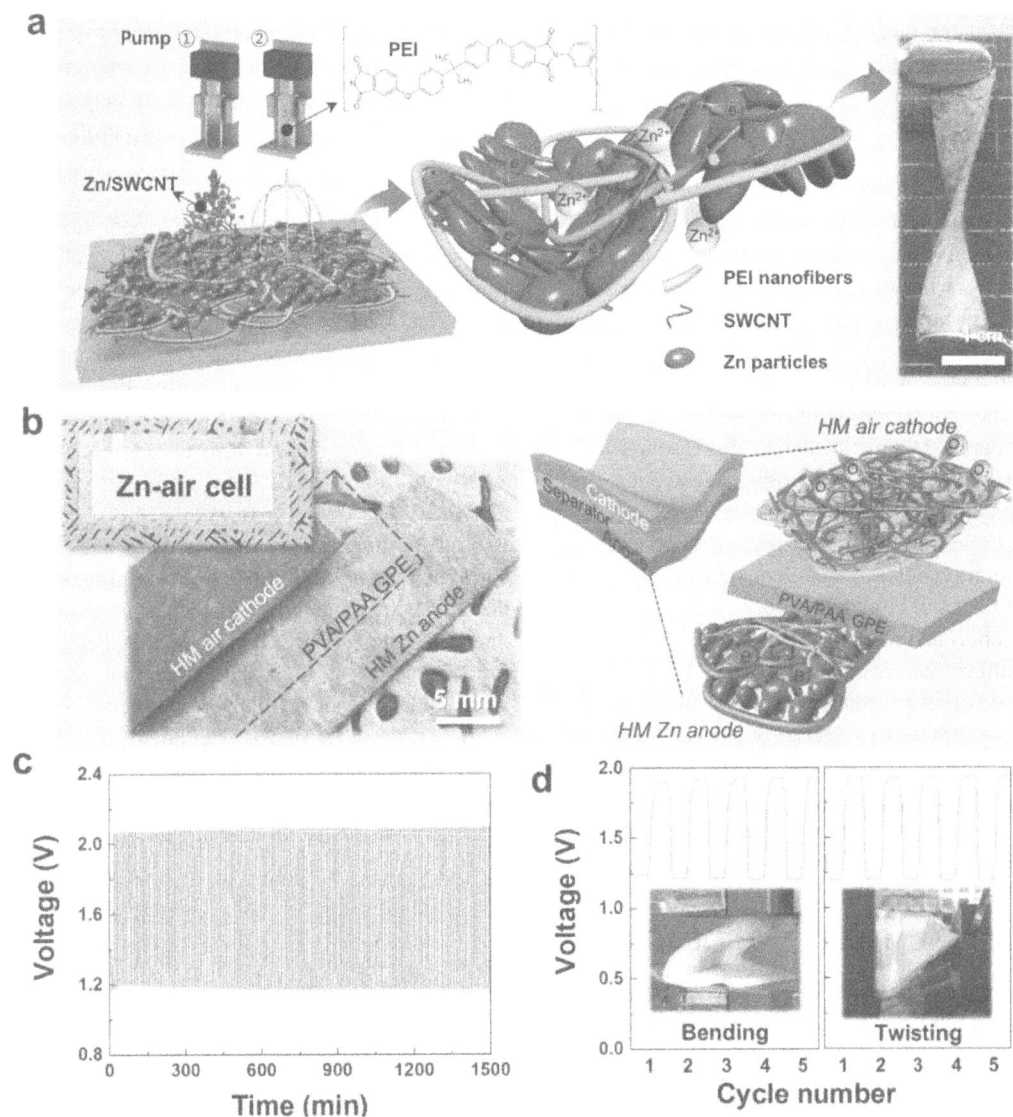

FIGURE 15.5   (a) Electrospinning/electrospraying methods for anode construction (HM-Zn); (b) construction of multifunctional Zn-air cells with HM structure; (c) galvanostatic discharge/charge cycling output at 0.5 mA square current for each 10-minute cycle period (5 minutes discharge and 5 minutes charge) during bending and twisting tests; (d) galvanostatic discharge/charge cycling performance during bending and twisting tests[55]. (Adapted with permission from Ref. [55]. Copyright (2018) American Chemical Society.)

were investigated to elucidate their electrochemical behavior. The structural electrodes supported by the frame under the deformation test displayed excellent electrical properties (Figure 15.5d). They had improved mechanical properties and electrochemical rechargeability[55].

Wearable electronics, paper-like products, and flexible biomedical healthcare equipment all have a sizable market share. To charge these devices, flexible energy storage is needed.

Capacitive cloud technology relies on stand-alone nanowires, which provide powerful energy sources for these lightweight electronics, which are one solution. The electrodes of superconductors are made up of heterogeneous coaxial nanowires with PEDOT as the shell and MnO_2 as the core material in a semi-cellular structure, which has higher speed and power than pure nanomaterials. Without any additives, these PEDOT coaxial nanowires were studied as anodes in a cell in both symmetric and asymmetric states in terms of charge, energy storage, and flexibility. The asymmetric cell outperformed the symmetric cell in terms of cycle capacity, speed, and energy density, according to the findings. Even at power densities above 850 watts per kilogram, the asymmetric device's electrode has an energy density of 8.9 watts per kilogram. This system is extremely adaptable, charging and discharging quickly while retaining 86% of its energy density even in its most versatile state. It has been demonstrated that the entire system has a total capacity of 0.26 F at a peak power of 1.7 volts, which is capable of supplying portable devices[56].

The largest volume changes of the electrode, the formation of the intermediate solution, the poor conductivity of ions, and the final discharge of lithium sulfides all contribute to the short cycle life and poor performance of lithium-sulfur batteries. Building a sulfur nanocomposite electrode with distinct lithium storage properties, mechanical amplification, ionic and electronic conductivity, and resulting in a solid and mixed ionic and electronic conductivity is an appealing way to solve these challenges. A new method for developing a sustainable cathode scaffold using sulfur, porous graphene oxide, and Mn_3O_4 nanoparticles in combination with polyaniline and sodium alginate was published in a recent study. This research could lead to a new strategy for creating durable mechanical conductive nanocomposite electrodes, a mix for elevated lithium-ion batteries which can be used to make versatile huge storage devices[57].

Conductive polymer hydrogels promise flexible electrode materials. However, hydrogels have limitations in terms of electrochemical performance or mechanical flexibility. By combining different forms of conductive polymers, polyaniline and PEDOT, via a phytic acid molecular connection, an approach to resolve this restriction has been established. This acid removes certain PSSs and changes the arrangement of PEDOT sequences from benzoic to quinoid. The generated gels are made up of a three-dimensional structure of PANI-coated PEDOT shells. The experimental results showed that this material has better mechanical properties than PEDOT hydrogel due to molecular interactions between PANI and PEDOT. A flexible solid-state supercapacitor based on PEDOT/PANI hydrogel with a high volumetric energy density of 0.25 mWh cm^{-3} with a power density of 107.14 mW cm^{-3} shows better properties compared to many previous solid-state capacitors based on PEDOT, hydrogels, and other conductive polymers[58].

15.10 SUMMARY AND OUTLOOK

Conductive polymers have various applications due to structural modifications, high conductivity, and unique electrochemical properties. Furthermore, these materials can be produced from large-scale biocompatible materials. Significant advancements in materials science, particularly their combination with nanostructures and the production of composite compounds, have opened up a plethora of options for manipulating morphology and preparing flexible polymer-based composites. Their spectrum can be expanded by incorporating them into emerging technologies such as organic and color-sensitive fuel cells, solar cells, electrochemical reactions, and environmental applications.

Theoretical approaches are needed to predict and explain conductive polymers' properties and control their stability to achieve industrial applications. Using these materials by adding mechanical properties and flexibility can expand their service and create technological changes shortly in various industry and medicine fields.

REFERENCES

1. Gómez IJ, Vázquez Sulleiro M, Mantione D, Alegret N. Carbon nanomaterials embedded in conductive polymers: A state of the art. *Polymers (Basel).* 2021;13(5):745. doi:https://doi.org/10.3390/polym13050745

2. Onggar T, Kruppke I, Cherif C. Techniques and processes for the realization of electrically conducting textile materials from intrinsically conducting polymers and their application potential. *Polymers (Basel).* 2020;12(12):1–46. doi:10.3390/polym12122867

3. Ehsani A, Parsimehr H, Nourmohammadi H, Safari R, Doostikhah S. Environment-friendly electrodes using biopolymer chitosan/poly ortho aminophenol with enhanced electrochemical behavior for use in energy storage devices. *Polym Compos.* 2019;40(12):4629–4637. doi:10.1002/pc.25330

4. Ehsani A, Mirtamizdoust B, Karimi F, Bigdeloo M, Parsimehr H. Influence of nanostructured VO-acetylacetonate coordination system with 2-(pyridin-4-ylmethylene) hydrazine-1-carbothioamide in pseudocapacitance performance of p-type conductive polymer composite film. *Plast Rubber Compos.* 2020;0(0):1–8. doi:10.1080/14658011.2020.1859875

5. Espinoza MSD, Poblete VH, Bernede JC, et al. Synthesis and optical behavior of PLED devices based on (PMMA)/(PAA)/Er(AP)6Cl3 complex and N,N′-didodecyl-3,4,9,10-perylene tetracarboxylic diimide composites. *Polym Bull.* 2013;70(10):2801–2814. doi:10.1007/s00289-013-0989-x

6. Zhan C, Yu G, Lu Y, Wang L, Wujcik E, Wei S. Conductive polymer nanocomposites: A critical review of modern advanced devices. *J Mater Chem C.* 2017;5(7):1569–1585. doi:10.1039/c6tc04269d

7. Pei Q. Light-emitting polymers. *Mater Matters.* 2007;2(3):26–28.

8. Sharma R, Alam F, Sharma AK, Dutta V, Dhawan SK. Role of zinc oxide and carbonaceous nanomaterials in non-fullerene-based polymer bulk heterojunction solar cells for improved cost-to-performance ratio. *J Mater Chem A.* 2015;3(44):22227–22238. doi:10.1039/c5ta06802a

9. Hadi A, Hashim A, Al-Khafaji Y. Structural, optical and electrical properties of PVA/PEO/SnO$_2$ new nanocomposites for flexible devices. *Trans Electr Electron Mater.* 2020;21(3):283–292. doi:10.1007/s42341-020-00189-w

10. Wu C, Li H, Yan Y, et al. Highly-stable organo-lead halide perovskites synthesized through green self-assembly process. *Sol RRL.* 2018;2(6):1800052. doi:10.1002/solr.201800052

11. Wang K, Wu C, Hou Y, Yang D, Priya S. Monocrystalline perovskite wafers/thin films for photovoltaic and transistor applications. *J Mater Chem A.* 2019;7(43):24661–24690. doi:10.1039/c9ta08823g

12. Mu G, Ding Z, Mu D, et al. Hierarchical void structured Si/PANi/C hybrid anode material for high-performance lithium-ion batteries. *Electrochim Acta.* 2019;300:341–348. doi:10.1016/j.electacta.2019.01.126

13. Xia X, Chao D, Qi X, et al. Controllable growth of conducting polymers shell for constructing high-quality organic/inorganic core/shell nanostructures and their optical-electrochemical properties. *Nano Lett.* 2013;13(9):4562–4568. doi:https://doi.org/10.1021/nl402741j

14. Li W, Chen R, Qi W, et al. Reduced graphene oxide/mesoporous ZnO NSs hybrid fibers for flexible, stretchable, twisted, and wearable NO$_2$ E-Textile gas sensor. *ACS Sensors.* 2019;4(10):2809–2818. doi:10.1021/acssensors.9b01509

15. Ramanujam J, Bishop DM, Todorov TK, et al. Flexible CIGS, CdTe and a-Si:H based thin film solar cells: A review. *Prog Mater Sci.* 2020;110:100619. doi:10.1016/j.pmatsci.2019.100619

16. Shi Y, Peng L, Ding Y, Zhao Y, Yu G. Nanostructured conductive polymers for advanced energy storage. *Chem Soc Rev.* 2015;44(19):6684–6696. doi:10.1039/c5cs00362h

17. Mohd Abdah MAA, Azman NHN, Kulandaivalu S, Sulaiman Y. Review of the use of transition-metal-oxide and conducting polymer-based fibres for high-performance supercapacitors. *Mater Des.* 2020;186:108199. doi:10.1016/j.matdes.2019.108199

18. Wang F, Kim HJ, Park S, Kee CD, Kim SJ, Oh IK. Bendable and flexible supercapacitor based on poly-pyrrole-coated bacterial cellulose core-shell composite network. *Compos Sci Technol.* 2016;128:33–40. doi:10.1016/j.compscitech.2016.03.012

19. Guo D, Lai L, Cao A, Liu H, Dou S, Ma J. Nanoarrays: Design, preparation and supercapacitor applications. *RSC Adv.* 2015;5(69):55856–55869. doi:10.1039/c5ra09453d

20. Zhang W, Feng P, Chen J, Sun Z, Zhao B. Electrically conductive hydrogels for flexible energy storage systems. *Prog Polym Sci.* 2019;88:220–240. doi:10.1016/j.progpolymsci.2018.09.001

21. Wu Q, Xu Y, Yao Z, Liu A, Shi G. Supercapacitors based on flexible graphene/polyaniline nanofiber composite films. *ACS Nano.* 2010;4(4):1963–1970. doi:10.1021/nn1000035

22. Xu, Y., Zhou, M. and Lei Y. Nanoarchitectured array electrodes for rechargeable lithium-and sodium-ion batteries. *Adv Energy Mater.* 2016;6(10):1502514. doi:https://doi.org/10.1002/aenm.201502514

23. Ellis BL, Knauth P, Djenizian T. Three-dimensional self-supported metal oxides for advanced energy storage. *Adv Mater.* 2014;26(21):3368–3397. doi:10.1002/adma.201306126

24. Iro ZS, Subramani C, Dash SS. A brief review on electrode materials for supercapacitor. *Int J Electrochem Sci.* 2016;11(12):10628–10643. doi:10.20964/2016.12.50

25. Dakshayini BS, Reddy KR, Mishra A, et al. Role of conducting polymer and metal oxide-based hybrids for applications in ampereometric sensors and biosensors. *Microchem J.* 2019;147(February):7–24. doi:10.1016/j.microc.2019.02.061

26. Yang J, Li XL, Zhou JW, Wang B, Cheng JL. Fiber-shaped supercapacitors: Advanced strategies toward high-performances and multi-functions. *Chinese J Polym Sci (English Ed.)* 2020;38(5):403–422. doi:10.1007/s10118-020-2389-7

27. Wu Y, Yuan W, Xu M, et al. Two-dimensional black phosphorus: Properties, fabrication and application for flexible supercapacitors. *Chem Eng J.* 2021;412:128744. doi:10.1016/j.cej.2021.128744

28. Wang J, Cui Y, Wang D. Design of hollow nanostructures for energy storage, conversion and production. *Adv Mater.* 2019;31(38):1801993. doi:10.1002/adma.201801993

29. Peng L, Fang Z, Zhu Y, Yan C, Yu G. Holey 2D nanomaterials for electrochemical energy storage. *Adv Energy Mater.* 2018;8(9):1–19. doi:10.1002/aenm.201702179

30. Pal RK, Kundu SC, Yadavalli VK. Fabrication of flexible, fully organic, degradable energy storage devices using silk proteins. *ACS Appl Mater Interfaces.* 2018;10(11):9620–9628. doi:10.1021/acsami.7b19309

31. Borenstein A, Hanna O, Attias R, Luski S, Brousse T, Aurbach D. Carbon-based composite materials for supercapacitor electrodes: A review. *J Mater Chem A.* 2017;5(25):12653–12672. doi:10.1039/c7ta00863e

32. Yan J, Li S, Lan B, Wu Y, Lee PS. Rational design of nanostructured electrode materials toward multifunctional supercapacitors. *Adv Funct Mater.* 2020;30(2):1–35. doi:10.1002/adfm.201902564

33. Nunes SP, Ruffmann B, Rikowski E, Vetter S, Richau K. Inorganic modification of proton conductive polymer membranes for direct methanol fuel cells. *J Memb Sci.* 2002;203(1–2):215–225. doi:10.1016/S0376-7388(02)00009-1

34. Liew CW, Ramesh S, Arof AK. A novel approach on ionic liquid-based poly(vinyl alcohol) proton conductive polymer electrolytes for fuel cell applications. *Int J Hydrog Energy.* 2014;39:2917–2928. doi:10.1016/j.ijhydene.2013.07.092

35. Jiang S, Yi B, Cao L, et al. Development of advanced catalytic layer based on vertically aligned conductive polymer arrays for thin-film fuel cell electrodes. *J Power Sources.* 2016;329:347–354. doi:10.1016/j.jpowsour.2016.08.098

36. Na SI, Kim SS, Jo J, Kim DY. Efficient and flexible ITO-free organic solar cells using highly conductive polymer anodes. *Adv Mater.* 2008;20(21):4061–4067. doi:10.1002/adma.200800338

37. Jeong S, Garnett EC, Wang S, et al. Hybrid silicon nanocone–Polymer solar cells. *Nano Lett.* 2012;12(6):2971–2976. doi:doi.org/10.1021/nl300713x

38. Dupont SR, Oliver M, Krebs FC, Dauskardt RH. Interlayer adhesion in roll-to-roll processed flexible inverted polymer solar cells. *Sol Energy Mater Sol Cells.* 2012;97:171–175. doi:10.1016/j.solmat.2011.10.012

39. Angmo D, Krebs FC. Flexible ITO-free polymer solar cells. *J Appl Polym Sci.* 2013;129(1):1–14. doi:10.1002/app.38854

40. Muhsin B, Roesch R, Gobsch G, Hoppe H. Flexible ITO-free polymer solar cells based on highly conductive PEDOT:PSS and a printed silver grid. *Sol Energy Mater Sol Cells.* 2014;130:551–554. doi:10.1016/j.solmat.2014.08.009

41. Wang J, Fei F, Luo Q, et al. Modification of the highly conductive PEDOT:PSS layer for use in silver nanogrid electrodes for flexible inverted polymer solar cells. *ACS Appl Mater Interfaces.* 2017;9(8):7834–7842. doi:10.1021/acsami.6b16341

42. Xiong SW, Zhang P, Xia Y, Zou Q, Jiang M-ying, Gai JG. Unique antimicrobial/thermally conductive polymer composites for use in medical electronic devices. *J Appl Polym Sci.* 2021;138(13):1–11. doi:10.1002/app.50113

43. Casado N, Hernández G, Sardon H, Mecerreyes D. Current trends in redox polymers for energy and medicine. *Prog Polym Sci.* 2016;52:107–135. doi:10.1016/j.progpolymsci.2015.08.003

44. Abidian MR, Ludwig KA, Marzullo TC, Martin DC, Kipke DR. Interfacing conducting polymer nanotubes with the central nervous system: Chronic neural recording using poly(3,4-ethylenedioxythiophene) nanotubes. *Adv Mater.* 2009;21(37):3764–3770. doi:10.1002/adma.200900887

45. Gelmi A, Ljunggren MK, Rafat M, Jager EWH. Influence of conductive polymer doping on the viability of cardiac progenitor cells. *J Mater Chem B.* 2014;2(24):3860–3867. doi:10.1039/c4tb00142g

46. Liang S, Zhang Y, Wang H, et al. Paintable and rapidly bondable conductive hydrogels as therapeutic cardiac patches. *Adv Mater.* 2018;30(23):1–10. doi:10.1002/adma.201704235

47. Blau A, Murr A, Wolff S, et al. Flexible, all-polymer microelectrode arrays for the capture of cardiac and neuronal signals. *Biomaterials*. 2011;32(7):1778–1786. doi:10.1016/j.biomaterials.2010.11.014

48. Xu M, Yadavalli VK. Flexible biosensors for the impedimetric detection of protein targets using silk-conductive polymer biocomposites. *ACS Sensors*. 2019;4(4):1040–1047. doi:10.1021/acssensors.9b00230

49. Kenry K, Liu B. Recent advances in biodegradable conducting polymers and their biomedical applications. *Biomacromolecules*. 2018;19(6):1783–1803. doi:10.1021/acs.biomac.8b00275

50. Nambiar S, Yeow JTW. Conductive polymer-based sensors for biomedical applications. *Biosens Bioelectron*. 2011;26(5):1825–1832. doi:10.1016/j.bios.2010.09.046

51. Shao Q, Niu Z, Hirtz M, et al. High-performance and tailorable pressure sensor based on ultrathin conductive polymer film. *Small*. 2014;10(8):1466–1472. doi:10.1002/smll.201303601

52. Jung YS, Jung WC, Tuller HL, Ross CA. Nanowire conductive polymer gas sensor patterned using self-assembled block copolymer lithography. *Nano Lett*. 2008;8(11):3777–3780. doi:10.1021/nl802099k

53. Wang T, Zhang Y, Liu Q, et al. A self-healable, highly stretchable, and solution processable conductive polymer composite for ultrasensitive strain and pressure sensing. *Adv Funct Mater*. 2018;28(7):1–12. doi:10.1002/adfm.201705551

54. Wu S, Peng S, Yu Y, Wang CH. Strategies for designing stretchable strain sensors and conductors. *Adv Mater Technol*. 2020;5(2):1–25. doi:10.1002/admt.201900908

55. Lee D, Kim HW, Kim JM, Kim KH, Lee SY. Flexible/rechargeable Zn-Air batteries based on multifunctional heteronanomat architecture. *ACS Appl Mater Interfaces*. 2018;10(26):22210–22217. doi:10.1021/acsami.8b05215

56. Duay J, Gillette E, Liu R, Lee SB. Highly flexible pseudocapacitor based on freestanding heterogeneous MnO$_2$/conductive polymer nanowire arrays. *Phys Chem Chem Phys*. 2012;14(10):3329–3337. doi:10.1039/c2cp00019a

57. Ghosh A, Manjunatha R, Kumar R, Mitra S. A facile bottom-up approach to construct hybrid flexible cathode scaffold for high-performance lithium-sulfur batteries. *ACS Appl Mater Interfaces*. 2016;8(49):33775–33785. doi:10.1021/acsami.6b11180

58. Yang Z, Ma J, Bai B, et al. Free-standing PEDOT/polyaniline conductive polymer hydrogel for flexible solid-state supercapacitors. *Electrochim Acta*. 2019;322:134769. doi:10.1016/j.electacta.2019.134769

LIST OF ABBREVIATIONS

BP	Black phosphorus
CCG	Chemically converted graphene
CP	Conductive polymer
CTS	Chitosan
ECs	Electrochemical capacitors
EES	Electrochemical energy storage
EML	Emitting layer
EQE	External quantum efficiency
FAO	United Nations Food and Agriculture Organization
FSCs	Flexible supercapacitors
GOx	Glucose oxidase
HC	High conductive
HTL	Hole transfer layer
ITO	Indium tin oxide
KOH	Potassium hydroxide
LEPs	Light-emitting polymers
MEH-PPV	Poly[2-methoxy-5-(2-ethyl hexyloxy)-1,4-phenylene vinylene]
NF	Nanofiber
OLED	Organic light-emitting diode
OPVs	Organic photovoltaic devices
P3HT	Poly(3-hexyl thiophene)
P3OT	Poly(3-octylthiophene)
PANI	Polyaniline

PAQS	Poly(anthraquinonyl sulfide)
PCBM	Phenyl-C61-butyric acid methyl ester
PEDOT	Poly(3,4-ethylene dioxythiophene)
PET	Polyethylene terephthalate
PFO	Poly(9,9-dioctylfluorene)
POAP	Poly ortho aminophenol
PPP	Poly(p-phenylene)
PPV	Poly(p-phenylene vinylene)
PPY	Polypyrrole
PSS	Polystyrene sulfonate
PTh	Polythiophene
SC	Solar cell
SPP	Sericin protein photoresist
TEMPO	2,2,6,6-tetramethyl-1-piperidinyloxy, free radical
VO(acac)	VO-acetylacetone
WF	Work function

16 Conducting Polymer Nanocomposites for Flexible Devices

Momath Lo[1], Modou Fall[1], Mohamed Lamine Sall[1],
Diariatou Gningue-Sall[1], Amadou Bélal Guèye[1,2],
Hanna J. Maria[1], and Sabu Thomas[2]

[1]Laboratory of Organic Physical Chemistry and Environmental Analyses (LCPOAE), Department of Chemistry, Faculty of Sciences and Techniques, Cheikh Anta Diop University, Dakar-Fann, Senegal

[2]School of Energy Materials and International and Inter University Center for Nanoscience and Nanotechnology (IIUCNN) and School of Chemical Sciences, Mahatma Gandhi University, Priyadarshini Hills, Kottayam, Kerala, India

CONTENTS

16.1 INTRODUCTION

The synthesis of conductive polymers (CPs) and their characterization are attracting increasing interest and are the subject of many researches because of their potential applications, especially in the field of energy. These nanostructured polymers require the presence of materials such as inorganic nanomaterials [1], graphene [2, 3], carbon nanotubes [4], and other CPs [5]. Among the CPs, polypyrrole (PPy), polyaniline (PANI), and polythiophenes (PThs) are the most popular synthesized materials [1–3].

The use of these conductive organic polymers is due to their low toxicity, flexibility, ability to improve electrical conductivity, increased stability, and ease of synthesis in various solvents. These CPs, synthesized either by the chemical route or by electropolymerization, exhibit good performance

DOI: 10.1201/9781003150374-16

in industrial applications, innocuousness, and good thermal stability. Improving the physical and chemical properties of CPs, such as controlling their morphology and changing thickness, broadens their fields of application. In this case, CPs generally need to be nanostructured and presented in nanocomposite forms of CPs (NCPs) in order to meet the ever-increasing energy needs in various technologies. These NCPs offer several advantageous features, including reducing the material size and providing applications in ultrathin screens. Additionally, the presence of heterojunctions at the nanoscale has shown high energy conversion efficiency in solar cells. The energy systems performance enhancement has become an essential research axis because of the ever-growing needs, especially in flexible electronics, smart devices for monitoring, and photovoltaic devices.

However, the selection of the flexible substrate for the material design is the main challenge to be faced when developing an energy-oriented platform. Indeed, some constituents on flexible substrates may lose their adhesions when stretched or may be detached from their substrates.

To overcome these problems, researchers are focusing on designing NCPs on flexible substrates to develop (i) performance comparable to that of rigid substrates, (ii) high stability materials, (iii) good electrochemical properties, (iv) superior mechanical performance, and (v) highly stretchability.

Here we review the recent trends in the evolution of NCPs applied in the energy field through flexible support. Additionally, we discuss the synthesis of CPs and the nanocomposite characterization along with the technique of depositing electroactive NCPs on flexible supports. The procedures used to make energy devices highly scalable and the basic building properties (flexible substrates) are also discussed along with the challenges and potential markets of NCPs for flexible device.

16.2   PREPARATION OF NANOCOMPOSITES OF CONDUCTIVE POLYMER

CPs, including PPy, PANI, and PTh (Scheme 16.1a), form a class of materials that are the subject of many studies in flexible devices. These organic polymers are electrically conductive through a system of conjugate bonds along the polymer chain. The manufacture of these CPs involves coupling and removal of protons. The proposed mechanism of PTh polymerization is shown in Scheme 16.1(b). However, it should also be noted that the complex polymerization mechanisms can be very different from a CP to another. A number of reports have described preparations for nanostructured CPs. Indeed, recent studies have shown that CPs that are flexible are often applied in the field of energy storage [6, 7], solar cells [8], and flexible electronics [9]. Monomers and CPs can be functionalized with different materials to tailor their properties. For this, various techniques of syntheses are used, including electrochemical, self-assembled, mixtures, chemical, and enzyme entrapment methods.

16.2.1   ELECTROCHEMICAL METHODS

Different electrochemical techniques have been used for the design of nanocomposites on electrode surfaces. The potentiodynamic, potentiostatic and galvanostatic methods are often employed [11]. These methods allow the homogeneous formation of a good quality nanocomposite on flexible electrode surfaces. Their interest is to regulate the size or the morphology of the materials [12]. For instance, PPy/MWCNT/cotton composite prepared by electrochemical polymerization exhibited outstanding flexibility, high conductivity, with very low resistance, and a cauliflower structure.

The electropolymerization process takes place in a three-electrode electrochemical cell from a solution containing an electrolytic salt and the selected monomer dissolved in the appropriate medium.

This electrochemical synthesis can be carried out in potentiostatic, galvanostatic, or potentiodynamic mode. It can be performed either in an organic medium [13] or a direct or reverse micellar medium [14, 15] and most often results in the formation of a polymer layer on the electrode.

This polymer formation procedure requires the control of certain parameters such as:

- the monomer concentration and the synthesized oligomers must have a lower oxidation potential than that of the monomer and be little or not soluble in the solvent used so that the polymerization process can continue

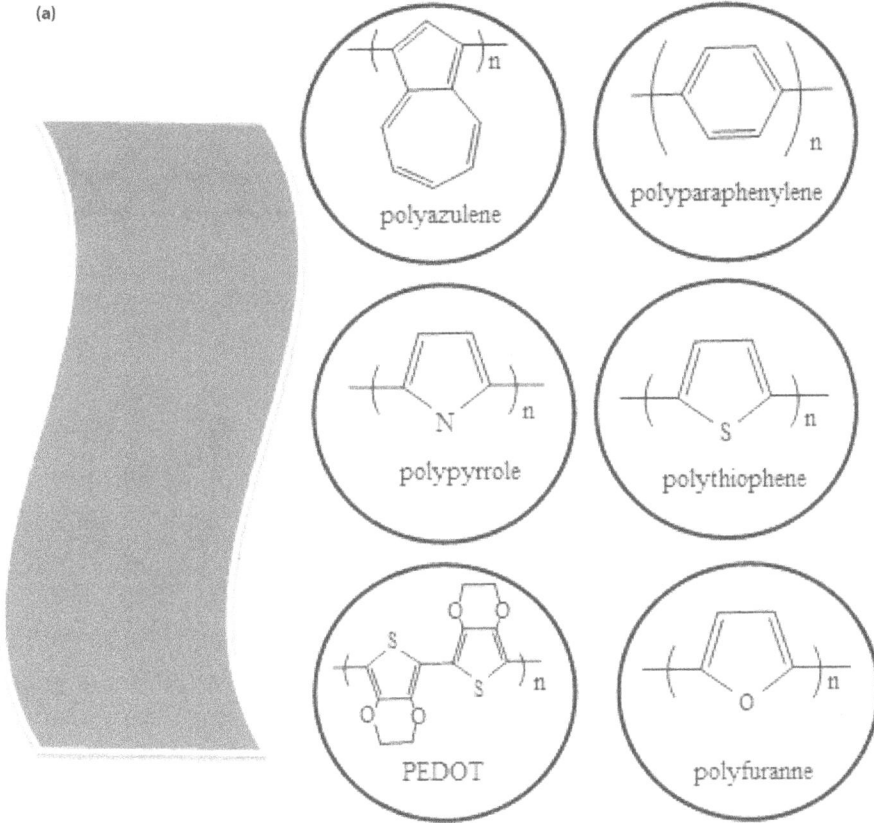

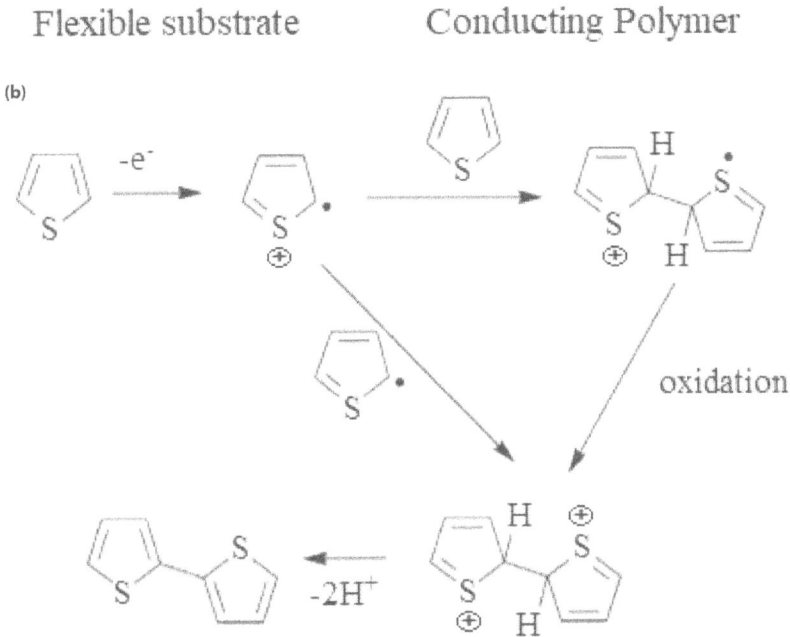

SCHEME 16.1 (a) Chemical structures of typical CPs. (b) Mechanism for thiophene polymerization.

(c)

Materials

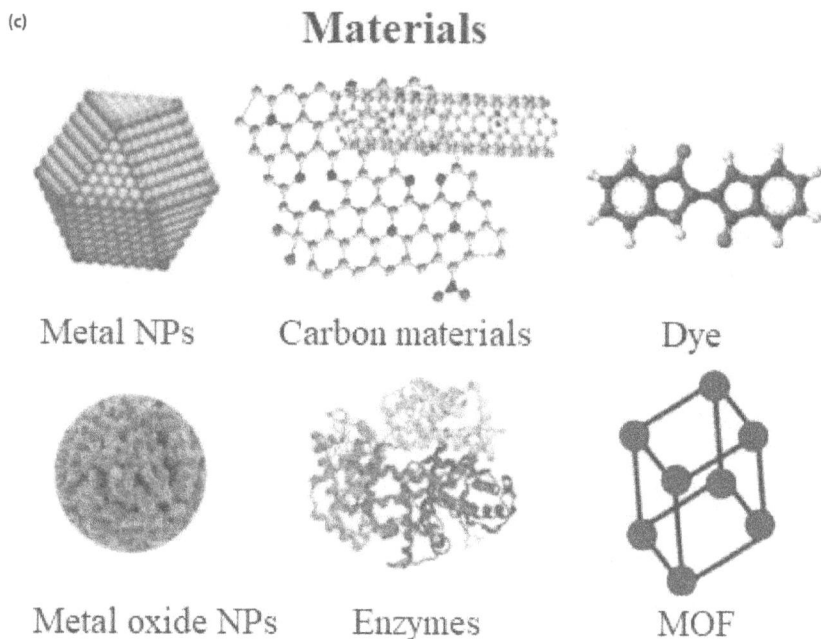

Metal NPs   Carbon materials   Dye

Metal oxide NPs   Enzymes   MOF

SCHEME 16.1 (Continued) (c) Different composite materials for nanostructured polymer. (Reproduced with permission from Ref. [10].)

- supporting-salt and its concentration
- solvent type (acidic, basic, or neutral character)
- the working electrode type
- the potential and current density values applied during the electropolymerization
- other used materials

The electrochemical method has several interests including the obtention of a large amount of polymer with the possibility of regulating the amount of charge deposited on the flexible electrode and obtaining doped, conductive, and relatively pure polymers.

In their works, Liu et al. [14] reported flexible cotton-based electrodes modified by a polypyrrole in the presence of carbon nanotubes by a two-step potentiostatic method. In the first step, in order to obtain conducting materials, flexible multiwall carbon nanotubes (MWCNT) scaffolds were made using 3D microporous cotton. Secondly, a potential was imposed for the electropolymerization of pyrrole doped with p-toluenesulfonic anions (TsO$^-$) on a conductive layer of cotton modified with MWCNT. The results showed the possibility of making textile composites that can be successfully applied in polymer series. Other authors [15] confirmed the interest in the techniques based on the mixture of PPy and carbon quantum dots (CQDs) deposited on a flexible ASSS substrate (Scheme 16.2). This material was synthesized by the potentiostatic method with the advantage of controlling the CQD/PPy composite layer for the formation of a porous material.

16.2.2 Chemical Method

The chemical polymerization needs a close interaction between the monomer and the oxidant. For the formation of the polymer, the oxidation potential of the oxidant should be higher than that of the monomer.

The advantage of this synthetic technique over the electrochemical method lies in obtaining a large amount of polymer in its doped and conductive form. Generally, the polymers obtained by this method are in the form of a fine powder, a coating on the electrode or accumulate on dispersed particles or on insulating substrates.

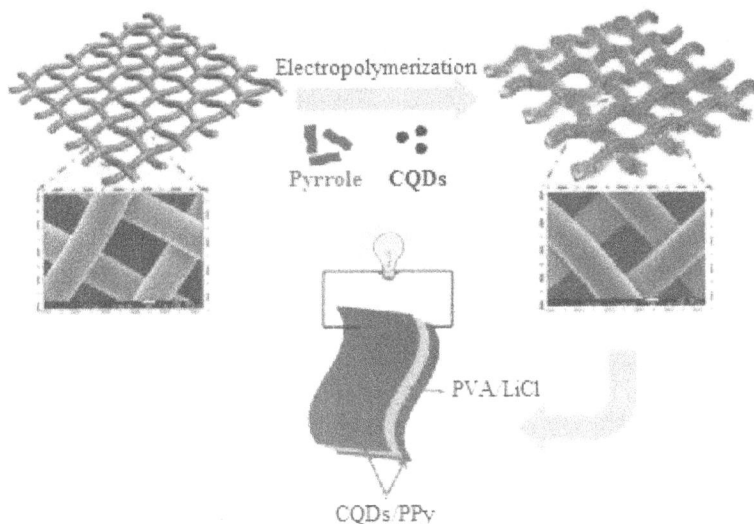

SCHEME 16.2 Procedure for the electrochemical synthesis of a CQDs/PPy composite on flexible ASSSs. (Reproduced with permission from Ref. [15].)

There are several chemical synthesis processes for nanocomposites of CP often leading to different morphological aspects of the polymer.

16.2.2.1 Synthesis in Solution: Powders

The chemical synthesis of nanocomposite of CPs requires an initiating or preliminary phase during which the monomer is oxidized into a radical-cation [16] under the action of an oxidant, such as ferric [Fe(III)] and cupric [(Cu(II)] salts or persulfate ammonium ($[NH_4]_2S_2O_8$). These radical-cations then couple resulting in a conductive black precipitate in powder form. The chemical synthesis of nanocomposite can be carried out in an organic solvent, in an aqueous medium, and even in a micellar medium.

16.2.2.2 "Layer by Layer" Deposition Method

The principle of this technique is based on successive soaking of a flexible substrate in a monomer solution, then in an oxidizing solution, and finally in rinsing solutions (Scheme 16.3). Like the

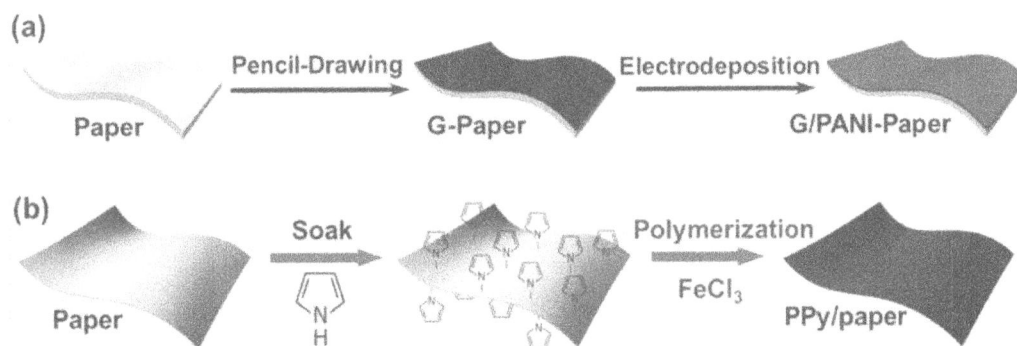

SCHEME 16.3 Procedure for manufacturing flexible cellulose paper-based material by in situ polymerization method. (a) G/PANI/cellulose paper. (b) RGO/PPy/cellulose paper. (Reproduced from Ref. [17] (open access).)

electrochemical method, its advantage is to allow deposition of polymers by layer-by-layer technique on metallic or insulating substrates and thus to be able to control their thickness. This process has already been successfully tested in the chemical polymerization of PPy or PANI for the design of new hybrid paper electrodes.

In Scheme 16.3, we show two main routes for the production of paper-based electrodes by the layer-by-layer deposition method. In the first method, a layer of graphite or other conductor is grafted with a pencil to activate the paper surface used to initiate the electropolymerization of PANI (Scheme 16.3a). The second route is a method of pre-dipping of the paper followed by polymerization. The paper is covered with a conductive layer of the polymer by monomers in situ polymerizations in a solution containing an oxidizing agent (Scheme 16.3b). In this case, the composite of RGO/PPy/cellulose, RPC was thus synthesized: surface modification of cellulose papers by in situ polymerization of Py using ferric ion as an oxidant, followed by dipping of the PPy/cellulose electrode in an aqueous solution containing GO in the presence of sodium borohydride ($NaBH_4$) as a reducing agent in order to obtain a conductive material RGO/PPy/cellulose. These results show that cellulose papers can be modified by conductive organic polymers exhibiting redox characteristics in an electrochemical probe. These materials could therefore be used to prepare flexible devices.

16.2.2.3 Vapor Phase Synthesis

This method is based on the vaporization of the monomer. In this technique, the monomer in the gaseous state is sent in the form of projectiles on a metal plate or an insulating substrate, previously impregnated in an oxidizing solution. By optimizing the application time of the vapors on the target, one can thus gradually improve the thickness of the polymer films. In the case of polypyrrole, Chen et al. [18] have successfully prepared a PEDOT/paper electrode by this method.

These authors reported a new very simple procedure for the development of flexible conductive electrodes based on cellulose paper. This is the use of the vapor phase polymerization (VPP) method to prepare an optimal PEDOT layer on cellulose paper (Scheme 16.4). This finest layer had a porous structure and exhibited very interesting energetic characteristics. Because of the flexibility of this material, the PEDOT/paper electrode exhibited high conductivity (78 S·cm^{-1}) and capacity (639 mF·cm^{-2}) at a current density of 1 mA·cm^{-2}.

Inspired by these advances, carbon cloth (CC) was chosen as a working electrode and the VPP method was used to fabricate a PEDOT nanofiber coating on it, in order to optimize the properties of the hybrids (Scheme 16.5). Here, vertically directed PEDOT-based nanofibers are deposited on

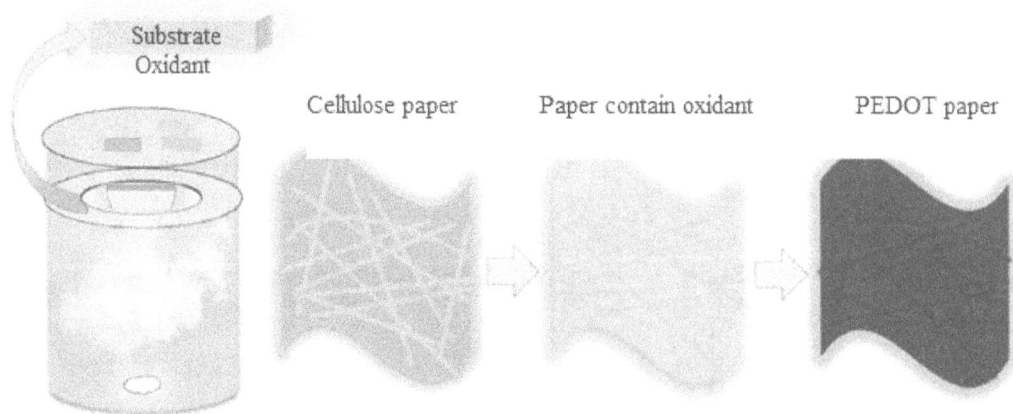

SCHEME 16.4 Procedure for the synthesis of PEDOT nanofibers on the CC electrode by vapor phase polymerization method. (Reproduced with permission from Ref. [18].)

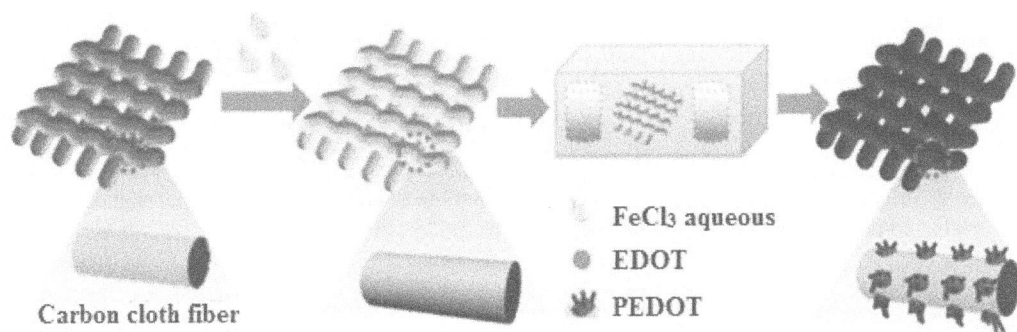

SCHEME 16.5 Schematic diagram of the fabrication of PEDOT/CC composites. (Reproduced with permission from Ref. [19].)

CC. The fabrication conditions are optimized to provide a good distribution of the PEDOT nanofiber film on the CC. Interestingly, the approach adopted by these authors allowed the development of flexible materials of PEDOT nanofibers on CC electrodes with a homogeneous layer distribution and strong adhesion on the CC surface. The PEDOT/CC electrode had good electronic characteristics, and it also achieved a hierarchical structure that allows the fast penetration of the electrolyte and could be successfully applied in flexible energy devices.

Flexible CCs were fabricated by directly using two PEDOT/CC electrodes wrapped in PET film.

16.3 APPLICATION OF NCPs FOR FLEXIBLE DEVICES IN THE ENERGY SECTOR

With the advancement of technology, the need for energy is becoming more and more increased. In this way, the scientific community is seeking to set up flexible devices that meet the requirements of high-performance portable electronics.

These flexible devices are often applied in lithium-ion batteries with relatively low power densities compared to supercapacitors. The application of these flexible devices in electronics is experiencing a new boom due to their important power density, portability, and wide applications.

These flexible materials are characterized by large specific surface area and a supple structure which enhances the electron transfer kinetics. Their capacity, energy, and power density can reach 86.8 mA·hg^{-1}, 59.7 Wh·kg^{-1}, and 447.8 W·kg^{-1}, respectively [20].

These materials are often in various forms. Metal electrode substrates are usually hydrophobic rigid surfaces with very high electrical properties [21, 22]. In addition, due to their flexibility powers, these substrates undergo a very scalable application process and their redox characteristics increase with the angle of twisting. Carbons are electrode materials that have reversible electrochemical characteristics [23, 24, 29–31]. They are one of the most widely used flexible materials due to their low cost and their surface expansions when modified and often used in energy storage [25, 26]. However, some of these types of materials require additional efforts, especially surface activation before use [27, 28]. Furthermore, carbon-based substrates have nowadays become alternative materials, thanks to compliance with their environmental standards (recyclability, biodegradability) [29–32]. Polymer materials are often composite films with strong interactions in their architecture and strong adhesion at the interface [33, 34]. These excellent properties provide good electrochemical and mechanical responses [35, 36]. These films are excellent candidates for flexible materials in the modern supercapacitor field [37, 38]. Other flexible materials such as sandpaper [39] are often applied in the field of flexible supercapacitors. We can also cite the case of cotton fibers [40], papers [41–43], which are large-scale flexible devices, thanks to their biodegradability properties and their availability.

These different materials synthesized by different methods are evaluated by their surface capacity (C_A) according to the following equation [44]:

$$C_A = \frac{\int I dV}{2S \times \Delta V \times v} \left(Fcm^{-2} \right) \tag{16.1}$$

The specific capacity (Cs) is one of the key parameters in the performance comparison of flexible materials and is calculated according to Equation (16.2) [45, 46]:

$$C_S = \frac{2 \times I \times \Delta t}{CV \times S} \left(F.cm^{-2} \right) \tag{16.2}$$

The performance of flexible devices can be evaluated by comparing the power density (P) and the energy density (E) according to the relationships below (Equations 16.3 and 16.4) [47]:

$$E = \frac{C \times \Delta V^2}{7200} \left(Wh.cm^{-2} \right) \tag{16.3}$$

$$P = \frac{3600 \times E}{\Delta t} \left(W.cm^{-2} \right) \tag{16.4}$$

To examine the cyclic stability (CS) of the material, the flexible electrode coated with nanocomposite of CP is scanned in a potential range for several cycles in an electrolytic medium. From a given number of cycles, no variation in intensity is observed. This shows that the material becomes practically stable. These results demonstrate the stability of the material obtained in an electrolytic medium. The parameter CS is defined as the ratio of the anodic peak current intensities of the first cycle (I_{pa}^1) and of the last cycle (no current variation) [48]:

$$CS = \frac{I_{pa}^1}{I_{pa}^\infty} \times 100 \tag{16.5}$$

16.3.1 Plastic Support

The fast development of the flexible electronics market has prompted the scientific world to develop flexible energetic plastic devices due to their availability and increased benefits. For this, several substrates have been developed, such as polyethylene naphthalate (PEN), polystyrene and polyimide (PI), parylene, nylon, PDMS, and polyethylene terephthalate (PET) in the field of soft electronics [27, 29]. Among these, flexible substrates based on PET and PEN are the most used, thanks to their electrical and thermal performances [49]. These substrates are transparent and inexpensive but are often used in temperature conditions below 150° C.

Table 16.1 summarizes the nanocomposite of CP onto flexible plastic for energy advanced found in the literature.

Alberto and Briand [50, 51] examined the role of the flexible plastic substrate. Their study focused on the application of flexible substrates in electronics. It is about improving the energy performance of PEDOT:PS by carbon nanotubes. This study showed a possibility of modification of transparent conductive electrodes (TCE) by conductive layers based on PEDOT:PS modified by carbon nanotubes using the photolithographic method. This synergistic effect exhibits a specific resistivity of about 0.26×10^{-4} Ω·m with optical transparency of 91.0% at 550 nm. These results show an improvement in mechanical and chemical properties when switching from the bare ITO electrode to the modified electrode (Figure 16.1).

TABLE 16.1

Nanocomposite of CP Deposed on Flexible Plastic Electrode and Their Electrochemical Performances

Materials	Polymerization-Type	Energetic Characteristics	Scope of the Work	References
CuNW–PEDOT:PSS/ PET	In situ polymerization	Sheet resistance: 15 $\Omega \cdot cm^{-1}$ Transparency: 76% (550 nm) Cyclic performance: 5000	Solar cells	[54]
PANI/PLA	Oxidation polymerization	Specific capacitance: 510.3 $F \cdot g^{-1}$ Cycling stability (2000): 111.5%	Energy storage	[55]
SWCNTs@ PEDOT:PSS/ITO	Spray coating	Sheetresistance: 89.1 $\Omega \cdot cm^{-1}$	Flexible electronics	[56]
P3HT-PCBM/ITO	Spin coated	CS (after 5000): 90% Conductivity: 4115 $S \cdot cm^{-1}$	Electrochromic device	[56]
GR-PANI/ITO	In situ polymerization	Specific capacitance: 267.2 $F \cdot cm^{-3}$	Energy storage	[53]
PEDOT:PSS:CFE/ PET	Slot-die printing	Conductivity: 4100 $S \cdot cm^{-1}$, Transmittance: 85% CS (after 5000): 90%	Solar Cells	[57]

Other modified TCE have been synthesized by Souza et al. [53]. They prepared place graphene/polyaniline nanocomposites on flexible ITO surface (GR-PANI/ITO-flex) and applied in the field of flexible energy storage (Figure 16.2a, b). This nanocomposite exhibited a strong material adherent to the ITO surface with attractive electrochemical properties (Figure 16.2c). In addition, this material withstands multiple deformations while maintaining its chemical structure and mechanical properties increase with the torsion. These characteristics also show that the redox couple of the nanocomposite and increases after twisting with a specific volumetric capacity of 95.5 $F \cdot cm^{-3}$, as do the charge and discharge curves (Figure 16.2d). These results have shown the possibility of constructing GR-PAN nanocomposites deposited on an ITO-flexible substrate in order to apply them as a new flexible supercapacitor.

In the same flexible substrate, a conductive layer was used by the combination of PEDOT and PSS for the purpose of fabricating perovskite solar modules (Figure 16.3a). In this case, to improve the conductivity of the material and increase the percentage of transmittance, shears were performed during the slot-die printing step (Figure 16.3b). This resulted in a printed electrode of PEDOT:PSS:CFE with a conductivity value of around 4100 S/cm and high flexibility (Figure 16.3c). Thanks to the shearing process, the transmittance of the PEDOT:PSS:CFE material increased and reached 85% at 550 nm. These results showed that it is possible to successfully construct perovskite solar cell (PSC) modules (Figure 16.3d) with very attractive mechanical properties (PCE = 12.5%) based on the modification of the transparent electrodes by conductive layers of the nanocomposite.

16.3.2 PAPER AND TEXTILE CARBON MATERIAL SUPPORT

Nowadays, with the advancement of technology, the development of new alternative flexible devices having the same properties has become unavoidable. It is in this context that flexible textile-based substrates were constructed by several authors. However, several approaches have been used for the development of flexible textile-based devices. One of the steps is to manufacture these flexible materials sensitive in the field of sensors such as polyaniline-nylon 6 composites [58]. The other

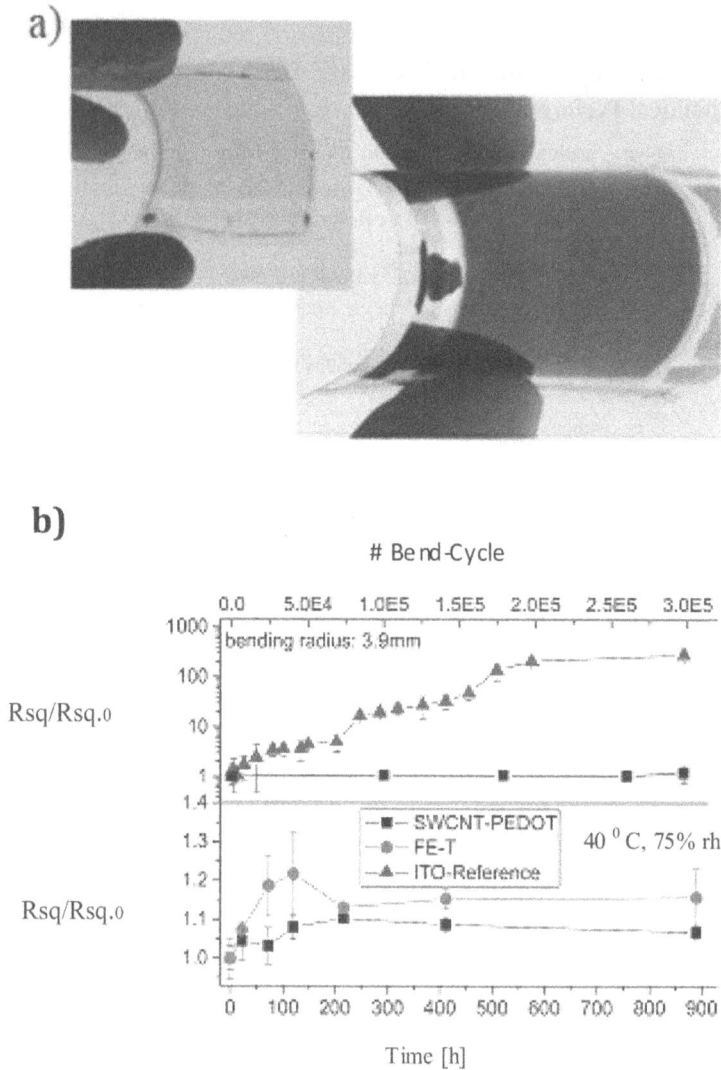

FIGURE 16.1   (a) Photographic images of a bare ITO electrode and an ITO electrode modified by SWCNT-PEDOT. (b) Improvement of the value of the relative resistance for flexible surfaces starting from bare ITO to ITO modified by SWCNT-PEDOT and depending on the number of twists. (Reproduced with permission from Ref. [52].)

step is to strengthen the architecture of flexible textile substrates for the design of flexible energy devices. Inspired by this method, Hang et al. modified cotton textile fibers with carbon nanotubes [59]. Also, Zheng et al. [60] synthesized hybrid nanocellulose electrodes based on polyaniline for high-performance supercapacitors. Their study demonstrates the central role of modified textiles in the field of new supercapacitor challenges.

Textile-based electrodes are experiencing remarkable growth in the energy field due to their availability, low cost, and ability to undergo several deformations. Among these electrodes, paper is one of the most widely used substrates due to biodegradability in environmental matrices and their possibility of being overlapped by layers of organic polymers. It is one of the oldest flexible products and is a great and promising alternative to flexible substrates. Their association with CPs provides materials with excellent mechanical and electrical properties and is a good candidate in the field

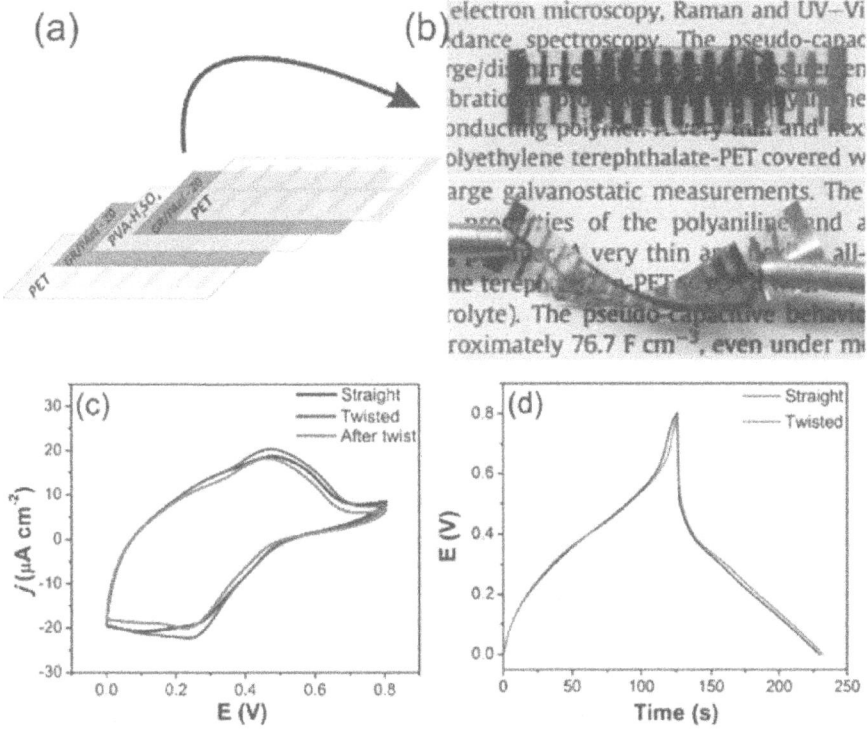

FIGURE 16.2 (a) Structure of the different modified PET surfaces; (b) twisted and straight flexible electrode of GR-PANI/PET nanocomposite; (c) cyclic voltammograms recorded in a potential range of 0–0.8 V of the working electrodes: GR-PANI/PET straight (black), GR-PANI/PET twisted, GR-PANI/PET after twist; (d) charge/discharge curves GR-PANI/PET straight, GR-PANI/PET twist. (Reproduced with permission from Ref. [53].)

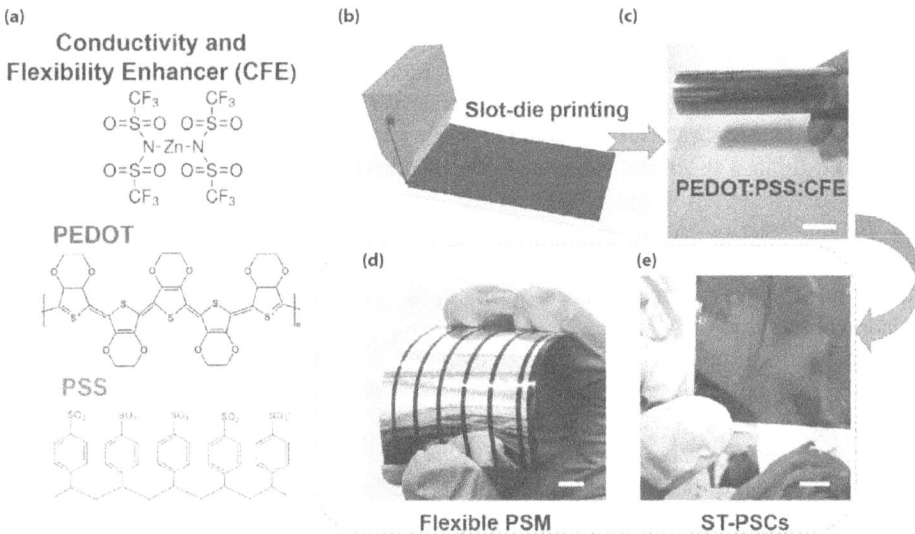

FIGURE 16.3 Manufacturing procedure of flexible perovskite solar cells. (a) Chemical structures of CFE, PEDOT and PSS compounds. (b) Diagram representing the printing phase of the flexible PEDOT electrode:PSS:CFE. (c) Diagram of the evolution of the printed electrode PEDOT:PSS:CFE. (d) Diagram illustrating the twisted PSCs. (e) Photographs of ST-PSC. (Reproduced with permission from Ref. [57].)

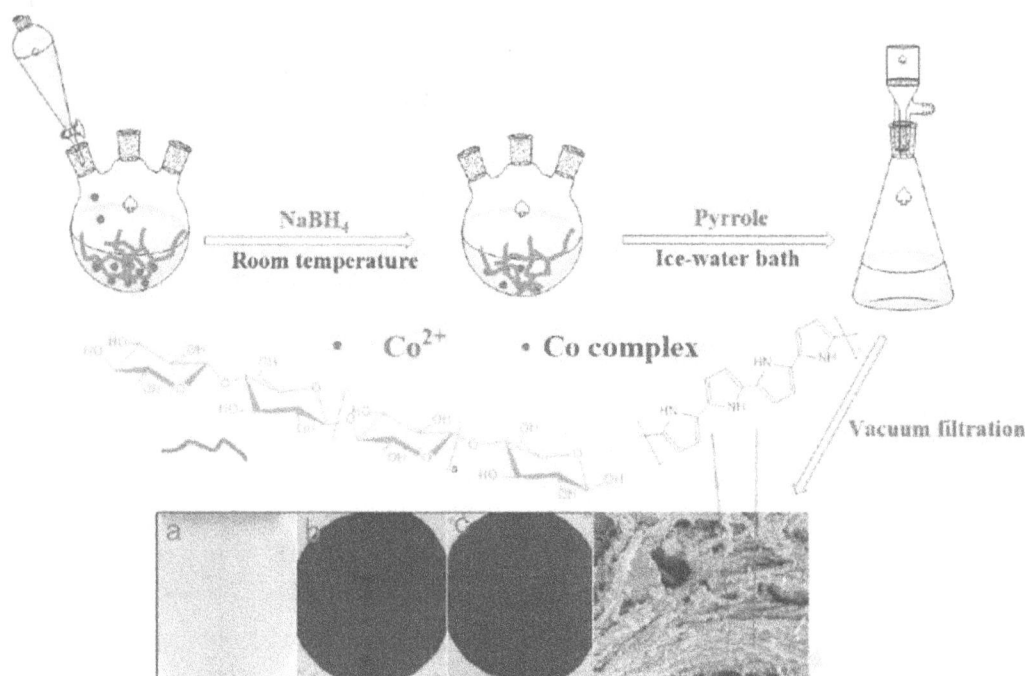

SCHEME 16.6 Procedure for developing a material based on cellulose fiber covered with a layer of PPy @ cobalt oxyhydroxide. (Reproduced with permission from Ref. [61].)

of energy storage. Previous research has confirmed this hypothesis by showing the improvement in the electrical properties of cellulose fibers coated with NCPs by the oxidative in situ polymerization method (Scheme 16.6). This method involves the preparation of a sensitive textile which can act both as a substrate and as a device layer in flexible electronics. In addition, Yang et al. [61] also developed flexible supercapacitors based on a cellulose fiber electrode modified with ternary polypyrrole@cobalt oxyhydroxide. They demonstrated in this experiment the possibility of constructing a nanocomposite via reduction in the liquid phase at room temperature. This involves covering cellulose fibers with an electroactive layer of polypyrrole@cobalt oxyhydroxide nanocomposite. This material exhibits highly resolved oxidation and reduction peaks with good electrical conductivity. In addition, it exhibits a specific capacity and good cyclic stability of the order of 571.3 F·g^{-1} at 0.2 A·g^{-1}, and 93.02%, respectively, after 1000 cycles. These results demonstrate the solid construction of a new dynamic energy-oriented flexible device based on flexible cellulose substrates reinforced with NCPs.

In this context, other researchers are working to develop flexible electrodes to design portable energy storage devices. For instance, Song et al. [62] reported on the high performance of flexible cotton-based electrodes coated with PANI modified with an optimal carbon ratio (PANI@carbon@textile) to ensure good surface homogeneity. PANI@carbon@textile composite synthesized by in situ polymerization method (Scheme 16.7) (assisted carbonization in H$_2$SO$_4$ medium). This synthesized flexible material shows excellent electrochemical and mechanical properties due to the symmetry of the meshes. It also has very high electrical performance due to its specific surface capacitance of 386.7 mF·cm^{-2}, a power density of 745 mW·m^{-2}, an electrochemical stability of over 70% beyond 5000 cycles (Figure 16.4). Due to its flexibility, the PANI@carbon@textile flexible substrate can undergo several twists at an angle varying from 0 to 180. This study demonstrates a remarkable material with a wide symmetrical supercapacitor application.

Another approach consists in electrochemically generating paper electrodes capable of meeting expectations in the field of flexible energy. Wang et al. could deposit polypyrrole nanocomposite on

SCHEME 16.7 Illustrative scheme for the development of flexible electrodes of modified carbon textiles PANI@carbon. (Reproduced with permission from Ref. [62].)

graphene paper. This elaborate paper-based electrode covered by a conductive layer is used in the field of flexible semiconductor supercapacitors (Figure 16.5).

In another study [2], different GrP-PPy were developed by cyclic voltammetry at numbers of cycles ranging from 1 to 15. These electrosynthesized electrodes were the subject of a comparison study using different electrochemical techniques (Figure 16.6). The voltammetry method shows that the synthesized GrP-PPy electrodes exhibit fairly strong oxidation and reduction peaks when they are subjected to a scan rate of 5 mV.s^{-1} in an electrolyte medium containing 0.1 M KCl and 5 mM $[Fe(CN)_6]^{3-/4-}$. This method shows the intensity of the oxidation and reduction peaks of the

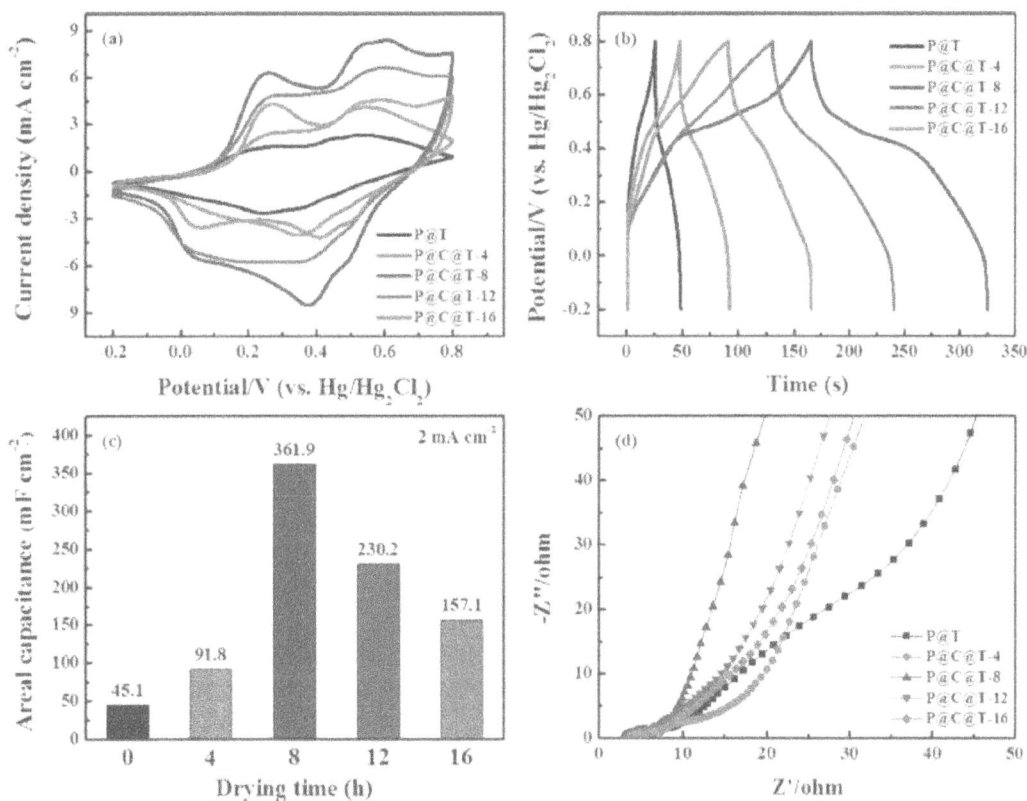

FIGURE 16.4 Electrochemical characterization of PANI@carbon@textile electrodes: (a) cyclic voltammograms of electrodes recorded in a potential range of 0.2–0.8; (b) charge/discharge curves; (c) specific surface capacities as a function of drying time; and (d) electrochemical impedance spectroscopy curve. (Reproduced with permission from Ref. [62].)

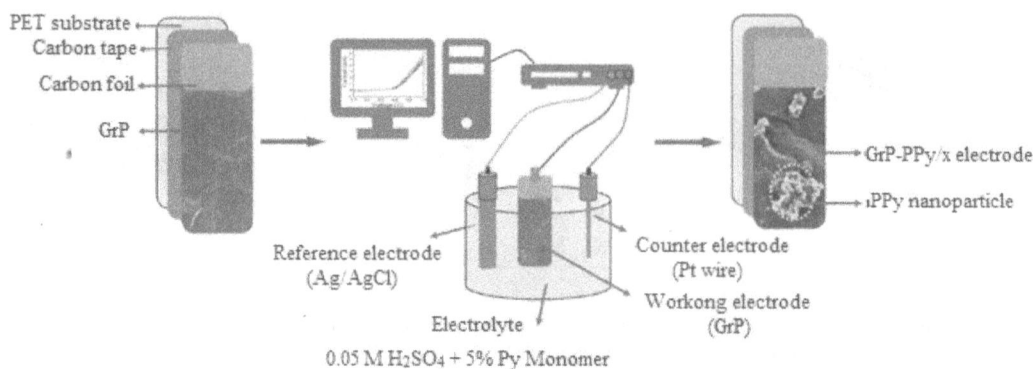

FIGURE 16.5 Procedure for manufacturing GrP-PPy electrodes by cyclic voltammetry method. (Reproduced with permission from Ref. [2].)

synthesized GrP-PPy electrode for 8 cycles (GrP-PPy/8-cycles) exhibits the best electrical properties compared to the others (Figure 16.6a). These results are confirmed by the application of these flexible electrodes in the field of all-solid-state supercapacitor (AASS) and show that the surface capacitance of the GrP-PPy/8-cycles electrode has the highest value of the order 128.9 mF·cm^{-2} to 0.1 mA.cm^{-2}. In addition, the optimal GrP-PPy/8-cycles electrode exhibited better cyclic stability and power density with, respectively, values of 180 mWh·cm^{-2} and 85% beyond 5000 cycles. Table 16.2 shows some of the important energetic characteristics of CP-based materials.

16.3.3 FLEXIBLE FREE-STANDING ELECTRODE

Nowadays, constraints related to environmental standards oblige the scientific community to develop biodegradable portable energy devices. For this, the design of a free-standing electrode has become an essential axis that aims to reduce the cost and use of toxic solvents in the field of manufacturing flexible devices of energy interest. These self-free-standing electrodes can be directly synthesized without the use of flexible electrodes or carbon substances to activate the surface. These electrodes are potential candidates in flexible electronics, thanks to their robustness, their high flexibility, their portability, and their ability to control the morphology of the surface. Free-standing electrodes are designed by several methods, of which the electrospinning method remains the most popular in flexible energy applications [65, 66]. And more recently, it has become more applicable and some free-standing electrodes categories like carbonaceous materials [39, 40], carbon nanotubes [67, 68], graphene [69, 70], and polymeric materials [71].

Table 16.3 summarizes free-standing electrodes, and fabrication methods for energy devices. Huang et al. [72] synthesized by coaxial electrospinning method flexible free-standing polymers based on sulfurized polyacrylonitrile. The polymer has been nanostructured by hollow tubular nanofibers (H-SPAN) to increase its contact surface and improve its charge transfer in the electrolyte medium. This nanostructured material is directly used in the field of lithium-sulfur batteries (Figure 16.7) and has excellent electrochemical properties. These results demonstrate the central role of H-SPAN in the constitution of lithium-sulfur batteries with a reversible capacity of 1250 mAh per g of sulfur. It has an excellent efficiency of 99% after 300 cycles and a high capacity of around 717 mAh per g of sulfur.

Other free-standing electrodes have been developed within the framework of portable flexible supercapacitors [73]. These authors synthesizing three-phase nanocomposites based on PPy in the presence of TiO$_2$ modified by graphene (TiO$_2$-G). Electrochemical studies have shown the synergistic effect of TiO$_2$ with an optimal amount of 14.6%, which shows good cyclic stability and improvement in current density (Table 16.3). This work shows the possibility of designing free-standing

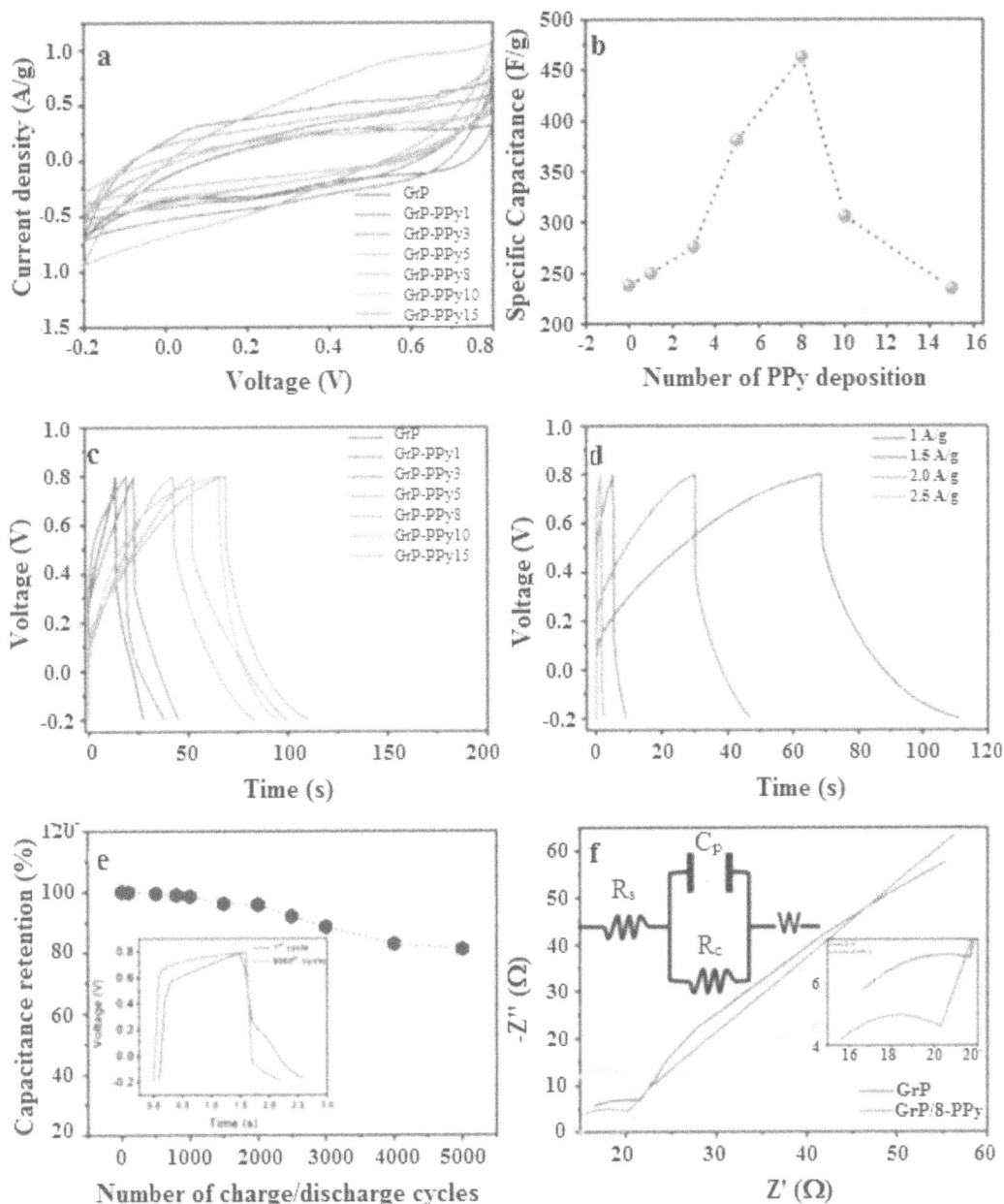

FIGURE 16.6 Electrochemical characteristics of the optimal GrP-PPy electrode. (a) Cyclic voltammogram of the bare electrode of GrP and GrP-PPy electrodes recorded in an electrolyte solution containing 1 M H$_2$SO$_4$ at v = 5 mV.s^{-1}; (b) specific capacity of GrP-PPy substrates as a function of the number of cycles during the electroplating; (c) charge and discharge curves of the electrodes prepared with 0–15 cycles during the electro-deposition at 3.5 A·g^{-1} current density; (d) charge and discharge curve of the optimal GrP-PPy electrode at different current density; (e) curve of retention capacitance as a function of the number of charge and discharge cycles; (f) spectroscopy curve of electrochemical impedance in a frequency range from 10 mHz to 100 kHz of the optimal electrode GrP-PPy in a solution containing 0.1 M KCl and 5 mM [Fe(CN)$_6$]$^{3-/4-}$. In inset, the equivalent electrical circuit. (Reproduced with permission from Ref. [2].)

TABLE 16.2

Nanocomposite of CP Deposed on Flexible Plastic Electrode and Their Electrochemical Performances

Materials (Polymerization Type)	Energetic Characteristics	Scope of the Work	Reference
PPy@CoOOH/cellulose fiber (in situ oxidation)	Specific capacitance: 571.3 F·g^{-1} (0.2 A·g^{-1}) CS (after 1000 cycles): 93%	Supercapacitor	[61]
PEDOT/cellulose (in situ oxidation)	Specific capacitance: 50.4 F·cm^{-2} (0.05 A·cm^{-3}) Energy density: 13.2 Wh·kg^{-1} Power density: 0.126 kW·kg^{-1} CS (after 1000 cycles): 90%	Energy-storage	[63]
PPy/porous cellulose (in situ oxidation)	Specific capacitance: 4117 mF·cm^{-2} (2 mA·cm^{-2})	Flexible supercapacitor	[64]
PPy /GrP (in situ electropolymerization)	Capacitance: 128.9 mF·cm^{-2} (0.1 mA·cm^{-2}) Energy density: 16.1 mWh·cm^{-2} Power density: 180 mW·cm^{-2} CS (after 5000 cycles): 85%	Flexible supercapacitor	[2]
PANI@carbon@textile (in situ polymerization)	Energy density: 35.8 mWh·m^{-2} Power density: 745 mW·m^{-2} CS (after 5000 cycles): 70%	Wearable energy storage	[62]

TiO_2-G-PPy materials for flexible supercapacitors with a capacity of 201.8 F·g^{-1} (100 mA·g^{-1}) and a cyclic stability of 76.5% after 100 cycles (Table 16.3).

Other applications of the free-standing and flexible electrodes as supercapacitors have been developed using the combination of carbon nanotubes (CNT) and poly (3,4-ethylenedioxythiophene) PEDOT. This CVD-synthesized nanocomposite has very attractive electrochemical and mechanical properties thanks to its spongy shape. It can be seen that the CNTs/PEDOT electrode shows a specific capacity of the order of 147 F·g^{-1} at 0.5 A·g^{-1} with an energy density of 12.6 Wh·kg^{-1}. This experiment shows a continuity in the design of free-standing electrodes, which can have remarkable cyclic stability of around 95% after 3000 cycles.

TABLE 16.3

A Summary of Free-Standing Electrodes and Energetic Characteristics and Fabrication Method for Energy Devices (Polymerization Type: Vacuum Infiltration, Oven Drying)

Materials	Energetic Characteristics	Scope of the Work	Reference
H-SPAN	Storage capacity: 1250 mAh·g^{-1} CS (after 300 cycles): 100%	Li/Na storage	[72]
TiO_2–G–PPy	Capacitance: 201.8 F·g^{-1} (100 mA·g^{-1}) CS (after 100 cycles): 76.5%	Supercapacitors	[73]
CNTs/PEDOT sponge electrode	Energy density of 12.6 Wh·kg^{-1} CS (after 3000 cycles): 95% Power density: 10.2 kW·kg^{-1}	Energy storage	[74]

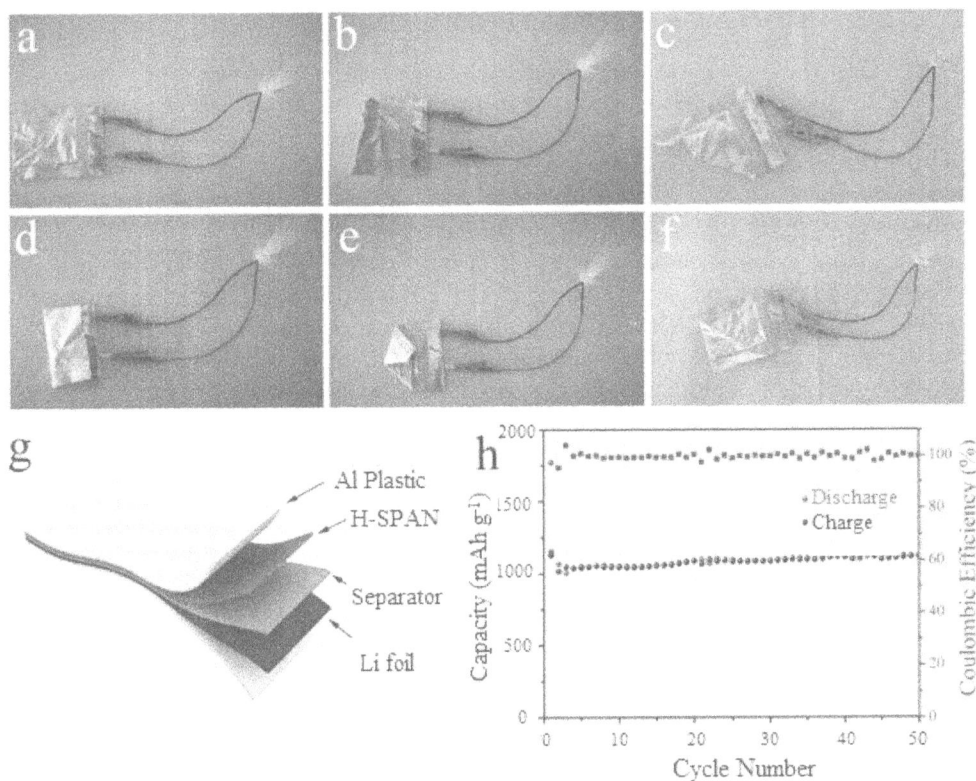

FIGURE 16.7 (a–f) Photographs of flexible pouch cell based on H-SPAN electrode in various folding angles to light up an LED illumination. (g) Schematic diagrams of as-assembled flexible pouch cell. (h) The cycle performance of pouch cell at 0.1 C current density. (Reproduced with permission from Ref. [72].)

16.4 MARKET POTENTIAL OF NCPs FOR FLEXIBLE DEVICE

Nanomaterials can fundamentally transfigure the market. Nanotechnology has played a key role in manufacturing since the end of the 1980s when nano-revolution occurred, which boosted the whole production of nanocomposites and yielded wealthy outcomes in the near future.

In the health sector, nanocomposites based on conducting polymers have largely boosted a new arena called drug delivery and molecular sieves for efficient transport technology [75]. We are witnessing a new healthcare revolution initiated by nanomaterials. For instance, various diseases (including tumors and cancers) can be healed by means of nanocomposites. The only limitation may be unknown side effects. PEDOT:PSS seems to be the most applied conducting polymer in the field of PEDOT:bioelectronics [76]. New PEDOT-based materials find applications in many other fields. Composites of CNTs and other CPs (polypyrrole and polyaniline) are used as actuators and components in energy storage and/or conversion devices. They are well conductive and cost-effective, and their properties can be enhanced by redox-active or polycharged aromatic anionic dopants. This is expected to enhance the dominance of the carbon nanotubes segment in the global market. Furthermore, CNTs segment controlled the polymer nanocomposites production in 2018: 35% of the total market [77]. Flexibility is among the major factors inducing the growth of this segment across the electronics industry. Also, semiconducting metal oxide nanomaterials appeared to be efficient sensors. They are more and more associated with CPs [78].

The demand for NCPs is expected to increase in the near future.

16.5 CONCLUSION AND FUTURE CHALLENGES

The chapter clearly allows to understand the different materials based on inorganic nanomaterials, graphene, carbon nanotubes, other CPs for engineering, NCPs and flexible electrodes, which provide important means of optimizing high-performance flexible devices based on NCPs. Research results have confirmed the interest of polymers nanocomposites, which gives rise to an improvement in electrochemical and mechanical properties of many devices, including cyclic stability, charge transfer, and greatly on the flexibility of the material.

Important work focusing on nanocomposites of conducting polymers for flexible devices for energy storage and conversion (supercapacitor and batteries) applications. In the last five years, the flexible materials based on NCPs are shifting to self-supporting electrodes. In addition, we also note that, in the majority of research, nanocomposites are more efficient on flexible storage. However, the energy performance of flexible devices developed today does not meet the expectations of the industrial world. In other words, the nano-structuring of CPs on flexible substrates certainly improves the potentials of these materials, however, their application is much more oriented in the fields of supercapacitors and energy storage. So that the orientation of research in the design of flexible devices applicable in the industrial field for commercialization still remains a major challenge. This nurtured ambition should boost the scientific community to develop other flexible devices that may vary the fields of application.

The applicability of NPCs in flexible energy devices require strong adhesion to the substrate and remain an issue in the durability of commercialized supercapacitors. To overcome these problems, the intervention of a thin adhesive layer could be a solution. For that, the use of the chemical interface of diazoniums or the silanization of surfaces could be a new method to develop commercial supercapacitors. In addition, improved charge transfers could consolidate the conductivity deficit in the electrolyte medium using copolymerization for the synthesis of flexible devices. These results could expand the fields of application of flexible devices through a wider window of potential.

ACKNOWLEDGMENTS

The authors of CAD University are grateful the [International Science Program (ISP), University of Uppsala (Sweden) under Grant to African Network of Electroanalytical Chemists [IPICS/ANEC] and to TWAS, The World Academy of Science for the Advancement of Science in developing countries under No. 16-499RG/CHE/AF/AC_G–FR3240293299.

REFERENCES

1. Yang J, Liu Y, Liu S, Li L, Zhang C, Liu T (2017) Conducting polymer composites: Material synthesis and applications in electrochemical capacitive energy storage. Mater Chem Front 1:251–268.
2. Hu R, Zhao J, Zhua G, Zheng J (2018) Fabrication of flexible free-standing reduced graphene oxide/polyaniline nanocomposite film for all-solid-state flexible supercapacitor. Electrochim Acta 261:151–159.
3. Guo X, Bai N, Tian Y, Gai L (2018) Free-standing reduced graphene oxide/polypyrrole films with enhanced electrochemical performance for flexible supercapacitors. J Power Sources 408:51–57.
4. Li H, Tang Z., Liu Z, Zhi C (2019) Evaluating flexibility and wearability of flexible energy storage devices. Joule 3:613–619.
5. Kayser LV, Lipomi DJ (2019) Stretchable conductive polymers and composites based on PEDOT and PEDOT:PSS. Adv Mater 31:1806133.
6. Chen S, Zhitomirsky I (2016) Strategies to optimize the capacitive behavior of polypyrrole electrodes. Mater Manuf Processes 31:2017–2022.
7. Bharti M, Singh A, Samanta S, Aswal DK (2018) Conductive polymers: Creating their niche in thermoelectric domain. Prog Mater Sci 93:270–310.
8. Shown I, Ganguly A, Chen L-C, Chen K-H (2015) Conducting polymer-based flexible supercapacitor. Energy Sci Eng 3:2–26.
9. Sun X, Zhang H, Zhou L, Huang X, Yu C (2018) Polypyrrole-coated zinc ferrite hollow spheres with improved cycling stability for lithium-ion batteries. Small 12:3732–3737.

10. Naveen MH, Gurudatt NG, Shim Y-B (2017) Applications of conducting polymer composites to electrochemical sensors: A review. Appl Mater Today 9:419–433.

11. Lo M, Ktari N, Gningue-Sall D, Madani A, Aaron SE, Aaron J-J, Mekhalif Z, Delhalle J, Chehimi MM (2020) Polypyrrole: A reactive and functional conductive polymer for the selective electrochemical detection of heavy metals in water. Emergent Mater 3:815–839.

12. Lo M, Seydou M, Bensghaïer A, Pires R, Gningue-Sall D, Aaron J.-J, Mekhalif Z, Delhalle J, Chehimi MM (2020) Polypyrrole wrapped carbon nanotube composite films coated on diazonium modified flexible ITO sheets for the electroanalysis of heavy metal ions. Sensors 20:580–598.

13. Debiemme-Chouvy C, Fakhry A, Pillier F (2018) Electrosynthesis of polypyrrole nano/micro structures using an electrogenerated oriented polypyrrole nanowire array as framework. Electrochim Acta 268:66–72.

14. Liu C, Cai Z, Zhao Y, Zhao H, Ge F (2016) Potentiostatically synthesized flexible polypyrrole/multiwall carbon nanotube/cotton fabric electrodes for supercapacitors. Cellulose 23:637–648.

15. Jian X, Yang H-M, Li J-G, Zhang E-H, Cao L-L, Liang Z-H (2017) Flexible all-solid-state high-performance supercapacitor based on electrochemically synthesized carbon quantum dots/polypyrrole composite electrode. Electrochim Acta 228:483–493.

16. Aggadi SE, Loudiyi N, Abbassi ZE, Hourch AE (2020) Electropolymerization of aniline monomer and effects of synthesis conditions on the characteristics of synthesized polyaniline thin films. Polym Chem 10:138–145.

17. Yao B, Zhang J, Kou T, Song Y, Liu T, Li Y (2017) Paper-based electrodes for flexible energy storage devices. Adv Sci 4:1700107–1700138.

18. Chen X, Jiang F, Jiang Q, Jia Y, Liu C, Liu G, Xu J, Duan X, Zhu C, Nie G, Liu P (2020) Conductive and flexible PEDOT-decorated paper as high performance electrode fabricated by vapor phase polymerization for supercapacitor. Colloids Surf, A 603:125173.

19. Zhao X, Dong M, Zhang J, Li Y, Zhang Q (2016) Vapor-phase polymerization of poly(3,4-ethylene dioxythiophene) nanofibers on carbon cloth as electrodes for flexible supercapacitors. Nanotechnology 27:385705.

20. Chen S, Ma L, Zhang K, Kamruzzaman M, Zhi C, Zapien JA (2019) A flexible solid-state zinc ion hybrid supercapacitors based on copolymer derived hollow carbon spheres. J. Mater Chem A7:7784–7790.

21. Yu JH, Xie FF, Wu ZC, Huang T, Wu JF, Yan DD, Huang CQ, Li L (2018) Flexible metallic fabric supercapacitor based on graphene/polyaniline composites. Electrochim Acta 259:968–974.

22. Yu PP, Zhao X, Huang ZL, Li YZ, Zhang QH (2014) Free-standing three-dimensional graphene and polyaniline nanowire arrays hybrid foams for high-performance flexible and lightweight supercapacitors. J Mater Chem A 2:14413–14420.

23. Lin YX, Zhang HY, Deng WT, Zhang DF, Li N, Wu QB, He CH (2018) In-situ growth of high-performance all-solid-state electrode for flexible supercapacitors based on carbon woven fabric/polyaniline/graphene composite. J Power Sources 384:278–286.

24. Choudhury A, Dey B, Mahapatra SS, Kim DW, Yang KS, Yang DJ (2018) Flexible and free-standing supercapacitor based on nanostructured poly(aminophenol)/carbon nanofiber hybrid mats with high energy and power densities. Nanotechnology 29:165401–165412.

25. Wang JJ, Dong LB, Xu CJ, Ren DY, Ma XP, Kang FY (2015) Polymorphous supercapacitors constructed from flexible three-dimensional carbon network/polyaniline/MnO_2 composite textiles. ACS Appl Mater Interfaces 8:10851–10859.

26. Chien HH, Liao CY, Hao YC, Hsu CC, Cheng IC, Yu IS, Chen JZ (2018) Improved performance of polyaniline/reduced-graphene-oxide supercapacitor using atmospheric-pressure-plasma-jet surface treatment of carbon cloth. Electrochim Acta 260:391–399.

27. Liu D, Du PC, Wei WL, Wang HX, Wang Q, Liu P (2018) Skeleton/skin structured (RGO/CNTs)@PANI composite fiber electrodes with excellent mechanical and electrochemical performance for all-solid-state symmetric supercapacitors. J Colloid Interface Sci 513:295–303.

28. Liu D, Wang HX, Du PC, Liu P (2017) Flexible and robust reduced graphene oxide/carbon nanoparticles/polyaniline (RGO/CNs/PANI) composite films: Excellent candidates as free-standing electrodes for high-performance supercapacitors. Electrochim Acta 233:201–209.

29. Liu D, Wang HX, Du PC, Liu P (2017) Independently double-crosslinked carbon nanotubes/polyaniline composite films as flexible and robust free-standing electrodes for high-performance supercapacitors. Carbon 122:761–774.

30. Miao FJ, Shao CL, Li XH, Wang KX, Lu N, Liu YC (2016) Electrospun carbon nanofibers/carbon nanotubes/polyaniline ternary composites with enhanced electrochemical performance for flexible solid-state supercapacitors. ACS Sustain Chem Eng 4:1689–1696.

31. Yu J, Lu WB, Pei SP, Gong K, Wang LY, Meng LH, Huang YD, Smith JP, Booksh KS, Li QW, Byun J-H, Oh Y, Yan YS, Chou T-W (2016) Omnidirectionally stretchable high-performance supercapacitor based on isotropic buckled carbon nanotube films. ACS Nano 10:5204–5211.

32. Zhang JT, Wang J, Yang JE, Wang YL, Chan-Park MB (2014) Three-dimensional macroporous graphene foam filled with mesoporous polyaniline network for high areal capacitance. ACS Sustain Chem Eng 2:2291–2296.

33. Pourjavadi A, Doroudian M, Ahadpour A, Pourbadiei B (2018) Preparation of flexible and free-standing graphene-based current collector via a new and facile self-assembly approach: Leading to a high-performance porous graphene/polyaniline supercapacitor. Energy 152:178–189.

34. Li WW, Gao FX, Wang XQ, Zhang N, Ma MM (2016) Strong and robust polyaniline-based supramolecular hydrogels for flexible supercapacitors. Angew Chem Int Ed 55:9196–9201.

35. Wang X, Yang CY, Jin J, Li XW, Cheng QL, Wang GC (2018) High-performance stretchable supercapacitors based on intrinsically stretchable acrylate rubber/MWCNTs@conductive polymer composite electrodes. J Mater Chem A 6:4432–4442.

36. Liu L, Luo S, Qing Y, Yan N, Wu YQ, Xie XF, Hu FY (2018) A temperature controlled, conductive PANI@CNFs/MEO2MA/PEGMA hydrogel for flexible temperature sensors. Macromol Rapid Commun 39:1700836.

37. Wang HX, Liu D, Du PC, Wei WL, Wang Q, Liu P (2017) Comparative study on polyvinyl chloride film as flexible substrate for preparing free-standing polyaniline-based composite electrodes for supercapacitors. J Colloid Interface Sci 506:572–581.

38. Wang HX, Liu D, Duan XJ, Du PC, Guo JS, Liu P (2016) Facile preparation of high strength polyaniline/polyvinyl chloride composite film as flexible freestanding electrode for supercapacitors. Mater Des 108:801–806.

39. Alcaraz-Espinoza JJ, de Oliveira HP (2018) Flexible supercapacitors based on a ternary composite of polyaniline/polypyrrole/graphite on gold coated sandpaper. Electrochim Acta 274:200–207.

40. Jin C, Wang HT, Liu YN, Kang XH, Liu P, Zhang JN, Jin LN, Bian SW, Zhu Q (2018) High-performance yarn electrode materials enhanced by surface modifications of cotton fibers with graphene sheets and polyaniline nanowire arrays for all-solid-state supercapacitors. Electrochim Acta 270:205–214.

41. Dong LB, Liang GM, Xu CJ, Ren DY, Wang JJ, Pan ZZ, Li BH, Kang FY, Yang QH (2017) Stacking up layers of polyaniline/carbon nanotube networks inside papers as highly flexible electrodes with large areal capacitance and superior rate capability. J Mater Chem A 5:19934–19942.

42. Wang HX, Liu D, Du PC, Liu P (2017) Flexible and robust amino-functionalized glass fiber filter paper/polyaniline composite films as free-standing tensile tolerant electrodes for high performance supercapacitors. Electrochim Acta 228:371–379.

43. Wang XJ, Kong DB, Zhang YB, Wang B, Li XL, Qiu TF, Song Q, Ning J, Song Y, Zhi LJ (2016) All-biomaterial supercapacitor derived from bacterial cellulose. Nanoscale 8:9146–9150.

44. Yadav N, Yadav N, Hashmi SA (2020) Ionic liquid incorporated, redox-active blend polymer electrolyte for high energy density quasi-solid-state carbon supercapacitor. J Power Sources 451:227771.

45. Oz A, Hershkovitz S, Belman N, Tal-Gutelmacher E, Tsur Y (2016) Analysis of impedance spectroscopy of aqueous supercapacitors by evolutionary programming: Finding DFRT from complex capacitance. Solid State Ion 288:311–314.

46. Laheäär A, Przygocki P, Abbas Q, Béguin F (2015) Appropriate methods for evaluating the efficiency and capacitive behavior of different types of supercapacitors. Electrochem Commun 60:21–25.

47. Cevik E, Bozkurt A (2021) Redox active polymer metal chelates for use in flexible symmetrical supercapacitors: Cobalt-containing poly(acrylic acid) polymer electrolytes. J Energy Chem 55:145–153.

48. Yuana YF, Chena Q, Zhua M, Caia GS, Guo SY (2021) Nano tube-in-tube CNT@void@TiO$_2$@C with excellent ultrahigh rate capability and long cycling stability for lithium-ion storage. J Alloys Compd 851:156795.

49. Alrammouz R, Podlecki J, Abboud P, Sorli B, Habchi R (2018) A review on flexible gas sensors: From materials to devices. Sens Actuators, A 284:209–231.

50. Alberto S, Wong WS (2009) Flexible Electronics: Materials and Applications (Electronic Materials: Science & Technology), Springer, New York.

51. Briand D, Molina-Lopez F, Quintero AV, Ataman C, Courbat J, Rooij NF (2011) Why going towards plastic and flexible sensors? Procedia Eng 25:8–15.

52. Fischer R, Gregori A, Sahakalkan S, Hartmann D, Büchele P, Tedde SF, Schmidt O (2018) Stable and highly conductive carbon nanotube enhanced PEDOT:PSS as transparent electrode for flexible electronics. Org Electron 62:351–356.

53. Souza VHR, Oliveira MM, Zarbin AJG (2017) Bottom-up synthesis of graphene/polyaniline nanocomposites for flexible and transparent energy storage devices. J Power Sources 348:87–93.
54. Chen J, Zhou W, Chen J, Fan Y, Zhang Z, Huang Z, Feng X, Mi B, Ma Y, Huang W (2015) Solution-processed copper nanowire flexible transparent electrodes with PEDOT:PSS as binder, protector and oxide-layer scavenger for polymer solar cells. Nano Res 8:1017–1025.
55. Wang Q, Wang H, Du P, Liu J, Liu D, Liu P (2019) Porous polylactic acid/carbon nanotubes/polyaniline composite film as flexible free-standing electrode for supercapacitors. Electrochim Acta 294:312–324.
56. Chaudhary A, Pathak DK, Tanwar M, Yogi P, Sagdeo PR, Kumar R (2019) Polythiophene–PCBM-based all-organic electrochromic device: Fast and flexible. ACS Appl Electron Mater 1:58–63.
57. Hu X, Meng X, Zhang L, Zhang Y, Cai Z, Huang Z, Su M, Wang Y, Li M, Li F, Yao X, Wang F, Ma W, Chen Y, Song Y, A (2019) Mechanically robust conducting polymer network electrode for efficient flexible perovskite solar cells. Joule 18:2205–2218.
58. Hong KH, Oh KW, Kang TJ (2004) Polyaniline–nylon 6 composite fabric for ammonia gas sensor. J Appl Polym Sci 92:37–42.
59. Han J-W, Kim B, Li J, Meyyappan M (2013) A carbon nanotube-based ammonia sensor on cotton textile. Appl Phys Lett 102:193104.
60. Zheng W, Lv R, Na B, Liu H, Jin T, Yuan D (2017) Nanocellulose-mediated hybrid polyaniline electrodes for high performance flexible supercapacitors. J Mat Chem A 5:12969–12976.
61. Yang S, Sun L, An X, Qian X (2020) Construction of flexible electrodes based on ternary polypyrrole@cobalt oxyhydroxide/cellulose fiber composite for supercapacitor. Carbohydr Polym 229:115455.
62. Song P, He X, Tao J, Shen X, Yan Z, Ji Z, Yuan A, Zhu G, Kong L (2021) H_2SO_4-assisted tandem carbonization synthesis of PANI@carbon@textile flexible electrode for high-performance wearable energy storage. Appl Surf Sci 535:147755.
63. Zhao D, Zhang Q, Chen WS, Yi X, Liu SX, Wang QW, Liu YX, Li J, Li XF, Yu HP (2017) Highly flexible and conductive cellulose-mediated PEDOT:PSS/MWCNT composite films for supercapacitor electrodes. ACS Appl Mater Interfaces 9:13213–13222.
64. Liu LM, Weng W, Zhang J, Cheng XL, Liu N, Yang JJ, Ding X (2016) Flexible supercapacitor with a record high areal specific capacitance based on a tuned porous fabric. J Mater Chem A 4:12981–12987.
65. Liu Q, Zhu J, Zhang L, Qiu Y (2018) Recent advances in energy materials by electrospinning. Renew Sustain Energy Rev 81:1825–1858.
66. Yoon J, Yang HS, Lee BS, Yu WR (2018) Recent progress in coaxial electrospinning: New parameters, various structures, and wide applications. Adv Mater 30:1704765.
67. Chou SL, Wang JZ, Chew SY, Liu HK, Dou SX (2008) Electrodeposition of MnO_2 nanowires on carbon nanotube paper as free-standing, flexible electrode for supercapacitors. Electrochem Commun 10:1724–1727.
68. Li P, Yang Y, Shi E, Shen Q, Shang Y, Wu S (2014) Core-double-shell, carbon nanotube@polypyrrole@MnO_2 sponge as freestanding, compressible supercapacitor electrode. ACS Appl Mater Interfaces 6:5228–5234.
69. De Oliveira HP, Sydlik SA, Swager TM (2013) Supercapacitors from free-standing polypyrrole/graphene nanocomposites. J Phys Chem C 117:10270–10276.
70. Jiang L, Lu X, Xu J, Chen Y, Wan G, Ding Y (2015) Free-standing microporous paper-like graphene films with electrodeposited PPy coatings as electrodes for supercapacitors. J Mater Sci Mater Electron 26:747–754.
71. Sultana I, Rahman MM, Li S, Wang J, Wang C, Wallace GG, Liu HK (2012) Electrodeposited polypyrrole (PPy)/para (toluene sulfonic acid) (pTS) free-standing film for lithium secondary battery application. Electrochim Acta 60:201–205.
72. Huang X, Liu J, Huang Z, Ke X, Liu L, Wang N, Liu J, Guo Z, Yang Y, Shi Z (2020) Flexible free-standing sulfurized polyacrylonitrile electrode for stable Li/Na storage. Electrochim Acta 333:135493.
73. Jiang L, Lu X, Xie C, Wan G, Zhang H, Youhong T (2015) Flexible, free-standing TiO_2–graphene–polypyrrole composite films as electrodes for supercapacitors. J Phys Chem C 119:3903–3910.
74. He X, Yang W, Mao X, Xu L, Zhou Y, Chen Y, Zhao Y, Yang Y, Xu J (2018) All-solid state symmetric supercapacitors based on compressible and flexible free-standing 3D carbon nanotubes (CNTs)/poly(3,4-ethylenedioxythiophene) (PEDOT) sponge electrodes. J Power Sources 376:138–146.
75. Ganachari SV, Viannie LR, Mogre P, Tapaskar RP, Yaradoddi JS (2019) Conducting polymer composite-based sensors for flexible electronics. Handbook of Ecomaterials, Springer, Switzerland, 1311–1341.
76. Conductive Polymers Market – Forecast (2021–2026). https://www.industryarc.com/Report/15567/conductive-polymers-market.html. Accessed 10 March 2021.

77. Acumen Research and Consulting: Polymer Nanocomposites Market Size Hit US$31.8 Bn By 2026. https://www.globenewswire.com/news-release/2020/02/20/1988215/0/en/Polymer-Nanocomposites-Market-Size-Hit-US-31-8-Bn-By-2026.html. Accessed 10 March 2021.
78. Fall B, Diaw AKD, Fall M, Sall ML, Lo M, Gningue-Sall D, Thotiyl MO, Maria HJ, Kalarikkal N, Thomas S (2021) Synthesis of highly sensitive rGO@CNT@Fe$_2$O$_3$/polypyrrole nanocomposite for the electrochemical detection of Pb^{2+}. Mater Today Commun 26:102005.

LIST OF ABBREVIATIONS

ASSS(s)	All-solid-state supercapacitor(s)
AgNWs	Silver Nanowires
C$_A$	Surface capacitance
CC	Carbon cloth
C-CNT(s)	Crosslinked multiwalled carbon nanotube(s)
CFE	Conductivity and flexibility enhancer
CNFs	Cellulose nanofibers CNFs
CNT(s)	Carbon nanotube(s)
Cp	pseudocapacity
CP(s)	Conducting polymer(s)
CQD	Carbon quantum dot
Cs	Specific capacitance
CS	Cyclic stability
CuNW(s)	Copper nanowire(s)
CV	Cyclic voltammetry
CVD	chemical vapor deposition
E	Energy density
ESR	Equivalent series resistance
FE-T	Heraeus Clevios™ Coating formulation of aqueous PEDOT/PSS dispersion
FTCE	Flexible transparent conductive electrodes
GCD	Galvanostatic charge/discharge
GO	Graphene oxide
GO	Graphene oxide
GrP	Graphene paper
H-SPAN	Sulfurized polyacrylonitrile film electrode with hollow tubular nanofibers
I	Discharge current (A)
Ipa	Anodic peak current intensity
ITO	Indium-tin-oxide
MEO$_2$MA	2-(2′-methoxyethoxy) ethyl methacrylate
MWCNT(s)	Multiwalled carbon nanotube(s)
NCPs	Nanocomposite of conducting polymers
OPD	Organic photodiode
P	Power density
P3HT	Poly(3-hexylthiophene)
PAC	Polyacetylene
PANI	Polyaniline
PCBM [6]	Phenyl-C$_{61}$-butyric acid methyl ester
PCE	Power conversion efficiency
PDMS	Polydimethylsiloxane
PEDOT	Polyethylenedioxythiophene
PEGMA	Poly(ethylene glycol) methyl ether methacrylate)
PEN	Polyethylene naphthalate

PET	Polyethylene terephthalate
PET	Polyethylene terephthalate
PI	Polyimide (PI)
PLA	Polylactic acid
PPP	Polyparaphenylene
Ppy	Polypyrrole
PTh	Polythiophene
Py	Pyrrole
PSCs	Perovskite solar cells
PSS	Poly(styrenesulfonate)
PTh	Polythiophene
Rc	Charge-transfer-resistance
RGO	Reduced graphene oxide
RPC	RGO/PPy/cellulose
Rs	Solution resistance
S	Surface area of the electrode material (cm^2)
ST	Semi-transparent
SWCNT(s)	Single walled carbon nanotubes(s)
TCE	Transparent conductive electrodes
TiO2–G–PPy	Titania–Graphene–Polypyrrole
TsO	p-toluenesulfonate
W	Warburg impedance
v	Scan rate (V.s^{-1}).
Δt	Discharge time (s),
ΔV	Scan potential window of the CV curve (V)

17 Conducting Polymers for Electrocatalysts

Saranya Narayanasamy and Jayapriya Jayaprakash
Department of Applied Science and Technology, Alagappa
College of Technology, Anna University, Chennai, India

CONTENTS

17.1 INTRODUCTION

Electrochemical devices used for the efficient production, storage, and utilization of clean energy from renewable resources have gained great interest among the scientific and industrial communities in recent times [1]. Electrolytic cells, Li-ion batteries, supercapacitors, solar cells, and fuel cells are some prominent electrochemical devices used for energy transformation and storage. Typically, they involve a few different types of typical half-cell reactions, such as the oxygen reduction reaction (ORR), hydrogen evolution reaction (HER), and oxygen evolution reaction (OER). For instance, electrochemical water splitting is an important electrochemical reaction that generates hydrogen (a clean energy source) through two half-reactions, HER (at the cathode) and four-electron OER at the anode [2]. Further, employing ORR at the air electrodes of electrochemical devices is considered an eco-friendly way toward H_2O_2 production and achieving hydrocarbon conversion in fuel cells (electrooxidation of formic acid, methanol, and ethanol). To translate this technology into business level, the low-cost electrocatalyst is the prerequisite. Such reactions are traditionally catalyzed by platinum-based (Pt-based) materials, other precious metals or their oxides, which have the disadvantages of low abundance and high cost. Therefore, finding ways to reduce the quantity of precious metals used or discovering alternate materials to develop new electrocatalysts with similar performance is crucial [3].

DOI: 10.1201/9781003150374-17

Electrocatalysts are materials used to speed up the electrochemical reactions that occur on the electrode surface or at the solid–liquid interfaces. They perform the task of decreasing the reaction energy barrier, thereby reducing the electrochemical overpotential required for the reaction to commence and proceed toward a specific pathway. An electrocatalyst must be conductive to aid electron transfer between the reactant, catalyst, and electrodes. To achieve a desired reaction, the catalyst should allow easy adsorption of the incoming reactant molecules on its surface, facilitate electron transfer between the molecule and the catalyst to form reaction intermediates, and mediate recombination among the reactive intermediates to produce the desired products. Further, product adsorption from the catalyst surface is also vital for the catalyst to regenerate for subsequent catalytic cycles.

Electrocatalysts applied in electrochemical reactions usually incorporate a heterogeneous catalyst. The most pronounced difference is that a heterogeneous catalyst must be insulating, while an electrocatalyst should possess excellent conductivity. Therefore, electrocatalysts and regular heterogeneous catalysts can be characterized by similar techniques; however, some special techniques are required for the conductive electrocatalysts.

For strongly uphill reactions that theoretically require large overpotentials, it is practically difficult to provide such high potentials and energies. In addition, poor product selectivity and Faradaic efficiency can impede specific half-reactions, such as ORR. Taking these challenges into consideration, the development of cost-effective and active electrocatalysts seems an urgent necessity. Electrocatalysts should have high electrical conductivity, large specific surface area, and electroactive properties. Thus, electrocatalysts are commonly characterized based on four parameters: selectivity, activity, overpotential, and stability. The major issue that hinders the implementation of these systems on a larger scale is their sluggish kinetics (high overpotential (η), $\eta_a > \eta_c$) of the oxygen production. OER is a four-electron-proton coupling reaction, while HER is a two electron-transfer reaction, and hence OER requires a higher potential to overcome the kinetic barrier and progress [4]. ORR follows both four-electron and two-electron pathways depending on the environmental conditions. The detailed mechanisms of ORR, HER, and OER with their environmental conditions are discussed in the following sections.

17.1.1 OXYGEN REDUCTION REACTION

In electrochemical devices with an aqueous electrolyte, the process of ORR at the cathode involves the following stages:

1. An oxygen molecule is diffused and adsorbed onto the surface of the electrocatalyst.
2. Adsorbed O_2 molecules gain electrons from the anode and splitting the O=O bonds.
3. Removal of the produced OH^- ions into the solution.

ORR can progress in two ways: (a) a highly efficient one-step four-electron pathway or (b) a sluggish two-electron pathway that involves two steps, which completely reduce the O_2 molecules to OH^- ions [5].

For the direct four-electron pathway, the reaction under alkaline conditions is as follows (Equation 17.1):

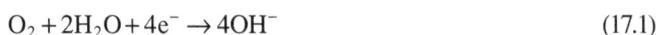

$$O_2 + 2H_2O + 4e^- \rightarrow 4OH^- \tag{17.1}$$

and that in an acidic (Equation 17.2) electrolyte is as follows:

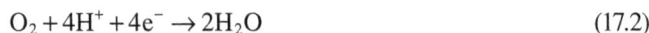

$$O_2 + 4H^+ + 4e^- \rightarrow 2H_2O \tag{17.2}$$

In the indirect two-electron process, after the initial reduction of two oxygen molecules in the alkaline solution, further reduction of H_2O_2 happens:

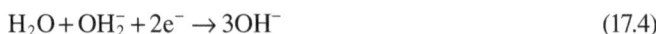

$$O_2 + H_2O + 2e^- \rightarrow HO_2^- + OH^- \tag{17.3}$$

$$H_2O + OH_2^- + 2e^- \rightarrow 3OH^- \tag{17.4}$$

In an acidic solution, the two-step-two-electron pathway proceeds as follows (Equations 17.5 and 17.6):

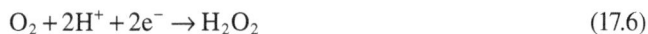

$$H_2O_2 + 2H^+ + 2e^- \rightarrow 2H_2O \tag{17.5}$$

$$O_2 + 2H^+ + 2e^- \rightarrow H_2O_2 \tag{17.6}$$

17.1.2 HYDROGEN EVOLUTION REACTION

Hydrogen production takes place at the cathode during the electrolytic water splitting reaction, and it is greatly dependent on environmental conditions.

In an acidic medium, HER proceeds as given below:

The first step is the Volmer reaction (a), in which hydrogen adsorption happens. Proton reduction on the surface forms H_{ad} intermediates (Equation 17.7).

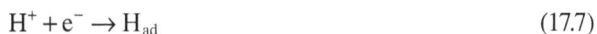

$$H^+ + e^- \rightarrow H_{ad} \tag{17.7}$$

Then, the HER can proceed by the Heyrovsky reaction (Equation 17.8) – Tafel pathway (Equation 17.9) to produce H_2.

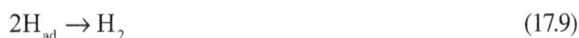

$$H^+ + e^- + H_{ad} \rightarrow H_2 \tag{17.8}$$

$$2H_{ad} \rightarrow H_2 \tag{17.9}$$

The two possible reaction steps in an alkaline medium are the Volmer reaction (Equation 17.10) and the Heyrovsky step (Equation 17.11), as shown in the following equations:

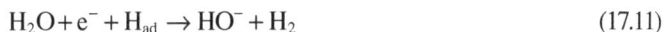

$$H_2O + e^- \rightarrow HO^- + H_{ad} \tag{17.10}$$

$$H_2O + e^- + H_{ad} \rightarrow HO^- + H_2 \tag{17.11}$$

Hu et al. [6] revealed that HER activity is dependent on hydrogen adsorption (H_{ad}) in alkaline media and maintains the balance between H_{ad} and hydroxyl adsorption (OH_{ad}). The free energy of hydrogen adsorption (ΔG_H), depends on the electronic configuration of the electrocatalysts, is widely accepted as the descriptor of HER activity. A moderate hydrogen binding energy is beneficial for the HER process. The volcano plot predicts that the electrocatalysts with superior HER activity have ΔG_{H^*} values close to zero. A very ΔG_{H^*} value indicates that the bonds between hydrogen and the catalyst surface are difficult to break, while a very small ΔG_{H^*} value leads to low production of H_2. Pt is the best HER catalyst in both alkaline and acidic media since it has the optimal hydrogen adsorption energy and shows the highest exchange current density [7]. HER activity is usually lower in alkaline when compared to acidic solutions because of the sluggish water dissociation step, which reduces the reaction rate by 2–3 orders of magnitude. However, alkaline electrolysis is preferred in a scaling-up. Therefore, electrocatalysts for high alkaline HER performance must be designed to bind to hydrogen species and dissociate water [8].

17.1.3 OXYGEN EVOLUTION REACTION

Oxygen evolution reaction is a half-reaction (at the anode) of the water-splitting reaction. The overall water splitting reaction is indicated by the following Equation (17.12),

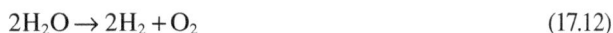

$$2H_2O \rightarrow 2H_2 + O_2 \tag{17.12}$$

Similar to ORR, OER is also environment-dependent and undergoes different pathways under acidic and alkaline conditions. The mechanism of the OER is well-established, and it is as shown below:

In an acidic medium, the cathode reaction is

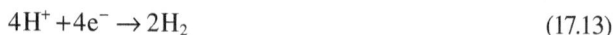

$$4H^+ + 4e^- \rightarrow 2H_2 \tag{17.13}$$

And the anode reaction is

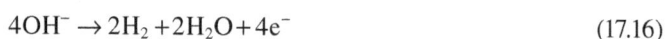

$$2H_2O \rightarrow O_2 + 4H^+ + 4e^- \tag{17.14}$$

In an alkaline medium, the cathode reaction is

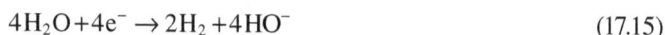

$$4H_2O + 4e^- \rightarrow 2H_2 + 4HO^- \tag{17.15}$$

And the anode reaction is

$$4OH^- \rightarrow 2H_2 + 2H_2O + 4e^- \tag{17.16}$$

Thus, electrocatalysts used for OER should have electroactive properties to boost the sluggish electrokinetic reaction. Though platinum is widely studied in this regard, the commercialization of electrochemical devices using platinum-based electrocatalysts is not economically feasible. Hence, alternate non-metal/metal-based electrocatalysts have been of keen interest in recent times. For developing non-metal-based electrocatalysts, conducting polymers (CPs) are suitable materials as they are electrochemically sensitive and electroactive.

In this chapter, the importance and basics of electrocatalysts suitable for various electrochemical reactions are discussed. The importance of CP-based electrocatalysts is highlighted by reviewing the synthesis methods, evaluation, and application of various CPs as electrocatalysts in different fields.

17.2   ROLE OF AN ELECTROCATALYST

Electrocatalysis refers to electrochemical redox reactions in the presence of an electrocatalyst, which lowers the kinetic energy barrier of the electrochemical reactions and drives them toward a specific pathway. Thus, an electrocatalyst is an electrochemically active catalyst with good charge transfer property that increases the rate of oxidation and reduction reactions in an electrochemical cell. The kinetics of OER/ORR/HER are often sluggish due to the high kinetic energy barriers even in the presence of such catalysts, limiting their performance and commercialization of productive electrochemical energy applications. The sluggish reaction kinetics of OER/ORR/HER is due to the high kinetic energy barriers. Thus, an electrocatalyst must be electronically conductive and possess an appropriate surface structure and abundant electrochemically active sites to be effective while also being stable against cycling [9]. The efficacy of the catalyst determines the overall performance of the electrochemical applications. The responsibilities of an electrocatalyst are described below:

1. Electrocatalysts are generally used for improving the reaction rate, i.e., the electrical current is produced by decreasing the activation energy required to initiate the desired electrochemical reaction and direct the reaction toward a specific pathway.
2. Unlike other catalysts, an electrocatalyst must be conductive to facilitate the electrode-electrolyte interaction and ensure fast reaction kinetics.
3. Electrocatalysts speed up electron transfer across the electrode and reactant species and/or facilitate their chemical transformation to form intermediates, and this process is termed as a half-reaction.
4. For a spontaneous reaction, the catalyst should permit the adsorption of reactants and desorption of products efficiently. Thus, the catalyst remains unaltered and can be regenerated for catalytic reaction cycles.

5. To achieve high efficiency in water splitting cells and fuel cells, an electrocatalyst should be able to effectively lower the energy barrier of the electrochemical reactions involving water, oxygen, and hydrogen.

17.3 PREREQUISITES OF AN ELECTROCATALYST

Electrocatalysts play a key role in establishing the overall performance of an electrochemical system. Thus, an efficient electrocatalyst material for electrochemical applications should possess the following properties:

- High intrinsic electrical conductivity (for easy diffusion of electrons).
- Large active surface (an increased number of reactive sites can lower the expense of catalyst loading).
- High catalytic activity.
- Long-term stability and durability (retention of catalytic performance over a long period and multiple cycles and corrosion resistance at high anodic potentials > 1.5 V).
- Superior chemical stability (consistent performance in different electrolytes and at various pH ranges).
- Excellent mechanical stability (compatible with adverse conditions, i.e., high temperature/pressure).
- Low overpotential at the current density of 10 mA/cm^2 in HER.
- A favorable structure with strong interactions between the catalyst particle and the support surface (Prevention of electrocatalyst detachment from the catalyst layer, which may lead to dissolution in the electrolyte and/or migration).
- Reasonable production cost with simple and environmentally friendly synthetic process.

17.4 MATERIAL AND SYNTHESIS OF CP-BASED ELECTROCATALYST

Conducting polymers are significant base materials for the successful development of electrocatalysts. Numerous CPs, including polypyrrole (PPY), polyaniline (PANI), poly(3,4-ethylenedioxythiophene) (PEDOT), and polythiophene (PTh), have been used. This chapter discusses the advances in this field and the benefits of using CPs that are easy to synthesize, such as PANI and PPy. Among them, PANI is the oldest known artificial CP and significant since it possesses high intrinsic electrical conductivity. PANI exhibits both insulating and electrically conductive behaviors based on the oxidation state and is found in three forms, namely emeraldine, pernigraniline, and leucoemeraldine. For example, for leucoemeraldine, x = 1 (LB(base), yellow) and, its structure is given below:

Leucoemeraldine (x = 1) 100% reduced form

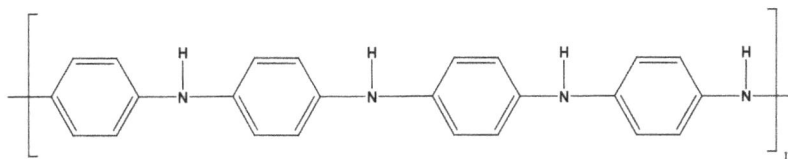

Pernigraniline (PG) (x = 0) (PGB(base), purple) and, its structure is given below:
Pernigraniline (x = 0) 100% oxidized form

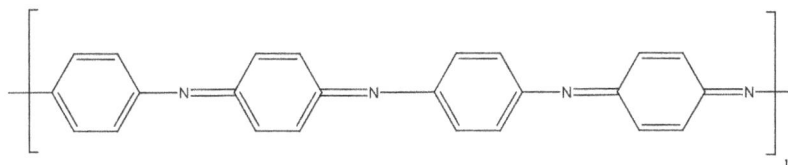

while emeraldine (EM) (x = 0.5) emeraldine salt (form (EMS), dark green and in (emeraldine base form (EMB), blue under basic and acidic conditions, respectively.

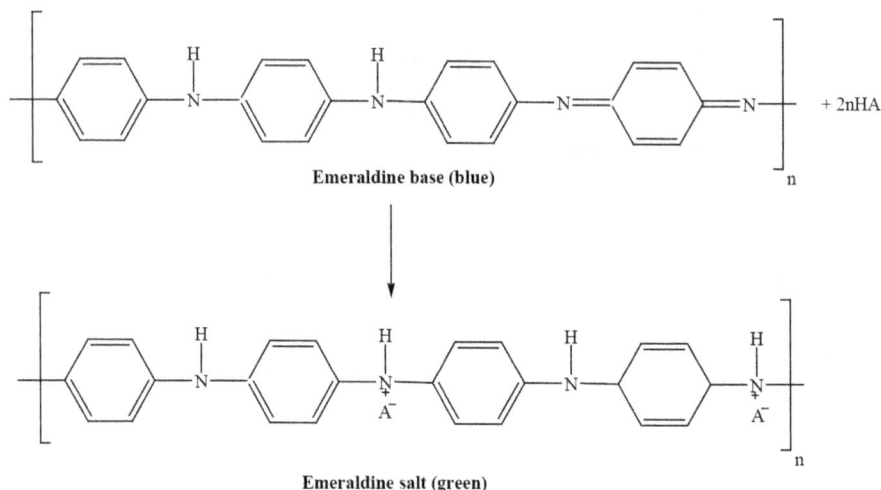

Emeraldine base (blue)

+ 2nHA

Emeraldine salt (green)

The presence of alternate saturated and unsaturated double bonds and nitrogen in the phenyl rings of PANI supports the different oxidation states, which influence its physical and chemical properties. The addition of doping agents enhances the electrical conductivity of PANI by delo- calizing π-bond electrons. Multidisciplinary applications are feasible because the structural and functional arrangements in PANI are customizable. PANI offers outstanding physicochemical and electrochemical properties, including effective morphological and thickness tunability, long life, flexibility, high porosity and hence improved surface area accessibility, excellent electrical conduc- tivity, and good ion-diffusion rate, effortless charge- transfer, and good cyclic retention. PANI is practically advantageous because of the (i) ease of synthesis, (ii) good stability under different envi- ronments, (iii) unique redox tunability, (iv) reasonable cost, (v) possibility of versatile applications (e.g., hydrogen storage, solar cells, anti-corrosive coatings, sensors, and diodes), and (vi) enhanced photocatalytic activity of the PANI-metal oxide composites.

Polypyrrole has also gained popularity in the recent past because of its outstanding electrical con- ductivity and redox potential, ease of processability, superior mechanical and magnetic tunability, and environmental sustainability. It has been investigated in a variety of applications, including electro- catalysts, sensors, anticorrosion coatings, electrochromic devices, batteries, and supercapacitors.

Conducting polymers like PANI and PPy are known for their electrochemical properties, which are comparable with those of metal- and carbon-based electrocatalysts [10]. They can be synthe- sized by chemical oxidative polymerization, electrochemical methods, interfacial polymerization, hydrothermal methods [11], and template methods. These CPs have gained focus because of their chemical structures that contain active sites or reactive groups, such as N-containing functional groups and π-conjugated structures, in their backbone. There are two main ways for preparing conductive polymers from monomers, namely chemical and electrochemical oxidative polymeriza- tion methods. However, chemical polymerization is not well-suited for large-scale manufacturing due to the high cost of oxidants and catalysts involved; further, the electrical conductivity decreases due to polymer agglomeration during chemical synthesis. Therefore, electrochemical polymeriza- tion is often preferred because of its ease and cost-effectiveness; it also produces cleaner films with desired thickness and morphology, besides better electrochemical conductivity and air stability. To date, potentiostatic (constant potential), galvanostatic (constant current), and potentiodynamic (cyclic voltammetry, CV) techniques have been developed to achieve aniline and pyrrole electropo- lymerization. Table 17.1 compares some CP-derived electrocatalysts synthesized using different polymerization approaches.

TABLE 17.1
Comparison of CPs as Electrocatalyst in Various Electrochemical Reactions

Electrocatalysts	Abbreviations	Polymer	Electrolytes	Catalyst Loading [mg cm^{-2}]	Tafel Slope [mV dec^{-1}]	Electrochemical Reaction	Polymerization	References
Carbon dots-Pt modified polyaniline nanosheets grown on carbon cloth (CC)	CDs/Pt-PANI	PANI	0.5 M H$_2$SO$_4$	8.1 μg/cm^2	41.7	HER	EP	[12]
Nitrogen-doped ACF	N-ACF	PPy	0.1 M KOH	0.118 mg cm^{-2}	–	ORR	CP	[13]
Cobalt oxide and polypyrrole coupled with a graphene nanosheet	Co$_3$O$_4$-PPy/GN	PPy	0.1 M KOH	0.2 mg·cm^{-2}	–	ORR	Ball milling	[14]
Co and N co-doped graphene networks	Co–N-GNWs	PANI	0.5 M H$_2$SO$_4$ aqueous	500 μg cm^{-2}	63	ORR	CP	[15]
N, P-co-doped mesoporous carbon	N, P-MC	PPy	0.1 M KOH	0.118 mg cm^{-2}	58	ORR	CP	[16]
Metal free-nitrogen-doped carbon	NC	PPy	0.1 M KOH	357 μg cm^{-2}	97	ORR	CP	[17]
Oxygen-doped carbonaceous polypyrrole nanotubes	OCPN	PPy	0.1 M KOH	4 mg	96	ORR	CP	[18]

Abbreviations: CP: chemical polymerization; EP: electrochemical polymerization; PANI: polyaniline; PPy: polypyrrole; ORR: oxygen reduction reaction; HER: hydrogen evolution reaction.

17.5 PERFORMANCE EVALUATION OF CP-BASED ELECTROCATALYSTS

An ideal catalyst must be highly catalytic with an abundance of expected active sites, cost-effective, remarkably stable, and highly porous for mass transport. Therefore, newly synthesized electrocatalysts are characterized for their structure, composition, and electrocatalytic features. It is also necessary to experimentally determine their physical and chemical properties. This section discusses the various techniques used for analyzing and evaluating the structural, morphological, and electrocatalytic properties of electrocatalysts.

The chemistry of electrocatalytic materials is analyzed by various spectroscopic techniques, such as Fourier transform infrared spectroscopy (FTIR), Raman spectroscopy, and nuclear magnetic resonance (NMR). Among these, FTIR reflects the chemical bonds and corresponding functional groups on the surface of the polymers. Raman spectroscopy is a dominant and non-destructive instrument that provides helpful information about the electronic and structural properties of the catalyst [19]. It is mainly used to analyze the carbon material–polymer chain interactions, which ultimately determine the properties of a polymeric composite material [20].

The electronic properties of an electrocatalyst can be analyzed by ultraviolet-visible spectroscopy (UV-Vis) and electron paramagnetic resonance (EPR). Brunauer-Emmett-Teller (BET) analysis is used to analyze the specific surface area of the materials, while Hg porosimetry gives information about the pore-size distribution. X-ray photo-electron spectroscopy (XPS) establishes the electrode surface elemental composition [21]. The crystallinity and structural characteristics can be determined by X-ray diffraction analysis (XRD). Morphological characterization can be performed by microscopic techniques like scanning electron microscopy (SEM) and transmission electron microscopy (TEM) along with elemental analysis, which gives a clear idea about the surface morphology. Thermal analysis is conducted to detect the physical or chemical changes that occur in a material during polymer synthesis from monomers. The techniques employed for this purpose include thermogravimetric analysis (TGA), differential scanning calorimetry (DSC), and differential thermal analysis (DTA) [22].

An in-depth evaluation of an electrocatalyst is mandatory prior to its exploitation in energy storage and conversion systems. The tested catalyst is usually considered stable when there is a stable or insignificant increase in the applied potential over the long term. However, (i) the oxidative degradation of an electrocatalyst, (ii) the surface of the catalysts blocked by the gas bubble evolution, and (iii) catalyst delaminated from the electrode surface may decrease the active catalyst surface area available to the electrolyte. In order to distinguish these effects and understand the mechanism underlying the degradation of electrocatalysts, electrochemical measurements such as electrochemical impedance spectroscopic analysis, cyclic voltammograms, potentiostatic and galvanostatic techniques has been devised. These techniques or methods are widely used to exemplify the electrocatalytic properties of CP-based catalysts in electrochemical reactions. The sequence in which these electrochemical techniques are to be in the analysis of an electrocatalyst was given by Maljusch et al. [23].

1. Electrochemical impedance spectroscopy (EIS) determines the solution resistance (R_s) and porous resistance (R_p).
2. Cyclic voltammetry evaluates the electrochemical active surface area (ECSA).
3. Linear sweep voltammetry (LSV) assesses the changes in the oxidation states.
4. Chronoamperometry or chronopotentiometry calculates the long-term stability.

The important features and the parameters involved in the evaluation of electrocatalysts are discussed below.

17.5.1 CYCLIC VOLTAMMETRY

Cyclic voltammetry is an electrochemical technique, which provides qualitative information about catalysts and their electrochemical reactions. This includes the electrochemical response and

(a)

(b)

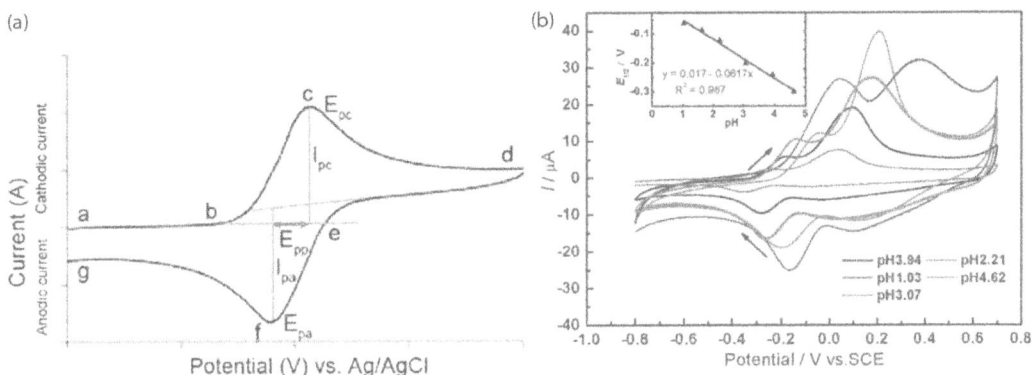

FIGURE 17.1 (a) Typical cyclic voltammogram. (Reproduced from Ref. [24]. Copyright (2016) RSC.) (b) CV of the anthraquinone-2-sulfonate/polyaniline (AQS/PANI) hybrid film-modified electrode at different pH values (Scan rate: 10 mV s^{-1}). Inset: the formal potential E$_{1/2}$ vs. pH. (Reproduced from Ref. [25]. Copyright (2011) RSC.)

catalytic activity of the catalysts pertaining to certain electrochemical reactions. In this chapter, we basically focus on the application of this technique for the evaluation of crucial electrocatalytic parameters.

CV can be used to evaluate electrode kinetics using a three-electrode (working, counter, and reference electrodes) system immersed in an electrolyte and connected to a dynamic potentiostat. The redox reaction that occurs at the interface between the electrolyte and CE is measured, and a resultant CV curve with the cathodic current (I_{pc}), anodic current (I_{pa}), cathodic peak potential (E_{pc}), and anodic peak potential (E_{pa}) is obtained, as shown in Figure 17.1a.

Let us consider the case of anthraquinone-2-sulfonate (AQS)/PANI modified electrode as proposed by Zhang and Yang (2011). Its electrochemical behavior in solutions of different pH (1.0 < pH < 5.0) was investigated by CV (Figure 17.1b), which produced two anionic and two cathodic peaks (the first pair corresponds to PANI and the second corresponds to the AQS$_{doping}$ redox process). The intrinsic redox peaks of PANI with anodic (I_{pa}) and cathodic (I_{pc}) peak currents showed a quasi-reversible pattern with a substantially low-current response. The half-wave potential of the redox reaction was calculated as the average value of the anodic (E_{pa}) and cathodic peak potentials (E_{pc}).

$$E_{1/2} = (E_{pa} + E_{pc})/2 \qquad (17.17)$$

The plot of E$_{1/2}$ (V) versus pH (inset of Figure 17.1b) was linear with a slope value corresponds to 61.7 mV pH^{-1}, which was comparable with the theoretical Nernstian value of 59.5 mV pH^{-1}, indicating a two electron-proton transfer (EPT).

The potentiodynamic polymerization of the aniline (AQS/PANI) film (Figure 17.2a) was achieved in an acidic medium by the electrochemical doping-dedoping-redoping technique [25]. The thickness of PANI in the prepared AQS/PANI film was determined based on the amount of charge (Q$_{PANI}$) required to switch from the LE form of PANI to the EM^{2+} form, according to Equation (17.18).

$$d = Q_{PANI}M_w/zFA_e\rho_a \qquad (17.18)$$

where M$_w$ is the molecular weight of aniline, ρ_a is the density of aniline (1.02 g cm^{-3}), A$_e$ is the electrode area, and z = 0.5 is the charge number of the aniline units in the polymer chains.

The influence of scan rate (v) on redox behavior can be studied using the CV technique (Figure 17.2e). The anodic (I_{pa}) and cathodic (I_{pc}) peak currents of the AQS$_{doping}$ redox process show linear

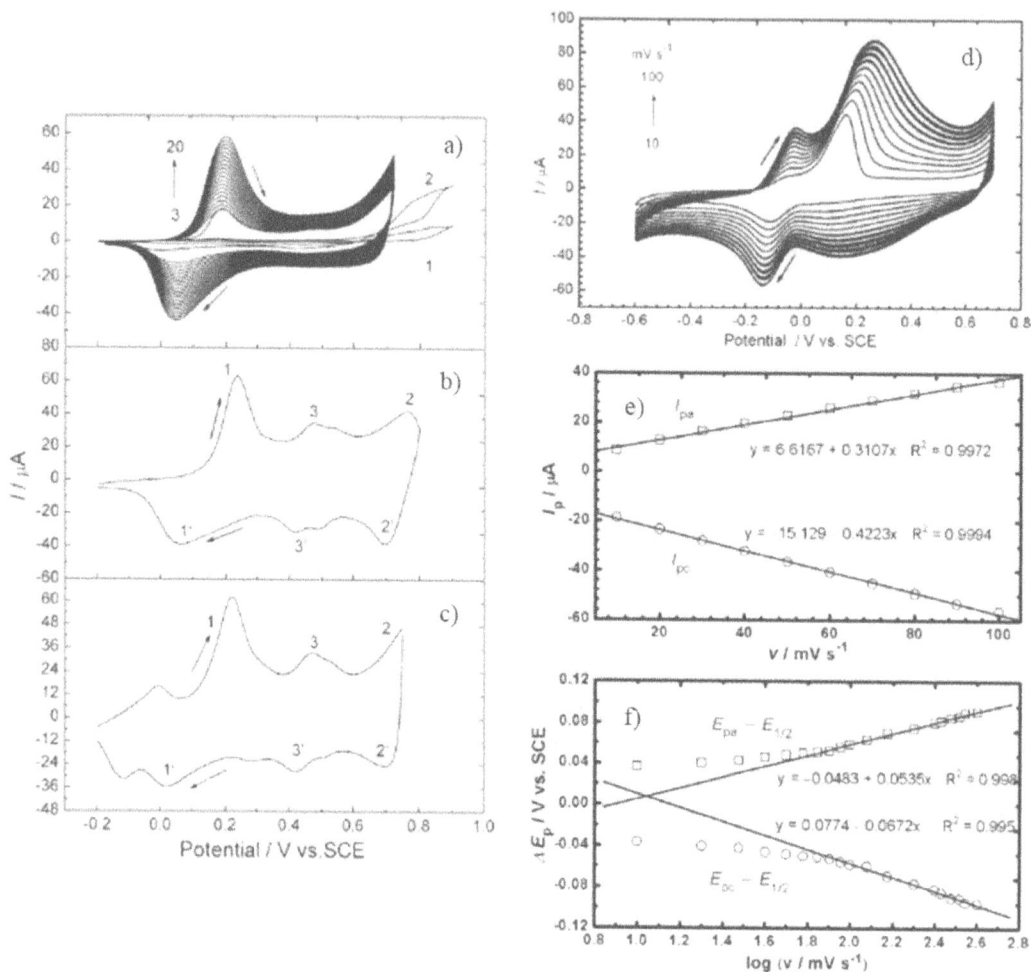

FIGURE 17.2 (a) Potentiodynamic polymerization of aniline; (b) stable cyclic voltammogram of PANI after electropolymerization; (c) potentiodynamic dedoping of emeraldine base-PANI at the scan rate of 10 mVs⁻¹; (d) cyclic voltammograms of the anthraquinone-2-sulfonate/polyaniline (AQS/PANI) hybrid film-modified electrode at different scan rates. Plots of (e) Ip vs. v and (f) ΔEp vs. log v for the AQS$_{doping}$ redox peak in potential range from –0.2 to 0 V. (Reproduced from Ref. [25]. Copyright (2011) RSC.)

dependence on scan rate (v) in the potential range from –0.2V to 0 V, which is typical of a surface redox process. This shows that theoretically, the redox reactions on the modified electrodes occur as a surface process at low scan rates and that the behavior shifts toward a diffusion-controlled process as the sweep rate increases. If the potentiodynamic kinetics of a redox process is diffusion-controlled, the peak current I_p can be calculated using the below formula:

$$Ip = 2.69 \times 10^5 A_e C_{AQS} n^{3/2} D^{1/2} v^{1/2} \tag{17.19}$$

where A_e is the electrode area, c_{AQS} is the AQS doping concentration, n is the number of electrons, D is the diffusion coefficient, and v is the scan rate.

Figure 17.2f shows that at high scan rates, there was a linear dependence of ΔEp on log v. The slope of the cathodic linear segment is given by 2.303 RT/α nF while that of anodic peaks is calculated as –2.303 RT/(1 – α)nF, where R (8.314 J mol⁻¹ K⁻¹) corresponds the molar gas constant and T represents the temperature. The α and n values obtained from the two slopes give information about the type of redox reaction and electron transfer process. With the AQS/PANI film, it was a quasi-reversible system ($\alpha = 0.44$ and $n = 2.03$).

The standard rate constant (k_s) can be estimated using the equation,

$$\log ks = \alpha \log(1-\alpha) + (1-\alpha)\log \alpha \qquad (17.20)$$

$$-\log\left(\frac{RT}{nFv}\right) - \left(\frac{\alpha\ (1-\alpha)nF\ \Delta Ep}{2.303\ RT}\right) \qquad (17.21)$$

17.5.2 ELECTROCHEMICAL ACTIVE SURFACE AREA

Electrochemical active surface area is a deciding factor of the activity of an electrocatalyst and is estimated based on its direct proportionality with the electrochemical double-layer capacitance. ECSA helps elucidate the catalytic and mechanistic behavior of the electrocatalyst. Since the catalytic activity is determined by both ECSA and the chemical activity of the catalyst, the chemical activity of different catalysts with different ECSAs is not comparable. Moreover, the electrochemical active area is basically different from the geometrical area of a flat electrode (Figure 17.3) [26]. Therefore, the specific activity, i.e., current density normalized to ECSA is critical for the comparison of the intrinsic activity of catalysts.

To evaluate the specific activity of a hybrid electrode, the capacitance for the double-layer (C_{dl}) is measured via CV at different scan rates in the potential range that allows only the absorption and desorption processes (No-Faradaic regions) but not the redox processes (charge transfer does not occur). Then, the C_{dl} is obtained by plotting $\Delta j = ja - jc$ against the scan rate, and the linear slope gives $2C_{dl}$ [27].

ECSA of active materials is calculated from the electrochemical double-layer capacitance (C_{dl}) of active materials, which is directly proportional to their electrochemically active surface area [28–30]. Thus, higher ECSA represents more effective active sites and higher intrinsic catalytic activity, which often leads to the enhancement of electrocatalytic activity [27, 29, 31].

ECSA value is calculated by

$$ECSA = C_{dl} / C_s \qquad (17.22)$$

where C_s is the specific capacitance of catalyst. The capacitance for the double-layer can also be used to identify the mechanical detachment of the catalyst layer from the inert support [23]. It is known that the non-Faradaic current density (j) must vary linearly with the scan rate (v), then the electrical double-layer capacitance is calculated by

$$C_{dl} = j / v \qquad (17.23)$$

The roughness factor is given by normalizing the double-layer capacitance of the electrode to the flat reference surface.

$$ECSA = C_{dl} / C_{dl.Ref} \qquad (17.24)$$

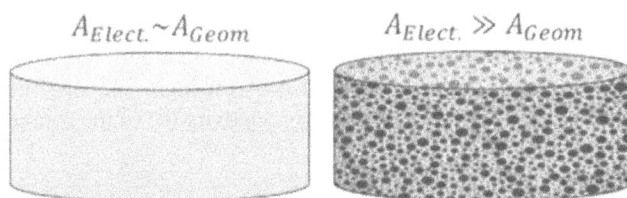

$A_{Elect.} \sim A_{Geom} \qquad A_{Elect.} \gg A_{Geom}$

FIGURE 17.3 Schematic illustration of the difference between the geometric surface area and electrochemically active surface area (nanostructured electrodes). (Reproduced from Ref. [26]. Copyright (2018) ACS.)

Then, the specific surface area of the electrode is the product of the flat reference surface area and roughness factor,

$$A_{Elect} = A_{Geom} \times ECSA \qquad (17.25)$$

Thus, the ECSA-normalized current density is calculated as

$$j_{ECSA} = j / ECSA \qquad (17.26)$$

Yang et al. correlated ECSA with ORR activity and showed that the sample with the highest ECSA produced the best result among all the tested samples [32]. They also evidenced that an enhanced ECSA can significantly play a vital role in decreasing the charge transfer resistance and thus improve the oxygen reduction rate.

17.5.3 ELECTROCHEMICAL IMPEDANCE SPECTROSCOPY

Electrochemical impedance spectroscopy is a powerful technique widely applied to identify the sources of polarization loss in electrochemical reactions within a short time. Basically, impedance is the measure of opposition to the flow of electrical current in a system. EIS studies performed with alternating current (AC) signals (~10 mV) at frequencies varying from 10 Hz to 1 MHz and reveal the distinct roles of ions and electron mobility in catalyst conductivity. As the mobility of electrons is faster compared with ions, electron transport is studied at low-frequency regions than those used to study ionic transportation. Besides, the amplitude of the applied AC voltage is so small that it does not perturb the equilibrium and hence enables the study of the steady-state of the system.

The Nyquist plot (the imaginary part Z′ vs. the real part Z″) and Bode plots (phase angle vs. frequency, |Z| vs. frequency) are some of the ways to plot impedance data. These plots depict a model based on the physicochemical properties of the catalyst. This impedance plot shows: (i) the high-frequency intercept, which indicates solution resistance (Rs), (ii) semicircle in the high-to-medium frequency region, the diameter of which represents the charge–transfer resistance (Rct) of the electrode/catalytic materials, and (iii) a straight line in the medium-to-low-frequency region, which illustrates ionic diffusion in the host material and resistance to electrolyte diffusion (Zw–Warburg impedance) at the electrode surface. Warburg impedance corresponds to mass-transport resistance that appears at the low-frequency region. The conduction occurs only through ionic diffusion, if the catalyst exhibits only Warburg impedance. The relaxation time (τ) is predicted from the maximum value in the semicircle of the Nyquist plot using the equation $τ = 1/2πf$. Therefore, low relaxation time (τ) denotes the fast charge-transfer property. Constant phase element (CPE) is usually employed to replace ideal capacitors owed to the inhomogeneity and roughness of the electrodes:

$$Z_{CPE} = T_{CPE} (jw)^{-n} \qquad (17.27)$$

where Z_{CPE} is the complex impedance of CPE, T_{CPE} and n are frequency-independent constants, and w is the angular frequency (w = 2πf, f is the frequency). n is related to the surface roughness of the electrode (usually between zero to one for a CPE, zero for a resistor, one for an ideal capacitor, and 0.5 for Warburg impedance/mass-transfer impedance). The solution resistance (R_S) and charge transfer resistance (R_{ct}) values of the catalyst are given by the high- and low-frequency intercepts of the real axis in the Nyquist plot, respectively. The conductivity (σ) of the catalyst is determined as

$$\sigma = \frac{L}{R_{ct} \ A} \qquad (17.28)$$

where L is the distance between the counter and working electrodes (cm), and A is the cross-sectional area of the electrocatalyst (cm²).

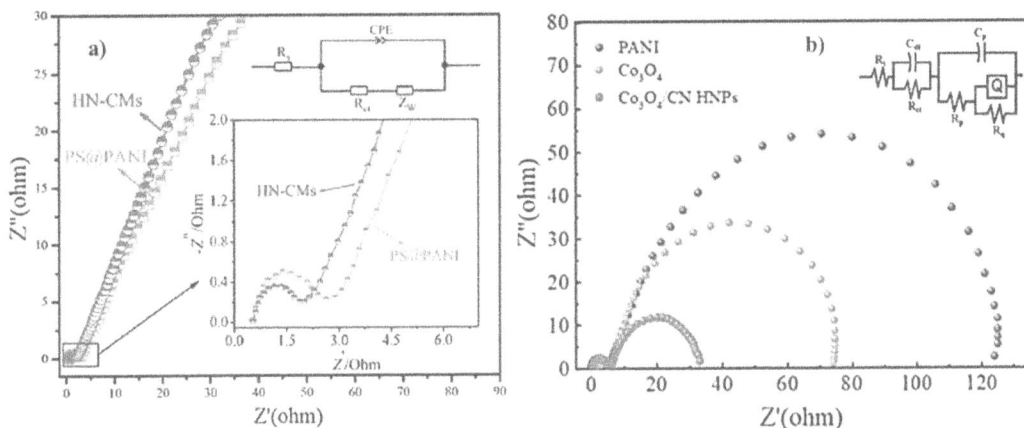

FIGURE 17.4 Nyquist plots of (a) polyvinylpyrrolidone (PVP)-functionalized polystyrene @polyaniline (PS@PANI) and hollow nitrogen-doped carbon microspheres (HN–CMs). (Reproduced from Ref. [35]. Copyright (2018) Elsevier). (b) PANI, Co_3O_4 and Co_3O_4/CN HNPs (defect-induced nitrogen-doped carbon-supported Co_3O_4 nanoparticles). Insets: Simulated equivalent circuits to fit the data. (Reproduced from Ref. [36]. Copyright (2020) Elsevier.)

EIS can measure the distribution of internal resistance, while it cannot estimate the overall internal resistance used in the maximum power density calculations [33]. In contrast, the polarization curve and cathode potential of the ORR catalysts used in microbial fuel cells do not only yield the overall internal resistance but also help compare the concentration overpotentials at high current outputs. A Nyquist plot with a smaller arc diameter suggests a decrease in R_{ct} and reduced contact resistance, which results in highly favorable reaction kinetics [30, 34]. Despite the advantages of EIS and the wide variety of data it furnishes, its major shortcoming is the ambiguity in selecting the equivalent circuit model. The equivalent circuit model R(Q(RW)) proposed to fit the Nyquist graphs of the prepared electrodes (Figure 17.4) in the three-electrode system and consists of an external solution resistance (R_s) connected in series respectively with a CPE, charge transfer resistance (R_{ct}) and Warburg diffusion resistance (W).

EIS can also be used to monitor the charge-transfer mechanism between the electrodes and electrolyte in interfacial reactions and thereby understand HER kinetics. It reveals the extent of electric resistance, which can depict the energy consumption pathways during the process. For this purpose, the charge transfer resistance (R_{ct}) associated with the electrochemical processes at the electrode/electrocatalyst interface is obtained. The R_{ct} value is proportional to the speed of the HER process and must ideally be very low when electron transport is highly enhanced, leading to superior HER activity [25].

We can consider the case of hollow nitrogen-doped carbon microspheres (HN-CMs) as an efficient electrochemical active material in an alkaline medium [35]. The Nyquist plot in Figure 17.4a, which shows lower solution resistances (R_s), less semi-circular diameters (R_{ct}), and a near-vertical line close to the real part of the impedance Nyquist plot, indicates the high electronic conductivity of the HN-CMs, thus signifying it as a proficient electrocatalytic material [35]. Furthermore, decreased charge-transport impedance and smaller contact resistance can favor reaction kinetics [30].

Chen et al. studied the charge-transfer mechanism by using EIS [36]. Of the two distinct capacitance arcs (Figure 17.4b) at the complicated plane in the complete frequency range of the plot, the one at high frequencies refers to the mass transfer resistance from the electrolyte to the exposed active sites, whereas that in the low-frequency region represents the resistance to charge transfer and equivalent resistance of the intermediates at the electrode–electrolyte interface. The former semicircle has a much shorter relaxation time than that of the latter, indicating that either the charge transfer resistance or the equivalent resistance of the intermediates was dominantly controlling in OER kinetics.

The kinetics of this OER process was studied by fitting the impedance results to the equivalent circuit of R_s $(C_{dl}$ $R_{ct})$ $(C_p$ $(R_p$ $(QR_q)))$ (Figure 17.4b, inset). R_s is the solution resistance; R_{dl} and C_{dl} are the polarization resistance and capacitance of the double layer, respectively; C_p is the double-layer capacitance at the electrode–electrolyte interface; R_p is the charge transfer resistance to the reaction across the electrode–electrolyte interface. R_q and Q are equivalent resistance and pseudo-capacitance, which reflect the adsorption of intermediates at the electrode, respectively. The fitted R_q was found to be relatively larger rather than R_p, revealing that the OER kinetics of these three electrocatalysts was mainly dependent on the equivalent resistance or the adsorption of intermediates. [36].

17.5.4 LINEAR SWEEP VOLTAMMETRY

In linear sweep voltammetry, the current is measured by varying the electrode potential at a constant rate. Linear sweep voltammograms (LSV) may reveal details regarding composition changes in the catalyst or changes in the oxidation state of the catalyst resulting from the prolonged application of high anodic potential.

The ORR process is practically a combination of the four-electron and two-electron reactions. Predominantly, the four-electron pathway is preferred for two reasons: it is more effective, and the two-electron process produces hydrogen peroxide, which may harm the electrochemical cell components. Thus, the number of electrons transferred is a crucial factor that determines the performance evaluation of ORR electrocatalysts. The electron transfer number (n) of an electrocatalyst (2 or 4) usually indicates the generation of H_2O or H_2O_2, respectively. Generally, the rotating ring-disk electrode (RRDE) and the Koutecký–Levich (K–L) equation are used to determine the electron transfer number based on the LSV data obtained at different speeds of the RRDE [37]. The polarization curves are obtained by sweeping at a potential (V) and a rotation speed (rpm) with different scan rates (mVs^{-1}). Furthermore, an RRDE analysis also provides insights into the electrochemical process of ORR based on parameters, including the starting potential, current densities, and overpotential of the catalyst.

The apparent electron number is determined from the slope of the K–L line, and it is defined by the Koutecký–Levich (K–L) equation.

$$\frac{1}{j(E)} = \frac{1}{j_k(E)} + \frac{1}{j_d} = \frac{1}{j_k} - \frac{1}{0.62 \cdot n \cdot F \cdot D(O_2)^{\frac{2}{3}} \cdot v^{-\frac{1}{6}} \cdot \omega^{\frac{1}{2}} \cdot c(O_2)} \tag{17.29}$$

where

  j(E) is the measured current density,
  $j_k(E)$ is the kinetic current density at a given electrode potential (E),
  j_d is the limiting diffusion current density,
  m is the kinematic viscosity of the solution (0.01 cm^2 s^{-1}),
  $D(O_2)$ is the diffusion coefficient of O_2 (1.9 × 10^{-5} cm^2 s^{-1}), and
  $c(O_2)$ is the dissolved O_2 concentration (1.2 × 10^{-6} mol cm^{-3}).

Let us consider the case of reduced graphene oxide (rGO) electrocatalysts modified PANI along with cobalt ferrite (CF) for ORR application in fuel cells [38]. The Koutecký –Levich (K–L) plots were plotted with the reciprocal rotating speed against reciprocal current density [28]. The number of electrons involved in ORR was calculated from the K–L plot (j^{-1} vs. $\omega^{-1/2}$) (Figure 17.5). It follows the first-order kinetics with respect to molecular oxygen, and the n value was in the order: rGO > PANI/rGO > CF/PANI/rGO. Based on Figure 17.5, the CF/PANI/rGO electrocatalyst can reduce the oxygen through a four-electron transfer pathway, and the order of ORR activity increased progressively with PANI and CF modifications [38].

The evaluation of electron transfer number is crucial for an electrocatalyst as it reveals information about catalyst selectivity and facilitates the optimization of material performance for specific electrochemical applications [39].

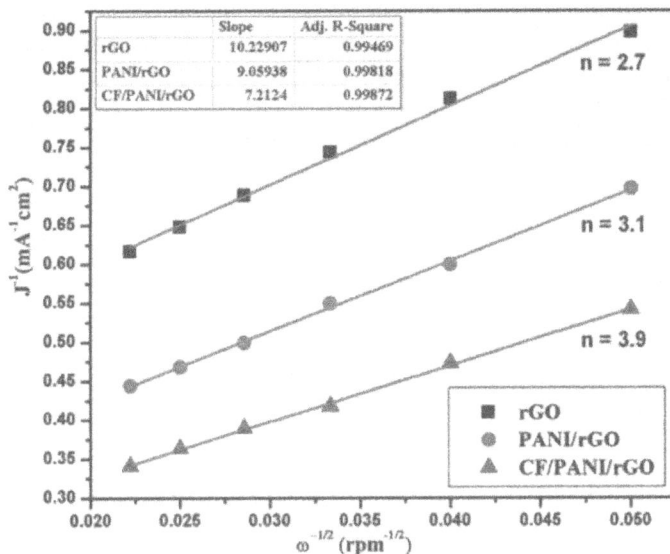

FIGURE 17.5 K–L plots of rGO, PANI/rGO, and CF/PANI/rGO electrocatalysts (–0.3 V). (Reproduced from Ref. [38]. Copyright (2015) Elsevier.)

The number of transferred electrons (n) in RRDE was calculated by

$$n = 4 \ I_{de} (I_{de} + I_{re} / \text{€})$$ (17.30)

$$\% \ H_2O_2 = 200 \ I_{re} / [\text{€}(I_{de} + I_{re} / \text{€})]$$ (17.31)

where I_{de} is the disk electrode current, I_{re} is the ring electrode current, and €(38%) is the collection efficiency of the ring current [37].

17.5.5 Tafel Analysis

Tafel analysis can be applied to understand the mechanism of electrocatalysis, as well as the intrinsic properties and electrocatalytic efficiency of the electrocatalyst in a particular electrochemical reaction, such as HER and ORR. This information is specifically obtained from two very important parameters, namely the Tafel slope (b) and exchange current density (i_o). The Tafel equation is used to calculate the Tafel slopes and thereby analyze the reaction kinetics [40]. LSV are transformed into a plot of η versus (log i) based on the Tafel equation given below:

$$\eta = b(\log j) + a$$ (17.32)

where $a = 2.303 \ RT \ \log j_0 / \alpha nF$ and $b = 2.303 \ RT / \alpha nF$,

η is the overpotential,
j is the current density, j_0 is the exchange current density,
b is the Tafel slope, and a is the intercept related to the exchange current density j_0 (A cm^{-2}),
R is the ideal gas constant, T represents the absolute temperature, n is the number of electrons transferred during the redox reaction, F is the Faraday constant, and α is the charge transfer coefficient.

The calculated slope value is termed the Tafel slope (b), which is a crucial parameter of an electrocatalyst and provides information regarding the rate-determining step (the slowest step) and possible reaction mechanism of the electrochemical reaction involved (HER). The Tafel slope of a catalyst is inversely proportional to its catalytic activity, which also indicates less activation energy required for the electron transfer process and lower activation losses [41]. Therefore, highly advanced electrocatalysts ideally involve very low overpotentials and exhibit significantly low Tafel slopes. Therefore, a high Tafel slope signifies a sharp rise in overpotential associated with low current density [42].

17.5.6 OVERPOTENTIAL (η)

The potential needed to drive the reaction at a certain current density, known as overpotential (η), determines the activity of an electrocatalyst in HER and OER. Meanwhile, the potential needed to overcome the intrinsic barriers, initiate the reaction and generates current is known as the onset potential. The high electrocatalytic activity of an electrocatalyst directly corresponds to a low overpotential value.

17.5.7 TURNOVER FREQUENCY

For an electrochemical reaction that involves a molecular catalyst in a homogeneous solution, the turnover frequency (TOF) is calculated as the ratio of moles of the product and moles of the catalyst present in the diffusion layer near the electrode per unit time. The TOF of an electrocatalyst in HER is defined as the mass of hydrogen produced per active site at a given time, and it is measured in H_2/s. TOF can be calculated as given below:

$$\text{Turn over frequency } TOF = \frac{jM}{zFm} \tag{17.33}$$

where
    M is the mass percentage,
    m is the mass/square centimeter,
    F is Faraday's constant, and
    z is the stoichiometric number of electrons, which is 2 for HER.

The greater the TOF value, the greater is the active sites in the electrocatalyst, and the higher is the hydrogen gas production at a given time, which indicates excellent intrinsic properties.

17.5.8 CHRONOAMPEROMETRY OR CHRONOPOTENTIOMETRY

Stability and durability are the most crucial electrocatalyst performance parameters considered during the commercialization process as they determine the ability of the electrocatalyst to sustain its original activity over a long period of time (stability) or repeated cycling (durability). In general, the electrolyte in a basic environment favors high catalyst stability, while acidic conditions hamper their performance. To access the durability of an electrocatalyst in different electrolytes, both long-term potential cycling and current-time responses at constant potentials are examined. If the polarization curves obtained from continuous CV at an increasing scanning rate over numerous cycles, practically overlay with the initial curve and show an insignificant cathodic current loss, the catalyst can be considered highly durable. The long-term stability is evaluated by chronopotentiometry (CP) or chronoamperometry (CA) performed over a period of time at a fixed current density and applied potential.

17.6 APPLICATION OF CP-BASED ELECTROCATALYSTS

The development of low-cost non-platinum group metal (non-PGM) catalysts for electrochemical devices has attracted much attention in recent years [17]. PANI and PPy are the widely used CPs in electrocatalysts. Even though these polymers have attractive benefits, including electrical and mechanical properties, they are widely used either as composite materials or heteroatom-doped/metal-doped materials [1, 43]. N-doped carbon is highly desirable because of its excellent tunability to form versatile designs and possessed high catalytic activity. Since CPs are rich in nitrogen, they are preferable precursors for heteroatom-doped pyrolyzed carbon materials [10, 18]. Improved surface area and heteroatom doping are the additional advantages of these materials, and they can hence be efficient alternatives to the Pt catalyst [22].

Structural changes that take place in N-doped carbon during carbonization between 500 and 800°C facilitate can [13]:

1. transform the nitrogen species in PPy into more active pyridinic-N groups and
2. improve the specific surface area and mesopores.

The surface structural changes in these materials are reflected by the number of electrons transferred in the redox reaction. Their enhanced ORR performance is attributed to the greater number of active sites and mesoporous structures [16]. Similar works have also shown better responses. For instance, PANI and glucose were used as precursors in a hydrothermal reaction to prepare nitrogen-doped carbon materials and activated by high-temperature carbonization, which resulted in the transfer of 4.2 electrons per molecule of O_2. The authors suggested that the ORR followed the mechanism of the four-electron pathway ($O_2 \rightarrow H_2O$) [44]. Higher N content usually represents enhanced ORR activity because more positive-charged C atoms are formed because of the highly electronegative neighboring N atoms [13, 45]. It has been found that the doping of element O is increased the ORR performance by the CPs since the co-doping of N and O would generate asymmetric spins and a better-charged structure [46]. Such doped atoms high or low electron-affinity (nitrogen, boron, etc.) create a charged site in the CPs, favoring O_2 adsorption. Thus, the B/N co-doped carbon materials are very attractive as they have unique electronic structures due to the synergistic coupling effect of the heteroatoms. For example, a boron and nitrogen co-doped porous carbon nanosheet was synthesized by chemical oxidative polymerization of aniline, 3-aminophenylboronic acid, and m-phenylenediamine using amino-functionalized graphene oxide (GO) as the template, generating GO-based PANI nanosheets activated with boronic acid (GO-CBP) [47]. Their ORR performance was comparable to that of Pt/C in terms of fuel cross-over effects, long-lasting durability, kinetic current density, and half-wave potential, as well as excellent methanol tolerance, owing to the large specific surface area, unique sheet nanostructure, and high content of doped heteroatom under alkaline conditions via a 4 e^- transfer mechanism [47–50]. Additionally, the asymmetric spin density produced by heteroatom doping in carbon provides active ORR sites. Doping of other heteroatoms (excluding N) has also been reported to encourage asymmetry in spin density and enhance the ORR activity of carbon [51]. Carbonization at inert atmosphere produced lesser yields than those treated under aerobic conditions since the oxygen atmosphere favors crosslinking and dehydration on carbonization. The small variations in the pore structure may be detected with a low amount of O_2 is used. Besides an increase in phenol-type groups, heat treatment in an oxygenated atmosphere leads to a greater number of N-functionalities at the edge sites (i.e., pyridine and pyrridone/pyrrole groups) compared with treatment in an N_2 atmosphere [1, 52]. Table 17.1 compares some CP-derived electrocatalysts used for different applications.

Metal-doped carbon materials developed from polymer precursors have been studied by various researchers. For instance, the ORR activity increased up to a concentration of cobalt loading, 1.0 wt%, and then further increase in the concentration of cobalt loading affects the O_2 reduction. This is due to metallic Co and cobalt oxide particles blocking the active sites of an electrocatalyst [53]. In another study, a simple pyrolysis of PANI fibers and cobalt(II) nitrate resulted in Co and N co-doped graphene networks (Co–N-GNWs) with an interconnected porous structure and high crystallinity with a high active surface area [15].

Electrical characteristics of the conductive porous structure have played a significant role in ORR, HER, and OER. The most active sites in PANI- and PPY-carbon derivatives are obtained during the heat treatment when the pyridine groups convert to edge-type quaternary N species in a zig-zag position [12]. The temperature used for heat treatment is crucial to achieving enhanced conductivity and increasing the active sites for the four-electron transfer pathway [14, 22].

17.7 CONCLUSION

Replacing the platinum metal catalyst with non-precious electrocatalyst for redox reactions is of keen interest in the last decades. In this view, this chapter highlights the importance, role, and applications of CPs. CPs are found to be efficient when they undergo pyrolysis at ambient conditions, which also allows for heteroatom- and metal-doping. The evaluation of electrocatalysts for their stability, durability, electrochemical surface area, and reaction kinetics is crucial, as discussed above. Various studies on CP-based electrocatalysts revealed that the content of nitrogen and oxygen groups almost invariable after pyrolysis are key determinants of the efficiency of the electrocatalyst. In conclusion, PANI and PPy are promising candidates for multifunctional electrochemical systems.

ACKNOWLEDGMENTS

One of the authors, N. Saranya wishes to acknowledge Council of Scientific and Industrial Research (CSIR-SRF), New Delhi and Anna University (ACRF), Chennai, for providing a research fellowship.

REFERENCES

1. J. Xu, S. Lu, H. Zhou, X. Chen, Y. Wang, C. Xiao, S. Ding, A highly efficient electrocatalyst derived from polyaniline@CNTs–SPS for the oxygen reduction reaction, Chem. Electro. Chem. 5 (2018) 195–200. doi:10.1002/celc.201700735.
2. S. Gupta, M.K. Patel, A. Miotello, N. Patel, Metal boride-based catalysts for electrochemical water-splitting: A review, Adv. Funct. Mater. 30 (2020). doi:10.1002/adfm.201906481.
3. M.D. Bhatt, J.Y. Lee, Advancement of platinum (Pt)-free (non-Pt precious metals) and/or metal-free (non-precious-metals) electrocatalysts in energy applications: A review and perspectives, Energy Fuels. 34 (2020) 6634–6695. doi:10.1021/acs.energyfuels.0c00953.
4. N.T. Suen, S.F. Hung, Q. Quan, N. Zhang, Y.J. Xu, H.M. Chen, Electrocatalysis for the oxygen evolution reaction: Recent development and future perspectives, Chem. Soc. Rev. 46 (2017) 337–365. doi:10.1039/c6cs00328a.
5. R.K. Gautam, A. Verma, Electrocatalyst materials for oxygen reduction reaction in microbial fuel cell, Elsevier B.V. (2019). doi:10.1016/b978-0-444-64052-9.00018-2.
6. C. Hu, L. Zhang, J. Gong, Recent progress made in the mechanism comprehension and design of electrocatalysts for alkaline water splitting, Energy Environ. Sci. 12 (2019) 2620–2645. doi:10.1039/c9ee01202h.
7. Y. Da, X. Li, C. Zhong, Y. Deng, X. Han, W. Hu, Advanced characterization techniques for identifying the key active sites of gas-involved electrocatalysts, Adv. Funct. Mater. 30 (2020) 1–28. doi:10.1002/adfm.202001704.
8. S. Wang, A. Lu, C.-J. Zhong, Hydrogen production from water electrolysis: Role of catalysts, Nano Converg. 8 (2021). doi:10.1186/s40580-021-00254-x.
9. C.L. Bentley, M. Kang, P.R. Unwin, Nanoscale surface structure-activity in electrochemistry and electrocatalysis, J. Am. Chem. Soc. (2018). doi:10.1021/jacs.8b09828.
10. S. Narayanasamy, J. Jayaprakash, Application of carbon-polymer based composite electrodes for microbial fuel cells, Rev. Environ. Sci. Biotechnol. 19 (2020) 595–620. doi:10.1007/s11157-020-09545-x.
11. X.L. Chen, L. Zhang, J.J. Feng, W. Wang, P.X. Yuan, D.M. Han, A.J. Wang, Facile solvothermal fabrication of polypyrrole sheets supported dendritic platinum-cobalt nanoclusters for highly efficient oxygen reduction and ethylene glycol oxidation, J. Colloid Interface Sci. 530 (2018) 394–402. doi:10.1016/j.jcis.2018.06.095.
12. Q. Dang, Y. Sun, X. Wang, W. Zhu, Y. Chen, F. Liao, H. Huang, M. Shao, Carbon dots-Pt modified polyaniline nanosheet grown on carbon cloth as stable and high-efficient electrocatalyst for hydrogen evolution in pH-universal electrolyte, Appl. Catal. B Environ. 257 (2019) 117905. doi:10.1016/j.apcatb.2019.117905.

13. A.C. Ramírez-Pérez, J. Quílez-Bermejo, J.M. Sieben, E. Morallón, D. Cazorla-Amorós, Effect of nitrogen-functional groups on the ORR activity of activated carbon fiber-polypyrrole-based electrodes, Electrocatalysis. 9 (2018) 697–705. doi:10.1007/s12678-018-0478-y.

14. G. Ren, Y. Li, Z. Guo, G. Xiao, Y. Zhu, L. Dai, L. Jiang, A bio-inspired Co_3O_4-polypyrrole-graphene complex as an efficient oxygen reduction catalyst in one-step ball milling, Nano Res. 8 (2015) 3461–3471. doi:10.1007/s12274-015-0844-5.

15. S. Peng, H. Jiang, Y. Zhang, L. Yang, S. Wang, W. Deng, Y. Tan, Facile synthesis of cobalt and nitrogen co-doped graphene networks from polyaniline for oxygen reduction reaction in acidic solutions, J. Mater. Chem. A. 4 (2016) 3678–3682. doi:10.1039/c5ta09615d.

16. Z. Zhang, J. Sun, M. Dou, J. Ji, F. Wang, Nitrogen and phosphorus codoped mesoporous carbon derived from polypyrrole as superior metal-free electrocatalyst toward the oxygen reduction reaction, ACS Appl. Mater. Interfaces. 9 (2017) 16236–16242. doi:10.1021/acsami.7b03375.

17. M. Yang, Y. Liu, H. Chen, D. Yang, H. Li, Porous N-doped carbon prepared from triazine-based polypyrrole network: A highly efficient metal-free catalyst for oxygen reduction reaction in alkaline electrolytes, ACS Appl. Mater. Interfaces. 8 (2016) 28615–28623. doi:10.1021/acsami.6b09811.

18. D. Xiao, J. Ma, C. Chen, Q. Luo, J. Ma, L. Zheng, X. Zuo, Oxygen-doped carbonaceous polypyrrole nanotubes-supported Ag nanoparticle as electrocatalyst for oxygen reduction reaction in alkaline solution, Mater. Res. Bull. 105 (2018) 184–191. doi:10.1016/j.materresbull.2018.04.030.

19. F. Papiya, A. Nandy, S. Mondal, P.P. Kundu, Co/Al_2O_3-rGO nanocomposite as cathode electrocatalyst for superior oxygen reduction in microbial fuel cell applications: The effect of nanocomposite composition, Electrochim. Acta. 254 (2017) 1–13. doi:10.1016/j.electacta.2017.09.108.

20. R.C.P. Oliveira, J. Milikić, E. Daş, A.B. Yurtcan, D.M.F. Santos, B. Šljukić, Platinum/polypyrrole-carbon electrocatalysts for direct borohydride-peroxide fuel cells, Appl. Catal. B Environ. 238 (2018) 454–464. doi:10.1016/j.apcatb.2018.06.057.

21. S. Li, S.-H. Ho, T. Hua, Q. Zhou, F. Li, J. Tang, Sustainable biochar as electrocatalysts for the oxygen reduction reaction in microbial fuel cells, Green Energy Environ. (2020). doi:10.1016/j.gee.2020.11.010.

22. J. Quílez-Bermejo, E. Morallón, D. Cazorla-Amorós, Oxygen-reduction catalysis of N-doped carbons prepared: Via heat treatment of polyaniline at over 1100°C, Chem. Commun. 54 (2018) 4441–4444. doi:10.1039/c8cc02105h.

23. A. Maljusch, O. Conradi, S. Hoch, M. Blug, W. Schuhmann, Advanced evaluation of the long-term stability of oxygen evolution electrocatalysts, Anal. Chem. 88 (2016) 7597–7602. doi:10.1021/acs.analchem.6b01289.

24. S. Wang, J. Tian, Recent advances in counter electrodes of quantum dot-sensitized solar cells, RSC Adv. 6 (2016) 90082–90099. doi:10.1039/c6ra19226b.

25. G. Zhang, F. Yang, Direct electrochemisty and electrocatalysis of anthraquinone-monosulfonate/polyaniline hybrid film synthesized by a novel electrochemical doping-dedoping-redoping method on pre-activated spectroscopically pure graphite surface, Phys. Chem. Chem. Phys. 13 (2011) 3291–3302. doi:10.1039/c0cp00608d.

26. D. Voiry, M. Chhowalla, Y. Gogotsi, N.A. Kotov, Y. Li, R.M. Penner, R.E. Schaak, P.S. Weiss, Best practices for reporting electrocatalytic performance of nanomaterials, ACS Nano. 12 (2018) 9635–9638. doi:10.1021/acsnano.8b07700.

27. T. Zhang, J. Du, P. Xi, C. Xu, Hybrids of cobalt/iron phosphides derived from bimetal-organic frameworks as highly efficient electrocatalysts for oxygen evolution reaction, ACS Appl. Mater. Interfaces. 9 (2017) 362–370. doi:10.1021/acsami.6b12189.

28. Y. Nie, J. Deng, S. Chen, Z. Wei, Promoting stability and activity of PtNi/C for oxygen reduction reaction via polyaniline-confined space annealing strategy, Int. J. Hydrogen Energy. (2019) 1–8. doi:10.1016/j.ijhydene.2019.01.125.

29. J. Shu, Q. Niu, N. Wang, J. Nie, G. Ma, Alginate derived Co/N doped hierarchical porous carbon microspheres for efficient oxygen reduction reaction, Appl. Surf. Sci. 485 (2019) 520–528. doi:10.1016/j.apsusc.2019.04.204.

30. J.T. Ren, Z.Y. Yuan, Bifunctional electrocatalysts of cobalt sulfide nanocrystals in situ decorated on N,S-codoped porous carbon sheets for highly efficient oxygen electrochemistry, ACS Sustain. Chem. Eng. 7 (2019) 10121–10131. doi:10.1021/acssuschemeng.9b01699.

31. Y. Shen, K. Dastafkan, Q. Sun, L. Wang, Y. Ma, Z. Wang, C. Zhao, Improved electrochemical performance of nickel-cobalt hydroxides by electrodeposition of interlayered reduced graphene oxide, Int. J. Hydrog. Energy. (2019). doi:10.1016/j.ijhydene.2018.12.098.

32. W. Yang, J. Li, L. Lan, Y. Zhang, H. Liu, Covalent organic polymers derived carbon incorporated with cobalt oxides as a robust oxygen reduction reaction catalyst for fuel cells, Chem. Eng. J. 390 (2020) 124581. doi:10.1016/j.cej.2020.124581.

33. S. Narayanasamy, J. Jayaprakash, Improved performance of *Pseudomonas aeruginosa* catalyzed MFCs with graphite/polyester composite electrodes doped with metal ions for azo dye degradation, Chem. Eng. J. 343 (2018) 258–269. doi:10.1016/j.cej.2018.02.123.

34. E. Fallah Talooki, M. Ghorbani, M. Rahimnejad, M. Soleimani Lashkenari, Investigating the effects of in-situ fabrication of a binder-free Co_3O_4-polyaniline cathode towards enhanced oxygen reduction reaction and power generation of microbial fuel cells, Synth. Met. 258 (2019) 116225. doi:10.1016/j.synthmet.2019.116225.

35. M. Hassan, D. Wu, X. Song, W. Qiu, Q. Mao, S. Ren, C. Hao, Polyaniline–derived metal–free hollow nitrogen–doped carbon microspheres as an efficient electrocatalyst for supercapacitors and oxygen reduction, J. Electroanal. Chem. 829 (2018) 157–167. doi:10.1016/j.jelechem.2018.09.051.

36. X. Chen, Y. Chen, X. Luo, H. Guo, N. Wang, D. Su, C. Zhang, T. Liu, G. Wang, L. Cui, Polyaniline engineering defect-induced nitrogen doped carbon-supported Co_3O_4 hybrid composite as a high-efficiency electrocatalyst for oxygen evolution reaction, Appl. Surf. Sci. 526 (2020) 146626. doi:10.1016/j.apsusc.2020.146626.

37. G. Lu, Z. Li, W. Fan, M. Wang, S. Yang, J. Li, Z. Chang, H. Sun, S. Liang, Z. Liu, Sponge-like N-doped carbon materials with Co-based nanoparticles derived from biomass as highly efficient electrocatalysts for the oxygen reduction reaction in alkaline media, RSC Adv. 9 (2019) 4843–4848. doi:10.1039/c8ra10462j.

38. K. Mohanraju, V. Sreejith, R. Ananth, L. Cindrella, Enhanced electrocatalytic activity of PANI and $CoFe_2O_4$/PANI composite supported on graphene for fuel cell applications, J. Power Sources. 284 (2015) 383–391. doi:10.1016/j.jpowsour.2015.03.025.

39. I.M. Minisy, N. Gavrilov, U. Acharya, Z. Morávková, C. Unterweger, M. Mičušík, S.K. Filippov, J. Kredatusová, I.A. Pašti, S. Breitenbach, G. Ćirić-Marjanović, J. Stejskal, P. Bober, Tailoring of carbonized polypyrrole nanotubes core by different polypyrrole shells for oxygen reduction reaction selectivity modification, J. Colloid Interface Sci. 551 (2019) 184–194. doi:10.1016/j.jcis.2019.04.064.

40. S. Das, R. Ghosh, P. Routh, A. Shit, S. Mondal, A. Panja, A.K. Nandi, Conductive MoS_2 quantum dot/polyaniline aerogel for enhanced electrocatalytic hydrogen evolution and photoresponse properties, ACS Appl. Nano Mater. 1 (2018) 2306–2316. doi:10.1021/acsanm.8b00373.

41. L. Xu, G. Zhang, J. Chen, G. Yuan, L. Fu, F. Yang, Prussian blue/graphene-modified electrode used as a novel oxygen reduction cathode in microbial fuel cell, J. Taiwan Inst. Chem. Eng. 58 (2016) 374–380. doi:10.1016/j.jtice.2015.06.013.

42. W. Xue, Q. Zhou, F. Li, B.S. Ondon, Zeolitic imidazolate framework-8 (ZIF-8) as robust catalyst for oxygen reduction reaction in microbial fuel cells, J. Power Sources. 423 (2019) 9–17. doi:10.1016/j.jpowsour.2019.03.017.

43. N. Saranya, J. Jayapriya, V. Ramamurthy, Unsaturated polyesters in microbial fuel cells and biosensors, in: S. Thomas, M. Hosur, C.J. Chirayil (Eds.), Unsaturated Polyester Resins: Fundamentals, Design, Fabrication, and Application, Elsevier Inc., 2019: pp. 557–578. doi:10.1016/b978-0-12-816129-6.00021-1.

44. X. Huang, X. Yin, X. Yu, J. Tian, W. Wu, Preparation of nitrogen-doped carbon materials based on polyaniline fiber and their oxygen reduction properties, Colloids Surf: A. 539 (2018) 163–170. doi:10.1016/j.colsurfa.2017.12.024.

45. A. Lahiri, G. Li, F. Endres, Highly efficient electrocatalytic hydrogen evolution reaction on carbonized porous conducting polymers, J. Solid State Electrochem. (2020). doi:10.1007/s10008-020-04577-3.

46. S. Xing, X. Yu, G. Wang, Y. Yu, Y. Wang, Y. Xing, Confined polyaniline derived mesoporous carbon for oxygen reduction reaction, Eur. Polym. J. 88 (2017) 1–8. doi:10.1016/j.eurpolymj.2017.01.011.

47. Y. Zhang, X. Zhuang, Y. Su, F. Zhang, X. Feng, Polyaniline nanosheet derived B/N co-doped carbon nanosheets as efficient metal-free catalysts for oxygen reduction reaction, J. Mater. Chem. A. 2 (2014) 7742–7746. doi:10.1039/c4ta00814f.

48. H. Jiang, Y. Zhu, Q. Feng, Y. Su, X. Yang, C. Li, Nitrogen and phosphorus dual-doped hierarchical porous carbon foams as efficient metal-free electrocatalysts for oxygen reduction reactions, Chem. Eur. J. 20 (2014) 3106–3112. doi:10.1002/chem.201304561.

49. F. Zhou, G. Wang, F. Huang, Y. Zhang, M. Pan, Polyaniline derived N- and O-enriched high surface area hierarchical porous carbons as an efficient metal-free electrocatalyst for oxygen reduction, Electrochim. Acta. 257 (2017) 73–81. doi:10.1016/j.electacta.2017.09.175.

50. A.M. Borges Honorato, M. Khalid, Q. Dai, L.A. Pessan, Nitrogen and sulfur co-doped fibrous-like carbon electrocatalyst derived from conductive polymers for highly active oxygen reduction catalysis, Synth. Met. 264 (2020) 116383. doi:10.1016/j.synthmet.2020.116383.

51. Q. Wei, G. Zhang, X. Yang, R. Chenitz, D. Banham, L. Yang, S. Ye, S. Knights, S. Sun, 3D porous Fe/N/C spherical nanostructures as high-performance electrocatalysts for oxygen reduction in both alkaline and acidic media, ACS Appl. Mater. Interfaces. 9 (2017) 36944–36954. doi:10.1021/acsami.7b12666.

52. J. Quílez-Bermejo, C. González-Gaitán, E. Morallón, D. Cazorla-Amorós, Effect of carbonization conditions of polyaniline on its catalytic activity towards ORR. Some insights about the nature of the active sites, Carbon N. Y. 119 (2017) 62–71. doi:10.1016/j.carbon.2017.04.015.

53. H.D. Sha, X. Yuan, L. Li, Z. Ma, Z.F. Ma, L. Zhang, J. Zhang, Experimental identification of the active sites in pyrolyzed carbon-supported cobalt-polypyrrole-4-toluenesulfinic acid as electrocatalysts for oxygen reduction reaction, J. Power Sources. 255 (2014) 76–84. doi:10.1016/j.jpowsour.2014.01.013.

18 Conducting Polymer-Based Microbial Fuel Cells

Charles Oluwaseun Adetunji[1], John Tsado Mathew[2],
Kshitij R.B. Singh[3], Abel Inobeme[4], Olugbemi T. Olaniyan[5],
Vanya Nayak[6], Jay Singh[7], and Ravindra Pratap Singh[6]

[1]Applied Microbiology, Biotechnology and Nanotechnology Laboratory,
Department of Microbiology, Edo University Iyamho,
Auchi, Edo State, Nigeria

[2]Department of Chemistry, Ibrahim Badamasi University Lapai,
Niger State, Nigeria

[3]Department of Chemistry, Government V.Y.T. Post Graduate
Autonomous College, Durg, Chhattisgarh, India

[4]Department of Chemistry, Edo University Iyamho,
Auchi, Edo State, Nigeria

[5]Laboratory for Reproductive Biology and
Developmental Programming, Department of Physiology,
Edo University Iyamho, Auchi, Edo State, Nigeria

[6]Department of Biotechnology, Indira Gandhi National Tribal University,
Amarkantak, Madhya Pradesh, India

[7]Department of Chemistry, Institute of Science,
Banaras Hindu University, Varanasi, Uttar Pradesh, India

CONTENTS

18.1 INTRODUCTION

Improvement in technology is continuously keen to improve the fossil fuel challenges in an eco-friendly way. A bacteriological fuel cell is such a supportable, green skill that could enable bio-electrochemical conversion. However, conducting polymers have shown auspicious materials for catalytic behavior, high conductivity, as well as exceptional electrochemical activity and, as a result, have become part of the highest demanding materials for use in a bacteriological fuel cell. This has been highly endorsed to amend electrodes and separators due to improved durability, high conductivity, and power density in the microbial fuel cell [1].

To improve the existing environmental pollution and energy crisis, ecofriendly and sustainable energy conversion and storage schemes are immediately needed. Owing to their promising catalytic activity, high conductivity, and exceptional electrochemical properties, conducting polymers have

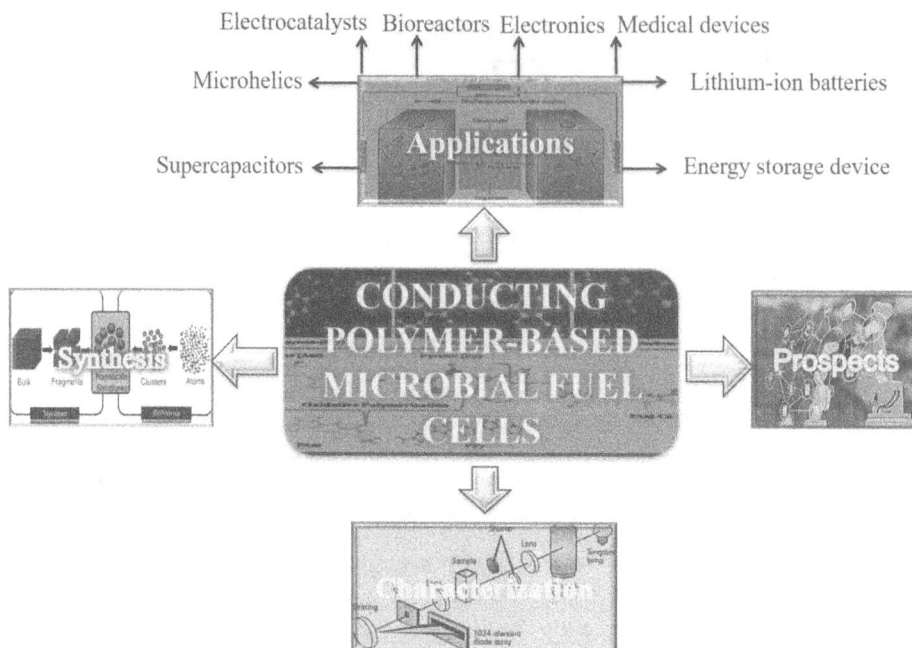

FIGURE 18.1 Illustration emphasizing details in this chapter.

been gaining strong consideration for application in electrochemical energy storage and conversion. The schemes engaged in increasing the electrochemical and electrocatalytic performances of conducting polymer-based materials are currently put into future research [2]. In recent times, carbon polymer-based composite electrodes for bacteriological fuel cells are essential together with a study of polymers and carbon materials. The demerits, merits, and performance of carbon polymer-based composites as cathode and anode materials for bacteriological fuel cells look into. The increase of complexes with surface and bulk modified electrode materials might overwhelm the main crisis posed by the low power density of bacteriological fuel cells. Fluorine- and nitrogen-containing polymers which polyaniline (PANI), polyacrylonitrile, polydopamine, polyacrylamide, and polytetrafluoroethylene, have been acknowledged as possible candidates for surface or bulk modification through the existence of redox-active species to improve the effectiveness and the biocompatibility of the decrease of the cathodes [3, 4]. Hence, this chapter briefly discusses the different synthesis and characterization techniques and utilities of conducting polymer-based microbial fuel cells and highlights their future aspects. Figure 18.1 shows a schematic illustration of the details discussed in the chapter.

18.2   SYNTHESIS AND CHARACTERIZATION OF CONDUCTING POLYMER-BASED MICROBIAL FUEL CELLS

Apart from traditional techniques for producing polymeric membranes through conductive polymers (lamination, irradiation, and phase inversion), an innovative method was established in recent times established-plasma procedures for plasma polymerization for plasma adjustment of skin surfaces. However, proton interchange membrane for membranes for alkaline fuel cells (AFCs) and proton exchange membrane fuel cells (PEMFCs) could be achieved by applying this method. The study improved an anode electrode through the carbon fiber-inserted bacteriological polyaniline/cellulose (BC/CF/PANI) nanocomposite for metal fuel cell uses. In this case, carbon fiber was infolded onto the bacteriological cellulose fibers system for the duration of the bacteriological cellulose synthesis. The BC/CF/PANI was achieved utilizing PANI polymerization on the bacteriological cellulose nanofibers as a framework. To synthesis, the electrode, Fourier-transform infrared spectroscopy, scanning electron microscopy (SEM), thermogravimetric analysis as well as X-ray diffraction were conducted [5].

A sequence of polymer-derived ceramic (PDC) proton exchange complex membranes utilizing high cation transport numbers, huge ion exchange capacity (IEC) values, and low oxygen dispersal measurements have been manufactured at numerous pyrolysis temperatures employing a pressing method. These constituents consisted of polysiloxane matrix diverse utilizing proton-conducting plasters, which are SiO_2 /$H_3PMo_{12}O_{40}$ and montmorillonite at diverse proportions [6].

In the case of platinum, nanoparticles attached over decreased (reduced graphene oxide [rGO]) and graphene oxide/conductive PANI complexes were exploited and synthesized as anode catalysts in bacteriological fuel cells. PANI links Pt and rGO nanoparticles using the π–π stacking/electrostatic interaction hydrogen/force Pt–N bond and bonding, separately, and improved the essential steadiness of PANI/rGO/Pt complex. The electrocatalytic recitals of PANI/rGO/Pt displayed the lower internal resistance and better oxidation current over the organized PANI/rGO and Pt/rGO composites as demonstrated through the electrochemical impedance and cyclic voltammetric methods, respectively. However, the number of active catalytic sites and high electrical conductivity, the organized PANI/rGO/Pt nanocomposite showed a concentrated bacteriological fuel cells' influence thickness of 2059 mW/m^2 the tangible life stability [7].

The TiO_2/PPy composite has been efficiently synthesized through chemical polymerization approaches as photopolymerization, in situ polymerizations, molecular imprinting polymerization (MIP), and electrochemical polymerization. However, other research work shows real in decreasing the posse gap energy, that proposes a rise in the establishment of photoexcited electron-hole pairs and, therefore, a development of the light absorption by the help of visible region (400–700 nm). Additionally, the doping of TiO_2/PPy through noble metals increases the departure of charges in the semi-conductor constituent part, constraining the recombination of photogenerated electron-hole pairs [8]. Further, the investigation indicated that PANI, PEDOT (poly(3,4-ethylenedioxythiophene)) as well as PPy (polypyrrole) microhelices were invented through *Spirulina platensis* as a biotemplate. However, the effective establishments of the conducting polymer microhelix were inveterate employing SEM. Raman, X-ray diffraction (XRD), and Fourier transform infrared spectroscopy (FTIR) was engaged to modify the molecular arrangements of the conducting polymer in microhelical forms [9].

18.3 APPLICATION OF CONDUCTING POLYMER-BASED MICROBIAL FUEL CELLS

Microbial fuel cells are an approach based on the bioelectrochemical process for naturally converting waste into energy by using the metabolic activity of microbes, and these microbes serve as biocatalysts. Further, the microbial fuel cells have enormous advantages (Figure 18.2) and can be termed green technology, which converts waste into useful without hampering the environment, and this approach uses completely sustainable materials to fabricate this tool [10]. Nowadays, conducting polymers are widely used in fabricating microbial fuel cells owing to the extraordinary properties (Figure 18.3) of conducting polymers; moreover, currently, many microbial fuel cells based on conducting polymers and different microorganisms (Table 18.1) are utilized for generating energy and all of them generate a great amount of energy. Further, the electronically conducting polymers (ECPs) are auspicious materials aimed at understanding high-performance fantastic capacitors. Thus, conducting polymers which consist of PANI, PPy, polyindole, and polythiophene, are characteristic sample materials for supercapacitors founded on the Faradaic procedure refer to as pseudocapacitance. In specific, conducting polymers is exclusively appropriate for assessing their affluence of flexibility and fabrication [11]. Further, some of the maximum stimulating naturally stirring geometries, microhelical arrangements have involved consideration depending on

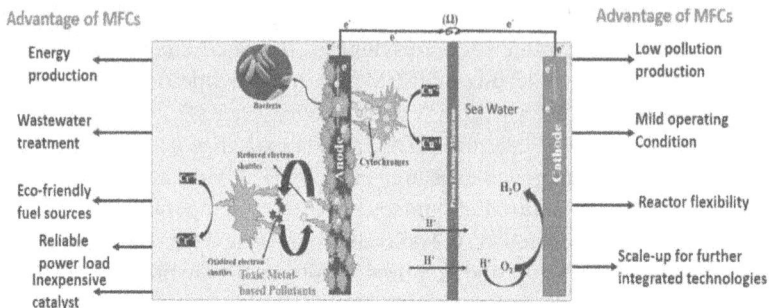

FIGURE 18.2 Diagrammatic illustration of general microbial fuel cells with their major advantages. (Reproduced with permission from Asim et al., 2020.)

their possible uses for fabricating microelectronic devices for biomedical utility. Conventional treating procedures for industrial microhelices are probably inadequate in mass productivity and cost, though *Spirulina* shows naturally acceptable microhelical systems that could be mass repeated at a tremendously low rate. Also, given the general usefulness of conducting polymers, it is absorbing to produce conducting polymer microhelices [12].

Carbon constituents like carbon nanotubes, graphene nanosheets, and carbon graphitic structures are widely applied as cares for electrocatalysts in fuel cells. Furthermore, conducting polymers that

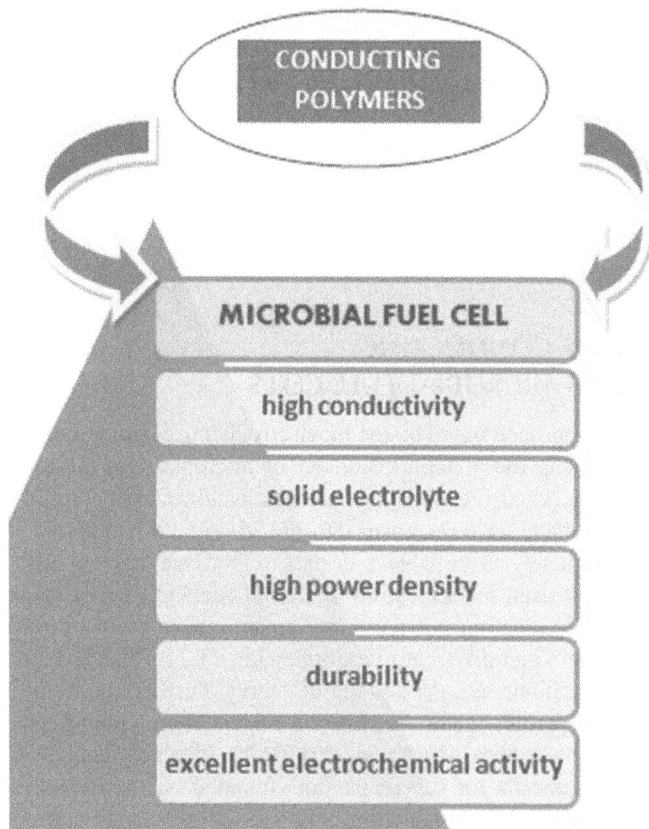

FIGURE 18.3 Properties accountable for the use of conducting polymers in microbial fuel cells.

TABLE 18.1
Power Density for Different Conducting Polymer-Based Microbial Fuel Cells and Strain of Microorganism Utilized

Microorganism	Anode Electrode Modification	Power Density (mWm^{-2})	References
Combined bacteria culture	Composite of polypyrrole and carbon nanotubes	1898	Tang et al. (2015)[a]
Shewanella oneidensis	Tartaric acid doped	490	Liao et al. (2015)[b]

[a] Tang X, Li H, Du Z 2015. "Conductive polypyrrole hydrogels and carbon nanotubes composite as an anode for microbial fuel cells". RSC Advances, 5(63). doi: 10.1039/C5RA06064H

[b] Liao Z, Sun J, Sun D 2015. "Enhancement of Power Production with Tartaric Acid Doped Polyaniline Nanowire Network Modified Anode in Microbial Fuel Cells". Bioresource Technology, 192, doi: 10.1016/j.biortech.2015.05.105

shown high chemical stability and ultra-high electrical conductivity have produced a concentrated investigation notice as catalyst sustenance aimed at polymer electrolyte membrane fuel cells and bacteriological fuel cells. Furthermore, metal oxides or metal catalysts might be immobilized on the functionalized polymer surface or the pure polymer to produce conducting polymer-based nanohybrids through enhanced catalytic stability as well as performance. However, metal oxides usually have an enormous surface area or/and permeable structures that showed exceptional synergistic effects through conducting polymers [13].

Improvement in technology is at all times insightful to improve the fossil fuel challenges in an eco-friendly system. A bacteriological fuel cell is supportable to green technology, which could enable bio-electrochemical changes. Therefore, conducting polymers have turned to be favorable materials for catalytic behavior, excellent electrochemical activity, and high conductivity, which have developed in the extreme demanding materials for the uses in a bacteriological fuel cell [14]. In the situation of conductive polymer membranes, they are carefully considered a result of the benefits obtainable through the fact that they purpose together as solid electrolytes and discerning separation obstructions for the classes obscure in electricity generation contained through the fuel cells. Major uses depended on production and attachment by the fuel cells arrangement of protons exchange membranes gotten through PTFE and per-sulfonic acid [15].

The emphasis on the enormous number of complex membranes depends on conductive polymers applied to construct large temperature proton interchange membrane fuel cells. Individually organic composite membranes due to the polymers through electric properties like sulfonated poly(ether ketone), sulfonated poly(p-phenylene), sulfonated poly(arylene ether sulfone), sulfonated polysulfone, sulfonated poly(sulfide ketone), sulfonated poly(aryl ether nitrile) as well as organic-inorganic composite membranes which consist of SiO_2/fluorinated polymer, Nafion/PTFE/zirconium phosphate, polyalkoxysilane/phosphotungstic acid and TiO_2/Nafion [16]. Further, four kinds of conductive polymers, which are copolymers poly(aniline-co-o-aminophenol) and polyaniline poly (aniline-1,8-diaminonaphthalene) as well as poly (aniline-co-2,4-diaminophenol), were used to transform carbon felt as the biocathodes in microbial fuel cells as well as aerobic abiotic cathodes. The copolymers through diverse functional groups outline had more special advantages in microbial fuel cells performance: PANOA and poly(aniline-co-2,4-diaminophenol) with –OH revealed less sensitivity to pH as well as dissolve oxygen change in cathode; poly(aniline-co-2,4-diaminophenol) and poly(aniline-1,8-diaminonaphthalene) through –NH$_3$ provided better attachment condition for biofilm that are capable their higher power output [17].

Studies have identified various applications of conducting polymer-based microbial like electrocatalysts through the utilization of carbon-based materials such as carbon nanotubes, graphene nanosheets, and carbon graphitic structures in fuel cells. These conducting polymer-based microbial cells display very significant high chemical stability and electrical conductivity for very important catalysts supporting system in the electrolyte membrane microbial fuel cells. Different microbes can be utilized as catalysts to convert metal/metal oxides into clean conducting polymers to produce nanohybrids functionalized polymer for enhanced stability and performance. Studies have shown that metallic oxides possess large surface areas, unique synergetic properties, and porous structures when combined with polymer-based microbial cells. These conducting polymer-based microbial cells are very stable, possess high bio/electro-catalyzing ability, and are environmentally friendly. Larissa Bach-Toledo et al. [18] reported that batteries and some supercapacitors are commonly utilized to develop electric vehicles and electronic devices due to their physicochemical properties like safety and reliability. The authors noted that these batteries and supercapacitors could be enhanced in performance through the combination of e electroactive materials in the construction processes. Conducting polymers-based microbial cells with transition metals oxides, carbon nanotubes, and graphene have been reported to improve nanorods, nanosheets, nanowires, nanoflowers, nanotubes, and nanofibers for batteries and supercapacitors. Through advanced technological development, environmentally friendly microbial fuel cells like green technology are sustainable approaches to replace fossil fuel problems and facilitate diverse bio-electrochemical conversion utilizing conducting polymer-based microbial fuel cells, which are now reliable, sustainable, and promising materials. Conducting polymer-based microbial cells possess high conductivity, power density, excellent electrochemical, efficient durability, high conductivity, and catalytic behavior. Conducting polymer-based microbial cells green technology is a promising approach for generating green electricity from biomass waste through an intense understanding of electrochemical processes for advanced power output generation and optimization [1]. Saranya and Jayapriya [3] reported that for the optimization and development of conducting polymer-based microbial cells green technology, carbon polymer-based composite electrodes for microbial fuel cells could be substituted with fluorine and nitrogen-based polymers, like polyacrylonitrile, polytetrafluoroethylene, PANI, polyacrylamide, and polydopamine, which possess large surface modification ability with active redox species for the improvement in effectiveness and biocompatibility and reduction in cathodes. The authors noted that these conducting polymer-based microbial cells green technology are bioreactors with the potential of converting chemical energy to electrical energy through biocatalytic reactions. Thus, a bacterium-catalyzed electrochemical device can produce current from the anode to the cathode surface. In fuel cell technology, microbial-based polymer membranes for energy transfer have become a subject of important research interest due to their biocompatibility, eco-friendly nature, energy production, consumption, and distribution approach. Kingshuk and Patit [19] reported that the alternative energy source for the next generation electronic and devices is the polymer electrolyte membrane fuel cells containing the microbial fuel cell. Several large available wastewater and sludge can be converted to electric energy through aromatic conducting polymers-based catalysts for enhanced distribution, anchoring catalysts, and dispersing electrode reactions. Over the past few decades, microbes have been revealed to have the capacity to generate an electric current through the utilization of extracellular electron transfer in the organic constituents. This is referred to as microbial fuel cells utilized in several applications like wastewater treatment, bioremediation, biosensors, desalination processes, and various renewable energy processes. The authors noted that microbial fuel cells involving sugar could be fermented into acetate, formate, butyrate, propionate, and lactate, with butyrate and acetate being the major acids in electricity generation. Again, Augustine et al. [20] reported that conducting polymers-based composites for batteries and supercapacitors like conducting polymer-based binary, quaternary, and ternary composites are utilized for several industrial applications in an energy storage system such as ionic lithium batteries, electronics, medical devices, and supercapacitors. Andriukonis et al. [21] reported that microbial biofuel cells and microbial amperometric biosensors are currently utilized for bioelectronics-based devices to increase the wider range of chemical fuels through efficient energy and charges transfer in the biofuel cell electrodes. The authors noted that improved charge transfer could be obtained utilizing polymers through immobilization of enzymes, optimal matrix, good cell performance, electron transfer enhancement, and cell leakage prevention.

18.4 CONCLUSION AND FUTURE RECOMMENDATION

Serious progress has been made in the development and utilization of conducting polymer-based composites across different industrial sectors. Several conducting polymer-based composites and materials are combined with nanoparticles representing a suitable candidate for electrocatalyst in microbial fuel cells. Even though more research work is still needed in this area, various efforts over the years have yielded positive responses in this regard. Thus, enhancement in microbial fuel cells' performance using polymer-based nanocomposites can scale up the charges transfer through immobilization of enzymes, optimal matrix, good cell performance, electron transfer enhancement, and cell leakage prevention. Furthermore, advanced simulation and modeling could be adopted to locate bioelectrochemical interaction by conducting polymer-based nanocomposites biofilms within the electrode. Thus in this chapter, we have highlighted recent advances in utilizing conducting polymer-based nanocomposites for energy applications, the synthesis and characterization of these materials using various techniques for flexible devices and novel electrode substrates, aiming at exploring low-cost, stable, and environmentally friendly materials with fascinating physiochemical properties.

ACKNOWLEDGMENTS

KRBS would like to express his gratitude for thanks to Professor A.K. Singh for providing constant support and guidance throughout this work. JS expresses his gratitude for the DST-INSPIRE faculty Fellowship, BHU (IoE grant), and UGC New Delhi to provide financial support. RPS is thankful to VC, IGNTU, Amarkantak, India, for providing constant financial support and motivating him to do good science.

REFERENCES

1. Rudra R, Pattanayak P, Kundu PP. 2019. "Conducting polymer-based microbial fuel cells. Enzymatic fuel cells". *Materials Research Foundations,* 44, 173–186. doi: http://dx.doi.org/10.21741/9781644900079-8

2. Wang J, Wang J, Kong Z, Lv K, Teng C, Zhu Y. 2017. "Conducting-polymer-based materials for electrochemical energy conversion and storage". *Advanced Materials,* 29(45), 1703044. doi:10.1002/adma.201703044

3. Narayanasamy S, Jayaprakash J. 2020. "Application of carbon-polymer based composite electrodes for Microbial fuel cells". *Reviews in Environmental Science and Bio/Technology,* https://doi.org/10.1007/s11157-020-09545-x.

4. Rossi R, Evans PJ, Logan BE. 2019. "Impact of flow recirculation and anode dimensions on performance of a large-scale microbial fuel cell". *Journal Power Sources,* 412, 294–300. doi.org/10.1016/j.jpowsour.2018.11.054

5. Trindade ECA, Antônio Ricardo RV, Brandes Letícia de Souza G, Neto A, Vargas VMM, Carminatti CA, de Oliveira D, Recouvreux S. 2020. "Carbon fiber-embedded bacterial cellulose/polyaniline nanocomposite with tailored for microbial fuel cells electrode". *Journal of Applied Polymer Science.* https://doi.org/10.1002/app.49036

6. Ahilan V, Wilhelm M, Rezwan K. 2018. "Porous polymer derived ceramic (PDC)-montmorillonite-$H_3PMo_{12}O_{40}/SiO_2$ composite membranes for microbial fuel cell (MFC) application". *Ceramics International,* 44, 16:19191–19199.

7. Kumar GG, Kirubaharan CJ, Udhayakumar S, Karthikeyan C, Nahm KS. 2014. "Conductive polymer/graphene supported platinum nanoparticles as anode catalysts for the extended power generation of microbial fuel cells". *Industrial & Engineering Chemistry Research,* 53(43), 16883–16893.

8. Amorim SM, Steffen G, de S Junior JMN. 2020. "Synthesis, characterization, and application of polypyrrole/TiO$_2$ composites in photocatalytic processes: A review." https://doi.org/10.1177/0967391120949489

9. Pu K, Ma Q, Cai W, Chen Q, Wang Y, Li F. 2018. "Polypyrrole modified stainless steel as high performance anode of microbial fuel cell". *Biochemical Engineering Journal*, 132: 255–261.

10. Yaqoob AA, Ibrahim MNM, Rodríguez-Couto S. 2020. "Development and modification of materials to build cost-effective anodes for microbial fuel cells (MFCs): An overview". *Biochemical Engineering Journal*, 164, 107779. https://doi.org/10.1016/j.bej.2020.107779

11. Magu T, Agobi A, Hitler L, Dass P. 2019. "A review on conducting polymers-based composites for energy storage application". *Journal of Chemical Reviews*, 1(1), 19–34. doi: 10.33945/SAMI/JCR.2019.1.1934

12. Hu X-Y, Ouyang J, Liu G-C, Gao M-J, Song L-B, Zang J, Chen W. 2018. "Synthesis and characterization of the conducting polymer micro-helix based on the *Spirulina* template". *Polymers*, 10(8), 882. https://doi.org/10.3390/polym10080882

13. Ghosh S, Das S, Mosquera MEG. 2020. "Conducting polymer-based nanohybrids for fuel cell application". *Polymers*, 12, 2993. https://doi.org/10.3390/polym12122993

14. Minteer S, Atanassov P, Luckarift HR, Johnson GR. 2012. "New materials for biological fuel cells". *Materials Today*, 15(4), 166–173. https://doi.org/10.1016/S1369-7021(12)70070-6

15. Fadhzir M, Kamaroddin A, Sabli N, Abdullah TAT. 2018. "Hydrogen production by membrane water splitting technologies". *Advances in Hydrogen Generation Technologies*. https://doi.org/10.5772/intechopen.76727

16. Dutta K, Kundu PP. 2014. "A review on aromatic conducting polymers-based catalyst supporting matrices for application in microbial fuel cells". *Polymer Reviews*, 54(3), 401–435. https://doi.org/10.1080/15583724.2014.881372

17. Li C, Ding L, Cui H, Ren H. 2012. "Application of conductive polymers in biocathode of microbial fuel cells and microbial community". *Bioresource Technology*, 116, 459–65. https://doi.org/10.1016/j.biortech.2012.03.115

18. Bach-Toledo L, Hryniewicz BM, Marchesi LF, Dall'Antonia LH, Vidotti M, Wolfart F. 2020. "Conducting polymers and composites nanowires for energy devices: A brief review". *Materials Science for Energy Technologies*, 3, 78–90.

19. Kingshuk D, Patit PK. 2014. "Application of polyaniline and polypyrrole, their derivatives and their nanostructures in direct methanol fuel cell technology". *Polymer Reviews*, 54(3), 401–435 https://doi.org/10.1080/15583724.2014.881372

20. Agobi AU, Louis H, Magu TO, Dass PM. 2019. "A review on conducting polymers-based composites for energy storage application". *Journal of Chemical Reviews*, 1(1), 19–34.

21. Andriukonis E, Celiesiute-Germaniene R, Ramanavicius S, Viter R, Ramanavicius A. 2021. "From microorganism-based amperometric biosensors towards microbial fuel cells". *Sensors*, 21, 2442. https://doi.org/10.3390/s21072442

19 Conducting Polymers as Membrane for Fuel Cells

Anuj Kumar[1], Ghulam Yasin[2],
Dipak Kumar Das[1], and Ram K. Gupta[3]
[1]Department of Chemistry, GLA University, Mathura, Uttar Pradesh, India
[2]Institute for Advance Study, Shenzhen University, Shenzhen, China
[3]Department of Chemistry, Kansas Polymer Research Center,
Pittsburg State University, Pittsburg, Kansas, USA

CONTENTS

19.1 INTRODUCTION

The fast-growing population demands economic growth, which in turn invites environmental imbalance and shortage of resources. These challenges need to be addressed for the sustenance of development across the globe [1]. Best on abundance, in the form of resources like natural gas, water and bio-materials, hydrogen can be considered as the best alternative to conventional fossil fuel [2]. In addition, hydrogen is also a clean source of energy capable of replacing petroleum products like diesel, kerosene, and petrol [3]. Therefore, fuel cells (FCs), in combination with water electrolyzers, can be considered as a clean and sustainable energy loop based on hydrogen energy that can meet the energy demand for modern society. FCs are a silent power resource free from carbon emission and noise pollution for the transformation of chemical energy into electrical energy. FCs have drawn huge attraction of scientific community due to their energy conversion efficiency, which largely depends on the proton exchange membrane (PEM) [4].

The membrane electrode assembly (MEA) is the heart of proton exchange membrane fuel cells (PEMFCs), comprising direct methanol fuel cells (DMFCs), and is responsible for the FC membrane efficiency. The materials, structure, and fabrication tools used in MEA components play a significant role in enhancing efficiency and optimization [5]. The catalyst membranes, for instance, are perhaps the most critical of the various aspects in PEMFCs because they are where the electrochemical reactions occur. An anode gas diffusion layer (GDL), an anode catalyst layer (ACL), a PEM, a cathode catalyst layer, and a cathode GDL are all contained in an MEA [6]. An ideal MEA would permit entire active catalytic sites in the catalyst layer to be available to the reactant (H_2 or O_2), protons, and electrons and would enable the effective removal of formed water from the catalyst layer and GDL [7]. Various catalyst layer/MEA structures and fabrication processes have been established during the last several years in an attempt to achieve the catalyst layer and MEA. As a consequence, while using various fabrication techniques [8], modifying the catalyst

DOI: 10.1201/9781003150374-19

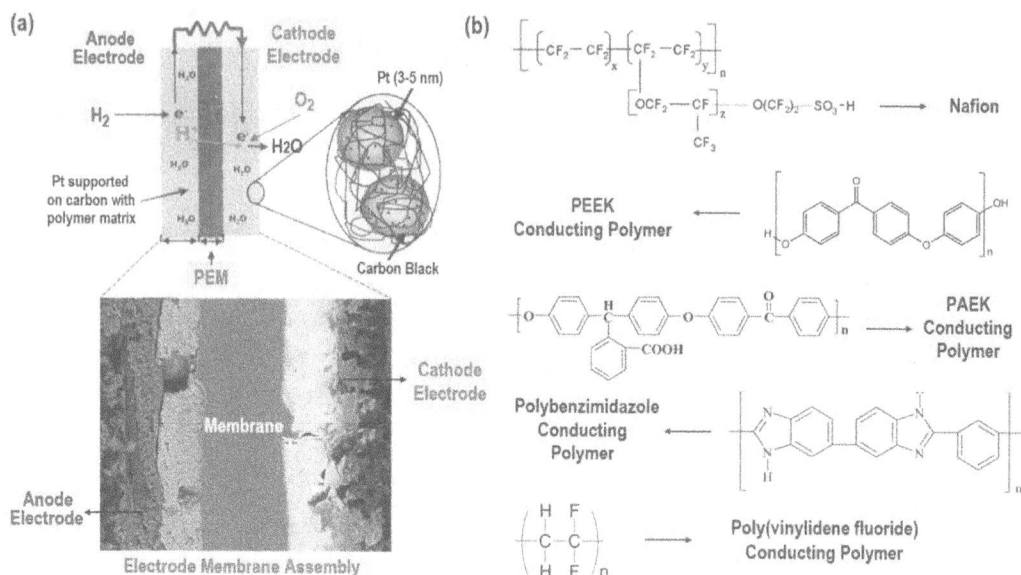

FIGURE 19.1 (a) Structure of FC along with TEM image of membrane electrode assembly. (Reproduced with permission from Ref. [9]. Copyright 2021, ACS Publishing Group.) (b) Chemical structures of various CPs used in fabrication of composite membranes.

layer architectures, and using specific membranes, MEA efficiency with advanced catalyst layers has been greatly improved. A systematic structure of FC along with a transmission electron microscopy (TEM) image of MEA is shown in Figure 19.1(a) [9].

A membrane is typically used to distinguish protons differentially by limiting fluid passage. Membrane assembly is a critical component of an FC that ensures its long life and good efficiency. Nevertheless, the proton transfer mechanism may be hampered by the membrane's intrinsic resistance that could be reduced by enhancing membrane conductivity. Therefore, researchers have been keenly engaged to find a suitable material for fabricating PEM for the practical application of FCs, DMFCs in particular, for several decades [10].

The most extensively used membranes are Dow XUSR©, FlemionR©R, poly(vinylidenefluoride) [11], poly(benzimidazole), sulfonatedpoly(etheretherketone) [12], Gore-TexR©, Gore-SelectR©, Aciplex-SR©, and NafionR©. Perfluorinated copolymeric backbone having –SO$_3$H hydrophilic functional groups, recognized as Nafion membrane possessing high proton conductivity, is the most widely utilized standard material for PEM, developed by DuPont laboratory. Tetrafluoroethylene (TFE) and a derivative of a perfluoro (alkyl vinylether) were firstly subjected to copolymerization with sulfonyl acid fluoride to blend Nafion©R derivatives. Nafion©R possesses exceptional ionic properties along with its thermal and mechanical stability due to the incorporation of perfluorovinyl ether (PFVE) groups terminated with –SO$_3$H groups onto a TFE [13]. A number of varieties of NafionR© (–105, –112, –115, –117, –1035, –1110, and –1135) having main difference in chain length and equivalent weight and membrane thickness are commercially available [14]. However, the limitation of high cost, alternation in pH, and sensitivity toward bio-fuels have restricted its application in FCs, which has inspired researchers to look for alternative membrane materials with enhanced H$^+$ conductivity and reduced methanol (MeOH) crossover, which has remained by and large elusive.

In this regard, various new membranes, like polyacrylonitrile [15], poly(hexafluoropropylene), poly(methylmethacrylate) [16], poly(styrenesulfonate) [17], polybenzimidazole, polypyrrole (Ppy) [18], poly(vinylidine fluoride) [19], polyaniline (PANI), and polythiophene (PTh) incorporated composite membranes were established and maintained their appropriateness for high thermal applications in DMFC. In this category are conducting polymers, capable of transferring electrons and facilitating the proton transfer channel in their composite membranes [20]. The chemical structures of various CPs used in the fabrication of conducting polymer membrane are illustrated in Figure 19.1(b).

CPs may play a significant role as material for the preparation of composite membranes, which are very similar to reported efficient membranes having good mechanical strength and elasticity. A membrane made of proton-conducting polymer is basically a tinny sheet of polymer having an extensive network of hydrophilic nanopores implanted in the hydrophobic structural domain. Moreover, a conducting polymer-based membrane offers a novel kind of advanced material due to its chemical properties being useful for the separation or transport of several chemical species.

This chapter deals with the fabrication of PEM using CPs, for example, PANI, and PPy and their derivatives, showing the significant electrocatalytic performance of DMFCs. Moreover, in this chapter, several critical points like H^+ conductivity, gas diffusion, methanol crossover, and thermal stability of conducting polymer-based membranes are highlighted for DMFCs. Further, the challenges and future prospects with PEM construction using CPs are also discussed.

19.2 FACTORS AFFECTING THE PERFORMANCE OF A DMFC

A variant form of PEMFCs is the DMFC. The DMFC uses lighter alcohols like MeOH or ethanol instead of hydrogen gas. In DMFCs, the anode is composed of mostly Pt/Ru/C catalyst and the MeOH solution (usually 1–2 M) in H_2O is fed in at the anode as a fuel. Only a small portion of MeOH is used as fuel continually being fed through to maintain the constant MeOH/H_2O ratio at the electrode. DMFCs operate at 60°C–90°C and have the same layout as PEMFCs, and the important condition for obtaining high power density is to increase the temperature of the cell, which eliminates the kinetic hindrance to methanol electro-oxidation; however, PEM dehydrates at temperatures above 100°C and develops resistivity problems [21]. Figure 19.2(a) shows the systematic structure, components, and chemistry of DMFCs.

The alcohols dissociate to electrons, H^+ ions, and carbon dioxide at the anode. The positively charged H^+ ions diffuse from anode to cathode through the PEM, and electrons migrate toward the cathode through an external circuit. The DMFCs are operated at lower temperatures and are thought to be the power source in the future for portable equipment like laptops, calculators, etc. At the cathode, electrons, H^+ ions, and oxygen (from the air) react with each other and form portable water. The reactions involved in DMFCs are shown below (Equations 19.1–19.3) [22].

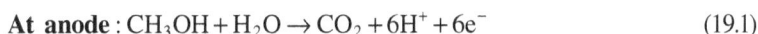

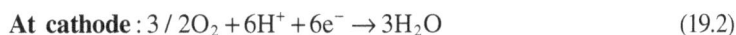

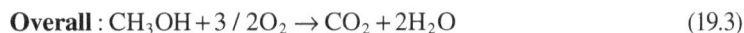

$$\textbf{At anode}: CH_3OH + H_2O \rightarrow CO_2 + 6H^+ + 6e^- \tag{19.1}$$

$$\textbf{At cathode}: 3/2O_2 + 6H^+ + 6e^- \rightarrow 3H_2O \tag{19.2}$$

$$\textbf{Overall}: CH_3OH + 3/2O_2 \rightarrow CO_2 + 2H_2O \tag{19.3}$$

Usually, Ru metal enhances the activity of Pt for the MeOH oxidation at anode because MeOH catalytic activity is low as compared to H_2. This occurs because OH forms at a lower voltage on Ru compared to Pt–OH, which is needed to oxidize intermediate adsorbents. However, Pt is a good catalyst for MeOH oxidation, but it leaves behind CO on the catalyst surface, blocking active sites for further oxidation, as shown in Figure 19.2(b) [23]. Therefore, Ru is added to help to remove CO from the active sites by oxidizing the CO into CO_2.

Thus, low operating temperatures, no dependency on electricity, refueling option, high durability, and the eco-friendly MeOH fuel having a specific energy density make DMFCs an efficient alternative to rechargeable batteries. Besides, several other factors such as simple construction, comparatively fast startup, and response to variations in the loading of DMFCs make it more preferable over traditional rechargeable energy devices. However, in spite of having several preferential qualities, DMFCs have two limitations of MeOH crossover and proton conductivity of used membrane, which hinder its practical and commercial applications. In this section, attempts have been made to identify the reasons for these limitations and their possible remedial aspects,

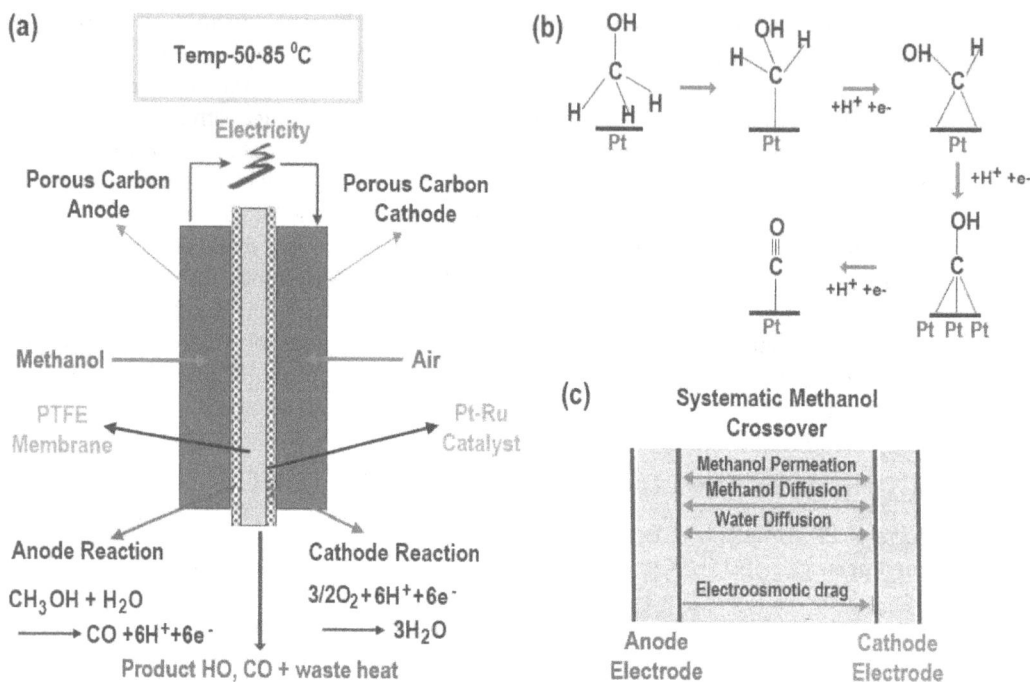

FIGURE 19.2 (a) Systematic structure, components, and chemistry of DMFCs, (b) the mechanism of MeOH oxidation and Co-poisoning at Pt catalyst, and (c) a systematic representation of MeOH crossover in DMFCs.

Generally, MeOH crossover results from the cumulative effects of electro-osmotic drag by H+-transport, diffusion by concentration gradient of MeOH, and convection between the electrodes, as shown in Figure 19.2(c) [24]. The H+ transportation from anode to cathode, while moving through the PEM, pushes the H_2O molecules along with it in the hydration sphere. As methanol gets dissolved in water and is applied as fuel, it is also carried through the PEM from anode to cathode, leading to MeOH crossover [25]. The MeOH crossover, a grave limitation due to undesired MeOH oxidation at the cathode, results in fuel loss; poisoning of the catalytic active sites on the cathode surface; and reduction in fuel efficiency, electrode potential, and current density. Almost 70% of the chemical energy of MeOH gets wasted due to the MeOH crossover and irreversible side reaction on the anode, as reported earlier [26]. The MeOH dilution may be an option to decrease the level of crossover; unfortunately, this action may result in an increase in cell stacked dimension proportionally, resulting in uncertain relation of dimensions and energy demands for portable uses. Additionally, MeOH crossover may bring sluggishness of oxygen reduction reaction at the cathode surface.

On the other hand, to have good H+ conductivity, PEM needs to get swollen significantly with water, which again invites high MeOH crossover [27]. Hence, attaining a PEM with suitable fabrication, satisfying high H+ conductivity, and minimum MeOH crossover is a practical challenge. Therefore, researchers across the globe are actively engaged to address the issues related to MeOH crossover to enhance the performance of DMFCs.

The linking between membrane permeability and MeOH crossover is quite complicated and is affected by the nature and morphological aspects of the material being used for PEM fabrication, PEM thickness, temperature, and methanol concentration [28]. It was observed that MeOH crossover was reduced with increasing membrane thickness due to the enhancement of flow resistance, resulting in a reduction in output current density. In the same way, MeOH crossover and H+ conductivity are favored with the increasing cell operating temperature. In addition, the increase in the feed of MeOH leads to an increment in current density as well as in MeOH crossover [29].

Furthermore, nanoscale segregation into hydrophobic (due to organic backbone of PEM) and hydrophilic (due to the cumulative effect of $-SO_3H$ groups) subphases upon exposure to water results due to the typical nanoarchitecture of PEMs. The need for MeOH-tolerant catalysts arises as mixed cathodic potential created due to the migration of MeOH through the PEM as well as ineffective counter-current flow of MeOH toward itself and CO_2 away from anodic active sites, which eventually results in scale-up to high power output as well as longer retention time [30].

Simultaneously CO_2 gas is captured in the pores of the anode, resulting in higher inducing MeOH and CO_2 crossover [31]. The increment in MeOH crossover leads to higher cell temperature, which initiates enhanced reaction kinetics in the electrodes. Moreover, NafionR©, the most widely used fabricating material for PEM, has severe limitations like MeOH and co-catalysts crossover, high expenditure, limiting operating temperature less than 80°C, and requirement of huge humidification [32]. In the next section, we will highlight some low-cost and efficient CPs that could be used as PEM materials.

19.3 RECENT ADVANCES IN THE USE OF CPs AS PEM MATERIALS

Usually, the NafionR© is the traditional fabrication material for various membranes; however, it has the limitations of MeOH crossover and effect on H^+ conductivity. In this regard, to address these issues, two approaches, namely (i) modification of NafionR© and (ii) development of ionic CPs, are being studied [33].

Modification of NafionR©: In this approach, attempts were made to minimize the MeOH crossover by blocking the channel by embedding the methanol-repellant species like aromatic moieties in the pores of NafionR© [34]. A similar approach successfully employed using aromatic CPs by several laboratories is discussed in this section.

Development of CPs Membrane: In this approach, alternative CPs and their composites are tried as suitable membrane constituents. CPs like PANI possess repeating imine groups, which are weak Lewis acid, and hence conducive for H^+ transportation. The location of the PANI layer within the PEM matrix determines the performance of PEM. Moreover, the high electronic conductivity of PANI makes it an important material for the fabrication of PEM. Usually, the resulting PEM is likely to be an insulator to conduct only H^+ ions. However, the direction of H^+ flow is from the anode to the cathode through the PEM; meanwhile, the electrons from the anode to the cathode via an external circuit make a complete loop of electricity generation [35].

As the structural material is a polymeric insulator, their blending with CPs thus reduces conductivities more than when using structural CPs. In this situation, if grafting of CPs is made on the insulating material surface, there is a minute possibility of adverse effects on the electronic properties of CPs. CP materials, at their oxidized state, are good conductors, but they behave perm selective toward cation on being neutral. In addition, the high chemical and electrochemical stability of CPs promote their application in corrosion-sensitive environments in FCs. However, an electronically conducting network should possess CP loading below the percolation threshold to avoid the short-circuit of electrodes in FCs [36]. The chemical structure of some very useful CPs is shown in Figure 19.3(a). Herein, we highlighted the applications of a few CPs as attractive materials for PEM construction.

19.3.1 PANI-BASED MEMBRANES

It is well known that H^+ passage by PEM proceeds following one of the mechanisms, namely the Grotthuss mechanism or vehicular mechanism. Although, the Grotthuss mechanism is reported to be more prevalent, in which excess H^+ or H^+ defects tunnel is constructed by the water molecule through an H-bonding network. The excess H^+ ions are generated due to the transfer of H^+ from H_3O^+ donor to any neighboring water acceptor leaving H_2O molecules [37]. This process, although faster than translational diffusion, is yet sluggish due to the lack of neighboring water acceptor sites as well as energy barriers in breaking the H-bonds.

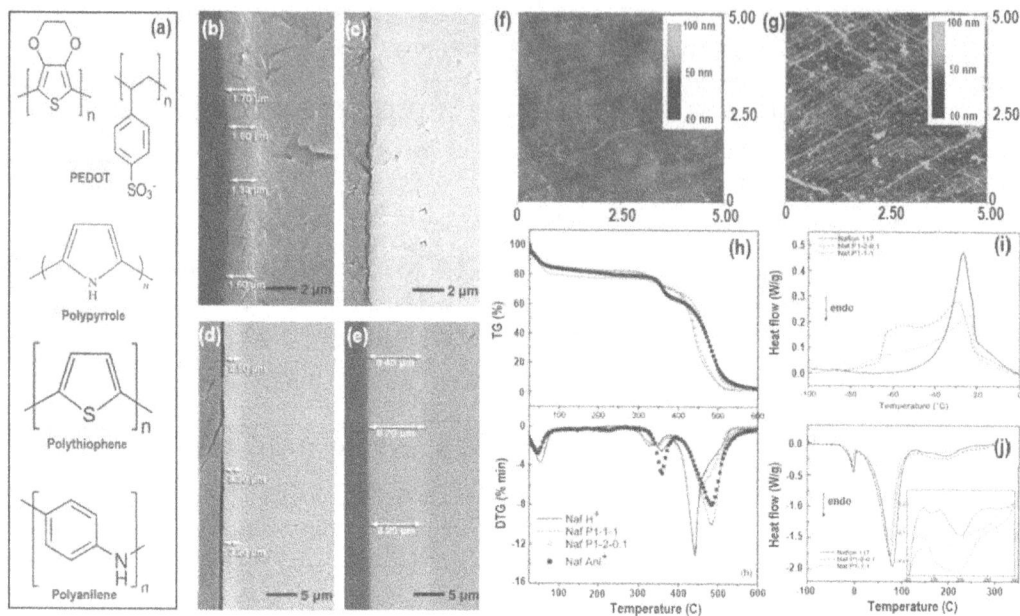

FIGURE 19.3 (a) The chemical structure of some very useful CPs like poly(3,4-ethylenedioxythiophene) (PEDOT), PPy, PTh, and PANI. Cross-section SEM images of (b) PANI/Nafion 117, (c) PPy/Nafion 117, (d) PANI/Nafion 117, and (e) PPy/Nafion 117. (Reproduced with permission from Ref. [40]. Copyright 2021, ACS Publishing Group.) The AFM images of (f) Nafion 117 and (g) Naf P1-1-1. (h) TGA and DTG curves of Nafion 117 in its protonated form (Naf H^+), Naf P1-1-1, Naf P1-2-0.1, and the differential scanning calorimetric curves at first cooling (i) and heating (j) curves of Nafion 117, Naf P1-2-0.1, and Naf P1-1-1 membranes. (Reproduced with permission from Ref. [41]. Copyright 2021, Elsevier Publishing Group.)

Whereas vehicular mechanism depends on the movement of excess H^+ via solvent on top of host molecules, the PEM containing SO_3H groups acts as an acid as on dissociation SO_3H groups provides $-SO_3^-$ and H^+ ions. The counter $-$ive ions get adsorbed into the pores of the nanostructure of the membrane while H^+ gets attached to water moieties. Now, it is understood that in equimolar acid-doped amine electrolyte, the Grotthuss mechanism is followed for H^+ conduction due to the deprotonation–protonation processes during H^+ transfer [38].

Moreover, the activation energy of H^+ transportation is affected partly by separating the distance between acid and amine sites. In order to improve the conductivity, a higher ratio of acid/amine groups in PEM (which reduces their distance) is therefore desirable [39]. In a study conducted at 120°C, it was found that phosphoric acid (H_3PO_4)-doped linear meta-PANI showed low humidity and H^+ conductivity. The higher humidity initiates the H^+ conductivity of the membrane in the Grotthuss mechanism and/or H^+ conduction phenomenon during vehicular mechanism. Moreover, in another investigation, the effect of the oxidation state of PANI on the physicochemical and transport properties of Nafion/PANI composite membranes were found [40] as supported by open-circuit voltage (OCV) values.

The Nafion/emeraldine salt oxidation form of PANI (Nafion-ES) also displayed the highest OCV, peak power density, and selectivity among all the membranes tried. Moreover, on electrochemical modification with PANI-ES, Nafion membrane showed the MeOH permeability reduction of 10 times at the cost of limited decrement in H^+ conductivity [41]. A larger reduction in MeOH permeability with simultaneous reduction of H^+ conductivity was observed during the modification of the membrane by three methods: at constant current and potential and under potential cycling. However, a huge reduction in MeOH permeability with a low decrease in H^+ conductivity at constant current was noticed as compared to other methods. Modified Nafion membrane is superior to pure Nafion membrane by 40% in terms of power density output [40].

Huang et al. [39] reported PANI-modified Nafion PEM, which showed 50% less MeOH permeability as compared to the PPy-modified counterpart with higher H^+ conductivity than the latter. Upon further modification by polymerizing aniline using ammonium peroxodisulfate, the Nafion membrane contained a higher proportion of oligomers, with most of PANI being formed inside the Nafion 117. The surface topography of Nafion indicated a change with heterogeneous PANI coating containing some aggregates having fibril-like structures. The AFM investigations suggested the presence of PANI covering the Nafion surface partially.

The PPy was connected poorly to SO_3H groups of Nafion membrane because of its weaker basic nature compared to PANI as the result of an experiment on differences between PANI/Nafion 117 and PPy/Nafion 117 composite membranes conducted by Schwenzer et al. [40]. Generally, PPy is hydrophobic and aggregated at the focal point of Nafion channels and does not get attached with $-SO_3H$ groups of hydrophilic wall as suggested by IR studies. On the other hand, PANI is found to form a strong bond with $-SO_3H$ groups along the side of Nafion pores and its surface, resulting in less obstructed water transport and H^+ conduction as compared to PPy. Figure 19.3(b–e) clearly indicates the morphological features of SEM images of PANI/Nafion 117 and PPy/Nafion 117 membranes, and the PPy coatings on Nafion appeared in SEM images.

Sophie Tan et al. [41] developed Nafion by chemical condensation of aniline with ammonium peroxodisulfate as the oxidizing agent. In Figure 19.3(f, g), the AFM illustrations of the untreated and treated Nafion (Naf P1-1-1) that demonstrate certain similarities, including the existence of linear and particulate frameworks, are given. Thermogravimetric analysis (TGA) and differential thermogravimetric (DTG) profiles of Nafion, Naf P1-1-1, and NafionP1-2-0.1 obtained in an air atmosphere are shown in Figure 19.3(h). The transitions in the Nafion DTG curve could be found at temperatures of 25°C–250°C, 290°C–390°C, 390°C–475°C, and 475°C–545°C, which correspond to the four degradation phases defined in the literature. Attributed to the prevalence of air, only the PTFE framework disintegrates at a low-temperature region. During aniline condensation on Nafion membranes, it is first reported that the modified membranes comprise less water (18%) than bare Nafion membranes (23%). The first cooling curves obtained with the Nafion and Nafion/PANI composite membranes are shown in Figure 19.3(i, j). Nafion has an exothermic peak of about –20°C, which is related to water crystallization. This crystallization temperature is consistent with literature values. It can be categorized as binding freezable because the crystallization temperature is below 0°C [42].

Further, the X-ray photoelectron spectroscopy (XPS) results indicated the well-characterized N1s signal having 399.6 eV, and 400.8 eV, 402.3 eV peaks corresponding to $-NH-$ and positively charged nitrogen, respectively. A comparative investigation of S_{2p} spectra of modified and unmodified Nafion membrane predicted shifting of binding energy toward a lower value of PANI, suggesting an interaction between $-SO_3^-$ and $-NH^+$ or $-NH_2^+$ functional groups. If the PANI layer covers the $-SO_3$ groups of Nafion membrane at its surface, S_{2p} intensity is lowered. The XPS spectra of O1s exhibit two peaks at 533.0 eV and 535.3 eV, which can be attributed to O of $-SO_3$ groups and ether oxygen, respectively. O1s peak intensity should have been very low as PANI does not contain any O-atom. However, curve fitting data of O1s signal indicated the presence of O-components at 534.1 eV and at 531.8 eV, which can be attributed to sulfonate oxygen compensated by the charged imine groups from PANI and to bicarbonate ions formation in the presence of air and water and their adsorption on PANI, respectively [43].

On the other hand, the integral adhesive nature of PANI also plays a crucial role as it has been reported that a PANI layer deposited on Nafion 117, Nafion 115, and Nafion 112 membranes displayed very high adhesion properties to the Nafion surface. Although uniformly spread over the entire surface of the membrane, the thickness of the coating was reported to be non-uniform across the surface. Methanol crossover is reported to be significantly more suppressed than the corresponding H^+ conductivity due to the penetration at a depth of 2% of the Nafion membrane as the polymer layer is of globular structure [44]. Further, the result of the study found PANI to increase the membrane's thermal stability by enhancing the hydration temperature and destructing the sulfonic group in Nafion, resulting in significant protection from degradation of the membrane [45]. Further, the morphological studies indicated that after 24 hours of polymerization, the coating thickness for PPy/Nafion 117 was in the range of 8.2 μm to 8.7 μm, greater than the coating thickness for the PANI/Nafion 117 prepared under similar experimental conditions. It can be concluded based on the above observations that aniline went through rapid polymerization on Nafion 117 surface. However, homopolymerization of PPy is reported to undergo rapidly as compared to PPy/Nafion 117 hetero-polymerization, as well as that of PANI homopolymerization.

The PANI and Nafion 112 membranes in combination displayed far better H^+ conductivity in comparison to Nafion 112, only at 60% relative humidity. Bhadra et al. [46] fabricated a self-cross-linked poly(benzimidazole-co-aniline), where aniline is reported to display the role of an extra basic unit (improving the H^+ conductivity) and cross-linker between benzimidazole moieties (enhanced the mechanical strength). Further, its modification via 45 wt% H_3PO_4 doping exhibited stress at break of 26±3 MPa and an H^+ conductivity of 0.17 Scm^{-1} at 120°C and 100% relative humidity [47].

19.3.2 PPy-Based Materials

The microbial FC utilizing PPy PEM is reported to have better performance in comparison to others, as internal polymerization offers more conductive azoles moieties to get attached to the PEM to improve its conductivity. Therefore, PPy can be utilized as conductive materials to fabricate MFC membranes to improve their overall performance. Recently, an in situ polymerized Pt-anchored pyrrole moieties-based Nafion matrix membrane was utilized in FCs. The analytical techniques suggested that the incorporation of positively charged pyrrole moieties played a crucial role in the linking of Pt nanoparticles (NPs) with the Nafion matrix through mutual interaction to bring morphological change in membrane surface, showing good physical and transport properties in FCs [48].

The results of another comparative study between PPy-modified Nafion 115 and Nafion 115 indicated better performance for the former under similar experimental conditions [49]. The improved performance may be because of decreased pore volume due to the presence of PPy in the pores. However, the interaction between SO_3H groups of Nafion and positively charged PPy may lead to a decrement in the mobile H^+ ions concentration, which eventually results in higher membrane and cell resistance. This drawback will nullify the advantages of decreased methanol crossover in a DMFC operating at higher current densities.

In contrast, upon conducting an experiment, Park et al. [50] suggested that PPy NPs were present near the surface of the Nafion membrane, not in the internal spaces, which favored the reduction in MeOH crossover (Figure 19.4a). Figure 19.4(b, c) depicts the SEM images of the prepared composites. Further, it was suggested that the interaction of polar phases of Nafion and the secondary ammonium groups of PPy was responsible for the enhancement in the mechanical and thermal properties of the composite Nafion/PPy membrane.

Xu et al. [51] reported the polymerization of pyrrole to take place at the boundary of hydrophilic and hydrophobic domains of Nafion 115 using solvents of high dielectric constants. Usually, the chemical polymerization results in the formation of PPy-modified Nafion 115 membranes with higher loading of oligomeric components when compared with electrochemical polymerization. It was also noticed that oxidants affect the penetration power of PPy, H_2O_2 being offered better results with reducing 30% in methanol permeability. The modified membrane, however, exhibited a minor reduction in H^+ conductivity as compared to Nafion 115 membrane. Moreover, the transport properties of this modified membrane are greatly affected by its morphology.

Moravcova et al. [52] reported a potential diffusion strategy to fix PPy on the membrane surface, separating the solutions of monomeric pyrrole and peroxodisulfate by Nafion membrane. The results indicated homogeneous morphology having strong adhesion and electrochemical activity. It was observed that PPy successfully restricted pyrrole permeation affecting the rate-determining polymerization of the secondary PPy layer. PPy NPs with a particle size of 20–40 nm attached to the Nafion membrane changed the morphology of the membrane significantly [53] due to the electrostatic interaction between SO_3H groups on Nafion and $-NH_2$ groups in PPy. These PPy NPs

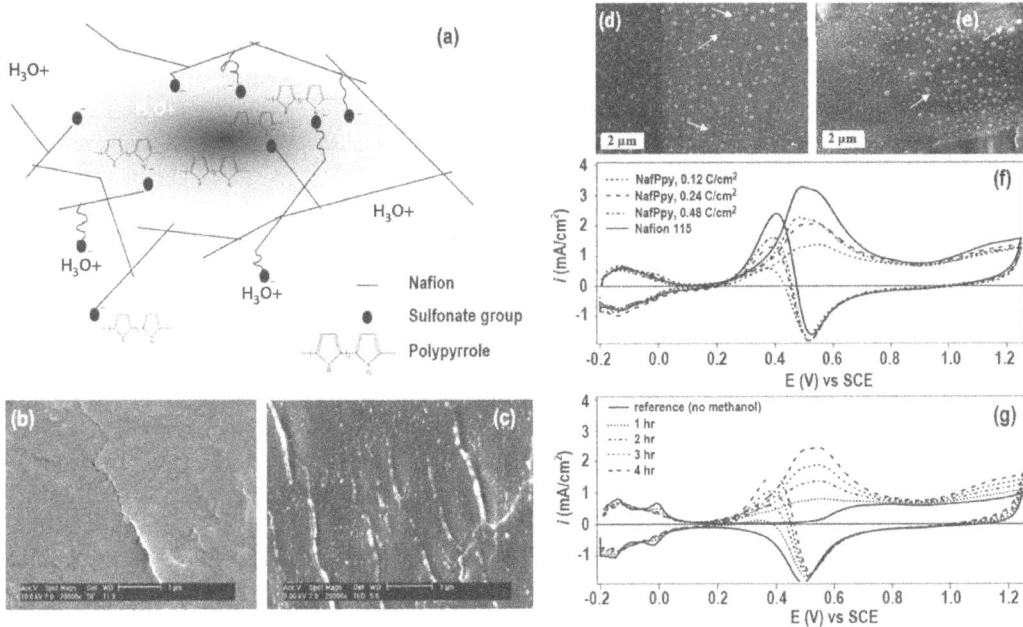

FIGURE 19.4 (a) Structure of the Nafion/polypyrrole composite membrane and SEM images of (b) Nafion and (c) N/P 002. (Reproduced with permission from Ref. [50]. Copyright 2021, Elsevier Publishing Group.). (d–e) SEM images of Nafion/Ppy membrane and the results for MeOH permeability tests for four different Nafion 115 membranes (f) and for the 0.12 C cm^{-2} Ppy membrane (g) (Reproduced with permission from Ref. [57]. Copyright 2021, Elsevier Publishing Group.)

influenced the Nafion backbone by obstructing its crystallization, affecting transportation as well as MeOH crossover [54]. Deposition of Pt catalyst on PPy-modified Nafion membranes through the reduction of Platonic chloride was developed by Li et al. [55]. The morphological characterization using SEM suggested a uniform depression of Pt NPs on the membrane surface.

Better selectivity for H$^+$ over methanol transportation, including efficient FC performance, was reported by Zhu et al. [56] using acidic and neutral Nafion 115 and Fe(III) and H$_2$O$_2$ as oxidants. The results indicated Fe(III) species to drastically the methanol crossover, but simultaneously, deposition of PPy on Nafion 115 membrane was also reported, which eventually enhanced membrane resistance leading to unsatisfactory performance due to poor bonding with the electrodes. However, H$_2$O$_2$, when used as an oxidant, displayed encouraging results through the eventual removal of PPy deposition. Moreover, the tetrabutylammonium form of Nafion exhibited low methanol permeability with locating the PPy on the Nafion structure. A reported study on PPy/PSS-modified Nafion membrane reveals that chemical polymerization produced more than ten times porosity when compared with electrochemical polymerization and, thereby, ten times more ionic conductivity.

The PPy, an aromatic-based CP, has been extensively used to fabricate PEMs for FCs. For example, Smit et al. [57] prepared the modified Nafion membranes by in situ electrodeposition of PPy within membrane pores and anode side so that MeOH crossover could be monitored in DMFCs. The modification of Nafion with PPy and studied by SEM (Figure 19.4d, e) suggested the presence of PPy in both Nafion surface as well as in its pores, having sizes of 100 nm and 700 nm, respectively. Specifically, small grains were observed on the membrane surface and pores while larger grains appeared outside the membrane, facilitating the reduction of methanol permeability [58]. An investigation was carried out to assess the effect of electrolytes on the output current density, which indicated the highest current density in MeOH/H$_2$O mixture followed by methanol alone, with the lowest being MeOH/H$_2$SO$_4$ mixture. In addition, with methanol alone, a peak appeared, which might be due to MeOH oxidation by PPy (Figure 19.4f, g).

Further, another class of CPs, sulfonated poly(ether ether ketones) (SPEEK) and poly(arylene ether ketones), attracted the concerned community to be used as potential PEM materials. The chemical modification of SPEEK by PPy produced a reduction in solution uptake and swelling ratio of SPEEK membrane [59]. Higher methanol concentration, however, brought increased value for both the parameters, which decreased subsequently. The modified membrane, however, is reported to undergo a decrease in mechanical stability in 10 vol% methanol above 60°C. The incorporation of PPy in the transport channel brought about a decrement in permeability and H^+ conductivity. However, the SPEEK PPy combination displayed a superior selectivity value than both pure SPEEK and pure Nafion 117 membranes individually.

Li et al. [60] fabricated $-CH_3$-substituted SPEEK membranes incorporating PPy, and the results showed that the diffusion coefficient for methanol was reduced significantly at 25°C due to the collaboration between PPy and SPEEK. Further, Xu et al. [61] prepared a few SPEEK/phosphotungstic acid (PWA)/PPy membranes by surface modification with PPy, which resulted in reduced methanol transport, maintaining a reasonable H^+ conductivity and better mechanical and thermal properties.

Lin et al. [62] fabricated a SPEEK membrane incorporating sulfonate and carboxyl groups via a layer-by-layer strategy. The morphological studies showed the surfaces of two bilayer and 10-bilayer PPy/PWA films, indicating the presence of PPy grains of diameter 50 nm to 700 nm. These PPy grains were smaller in size and lesser in number as compared to the ten-bilayer counterpart, and the methanol diffusion value was one order of magnitude lower than pure SPEEK-C and pure Nafion 117 membranes.

19.3.3 OTHER CONDUCTING POLYMERS

Poly(3,4-ethylenedioxythiophene) (PEDOT), a member of PTh class, is a widely used CP-based membrane material. Chemically modified Nafion 117 using polymerized 3,4-ethylenedioxythiophene showed Arrhenius-type methanol permeability dependency and H^+ conductivity. The SEM images of the membrane surface for the Nafion 117 membrane are shown in Figure 19.5(a, b). This

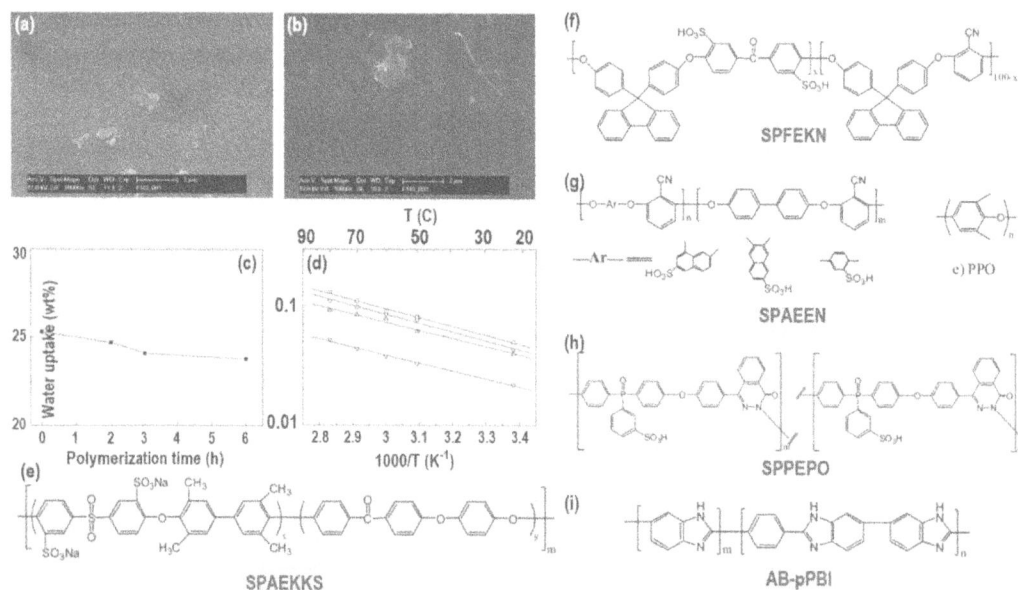

FIGURE 19.5 SEM images of the membrane surface for (a) Nafion 117 membrane, (b) PEDOT-modified Nafion membranes of polymerization after 2 h, (c) water uptake of PEDOT-modified Nafion membranes, and (d) proton conductivity of PEDOT-modified Nafion membranes as a function of temperature. (Reproduced with permission from Refs. [63, 64]. Copyright 2021, Elsevier Publishing Group.) (e–i) The chemical structure of SPAEKS, SPFEKN, SPAEEN, SPPEPO, and AB-pPBI CPs [66].

modified membrane displayed higher methanol permeation than the unmodified membrane with similar activation energy for H+ migration. At 60°C, methanol permeability of the modified membrane underwent a reduction from 72% to 30%, and it displayed a maximum power density of 48.4 mWcm^{-2} compared to the unmodified Nafion membrane (37 mWcm^{-2}) under an identical situation (Figure 19.5c, d) [63, 64].

Li et al. [65] also reported a similar reduction in methanol permeability (about 75%) of PEDOT-modified Nafion 117 composite membrane in comparison to unmodified Nafion 117 with no significant variation in H+ conductivity. Higgins et al. [66] fabricated PEDOT coated Nafion 117 and PSS-grafted poly(vinylidene fluoride) (PVdF) composite membranes and reported the PEDOT layer to have tightly attached toward both the membranes with high H+ conductivity. Besides, PVdF, a potential conducting polymer, can be suitably manipulated with Nafion to fabricate polymer electrolyte membranes.

Furthermore, several CPs like sulfonated poly(aryl ether ketone) (SPAEKS), sulfonated poly(fluorenyl ether ketone nitrile) (SPFEKN), sulfonated poly(arylene ether ether nitrile) (SPAEEN), sulfonated poly(phthalazinone ether phosphine oxide) (SPPEPO), and a copolymer of poly(2,5-benzimidazole) (AB-pPBI) have been to be very useful materials to fabricate various Nafion-based membranes [64]. The chemical structures of SPAEKS, SPFEKN, SPAEEN, SPPEPO, and AB-pPBI CPs are shown in Figure 19.5(e–i).

19.4 CONCLUSIONS AND FUTURE PROSPECTS

The PEMFCs have been described as renewable energy conversion devices devoid of C-emission and have attracted the scientific community working in this field due to their high proficiency and high energy density, as well as their potential application in transportation. However, FCs still have the challenges of high cost of noble metal (Pt)-based catalyst materials and membrane and catalyst layer designs. Therefore, low-cost materials and membrane and catalyst layer designs are highly sought for the development of sustainable commercialization of PEMFCs.

In this chapter, efforts have been made to cumulatively highlight the significant outcomes of several outstanding works on CPs being utilized as potential fabricating materials for polymer electrolyte membranes. To sum up, the significant findings are presented in terms of (i) designs and modeling of polymer electrolyte membrane for the DMFCs; (ii) fundamentals of PEM function, catalyst deterioration and diagnosis, and techniques for reducing failure modes; (iii) incorporation of CPs in PEM to bring down MeOH crossover significantly with negligible negotiation of H+ conductivity; (iv) CPs having pores successfully initiating the diffusion of gases out of MEA; (v) CPs-based membrane displaying higher overall stability (due to higher thermal stability CPs); and (vi) making it cost effective by replacing costly Nafion through CPs for PEM fabrication.

Besides, to account for the solution of the crossover dilemma, diverse relationships that are involved can be significantly improved to obtain better outcomes by integrating sufficient functional classes into the CP architecture. Furthermore, the following options can also be considered as promising strategies: (a) altering the thickness of polymer membrane (because a thinner membrane may decrease the ion flow resistance, and thus the MeOH crossover); (b) adopting Cs+ doping in membranes (Cs+ can alter the microstructure of the membrane, lowering the conductivity because Cs+ has lower hydration energy and a poor affinity to water); (c) using MeOH-tolerant cathodes like $Mo_2Ru_5S_5$ cluster (which can lower oxygen reduction activity as well as deterioration) to address crossover issues.

Summing up, it is important to connect cross-linkable units to potential PEMs and/or strengthen them at the same time in order to increase swelling and longevity. Expecting a single form of the membrane to address the needs of all industrial, stationary, and portable FC technologies is unrealistic. Nevertheless, it is crucial to recognize the efforts of experts from different fields, such as polymer chemists, developers, scientists, electrochemists, and engineering technicians, which are critical to improving performance and ensuring the future of DMFCs based on conducting polymer-based membranes.

REFERENCES

1. A. Kumar, V.K. Vashistha, D.K. Das, Recent Development on Metal Phthalocyanines Based Materials for Energy Conversion and Storage Applications, Coordination Chemistry Reviews, (2020) 213678.
2. Y. Zhong, S. Wang, M. Li, J. Ma, S. Song, A. Kumar, H. Duan, Y. Kuang, X. Sun, Rational Design of Copper-based Electrocatalysts and Electrochemical Systems for CO_2 Reduction: From Active Sites Engineering to Mass Transfer Dynamics, Materials Today Physics, (2021) 100354.
3. D. Yu, A. Kumar, T.A. Nguyen, M.T. Nazir, G. Yasin, High-Voltage and Ultrastable Aqueous Zinc–Iodine Battery Enabled by N-Doped Carbon Materials: Revealing the Contributions of Nitrogen Configurations, ACS Sustainable Chemistry & Engineering, 8 (2020) 13769–13776.
4. M. Ma, A. Kumar, D. Wang, Y. Wang, Y. Jia, Y. Zhang, G. Zhang, Z. Yan, X. Sun, Boosting the Bifunctional Oxygen Electrocatalytic Performance of Atomically Dispersed Fe Site via Atomic Ni Neighboring, Applied Catalysis B: Environmental, 274 (2020) 119091.
5. H. Yang, W. Lee, B. Choi, Y. Ko, S. Yi, W. Kim, Self-Humidifying Pt-C/Pt-TiO_2 Dual-Catalyst Electrode Membrane Assembly for Proton-Exchange Membrane Fuel Cells, Energy, 120 (2017) 12–19.
6. F. Yang, J. Ma, X. Zhang, X. Huang, P. Liang, Decreased Charge Transport Distance by Titanium Mesh-Membrane Assembly for Flow-Electrode Capacitive Deionization with High Desalination Performance, Water Research, 164 (2019) 114904.
7. M. Wang, M. Chen, Z. Yang, G. Liu, J.K. Lee, W. Yang, X. Wang, High-Performance and Durable Cathode Catalyst Layer with Hydrophobic C@ PTFE Particles for Low-Pt Loading Membrane Assembly Electrode of PEMFC, Energy Conversion and Management, 191 (2019) 132–140.
8. I. Vincent, A. Kruger, D. Bessarabov, Development of Efficient Membrane Electrode Assembly for Low Cost Hydrogen Production by Anion Exchange Membrane Electrolysis, International Journal of Hydrogen Energy, 42 (2017) 10752–10761.
9. K. Hengge, C. Heinzl, M. Perchthaler, D. Varley, T. Lochner, C. Scheu, Unraveling Micro-and Nanoscale Degradation Processes During Operation of High-Temperature Polymer-Electrolyte-Membrane Fuel Cells, Journal of Power Sources, 364 (2017) 437–448.
10. M. Noroozifar, Z. Yavari, M. Khorasani-Motlagh, T. Ghasemi, S.-H. Rohani-Yazdi, M. Mohammadi, Fabrication and Performance Evaluation of a Novel Membrane Electrode Assembly for DMFCs, RSC Advances, 6 (2016) 563–574.
11. G. Merle, S.S. Hosseiny, M. Wessling, K. Nijmeijer, New Cross-Linked PVA Based Polymer Electrolyte Membranes for Alkaline Fuel Cells, Journal of Membrane Science, 409 (2012) 191–199.
12. L. Li, J. Zhang, Y. Wang, Sulfonated Poly (Ether Ether Ketone) Membranes for Direct Methanol Fuel Cell, Journal of Membrane Science, 226 (2003) 159–167.
13. M. Yoshitake, Y. Kunisa, E. Endoh, E. Yanagisawa, Solid Polymer Type Fuel Cell and Production Method Thereof, Google Patents, 2005.
14. K. Dutta, P. Kumar, S. Das, P.P. Kundu, Utilization of Conducting Polymers in Fabricating Polymer Electrolyte Membranes for Application in Direct Methanol Fuel Cells, Polymer Reviews, 54 (2014) 1–32.
15. D. Satolli, M. Navarra, S. Panero, B. Scrosati, D. Ostrovskii, P. Jacobsson, I. Albinsson, B.-E. Mellander, Macro-and Microscopic Properties of Nonaqueous Proton Conducting Membranes Based on PAN, Journal of the Electrochemical Society, 150 (2003) A267.
16. H. Ericson, C. Svanberg, A. Brodin, A. Grillone, S. Panero, B. Scrosati, P. Jacobsson, Poly (Methyl Methacrylate)-Based Protonic Gel Electrolytes: A Spectroscopic Study, Electrochimica Acta, 45 (2000) 1409–1414.
17. P. Kundu, B.T. Kim, J.E. Ahn, H.S. Han, Y.G. Shul, Formation and Evaluation of Semi-IPN of Nafion 117 Membrane for Direct Methanol Fuel Cell: 1. Crosslinked Sulfonated Polystyrene in the Pores of Nafion 117, Journal of Power Sources, 171 (2007) 86–91.
18. G. Li, P.G. Pickup, Ion Transport in a Chemically Prepared Polypyrrole/Poly (Styrene-4-Sulfonate) Composite, The Journal of Physical Chemistry B, 103 (1999) 10143–10148.
19. H.J. Kim, H.J. Kim, Y.G. Shul, H.S. Han, Nafion–Nafion/Polyvinylidene Fluoride–Nafion Laminated Polymer Membrane for Direct Methanol Fuel Cells, Journal of Power Sources, 135 (2004) 66–71.
20. A. Han, B. Wang, A. Kumar, Y. Qin, J. Jin, X. Wang, C. Yang, B. Dong, Y. Jia, J. Liu, Recent Advances for MOF-derived Carbon-Supported Single-Atom Catalysts, Small Methods, 3 (2019) 1800471.
21. T. Schultz, U. Krewer, T. Vidaković, M. Pfafferodt, M. Christov, K. Sundmacher, Systematic Analysis of the Direct Methanol Fuel Cell, Journal of Applied Electrochemistry, 37 (2007) 111–119.
22. K.M. McGrath, G.S. Prakash, G.A. Olah, Direct Methanol Fuel Cells, Journal of Industrial and Engineering Chemistry, 10 (2004) 1063–1080.
23. J. Han, H. Liu, Real Time Measurements of Methanol Crossover in a DMFC, Journal of Power Sources, 164 (2007) 166–173.

24. N. Nakagawa, K. Sekimoto, M.S. Masdar, R. Noda, Reaction Analysis of a Direct Methanol Fuel Cell Employing a Porous Carbon Plate Operated at High Methanol Concentrations, Journal of Power Sources, 186 (2009) 45–51.

25. M. Ahmed, I. Dincer, A Review on Methanol Crossover in Direct Methanol Fuel Cells: Challenges and Achievements, International Journal of Energy Research, 35 (2011) 1213–1228.

26. A. Heinzel, V. Barragan, A Review of the State-of-the-Art of the Methanol Crossover in Direct Methanol Fuel Cells, Journal of Power Sources, 84 (1999) 70–74.

27. S.J. Zaidi, Research Trends in Polymer Electrolyte Membranes for PEMFC, Polymer Membranes for Fuel Cells, Springer 2009, pp. 7–25.

28. W. Yang, S. Chou, C. Shu, Effect of Current-Collector Structure on Performance of Passive Micro Direct Methanol Fuel Cell, Journal of Power Sources, 164 (2007) 549–554.

29. A. Bauer, E.L. Gyenge, C.W. Oloman, Direct Methanol Fuel Cell with Extended Reaction Zone Anode: PtRu and PtRuMo Supported on Graphite Felt, Journal of Power Sources, 167 (2007) 281–287.

30. A. Kumar, R. Randhir, Surface-Modified Electrodes: Oxidative Electropolymerisation of Macrocyclic Complexes of Fe (III) and Ni (II), Analytical & Bioanalytical Electrochemistry, 8 (2016) 382–396.

31. J. Sauk, J. Byun, H. Kim, Grafting of Styrene on to Nafion Membranes using Supercritical CO_2 Impregnation for Direct Methanol Fuel Cells, Journal of Power Sources, 132 (2004) 59–63.

32. P. Costamagna, C. Yang, A.B. Bocarsly, S. Srinivasan, Nafion® 115/Zirconium Phosphate Composite Membranes for Operation of PEMFCs above 100 C, Electrochimica Acta, 47 (2002) 1023–1033.

33. K.-D. Kreuer, Proton Conductivity: Materials and Applications, Chemistry of Materials, 8 (1996) 610–641.

34. Z. Yang, D.H. Coutinho, R. Sulfstede, K.J. Balkus Jr, J.P. Ferraris, Proton Conductivity of Acid-Doped Meta-Polyaniline, Journal of Membrane Science, 313 (2008) 86–90.

35. E. Riva, M. Mariani, Analysis of the Effects of Different Carbonaceous Phases in the Microporous Layer Composition on the Reliability and Durability of PEM Fuel Cells, School of Industrial and Information Engineering, Italy (2018).

36. J.A. Wrubel, Y. Chen, Z. Ma, T.G. Deutsch, Modeling Water Electrolysis in Bipolar Membranes, Journal of The Electrochemical Society, 167 (2020) 114502.

37. A. Krimalowski, D. Rosenbach, H. Burchardt-Tofaute, M. Thelakkat, Nanostructured Functional Polymers for Electrical Energy Storage, Nano MD, 2019 (2019) 32.

38. B.G. Choi, H. Park, H.S. Im, Y.J. Kim, W.H. Hong, Influence of Oxidation State of Polyaniline on Physicochemical and Transport Properties of Nafion/Polyaniline Composite Membrane for DMFC, Journal of Membrane Science, 324 (2008) 102–110.

39. Q. Huang, Q. Zhang, H. Huang, W. Li, Y. Huang, J. Luo, Methanol Permeability and Proton Conductivity of Nafion Membranes Modified Electrochemically with Polyaniline, Journal of Power Sources, 184 (2008) 338–343.

40. B. Schwenzer, S. Kim, M. Vijayakumar, Z. Yang, J. Liu, Correlation of Structural Differences between Nafion/Polyaniline and Nafion/Polypyrrole Composite Membranes and Observed Transport Properties, Journal of Membrane Science, 372 (2011) 11–19.

41. S. Tan, D. Belanger, Characterization and Transport Properties of Nafion/Polyaniline Composite Membranes, The Journal of Physical Chemistry B, 109 (2005) 23480–23490.

42. A. Hsini, Y. Naciri, M. Laabd, M. El Ouardi, Z. Ajmal, R. Lakhmiri, R. Boukherroub, A. Albourine, Synthesis and Characterization of Arginine-Doped Polyaniline/Walnut Shell Hybrid Composite with Superior Clean-Up Ability for Chromium (VI) from Aqueous Media: Equilibrium, Reusability and Process Optimization, Journal of Molecular Liquids, 316 (2020) 113832.

43. P. Karthikeyan, S.S. Elanchezhiyan, J. Preethi, S. Meenakshi, C.M. Park, Mechanistic Performance of Polyaniline-Substituted Hexagonal Boron Nitride Composite as a Highly Efficient Adsorbent for the Removal of Phosphate, Nitrate, and Hexavalent Chromium Ions from an Aqueous Environment, Applied Surface Science, 511 (2020) 145543.

44. I.Y. Sapurina, M. Kompan, V. Malyshkin, V. Rosanov, J. Stejskal, Properties of Proton-Conducting Nafion-Type Membranes With Nanometer-Thick Polyaniline Surface Layers, Russian Journal of Electrochemistry, 45 (2009) 697–706.

45. X. Wei, Q. Liu, H. Zhang, J. Liu, R. Chen, R. Li, Z. Li, P. Liu, J. Wang, Rapid and Efficient Uranium (VI) Capture by Phytic Acid/Polyaniline/FeOOH Composites, Journal of Colloid and Interface Science, 511 (2018) 1–11.

46. S. Bhadra, N.H. Kim, J.H. Lee, A New Self-Cross-Linked, Net-Structured, Proton Conducting Polymer Membrane for High Temperature Proton Exchange Membrane Fuel Cells, Journal of Membrane Science, 349 (2010) 304–311.

47. J. Yang, P.K. Shen, J. Varcoe, Z. Wei, Nafion/Polyaniline Composite Membranes Specifically Designed to Allow Proton Exchange Membrane Fuel Cells Operation at Low Humidity, Journal of Power Sources, 189 (2009) 1016–1019.

48. P. Sengodu, Conducting Polymers/Inorganic Nanohybrids for Energy Applications, Polymer-Engineered Nanostructures for Advanced Energy Applications, Springer 2017, pp. 365–417.

49. E.B. Easton, B.L. Langsdorf, J.A. Hughes, J. Sultan, Z. Qi, A. Kaufman, P.G. Pickup, Characteristics of Polypyrrole/Nafion Composite Membranes in a Direct Methanol Fuel Cell, Journal of the Electrochemical Society, 150 (2003) C735.

50. H.S. Park, Y.J. Kim, W.H. Hong, H.K. Lee, Physical and Electrochemical Properties of Nafion/Polypyrrole Composite Membrane for DMFC, Journal of Membrane Science, 272 (2006) 28–36.

51. F. Xu, C. Innocent, B. Bonnet, D. Jones, J. Roziere, Chemical Modification of Perfluorosulfo-nated Membranes with Pyrrole for Fuel Cell Application: Preparation, Characterisation and Methanol Transport, Fuel Cells, 5 (2005) 398–405.

52. S. Moravcova, Z. Cílová, K. Bouzek, Preparation of a Novel Composite Material Based on a Nafion® Membrane and Polypyrrole for Potential Application in a PEM Fuel Cell, Journal of Applied Electrochemistry, 35 (2005) 991–997.

53. H. Park, Y. Kim, W.H. Hong, Y.S. Choi, H. Lee, Influence of Morphology on the Transport Properties of Perfluorosulfonate Ionomers/Polypyrrole Composite Membrane, Macromolecules, 38 (2005) 2289–2295.

54. H. Park, Y. Kim, Y.S. Choi, W.H. Hong, D. Jung, Surface Chemistry and Physical Properties of Nafion/Polypyrrole/Pt Composite Membrane Prepared by Chemical In Situ Polymerization for DMFC, Journal of Power Sources, 178 (2008) 610–619.

55. L. Li, Y. Zhang, J.-F. Drillet, R. Dittmeyer, K.-M. Jüttner, Preparation and Characterization of Pt Direct Deposition on Polypyrrole Modified Nafion Composite Membranes for Direct Methanol Fuel Cell Applications, Chemical Engineering Journal, 133 (2007) 113–119.

56. J. Zhu, R.R. Sattler, A. Garsuch, O. Yepez, P.G. Pickup, Optimisation of Polypyrrole/Nafion Composite Membranes for Direct Methanol Fuel Cells, Electrochimica Acta, 51 (2006) 4052–4060.

57. M. Smit, A. Ocampo, M. Espinosa-Medina, P. Sebastian, A Modified Nafion Membrane with In Situ Polymerized Polypyrrole for the Direct Methanol Fuel Cell, Journal of Power Sources, 124 (2003) 59–64.

58. R. Escudero-Cid, M. Montiel, L. Sotomayor, B. Loureiro, E. Fatás, P. Ocón, Evaluation of Polyaniline-Nafion® Composite Membranes for Direct Methanol Fuel Cells Durability Tests, International Journal of Hydrogen Energy, 40 (2015) 8182–8192.

59. S. Xue, G. Yin, Proton Exchange Membranes Based on Modified Sulfonated Poly (Ether Ether Ketone) Membranes with Chemically In Situ Polymerized Polypyrrole, Electrochimica Acta, 52 (2006) 847–853.

60. X. Li, C. Liu, D. Xu, C. Zhao, Z. Wang, G. Zhang, H. Na, W. Xing, Preparation and Properties of Sulfonated Poly (Ether Ether Ketone)s (SPEEK)/Polypyrrole Composite Membranes for Direct Methanol Fuel Cells, Journal of Power Sources, 162 (2006) 1–8.

61. D. Xu, G. Zhang, N. Zhang, H. Li, Y. Zhang, K. Shao, M. Han, C.M. Lew, H. Na, Surface Modification of Heteropoly Acid/SPEEK Membranes by Polypyrrole with a Sandwich Structure for Direct Methanol Fuel Cells, Journal of Materials Chemistry, 20 (2010) 9239–9245.

62. H. Lin, C. Zhao, W. Ma, H. Li, H. Na, Layer-by-Layer Self-Assembly of In Situ Polymerized Polypyrrole on Sulfonated Poly (Arylene Ether Ketone) Membrane with Extremely Low Methanol Crossover, International Journal of Hydrogen Energy, 34 (2009) 9795–9801.

63. L. Li, Y. Zhang, Chemical Modification of Nafion Membrane with 3,4-Ethylenedioxythiophene for Direct Methanol Fuel Cell Application, Journal of Power Sources, 175 (2008) 256–260.

64. H. Zhang, P.K. Shen, Recent Development of Polymer Electrolyte Membranes for Fuel Cells, Chemical Reviews, 112 (2012) 2780–2832.

65. L. Li, J.-F. Drillet, R. Dittmeyer, K. Jüttner, Formation and Characterization of PEDOT-Modified Nafion 117 Membranes, Journal of Solid State Electrochemistry, 10 (2006) 708–713.

66. S.J. Higgins, K.V. Lovell, R.G. Rajapakse, N.M. Walsby, Grafting and Electrochemical Characterisation of Poly-(3,4-Ethylenedioxythiophene) Films, on Nafion and on Radiation-Grafted Polystyrenesulfonate–Polyvinylidene Fluoride Composite Surfaces, Journal of Materials Chemistry, 13 (2003) 2485–2489.

20 Synthesis and Characterization of Poly(zwitterionic) Structures for Energy Conversion and Storage

Adrian Olejnik[1,2], Katarzyna Grochowska[1], and Katarzyna Siuzdak[1]

[1]Centre for Plasma and Laser Engineering, The Szewalski Institute of Fluid-Flow Machinery, Polish Academy of Sciences, Gdańsk, Poland

[2]Gdańsk University of Technology, Gdańsk, Poland

CONTENTS

20.1 INTRODUCTION

International Union of Pure and Applied Chemistry defines zwitterions (ZI) as "neutral compounds having formal unit electrical charges of opposite sign" (Chemistry (IUPAC) 2021). In other words, zwitterions are molecules possessing an equal number of functional groups bearing positive and negative charges with zero net charge. Typical examples of zwitterions are given in Figure 20.1.

Aminoacids (e.g., glycine) in the environment, where pH is equal to their isoelectric point, are zwitterions. In this case, the positive charge is provided by protonated ammonium group and

DOI: 10.1201/9781003150374-20

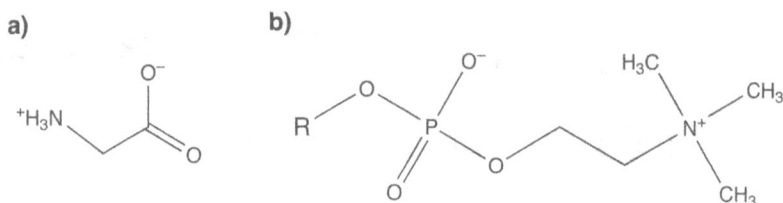

FIGURE 20.1 Exemplar zwitterionic molecules: (a) glycine and (b) phosphatydylcholine derivative (R contains glycerol and fatty acids moieties).

negative charge is ensured by the deprotonated carboxylic group. Zwitterions are abundant in biochemistry, especially as a building block of lipid bilayers of cells, where they fulfill several roles including compartmentalization and transport of different compounds through the membranes (Laschewsky and Rosenhahn 2019).

Besides single molecular zwitterions, there are also polymeric zwitterions (PZ). Further in the text, ZI will correspond to single molecular zwitterions and PZ to polymeric zwitterions. According to the terminology defined by Laschewsky, polyzwitterions are polymers, which "bear, within their structural repeating unit, the same amount of cationic and anionic groups" (Laschewsky 2014). In other words, the structural units of PZ are zwitterions and the overall electric charge is zero. Some authors distinguish strong PZ, where every unit has two oppositely charged ionic groups, and weak PZ, where only a fraction of units bear two ionic groups. (Kharlampieva, Izumrudov, and Sukhishvili 2007) Standard cationic groups in PZ are ammoniums and imidazoliums, while anionic groups are sulfanate, carboxylate, and phosphate groups. The most abundant class of synthetic PZ are poly(methacrylates) such as poly(carboxybetaine methacrylate) (PCBMA), poly(sulfobetaine methacrylate) (PSBMA), or poly(phosphoryl choline methacrylate) (PMPC) derivatives. Monomers corresponding to those polymers are shown in Figure 20.2.

Monomers might differ by the n number, which is the number of methylene groups that separate the cationic and anionic moieties (Figure 20.2), i.e., a spacer. The length of a spacer is crucial and

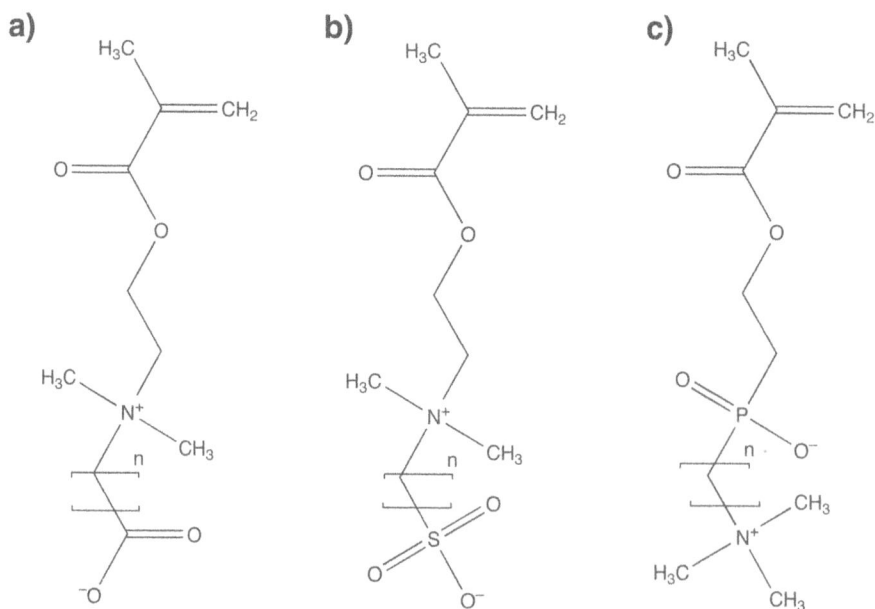

FIGURE 20.2 Three canonical monomers used for synthesis of polyzwitterions: (a) CBMA, (b) SBMA, (c) MPC. Value of n lies typically between 1 and 3 methylene groups.

determines the strength of intermolecular interactions of polyzwitterions (Li et al. 2019), which as a consequence determine their physiochemical properties and application potential. A noble theoretical chemist Mark Ratner in his memorable seminar in 2004, said, "The chemistry of XX-th century was about intramolecular interactions; the chemistry of XXI century will be about intermolecular interactions." Therefore, Section 20.2 focuses on explaining those interactions and physicochemical properties of polyzwitterions and relating them to the experimentally observed phenomena.

It can be easily seen that monomers in Figure 20.2 are methacrylates having C=C bond, being susceptible to radical polymerization. Typical routes of synthesis involve classical free-radical polymerization or controlled polymerization methods such as Atom Transfer Radical Polymerization (ATRP). By means of controlled polymerizations, a wide variety of zwitterionic structures have already been obtained, including thin layers, brushes, star-shaped colloidal polymers, crosslinked coatings, or hydrogels (Laschewsky and Rosenhahn 2019). Therefore, Section 20.3 gives an overview of synthetic methods, which are commonly used in transforming ZI monomers into PZ and depositing PZ on surfaces as well as recent trends in this tremendously fast-developing field. Finally, some recent applications of polyzwitterions in the energy field will be thoroughly described with particular attention put toward supercapacitors (SCs), batteries, solar cells (SoC), and organic light-emitting diodes (OLEDs).

20.2 INTERMOLECULAR INTERACTIONS AND PHYSIOCHEMICAL PROPERTIES OF ZWITTERIONS

The molecular structure of ZI and PZ is a source of their inherent properties on both microscopic and macroscopic levels. From the nanoscale perspective, they exhibit strong intramolecular and intermolecular interactions. The physical nature of those interactions can be coulombic (electrostatic), dipole-dipole, dipole-multipole, or dipole-induced dipole (van der Waals) attractions repulsions. However, due to the high dipole moments imposed by anionic and cationic groups, these interactions can have a tremendous impact on the larger scale effects yielding unique macroscopic properties (Laschewsky and Rosenhahn 2019; Schlenoff 2014).

20.2.1 ANTIFOULING PROPERTY AND HYDRATION STRUCTURE OF ZWITTERIONS

The most famous property of ZI and PZ is the capability of reducing the adsorption of biomolecules (e.g., proteins) and whole cells onto their surface (antifouling). This property opened a vast area of applications involving biocompatible coatings (Vaterrodt et al. 2016), membranes for ultrafiltration (Zhou et al. 2014), sensing of biomolecules (Yang et al. 2011), and many more. Antifouling materials have several crucial properties in common. Their net electrostatic charge on exposure to a solution should be zero and they should be well hydrated. ZI and PZ fulfill both requirements perfectly. The detailed mechanism of antifouling on a molecular level is a complicated issue and was thoroughly discussed by Schlenoff (2014). Understanding the two intuitive arguments is worth discussing because they play a major role in the energy applications of ZI and PZ.

Let us assume that some surface is modified with ZI molecules or polymeric chains of PZ. If the structure of hydrogen bonds in the water around the surface is the same as in the bulk solution, the adsorption of foreign molecules on that surface is energetically unfavorable. That is because the cleavage of hydrogen bonds during adsorption requires some amount of free energy. This energy is typically granted from increased entropy of ordered water dipoles oriented toward a single charge (Figure 20.3a). However, in the case of ZI or PZ, there is no order so that no free energy can be granted. Therefore, a strong not-ordered hydration shell leads to the repulsion of proteins, which "do not see" the zwitterionic surface.

The second argument invokes the fact that the releasing of counterions facilitates adsorption. Negatively charged surface with positive counterions can induce dipole moments in the protein so that its partial positive charge is directed toward the surface. Then the protein approaches the

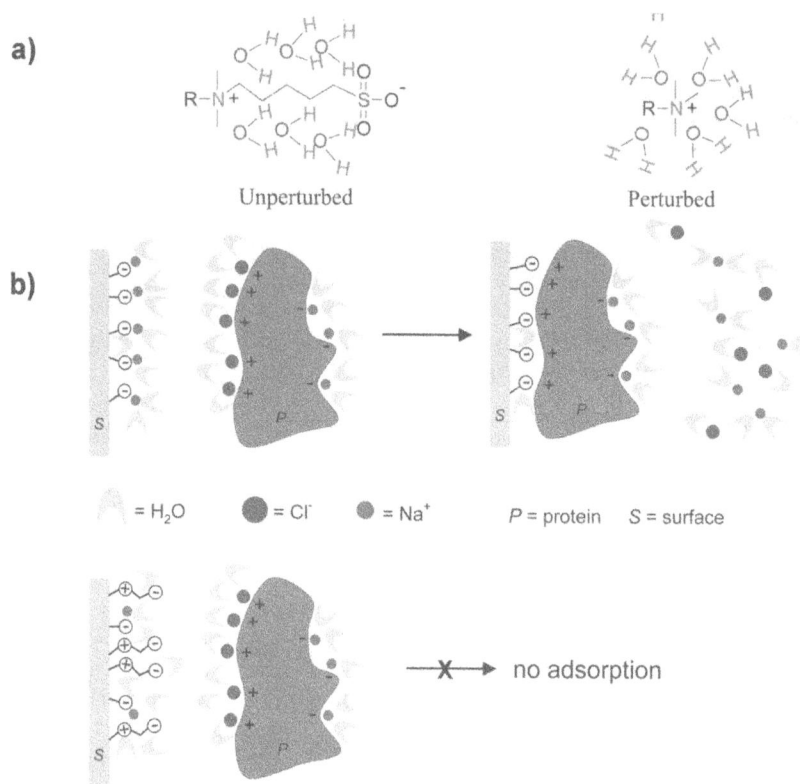

FIGURE 20.3 (a) Hydrogen bonding structure, (b) antifouling mechanism on the molecular level. (Copyright Schlenoff 2014.)

surface and destroys the ordering of counterions providing an increase of entropy of the system (entropic gain). An increase in entropy contributes to the decrease of free energy and adsorption is thermodynamically favorable. On the other hand, zwitterionic surfaces are well hydrated and have a very low amount of counterions, in contrast to polycations and polyanions. Therefore, no free energy can be granted from entropic gain and adsorption does not occur.

On the phase boundary between the surface and electrolyte, there is a layer of dipoles (regardless of whether they come from ZI or PZ). This layer is beneficial not only in terms of reducing adsorption but also facilitates charge transfer between phases in solar cells and OLEDs. The common observations are that the zwitterionic coatings between electrodes and active layer change with the work function of electrodes (up to 0.9 eV) so that the optimal alignment of the bands is obtained (Page et al. 2015, Zhu 2017, Sun et al. 2012a). Intuitively, the dipole layer causes a lowering of the energy difference between phases due to the intermolecular interactions. The results of zwitterionic modifications lead to the increase of solar cell and OLED performance and their long-term stability. Those issues will be discussed thoroughly in Section 20.4 dedicated to the ZI applications.

20.2.2 ANTIPOLYELECTROLYTE EFFECT AND SENSITIVITY TO SALTS

While ZI is just a single molecule, PZ is a polymer that can have different conformations depending on many factors. The nearest environment involving water, molecules, and ions is a crucial factor determining the physical structure of PZ when exposed to a solution. Let us elucidate this topic more deeply. The separation of charges caused by the chemical structure exhibits large dipole moments. Therefore, PZ remains highly hydrated due to dipole–dipole interactions with water. This effect is especially pronounced for brushed structures (Higaki, Kobayashi, and Takahara 2020). In

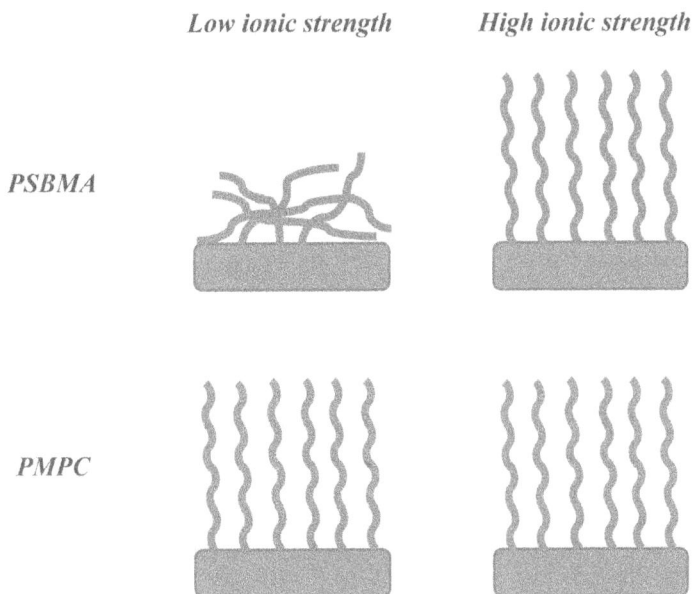

FIGURE 20.4 Graphical comparison of PSBMA and PMPC brush conformations in low and high salts environments.

salt-free water, PSBMA tends to aggregate due to large intramolecular interactions and ion-pairing (Laschewsky and Rosenhahn 2019). PSBMA brushes easily collapse when no external electrolyte is provided. However, when the salt is introduced to the solution, ion pairs dissociate due to the charge screening effect induced by interactions with salt ions. In consequence, the chain conformations change so that the brushes are extended and their height increases. This is observed as swelling of the brushes and can be measured with Atomic Force Microscopy in tapping mode (Xiang, Xu, and Leng 2020). Change of chain conformation in response to ions is known as the antipolyelectrolyte effect and can be studied by small-angle X-ray scattering (SAXS), light and neutron scattering experimental techniques (Kobayashi, Ishihara, and Takahara 2014).

In contrast to PSBMA, PMPC usually exhibits a weaker antipolyelectrolyte effect and is completely soluble in water. Additionally, the height of PMPC brushes is almost independent of the ionic strength and the conformations are elongated (Figure 20.4). This can be attributed to the higher affinity of sulfonate than phosphate to create ion pairs with the quaternary ammonium groups. Just for comparison, it is worth mentioning that the salt behavior of PCBMA is somewhere in between PSBMA and PMPC. There is also a large role of ion specificity in the hydration of polyzwitterions and modulation of antipolyelectrolyte effect. Less hydrated chaotropic anions such as thiocyanate, SCN^- and ClO_4^- promote dissolution of PSBMA, while heavily hydrated anions such as SO_4^{2-} and OH^- supports creating PSBMA aggregates (Higaki, Kobayashi, and Takahara 2020).

Salt behavior of PZ is a complex phenomenon and there are several inconsistencies about those issues in experimental data. However, one can withdraw some heuristic practical implications and employ them in energy applications. For example, if PZ chains are free in solution, the antipolyelectrolyte effect causes an increase in polymer solubility while ionic strength increases. If, however, they are tethered to the surface or they are in the form of gel, they have a very high level of hydration which increases with salt content. This is beneficial when polyzwitterions are applied as a polymer electrolyte for supercapacitors or batteries because significant concentration of ions can be accumulated between capacitor plates therefore enhancing the energy density. Besides the unique salination behavior, conformational changes can also occur due to the external electric field. Due to the presence of high dipole moments, PZ can exhibit a high level of self-ordering after the field is turned on. Therefore, fast ion-transport channels are created and the performance of the supercapacitor is enhanced. This was shown using (X-ray absorption near-edge spectroscopy (XANES) measurements (Peng et al. 2016).

20.3 OVERVIEW OF PZ AND ZI STRUCTURES AND SYNTHETIC APPROACHES

To guide the reader through the forthcoming discussion about the applications of zwitterions, a short overview of the different architectures of PZ and ZI will be conducted. Molecular structures of different ZI molecules can be very diverse depending on the choice of cationic and anionic functional groups and their placement in the molecule. Many combinations can be realized using various organic chemistry approaches and depending on the requirements of different applications. Possibilities in this field are limitless and their discussion is beyond the scope of this chapter. However, the most commonly used ZI are methacrylates that are further polymerized into PZ. Both PZ and ZI can be anchored onto the desired surface in many ways, as we describe them soon. There are several methods of obtaining PZ, which are most commonly radical polymerization techniques. Therefore, ideas of controlled radical polymerization and photopolymerization will also be briefly discussed.

20.3.1 PZ Architectures

Let us now discuss the possible architectures of PZ. In this chapter, we are mostly interested in surfaces modified with PZ or free-standing PZ materials, e.g., polymer electrolytes for batteries and supercapacitors. Several approaches to surface modification are schematically depicted in Figure 20.5. They include layers, brushes, and crosslinked coatings.

One of the simplest polymer coating techniques is layer-by-layer deposition by, e.g., evaporation of the solvent. However, PZ is generally not prone to physically adsorb onto surfaces spontaneously. This phenomenon is actually beneficial in most applications because it means that other objects do not adsorb onto PZ as well. It is worth noting that below the pKa of carboxylate/sulfonate group, PCBMA/PSBMA is no longer zwitterionic because those groups are protonated. Then, in strongly acidic media, they can spontaneously adsorb, but when pH goes back to neutral, they desorb. Moreover, only very thin films can be obtained this way (Kharlampieva, Izumrudov, and Sukhishvili 2007). Therefore, layer-by-layer deposition is typically accompanied by other materials. Although those layers are relatively easy to obtain, they have a certain window of stability in terms of pH (protonation or deprotonation of carboxylic groups) and salt concentration (antipolyelectrolyte effect), outside which they become soluble.

PZ brushes, however, can be deposited in a variety of environments (organic, aqueous) and in neutral pH and maintain perfect stability. One crucial requirement although is a covalent attachment of the polymerization initiators onto the desired surface. A very promising strategy of covalent

Layers   *Brushes*   *Crosslinked coatings*

FIGURE 20.5 Simplified topologies of PZ tethered to the surface.

attachment is via the employment of polydopamine chemistry. Polydopamine (PDA) is a bioinspired polymer obtained through electropolymerization or oxidative polymerization of dopamine onto the desired surface. Due to its strong intermolecular interactions (π-π stacking, π-cation interactions, amine-catechol interplay, etc.), it has a spectacularly good adhesion toward different materials: metals, metal oxides, polymers, and other organic materials (Lee, Park, and Lee 2020). Additionally, the catechol groups of PDA are prone to nucleophilic addition/substitution that can be utilized in the covalent attachment of initiators. Synthesis of brushes is realized by the controlled radical polymerization and a prominent example of this technique ATRP will be discussed soon. Alternatively, whole ZI or PZ molecules can be tethered to the surface.

PDA can be utilized in the synthesis of crosslinked PZ as well. Zhou obtained hybrid crosslinked PDA/PSBMA membranes with enhanced antifouling properties and washing stability by simple one-pot deposition onto porous polypropylene substrate (Zhou et al. 2014). Simultaneous deposition of PSBMA and polymerization of dopamine in slightly alkaline pH lead to stable porous polyzwitterionic membranes. Photopolymerization is also a universal tool for obtaining PZ crosslinked hydrogels and electrolytes not only in free-standing form but also as elements of functional materials, e.g., biosensors (Yang et al. 2011).

20.3.2 Atom Transfer Radical Polymerization

ATRP is a special type of radical polymerization that involves the catalytic activity of copper complexes to create radicals; complexes of other transition metals are also possible, although copper is the most typically used one. To perform ATRP, one needs several ingredients. The first of them is an initiator containing at least one halogen atom. Typically these are tertiary alkyl bromides, although most kinds of halogens are sufficient. Two commonly used examples, PEBr (phenyl ethylbromide) and EtBriB (ethyl bromoisobutyrate) are given in Figure 20.6. Then, a copper complex composed of copper salt and a ligand that is stable and soluble in the desired solvent is required. Two examples of ligands – bipyridine and Me_6TREN (tris[2-(methylamino)ethyl]amine) (Matyjaszewski 2012) – are also provided in Figure 20.6.

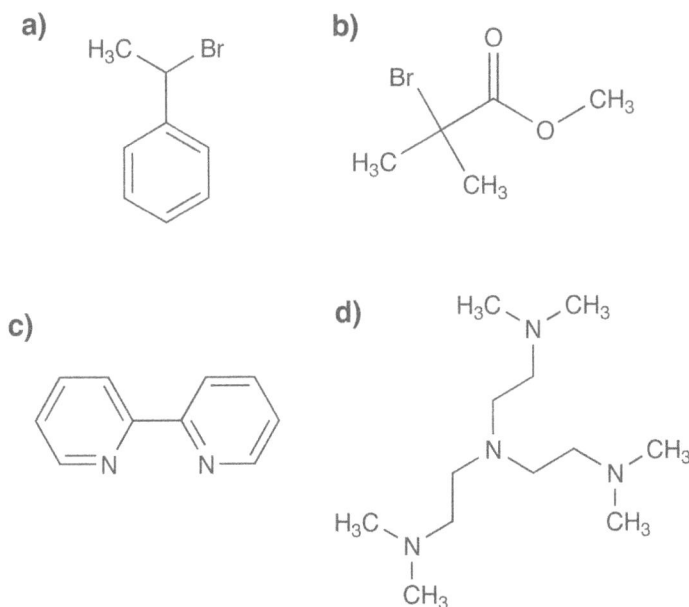

FIGURE 20.6 Typically used ATRP initiators: (a) PEBr, (b) EtBriB and ligands for ATRP copper catalyst, (c) bipyridine, and (d) Me_6TREN.

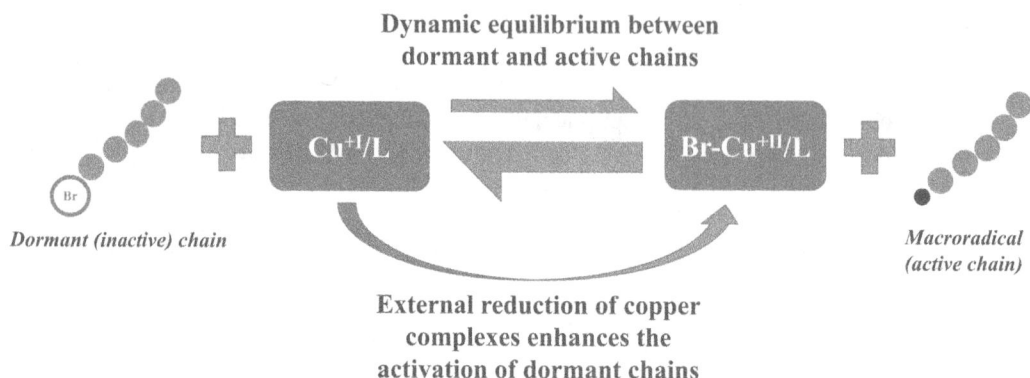

FIGURE 20.7   A simplified scheme of ATRP procedure.

Solvents used in ATRP vary between organic compounds such as dimethylformamide (DMF), MeCN, simple aliphatic alcohols, and water. The selection of the solvent is strictly dependent on the monomer used because of the solubility criterion. The general idea of ATRP is the following (Figure 20.7). Firstly, some amount of Cu(I) complex is generated (approaches to the generation of active complexes will be covered soon). Then, there is a halogen atom transfer from Br-terminated activator to Cu(I) complex, which in turn creates the first radical and oxidizes copper to Br-Cu(II). There is an equilibrium between the inactive (dormant) chains of the polymer and active macroradicals, but it is strongly shifted toward the dormant form. In this scenario, the macroradical is allowed to propagate for some time, but it is quickly turned into a dormant form again. By applying an external reduction power, Br-Cu(II) is transformed back to Cu(I) that leads to the possibility of another activation of the yet dormant chain. This property allows precise control of molecular mass and results in a relatively small polydispersity, i.e., the distribution of molecular mass is relatively narrow. It is a real power underlying ATRP. This enables a researcher to make polymer chains of defined length and topology that are tailored for each one's needs by manipulating reaction times and concentrations.

One can find an abbreviation SI-ATRP which stands for Surface Initiated ATRP, where initiator is permanently attached (most often covalently) to some surface. In this approach, one can obtain copolymers and brushes. The typical initiator used in the synthesis of PZ brushes is bromoisobutyryl bromide (BiBB) and its structure is similar to EtBriB (Figure 20.8). This approach was applied

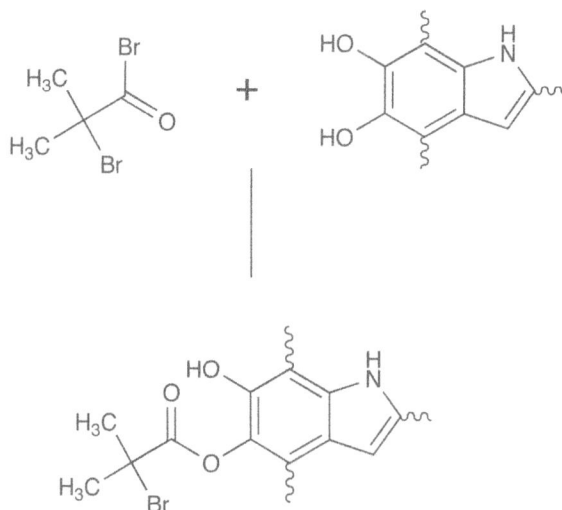

FIGURE 20.8   Covalent attachment of BiBB initiator to the surface of polydopamine.

by Liu et al. (2017) in the synthesis of PZ brushes for composite membranes. There are also plenty of other chemistries allowing covalent attachment of ATRP initiator. One typically found is the modification of gold with compounds terminated by bromine from one side and thiols from the other side. Thiol is attached to gold and the other end of the molecule containing bromine serves as SI-ATRP initiation site (Hu et al. 2015). Controlled radical polymerizations allow to manipulate the topology of polymers and therefore tailor the properties of polyzwitterions such as thermo, pH, and salt responsiveness.

20.3.3 PHOTOPOLYMERIZATION

Photopolymerization is a widely used technique on both industrial and laboratory scales as well as in advanced technology such as microlithography (Free Radical Photopolymerization of Multifunctional Monomers 2016). Basically, it is another variant of free radical polymerization that radicals are created through the interaction of the precursor with light. Reaction mixture designed for photopolymerization should include four primary components: monomer, curing (crosslinking) agent, solvent, and photoinitiator. Most typically used monomers for photopolymerization are substituted acrylates and methacrylates so that created radicals are resonance stabilized. The chemical structure of the commonly used zwitterionic monomers, PSBMA, PCBMA, and PMPC, make them suitable for photopolymerization. Curing agents are compounds that have two or more C=C bonds, e.g., EGDMA (ethylene glycol dimethacrylate, see Figure 20.9a). The presence of multiple C=C bonds causes the emergence of 3D crosslinked topology during polymerization. Thus hydrogels and crosslinked coatings are easily synthesized. Unless the crosslinking agent is present, only straight polymeric chains are created. The selection of the solvent depends on the choice of the monomer and curing agent. In mixtures of zwitterionic monomers, water is usually used as a solvent (Yang et al. 2011).

An exemplar photoinitiator molecule is 2-hydroxy-2-methylpropiophenone (Yang et al. 2011) (Figure 20.9b). Photoinitiators can have different chemistry, but they have one aspect in common. They should have high absorption coefficients for the desired wavelengths. When those molecules are subjected to proper wavelengths, their electrons are excited to higher electronic levels. When this electron relaxes back to lower level, it can create enough thermal (vibrational) energy to cause homolysis of a particular bond and create radicals.

If one needs to obtain high molecular mass macromolecules, photopolymerization should be performed in an oxygen-free environment. Oxygen molecules cause quenching of excited states and react with already existing radicals limiting the number of free radicals and therefore reducing chain lengths. Unfortunately, those reactions are around a million times faster than the speed of chain propagation. (Free Radical Photopolymerization of Multifunctional Monomers 2016) However, there are several solutions to this issue. Performing synthesis in a neutral gas atmosphere is usually easily done on a laboratory scale, but is difficult for industrial applications. Another idea is the application of a physical barrier between the light source and the reaction mixture that blocks the diffusion of oxygen. Last but not least, using the high intensity of light, especially originating from laser sources, creates such a high number of radicals that oxygen inhibition interrupts photopolymerization only to a small extent (Malinauskas 2013).

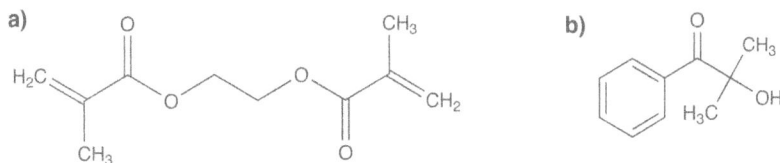

FIGURE 20.9 Exemplar structures of (a) curing agent EGDMA, (b) photoinitiator 2-hydroxy-2-methylpropiophenone (Benacure 1173).

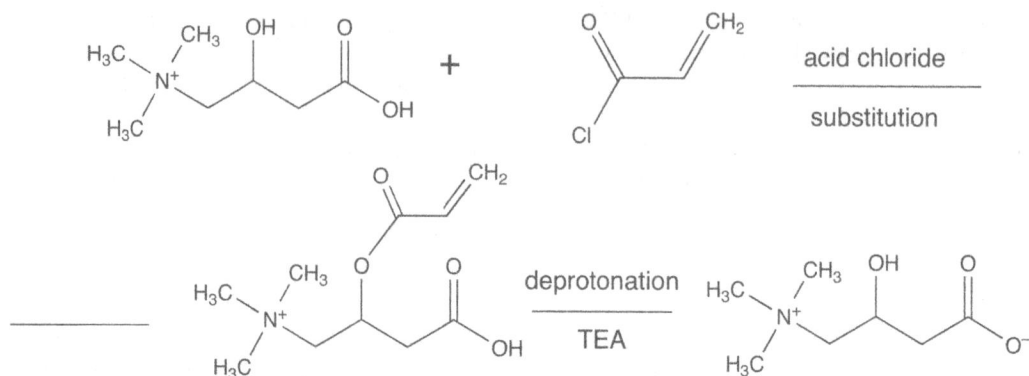

FIGURE 20.10 Synthetic route toward L-carnitine-based ZI monomer.

20.3.4 Recent Trends in the Synthesis of ZI and PZ

Increasing the structural diversity of ZI monomers, polymers, and materials is of paramount importance in the field as the development of sustainable materials from renewable resources is one of the key needs of modern science. This universal concept also applies to PZ, especially because the most common monomers do not come from renewable feedstocks. Wang et al. (2017) proposed a single-step synthetic approach toward creating a monomer based on L-carnitine, a naturally occurring amino acid derivative (Figure 20.10). Briefly, L-carnitine was subjected to a nucleophilic substitution reaction with acid chloride containing acrylate double bond. Then subsequently, deprotonation of carboxyl groups was achieved by triethylene amine (TEA) resulting in zwitterionic L-carnitine acrylate.

Based on the obtained monomer, PZ brushes on gold substrate and hydrogels were synthesized by ATRP and photopolymerization techniques, respectively. Those materials exhibited antifouling properties similar to the commercially available ones, which is a promising result opening new opportunities toward bio-based PZ.

Poly(zwitterions) can also be obtained through postpolymerization procedures, i.e., reaction with already synthesized polymers by introducing new functionalities. Scott et al. (2020) synthesized phosphonium-based PZ according to the following protocol (Figure 20.11). Firstly, DPPS (4-diphenylphosphino) styrene was copolymerized with DEGMEMA (diethylene glycol methyl ether methacrylate) to obtain random copolymers. Then, the structural units containing phosphine were subjected to sultone alkylation (see Figure 20.11) to obtain zwitterionic moieties on the polymeric

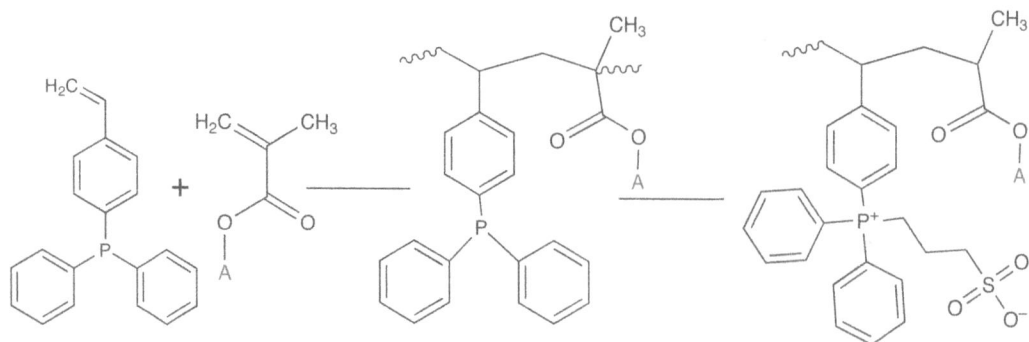

FIGURE 20.11 Synthetic scheme of phosphonium-based polyzwitterions. A stands for oligo(ethylene glycol) chain.

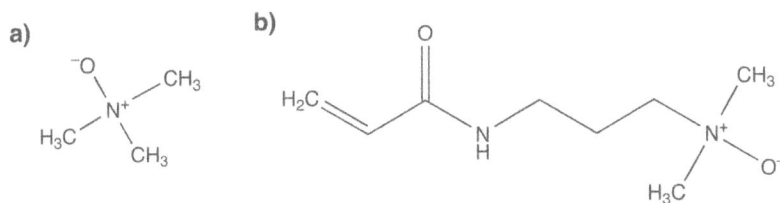

FIGURE 20.12 Zwitterionic (a) trimethylammonium N–oxide, TMAO original molecule, (b) TMAO monomer.

chains. Similarly, polycations were also synthesized using n-propyl bromide as a modifying agent for comparison purposes.

Besides introducing a new chemistry toward the synthesis of PZ, several interesting properties of zwitterated materials were observed during tensile testing. Solvent-casted phosphine-based PZ exhibits markedly higher tensile strength due to physical crosslinking than its polycation analogs. This was attributed to the physical crosslinking caused by attractive intermolecular interactions of PZ. Large dipole moments enforced by the molecular structure of the polymer lead to ionic associations that reinforce the mechanical structure of the material. The existence of those ionic aggregates was confirmed by SAXS. As a natural consequence of crosslinking, the elasticity of the polymer decreased and classification temperature T_g increased.

The chemical structure and morphology of biomaterials are the results of billions of years of evolution. With limited resources and a miserably finite lifetime, living organisms shaped their building structures with incredible complexity and unique physiochemical properties (Zhao et al. 2014). Many modern materials such as electroactive polymers, photonic crystals, polydopamine, and many more are inspired by this complexity. An interesting example of grasping the patterns of nature was performed by Li et al. (2019). Trimethylammonium N–oxide (TMAO) is a zwitterionic osmolyte commonly present in saltwater fishes. The uniqueness of this molecule is due to the length of a spacer between the positively charged ammonium group and negatively charged oxygen being just the length of a single bond (Figure 20.12).

Li et al. (2019) synthesized a ZI monomer analogous to TMAO and used it to make hydrogels and brush coatings for different surfaces such as gold and protein carriers. These materials were subjected to a very hard test of antifouling properties both in vitro and in vivo. Results show that this material exhibits ultralow protein adsorption and a decrease in immunogenicity of protein carriers. These ones confirm that the shorter length of a spacer between ionic groups in PZ, the better antifouling performance achieved.

20.4 APPLICATIONS OF ZWITTERIONS IN THE FIELDS OF ENERGY STORAGE AND ENERGY CONVERSION

With the large development of the materials dedicated to energy conversion and storage, still many improvements have to be done to increase their efficiency, lower the cost of the device, and make the processing route simpler that facile the whole commercialization process. The most typical approach regards the synthesis of new materials, modification of those already adopted (e.g., doping with metal/non-metal atoms, adding new functionalities) and the usage of structures that were successfully applied in other research areas, e.g., a transfer of the material that was utilized in the field of biosensing to the supercapacitors: batteries, solar cells, and organic light-emitting diodes. Below, the specific applications of zwitterion molecules in solar cells, organic light-emitting diodes, supercapacitors, and batteries are presented. We describe some exemplary ZI nanostructures and demonstrate how their dual nature affects the performance of the whole device.

20.4.1 SUPERCAPACITORS

Supercapacitors are devices that combine the electrochemical properties of both batteries and capacitors. Like batteries, they can store a large amount of energy during charging state and similar to capacitors they are capable of discharging high power density in a very short time. In other words, they fill the void between capacitors and rechargeable batteries. The concept of supercapacitor assumes tremendously high capacitance with an almost unlimited charge/discharge cycle life, even above 100 thousand cycles. Due to that fact, they have attracted a lot of attention. It should be kept in mind that basically supercapacitors can be divided into three classes: pseudocapacitors, electric double-layer capacitors (EDLC), and hybrid capacitors that are based on their cell pattern or storage mechanism. In the case of pseudocapacitors, the pseudocapacitance is achieved by Faradaic reactions and the storage mechanism is electrochemical. In EDLC, the separation of charge occurs in the double layer and the charge storage is realized electrostatically. Lastly, in hybrid capacitors, the synergistic effect of both EDLC and pseudocapacitor is exploited and the device is composed of electrodes exhibiting electrostatic (one electrode) or electrochemical capacitance (another electrode). Of course, one may not forget about electrolyte that ensures the ionic conductive connection between the two electrodes. Therefore, the electrolyte influences the operating voltage, temperature range, and capacitance of the SCs. Due to the inflammability, wide potential window and high ionic conductivity, ionic liquids are a promising type of electrolyte (Pan et al. 2020). They can be classified into three categories: aprotic, protic, and zwitterionic. Interestingly, aprotic ionic liquids are considered to be suitable for batteries and supercapacitors, protonic ones – for fuel cells, while zwitterionic ones – for ionic-liquid-based membranes (Armand et al. 2009). Nevertheless, in this section, we will show the possible application of zwitterions in supercapacitors.

Afrifah et al. (2020) combined poly (2-acrylamido-2-methylpropane sulfonic acid) and silica–sulfobetaine silane zwitterion to be used in SCs. The combination of those two substances resulted in increased ionic conductivity as well as in improved electrochemical stability and cycling. When electrolyte contained zwitterion, the specific capacity at 10 mVs^{-1} was established to be 321 Fg^{-1} which was about twice higher than the one obtained without zwitterion. Moreover, the increase of capacitance was observed with the higher amount of zwitterion and nearly rectangular shape of the cyclic voltammetry curve at 100 mVs^{-1} indicated almost ideal capacitor behavior – see Figure 20.13. What is worth mentioning, the SC with zwitterion containing electrolyte exhibited excellent cycling

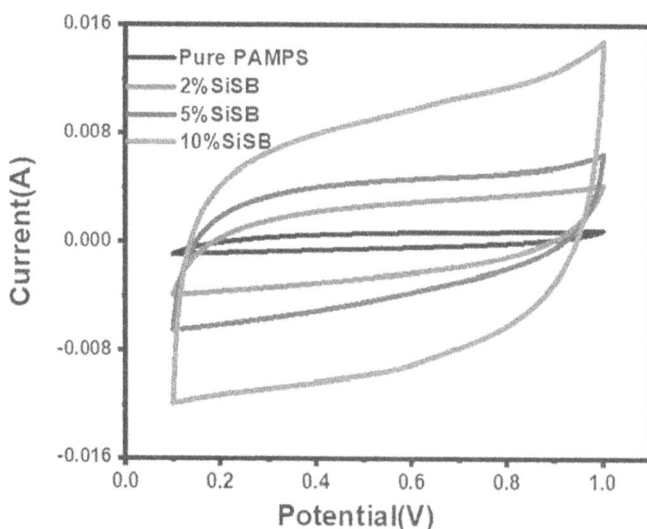

FIGURE 20.13 Cyclic voltammetry of poly(2-acrylamido-2-methylpropane sulfonic acid) (PAMPS) with different amounts of silica–sulfobetaine silane zwitterion (SiSB). (Copyright Afrifah et al. 2020.)

stability, namely nearly 100% after 5000 charging/discharging cycles comparing to the one without zwitterion – 73%. Enhancement of the performance through introducing zwitterions can be explained in the following way. Zwitterions have the capability of self-ordering when an external electric field is applied (Peng et al. 2016). During charge–discharge of the supercapacitor, the electric field causes directional changes in chain conformation leading to the creation of fast ion transport channels for both cations and anions.

The same idea of ion-transport channels was implemented by Peng et al. (2016). They worked on the zwitterionic gel electrolyte that can be used in graphene-based solid-state supercapacitors. In this case, the poly(propylsulfonate dimethylammonium propylmethacrylamide) was used. The volume capacitance of 300.8 F cm^{-3} at 0.8 A cm^{-3} was achieved, and the authors claimed that this value is the best among previously reported ones for graphene-based solid-state SCs. Moreover, it should be underlined that the capacitance retention at the level of 103% was maintained after 10,000 charge–discharge cycles.

In the work of Hyeon et al. (2019), the quasi-solid-state supercapacitor with zwitterionic gel electrode with a dual redox additive was studied. Ethyl viologen (EV) dibromide was chosen to be incorporated into polyzwitterion charged polymer (poly([2-(methacryloyloxy)ethyl]dimethyl-(3-sulfopropyl)ammonium hydroxide). It should be emphasized that the gel polymer served not only as an electrolyte but also as a separator placed between the electrodes. Moreover, added ethyl viologen dibromide was employed both as a cationic and anionic redox additive. Thanks to that, redox additive can increase the electrical performance of the assembled SC. The fabricated SC exhibited an ionic conductivity of 20 mS cm^{-1} and a specific capacitance of 677 F g^{-1}. After 2000 cycles, the device maintained 66% of the initial capacitance (Figure 20.14). The authors indicated that although the cyclic stability is poor, the retained energy density was excellent – 327.11 Wh kg^{-1}.

Another example of a supercapacitor that contains polyzwitterion was presented by Diao et al. (2020). The polymer gel was fabricated through copolymerization of acrylamide, SBMA zwitterionic monomer, 2-acrylamido-2-methyl-1-propanesulfonic acid anionic monomer, and vinyl silica nanoparticles. The authors reported an ionic conductivity of 3.4 S m^{-1}, capacitance of 60.6 F g^{-1} at 0.5 A g^{-1}, and capacitance retention of ca. 98% after 1000 cycles.

The last example in this section showing the flexible supercapacitor that contains zwitterion was described by Han et al. (2020). The authors introduced acroleic acid and acrylamide with ZnSO$_4$ electrolyte through free radical polymerization while sodium alginate served as a natural polymer

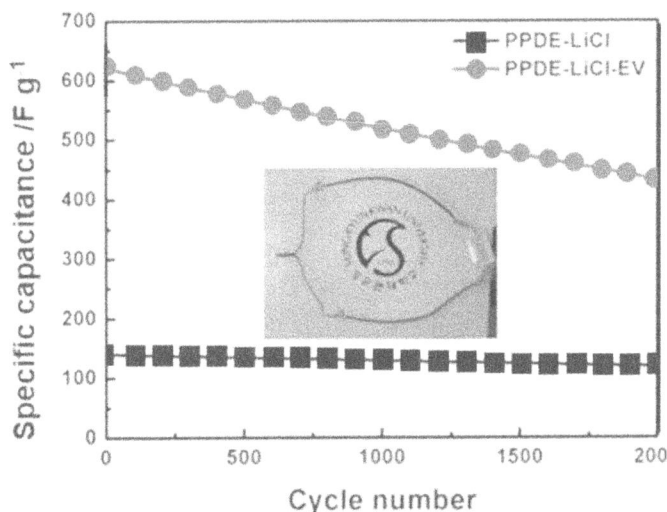

FIGURE 20.14 Capacitance retention of the supercapacitor cell with and without EV in the electrolyte during 2000 cycles. (Copyright Hyeon et al. 2019.)

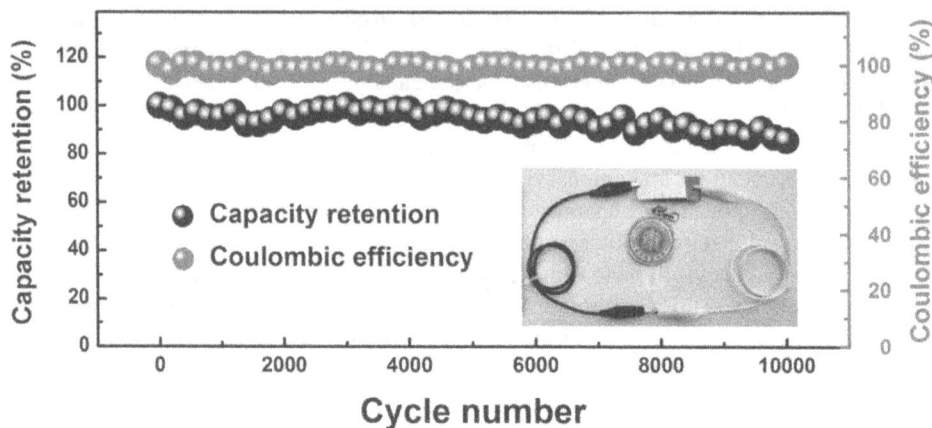

FIGURE 20.15 Cycling performance of solid-state zinc-ion hybrid supercapacitor. Inset shows the LED powered by SC. (Copyright Han et al. 2020.)

backbone. The prepared solid-state zinc-ion hybrid supercapacitor was characterized by a high maximum energy density of 286.6 Wh kg^{-1} at the 220 W kg^{-1} power density. Moreover, it retained capacity at the level of 95.4% after 2000 cycles at 2 A g^{-1} and 85.9% after 10,000. It also exhibited a wide and stable window of 2.4 V. What is of key importance, the possible application was also verified. Namely, it was found out that the SC was able to light up a green LED (Figure 20.15) and power an electrical watch (Figure 20.16) for at least 168 hours.

20.4.2 RECHARGEABLE BATTERIES

The battery is a device to power electrical tools, gadgets, or even some machines. It can be said that the modern world strongly depends on rechargeable energy store and researches on this topic are of key importance especially in terms of low-carbon society. Due to the best energy density and long shelf-life, lithium-ion batteries are the most popular. Nevertheless, rechargeable batteries suffer from such problems as leakage of safety due to the usage of liquid electrolytes. However, researchers are working on the development of the ideal electrolyte and one of the solutions is the introduction of zwitterions as they are able to facilitate the salt dissociation and dissociated ions migration (Ohno, Yoshizawa-Fujita, and Kohno 2018). Below, we will show a few examples of the application of zwitterions in electrolytes for batteries.

Lu et al. (2020) proposed a strategy to enhance the conducting properties of a single lithium-ion solid conducting polymer by introducing zwitterions that serve as synergistic ion dissociator and

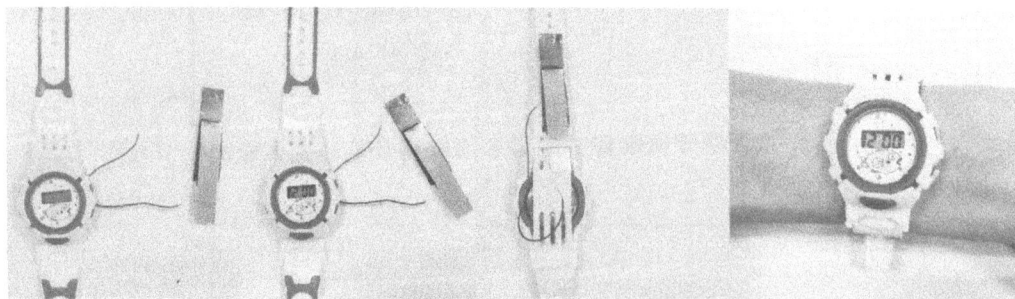

FIGURE 20.16 Photographs of an electrical watch powered by solid-state zinc-ion hybrid supercapacitor. (Copyright Han et al. 2020.)

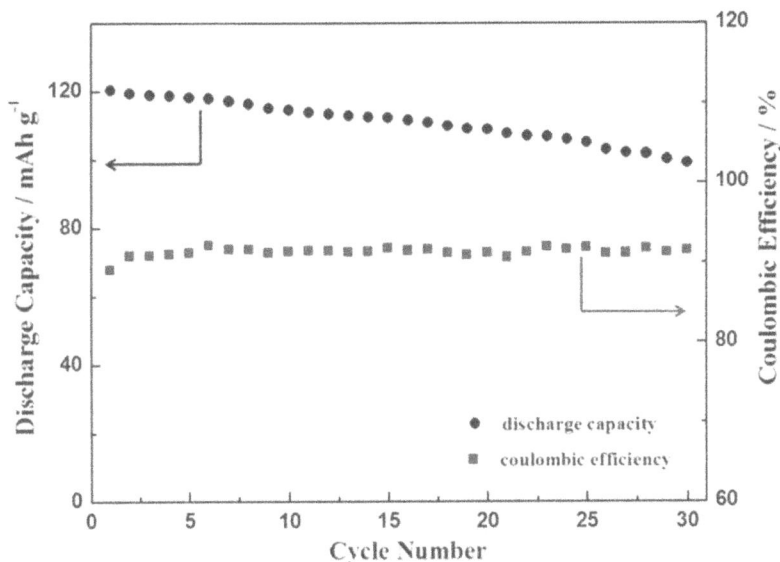

FIGURE 20.17 Cycle performance of LiFePO$_4$ half-cell with PIL/PC film as electrolyte. (Copyright Lu et al. 2017.)

plasticizer to reduce the poly(ethylene oxide) crystallinity. After adding the zwitterions, the ionic conductivity of the electrolyte can be increased by more than one magnitude. Those results are attributed to the strong intermolecular interactions between zwitterions and single lithium conductors as well as presumably the creation of fast ion transport channels under the applied electric field. Moreover, the LiPO$_4$ half-cell was constructed and it showed decent cycling performance. The initial discharge capacity was equal to 120.8 mAh g^{-1} at 80°C and no significant drop was observed after 50 cycles.

Lu et al. (2017) studied flexible polymer electrolyte films based on a lithium-containing zwitterionic poly(ionic liquid) (PIL) with and without the addition of propylene carbonate (PC). Such electrolytes exhibited good thermal stability and high electrochemical ones. The Li/LiFePO$_4$ half-cell with PIL and PC showed an initial discharge capacity of 120.4 mAh g^{-1} at 30°C (Figure 20.17). Nevertheless, the discharge capacity slowly decreased to 99.2 mAh g^{-1} in the 30th cycle. However, the authors concluded that the obtained data suggest that zwitterion-based polymer electrolytes offer a prospect in batteries.

Zwitterion-type molecule was also used as an additive to the electrolyte in vanadium redox flow battery (VRFB) (Wei et al. 2020). The influence of pyridinium propyl sulfobetaine (PPS) with negatively charged sulfonic group and positively charged pyridine group on electrochemical performance and thermal stability of positive electrolyte was investigated. The cells containing PPS showed discharge capacity retention of ca.76% after 100 cycles that is over 5% higher than that demonstrated by the cell without zwitterion. Moreover, the working mechanism of PPS on the electrode/electrolyte interface was proposed. The presence of PPS results in hampering of tight Helmholtz layer formation and in lowering down the voltage drop as PPS can be adsorbed at the electrode surface (Figure 20.18). Therefore, the overpotential is decreased and in turn it yields better reversibility for VO^{2+}/VO$_2^+$ redox reaction. This is yet another way of improving the performance of VRFB together with modification of VRFB membranes using zwitterions described in a previous section in the work of Dai et al. (2018).

Leng et al. (2020) investigated Zn metal batteries (ZMB) with polyzwitterionic hydrogel electrolyte (PZHE) of ionic conductivity of 33 mS cm^{-1}. The long cyclic life up to 3500 h was achieved for symmetric ZMBs containing PZHE, and the heavily loaded VS$_2$ cathode exhibited capacity

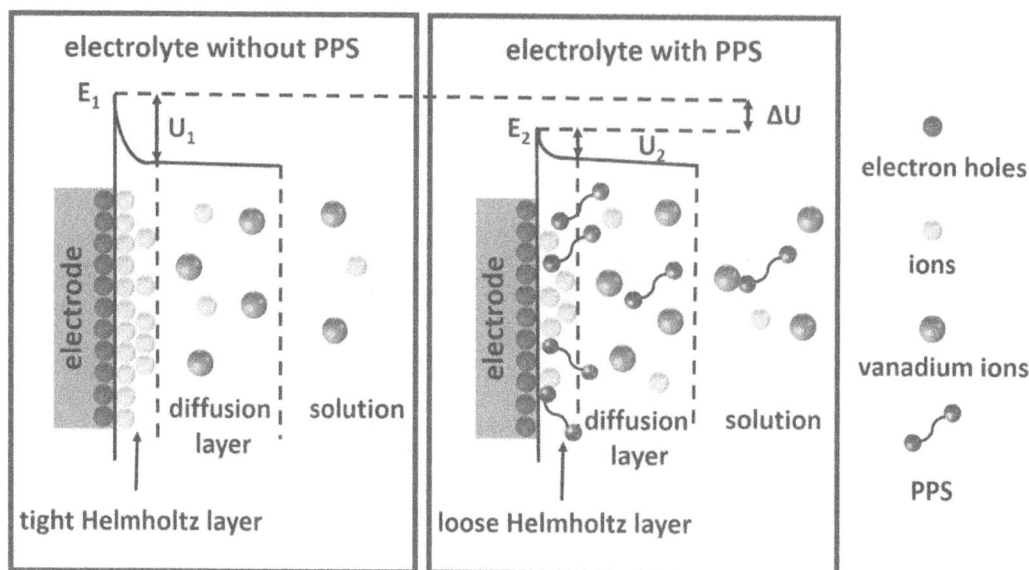

FIGURE 20.18 The influence of PPS presence on the electrode/electrolyte interface. E_1, E_2 – practical reaction potential for the V(IV)/V(V) redox couple on the electrode; U_1, U_2 – voltage drop.

retention of 83% after 500 cycles. In the case of Zn/PHZE/MnO$_2$ cells, the specific capacity of 200 mAh g^{-1} with the coulombic efficiency of above 99% after 600 cycles was obtained. Moreover, cutting and soaking, low-temperature anti-icing, and self-healing tests were performed to show the superior performance of ZMBs.

20.4.3 SOLAR CELLS

Solar cells are devices enabling the conversion of solar light into electricity. They are classified into four generations (Khatibi, Astaraei, and Ahmadi 2019) depending on the technological development and the materials used for their fabrications. The silicon-based solar cells (SiSoC), containing single and multicrystalline, are still the most common SoC on the market and are considered as a first generation. Due to the high amount of used material (thick Si layer) affecting the cost of solar cells, the second generation was introduced. Those devices were characterized by the significantly reduced thickness of an active layer, even to few nanometers, e.g., CuTe, CIGS (copper, indium, gallium, diselenide), and CZTS (copper, zinc, tin, sulfide). The rapid development in nanotechnology, especially in materials engineering and nanoprocessing, impacts further achievements also in SoC. Thereby the third generation arose. In this group, dye-sensitized solar cells, perovskite solar cells (PSC), organic solar cells (OCS), photoelectrochemical, and quantum dot-based cells are involved. In the latest fourth generation, the most advanced hybrid materials, where inorganic crystals are combined with polymer matrix, are applied. What should be highlighted is that the whole device can be flexible and therefore can take different shapes. There are already constructed OSC that can reach over a dozen percent of photon conversion efficiency (PCE), e.g., even above 21% (Wang et al. 2021).

According to the literature, one of the widely used approaches within OSC targeted to reach efficiency on the level of SiSoC, is based on the development within the photoactive layer. However, it turns out that the whole device architecture also includes the chemical character of interlayers playing an important role in the charge transport from the photoactive layer to the current collector. Such films deposited between the active layer and the substrate minimizes the energy level mismatch arising due to the difference in the working function of the contact and the highest occupied molecular orbital/the lowest unoccupied molecular orbital (HOMO/LUMO) positions (see

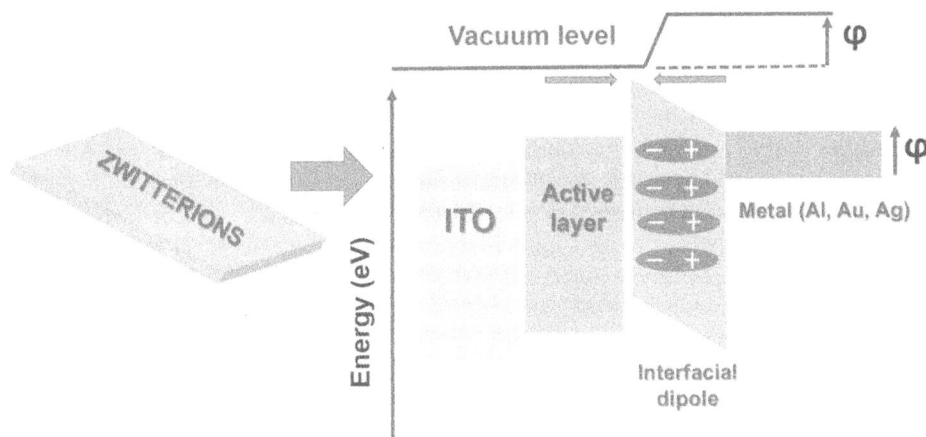

FIGURE 20.19 The schematic representation of the mechanism responsible for the improvement of the bandgap alignment in OSC via application of a molecule exhibiting zwitterion nature. ITO stands for indium tin oxide, while φ stays for working function.

Figure 20.19). As promising candidates, various molecules exhibiting zwitterionic character were proposed (Xiao et al. 2020, Hu et al. 2021).

20.4.3.1 Small Molecule Zwitterions for OSCs

In 2012, zwitterions were used by Sun et al. (2012b) in polymer solar cells for the first time, acting as an electrolyte interlayer. The formation of rhodamine 101 layer on Al contact can be realized via solution processing or thermal deposition. Introduction of rhodamine 101 into the cell based on fullerene derivatives (PCBM) compared to bare Ca/Al cathode results in PCE improvement from 4.57% to 6.15%. This change is explained by the lowering of the work function of the ZI-containing cathode. However, a similar impressive enhancement has not been achieved for cells with other active materials due to the difference in reactivity at the active layer/contact interface. The same compound was used in solar cells of inverted structures that are regarded as more stable compared to the typical OSCs (Sun et al. 2012a). In this case, the deposition of zwitterionic molecules leads to better performance via improving the charge transport at the boundary between the active layer and contact owing to the reduced work function of ITO.

After preliminary work with rhodamine as ZI molecule, perylene diimide (PDI) was included in solar cells. PDI was selected due to its high electron affinity, the opportunity to facile modifications, stability upon irradiation, and high conductivity. The elevated conductivity results from the π-type interactions characterized by self-organization, while the particular value depends on the layer thickness (Yue et al. 2012). As an alternative, PDI containing amino N-oxide (PDINO) (Zhang et al. 2014) has been proposed, exhibiting conductivity on the same level irrespectively of the film thickness. The device where such interlayer was applied reached an efficiency of 8.24% and thus, doubled the performance of the cell with a conventional structure. Explanation of such enhancement on the molecular level is based on the very short spacer length of the metal | active layer interphase. This is yet another argument confirming the importance of intermolecular dipolar interactions of zwitterions and their still undiscovered application potential.

Further development of interfacial materials concerns the fabrication of naphthalene diimide N-oxide (NDIO) for cathode formation (Zhao et al. 2015). This compound exhibits good solubility in water, high transparency in the visible waverange, as well as well-matched energy levels. It was shown that NDIO could decrease the work function of ITO to 3.6 eV. The application of NDIO together with the active layer of (PBDT-TS1)/$PC_{71}BM$ ((6,6)-phenyl-C71-butyric acid methyl ester) results in an efficiency of 9.51% that is two times higher compared to the device without ZI content.

Similar effect was achieved when the other active layer was used in SoC construction, namely, PBDTT-EFT/N2200 (2,6-bis(trimethyltin)-4,8-bis(5-(2-ethylhexyl)thiophen-2-yl)benzo[1,2-b:4,5-b]dithiophene/Poly{[N,N'-bis(2-octyldodecyl)-naphthalene-1,4,5,8-bis(dicarboximide)-2,6-diyl]-alt-5,5'-(2,2'-bithiophene)}) (Zhao et al. 2015). According to the presented results, NDIO presence reduces recombination rate and simultaneously the improved ohmic contact was ensured. The decrease of the work function was ascribed to the formation of molecular dipoles, while the smooth, uniform morphology of NDIO affected the effective collection of charge. Again, the NDIO has only one bond long spacer between charges, providing a very large net intermolecular interaction, and therefore a significant improvement in the cell performance.

Other groups of ZI, isoindigo derivatives were also successfully implemented in SoC. This class of materials exhibits a deficiency of electrons and is typically used for the synthesis of donor-acceptor molecules. The derivatives of isoindigo dedicated for solar cells, are composed of pyridinium and sulfonate moieties (Han et al. 2016). This structure shows excellent film formation and electron transfer abilities. The investigated interlayers exhibit absorption shifted toward visible light and proceed to deep LUMO levels, and finally the device can reach an efficiency of even 9.12%. Spectra obtained by the irradiation of a material with a beam of X-rays, indicated that polar functional groups were accumulated on the surface of the interlayer, forming the molecular dipole and enhanced electron extraction.

Other approaches presented by Wang et al. (2016) concerned the usage of star-shaped ZI. The derivative of triphenylamine acted as a core and different polar moieties (e.g., quaternary ammonium ions, amino N-oxides, sulfobetaines) as the arms of the star. Due to the simple structures of those molecules (TFN, TFB, TFO, TFS, see Figure 20.20), the processing cost can be reduced and thus the increase of technology readiness level can be achieved. As shown in Figure 20.20, TFB, TFO, and TFS are quaternary ammonium ions, amino N-oxide and sulfobetaine ion derived from the neutral TFN molecule, respectively.

Listed compounds absorb light in a similar waverange of 300–450 nm because of the same π-conjugated structure and have almost the same position of HOMO and LUMO levels. Interestingly, in the cases of TFS and TFO, the presence of zwitterion does not enhance the performance more than the other listed compounds. Cell characteristics are rather similar to other molecules used in this work, such as cationic TFN and neutral TFB. The improvement of PCE is rather attributed to the star-shaped topology in the described work.

Among listed molecules, the usage of TFB results in the highest device efficiency up to 10.1% with open-circuit voltage (V_{OC}) of 0.78 V and short circuit current (Jsc) of 17.57 mA cm^{-2}. Such a high value of short circuit current density results from the much increased mobility of charge carriers while superb Ohmic contact with the active layer improves charge transfer at the interface.

The other group that gained attention and can be used to construct an active layer are molecules characterized by the twisted geometrical configuration being responsible for diminishing the π–π interactions. Those π–π interactions cause aggregation and thereby can be characterized by aggregation-induced emission (AIE), and in consequence, a quenching effect can be achieved. The combination of AIE luminogen and tetraphenylethylene (TPE) as a cathode material results in the highly efficient solar cell with PCE reaching 8.94% that is much higher than the conventional device (PCE = 7.31%) (Wang et al. 2017).

20.4.3.2 Polyzwitterions for OSC

Further development in ZI in application in SoC concerns the application of polymeric molecules. The first approach was based on the thiophenes (Liu et al. 2013) containing methylene or butylene units. When such material was inserted between an active layer and a cathode in the case of PTB7:PC$_{71}$BM (poly[[4,8-bis[(2-ethyl-hexyl)oxy]benzo[1,2-b:4,5-b0]dithiophene-2,6-diyl][3-fluoro-2-(2-ethylhexy)carbonyl]thieno[3,4-b]thiophenediyl]]: [6,6]-phenyl-C71-butyric acid methyl ester) and Al/Ag, the photoconversion efficiency grows rapidly from nearly 1% to even 5.8% for Ag and from nearly 4.4% to 7.4% for Al electrode. As already mentioned and schematically shown in

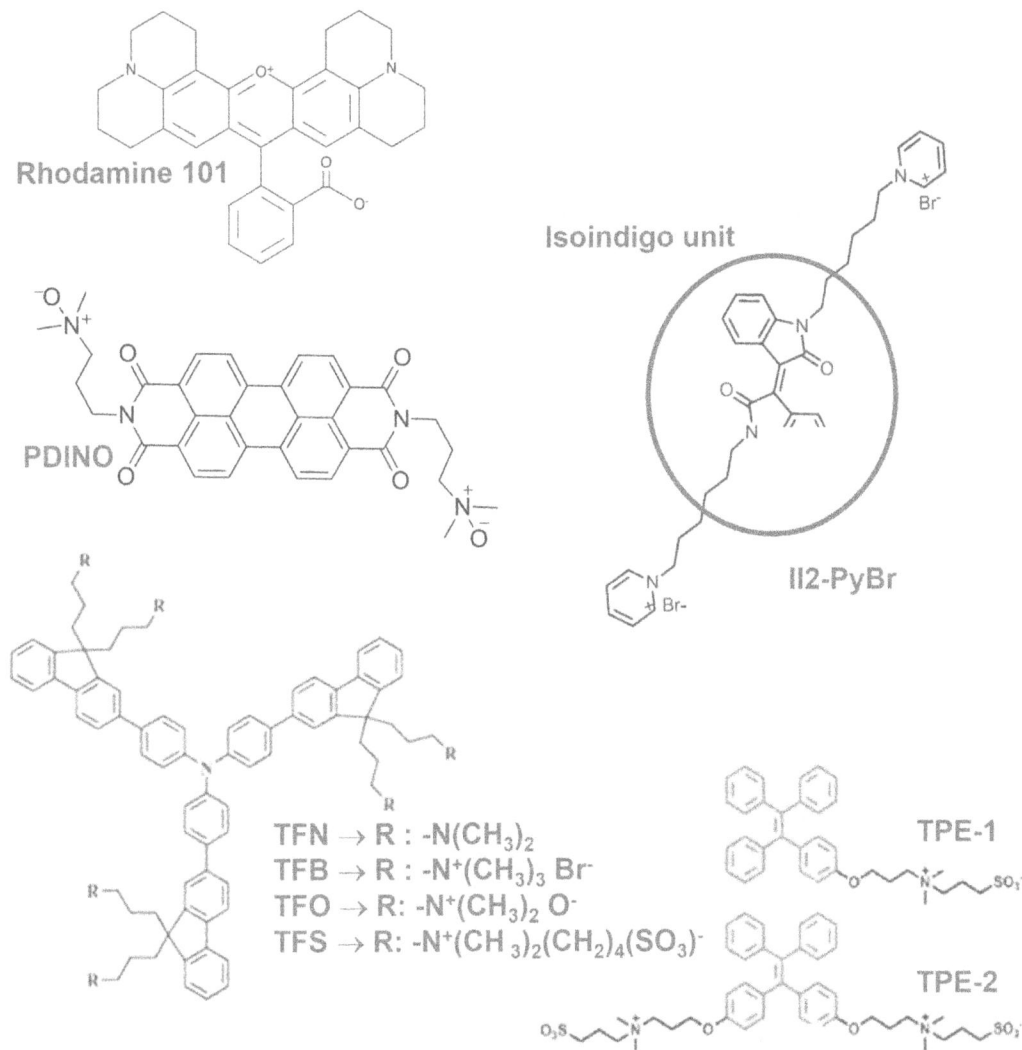

FIGURE 20.20 The structures of different ZI applied in smart organic devices (R – substituent).

Figure 20.19, ZI induces dipole layer formation at the interface. Among the group of PSBMA based zwitterions, the performance was verified depending on the alkyl chain. The longer the chain, the higher photoconversion efficiency. To enrich a group of conjugate ZI, the series of compounds based on the diketopyrrolopyrrole and isoindigo backbone, and spirobiindane bisphenol as side chains were elaborated (Page et al. 2015). Despite the numerous reports showing the improved efficiency of the device where polymeric ZI was applied, the low electron mobility became problematic. Of course, some solutions for this issue were proposed, e.g., via the nucleophilic opening of the ring with PDI or fullerene diamines.

20.4.3.3 Zwitterion Materials for Perovskite Solar Cells

As has been mentioned, perovskite solar cells attracted a lot of attention and currently many efforts are undertaken to put this technology into the worldwide market (Park 2015). According to the recent approaches, the whole processing technology can be regarded as a cost-effective alternative compared to well-known SiSoC. However, still even some slight changes in the construction and application of some novel molecules can affect the overall photostability and photoconversion efficiency. As for high impact, the following criteria should be fulfilled: the wide absorption from the visible toward near-infrared light, high charge mobility, and their long diffusion length, dielectric constant as well as the convenient fabrication process. Among many additional achievements, the improvement of the charge transfer at the interface was also explored as was in the case of organic solar cells. To improve the efficiency of PSC, different metal oxides (e.g., TiO_2, NiO_x, VO_x, ZnO) and metal salts (e.g., LiF) form an additional film at the active layer/contact interface have been proposed so far. However, the application of that structures in the PSC makes the fabrication route complicated and still some simplification is needed.

Following the recent development in OSC via the introduction of ZI in the interfacial region, a similar approach was applied toward PSC. In works dedicated to the use of ZI in perovskite solar cells, zwitterionic PSBMA was introduced as an interlayer (Chen et al. 2019). After modification via the deposition of ultrathin PSBMA film (3 nm), the photoconversion efficiency reached almost 19.16%, while it equals 17.31% without interlayer. It should be underlined that the materials are simply deposited via spin-coating, and therefore in the real fabrication of the ZI film can be replaced by ink-printing.

Moreover, interfacial layers both inorganic and organic ones, can be combined and used together in the PSC. This route was applied by Sun et al. (2015), that proposed LiF and rhodamine 101 placed at the interface between PCBM active layer and Ag contact. The conversion was enhanced by 2.1% and reached 13.2%, while when only ZI was applied, the efficiency was lowered by 0.1%. It should be noted that the presence of rhodamine 101 does not change the absorbance spectrum of the whole device, but the stability of PSBM is preserved while its surface is smooth and uniform.

20.4.4   ZI AND PZ FOR ORGANIC LIGHT-EMITTING DIODES

Organic light-emitting diodes have gained lastly also enormous attention since their potential application in currently very popular wearable devices and multifunctional smart systems (Zou et al. 2020). The light emission originates from small organic molecules and semiconducting polymers. Among others, currently the great challenge concerns achieving efficient, pure light emission with the narrow waverange as well as long-term stability. This goal can be reached by introducing a stack of thin layers where each of them has particular function, namely, hole and electron transporting layer (HTL and ETL, respectively), emitting layer, and contacts. Similar to the SoC, the charge transfer between those films plays a crucial role and is directly related to the alignment of energy levels. Following that, the efficient charge injection to the emitter is very important. Therefore, the introduction of the additional interface between ETL and both contacts can improve the overall performance taking advantage from the bipolar nature of zwitterions.

Similar to SoC, also in the case of OLED small molecules of ZI were introduced. The new class of ZI, being the derivative of tetrakis(1-imidazolyl)borate, was applied to the solution-based fabrication of organic light-emitting diode (Li et al. 2009). It was found that the improvement of the OLED efficiency is related to the length of the alkyl chain. As the chain in ZI is elongated, both the efficiency and current density increase while the value of onset potential is reduced. Other groups of ZI contained the core of (4,4′-sulfonylbis(4,1-phenylene)bis(oxy))bis(ethane-2,1-diyl) (Min et al. 2013). Those kinds of structures used for the interface formation were found to stabilize emission. Taking into account the approach of ZI application in SoC, PSBMA was introduced to organic diodes in both conventional and inverted construction. The performance was ten times enhanced comparing to the device without PSBMA, indicating balanced electron and hole current. Regarding the value of V_{OC}, the decrease was recorded being an effect of the superior charge injection. ZI can also take the role of the emitting layer, as is in the case of the iridium(III) complex, $(piq)_2Ir^+[BP(O-)(OH)]$. The derivative of this compound was used together with the polymer forming PVK:PBD (olyN-vinylcarbazole with 2-tert-butylphenyl-5-biphenyl-1,3,4-oxadiazole) mixture. The prepared device exhibited high luminescence efficiency of 12.62 cd/A with the reduced exciton quenching. Also, ZI based on Ir(III) containing complex with iridium core and N,N′-heteroatomic ligand with units possessing negative charge was applied. Such structures provide a very bright blue color.

20.5 CONCLUSIONS

Zwitterionic materials are commonly used in anti-fouling and anti-adsorption applications. Yet, they attract increasing attention from different fields such as energy storage and conversion. Their molecular structure leads to several unique properties that make them perfect candidates for inter-layers in solar cells and organic light-emitting diodes, electrolytes in supercapacitors as well as separators in batteries.

Strong dipole moments of ZI or PZ created at the phase boundary between the active layer and electrodes in solar cells and OLED lead to better alignment of bands and facilitate charge transfer. The same property that leads to anti-fouling allows to increase the performance and long-term stability of the solar cells and OLED. Strong hydration of zwitterionic electrolytes and the ability to change PZ conformations in the presence of salts or electric fields play a major role in enhancing both the energy and power density of supercapacitors. Although polyzwitterions are not conductive themselves, they can greatly enhance the performance of existing energy storage and energy conversion solutions. We believe, however, that there is still a vastly unexplored area involving both fundamental physiochemical properties of zwitterions and polyzwitterions as well as their application in many devices in different fields including energy storage and conversion.

REFERENCES

Afrifah, Vera Afumaa, Isheunesu Phiri, Louis Hamenu, Alfred Madzvamuse, Kwang Se Lee, and Jang Myoun Ko. 2020. "Electrochemical Properties of Poly(2-Acrylamido-2-Methylpropane Sulfonic Acid) Polyelectrolyte Containing Zwitterionic Silica Sulfobetaine for Supercapacitors." *Journal of Power Sources* 479 (December): 228657. doi:10.1016/j.jpowsour.2020.228657.

Armand, Michel, Frank Endres, Douglas R. MacFarlane, Hiroyuki Ohno, and Bruno Scrosati. 2009. "Ionic-Liquid Materials for the Electrochemical Challenges of the Future." *Nature Materials* 8 (8): 621–29. doi:10.1038/nmat2448.

Chemistry (IUPAC), The International Union of Pure and Applied. 2021. "IUPAC – Zwitterionic Compounds/Zwitterions (Z06752)." Accessed March 23. doi:10.1351/goldbook.Z06752.

Chen, Qiaoyun, Ligang Yuan, Ruomeng Duan, Peng Huang, Jianfei Fu, Hui Ma, Xiaocheng Wang, Yi Zhou, and Bo Song. 2019. "Zwitterionic Polymer: A Facile Interfacial Material Works at Both Anode and Cathode in *P-i-n* Perovskite Solar Cells." *Solar RRL* 3 (9): 1900118. doi:10.1002/solr.201900118.

Dai, Jicui, Yichao Dong, Cong Yu, Yaxin Liu, and Xiangguo Teng. 2018. "A Novel Nafion-g-PSBMA Membrane Prepared by Grafting Zwitterionic SBMA onto Nafion via SI-ATRP for Vanadium Redox Flow Battery Application." *Journal of Membrane Science* 554 (May): 324–30. doi:10.1016/j.memsci.2018.03.017.

Diao, Wenjing, Linlin Wu, Xiaofeng Ma, Lei Wang, Ximan Bu, Wei Ni, Xinfeng Yang, and Ying Fang. 2020. "Reversibly Highly Stretchable and Self-healable Zwitterion-Containing Polyelectrolyte Hydrogel with High Ionic Conductivity for High-Performance Flexible and Cold-resistant Supercapacitor." *Journal of Applied Polymer Science* 137 (34): 48995. doi:10.1002/app.48995.

"Free Radical Photopolymerization of Multifunctional Monomers." 2016. *Three-Dimensional Microfabrication Using Two-Photon Polymerization*, William Andrew Publishing, 62–81. doi:10.1016/B978-0-323-35321-2.00004-2.

Han, Jianxiong, Youchun Chen, Weiping Chen, Chengzhuo Yu, Xiaoxian Song, Fenghong Li, and Yue Wang. 2016. "High Performance Small-Molecule Cathode Interlayer Materials with D-A-D Conjugated Central Skeletons and Side Flexible Alcohol/Water-Soluble Groups for Polymer Solar Cells." *ACS Applied Materials & Interfaces* 8 (48): 32823–32. doi:10.1021/acsami.6b10900.

Han, Lu, Hailong Huang, Xiaobin Fu, Junfeng Li, Zhongli Yang, Xinjuan Liu, Likun Pan, and Min Xu. 2020. "A Flexible, High-Voltage and Safe Zwitterionic Natural Polymer Hydrogel Electrolyte for High-Energy-Density Zinc-Ion Hybrid Supercapacitor." *Chemical Engineering Journal* 392 (July): 123733. doi:10.1016/j.cej.2019.123733.

Higaki, Yuji, Motoyasu Kobayashi, and Atsushi Takahara. 2020. "Hydration State Variation of Polyzwitterion Brushes through Interplay with Ions." *Langmuir* 36 (31): 9015–24. doi:10.1021/acs.langmuir.0c01672.

Hu, Yichuan, Guang Yang, Bo Liang, Lu Fang, Guanglong Ma, Qin Zhu, Shengfu Chen, and Xuesong Ye. 2015. "The Fabrication of Superlow Protein Absorption Zwitterionic Coating by Electrochemically Mediated Atom Transfer Radical Polymerization and Its Application." *Acta Biomaterialia* 13 (February): 142–49. doi:10.1016/j.actbio.2014.11.023.

Hu, Zhao, Biao Yang, Jingsheng Miao, Tingting Li, Muhammad Umair Ali, Chaoyi Yan, Osamu Goto, et al. 2021. "π-Conjugated Zwitterion for Dual-Interfacial Modification in High-Performance Perovskite Solar Cells." *Chemical Engineering Journal* 416 (July): 129153. doi:10.1016/j.cej.2021.129153.

Hyeon, Suh-Eun, Jung Yong Seo, Jong Wook Bae, Woo-Jae Kim, and Chan-Hwa Chung. 2019. "Faradaic Reaction of Dual-Redox Additive in Zwitterionic Gel Electrolyte Boosts the Performance of Flexible Supercapacitors." *Electrochimica Acta* 319 (October): 672–81. doi:10.1016/j.electacta.2019.07.043.

Kharlampieva, Eugenia, Vladimir A. Izumrudov, and Svetlana A. Sukhishvili. 2007. "Electrostatic Layer-by-Layer Self-Assembly of Poly(Carboxybetaine)s: Role of Zwitterions in Film Growth." *Macromolecules* 40 (10): 3663–68. doi:10.1021/ma062811e.

Khatibi, Ali, Fatemeh Razi Astaraei, and Mohammad Hossein Ahmadi. 2019. "Generation and Combination of the Solar Cells: A Current Model Review." *Energy Science & Engineering* 7 (2): 305–22. doi:10.1002/ese3.292.

Kobayashi, Motoyasu, Kazuhiko Ishihara, and Atsushi Takahara. 2014. "Neutron Reflectivity Study of the Swollen Structure of Polyzwitterion and Polyeletrolyte Brushes in Aqueous Solution." *Journal of Biomaterials Science, Polymer Edition* 25 (14–15): 1673–86. doi:10.1080/09205063.2014.952992.

Laschewsky, André. 2014. "Structures and Synthesis of Zwitterionic Polymers." *Polymers* 6 (5): 1544–1601. doi:10.3390/polym6051544.

Laschewsky, André, and Axel Rosenhahn. 2019. "Molecular Design of Zwitterionic Polymer Interfaces: Searching for the Difference." *Langmuir* 35 (5): 1056–71. doi:10.1021/acs.langmuir.8b01789.

Lee, Haesung A., Eunsook Park, and Haeshin Lee. 2020. "Polydopamine and Its Derivative Surface Chemistry in Material Science: A Focused Review for Studies at KAIST." *Advanced Materials* 32 (35): 1907505. doi:https://doi.org/10.1002/adma.201907505.

Leng, Kaitong, Guojie Li, Jingjing Guo, Xinyue Zhang, Aoxuan Wang, Xingjiang Liu, and Jiayan Luo. 2020. "A Safe Polyzwitterionic Hydrogel Electrolyte for Long-Life Quasi-Solid State Zinc Metal Batteries." *Advanced Functional Materials* 30 (23): 2001317. doi:10.1002/adfm.202001317.

Li, Bowen, Priyesh Jain, Jinrong Ma, Josh K. Smith, Zhefan Yuan, Hsiang-Chieh Hung, Yuwei He, et al. 2019. "Trimethylamine *N*-Oxide–Derived Zwitterionic Polymers: A New Class of Ultralow Fouling Bioinspired Materials." *Science Advances* 5 (6): eaaw9562. doi:10.1126/sciadv.aaw9562.

Li, Huaping, Yunhua Xu, Corey V. Hoven, Chunzeng Li, Jung Hwa Seo, and Guillermo C. Bazan. 2009. "Molecular Design, Device Function and Surface Potential of Zwitterionic Electron Injection Layers." *Journal of the American Chemical Society* 131 (25): 8903–12. doi:10.1021/ja9018836.

Liu, Caihong, Jongho Lee, Jun Ma, and Menachem Elimelech. 2017. "Antifouling Thin-Film Composite Membranes by Controlled Architecture of Zwitterionic Polymer Brush Layer." *Environmental Science & Technology* 51 (4): 2161–69. doi:10.1021/acs.est.6b05992.

Liu, Feng, Zachariah A. Page, Volodimyr V. Duzhko, Thomas P. Russell, and Todd Emrick. 2013. "Conjugated Polymeric Zwitterions as Efficient Interlayers in Organic Solar Cells." *Advanced Materials* 25 (47): 6868–73. doi:10.1002/adma.201302477.

Lu, Fei, Xinpei Gao, Aoli Wu, Na Sun, Lijuan Shi, and Liqiang Zheng. 2017. "Lithium-Containing Zwitterionic Poly(Ionic Liquid)s as Polymer Electrolytes for Lithium-Ion Batteries." *The Journal of Physical Chemistry C* 121 (33): 17756–63. doi:10.1021/acs.jpcc.7b06242.

Lu, Fei, Gaoran Li, Yang Yu, Xinpei Gao, Liqiang Zheng, and Zhongwei Chen. 2020. "Zwitterionic Impetus on Single Lithium-Ion Conduction in Solid Polymer Electrolyte for All-Solid-State Lithium-Ion Batteries." *Chemical Engineering Journal* 384 (March): 123237. doi:10.1016/j.cej.2019.123237.

Malinauskas, Mangirdas. 2013. "Ultrafast Laser Nanostructuring of Photopolymers: A Decade of Advances." *Physics Reports* 533:1–31.

Matyjaszewski, Krzysztof. 2012. "Atom Transfer Radical Polymerization (ATRP): Current Status and Future Perspectives." *Macromolecules* 45 (10): 4015–39. doi:10.1021/ma3001719.

Min, Chao, Changsheng Shi, Wenjun Zhang, Tonggang Jiu, Jiangshan Chen, Dongge Ma, and Junfeng Fang. 2013. "A Small-Molecule Zwitterionic Electrolyte without a π-Delocalized Unit as a Charge-Injection Layer for High-Performance PLEDs." *Angewandte Chemie International Edition* 52 (12): 3417–20. doi:10.1002/anie.201209959.

Ohno, Hiroyuki, Masahiro Yoshizawa-Fujita, and Yuki Kohno. 2018. "Design and Properties of Functional Zwitterions Derived from Ionic Liquids." *Physical Chemistry Chemical Physics* 20 (16): 10978–91. doi:10.1039/C7CP08592C.

Page, Zachariah A., Feng Liu, Thomas P. Russell, and Todd Emrick. 2015. "Tuning the Energy Gap of Conjugated Polymer Zwitterions for Efficient Interlayers and Solar Cells." *Journal of Polymer Science Part A: Polymer Chemistry* 53 (2): 327–36. doi:10.1002/pola.27349.

Pan, Shanshan, Meng Yao, Jiahe Zhang, Bosen Li, Chunxian Xing, Xianli Song, Peipei Su, and Haitao Zhang. 2020. "Recognition of Ionic Liquids as High-Voltage Electrolytes for Supercapacitors." *Frontiers in Chemistry* 8 (May): 261. doi:10.3389/fchem.2020.00261.

Park, Nam-Gyu. 2015. "Perovskite Solar Cells: An Emerging Photovoltaic Technology." *Materials Today* 18 (2): 65–72. doi:10.1016/j.mattod.2014.07.007.

Peng, Xu, Huili Liu, Qin Yin, Junchi Wu, Pengzuo Chen, Guangzhao Zhang, Guangming Liu, Changzheng Wu, and Yi Xie. 2016. "A Zwitterionic Gel Electrolyte for Efficient Solid-State Supercapacitors." *Nature Communications* 7 (1): 11782. doi:10.1038/ncomms11782.

Schlenoff, Joseph B. 2014. "Zwitteration: Coating Surfaces with Zwitterionic Functionality to Reduce Nonspecific Adsorption." *Langmuir* 30 (32): 9625–36. doi:10.1021/la500057j.

Scott, Philip J., Glenn A. Spiering, Yangyang Wang, Zach D. Seibers, Robert B. Moore, Rajeev Kumar, Bradley S. Lokitz, and Timothy E. Long. 2020. "Phosphonium-Based Polyzwitterions: Influence of Ionic Structure and Association on Mechanical Properties." *Macromolecules* 53 (24): 11009–18. doi:10.1021/acs.macromol.0c02166.

Sun, Kuan, Jingjing Chang, Furkan Halis Isikgor, Pengcheng Li, and Jianyong Ouyang. 2015. "Efficiency Enhancement of Planar Perovskite Solar Cells by Adding Zwitterion/LiF Double Interlayers for Electron Collection." *Nanoscale* 7 (3): 896–900. doi:10.1039/C4NR05975A.

Sun, Kuan, Baomin Zhao, Amit Kumar, Kaiyang Zeng, and Jianyong Ouyang. 2012a. "Highly Efficient, Inverted Polymer Solar Cells with Indium Tin Oxide Modified with Solution-Processed Zwitterions as the Transparent Cathode." *ACS Applied Materials & Interfaces* 4 (4): 2009–17. doi:10.1021/am201844q.

Sun, Kuan, Baomin Zhao, Vajjiravel Murugesan, Amit Kumar, Kaiyang Zeng, Jegadesan Subbiah, Wallace W. H. Wong, David J. Jones, and Jianyong Ouyang. 2012b. "High-Performance Polymer Solar Cells with a Conjugated Zwitterion by Solution Processing or Thermal Deposition as the Electron-Collection Interlayer." *Journal of Materials Chemistry* 22 (45): 24155. doi:10.1039/c2jm35221d.

Vaterrodt, Anne, Barbara Thallinger, Kevin Daumann, Dereck Koch, Georg M. Guebitz, and Mathias Ulbricht. 2016. "Antifouling and Antibacterial Multifunctional Polyzwitterion/Enzyme Coating on Silicone Catheter Material Prepared by Electrostatic Layer-by-Layer Assembly." *Langmuir* 32 (5): 1347–59. doi:10.1021/acs.langmuir.5b04303.

Wang, Zhenguo, Zuojia Li, Xiaopeng Xu, Ying Li, Kai Li, and Qiang Peng. 2016. "Polymer Solar Cells Exceeding 10% Efficiency Enabled via a Facile Star-Shaped Molecular Cathode Interlayer with Variable Counterions." *Advanced Functional Materials* 26 (26): 4643–52. doi:10.1002/adfm.201504734.

Wang, Can, Zhiyang Liu, Mengshu Li, Yujun Xie, Bingshi Li, Shuo Wang, Shan Xue, et al. 2017. "The Marriage of AIE and Interface Engineering: Convenient Synthesis and Enhanced Photovoltaic Performance." *Chemical Science* 8 (5): 3750–58. doi:10.1039/C6SC05648B.

Wang, Xiao, Kasparas Rakstys, Kevin Jack, Hui Jin, Jonathan Lai, Hui Li, Chandana Sampath Kumara Ranasinghe, et al. 2021. "Engineering Fluorinated-Cation Containing Inverted Perovskite Solar Cells with an Efficiency of >21% and Improved Stability towards Humidity." *Nature Communications* 12 (1): 52. doi:10.1038/s41467-020-20272-3.

Wang, Wei, Jianhai Yang, Ershuai Zhang, Yang Lu, and Zhiqiang Cao. 2017. "L-Carnitine Derived Zwitterionic Betaine Materials." *Journal of Materials Chemistry B* 5 (44): 8676–80. doi:10.1039/C7TB02431B.

Wei, Xianli, Suqin Liu, Jue Wang, Zhen He, Kuangmin Zhao, Yuliang Yang, Bingjun Liu, Rongjiao Huang, and Zhangxing He. 2020. "Boosting the Performance of Positive Electrolyte for VRFB by Employing Zwitterion Molecule Containing Sulfonic and Pyridine Groups as the Additive." *Ionics* 26 (6): 3147–59. doi:10.1007/s11581-020-03481-0.

Xiang, Yuan, Rong-Guang Xu, and Yongsheng Leng. 2020. "Molecular Understanding of Ion Effect on Polyzwitterion Conformation in an Aqueous Environment." *Langmuir* 36 (26): 7648–57. doi:10.1021/acs.langmuir.0c01287.

Xiao, Ke, Renxing Lin, Qiaolei Han, Yi Hou, Zhenyuan Qin, Hieu T. Nguyen, Jin Wen, et al. 2020. "All-Perovskite Tandem Solar Cells with 24.2% Certified Efficiency and Area over 1 cm² Using Surface-Anchoring Zwitterionic Antioxidant." *Nature Energy* 5 (11): 870–80. doi:10.1038/s41560-020-00705-5.

Yang, Wei, Hong Xue, Louisa R. Carr, Joseph Wang, and Shaoyi Jiang. 2011. "Zwitterionic Poly(Carboxybetaine) Hydrogels for Glucose Biosensors in Complex Media." *Biosensors and Bioelectronics* 26 (5): 2454–59. doi:10.1016/j.bios.2010.10.031.

Yue, Wan, Aifeng Lv, Jing Gao, Wei Jiang, Linxiao Hao, Cheng Li, Yan Li, et al. 2012. "Hybrid Rylene Arrays via Combination of Stille Coupling and C–H Transformation as High-Performance Electron Transport Materials." *Journal of the American Chemical Society* 134 (13): 5770–73. doi:10.1021/ja301184r.

Zhang, Zhi-Guo, Boyuan Qi, Zhiwen Jin, Dan Chi, Zhe Qi, Yongfang Li, and Jizheng Wang. 2014. "Perylene Diimides: A Thickness-Insensitive Cathode Interlayer for High Performance Polymer Solar Cells." *Energy & Environmental Science* 7 (6): 1966. doi:10.1039/c4ee00022f.

Zhao, Ning, Zhen Wang, Chao Cai, Heng Shen, Feiyue Liang, Dong Wang, Chunyan Wang, et al. 2014. "Bioinspired Materials: From Low to High Dimensional Structure." *Advanced Materials* 26 (41): 6994–7017. doi:https://doi.org/10.1002/adma.201401718.

Zhao, Kang, Long Ye, Wenchao Zhao, Shaoqing Zhang, Huifeng Yao, Bowei Xu, Mingliang Sun, and Jianhui Hou. 2015. "Enhanced Efficiency of Polymer Photovoltaic Cells via the Incorporation of a Water-Soluble Naphthalene Diimide Derivative as a Cathode Interlayer." *Journal of Materials Chemistry C* 3 (37): 9565–71. doi:10.1039/C5TC02172C.

Zhou, Rong, Peng-Fei Ren, Hao-Cheng Yang, and Zhi-Kang Xu. 2014. "Fabrication of Antifouling Membrane Surface by Poly(Sulfobetaine Methacrylate)/Polydopamine Co-Deposition." *Journal of Membrane Science* 466 (September): 18–25. doi:10.1016/j.memsci.2014.04.032.

Zhu, Liping. 2017. "Cathode Modification in Planar Hetero-Junction Perovskite Solar Cells through a Small-Molecule Zwitterionic Carboxylate." *Organic Electronics*, 48: 204–210.

Zou, Shi-Jie, Yang Shen, Feng-Ming Xie, Jing-De Chen, Yan-Qing Li, and Jian-Xin Tang. 2020. "Recent Advances in Organic Light-Emitting Diodes: Toward Smart Lighting and Displays." *Materials Chemistry Frontiers* 4 (3): 788–820. doi:10.1039/C9QM00716D.

21 High-Performance Conducting Polymer Nanocomposites for EMI Shielding Applications

Avinash R. Pai[1], Gopika G. Nair[2],
Preema C. Thomas[2], and Sabu Thomas[1]
[1]International and Inter University Center for Nanoscience and Nanotechnology, (IIUCNN), and School of Chemical Sciences, Mahatma Gandhi University, Kottayam, Kerala, India
[2]Department of Physics, CMS College (Autonomous) Kottayam, Kerala, India

CONTENTS

21.1 INTRODUCTION

Electromagnetic pollution is an aftermath of the excessive use of electromagnetic (EM) radiations used to power electronic modules that function in a broad spectrum of microwave region. The intermixing of EM radiations often leads to a phenomenon known as electromagnetic interference, which disrupts the performance of electronic devices and is also envisaged to have adverse effects on human health. EM pollution is an aftereffect of the increasing use of microwave radiation in telecommunication and wireless technologies. One of the main consequences of EM pollution is the huge exposure of humans to electric/magnetic fields generated from various electronic devices such as mobile phones, Bluetooth, Wi-Fi, etc. [1]. Prolonged exposure can damage the overall metabolism of the human body [2]. To overcome this issue, efficient EMI shielding materials need to be developed to resist EM waves and suppress them from excess leakage to nearby fields. There are three different mechanisms for shielding EM radiations: shielding by reflection, absorption and

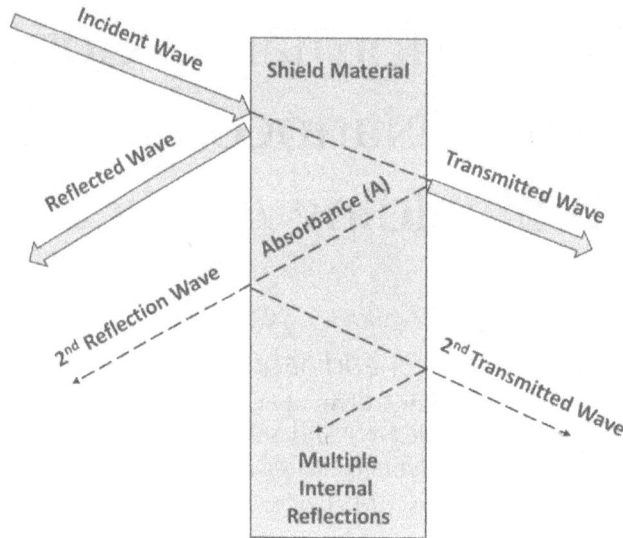

FIGURE 21.1 Basic EMI shielding mechanism viz absorption, reflection and multiple internal reflection of EM waves.

multiple internal reflection as denoted in Figure 21.1 [3]. In the reflection mechanism, the shielding occurs by the reflection of incoming EM waves due to the impedance mismatch between the free space and shielding material. The absorption of incident EM wave is ascribed due to the current developed within the material due to ohmic/magnetic hysteresis losses. EMI shielding by multiple internal reflection occurs mainly in porous materials where the incident EM waves get scattered inside the cellular structure and thereby get dissipated as heat energy. Among the three mechanisms shielding by absorption is the most effective and desired method for attenuation of EM energy [4]. In practice, thin metallic sheets were used as shielding materials that only reflect and scatter the EM radiations in all directions. Due to the difficulties in processing, high density, poor corrosion resistance and lack of flexibility of metallic shields, they are barely used nowadays. For many high-end applications like military, defense and aerospace, lightweight shielding materials with low thickness, good flexibility coupled with strong microwave absorption over a wide frequency range are considered ideal EMI shielding materials

21.2 EMI SHIELDING EFFECTIVENESS

EMI shielding effectiveness (EMI SE) is defined as the capacity of a material to attenuate electromagnetic signals incident on its surface. More specifically, it can be defined as the ratio between the strength of the incident wave to that of the transmitted wave.

$$SE_P = 10\log\left(P_{in} / P_{out}\right) \tag{21.1}$$

$$SE_E = 10\log\left(E_{in} / E_{out}\right) \tag{21.2}$$

$$SE_H = 10\log\left(H_{in} / H_{out}\right) \tag{21.3}$$

Equations (21.1–21.3) represent the SE in terms of plane wave strength (P), electric field (E) and magnetic field (H), respectively [5, 6]. The total shielding effectiveness includes the contribution from different shielding mechanisms, namely reflection (SE_R), absorption (SE_A) and multiple internal reflection (SE_M).

Thus the equation for total shielding effectiveness is given as

$$SE_T = SE_R + SE_A + SE_M \tag{21.4}$$

21.2.1 EMI Shielding Mechanisms

The term EMI shielding describes the process of blocking undesired EM radiations from any device. The materials used for this purpose are referred to as EMI shielding materials. To block the incident EM radiations from getting transmitted, the shielding material should be capable of reflecting or absorbing them effectively. The properties of the shielding material and the nature of the EM radiation determine the shielding mechanism of a particular EM shield [6].

21.2.1.1 Shielding by Reflection

Reflection is the fundamental shielding mechanism that basically depends on the mismatch in impedance of the incident radiation and the shielding material [6]. For reflecting the incident EM waves, the material should consist of mobile charge carriers (electrons/holes), which can interact with the electric/magnetic fields of the incident EM energy. For a material to possess superior shielding by reflection, it should have a minimum conductivity of 1 S/cm [7]. Materials with high conductivity are best suitable for shielding by reflection. Metals are good examples for this type of EM shield due to the abundance of free electrons [7, 8]. The loss due to reflection is given by the Equation (21.5),

$$SE_R\,(dB) = -10\log_{10}\left(\frac{\sigma_T}{16\omega\varepsilon_0\,\mu_r}\right) \tag{21.5}$$

σ_T is the total conductivity of the material, ε_0 is the absolute permittivity and μ_r is the relative permeability.

21.2.1.2 Shielding by Absorption

The absorption of EM waves is the most effective mechanism for EMI shielding. Shielding by absorption is primarily due to the interaction with electric or magnetic dipoles present in the shield material. When the dipole interacts with the radiation, it will start to oscillate. The energy for the oscillation is obtained from the radiation and is dissipated as heat energy. Thus, no radiation is reflected back and does not give rise to secondary pollution. This is the reason for considering shielding by absorption much superior to shielding by reflection. The materials with a high dielectric constant can be suitable EM absorbing materials since they have sufficient electric dipoles [9]. The shielding by absorption only depends on the physical properties of the shield material and it is independent of the source of the field. When the EM wave gets transmitted through the shielding material, its strength will decrease exponentially. At a particular distance inside the material, its strength gets reduced to 1/e of its initial strength and is referred to as the skin depth (δ) of the material. The absorption loss is proportional to the thickness of the shield material, permeability, conductivity and frequency on the incident EM wave [10–12]. The loss due to absorption can be expressed by the Equation (21.6),

$$SE_A\,(dB) = -20\frac{t}{\delta}\log_{10}e \tag{21.6}$$

where δ is the skin depth and t is the thickness of the shield material.

21.2.1.3 Shielding by Multiple Internal Reflections

Attenuation of the EM waves by its reflection from different surfaces or interfaces of a shielding material is referred to as shielding by multiple internal reflections. The material should have a large surface area or a highly porous nature for multiple internal reflections to occur. Normally composite materials with one or more fillers have a high surface area. The shielding effect due to this phenomenon can be neglected when the distance between the reflecting surfaces/interfaces is large when compared to the skin depth of the shield [10–12]. Loss due to multiple internal reflections is described by the Equation (21.7),

$$SE_M = 20\log_{10}\left(1-e^{-2t/\delta}\right) = 20\log_{10}\left|\left(1-10^{\frac{-SE_A}{10}}\right)\right| \tag{21.7}$$

From the equation, we can analyze that SE_M depends on the loss due to absorption (SE_A). For materials with high absorption loss, the loss due to multiple reflection can be neglected since the wave will get attenuated before it reaches the second surface of the shield material. Practically we consider that for $SE_A > 10$ dB, SE_M is negligible [12].

21.3   EMI SHIELDING MEASUREMENT TECHNIQUES

Network analyzers are the standard instruments used to measure the EMI shielding effectiveness of samples experimentally. Network analyzers are classified into two categories, namely scalar network analyzer (SNA) and vector network analyzer (VNA). SNA measures only the amplitude of the electrical signal, while VNA can measure both amplitude and phase of the electrical signal. VNA is preferred over SNA since it can measure complex quantities such as complex permittivity and complex permeability.

In a two-port VNA instrument, the incident and transmitted waves are represented mathematically in terms of S-parameters. The S-parameters are S_{11} (or S_{22}) and S_{12} (or S_{21}). In this, S_{11} indicates the reflection from the material under test (MUT) and S_{21} indicates the transmission through MUT. The coefficient of shielding mechanism is connected to these parameters as [13]

$$\text{Reflection coefficient } R = |S_{11}|^2 = |S_{22}|^2 \tag{21.8}$$

$$\text{Transmission coefficient } T = |S_{12}|^2 = |S_{21}|^2 \tag{21.9}$$

$$\text{Absorption coefficient } A = (1-R-T) \tag{21.10}$$

In addition, VNA can also be used to estimate the shielding effectiveness due to reflection (SE_R) and shielding effectiveness due to absorption (SE_A). If the MUT is uniform in nature, we can consider that $S_{11} = S_{22}$ and $S_{12} = S_{21}$.

Shielding effectiveness of the material can be expressed as [14]

$$SE_T = 10\log_{10}\left(\frac{1}{T}\right) = 10\log_{10}\left(\frac{1}{|S_{21}|^2}\right) \tag{21.11}$$

$$SE_R = 10\log_{10}\left(\frac{1}{1-R}\right) = 10\log_{10}\left(\frac{1}{1-|S_{11}|^2}\right) \tag{21.12}$$

$$SE_A = 10\log_{10}\left(\frac{1-R}{T}\right) = 10\log_{10}\left(\frac{1-|S_{21}|^2}{|S_{21}|^2}\right) \tag{21.13}$$

The coefficients provide information about the quantity of radiation lost by the mechanisms of reflection and absorption.

A two-port network analyzer will emit EM radiation with intensity I from each side in the required frequency range. Also, it analyzes the radiation which is reflected from and transmitted through the MUT. There are three main configurations for analyzing the solid samples: waveguide, coaxial line and free space arrangements.

21.3.1 Waveguide Method

Waveguide method consists of a rectangular section. The sample is placed at a particular distance from the waveguide end. The size of the waveguide will change in accordance with the range of frequency for which it is designed. As the frequency increases, the size of the waveguide will decrease. In this system, the preparation of the sample is comparatively easy. But this method can be used for only small ranges of frequency, which is a limitation. So if we want to measure high frequency, we have to use multiple waveguides [14].

Advantages of this method include

- Easy manufacture of the sample by assuring good mechanical contact between the sample and the sample holder. This will help to avoid errors in the measurement by avoiding microwave leakage.
- Possibility of measuring dielectric properties of the sample along the two main directions of electric field which is useful for the analysis of composite materials.

Disadvantages of this method include

- The measurable range of frequency is small and hence several waveguides are required to measure high frequencies.
- The dimension of the sample holder is different for different waveguides, which increases the possibility of air gap and error in the measurement.
- A new calibration with a suitable calibration kit is required for each waveguide which introduces new errors.
- The propagation of EM waves in the waveguide is either in transverse electric mode (TE) or transverse magnetic mode (TM).

21.3.2 Coaxial Line Method

Coaxial line method consists of an inner and outer conductor. The sample is placed in between these two conductors at a particular distance from the ends. In this configuration, the sample must be in the form of a rectangular toroid. High-frequency measurements can be taken on the same sample in this method. Some difficulties are there in this method which are, the thickness of the toroid should be very precise and also there should not be any air gaps in the sample otherwise there are chances for error to occur [13, 14].

Advantages of this method are

- Comparatively extended range of frequency and thus one sample material can be used to measure a large band of frequency.
- Here TEM mode is used. This enables easy comparison with the SE measurement in free space.
- Calibration is done only once. Dielectric properties can be analyzed from single measurements of the scattering parameters

The disadvantages are mainly the complications in the preparations of the samples.

21.3.3 FREE SPACE METHOD

Free space method is a noncontact measurement method. It consists of two opposite antennae, flanking the sample. Its main advantage is that this can be used for a wide range of frequencies. Also it is possible to change the angle of incidence of the radiation on MUT. But the dimension of the sample must be comparatively large. For instance, to measure a few GHz, the sample should have a dimension of 30–40 cm [13].

21.4 POLY(ANILINE) (PANI)-BASED EMI SHIELDING MATERIALS

Polyaniline (PANI) is one of the most studied intrinsically conducting polymer (ICP) due to its outstanding electronic and optical properties. PANI exists in three different redox forms, namely emeraldine (E), leucoemeraldine (L) and pernigraniline (P). Among these, only emeraldine salt of PANI is electrically conductive and is considered useful for designing sensors and EMI shielding materials. Figure 21.2 shows the structure of different forms of PANI [15].

PANI is most widely synthesized by employing *in situ* electro-oxidative polymerization of aniline monomer onto surfaces like fibers and fabrics to form robust and conductive composites. Seema Joon et al., fabricated flexible and free-standing polyaniline-carbon fiber (PANI-CF) composite via *in situ* emulsion polymerization technique in the presence of β-naphthalene sulfonic acid (NSA) which acts as a dopant as well as a surfactant. Phenolic resin was employed as a binder into which polyaniline/carbon nanofibers (PANI/CNF) composite was incorporated and homogenized to form a uniform suspension. The solution was later vacuum filtered to form thin sheets and compression molded at 100°C to form free-standing PANI-CF sheets as depicted in Figure 21.3(a). These composite sheets demonstrated high electrical conductivity of ~1.02 S/cm with improved thermal stability and mechanical properties. The maximum EMI SE value of 31.9 dB was recorded at 12.4 GHz and 1.5 mm sheet thickness with an absorption-dominated shielding mechanism [16]. The possible effects for high SE value and its mechanism are schematically represented in Figure 21.3(b).

The microwave absorbing property of any material is highly correlated with its microstructure and can be boosted by incorporating magnetic nanoparticles (NPs). Weidong Zhang et al., fabricated 1D molybdenum–amine complex with 1D nanowire-like structure. PANI was later coated

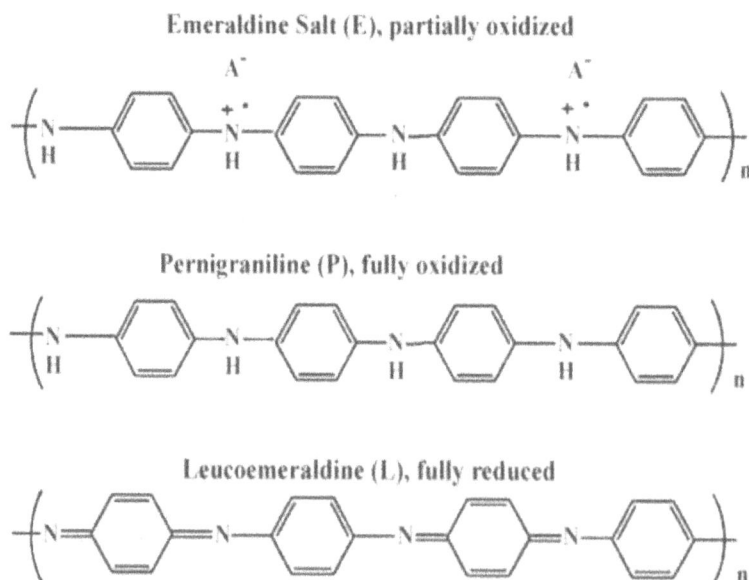

FIGURE 21.2 The various oxidation states of polyaniline. Reproduced with permission from Ref [15].

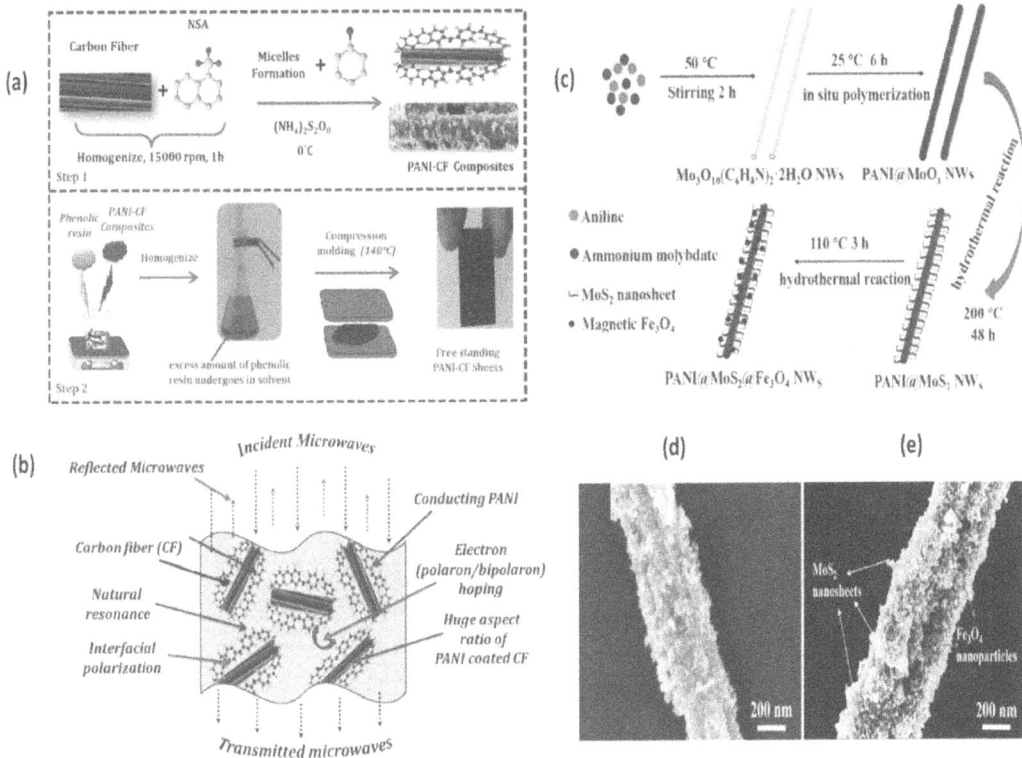

FIGURE 21.3 (a) Step 1: Schematic for the synthesis of PANI-CF composite and Step 2: Fabrication of phenolic resin binded PANI/CF sheets by compression molding. (b) Schematic representation of shielding mechanism in PANI/CF composite. (c) Synthesis and growth mechanism of PANI/MoS$_2$/Fe$_3$O$_4$ NWs. (d, e) SEM images of PANI/MoS$_2$/Fe$_3$O$_4$ NWs. Reproduced with permission from Refs. [16] and [17].

onto these nanowires (NWs) by *in situ* polymerization to form PANI/MoO$_x$ NWs. Subsequently, MoS$_2$ nanosheets were self-assembled onto the NWs and decorated with Fe$_3$O$_4$ by the hydrothermal reaction to form PANI/MoS$_2$/Fe$_3$O$_4$ NWs as shown in Figure 21.3(c). From the SEM images (see Figure 21.3(d, e)), the presence of MoS$_2$ nanosheets and Fe$_3$O$_4$ NPs is very much visible. These heterostructures of PANI and magnetic NPs showcased a minimum R$_L$ value of −49.7 dB (>99.99% microwave absorption) at 1.3 mm thickness [17]. Interestingly, a broad microwave absorption bandwidth (R$_L$ < −10dB) of 6.4 GHz was observed at 10.2–16.7 GHz for these nanostructures, which makes them as favorable materials for high-performance EM absorption.

Nanocellulose is one of the most versatile plant-based nanofibers that possess excellent processability and mechanical properties. Due to the poor processability of ICPs, it is often used in conjunction with nanocellulose to fabricate robust and flexible nanopapers and ultra-lightweight aerogels for microwave absorption and EMI shielding applications [18].

Aniline monomer can be *in situ* polymerized onto the surface of cellulose nanofibers (CNFs) to form stable PANI/CNF suspensions or hydrogels. Gopakumar et al., converted these suspensions into flexible nanopapers by using vacuum filtration followed by hot pressing to form highly conductive PANI/CNF papers, as shown in Figure 21.4(a). These nanopapers exhibited a high EMI SE of −23 dB at 8.2 GHz and 1 mm paper thickness with a predominant absorption of EM energy (ca. 87%) as the prime shielding mechanism (see Figure 21.4(b)) [19]. The formation of a uniform coating of PANI on CNF was achieved due to the formation of intermolecular H bonding and is envisaged for its high electrical conductivity and EMI shielding performance. As an extension to this work, Pai et al. adopted a similar strategy wherein, PANI/CNF hydrogels were freeze-dried at

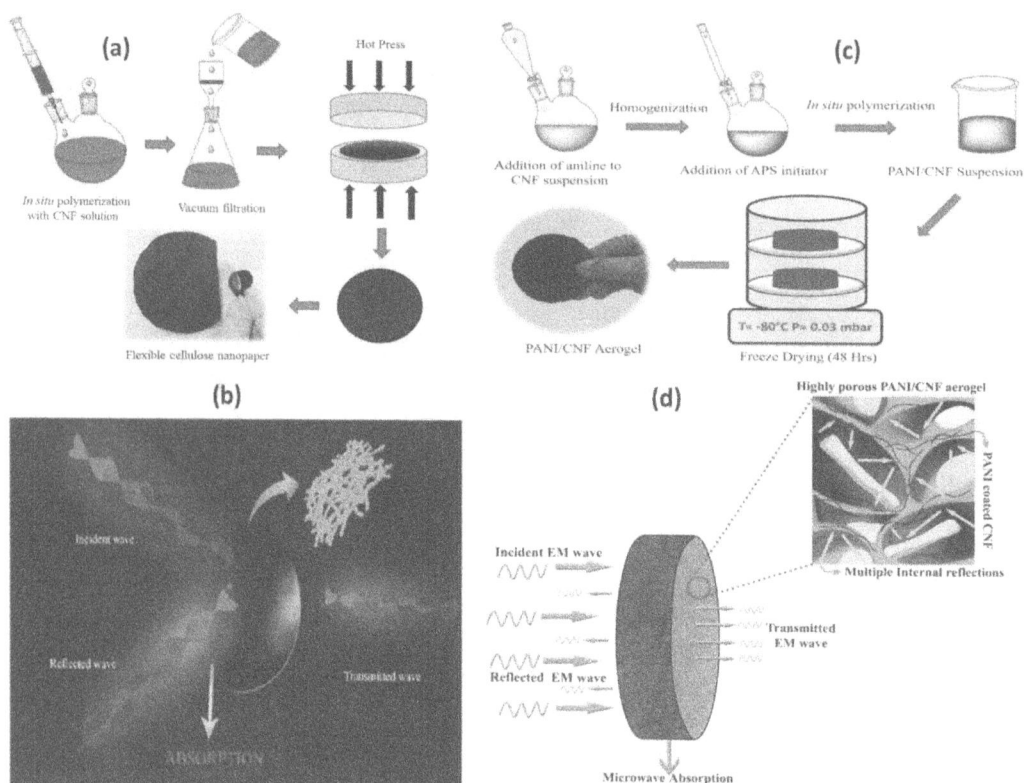

FIGURE 21.4 (a) Schematic demonstrating the fabrication of flexible PANI/CNF nanopaper. (b) The possible EMI shielding mechanism in PANI/CNF paper. (c) The fabrication process of PANI/CNF aerogels. (d) A schematic depicting plausible shielding mechanism for EM radiation trapped within PANI/CNF aerogel. Reproduced with permission from Refs. [19] and [20].

−85°C/0.05 mbar pressure for 48 hours to form highly porous and conductive aerogels, as shown in Figure 21.4(c). These porous EMI shields showcased a maximum shielding value of −32 dB with a superior specific EMI SE of 1667 dB.cm³· g⁻¹. The PANI/CNF aerogels demonstrated higher microwave absorption (95%) when compared to its nanopapers at a similar concentration. Due to the high porosity (98.6%) in aerogels, the EM waves that enter the bulk of the aerogel get trapped within the porous structure and undergo multiple internal scattering within the pores and get dissipated as heat energy [20]. A possible microwave absorption phenomenon in PANI/CNF aerogels is illustrated in Figure 21.4(d). This makes PANI/CNF aerogels and nanopapers as sustainable candidates for suppressing EM pollution by absorption of excess EM energy emanating from electronic devices.

21.5 POLY(PYRROLE) (PPy)-BASED EMI SHIELDING MATERIALS

Poly(pyrrole) (PPy) is one of the most extensively studied π-electron conjugated conducting polymers, due to its facile synthesis, high electrical conductivity, stability in oxidized form and excellent redox properties. In the oxidized state, PPy is a positively charged conducting polymer and over oxidation usually deteriorates its electrical conductivity. It is also insulating and mechanically weak in its neutral state. Due to its excellent electronic properties, PPy has proven to be a prospective candidate for technological applications such as supercapacitors, strain sensors, batteries, anti-static coatings for fabrics, EMI shielding, etc. One main advantage of PPy is that, unlike PANI, it can be synthesized in an aqueous medium. The intrinsic properties of PPy are also highly dependent on the polymerization conditions like monomer: oxidant ratio, surfactants, etc. Chemical polymerization method is used mainly to synthesize polypyrrole, as shown in Figure 21.5 [21]. This method

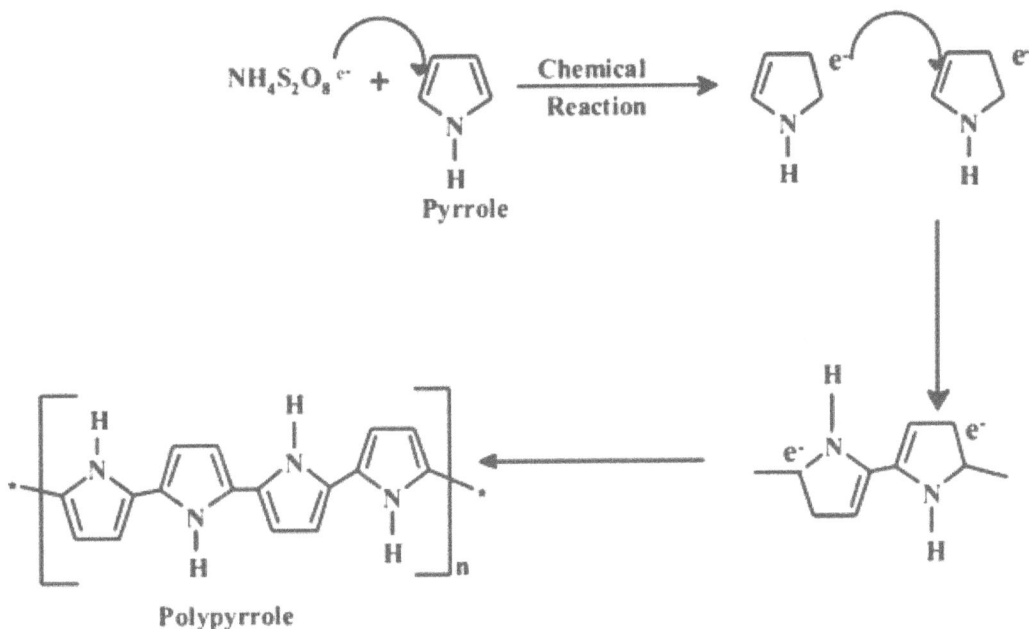

FIGURE 21.5 Reaction mechanism of polypyrrole. Reproduced with permission from Ref. [21].

involves the oxidation of the pyrrole monomer in the aqueous medium, in the presence of oxidizing agents like ammonium persulfate (APS) and anionic surfactants like sodium bis (2-ethylhexyl) sulfosuccinate, dodecyl benzene sulfonic acid (DBSA) and its sodium salt and sodium dodecyl sulfate.

The formation of a long order π conjugated network of PPy onto the surface of cellulose nanofiber has attracted much attention for designing efficient microwave absorbers. This imparts high electrical conductivity and the electrical dipoles present in PPy can interact with the incident EM energy. Gopakumar et al. employed this technique to fabricate flexible papers of PPy/CNF via *in situ* polymerization technique. Initially, pyrrole monomer was homogenized in CNF suspension and APS (oxidant) was added dropwise to initiate polymerization. The reaction was performed at 0–5°C for 3–5 hours and the unreacted pyrrole monomers were washed out from the suspension using acetone and distilled water. This suspension was then subjected to a vacuum filtration process and hot-pressed at 70°C to form flexible PPy/CNF papers. The hydrogen bonding interaction between PPy and CNF is envisaged for its high electrical conductivity (0.21 S/cm) and high shielding value of −22 dB with 89% microwave absorption at 1 mm paper thickness [22].

Coating of PPy on industrial textiles and fabrics has been widely used for making smart textiles, wearable sensors and EMI shielding materials, etc. This ideology was adapted by Pragati Gahlout et al. to fabricate a series of electrically conductive fabrics from industrially relevant fabrics such as cotton, nylon, polyester and lycra. These fabrics were immersed in pyrrole and dopant (sodium lauryl sulfate) solution followed by $FeCl_3$ oxidant solution to complete the polymerization. This impregnation technique was repeated four times to enhance the electrical conductivity and EMI shielding performance. It is noteworthy that lycra fabrics demonstrated the highest conductivity ($\sim 3.92 \times 10^{-1}$ S/cm) and EMI shielding value (−26 dB) among all the fabrics examined [23]. This opens new horizons for developing higher performance EMI shielding fabrics for smart wearable applications in defense and military sectors.

21.6 POLY(THIOPHENE) (PTh)-BASED EMI SHIELDING MATERIALS

Due to the high stability in its (un)doped states, excellent electrical conductivity and ease of structural modification make polythiophene and its derivatives as one of the potential materials for electronic applications like resistive memory devices, field-effect transistors, LEDs, biosensors, solar

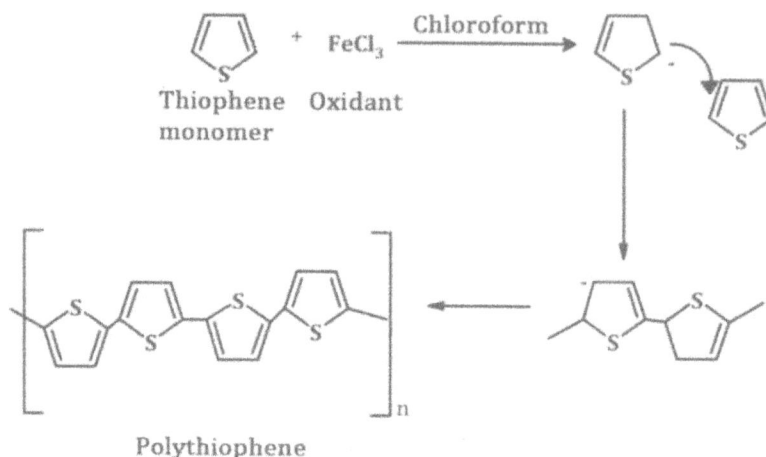

FIGURE 21.6   Reaction mechanism of polythiophene. Reproduced with permission from Ref. [25].

cells, batteries and supercapacitors [24]. In general, there are three major routes to polymerize thiophene, namely electro-polymerization, chemical oxidative polymerization and metal-catalyzed coupling reactions. Polythiophene can be synthesized with different oxidizing agents, such as ferric chloride, copper perchlorate hydrate, iron perchlorate hydrate and ferric chloride hydrate [24]. Figure 21.6 shows a typical reaction scheme for the synthesis of PTh.

The ability to tune the electrical properties makes PTh an ideal candidate for EMI shielding applications. Moreover, the magnetic permeability of PTh composites can also be enhanced by incorporating magnetic fillers like Ni, Co, Mn and ferrites. Erdoğan et al. fabricated conductive PTh/poly(ethyleneterephthalate) composite fibers for EMI shielding application. These composite fibers were fabricated by *in situ* polymerization of thiophene monomer on the surface of polyethylene terephthalate (PET) fibers using $FeCl_3$ as an oxidant in acetonitrile solution. A maximum shielding value of 21 dB was achieved at 20 MHz with absorption dominant shielding as the major factor for attenuation [26]. Such fabrics can also find applications in other sectors like wearable sensors, anti-static fabrics and energy storage devices.

PTh/magnetic hybrids are widely used to improve the microwave absorption characteristics in conducting polymer nanocomposites. Sajid Iqbal et al. synthesized barium ferrite-polythiophene ($BaFe_{12}O_{19}$/PTh) nanocomposite. A high EMI SE value of −43.27 dB was obtained at 11.65 GHz and is envisaged due to the synergistic effect of both dielectric and magnetic losses in these nanocomposites [27]. Doping of ferrites with metals is another approach used to enhance microwave absorption in ICP nanocomposites. In another work, Sajid Iqbal et al. doped bismuth into barium ferrite and *in situ* polymerized with thiophene monomer at various concentrations to form bismuth-doped barium ferrite/PTh nanocomposites. These hybrid materials exhibit a maximum EMI SE value of −47.12 dB at 9.08 GHz with enhanced microwave absorption characteristics in the X band region [28].

21.7   POLY(3,4-ETHYLENEDIOXYTHIOPHENE):POLY(STYRENESULF ONATE) (PEDOT:PSS)-BASED EMI SHIELDING MATERIALS

PEDOT:PSS is a polyelectrolyte that consists of a positively charged conjugated PEDOT part and PSS polymer anions component with a negative charge. The PSS part is incorporated to stabilize the conjugated PEDOT cations and to induce solubility in few polar organic solvents as well as water [29]. This technique is a significant breakthrough that has greatly improved the processability of conjugated polymers and makes PEDOT:PSS unique amongst the family of ICPs. High electrical conductivity and transparency are the most fascinating properties of PEDOT:PSS and has the potential to replace indium tin oxide (ITO) in transparent conductive electrodes, touch screens and other flexible organic devices. The electrical conductivity of these conjugated polymers can be as high as 4600 S/cm, which has also gathered much attention for its use as higher performance EMI shielding materials [30]. Sabyasachi Ghosh et al. fabricated PEG/PEDOT:PSS treated cotton fabric by facile dipping technique. Cotton fabrics were pretreated with NaOH solution and then dipped into PEG/PEDOT:PSS solution and dried at 70°C. At 20 dipping cycles, these fabrics demonstrated high electrical conductivity of 51 S/cm and a maximum EMI SE value of 46.80 dB [31]. Conductive hybrids of PEDOT:PSS with other nanofillers such as graphene, CNT and their analogues have been studied for their microwave absorption characteristics. Nidhi Agnihotri et al. prepared graphene nanoplatelet (GNP)/PEDOT:PSS nanocomposites by *in situ* polymerization of EDOT monomers in the presence of calculated amounts of GNPs. These hybrid nanocomposites exhibited a maximum shielding of 70 dB at 0.8 mm thickness and high electrical conductivity of 6.84 S/cm [32]. It is noteworthy that the main contributing factor towards the shielding is absorption of EM waves which is highly desired for high-end technological applications such as aerospace, military and defense sectors.

Ultra-thin PEDOT:PSS films can be made water-insoluble via crosslinking reaction. This makes crosslinked films useful in exterior applications where it comes in contact with moisture and humidity. Pritom J. Bora et al. fabricated crosslinked ultra-thin conductive films of PEDOT:PSS by dropcasting on a PET film. Divinylsulfone (DVS) was used as a crosslinking agent to form robust films that are water-insoluble. A 9 μm thickness film exhibited an average EMI SE of 40 dB and a maximum electrical conductivity of 769 S/cm with an absorption dominant shielding mechanism [33].

21.8 MXene/ICP HYBRIDS FOR EMI SHIELDING AND MICROWAVE ABSORPTION

In 2011, Yury Gogotsi and coworkers at Drexel University, USA discovered a class of novel 2D transition metal carbides, nitrides and/or carbonitrides known as MXenes. They have a general formula of $M_{n+1}X_nT_x$, where M is an early transition metal like Ti, Zr, V, Ta, Nb, or Mo, X represents C and/or N, and Tx is the functional groups present on the MXene surface [34]. MXene surface chemistry can be altered by introducing surface functional groups such as –F, −OH, or =O. This imparts hydrophilicity and ease of processability for the fabrication of MXene composites in an aqueous solution. Owing to the exceptional electrical conductivity (over 4600 S/cm), large surface areas and low densities, MXenes are one of the frontrunners for EMI shielding application [4]. In order to investigate the hybrid effect, MXenes are combined with ICPs for tuning the microwave absorption characteristics and EMI shielding ability for high-end technological applications like aerospace, military and defense applications.

Ruiting Liu et al. fabricated ultrathin, flexible and freestanding $Ti_3C_2T_x$/(PEDOT:PSS) films with a brick-and-mortar like structure using a facile vacuum-assisted filtration process, as shown in Figure 21.7(a). These composite films showcased a maximum SE value of 42.10 dB @ 11.1 μm thickness with excellent conductivity of 340.5 S/cm (see Figure 21.7(b, c)) [35]. It is worth mentioning that the absorption of EM waves was the prime shielding mechanism in these hybrid nanocomposites. Conducting polymers like PANI can also be combined with MXenes for achieving better microwave absorption and shielding performance. Guang Yin et al. prepared carbon fabric impregnated with MXene and PANI using a layer by layer assembly technique to form conductive and flexible PANI/MXene/CF fabrics. Initially, precursor Ti_3AlC_2 powder was etched using LiF to form exfoliated 2D (Ti_3C_2) MXene sheets. These MXene sheets were later deposited onto the carbon fabric surface via dip-coating method and followed by *in situ* polymerization of aniline monomer, which eventually forms a coating of PANI on these fabrics. A high EM SE of 26.0 dB was achieved at 0.55 mm fabric thickness and showcased superior electrical conductivity of 24.57 S/m with a predominant absorption as the prime shielding mechanism in the X band region [36].

Incorporating magnetic particles into MXenes has enhanced the overall microwave absorption characteristics of MXene hybrid composites due to enhancement in magnetic losses. Zhen Wang et al. synthesized magnetic hybrids of MXene by adding Fe_3O_4/PANI at different weight ratios. This

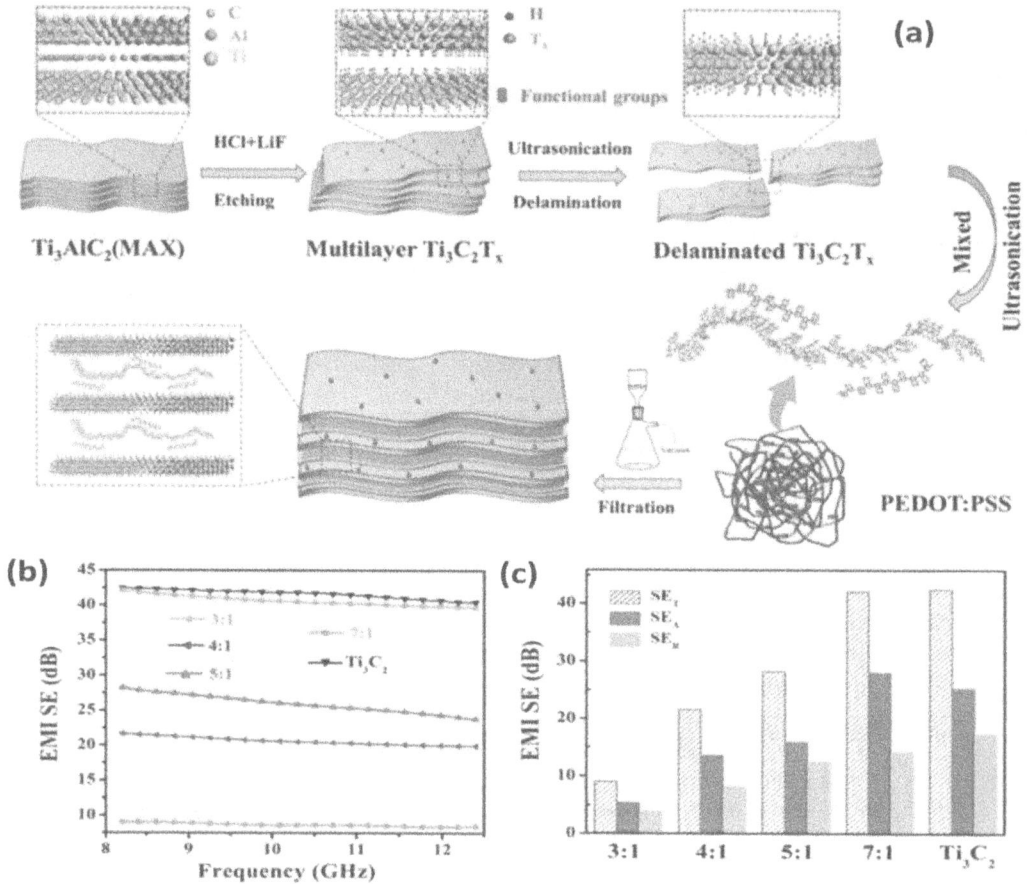

FIGURE 21.7 (a) Schematic representing the assembly protocol for $Ti_3C_2T_x$ MXene/PEDOT:PSS hybrid films. (b) EMI SE of neat $Ti_3C_2T_x$ MXene film and its polymeric composite films at varying ratios. (c) Total shielding value (SE_T), absorption (SE_A), reflection (SE_R) components of pristine $Ti_3C_2T_x$ MXene film and its polymeric composite films at varying ratios. Reproduced with permission from Ref. [35].

MXene/Fe_3O_4/PANI solution was then vacuum filtered to form flexible nanocomposite films. This ultrathin composite film (12.1 μm thick) with $Ti_3C_2T_x$ nanosheets to Fe_3O_4@PANI weight ratio of 12:5 demonstrated a maximum shielding value of 58.8 dB. This value can be enhanced up to 62 dB by increasing the film thickness to 16.7 μm and the concentration of MXene in the nanocomposite. These composite films also showcased superior microwave absorption of EM radiations, opening up its use for higher performance electronic devices [37].

21.9   CONCLUSION AND FUTURE PROSPECTS

To date, numerous combinations of ICPs like PANI, PPy, PTh, PEDTO:PSS and their hybrid combinations with nanofillers and magnetic NPs have been extensively studied for EMI shielding applications. The tunable electrical conductivity of ICPs makes them ideal candidates for fabricating higher-performance EM shielding materials and some of the works reported in the literature are highlighted in Table 21.1. Both the permittivity and permeability values of these nanocomposites are crucial parameters for achieving high microwave absorption and shielding performance. These ICP hybrid composites convert the incident EM radiations into heat energy through dielectric and magnetic losses which arises due to the conductivity and magnetic resonance. Nevertheless, ICP nanocomposites, unlike

TABLE 21.1

Comparison of EMI Shielding Performance of Various ICP Composites Reported in the Literature

Sr No.	Polymer Composite (Matrix/Filler)	Thickness (mm)	Frequency (GHz)	SE_R (dB)	SE_A (dB)	SE_T (dB)	Ref
1.	PANI/MWCNT/PS	–	8.2–12.4	–4.6	–18.7	–23.3	[38]
2.	PANI-MnO$_2$	169 µm	12.4–18	–10	–29	–39	[39]
3.	PANI/PFO	3	8–18	–4	–20	–24	[40]
4.	PTh/SrF	1	15.73	–7.46	–24.18	–31.64	[41]
5.	CF-PTh/graphene	–	12.4–18	–10	–20	–30	[42]
6.	PPy-MWCNT/ polyurethane	3	8.2–12.4	–5.1	–40.9	–46	[43]
7.	CF/PPy/Fe$_3$O$_4$	2	14–15	–10.2	–61.8	–72	[44]
8.	PEDOT:PSS/ graphene-CNT-Fe$_2$O$_3$	0.6	8–12	–10	–120	–130	[45]
9.	PANI/Ni-Co-Fe-P/PI	0.2	12.1	–5.7	–63.70	–69.40	[46]
10.	Lycra/PPy	0.3	8–12	–5.57	–13.13	–18.7	[23]
11.	Cotton/PEG-treated PEDOT:PSS	0.38	8–12	–10.62	–36.17	–46.79	[31]
12.	CDCA/PPy/Fe$_3$O$_4$ aerogel	2	8–12	–7.0	32.4	–39.4	[47]
13.	Ti3C2Tx/Fe$_3$O$_4$/PANI	16.7 µm	8–12	–26	–36	–62	[37]
14.	PPy/BST/RGO/Fe$_3$O$_4$	–	8.2–12	–5.83	–42.75	–48.58	[48]
15.	GF/PEDOT:PSS	–	8–12	–13	78.9	–91.9	[49]

metallic surfaces, microwave absorption predominates reflection mechanism, which is very much essential to curb EM pollution. Reflection of EM wave tends to create further interference with nearby circuitry which deteriorates the lifespan and performance of electronic devices. The combination of ICPs with MXene and their magnetic hybrids can be foreseen as the area where tremendous improvements in microwave absorption and shielding performance can be achieved for future cutting edge technologies like 5G, artificial intelligence (AI), Internet of Things (IoT), etc.

REFERENCES

1. Röösli, M.; Rapp, R.; Braun-Fahrländer, C. Radio and microwave frequency radiation and health–an analysis of the literature. *Gesundheitswesen* **2003**, *65*, 378–392, doi:10.1055/s-2003-40311.
2. Ozdemir, F.; Kargi, A. Electromagnetic waves and human health. Electromagnetic Waves. IntechOpen, London, UK. **2011**.
3. Singh, A.P.; Mishra, M; Dhawan, S.K. Conducting multiphase magnetic nanocomposites for microwave shielding application. *Nanomagnetism* **2014**, *2*, 246–277.
4. Shahzad, F.; Alhabeb, M.; Hatter, C.B.; Anasori, B.; Hong, S.M.; Koo, C.M.; Gogotsi, Y. Electromagnetic interference shielding with 2D transition metal carbides (MXenes). *Science* **2016**, *353*, 1137–1140, doi:10.1126/science.aag2421.
5. Jiang, D.; Murugadoss, V.; Wang, Y.; Lin, J.; Ding, T.; Wang, Z.; Shao, Q.; Wang, C.; Liu, H.; Lu, N.; et al. Electromagnetic interference shielding polymers and nanocomposites—a review. *Polymer Reviews* **2019**, *59*, 280–337.
6. Jagatheesan, K.; Ramasamy, A.; Das, A.; Basu, A. Electromagnetic shielding behaviour of conductive filler composites and conductive fabrics—a review. *Indian Journal of Fibre and Textile Research* **2014**, *39*, 329–342.

7. Lai, K.; Sun, R.J.; Chen, M. yu; wu, H.; Zha, an X. Electromagnetic shielding effectiveness of fabrics with metallized polyester filaments. *Textile Research Journal* **2007**, *77*, 242–246, doi:10.1177/0040517507074033.

8. Roh, J.S.; Chi, Y.S.; Kang, T.J.; Nam, S.W. Electromagnetic shielding effectiveness of multifunctional metal composite fabrics. *Textile Research Journal* **2008**, *78*, 825–835, doi:10.1177/0040517507089748.

9. Sankaran, S.; Deshmukh, K.; Ahamed, M.B.; Khadheer Pasha, S.K. Recent advances in electromagnetic interference shielding properties of metal and carbon filler reinforced flexible polymer composites: A review. *Composites Part A: Applied Science and Manufacturing* **2018**, *114*, 49–71.

10. Abdi, M.M.; Kassim, A.B.; Ekramul Mahmud, H.N.M.; Yunus, W.M.M.; Talib, Z.A. Electromagnetic interference shielding effectiveness of new conducting polymer composite. *Journal of Macromolecular Science, Part A: Pure and Applied Chemistry* **2010**, *47*, 71–75, doi:10.1080/10601320903399834.

11. Sambyal, P.; Singh, A.P.; Verma, M.; Farukh, M.; Singh, B.P.; Dhawan, S.K. Tailored polyaniline/barium strontium titanate/expanded graphite multiphase composite for efficient radar absorption. *RSC Advances* **2014**, *4*, 12614–12624, doi:10.1039/c3ra46479b.

12. Singh, A.P.; Garg, P.; Alam, F.; Singh, K.; Mathur, R.B.; Tandon, R.P.; Chandra, A.; Dhawan, S.K. Phenolic resin-based composite sheets filled with mixtures of reduced graphene oxide, γ-Fe_2O_3 and carbon fibers for excellent electromagnetic interference shielding in the X-band. *Carbon* **2012**, *50*, 3868–3875, doi:10.1016/j.carbon.2012.04.030.

13. Micheli, D.; Pastore, R.; Vricella, A.; Delfini, A.; Marchetti, M.; Santoni, F. Electromagnetic characterization of materials by vector network analyzer experimental setup. In *Spectroscopic Methods for Nanomaterials Characterization*. Elsevier. **2017**; Vol. 2, pp. 195–236.

14. Berber, M.; Inas Hazzaa Hafez, E. *Carbon Nanotubes—Current Progress of Their Polymer Composites*. IntechOpen. **2016**.

15. Jamadade, V.S.; Dhawale, D.S.; Lokhande, C.D. Studies on electrosynthesized leucoemeraldine, emeraldine and pernigraniline forms of polyaniline films and their supercapacitive behavior. *Synthetic Metals* **2010**, *160*, 955–960, doi:10.1016/j.synthmet.2010.02.007.

16. Joon, S.; Kumar, R.; Singh, A.P.; Shukla, R.; Dhawan, S.K. Fabrication and microwave shielding properties of free standing polyaniline-carbon fiber thin sheets. *Materials Chemistry and Physics* **2015**, *160*, 87–95, doi:10.1016/j.matchemphys.2015.04.010.

17. Zhang, W.; Zhang, X.; Zheng, Y.; Guo, C.; Yang, M.; Li, Z.; Wu, H.; Qiu, H.; Yan, H.; Qi, S. Preparation of polyaniline@MoS_2@Fe_3O_4 nanowires with a wide band and small thickness toward enhancement in microwave absorption. *ACS Applied Nano Materials* **2018**, *1*, 5865–5875, doi:10.1021/acsanm.8b01452.

18. Pai, A.R.; Paoloni, C.; Thomas, S. Nanocellulose-based sustainable microwave absorbers to stifle electromagnetic pollution. In *Nanocellulose Based Composites for Electronics*. Elsevier. **2021**, pp. 237–258.

19. Gopakumar, D.A.; Pai, A.R.; Pottathara, Y.B.; Pasquini, D.; Carlos De Morais, L.; Luke, M.; Kalarikkal, N.; Grohens, Y.; Thomas, S. Cellulose nanofiber-based polyaniline flexible papers as sustainable microwave absorbers in the X-band. *ACS Applied Materials and Interfaces* **2018**, *10*, 20032–20043, doi:10.1021/acsami.8b04549.

20. Pai, A.R.; Binumol, T.; Gopakumar, D.A.; Pasquini, D.; Seantier, B.; Kalarikkal, N.; Thomas, S. Ultrafast heat dissipating aerogels derived from polyaniline anchored cellulose nanofibers as sustainable microwave absorbers. *Carbohydrate Polymers* **2020**, *246*, 116663, doi:10.1016/j.carbpol.2020.116663.

21. Dubal, D.P.; Patil, S. V.; Jagadale, A.D.; Lokhande, C.D. Two step novel chemical synthesis of polypyrrole nanoplates for supercapacitor application. *Journal of Alloys and Compounds* **2011**, *509*, 8183–8188, doi:10.1016/j.jallcom.2011.03.080.

22. Gopakumar, D.A.; Pai, A.R.; Pottathara, Y.B.; Pasquini, D.; de Morais, L.C.; Khalil H.P.S., A.; Nzihou, A.; Thomas, S. Flexible papers derived from polypyrrole deposited cellulose nanofibers for enhanced electromagnetic interference shielding in gigahertz frequencies. *Journal of Applied Polymer Science* **2021**, *138*, 50262, doi:10.1002/app.50262.

23. Gahlout, P.; Choudhary, V. Microwave shielding behaviour of polypyrrole impregnated fabrics. *Composites Part B: Engineering* **2019**, *175*, 107093, doi:10.1016/j.compositesb.2019.107093.

24. Kaloni, T.P.; Giesbrecht, P.K.; Schreckenbach, G.; Freund, M.S. Polythiophene: From fundamental perspectives to applications. *Chemistry of Materials* **2017**, *29*, 10248–10283.

25. Kulkarni, G.; Kandesar, P.; Velhal, N.; Phadtare, V.; Jatratkar, A.; Shinde, S.K.; Kim, D.Y.; Puri, V. Exceptional electromagnetic interference shielding and microwave absorption properties of room temperature synthesized polythiophene thin films with double negative characteristics (DNG) in the Ku-band region. *Chemical Engineering Journal* **2019**, *355*, 196–207, doi:10.1016/j.cej.2018.08.114.

26. Erdoğan, M.K.; Karakişla, M.; Saçak, M. Preparation, characterization and electromagnetic shielding effectiveness of conductive polythiophene/poly(ethylene terephthalate) composite fibers. *Journal of Macromolecular Science, Part A: Pure and Applied Chemistry* **2012**, *49*, 473–482, doi:10.1080/10601325.2012.676896.

27. Iqbal, S.; Shah, J.; Kotnala, R.K.; Ahmad, S. Highly efficient low cost EMI shielding by barium ferrite encapsulated polythiophene nanocomposite. *Journal of Alloys and Compounds* **2019**, *779*, 487–496, doi:10.1016/j.jallcom.2018.11.307.

28. Iqbal, S.; Khatoon, H.; Kotnala, R.K.; Ahmad, S. Bi-doped barium ferrite decorated polythiophene nanocomposite: Influence of Bi-doping on structure, morphology, thermal and EMI shielding behavior for X-band. *Journal of Materials Science* **2020**, *55*, 15894–15907, doi:10.1007/s10853-020-05134-z.

29. Yano, H.; Kudo, K.; Marumo, K.; Okuzaki, H. Fully soluble self-doped poly(3,4-ethylenedioxythiophene) with an electrical conductivity greater than 1000 S cm^{-1}. *Science Advances* **2019**, *5*, eaav9492, doi:10.1126/sciadv.aav9492.

30. Worfolk, B.J.; Andrews, S.C.; Park, S.; Reinspach, J.; Liu, N.; Toney, M.F.; Mannsfeld, S.C.B.; Bao, Z. Ultrahigh electrical conductivity in solution-sheared polymeric transparent films. *Proceedings of the National Academy of Sciences of the United States of America* **2015**, *112*, 14138–14143, doi:10.1073/pnas.1509958112.

31. Ghosh, S.; Ganguly, S.; Remanan, S.; Das, N.C. Fabrication and investigation of 3D tuned PEG/PEDOT:PSS treated conductive and durable cotton fabric for superior electrical conductivity and flexible electromagnetic interference shielding. *Composites Science and Technology* **2019**, *181*, 107682, doi:10.1016/j.compscitech.2019.107682.

32. Agnihotri, N.; Chakrabarti, K.; De, A. Highly efficient electromagnetic interference shielding using graphite nanoplatelet/poly(3,4-ethylenedioxythiophene)-poly(styrenesulfonate) composites with enhanced thermal conductivity. *RSC Advances* **2015**, *5*, 43765–43771, doi:10.1039/c4ra15674a.

33. Bora, P.J.; Anil, A.G.; Vinoy, K.J.; Ramamurthy, P.C. Outstanding absolute electromagnetic interference shielding effectiveness of cross-linked PEDOT:PSS film. *Advanced Materials Interfaces* **2019**, *6*, 1901353, doi:10.1002/admi.201901353.

34. Naguib, M.; Mochalin, V.N.; Barsoum, M.W.; Gogotsi, Y. 25th anniversary article: MXenes: A new family of two-dimensional materials. *Advanced Materials* **2014**, *26*, 992–1005, doi:10.1002/adma.201304138.

35. Liu, R.; Miao, M.; Li, Y.; Zhang, J.; Cao, S.; Feng, X. Ultrathin biomimetic polymeric $Ti_3C_2T_x$ MXene composite films for electromagnetic interference shielding. *ACS Applied Materials and Interfaces* **2018**, *10*, 44787–44795, doi:10.1021/acsami.8b18347.

36. Yin, G.; Wang, Y.; Wang, W.; Yu, D. Multilayer structured PANI/MXene/CF fabric for electromagnetic interference shielding constructed by layer-by-layer strategy. *Colloids and Surfaces A: Physicochemical and Engineering Aspects* **2020**, *601*, 125047, doi:10.1016/j.colsurfa.2020.125047.

37. Wang, Z.; Cheng, Z.; Xie, L.; Hou, X.; Fang, C. Flexible and lightweight $Ti_3C_2T_x$ MXene/Fe_3O_4@PANI composite films for high-performance electromagnetic interference shielding. *Ceramics International* **2021**, *47*, 5747–5757, doi:10.1016/j.ceramint.2020.10.161.

38. Saini, P.; Choudhary, V. Enhanced electromagnetic interference shielding effectiveness of polyaniline functionalized carbon nanotubes filled polystyrene composites. *Journal of Nanoparticle Research* **2013**, *15*, 1–7, doi:10.1007/s11051-012-1415-2.

39. Bora, P.J.; Vinoy, K.J.; Ramamurthy, P.C.; Madras, G. Electromagnetic interference shielding efficiency of MnO_2 nanorod doped polyaniline film. *Materials Research Express* **2017**, *4*, 025013, doi:10.1088/2053-1591/aa59e3.

40. Choudhary, H.K.; Pawar, S.P.; Bose, S.; Sahoo, B. EMI shielding performance of lead hexaferrite/polyaniline composite in 8–18 GHz frequency range. In *Proceedings of the AIP Conference Proceedings*; **2018**; Vol. *1953*, p. 120061.

41. Iqbal, S.; Khatoon, H.; Kotnala, R.K.; Ahmad, S. Mesoporous strontium ferrite/polythiophene composite: Influence of enwrappment on structural, thermal, and electromagnetic interference shielding. *Composites Part B: Engineering* **2019**, *175*, 107143, doi:10.1016/j.compositesb.2019.107143.

42. Bhardwaj, P.; Grace, A.N. Antistatic and microwave shielding performance of polythiophene-graphene grafted 3-dimensional carbon fibre composite. *Diamond and Related Materials* **2020**, *106*, 107871, doi:10.1016/j.diamond.2020.107871.

43. Gahlout, P.; Choudhary, V. EMI shielding response of polypyrrole-MWCNT/polyurethane composites. *Synthetic Metals* **2020**, *266*, 116414, doi:10.1016/j.synthmet.2020.116414.

44. Liu, C.; Liao, X. Collagen fiber/Fe_3O_4/polypyrrole nanocomposites for absorption-type electromagnetic interference shielding and radar stealth. *ACS Applied Nano Materials* **2020**, *3*, 11906–11915, doi:10.1021/acsanm.0c02472.

45. Lee, S.H.; Kang, D.; Oh, I.K. Multilayered graphene-carbon nanotube-iron oxide three-dimensional heterostructure for flexible electromagnetic interference shielding film. *Carbon* **2017**, *111*, 248–257, doi:10.1016/j.carbon.2016.10.003.

46. Wang, Y.; Wang, W.; Ding, X.; Yu, D. Multilayer-structured Ni-Co-Fe-P/polyaniline/polyimide composite fabric for robust electromagnetic shielding with low reflection characteristic. *Chemical Engineering Journal* **2020**, *380*, 122553, doi:10.1016/j.cej.2019.122553.

47. Wan, C.; Li, J. Synthesis and electromagnetic interference shielding of cellulose-derived carbon aerogels functionalized with α-Fe$_2$O$_3$ and polypyrrole. *Carbohydrate Polymers* **2017**, *161*, 158–165, doi:10.1016/j.carbpol.2017.01.003.

48. Sambyal, P.; Dhawan, S.K.; Gairola, P.; Chauhan, S.S.; Gairola, S.P. Synergistic effect of polypyrrole/BST/RGO/Fe$_3$O$_4$ composite for enhanced microwave absorption and EMI shielding in X-band. *Current Applied Physics* **2018**, *18*, 611–618, doi:10.1016/j.cap.2018.03.001.

49. Wu, Y.; Wang, Z.; Liu, X.; Shen, X.; Zheng, Q.; Xue, Q.; Kim, J.K. Ultralight graphene foam/conductive polymer composites for exceptional electromagnetic interference shielding. *ACS Applied Materials and Interfaces* **2017**, *9*, 9059–9069, doi:10.1021/acsami.7b01017.

22 Challenges and Future Lookout of Conductive Polymers

Hamidreza Parsimehr and Amir Ershad Langroudi
Color and Surface Coating Group, Polymer Processing Department,
Iran Polymer and Petrochemical Institute (IPPI), Tehran, Iran

CONTENTS

22.1 INTRODUCTION

Most of the organic compounds are electric insulators. Conductive polymers (CPs) are polymers that contain conductivity from the resonance of π electrons [1]. Tremendous efforts have been made to develop the CPs and their applications in the last two centuries [2]. Polyaniline (PANI), is the first CP, was developed in the early 1860s [3]. Afterward, the polycyclic aromatic compounds and semi-conducting charge-transfer complexes were rapidly developed [4]. Numerous CPs have been developed in recent decades to enhance the important applications of the CPs. The most common CPs are PANI, polypyrrole (PPY), poly(P-phenylene) (PPP), poly(P-phenylenevinylene) (PPV), polythiophene (PTH), and poly(3,4-ethylene dioxythiophene) (PEDOT), which are illustrated in Figure 22.1. The other common conductive derivatives that considered as CPs are polythiophene-vinylene (PTV), polypyrrole-poly(styrene sulfonate) (PPY-PSS), poly P-phenylene-sulfide (PPS), polyacetylene (PAC), poly(2,5-thienylenevinylene) (PTV), poly(3-alkyl thiophene) (PAT), poly(p-phenylene-terephthalamide) (PPTA), poly(isothianaphthene) (PITN), poly(α-naphthylamine) (PNA), polyisoprene (PIP), and poly(p-phenylene-terephthalamide) (PPTA).

The unique application of the different CPs leads to extensive investigations in developing the CPs in the future. All of these advanced applications have originated owing to the presence of π bonds in the CPs [5]. One of the most significant properties of the CPs is their flexibility. Therefore, tremendous efforts have been made to fabricate advanced flexible devices and nanocomposites from CPs [6, 7].

DOI: 10.1201/9781003150374-22

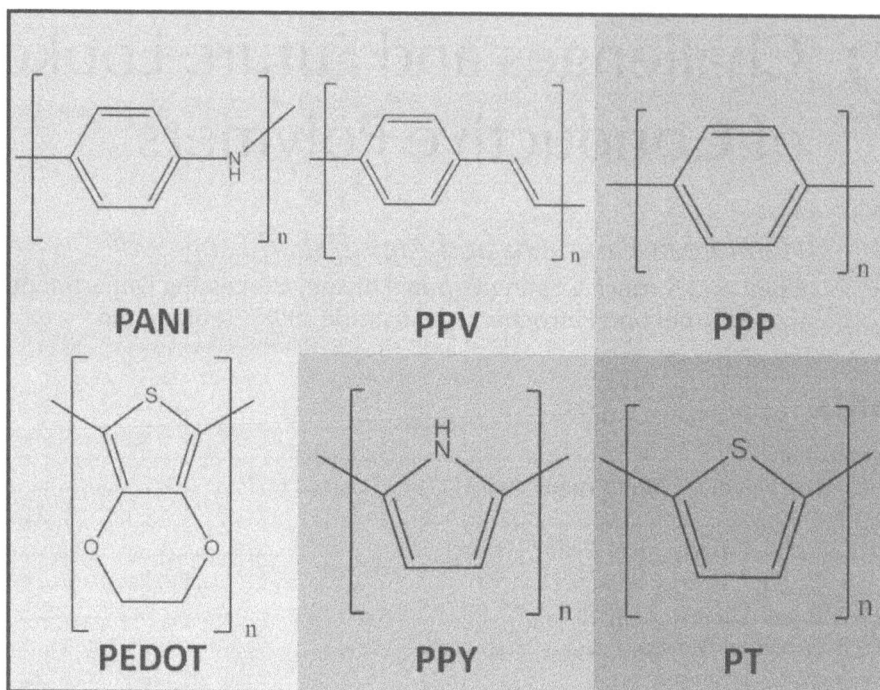

FIGURE 22.1   The most common conductive polymers (CPs).

22.2   CONDUCTIVE POLYMERS' MAJOR APPLICATIONS

The CPs have numerous applications in materials science [8]. The most significant applications of the CPs include energy (harvesting and storage), biomedical applications, sensors, and environmental remediation have been rapidly developed in recent years. These important applications have determined the valuable potential of the CPs in materials science. Numerous advanced devices and substances have been obtained by using CPs. Advanced applications of CPs stem from four significant properties: flexibility [9, 10], transparency [11, 12], stretchability [13, 14], and wearability [15, 16]. These properties are the reasons for the swift development of the CPs' applications in recent years. Also, novel investigations have focused on consolidating these four properties [17–19].

22.2.1   Energy

Energy is considered one of the most important issues of the modern era. Several aspects of humans living on the earth depended on energy. The rapid development of industries and societies in recent years stems from the advent of high-performance energy sources, however, they are polluting the environment too. Therefore, green renewable energy resources have been rapidly developed in recent years to reduce the oil dependence of the industries. Most of the novel investigations on materials science are focused on preparing novel materials [20] or energy [21] based on green chemistry principles [22]. Energy has two major aspects. These vital aspects have originated from two important questions: first, that of how to produce energy appropriately and second, that of how to store energy appropriately. The first important aspect of energy is the production of energy or energy harvesting, and the second is energy storage. These two major aspects of energy are illustrated in Figure 22.2.

22.2.1.1   Energy Harvesting Devices

Energy production (energy harvesting) is considered a crucial issue in the current era. The massive energy production through non-renewable hydrocarbons like oil and coal leads to extensive damages to the environment. This severe environmental pollution has several negative consequences, such as

FIGURE 22.2 The major aspects of energy.

global warming, ozone layer depletion, and flora/fauna species extinction [23]. Therefore, the novel methods of energy production based on green chemistry technology have been widely investigated in recent decades to reduce the environmental pollution arising due to energy production procedures. Also, the contemporary methods of energy production have considerable costs. Therefore, novel methods have been developed to reduce production costs and environmental pollution. Several novel methods have been developed to provide renewable benign energy instead of non-renewable energy. Solar cells and fuel cells as renewable energy harvesting devices have been swiftly developed in recent decades to provide renewable benign energy for different purposes. Numerous investigations have been developed to improve the solar cells [24, 25] and fuel cells [26, 27] through CPs. The microscopic images of CP-based solar cells from PEDOT:PSS are illustrated in Figure 22.3. Also, the microscopic images of CP-based microbial fuel cells from polyaniline are illustrated in Figure 22.4.

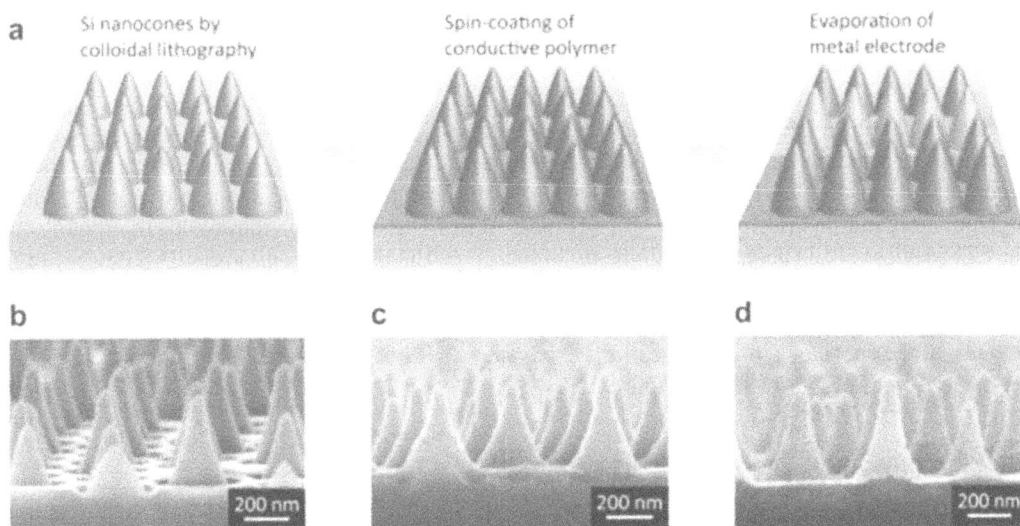

FIGURE 22.3 Conductive polymer-based solar cell. (a) Schematic illustration of the fabrication process for Si nanocone/polymer solar cell. SEM images of the cross-sectional view of Si nanocones. (b) After nanosphere lithography and RIE. (c) After a spin-coating of PEDOT:PSS with a spin speed of 4000 rpm. (d) After evaporation of a metal electrode, Au, with a thickness of 15 nm. Copyright (2012) American Chemical Society [28].

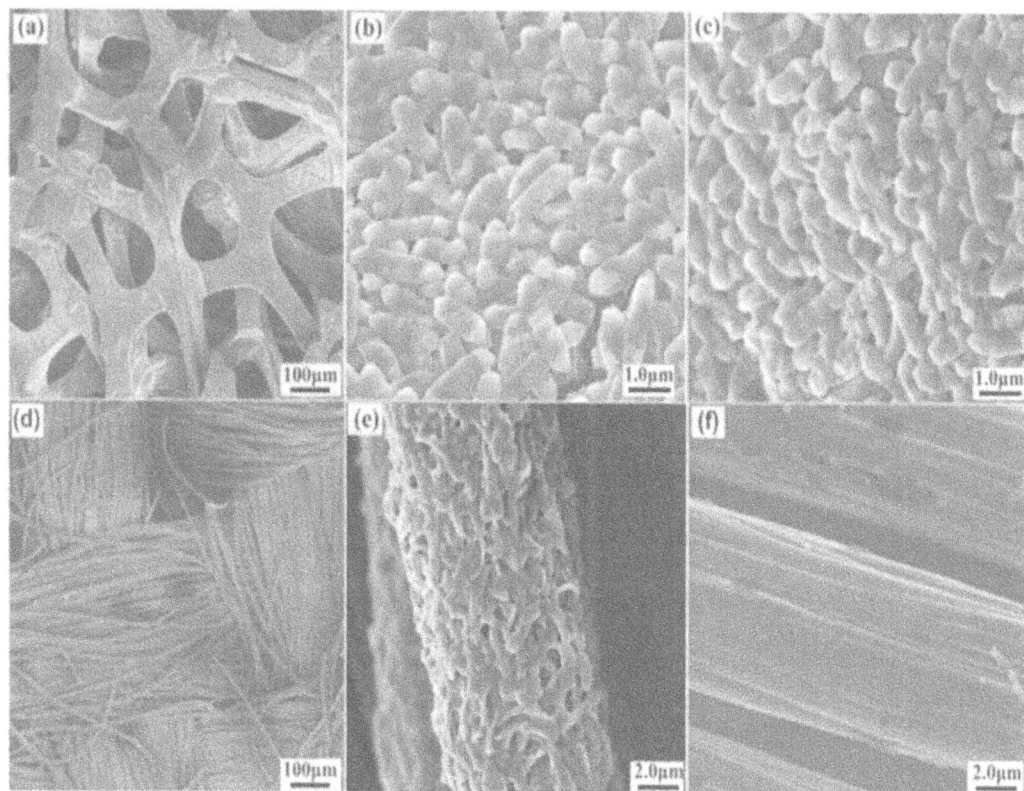

FIGURE 22.4 Conductive polymer-based microbial fuel cell. SEM images of graphene/polyaniline (a–c) and carbon cloth (d–f) electrodes after 60 h incubation in MFC with *S. oneidensis* MR-1 cell suspension. With higher magnification, images (b) and (e) were taken at the electrode surface, while images (c) and (f) were focused on the electrode interior. Copyright (2012) American Chemical Society [29].

22.2.1.2 Energy Storage Devices

The second significant aspect of energy is energy storage. One of the best methods for storing energy is electrochemical energy storage (EES) devices, including rechargeable batteries, supercapacitors, and hybrid EES devices. Using the EES devices provides several significant advantages such as environmental protection and reduction of costs. Therefore, tremendous efforts have been accomplished to develop EES devices [30–33]. Various organic [34–36], inorganic [37–39], and organic–inorganic hybrid materials [40, 41] have been investigated for improving the quality of the EES devices. The CPs have a significant influence on the development of the EES devices [42, 43]. For example, advanced flexible EES devices [44], wearable EES devices [15], stretchable EES devices [13], and transparent EES devices with CPs [45] have been rapidly developed in recent years because of the enormous importance of these advanced EES devices. A self-charging supercapacitor made from CP is illustrated in Figure 22.5.

22.2.2 Biomedical Applications

One of the most significant concepts of CPs' applications is the biomedical applications of the CPs. There are several significant biomedical applications of CPs, such as modified natural polymers for conductivity (e.g., chitosan [47], collagen [48], fibroin [49], and other polymers with 3D network structures), CP-based 3D scaffolds [50], cardiovascular disorders [51], tissue engineering [52], regenerative medicine [53], electrospinning [54], neural system [55], and wound healing [56].

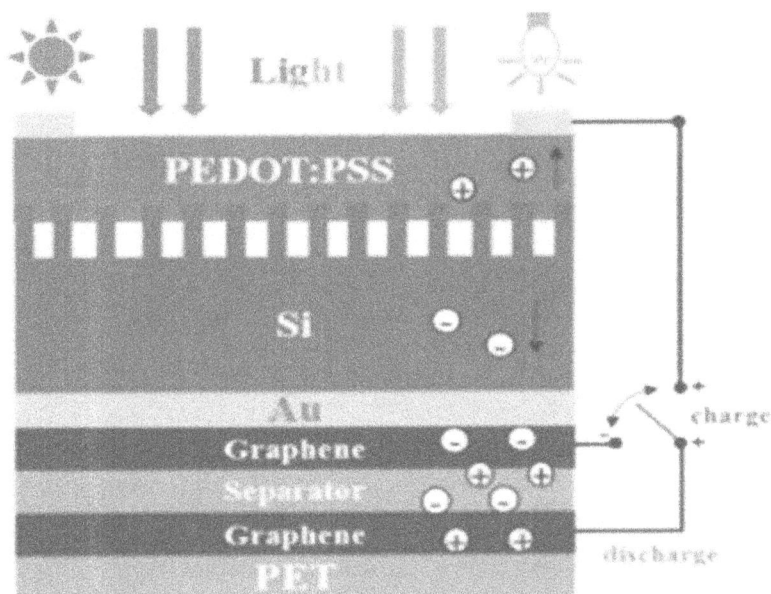

FIGURE 22.5 Schematic illustration of the structure of the self-charging power pack consisting of a silicon nanowire array/conductive polymer hybrid solar cell and a laser-scribed graphene supercapacitor. Copyright (2018) American Chemical Society [46].

These advanced applications have enormous importance in medicine. Hence numerous biomedical applications of the CPs due to increasing investigations on the biomedical applications of the CPs.

22.2.3 DETECTION DEVICES

22.2.3.1 Sensors

The sensors have been defined as devices to convert physical quantity input into functionally related output, usually in the form of optical or electrical signals that can be detected or read either by electronic instruments or by human users. Numerous investigations have been accomplished in the last 20 years to develop CP-based sensors because of the huge potential of CPs as sensors and biosensors. For example, CPs have been widely used to prepare acetic acid vapor-sensor [57], aliphatic alcohols-sensor [58], ammonia-sensor [59], chloroform-vapor sensor [60], glucose-sensor [61], humidity-sensor [62], methanol-sensor [63], NO_2-sensor [64], and pesticide-sensor [65]. These sensors have wide applications in materials science and industries. A highly sensitive glucose sensor based on Pt nanoparticle/PANI hydrogel heterostructures is illustrated in Figure 22.6.

22.2.3.2 Biosensors

Also, CPs have been widely used to prepare biosensors [66]. Biosensors contain bioreceptor, physicochemical detectors, transducers, and biological components. A biomolecule, such as an enzyme, antibody, nucleic acid, aptamer, or cell, capable of identifying or detecting the target analyte is used as the bioreceptor. These sensors have such advantages as high selectivity to the target analyte mainly due to the specific interaction of the bioreceptor present in their structure with the target analyte that is recognized as the biorecognition. For example, novel biosensors such as acetylcholinesterase-biosensor [67], DNA-biosensor [68], enzyme-biosensor [69], and protein-biosensor [70] have been rapidly developed in recent years because of the significance of biorecognition in modern life. The schematic depiction of the glucose biosensor from PANI is illustrated in Figure 22.7.

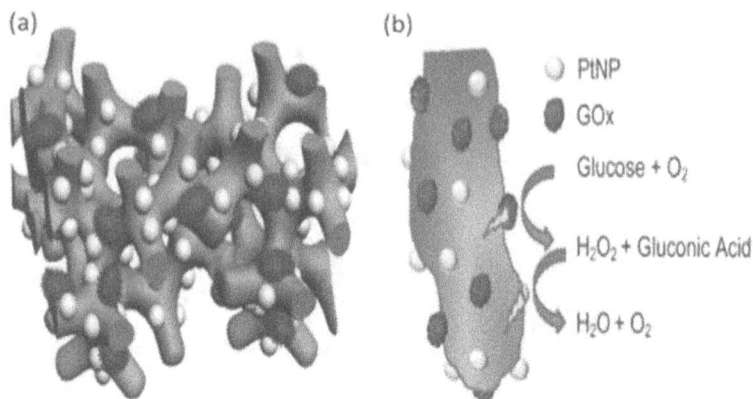

FIGURE 22.6 (a) Schematic representation of the 3D heterostructure of the Pt/PANI hydrogel, in which the PANI hydrogel acts as a matrix for the immobilization of the GOx enzyme and homogeneous loading of PtNPs. (b) A 2D scheme showing the reaction mechanism of the glucose sensor based on the Pt/PANI hydrogel heterostructure. Copyright (2013) American Chemical Society [61].

22.2.3.3 Actuators

The actuators are the part of devices that are responsible for moving and controlling a mechanism or process. The actuators have required control signals and energy to perform the procedure. Different actuators have been widely developed in recent decades [72]. CPs have been widely used as actuators in recent years because of their stimuli-responsiveness feature [73]. The electrochemical actuators based on CPs have been rapidly developed in recent years [74]. The unique performance of the CP-based actuators is due to increasing attention to these advanced actuators [75, 76].

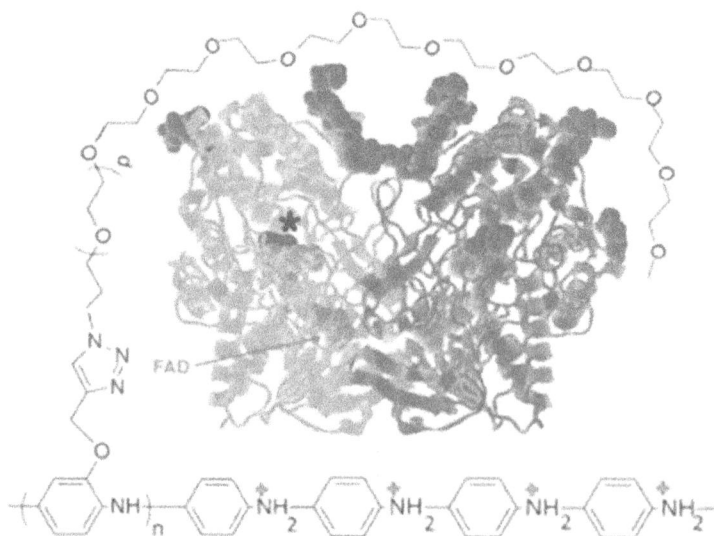

FIGURE 22.7 Schematic depiction of electrostatic interactions between GOx and PA-g-PEG. Bottom: polyanion–polycation interaction. Arithmetic signs represent full charges at pH 6.5. Top: oligosaccharide-PEG hydrogen bond network. Magenta-colored parts indicate enzyme-oligosaccharide units. Copyright (2017) American Chemical Society [71].

22.2.4 ENVIRONMENTAL REMEDIATION

In recent decades, increasing environmental pollution is the reason for severe damage to the environment. Therefore tremendous efforts have been accomplished to reduce environmental pollution. The absorption/adsorption of hazardous materials by absorbent as a practical method to reduce environmental pollution has been widely developed in recent years. The polymeric absorbents have an appropriate potential to absorb hazardous materials such as heavy metals like mercury, lead, and cadmium. Recent investigations have been determined the huge potential of the CPs as polymeric absorbents. Hence the CPs have been widely used as absorbents in recent investigations to absorb and remove environmental pollutions especially hazardous heavy metals. The hierarchical composite polyaniline-(electrospun polystyrene) fibers for absorption of Hg(II), Cd(II), Pb(II), Cr(VI), and Cu(II) ions are illustrated in Figure 22.8.

CPs have improved identify, quantify, and eliminate pollutants in various environmental modifications. CPs can be used as adsorbents, ion exchangers, filters, and membranes to remove contaminants in complex or redox reactions. The adsorption and release of pollutants, doping/dedoping, reduction, and oxidation electrocatalytic reactions are the fundamental mechanisms employed by CPs or composites in the presence of a potential field. Charged pollutants may be removed from aqueous solutions or liquid wastes by combining them with thin layers of charged or uncharged CP. Metal ion complexes can be generated in uncharged CPs by functional groups containing nitrogen or nitrogen in the polymer ring of composites and nanofibers. Loading/unloading cycles in charged CPs are performed depending on the need for charge carriers. CPs absorb charged contaminants using an electric charge with the opposite polarity of the polluting particles' charge. CPs remove

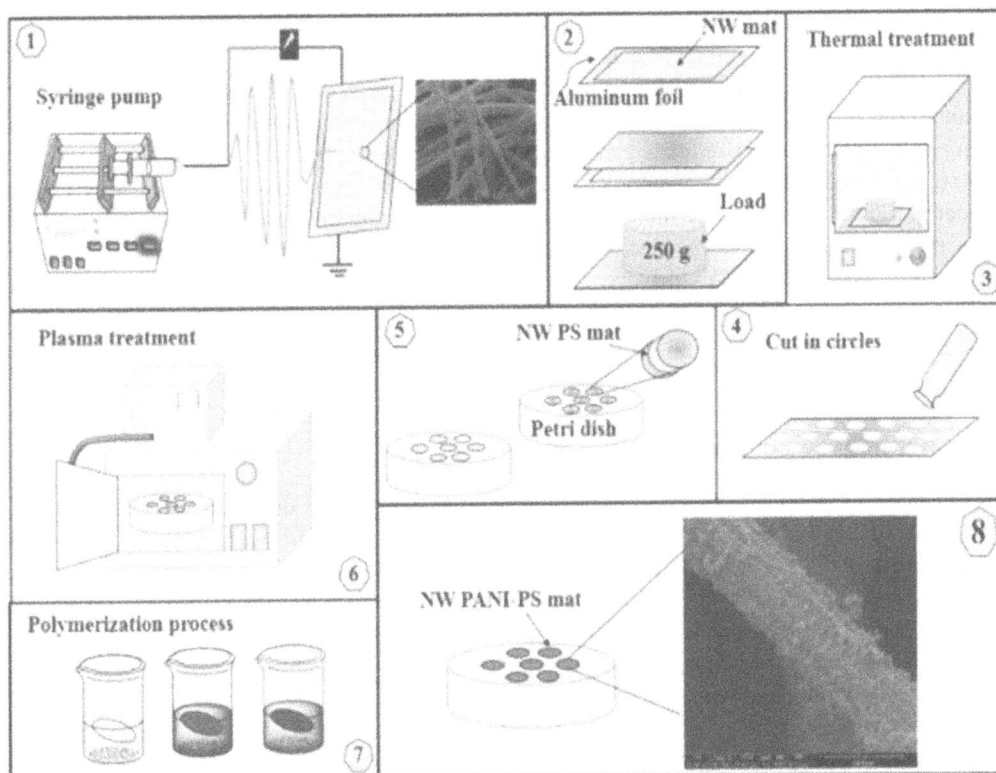

FIGURE 22.8 Schematic representation of nanowires (NW) PANI-PS mat production process. (1). The sequence of steps for the treatment of the NW PS mats (2–6). Polymerization process (7) and final NW PANI-PS mats (8). Copyright (2015) American Chemical Society [77].

charged contaminants with the help of some electrochemical ion exchange. A positively charged CP is reduced to neutral by removing anionic contaminants. Throughout this reduction phase, the amount of anions released is determined by the surface morphology as well as the reduction potential. CP is often used for electron transfer because it is more stable and hydrophobic than monolayers of adsorbent surface or covalently bonded catalysts in terms of stability and hydrophobicity. Adsorption, mineral removal, and photolytic methods all require more energy to extract pollutants.

Polymers are subjected to a variety of environmental modifications. Dye is absorbed from an aqueous medium by an environmentally friendly composite containing PANI and a biopolymer containing chitosan. Since pollutants are often present in low concentrations in the environment, using CPs on 3D porous materials, such as graphite or carbon, speeds up and improves the reaction's performance. Due to its low-density structure and large surface area, the use of PANI hollow nanospheres has a tremendous practical perspective in this regard.

Precipitators can selectively eliminate cationic, anionic, aromatic, and polyaromatic hydrocarbons. Dendrimeric structures can selectively eliminate contaminant species based on size, load, or shape. The pH factor can play an important role in the removal process as some dendrimer groups may control the uptake or desorption of chemicals in this way or Lewis acid–base interactions. Mass transfer and selection are preferred by dendrimers (3D materials). Physical adsorption is usually the preferred method to remove unwanted minor molecule gases such as H_2S, NO_x, CO_x, and SO_2. However, eliminating these gases in moisture's presence is significantly weakened. One solution can be reactive absorption. PANI graphite oxide composite could be used for such applications.

Furthermore, due to its high specific surface area and recycled capacity, fine-grained, porous S-doped carbon from graphene oxide reduced polythiophene layer can be used to selectively remove carbon dioxide from its gaseous mixture with hydrogen, nitrogen, and methane. Dechlorination is an alternative method for the modification of halogenated harmful organic matter. Therefore, removing halogen atoms with a CP can be a good strategy. Electrochemical halogen removal can be done in stages. The reduction of 3,3,4,4-tetrachlorobiphenyl in the presence of hydrogen using a CP containing zero-valent Pd nanoparticles in a PPy medium is an example of dechlorination. The toxicity of this compound is reduced by removing chlorine atoms.

Titanium dioxide is a popular photocatalyst that comes in a variety of crystal structures. The polymer substrate appears to be the most promising of the numerous substrates tested to support photocatalysts because of its many advantages, including low cost, stability, good chemical and mechanical properties, and low density. By creating a synergistic effect, the presence of WO_3 in its composites with titanium dioxide in PANI and PPy substrates causes better charge separation, higher porosity, and expansion of the optical reaction amplitude. Other inorganic oxides are used in addition to titanium dioxide's photocatalytic properties.

22.3 PROGRESS IN ADVANCED APPLICATIONS

The advanced applications of CPs are related to the consolidation of other advanced features. For example, flexibility and stretchability should be merged in one device to obtain higher performance. Therefore, the novel advanced applications on the CPs have been focused on consolidating other advanced properties in CPs, such as flexibility, wearability, stretchability, transparency, biocompatibility, and stimuli-responsiveness. The future of advanced CPs belongs to some devices and compounds that contained more practical properties (flexibility, wearability, stretchability, transparency, biocompatibility, and stimuli-responsiveness).

22.4 STRUCTURE–PROPERTIES CORRELATION

The novel investigations have been determined the huge importance of the specific structures on the CPs applications. CPs with specific structures including crosslink [78], core-shell [79], hierarchical [80], and self-assembly [81] have been rapidly developed in recent years to improve the CPs' applications. Also, the combination of these specific structures in CPs has been rapidly developed in recent years to provide higher performances. Therefore, future investigations on CPs focus on these specific structures to enhance the CPs' applications.

22.5 CHALLENGES AND FUTURE LOOKOUT

The future of the CPs belongs to the multifunctional compounds and devices that contained several properties altogether. The consolidation of advanced properties in novel CPs, such as flexibility, wearability, stretchability, transparency, biocompatibility, and stimuli-responsiveness, leads to emerging novel CP-based composites. This is the most significant stage in developing the CPs. The multi-responsive CPs, as one of the most advanced CPs, have been rapidly developed in recent years. These advanced CPs have a responsive ability to different external stimuli [82, 83]. Also, using low-cost materials such as biochar in CP-based composites has dramatically reduced the CP nanocomposites' cost. Therefore, the novel CP nanocomposites should be low-cost to increase their practical applications. Finally, the preparation of CPs, CP-based nanocomposites, and CP-based devices should be performed based on the green chemistry priorities. All chemical industries have been tried to reduce their chemical pollutions, and also all societies have developed strict regulations to control and reduce environmental pollutions. Therefore, the development of CPs should be conformed to environmental protection regulations.

22.6 CONCLUSION

The wide applications of CPs are due to numerous investigations on the development of the CPs. Several useful applications have been obtained from CPs, such as flexibility, wearability, stretchability, transparency, biocompatibility, and stimuli-responsiveness. The novel investigations have been focused on merging different properties of the CPs to fabricate the advanced CPs that contained more than one feature simultaneously. Also, the novel investigations should be focused on economic priorities and environmental priorities to provide low-cost and benign CP-nanocomposites. Therefore the future of CPs belongs to some low-cost and benign advanced CPs that simultaneously contained several useful properties.

REFERENCES

1. Nezakati T, Seifalian A, Tan A, Seifalian A (2018) Conductive Polymers: Opportunities and Challenges in Biomedical Applications. Chem Rev 118:6766–6843. https://doi.org/10.1021/acs.chemrev.6b00275
2. Hush N (2003) An Overview of the First Half-Century of Molecular Electronics. Ann N Y Acad Sci 1006:1–20. https://doi.org/10.1196/annals.1292.016
3. Geniès E, Boyle A, Lapkowski M, Tsintavis C (1990) Polyaniline: A Historical Survey. Synth Met 36:139–182. https://doi.org/10.1016/0379-6779(90)90050-U
4. Dabestani R, Reszka K, Sigman M (1998) Surface Catalyzed Electron Transfer from Polycyclic Aromatic Hydrocarbons (PAH) to Methyl Viologen Dication: Evidence for Ground-State Charge Transfer Complex Formation on Silica Gel. J Photochem Photobiol A Chem 117:223–233. https://doi.org/10.1016/S1010-6030(98)00327-X
5. Grancarić A, Jerković I, Koncar V, et al. (2018) Conductive Polymers for Smart Textile Applications. J Ind Text 48:612–642. https://doi.org/10.1177/1528083717699368
6. Hu W, Chen S, Yang Z, et al. (2011) Flexible Electrically Conductive Nanocomposite Membrane Based on Bacterial Cellulose and Polyaniline. J Phys Chem B 115:8453–8457. https://doi.org/10.1021/jp204422v
7. Chang CM, Hu ZH, Lee TY, et al. (2016) Biotemplated Hierarchical Polyaniline Composite Electrodes with High Performance for Flexible Supercapacitors. J Mater Chem A 4:9133–9145. https://doi.org/10.1039/c6ta01781a
8. Zhan C, Yu G, Lu Y, et al. (2017) Conductive Polymer Nanocomposites: A Critical Review of Modern Advanced Devices. J Mater Chem C 5:1569–1585. https://doi.org/10.1039/C6TC04269D

9. Miao F, Shao C, Li X, et al. (2015) Flexible Solid-State Supercapacitors Based on Freestanding Electrodes of Electrospun Polyacrylonitrile@polyaniline Core-Shell Nanofibers. Electrochim Acta 176:293–300. https://doi.org/10.1016/j.electacta.2015.06.141

10. Xiang X, Zhang W, Yang Z, et al. (2016) Smart and Flexible Supercapacitor Based on a Porous Carbon Nanotube Film and Polyaniline Hydrogel. RSC Adv 6:24946–24951. https://doi.org/10.1039/C6RA00705H

11. Jamdegni M, Kaur A (2020) Highly Efficient Dark to Transparent Electrochromic Electrode with Charge Storing Ability Based on Polyaniline and Functionalized Nickel Oxide Composite Linked Through a Binding Agent. Electrochim Acta 331:135359. https://doi.org/10.1016/j.electacta.2019.135359

12. Cheng Y, Ren X, Duan L, Gao G (2020) A Transparent and Adhesive Carboxymethyl Cellulose/Polypyrrole Hydrogel Electrode for Flexible Supercapacitors. J Mater Chem C 8:8234–8242. https://doi.org/10.1039/D0TC01039A

13. Xie Y, Liu Y, Zhao Y, et al. (2014) Stretchable All-Solid-State Supercapacitor with Wavy Shaped Polyaniline/Graphene Electrode. J Mater Chem A 2:9142–9149. https://doi.org/10.1039/c0xx00000x

14. Hao G-P, Hippauf F, Oschatz M, et al. (2014) Stretchable and Semitransparent Conductive Hybrid Hydrogels for Flexible Supercapacitors. ACS Nano 8:7138–7146. https://doi.org/10.1021/nn502065u

15. An T, Ling Y, Gong S, et al. (2019) A Wearable Second Skin-Like Multifunctional Supercapacitor with Vertical Gold Nanowires and Electrochromic Polyaniline. Adv Mater Technol 4:1800473. https://doi.org/10.1002/admt.201800473

16. Gong S, Lai D, Wang Y, et al. (2015) Tattoolike Polyaniline Microparticle-Doped Gold Nanowire Patches as Highly Durable Wearable Sensors. ACS Appl Mater Interfaces 7:19700–19708. https://doi.org/10.1021/acsami.5b05001

17. Cai J, Zhang C, Khan A, et al. (2018) Highly Transparent and Flexible Polyaniline Mesh Sensor for Chemiresistive Sensing of Ammonia Gas. RSC Adv 8:5312–5320. https://doi.org/10.1039/C7RA13516E

18. Dudem B, Mule A, Patnam H, Yu J (2019) Wearable and Durable Triboelectric Nanogenerators via Polyaniline Coated Cotton Textiles as a Movement Sensor and Self-Powered System. Nano Energy 55:305–315. https://doi.org/10.1016/j.nanoen.2018.10.074

19. Kundu S, Majumder R, Ghosh R, et al. (2020) Relative Humidity Sensing Properties of Doped Polyaniline-Encased Multiwall Carbon Nanotubes: Wearable and Flexible Human Respiration Monitoring Application. J Mater Sci 55:3884–3901. https://doi.org/10.1007/s10853-019-04276-z

20. Parsimehr H, Pazokifard S (2021) Ambient Temperature Cross-Linkable Acrylic Latexes: Effect of Cross-Link Density, Glass Transition Temperature and Application Temperature Difference on Mechanical Properties. Polym Bull https://doi.org/10.1007/s00289-020-03515-7

21. Najafi G, Ghobadian B, Yusaf TF (2011) Algae as a Sustainable Energy Source for Biofuel Production in Iran: A Case Study. Renew Sustain Energy Rev 15:3870–3876. https://doi.org/10.1016/j.rser.2011.07.010

22. Anastas P, Eghbali N (2010) Green Chemistry: Principles and Practice. Chem Soc Rev 39:301–312. https://doi.org/10.1039/b918763b

23. van den Hove S, Le Menestrel M, de Bettignies H-C (2002) The Oil Industry and Climate Change: Strategies and Ethical Dilemmas. Clim Policy 2:3–18. https://doi.org/10.3763/cpol.2002.0202

24. Hau S, Yip H-L, Jen A (2010) A Review on the Development of the Inverted Polymer Solar Cell Architecture. Polym Rev 50:474–510. https://doi.org/10.1080/15583724.2010.515764

25. Tai Q, Chen B, Guo F, et al. (2011) In Situ Prepared Transparent Polyaniline Electrode and Its Application in Bifacial Dye-Sensitized Solar Cells. ACS Nano 5:3795–3799. https://doi.org/10.1021/nn200133g

26. Nunes S, Ruffmann B, Rikowski E, et al. (2002) Inorganic Modification of Proton Conductive Polymer Membranes for Direct Methanol Fuel Cells. J Memb Sci 203:215–225. https://doi.org/10.1016/S0376-7388(02)00009-1

27. Khilari S, Pandit S, Varanasi J, et al. (2015) Bifunctional Manganese Ferrite/Polyaniline Hybrid as Electrode Material for Enhanced Energy Recovery in Microbial Fuel Cell. ACS Appl Mater Interfaces 7:20657–20666. https://doi.org/10.1021/acsami.5b05273

28. Jeong S, Garnett E, Wang S, et al. (2012) Hybrid Silicon Nanocone–Polymer Solar Cells. Nano Lett 12:2971–2976. https://doi.org/10.1021/nl300713x

29. Yong Y-C, Dong X-C, Chan-Park M, et al. (2012) Macroporous and Monolithic Anode Based on Polyaniline Hybridized Three-Dimensional Graphene for High-Performance Microbial Fuel Cells. ACS Nano 6:2394–2400. https://doi.org/10.1021/nn204656d

30. Ehsani A, Parsimehr H (2020) Electrochemical Energy Storage Electrodes via Citrus Fruits Derived Carbon: A Minireview. Chem Rec 20:820–830. https://doi.org/10.1002/tcr.202000003

31. Ehsani A, Parsimehr H (2020) Electrochemical Energy Storage Electrodes from Fruit Biochar. Adv Colloid Interface Sci 102263. https://doi.org/10.1016/j.cis.2020.102263

32. Parsimehr H, Ehsani A (2020) Algae-Based Electrochemical Energy Storage Devices. Green Chem https://doi.org/10.1039/D0GC02246B

33. Parsimehr H, Ehsani A (2020) Corn-Based Electrochemical Energy Storage Devices. Chem Rec 20:1163–1180. https://doi.org/10.1002/tcr.202000058

34. Kowsari E, Ehsani A, Dashti Najafi M (2017) Electrosynthesis and Pseudocapacitance Performance of Ionic Liquid—Cr (η6-C6H5) Complex Functionalized Reduced Graphene Oxide/Poly Ortho Aminophenol Nanocomposite Film. J Colloid Interface Sci 504:507–513. https://doi.org/10.1016/j.jcis.2017.05.117

35. Kowsari E, Ehsani A, Dashti Najafi M, Bigdeloo M (2018) Enhancement of Pseudocapacitance Performance of p-type Conductive Polymer in the Presence of Newly Synthesized Graphene Oxide-hexamethylene Tributylammonium Iodide Nanosheets. J Colloid Interface Sci 512:346–352. https://doi.org/10.1016/j.jcis.2017.10.076

36. Ehsani A, Kowsari E, Dashti Najafi M, et al. (2017) Enhanced Pseudocapacitive Performance of Electroactive p-type Conductive Polymer in the Presence of 1-octadecyl-3-methylimidazolium Bromide. J Colloid Interface Sci 503:10–16. https://doi.org/10.1016/j.jcis.2017.05.008

37. Mohammad Shiri H, Ehsani A, Shabani Shayeh J (2015) Synthesis and Highly Efficient Supercapacitor Behavior of a Novel Polypyrrole/Ceramic Oxide Nanocomposite Film. RSC Adv 5:91062–91068. https://doi.org/10.1039/C5RA19863A

38. Naseri M, Fotouhi L, Ehsani A, Shiri HM (2016) Novel Electroactive Nanocomposite of POAP for Highly Efficient Energy Storage and Electrocatalyst: Electrosynthesis and Electrochemical Performance. J Colloid Interface Sci 484:308–313. https://doi.org/10.1016/j.jcis.2016.08.071

39. Salehifar N, Shayeh JS, Ranaei Siadat SO, et al. (2015) Electrochemical Study of Supercapacitor Performance of Polypyrrole Ternary Nanocomposite Electrode by Fast Fourier Transform Continuous Cyclic Voltammetry. RSC Adv 5:96130–96137. https://doi.org/10.1039/C5RA18694C

40. Boorboor Ajdari F, Kowsari E, Ehsani A (2018) P-type Conductive Polymer/Zeolitic Imidazolate Framework-67 (ZIF-67) Nanocomposite Film: Synthesis, Characterization, and Electrochemical Performance as Efficient Electrode Materials in Pseudocapacitors. J Colloid Interface Sci 509:189–194. https://doi.org/10.1016/j.jcis.2017.08.098

41. Naseri M, Fotouhi L, Ehsani A, Dehghanpour S (2016) Facile Electrosynthesis of Nano Flower Like Metal-organic Framework and Its Nanocomposite with Conjugated Polymer as a Novel and Hybrid Electrode Material for Highly Capacitive Pseudocapacitors. J Colloid Interface Sci 484:314–319. https://doi.org/10.1016/j.jcis.2016.09.001

42. Ehsani A, Parsimehr H, Nourmohammadi H, et al. (2019) Environment-friendly Electrodes Using Biopolymer Chitosan/Poly Ortho Aminophenol with Enhanced Electrochemical Behavior for Use in Energy Storage Devices. Polym Compos 40:4629–4637. https://doi.org/10.1002/pc.25330

43. Ehsani A, Mirtamizdoust B, Karimi F, et al. (2020) Influence of Nanostructured VO-Acetylacetonate Coordination System with 2-(pyridin-4-ylmethylene) Hydrazine-1-carbothioamide in Pseudocapacitance Performance of p-Type Conductive Polymer Composite Film. Plast Rubber Compos https://doi.org/10.1080/14658011.2020.1859875

44. Horng Y-Y, Lu Y-C, Hsu Y-K, et al. (2010) Flexible Supercapacitor Based on Polyaniline Nanowires/Carbon Cloth with Both High Gravimetric and Area-Normalized Capacitance. J Power Sources 195:4418–4422. https://doi.org/10.1016/j.jpowsour.2010.01.046

45. Ge J, Cheng G, Chen L (2011) Transparent and Flexible Electrodes and Supercapacitors Using Polyaniline/Single-walled Carbon Nanotube Composite Thin Films. Nanoscale 3:3084–3088. https://doi.org/10.1039/c1nr10424a

46. Liu H, Li M, Kaner R, et al. (2018) Monolithically Integrated Self-Charging Power Pack Consisting of a Silicon Nanowire Array/Conductive Polymer Hybrid Solar Cell and a Laser-Scribed Graphene Supercapacitor. ACS Appl Mater Interfaces 10:15609–15615. https://doi.org/10.1021/acsami.8b00014

47. Mobarak N, Ahmad A, Abdullah M, et al. (2013) Conductivity Enhancement via Chemical Modification of Chitosan Based Green Polymer Electrolyte. Electrochim Acta 92:161–167. https://doi.org/10.1016/j.electacta.2012.12.126

48. Bonfrate V, Manno D, Serra A, et al. (2017) Enhanced Electrical Conductivity of Collagen Films Through Long-Range Aligned Iron Oxide Nanoparticles. J Colloid Interface Sci 501:185–191. https://doi.org/10.1016/j.jcis.2017.04.067

49. Xia Y, Yun Lu (2008) Fabrication and Properties of Conductive Conjugated Polymers/Silk Fibroin Composite Fibers. Compos Sci Technol 68:1471–1479. https://doi.org/10.1016/j.compscitech.2007.10.044

50. Hardy J, Lee J, Schmidt C (2013) Biomimetic Conducting Polymer-Based Tissue Scaffolds. Curr Opin Biotechnol 24:847–854. https://doi.org/10.1016/j.copbio.2013.03.011

51. Gelmi A, Ljunggren M, Rafat M, Jager E (2014) Influence of Conductive Polymer Doping on the Viability of Cardiac Progenitor Cells. J Mater Chem B 2:3860–3867. https://doi.org/10.1039/C4TB00142G

52. Balint R, Cassidy N, Cartmell S (2014) Conductive Polymers: Towards a Smart Biomaterial for Tissue Engineering. Acta Biomater 10:2341–2353. https://doi.org/10.1016/j.actbio.2014.02.015

53. Antoniadou E, Cousins B, Seifalian A (2010) Development of Conductive Polymer with Carbon Nanotubes for Regenerative Medicine Applications. In: 2010 Annual International Conference of the IEEE Engineering in Medicine and Biology. IEEE, pp 815–818

54. Wang X-X, Yu G-F, Zhang J, et al. (2021) Conductive Polymer Ultrafine Fibers via Electrospinning: Preparation, Physical Properties and Applications. Prog Mater Sci 115:100704. https://doi.org/10.1016/j.pmatsci.2020.100704

55. Nezakati T, Tan A, Lim J, et al. (2019) Ultra-Low Percolation Threshold POSS-PCL/Graphene Electrically Conductive Polymer: Neural Tissue Engineering Nanocomposites for Neurosurgery. Mater Sci Eng C 104:109915. https://doi.org/10.1016/j.msec.2019.109915

56. Qu J, Zhao X, Liang Y, et al. (2019) Degradable Conductive Injectable Hydrogels as Novel Antibacterial, Anti-oxidant Wound Dressings for Wound Healing. Chem Eng J 362:548–560. https://doi.org/10.1016/j.cej.2019.01.028

57. Turemis M, Zappi D, Giardi MT, et al. (2020) ZnO/Polyaniline Composite Based Photoluminescence Sensor for the Determination of Acetic Acid Vapor. Talanta 211:120658. https://doi.org/10.1016/j.talanta.2019.120658

58. Athawale A, Kulkarni M (2000) Polyaniline and Its Substituted Derivatives as Sensor for Aliphatic Alcohols. Sensors Actuators B Chem 67:173–177. https://doi.org/10.1016/S0925-4005(00)00394-4

59. Jin Z, Su Y, Duan Y (2001) Development of a Polyaniline-Based Optical Ammonia Sensor. Sensors Actuators B Chem 72:75–79. https://doi.org/10.1016/S0925-4005(00)00636-5

60. Sharma S, Nirkhe C, Pethkar S, Athawale A (2002) Chloroform Vapour Sensor Based on Copper/Polyaniline Nanocomposite. Sensors Actuators B Chem 85:131–136. https://doi.org/10.1016/S0925-4005(02)00064-3

61. Zhai D, Liu B, Shi Y, et al. (2013) Highly Sensitive Glucose Sensor Based on Pt Nanoparticle/Polyaniline Hydrogel Heterostructures. ACS Nano 7:3540–3546. https://doi.org/10.1021/nn400482d

62. Zeng F-W, Liu X-X, Diamond D, Lau K (2010) Humidity Sensors Based on Polyaniline Nanofibres. Sensors Actuators B Chem 143:530–534. https://doi.org/10.1016/j.snb.2009.09.050

63. Athawale A, Bhagwat S, Katre P (2006) Nanocomposite of Pd–Polyaniline as a Selective Methanol Sensor. Sensors Actuators B Chem 114:263–267. https://doi.org/10.1016/j.snb.2005.05.009

64. Xie D, Jiang Y, Pan W, et al. (2002) Fabrication and Characterization of Polyaniline-Based Gas Sensor by Ultra-Thin Film Technology. Sensors Actuators B Chem 81:158–164. https://doi.org/10.1016/S0925-4005(01)00946-7

65. Ivanov A, Lukachova L, Evtugyn G, et al. (2002) Polyaniline-Modified Cholinesterase Sensor for Pesticide Determination. Bioelectrochemistry 55:75–77. https://doi.org/10.1016/S1567-5394(01)00163-3

66. Dhand C, Das M, Datta M, Malhotra B (2011) Recent Advances in Polyaniline Based Biosensors. Biosens Bioelectron 26:2811–2821. https://doi.org/10.1016/j.bios.2010.10.017

67. Somerset V, Klink M, Baker P, Iwuoha E (2007) Acetylcholinesterase-Polyaniline Biosensor Investigation of Organophosphate Pesticides in Selected Organic Solvents. J Environ Sci Heal Part B 42:297–304. https://doi.org/10.1080/03601230701229288

68. Chang H, Yuan Y, Shi N, Guan Y (2007) Electrochemical DNA Biosensor Based on Conducting Polyaniline Nanotube Array. Anal Chem 79:5111–5115. https://doi.org/10.1021/ac070639m

69. Gerard M, Malhotra B (2005) Application of Polyaniline as Enzyme Based Biosensor. Curr Appl Phys 5:174–177. https://doi.org/10.1016/j.cap.2004.06.016

70. Chowdhury AD, Gangopadhyay R, De A (2014) Highly Sensitive Electrochemical Biosensor for Glucose, DNA and Protein Using Gold-Polyaniline Nanocomposites as a Common Matrix. Sensors Actuators, B Chem 190:348–356. https://doi.org/10.1016/j.snb.2013.08.071

71. Bicak T, Gicevičius M, Gokoglan T, et al. (2017) Simultaneous and Sequential Synthesis of Polyaniline-g-poly(ethylene glycol) by Combination of Oxidative Polymerization and CuAAC Click Chemistry: A Water-Soluble Instant Response Glucose Biosensor Material. Macromolecules 50:1824–1831. https://doi.org/10.1021/acs.macromol.7b00073

72. Hong C-H, Ki S-J, Jeon J-H, et al. (2013) Electroactive Bio-Composite Actuators Based on Cellulose Acetate Nanofibers with Specially Chopped Polyaniline Nanoparticles Through Electrospinning. Compos Sci Technol 87:135–141. https://doi.org/10.1016/j.compscitech.2013.08.006

73. Gao H, Zhang J, Yu W, et al. (2010) Monolithic Polyaniline/Polyvinyl Alcohol Nanocomposite Actuators with Tunable Stimuli-Responsive Properties. Sensors Actuators B Chem 145:839–846. https://doi.org/10.1016/j.snb.2010.01.066

74. García-Gallegos J, Martín-Gullón I, Conesa J, et al. (2016) The Effect of Carbon Nanofillers on the Performance of Electromechanical Polyaniline-Based Composite Actuators. Nanotechnology 27:015501. https://doi.org/10.1088/0957-4484/27/1/015501

75. Molberg M, Crespy D, Rupper P, et al. (2010) High Breakdown Field Dielectric Elastomer Actuators Using Encapsulated Polyaniline as High Dielectric Constant Filler. Adv Funct Mater 20:3280–3291. https://doi.org/10.1002/adfm.201000486

76. Okuzaki H, Takagi S, Hishiki F, Tanigawa R (2014) Ionic Liquid/Polyurethane/PEDOT:PSS Composites for Electro-Active Polymer Actuators. Sensors Actuators B Chem 194:59–63. https://doi.org/10.1016/j.snb.2013.12.059

77. Alcaraz-Espinoza J, Chávez-Guajardo A, Medina-Llamas J, et al. (2015) Hierarchical Composite Polyaniline–(Electrospun Polystyrene) Fibers Applied to Heavy Metal Remediation. ACS Appl Mater Interfaces 7:7231–7240. https://doi.org/10.1021/acsami.5b00326

78. Guo H, He W, Lu Y, Zhang X (2015) Self-Crosslinked Polyaniline Hydrogel Electrodes for Electrochemical Energy Storage. Carbon N Y 92:133–141. https://doi.org/10.1016/j.carbon.2015.03.062

79. Zhang B, Du Y, Zhang P, et al. (2013) Microwave Absorption Enhancement of Fe_3O_4/Polyaniline Core/Shell Hybrid Microspheres with Controlled Shell Thickness. J Appl Polym Sci 130:1909–1916. https://doi.org/10.1002/app.39332

80. Xiong P, Huang H, Wang X (2014) Design and Synthesis of Ternary Cobalt Ferrite/Graphene/Polyaniline Hierarchical Nanocomposites for High-Performance Supercapacitors. J Power Sources 245:937–946. https://doi.org/10.1016/j.jpowsour.2013.07.064

81. Wu J, Zhang Q, Wang J, et al. (2018) A Self-Assembly Route to Porous Polyaniline/Reduced Graphene Oxide Composite Materials with Molecular-Level Uniformity for High-Performance Supercapacitors. Energy Environ Sci 11:1280–1286. https://doi.org/10.1039/C8EE00078F

82. Döbbelin M, Tena-Zaera R, Marcilla R, et al. (2009) Multiresponsive PEDOT-Ionic Liquid Materials for the Design of Surfaces with Switchable Wettability. Adv Funct Mater 19:3326–3333. https://doi.org/10.1002/adfm.200900863

83. Marina S, Mantione D, ManojKumar K, et al. (2018) New Electroactive Macromonomers and Multi-Responsive PEDOT Graft Copolymers. Polym Chem 9:3780–3790. https://doi.org/10.1039/C8PY00680F

Index

For Product Safety Concerns and Information please contact our EU
representative GPSR@taylorandfrancis.com
Taylor & Francis Verlag GmbH, Kaufingerstraße 24, 80331 München, Germany